SUBATOMIC PHYSICS

Third Edition

SUBATOMIC PHYSICS

Third Edition

Ernest M Henley
Alejandro García
University of Washington, USA

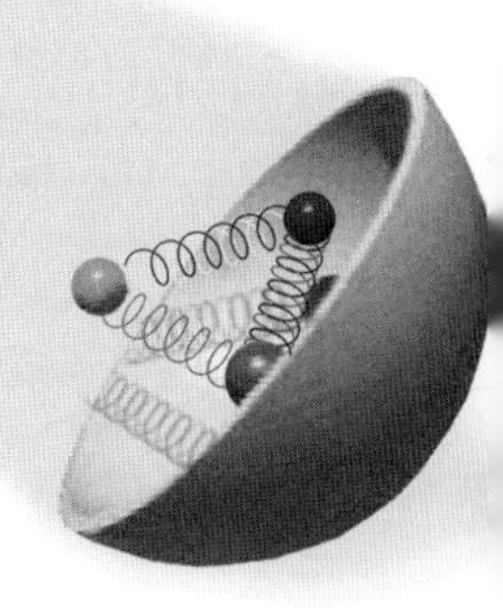

NEW JERSEY • LONDON • SINGAPORE • BEIJING • SHANGHAI • HONG KONG • TAIPEI • CHENNAI

Published by

World Scientific Publishing Co. Pte. Ltd.
5 Toh Tuck Link, Singapore 596224
USA office: 27 Warren Street, Suite 401-402, Hackensack, NJ 07601
UK office: 57 Shelton Street, Covent Garden, London WC2H 9HE

SUBATOMIC PHYSICS (3rd Edition)

ISBN-13 978-981-270-056-8

Printed in the U.S.A.

To Elaine and Viviana

Acknowledgments

In writing the present book, we have had generous help from students, friends and colleagues who have cleared up questions, provided us with data and/or figures, and have made helpful comments on parts of the manuscript.

We thank particularly Eric Adelberger, Hans Bichsel, Owen Biesel, Rick Casten, John Cramer, Kees de Jager, Paolo Gondolo, Paul Grannis, Wick Haxton, Craig Hogan, Seth Hoedl, David Kaplan, Joseph Kapusta, Kevin Lesko, Henry Lubatti, Augusto Machiavelli, Dan Melconian, Gerald Miller, Berndt Müller, John Negele, Ann Nelson, Mal Ruderman, Martin Savage, Anne Sallaska, Hendrik Schatz, Steve Sharpe, Reinhard Schumacher, Sky Sjue, Dick Seymour, Kurt Snover, Dam Thanh Son, Stephanie Stattel, Smarajit Triambak, Bob Tribble, Raju Venugopalan, John Wilkerson.

The figures from journals of the American Physical Society are reprinted with courtesy of the American Physical Society, which holds the copyrights; those from Physics Letters and Nuclear Physics are courtesy of Elsevier Publishing Company.

Preface to the First Edition

Subatomic Physics, the physics of nuclei and particles, has been one of the frontiers of science since its birth in 1896. From the study of the radiations emitted by radioactive nuclei to the scattering experiments that point to the presence of subunits in nucleons, from the discovery of the hadronic interactions to the realization that the photon possesses hadronic (strong) attributes, and that weak and electromagnetic forces may be intimately related, subatomic physics has enriched science with new concepts and deeper insights into the laws of nature.

Subatomic Physics does not stand isolated; it bears on many aspects of life. Ideas and facts emerging from studies of the subatomic world change our picture of the macrocosmos. Concepts discovered in subatomic physics are needed to understand the creation and abundance of the elements, and the energy production in the sun and the stars. Nuclear power may provide most of the future energy sources. Nuclear bombs affect national and international decisions. Pion beams have become a tool to treat cancer. Tracer and Mössbauer techniques give information about structure and reactions in solid state physics, chemistry, biology, metallurgy, and geology.

Subatomic Physics, because it reaches into so many areas, should not only be accessible to physicists, but also to other scientists and to engineers. The chemist observing the Mössbauer effect, the geologist using a radioactive dating method, the physician injecting a radioactive isotope, or the nuclear engineer designing a power plant have no immediate need to understand isospin or inelastic electron scattering. Nevertheless, their work may be more satisfying and they may be able to find new connections if they have a grasp of the basic principles of subatomic physics. While the present book is mainly intended as an introduction for physicists, we hope that it will also be useful to other scientists and to engineers.

Subatomic Physics deals with all entities smaller than the atom; it combines nuclear and particle physics. The two fields have many concepts and features in common. Consequently, we treat them together and attempt to stress unifying ideas, concepts and currently unsolved problems. We also show how subatomic physics is involved in astrophysics. The level of presentation is aimed at the senior undergraduate or first-year graduate student who has some understanding of electromagnetism, special relativity, and quantum theory. While many aspects of subatomic physics can be elucidated by hand waving and analogies, a proper understanding requires equations. One of the most infuriating sentences in textbooks is "It can be shown..." We would like to avoid this sentence but it is just not possible. We include most derivations but use equations without proof in two situations. Many of the equations from other fields will be quoted without derivation in order to save space and time. The second situation arises when the proper tools, for instance Dirac theory of field quantization, are too advanced. We justify omission in both situations by an analogy. Mountain climbers usually like to reach the unexplored parts of a climb quickly rather than spend days walking through familiar terrain. Quoting equations from quantum theory and electrodynamics corresponds to reaching the starting point of an adventure by car or cable car. Some peaks can only by reached by difficult routes. An inexperienced climber, not yet capable of mastering such a route, can still learn by watching from a safe place. Similarly, some equations can only be reached by difficult derivations, but the reader can still learn by exploring the equations without following their derivations. Therfore, we will quot some relations without proof, but we will try to make the result plausible and to explore the physical consequences. Some more difficult parts will be denoted with bullets (•); these parts can be omitted on first reading.

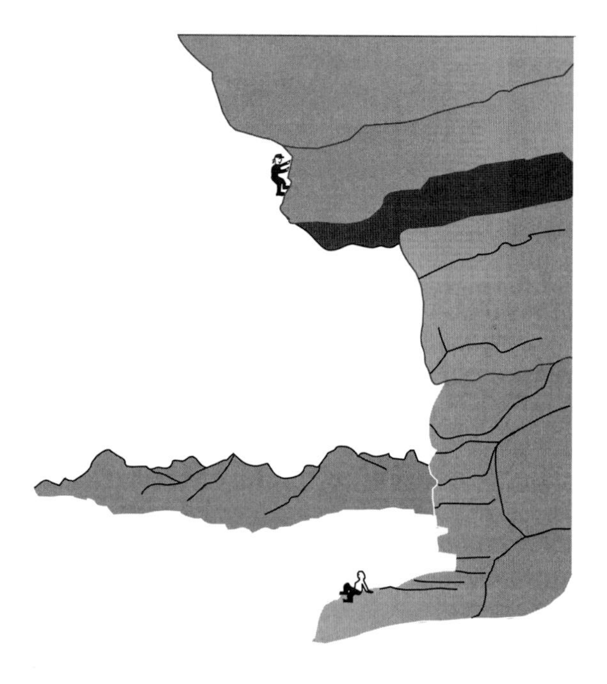

Preface to the Third Edition

Subatomic Physics has continued to make rapid strides since the 2nd. Edition was published in 1991 (by Prentice-Hall). New particles have been found; the distributions of electric charge and magnetism within the proton have been found to be significantly different; neutrinos have been found to have masses and undergo oscillations, and the standard model needs to be accordingly modified; CP violation has been established to be compatible with the Cabibbo-Kobayashi-Maskawa matrix; chiral and effective field theories have been developed, lattice QCD has made enormous strides. Nuclear structure far from the region of stability has started to been studied, relativistic heavy ions have opened new doors and understanding, and astrophysics and cosmology have provided us with a much improved understanding of the world around us. Data has become much more precise. Although there is a perception that physics has changed from being a unified science to a series of subfields that ignore each other, here we find the opposite: in the last twenty years there has been much progress at the intersection between atomic, nuclear, particle, and astro physics.

In the new edition we have updated all the material trying to expose the excitment that we feel about progress in the last two decades. We have reorganized chapters to make the material more clear, we have written new sections where new discoveries justified it, and we have trimmed parts of the 2nd Edition to allow us to incorporate new material. We have included new problems and, on the basis of comments we have received on the previous editions, we have starred problems which require the student to find library material. Overall there is more material in this edition than in the previous ones and we do realize that this is too much to be covered in a single quarter or semester. We nevertheless believe that this gives some freedom for the instructor to concentrate on the areas of choice. In addition, it gives the students the possibility of using the additional material to explore it on their own.

Hans Frauenfelder, who was one of the authors of the first two editions (1976, 1991) has been out of the field long enough to ask not to participate in the present work.

General Bibliography

The reader of the present book is expected to have some understanding of electromagnetism, special relativity, and quantum theory. We shall quote many equations from these fields without proof, but shall indicate where derivations can be found. the books listed here are referred to in the text by the name of the author.

Electrodynamics J.D. Jackson, *Classical Electrodynamics*, 3rd edition, Wiley, New York, 1999. Jackson's book is not an undergraduate text, but it is beautifully written and provides an exceptionally lucid treatment of classical electrodynamics. An alternative textbook undergraduates are more familiar with is D.J. Griffiths, *Introduction to Electrodynamics*, 3rd edition, Prentice Hall, NJ, 1999.

Modern Physics P.A. Tipler and R.A. Llewellyn, *Modern Physics*, 4th edition, W.H. Freeman and Co., New York, 2002. This book gives most of the needed background in special relativity, quantum mechanics, and atomic theory. An alternative is R. Eisberg, R. Resnick, *Quantum Physics of Atoms, Molecules, Solids, Nuclei, and Particles*, John Wiley & Sons, NY, 1985.

Quantum Mechanics E. Merzbacher, *Quantum Mechanics*, Wiley, New York, 3rd Edition, 1998; R. Shankar, *Principles of Quantum Mechanics*, 2nd edition, Springer Science, 1994; D.J. Griffiths, *Introduction to Quantum Mechanics*, 3rd edition, Pearson Prentice Hall, 2005. R.P. Feynman, R.B. Leighton, and M. Sands, *The Feynman Lectures on Physics*, Addison-Wesley, Reading, MA, 1965.

Mathematical Physics G.B. Arfken and H.J. Weber, *Mathematical Methods of Physicists*, 5th edition, Harcourt Acad. Press, San Diego (2001); or J. Mathews and R.L. Walker, *Mathematical Methods of Physics*, Benjamin Reading, MA, 1964, 1970, are easy-to-read books that cover the mathematical tools needed.

Data In the textbook we make extensive reference to data that has been evaluated by the Particle Data Group which we will refer to as 'PDG'. Their last publication is W.-M. Yao et al., *J. Phys. G* **33**, 1 (2006) and the data can be found online at http://pdg.lbl.gov/. For nuclear structure we refer to the National Nuclear Data Center at Brookhaven, online at http://www.nndc.bnl.gov/.

Miscellaneous Finally we should like to say that physics, despite its cold appearance, is an intensely human field. Its progress depends on hard-working people. Behind each new idea lie countless sleepless nights and long struggles for clarity. Each major experiment involves strong emotions, often bitter competition, and nearly always dedicated collaboration. Each new step is bought with disappointments; each new advance hides failures. Many concepts are connected to interesting stories and sometimes funny anecdotes. A book like this one cannot dwell on these aspects, but we add a list of books related to subatomic physics that we have read with enjoyment.

L. Fermi, *Atoms in the Family*, University of Chicago Press, Chicago, 1954.

L. Lamont, *Day of Trinity*, Atheneum, New York, 1965.

R. Moore, *Niels Bohr*, A.A. Knopf, New York, 1966.

V.F. Weisskopf, *Physics in the Twentieth Century: Selected Essays*, MIT Press, Cambridge, 1972.

G. Gamow, *My World Line*, Viking, New York, 1970.

E. Segre, *Enrico Fermi, Physicist*, University of Chicago Press, Chicago, 1970.

M. Oliphant, *Rutherford Recollections of the Cambridge Days*, Elsevier, Amsterdam, 1972.

W. Heisenberg, *Physics and Beyond; Encounters and Conversations*, Allen and Unwin, London, 1971.

R. Jungk, *The Big Machine*, Scribner, New York, 1968.

P.C.W. Davies, *The Forces of Nature*, Cambridge University Press, Cambridge, 1979.

E. Segre, *From X Rays to Quarks*, Freeman, San Francisco, 1980.

Y. Nambu, *Quarks*, World Sci., Singapore, 1981.

P. Davies, *Superforce*, Simon & Schuster, New York, 1984.

F. Close, *The Cosmic Onion*, American Institute of Physics, New York, 1983.

R.P. Feynman, *Quantum Electrodynamics*, Princeton University Press, Princeton, 1985.

H.R. Pagels, *Perfect Symmetry*, Simon & Schuster, New York, 1983.

A. Zee, *Fearful Symmetry*, MacMillan Publishing Co., New York, 1986.

R.E. Peierls, *Atomic Histories*, American Institute of Physics, New York, 1997.

F. Close, *Lucifer's Legacy*, Oxford University Press, Oxford, 2000.

F. Close, M. Marten, and C. Sutton, *A Journey to the Heart of Matter*, Oxford University Press, Oxford, 2002.

K.S. Thorne, *Black Holes and Time Warps: Einstein's outrageous legacy*, W.W. Norton, New York, 1994.

S. Weinberg, *The Discovery of Subatomic Particles*, Cambridge University Press, Cambridge, New York, 2003.

Contents

Chapter 1

Background and Language

> Human existence is based on two pillars: compassion and curiosity. Compassion without curiosity is ineffective. Curiosity wihtout compassion is inhuman.
>
> Victor F. Weisskopf

The exploration of subatomic physics started in 1896 with Becquerel's discovery of radioactivity; since then it has been a constant source of surprises, unexpected phenomena, and fresh insights into the laws of nature.

In this first chapter we shall describe the orders of magnitude encountered in subatomic physics, define our units, and introduce the language needed for studying subatomic phenomena.

1.1 Orders of Magnitude

Subatomic physics is distinguished from all other sciences by one feature: it is the playground of three different interactions, and two of them act only when the objects are very close together. Biology, chemistry, and atomic and solid-state physics are dominated by the long-range electromagnetic force. Phenomena in the universe are ruled by two long-range forces, gravity and electromagnetism. Subatomic physics, however, is a subtle interplay of three interactions—the strong, the electromagnetic, and the weak—and the strong and the weak vanish at atomic and larger distances. The strong (or hadronic, or nuclear) force holds nuclei together; its range is very short, but it is strong. The weak interaction has an even shorter range. At this point *strong, weak*, and *short range* are just names, but we shall become familiar with the forces as we go along.

Figures 1.1, 1.2, and 1.3 give an idea of the orders of magnitude involved in the various phenomena. We present them here without discussion; they speak for themselves.

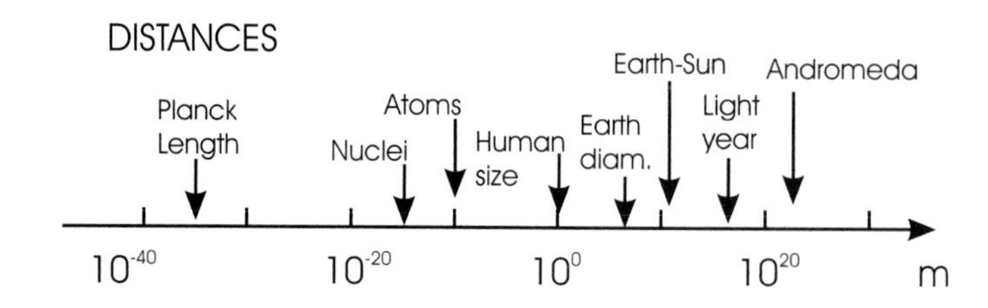

Figure 1.1: Typical distances. The region below about 10^{-18}m is unexplored. It is unknown if new forces and new phenomena appear.

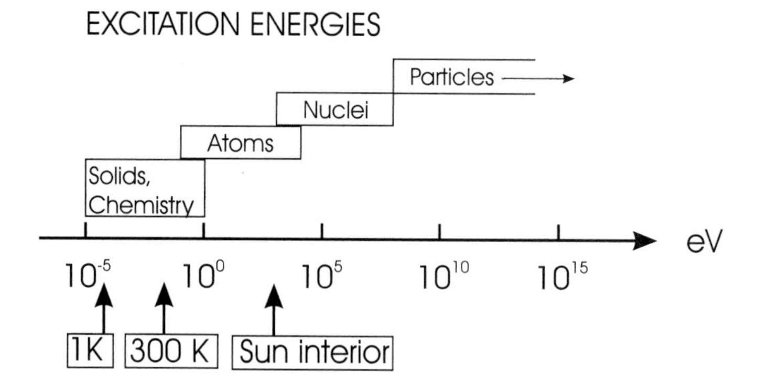

Figure 1.2: Range of excitation energies. The temperatures corresponding to the energies are also given.

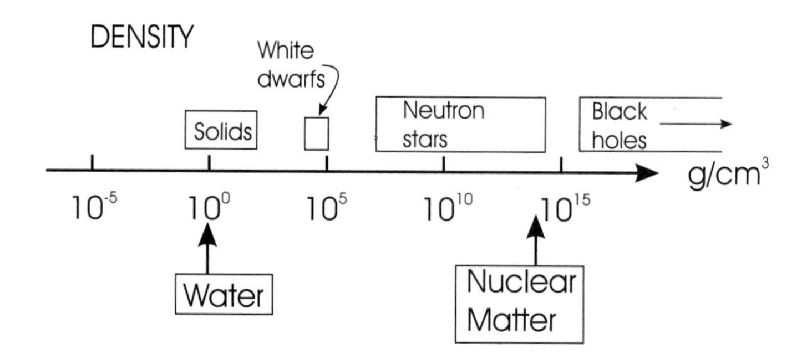

Figure 1.3: Range of densities.

Table 1.1: BASIC UNITS. c is the velocity of light.

Quantity	Unit	Abbreviation
Length	Meter	m
Time	Second	sec or s
Energy	Electron volt	eV
Mass		eV/c^2
Momentum		eV/c

Table 1.2: PREFIXES FOR POWERS OF 10.

Power	Name	Symbol	Power	Name	Symbol
10^1	Deca	da	10^{-1}	Deci	d
10^2	Hecto	h	10^{-2}	Centi	c
10^3	Kilo	k	10^{-3}	Milli	m
10^6	Mega	M	10^{-6}	Micro	μ
10^9	Giga	G	10^{-9}	Nano	n
10^{12}	Tera	T	10^{-12}	Pico	p
10^{15}	Peta	P	10^{-15}	Femto	f
10^{18}	Exa	E	10^{-18}	Atto	a

1.2 Units

The *basic units* to be used are given in Table 1.1. The prefixes defined in Table 1.2 give the decimal fractions or multiples of the basic units. As examples, 10^6 eV = MeV, 10^{-12} sec = psec, and 10^{-15}m = fm. The last unit, femtometer, is often also called *Fermi*, and it is extensively used in particle physics. The introduction of the electron volt as an energy unit requires a few words of justification. One eV is the energy gained by an electron if it is accelerated by a potential difference of 1 V (volt):

$$\begin{aligned} 1 \text{ eV} &= 1.60 \times 10^{-19} \text{ C (coulomb)} \times 1 \text{ V} \\ &= 1.60 \times 10^{-19} \text{ J (joule)} \\ &= 1.60 \times 10^{-12} \text{ erg.} \end{aligned} \tag{1.1}$$

The electron volt (or any decimal multiple thereof) is a convenient energy unit because particles of a given energy are usually produced by acceleration in electromagnetic fields. To explain the units for mass and momentum we require one of the most important equations of special relativity, connecting total energy E, mass m, and momentum $\boldsymbol{p}$ of a free particle[(1)]:

$$E^2 = p^2c^2 + m^2c^4. \tag{1.2}$$

[1]Tipler & Llewellyn, Eq. (2-31), or Jackson, Eq. (11.55).

This equation states that the total energy of a particle consists of a part independent of the motion, the rest energy mc^2, and a part that depends on the momentum. For a *particle without mass*, Eq. (1.2) reads

$$E = pc; \tag{1.3}$$

on the other hand, for a *particle at rest*, the famous relation

$$E = mc^2 \tag{1.4}$$

follows. These equations make it clear why the units eV/c^2 for mass and eV/c for momentum are convenient. For instance, if the mass and energy of a particle are known, then the momentum in eV/c follows immediately from Eq. (1.2). In the previous equations, we have denoted a vector by $\boldsymbol{p}$ and its magnitude by p. In equations where we require electromagnetic quantities we shall use *Gaussian units.* Gaussian units are used by Jackson and his Appendix 4 gives clear prescriptions for the conversion from Gaussian to mks units.

1.3 Special Relativity, Feynman Diagrams

In our discussions we shall use concepts and equations from electrodynamics, special relativity, and quantum mechanics. The fact that we need some *electrodynamics* is not surprising. After all, most particles and nuclei are charged; their mutual interaction and their behavior in external electric and magnetic fields are governed by Maxwell's laws.

The fact that the theory of *special relativity* is essential can be seen most clearly from two features. First, subatomic physics involves the creation and destruction of particles, or, in other words, the change of energy into matter and vice versa. If the matter is at rest, the relation between energy and matter is given by Eq. (1.4); if it is moving, Eq. (1.2) must be used. Second, the particles produced by modern accelerators move with velocities that are close to the velocity of light, and nonrelativistic (Newtonian) mechanics does not apply. Consider two coordinate systems, K and K'. System K' has its axes parallel to those of K but is moving with a velocity v in the positive z direction relative to K. The connection between the coordinates (x', y', z', t') of system K' and $(x,\ y,\ z,\ t)$ of K is given by the *Lorentz transformation,*[2]

$$\begin{aligned} x' &= x, \quad y' = y, \\ z' &= \gamma(z - vt), \\ t' &= \gamma\left(t - \frac{\beta}{c}z\right), \end{aligned} \tag{1.5}$$

[2] Tipler & Llewellyn, Eq. (1-20); Jackson, Eq. (11.16).

where

$$\gamma = \frac{1}{(1-\beta^2)^{1/2}}, \quad \beta = \frac{v}{c}. \tag{1.6}$$

Momentum and velocity are connected by the relation

$$\boldsymbol{p} = m\gamma\boldsymbol{v}. \tag{1.7}$$

Squaring this expression and using Eqs. (1.2) and (1.6) yields

$$\beta \equiv \frac{v}{c} = \frac{pc}{E}. \tag{1.8}$$

As one application of the Lorentz transformation to subatomic physics, consider the muon, a particle that we shall encounter often. It is basically a heavy electron with a mass of 106 MeV/c^2. While the electron is stable, the muon decays with a mean life τ:

$$N(t) = N(0)e^{-t/\tau},$$

where $N(t)$ is the number of muons present at time t. If $N(t_1)$ muons are present at time t_1, only $N(t_1)/e$ are still around at time $t_2 = t_1 + \tau$. The mean life of a muon *at rest* has been measured as 2.2μ sec. Now consider a muon produced at the FNAL (Fermi National Accelerator Laboratory) accelerator with an energy of 100 GeV. If we observe this muon in the laboratory, what mean life τ_{lab} do we measure? Nonrelativistic mechanics would say 2.2μ sec. To obtain the correct answer, the Lorentz transformation must be used. In the muon's rest frame (unprimed), the mean life is the time interval between the two times t_2 and t_1 introduced above, $\tau = t_2 - t_1$. The corresponding times, t'_2 and t'_1, in the laboratory (primed) system are obtained with Eq. (1.5) and the *observed* mean life $\tau_{\text{lab}} = t'_2 - t'_1$ becomes

$$\tau_{\text{lab}} = \gamma\tau.$$

With Eqs. (1.6) and (1.8), the ratio of mean lives becomes

$$\frac{\tau_{\text{lab}}}{\tau} = \gamma = \frac{E}{mc^2}. \tag{1.9}$$

With $E = 100$ GeV, $mc^2 = 106$ MeV, $\tau_{\text{lab}}/\tau \approx 10^3$. The mean life of the muon observed in the laboratory is about 1000 times longer than the one in the rest frame (called proper mean life).

Although we will not use relativistic notation (e.g., four-vectors) very often, we introduce it here for convenience. The quantity $A \equiv A_\mu = (A_0, \boldsymbol{A})$ is called a four-vector if it transforms under a Lorentz transformation like $(ct, \boldsymbol{x})$. The time component is A_0. The scalar product of two four vectors A and B is defined as

$$A \cdot B = \sum_{\mu,\nu=0}^{3} g_{\mu,\nu} A_\mu B_\nu = A_0 B_0 - \boldsymbol{A} \cdot B, \tag{1.10}$$

with $g_{00} = 1, g_{ii} = -1$ ($i = x, y, z$ or $1, 2, 3$) and $g_{\mu,\nu} = 0$ for $\mu \neq \nu$. Such a scalar product is a Lorentz scalar; it remains constant or invariant under a Lorentz transformation. The four-vectors that occur most often are

$$\begin{array}{ll} \text{time–space} & x_\mu = (ct, \boldsymbol{x}), \\ \text{four-momentum} & p_\mu = (\frac{E}{c}, \boldsymbol{p}), \\ \text{four-current} & j_\mu = (c\rho, \boldsymbol{j}), \\ \text{four-potential} & A_\mu = (A_0, \boldsymbol{A}), \\ \text{four-gradient} & \nabla_\mu = \left(\frac{1}{c}\frac{\partial}{\partial t}, -\boldsymbol{\nabla}\right). \\ \text{(note the sign)} & \end{array} \tag{1.11}$$

Relativistic kinematics are introduced in Section 2.6. In order to transform energies and momenta from one frame of reference to another, it is helpful to use a relativistic invariant of the above type. For example, in the collision of particles a and b, we have

$$(p_{a\mu}c + p_{b\mu}c)^2 = (E_a + E_b)^2 - (\boldsymbol{p}_a c + \boldsymbol{p}_b c)^2 = M_{ab}^2 c^4, \tag{1.12}$$

with M_{ab} an invariant.

Quantum mechanics was forced on physics because of otherwise unexplained properties of atoms and solids. It is therefore not surprising that subatomic physics also requires quantum mechanics for its description. Indeed the existence of quantum levels and the occurrence of interference phenomena in subatomic physics make it clear that quantum phenomena occur. But will the knowledge gained from atomic physics be sufficient? The dominant features of atoms can be understood without recourse to relativity, and nonrelativistic quantum mechanics describes nearly all atomic phenomena well. In contradistinction, subatomic physics cannot be explained without relativity, as outlined above. It is therefore to be expected that nonrelativistic quantum mechanics is inadequate. An example of its failure can be explained simply: assume a particle described by a wave function $\psi(\boldsymbol{x}, t)$. The normalization condition[3]

$$\int_{-\infty}^{+\infty} \psi^*(\boldsymbol{x}, t)\psi(\boldsymbol{x}, t) d^3x = 1 \tag{1.13}$$

states that the particle must be found somewhere at all times. However, the *creation* and *destruction* of particles is a phenomenon that occurs frequently in subatomic physics. A spectacular example is shown in Fig. (1.4). On the left-hand side, a bubble chamber picture is reproduced. (Bubble chambers will be discussed in Section 4.4.) On the right-hand side, the important tracks in the bubble chamber are redrawn and identified. We shall describe the various particles in Chapter 5. Here we just assume that particles with the names indicated in Fig. 1.4 exist and do not worry about their properties. The figure then tells the following story. A

[3] The integral should properly be written as $\iiint d^3x$. Following custom, we write only one of the three integrals.

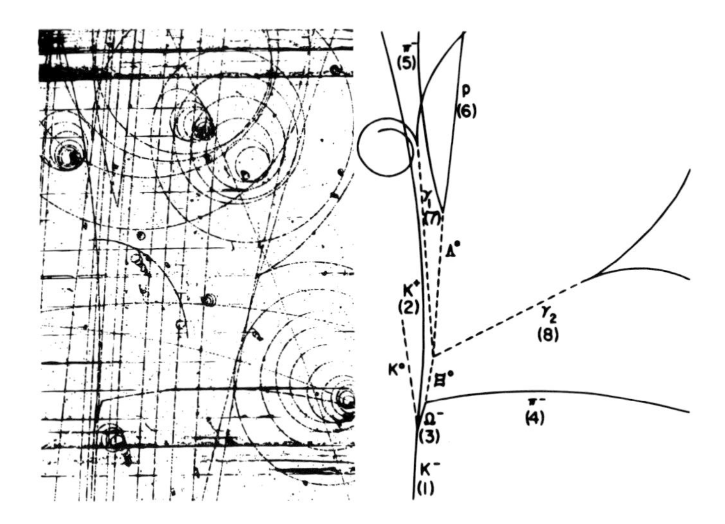

Figure 1.4: Liquid hydrogen bubble chamber picture. This photograph and the tracing at right show the production and the decay of many particles. Part of the story is told in the text. [Courtesy Brookhaven National Laboratory, where the photograph was taken in 1964.]

K^-, or negative kaon, enters the bubble chamber from below. The bubble chamber is filled with hydrogen and the only particle with which the kaon can collide with appreciable probability is the nucleus of the hydrogen atom, namely the proton. The negative kaon indeed collides with a proton and produces a positive kaon, a neutral kaon, and an *omega minus*. The Ω^- decays into a Ξ^0 and a π^-, and so forth. The events shown in Fig. 1.4 make the essential point forcefully: particles are created and destroyed in physical processes. Without special relativity, these observations cannot be understood. Equally strongly, Eq. (1.12) cannot be valid since it states that the total probability of finding the particle described by ψ must be independent of time. Nonrelativistic quantum mechanics cannot describe the creation and destruction of particles.[4]

We need at least a language to describe these phenomena. Such a language exists and is used universally. It is the method of Feynman diagrams or graphs. The diagrams, which are a pictorial representation of particle interactions, have a more sophisticated use than would appear from the way we describe them here. Arrows indicate the time sense. Energy, momentum, and charge are conserved at vertices. Lines entering a Feynman diagram indicate initial state free particles and those leaving it are final state free particles. The Feynman graphs for two of the processes contained in Fig. 1.4 are given in Fig. 1.5. The first one describes the decay of a lambda (Λ^0) into a proton and a negative pion, and the second one

[4] The theorem that nonrelativistic quantum mechanics cannot describe unstable elementary particles was proved by Bargmann. The proof can be found in Appendix 7 of F. Kaempffer, *Concepts in Quantum Mechanics*, Academic Press, New York, 1965. The appendix is entitled "If Galileo Had Known Quantum Mechanics."

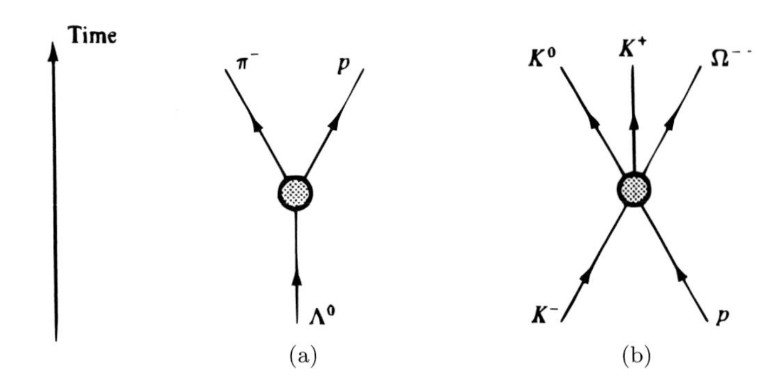

Figure 1.5: Feynman diagrams for (a) the decay $\Lambda^\circ \to p\pi^-$ and (b) the reaction $K^-p \to K^\circ K^+\Omega^-$.

the collision of a negative kaon and a proton, giving rise to a neutral and a positive kaon and an omega minus. In both diagrams, the interaction is drawn as a "blob" to indicate that the exact mechanism remains to be explored. In the following chapters we shall use Feynman diagrams often and explain more details as we need them.

1.4 References

Special relativity is treated in many books, and every teacher and reader has his favorites. Good first introductions can be found in the *Feynman Lectures*, Vol. I, Chapters 15–17. A concise and complete exposition is given in Jackson, Chapters 11 and 12. These two chapters form an excellent base for all applications to subatomic physics. Some more recent useful references are W Rindler, *Introduction to Special Relativity*, 2nd Ed., Clarendon Press, Oxford, 1991; R.P. Feynman, *Six Not-so-Easy Pieces*, Addison Wesley, Reading, MA, 1997; J.B. Kogut, *Introduction to Relativity*, Harcourt/Academic Press, San Diego 2001; W.S.C. Williams, *Introductory Special Relativity*, Taylor and Francis, London, 2002.

Books on quantum mechanics have already been listed at the end of the Preface. However, a few additional remarks concerning Feynman diagrams are in order here. There is no place where Feynman diagrams can be learned without effort. Relatively gentle introductions can be found in

R.P. Feynman, *Theory of Fundamental Processes*, Benjamin, Reading, Mass., 1962.

F. Mandl, *Introduction to Quantum Field Theory*, Wiley-Interscience, New York, 1959.

J.M. Ziman, *Elements of Advanced Quantum Theory*, Cambridge University Press, Cambridge, 1969.

K. Gottfried and V.F. Weisskopf, *Concepts in Particle Physics*, Oxford University Press, New York, Vol. I, 1984, Vol. II, 1986.

Problems

1.1. * Use what information you can find to get a number characterizing the strength of each of the four basic interactions. Justify your numbers.

1.2. Discuss the range of each of the four basic interactions.

1.3. List a few important processes for which the electromagnetic interaction is essential.

1.4. For what cosmological and astrophysical phenomena is the weak interaction essential?

1.5. It is known that the muon (the heavy electron, with a mass of about 100 MeV/c^2) has a radius that is smaller than 0.1 fm. Compute the minimum density of the muon. Where would the muon lie in Fig. 1.3? What problems does this crude calculation raise?

1.6. Verify Eq. (1.8).

1.7. Verify Eq. (1.9).

1.8. Consider a pion with a kinetic energy of 200 MeV. Find its momentum in MeV/c.

1.9. A proton is observed to have a momentum of 5 MeV/c. Compute its kinetic energy in MeV.

1.10. For a certain experiment, kaons with a kinetic energy of 1 GeV are needed. They are selected with a magnet. What momentum does the magnet have to select?

1.11. Find two examples where special relativity is essential in subatomic physics.

1.12. How far does a beam of muons with kinetic energy of

(a) 1 MeV,

(b) 100 GeV

travel in empty space before its intensity is reduced to one half of its initial value?

1.13. Repeat Problem 1.12 for charged and for neutral pions. Also repeat for an intensity reduction to one half of its initial value.

1.14. Which subatomic phenomena exhibit quantum mechanical interference effects?

1.15. If the strong and weak forces are assumed to be approximately constant over 1 fm, find the order of magnitudes for

$$F_{\mathrm{h}} : F_{\mathrm{em}} : F_{\mathrm{weak}} : F_{\mathrm{gravit}}$$

for two protons that are 1 fm apart. Use any physical knowledge or arguments at your disposal to obtain the desired ratios.

Part I

Tools

One of the most frustrating experiences in life is to be stranded without proper tools. The situation can be as simple as being in the wilderness with a broken shoe strap but no wire or knife. It can be as simple as having a leaking radiator hose in Death Valley and no tape to fix it. In these instances we at least know what we miss and what we need. Confronted with the mysteries of subatomic physics, we also need tools and we often do not know what is required. However, during the past century, we have learned a great deal, and many beautiful tools have been invented and constructed. We have accelerators to produce particles, detectors to see them and to study their interactions, instruments to quantify what we observe, and computers to evaluate the data. In the following three chapters we sketch some important tools.

Chapter 2

Accelerators

2.1 Why Accelerators?

Accelerators cost a lot of money. What can they do? Why are they crucial for studying subatomic physics? As we proceed through various fields of subatomic physics, these questions will be answered. Here we shall simply point out a few of the important aspects.

Accelerators produce beams of charged particles with energies ranging from a few MeV to several TeV. Intensities can be as high as 10^{17} particles/sec, and the beams can be concentrated onto targets of a few mm^2 or less in area. The particles that are most often used as primary projectiles are protons and electrons.

Two tasks can be performed well only by accelerators, namely the production of new particles and new states, and the investigation of the detailed structure of subatomic systems. Consider, first, particles and nuclei. Only very few stable particles exist in nature—the proton, the electron, the neutrino, and the photon. Only a limited number of nuclides are available in terrestrial matter, and they are usually in the ground state. To escape the narrow limitations of what is naturally available, new states must be produced artificially. To create a state of mass m, we need at least the energy $E = mc^2$. Very often, considerably more energy is required, as we shall find out. So far, no limit on the mass of new particle states has been found, and we do not know if one exists. It is suspected that the Planck mass, $(\hbar c/G_g)^{1/2} = 1.22 \times 10^{28}\ \text{eV}/c^2$ may set a limit; here G_g is the gravitational constant. Clearly, higher energies are a prerequisite to finding out.

High energies are not only needed to produce new states; they are also essential in finding out details concerning the structure of subatomic systems. It is easy to see that the particle energy has to be higher as the dimension to be looked at becomes smaller. The de Broglie wavelength of a particle with momentum p is given by

$$\lambda = \frac{h}{p}, \tag{2.1}$$

where h is Planck's constant. In most expressions, we shall use the *reduced* de Broglie wavelength,

$$\lambda\!\!\!^{-} = \frac{\lambda}{2\pi} = \frac{\hbar}{p}, \tag{2.2}$$

where h-bar, or Dirac's $\hbar$, is

$$\hbar = \frac{h}{2\pi} = 6.5821 \times 10^{-22} \text{ MeV sec.} \tag{2.3}$$

As is known from optics, in order to see structural details of linear dimensions d, a wavelength comparable to, or smaller than, d must be used:

$$\lambda\!\!\!^{-} \le d. \tag{2.4}$$

The momentum required then is

$$p \ge \frac{\hbar}{d}. \tag{2.5}$$

To see small dimensions, high momenta and thus high energies are needed. As an example, we consider $d = 1$ fm and protons as a probe. We shall see that a nonrelativistic approximation is permitted here; the minimum kinetic energy of the protons then becomes, with Eq. (2.5),

$$E_{\text{kin}} = \frac{p^2}{2m_p} = \frac{\hbar^2}{2m_p d^2}. \tag{2.6}$$

It is straightforward to insert the constants $\hbar$ and m_p (see PDG.) However, we shall use this example to compute E_{kin} in a more roundabout but also more convenient way: Express as many quantities as possible as dimensionless ratios. E_{kin} has the dimension of an energy, as does $m_p c^2 = 938$ MeV. The kinetic energy is consequently rewritten as a ratio:

$$\frac{E_{\text{kin}}}{m_p c^2} = \frac{1}{2d^2}\left(\frac{\hbar}{m_p c}\right)^2.$$

The quantity in parentheses is just the Compton wavelength of the proton

$$\lambda\!\!\!^{-}_p = \frac{\hbar}{m_p c} = \frac{\hbar c}{m_p c^2} = \frac{197.3 \text{ MeV fm}}{938 \text{ MeV}} = 0.210 \text{ fm} \tag{2.7}$$

so that the kinetic energy is given by

$$\frac{E_{\text{kin}}}{m_p c^2} = \frac{1}{2}\left(\frac{\lambda\!\!\!^{-}_p}{d}\right)^2 = 0.02. \tag{2.8}$$

The combination $\hbar c$ will be found very useful throughout the text. The kinetic energy required to see linear dimensions of the order of 1 fm is about 20 MeV. Since this kinetic energy is much smaller than the rest energy of the nucleon, the nonrelativistic approximation is justified. Nature does not provide us with intense particle beams of such energies; they must be produced artificially. (Cosmic rays contain particles with much higher energies, but the intensity is so low that only very few problems can be attacked in a systematic way.)

The common way to produce a particle beam of high energy is to accelerate charged particles in an electric field. The force exerted on a particle of charge q by an electric field $\boldsymbol{E}$ is

$$\boldsymbol{F} = q\boldsymbol{E}. \tag{2.9}$$

In the simplest accelerator, two grids with a potential difference V at a distance d (Fig. 2.1), the average field is given by $|\boldsymbol{E}| = V/d$, and the energy gained by the particle is

$$E = Fd = qV. \tag{2.10}$$

Of course, the system must be placed in a vacuum; otherwise the accelerated particles will collide with air molecules and continuously lose much of the acquired energy. Figure 2.1. therefore includes a vacuum pump. Moreover, an ion source is also indicated—it produces the charged particles. These elements—particle source, accelerating structure, and vacuum pump—appear in every accelerator.

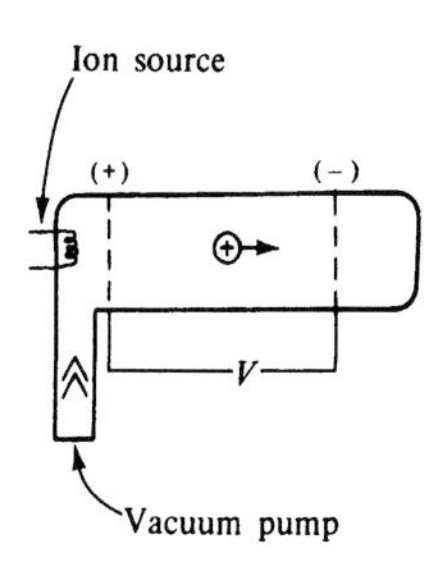

Figure 2.1: Prototype of the simplest accelerator.

Can particle beams of 20 MeV be reached with simple machines as sketched in Fig. 2.1? Anyone who has played with high voltages knows that such an approach is not easy. At a few kV, voltage breakdowns can occur and it requires experience to exceed even 100 kV. Indeed, it has taken considerable ingenuity and work to bring *electrostatic generators* to the point where they can produce particles of charge $|e|$ with energies of the order of tens of MeV.

However, it is impossible to achieve energies that are orders of magnitude higher, no matter how sophisticated the electrostatic generator. A new idea is needed, and such an idea was found—successive application of a given voltage to the same particle. Actually, a few times during the long road to the giant accelerators of today it looked as though the maximum accelerator energy had been reached. However, every apparently unsurmountable difficulty was overcome by an ingenious new approach.

We shall discuss only three types of accelerators: the electrostatic generator, the linear accelerator, and the synchrotron.

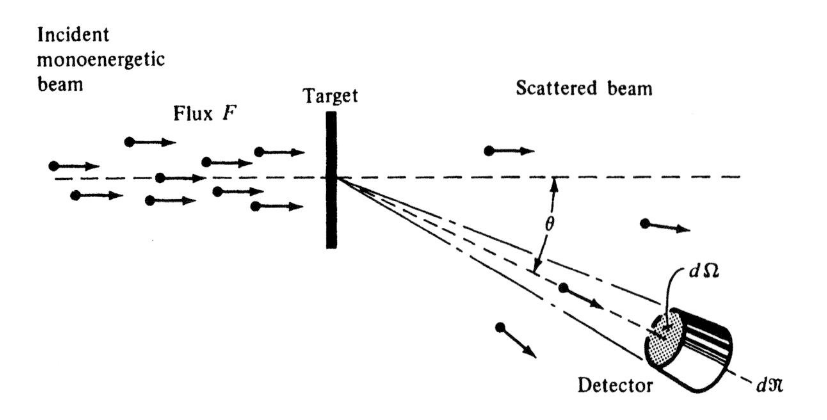

Figure 2.2: An incident monoenergetic beam is scattered by a target; the counter observing the scattered particles makes an angle θ with respect to the incident beam direction, subtends a solid angle $d\Omega$, and records $d\mathcal{N}$ particles per unit time.

2.2 Cross Sections and Luminosity

Before we describe accelerators we need to understand two quantities that are of interest to describe their power. Collisions are the most important processes used to study structure in subatomic physics. The behavior of a collision is usually expressed in terms of a cross section. To define cross section, a monoenergetic particle beam of well-defined energy is assumed to impinge on a target (Fig. 2.2). The flux F of the incident beam is *defined* as the number of particles crossing a unit area perpendicular to the beam per unit time. If the beam is uniform and contains n_i particles per unit volume, moving with velocity v with respect to the stationary target, the flux is given by

$$F = n_i v. \tag{2.11}$$

In most calculations, the number of incident particles is normalized to *one particle per volume* V. The number n_i is then equal to $1/V$. Particles scattered by the target are observed with a counter that detects all particles scattered by an angle θ into the solid angle $d\Omega$. The number $d\mathcal{N}$ recorded per unit time is proportional to the incident flux F, the solid angle $d\Omega$, and the number N of independent scattering centers in the target that are intercepted by the beam[1]:

$$d\mathcal{N} = FN\sigma(\theta)d\Omega. \tag{2.12}$$

The coefficient of proportionality is designated by $\sigma(\theta)$; it is called the *differential scattering cross section*, and we also write

$$\sigma(\theta)d\Omega = d\sigma(\theta) \qquad \text{or} \qquad \sigma(\theta) = \frac{d\sigma(\theta)}{d\Omega}. \tag{2.13}$$

[1]It is assumed here that each particle scatters at most once in the target and that each scattering center acts independently of each other one.

The total number of particles scattered per unit time is obtained by integrating over all solid angles,

$$\mathcal{N}_s = FN\sigma_{\text{tot}}, \tag{2.14}$$

where

$$\sigma_{\text{tot}} = \int \sigma(\theta)d\Omega \tag{2.15}$$

is called the *total scattering cross section.* Equation (2.14) shows that the total cross section has the dimension of an area, and it is customary to quote subatomic cross sections in barns (b) or decimal fractions of barns, where

$$1 \text{ b} = 10^{-24}\text{cm}^2 = 100 \text{ fm}^2.$$

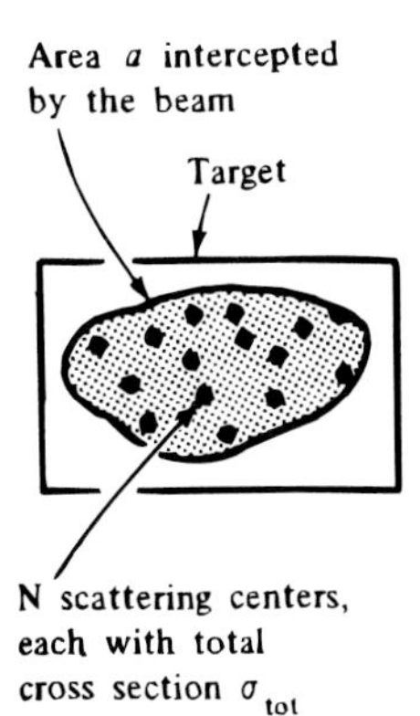

Figure 2.3: An area a of the target is struck by the incident beam. The area a contains N scattering centers, each with cross section σ_{tot}.

The significance of σ_{tot} can be understood by computing the fraction of particles that are scattered. Figure 2.3 represents the target seen in the beam direction. The area a intercepted by the beam contains N scattering centers. The total number of incident particles per unit time is given by

$$\mathcal{N}_{\text{in}} = Fa;$$

the total number of scattered particles is given by Eq. (2.14) so that the ratio of scattered to incident particle numbers is

$$\frac{\mathcal{N}_s}{\mathcal{N}_{\text{in}}} = \frac{N\sigma_{\text{tot}}}{a}. \tag{2.16}$$

The interpretation of this relation is straightforward: if no multiple scattering events occur, then the fraction of particles scattered is equal to the effective fraction of the total area occupied by scattering centers. $N\sigma_{\text{tot}}$ consequently must be the total area of all scattering centers and σ_{tot} the area of one scattering center. We stress that σ_{tot} is the area effective in scattering. It depends on the type and energy of the particles and is only occasionally equal to the actual geometrical area of the scattering center.

Finally, we note that if n is the number of scattering centers per unit volume, d the target thickness, and a the area intercepted by the beam, N is given by

$$N = and.$$

If the target consists of nuclei with atomic weight A and has a density ρ, n is given by

$$n = \frac{N_0 \rho}{A}, \tag{2.17}$$

where $N_0 = 6.0222 \times 10^{23}$ mole^{-1} is Avogadro's number.

Equation (2.14) describes the number, $\mathcal{N}_s$, of events per unit time in a *fixed target experiment*, where the incident beam impinges on a stationary target. Since the number N of scattering centers in a solid or liquid target is very large, $\mathcal{N}_s$ is measurable even for processes with small cross sections. We have, however, shown in Section 2.7 that the energy available in the c.m. is limited in fixed target experiments. In *colliding beam experiments* (Section 2.8), high energies can be obtained, but the number of scattering events becomes much smaller. The number of events per unit time is characterized by the *luminosity* $\mathcal{L}$, defined as the number of events per unit cross section that take place at a single beam encounter region per unit time. In the simplest situation, each colliding beam contains a single bunch, the bunches collide head-on, and each beam is uniform over an area A. If the bunches collide with a frequency f and if bunch i contains N_i particles, the luminosity in the interaction region of beams 1 and 2 is given by

$$\mathcal{L} = \frac{\mathcal{N}_s}{\sigma_{\text{tot}}} = \frac{N_1 N_2 f}{A}. \tag{2.18}$$

As an example the design luminosity for the Large Hadron Collider at CERN is $\approx 10^{34}$ cm^{-2} sec^{-1}.

2.3 Electrostatic Generators (Van de Graaff)

It is difficult to produce a very high voltage directly, for instance, by a combination of transformer and rectifier. In the Van de Graaff generator,(2) the problem is circumvented by transporting a charge Q to one terminal of a condenser C; the resulting voltage,

$$V = \frac{Q}{C}, \tag{2.19}$$

is used to accelerate the ions. The main elements of a Van de Graaff generator are shown in Fig. 2.4. Positive charges are sprayed onto an insulating charging belt by using a voltage of about 20–30 kV. The positive charge is carried to the terminal by the motor-driven belt; it is collected there by a set of needles and travels to the terminal surface. Positive ions (protons, deuterons, etc.) are produced in the ion source and are accelerated in the evacuated accelerating column. The beam emerging from the column is usually deflected by a magnet onto the target. If the entire system is placed in air, voltages of up to about a few MV can be reached before artificial lightning discharges the terminal. If the system is placed in a pressure tank

[2]R. J. Van de Graaff, *Phys. Rev.* **38**, 1919A (1931); R. J. Van de Graaff, J. G. Trump, and W. W. Buechner, *Rep. Prog. Phys.* **11**, 1 (1948).

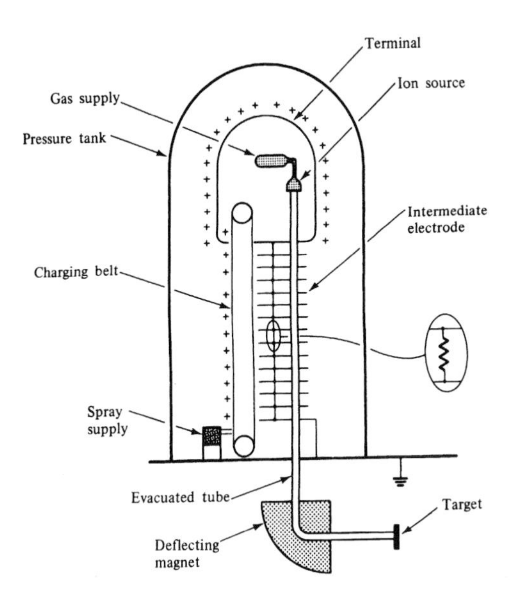

Figure 2.4: Schematic diagram of a Van de Graaff generator.

filled with an inert gas (N_2, CO_2, SF_6 at $\sim$ 15 atm are used) voltages of up to 20 MV can be obtained.

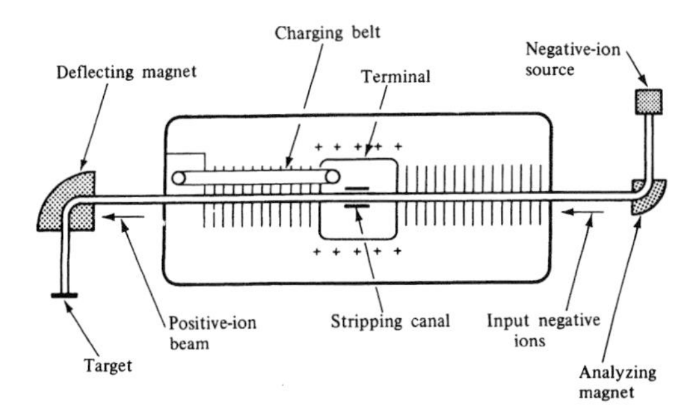

Figure 2.5: Tandem Van de Graaff. Negative ions are first accelerated to the central terminal. There they are stripped of their electrons and accelerated as positive ions to the target.

Twice the maximum voltage can be utilized in *tandem* machines, sketched in Fig. 2.5. Here, the terminal is in the middle of a long high-pressure tank; the ion source is at one end and it produces negative ions, for instance H^-. These ions are accelerated toward the central terminal where they are stripped of their two electrons by passage through a foil or a gas-containing canal. The positive ions now accelerate away from the terminal and again acquire energy. The total energy

Photo 1: The tandem Van de Graaff accelerator at the University of Washington, Seattle, WA.

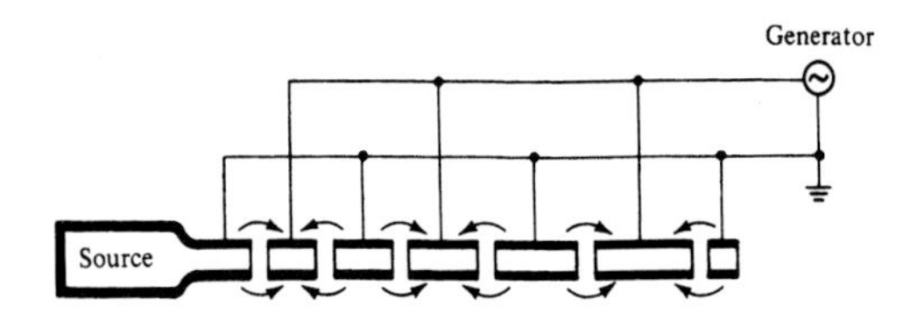

Figure 2.6: Drift tube linac. The arrows at the gaps indicate the direction of the electric field at a given time.

gain is therefore twice that of a single-stage machine. Photo 1 shows the Tandem accelerator at the University of Washington.

Van de Graaff generators in various energy and price ranges can be obtained commercially, and they are ubiquitous. They have a high beam intensity (up to 100 μA); this beam can be continuous and well collimated and the output energy is well stabilized (± 10 keV). Until the end of the last century, they were the workhorses of nuclear structure research and some are still in use. However, their present maximum energy is limited to about 30–40 MeV for protons, and they can therefore not be used in elementary particle research.

2.4 Linear Accelerators (Linacs)

To reach very high energies, particles must be accelerated many times over. Conceptually the simplest system is the linear accelerator,[3] sketched in Fig. 2.6.

[3]R. Wideröe, *Arch. Elektrotech.* **21**, 387 (1928); D. H. Sloan and E. O. Lawrence, *Phys. Rev.* **38**, 2021 (1931).

A series of cylindrical tubes are connected to a high-frequency oscillator. Successive tubes are arranged to have opposite polarity. The beam of particles is injected along the axis. Inside the cylinders the electric field is always zero; in the gaps it alternates with the generator frequency. Consider now a particle of charge e that crosses the first gap at a time when the accelerating field is at its maximum. The length L of the next cylinder is so chosen that the particle arrives at the next gap when the field has changed sign. It therefore again experiences the maximum accelerating voltage and has already gained an energy $2\ eV_0$. To achieve this feat, L must be equal to $\frac{1}{2}vT$, where v is the particle velocity and T the period of the oscillator. Since the velocity increases at each gap, the cylinder lengths must increase also. For electron linacs, the electron velocity soon approaches c and L tends to $\frac{1}{2}cT$. The drift-tube arrangement is not the only possible one; electromagnetic waves propagating inside cavities can also be used to accelerate the particles. In both cases large rf power sources are required for the acceleration, and enormous technical problems had to be solved before linacs became useful machines.

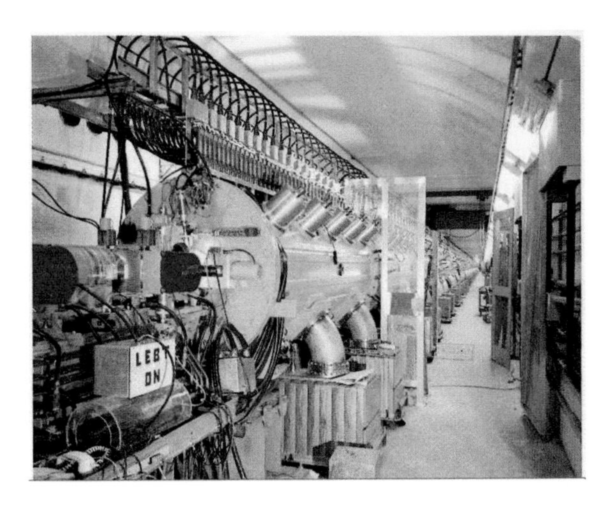

Photo 2: A view of the linac at RHIC. Its purpose is to provide currents of up to 35 milliamperes of protons at energies $\approx$ 200 MeV for injection in a synchrotron for further acceleration. The basic components of the linac include a radiofrequency quadropole pre-injector, and nine accelerator radiofrequency cavities spanning the length of a 150 meters tunnel (shown above.) [Courtesy of Brookhaven National Lab.]

At present, Stanford has an *electron linac* that is 3 km ("2 miles") long and produces electrons of 50 GeV energy. A *proton linac* of 800 MeV energy with a beam current of 1 mA, a so-called meson factory, was constructed at Los Alamos. It is now primarily used to bombard targets made of neutron-rich elements and produce neutrons that are subsequently used to study properties of materials. The Relativistic Heavy Ion Collider (RHIC) at Brookhaven has as one of its components a linac (see Photo 2) and the planned Rare Isotope Accelerator (RIA) will produce large numbers of rare isotopes by bombarding a variety of targets with beams of stable ions accelerated with a linac.

2.5 Beam Optics

In the description of linacs we have swept many problems under the rug, and we shall leave most of them there. However, one question must occur to anyone thinking about a machine that is a few km in length: How can the beam be kept well collimated? The beam of a flashlight, for instance, diverges, but it can be refocused with lenses. Do lenses for charged particle beams exist? Indeed they do, and we shall discuss here some of the elementary considerations, using the analogy to ordinary optical lenses. In light optics, the path of a monochromatic light ray through a system of thin lenses and prisms can be found easily by using geometrical optics.[4] Consider, for instance, the combination of a positive and a negative thin lens, with equal focal lengths f and separated by a distance d (Fig. 2.7). This combination is *always* focusing, with an overall focal length given by

$$f_{\text{comb}} = \frac{f^2}{d}. \tag{2.20}$$

In principle one could use electric or magnetic lenses for the guidance of charged particle beams. The electric field strength required for the effective focusing of high-energy particles is, however, impossibly high, and only magnetic elements are used.

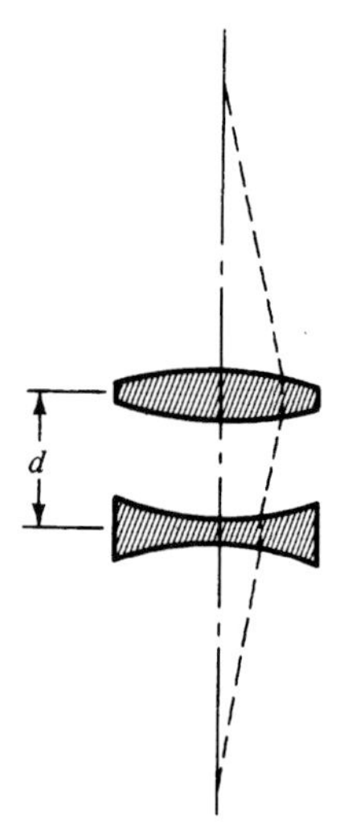

Figure 2.7: The combination of a focusing and a defocusing thin lens with equal focal lengths is always focusing.

The deflection of a monochromatic (monoenergetic) beam by a desired angle, or the selection of a beam of desired momentum, is performed with a *dipole magnet*, as shown in Fig. 2.8. The radius of curvature, ρ, can be computed from the *Lorentz equation*,[5] which gives the force $\boldsymbol{F}$ exerted on a particle with charge q and velocity $\boldsymbol{v}$ in an electric field $\boldsymbol{E}$ and a magnetic field $\boldsymbol{B}$:

$$\boldsymbol{F} = q\left(\boldsymbol{E} + \frac{1}{c}\boldsymbol{v} \times \boldsymbol{B}\right). \tag{2.21}$$

[4] See, for instance, E. Hecht, *Optics*, 4th. Ed., Addison-Wesley, San Francisco, 2002.
[5] Jackson, Eq. (6.113).

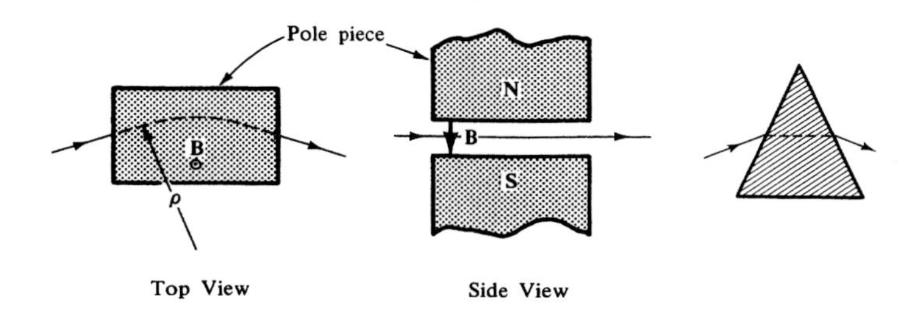

Figure 2.8: Rectangular dipole magnet. The optical analog is a prism, shown at the right.

The force is normal to the trajectory. For the normal component of the force, Newton's law, $\boldsymbol{F} = d\boldsymbol{p}/dt$, and Eq. (1.7) give

$$F_n = \frac{pv}{\rho}, \tag{2.22}$$

so that with Eq. (2.21) the radius of curvature becomes [6]

$$\rho = \frac{pc}{|q|\boldsymbol{B}}. \tag{2.15}$$

Problems arise when a beam should be focused. Figure 2.8 makes it clear that an ordinary (dipole) magnet bends particles only in one plane and that focusing can be achieved only in this plane. No magnetic lens with properties analogous to that of an optical focusing lens can be designed, and this fact stymied physicists for many years. A solution was finally found in 1950 by Christofilos and independently by Courant, Livingston, and Snyder in 1952.[7] The basic idea of the so-called strong focusing can be explained simply by referring to Fig. 2.7: If focusing and defocusing elements of equal focal lengths are alternated, a net focusing effect occurs. In beam transport systems, strong focusing is most often achieved with *quadrupole magnets.* A cross section through such a magnet is shown in Fig. 2.9. It consists of four poles; the field in the center vanishes and the magnitude of the field increases from

[6] Equation (2.15) is given in Gaussian units, where the unit for B is 1 G, and the unit of potential is 1 stat $V = 300$V. To compute ρ for a particle with unit charge ($|q| = e$), express pc in eV; then Eq. (2.15) yields

$$B(\text{Gauss}) \times \rho(\text{cm}) = \frac{V}{300}. \tag{2.23a}$$

As an example, consider an electron with a kinetic energy of 1 MeV; pc follows from Eq. (1.2) as

$$pc = (E_{\text{kin}}^2 + 2E_{\text{kin}}mc^2)^{1/2} = 1.42 \times 10^6\text{eV}.$$

V then is 1.42×10^6V and $B\rho = 4.7 \times 10^3$G cm. Equation (2.15) can also be rewritten in mks units, where the unit of B is

$$1 \text{ T (Tesla)} = 1 \text{ Wb (Weber)m}^{-2} = 10^4\text{G}.$$

[7] E. D. Courant, M. S. Livingston, and H. S. Snyder, *Phys. Rev.* **88**, 1190 (1952).

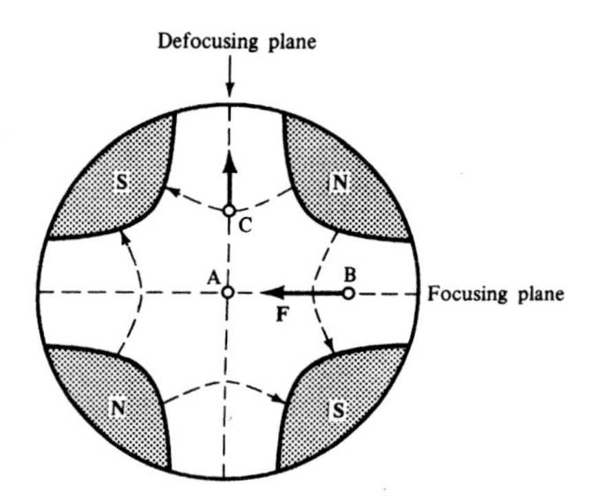

Figure 2.9: Cross section through a quadrupole magnet. Three positive particles enter the magnet parallel to the central symmetry axis at points A, B, and C. The particle at A is not deflected, B is pushed toward the center, and C is deflected outward.

the center in all directions. To understand the operation of a quadrupole magnet, consider three positive particles, going into the magnet at the points denoted by A, B, and C. Particle A in the center is not deflected; the Lorentz force, Eq. (2.21), pushes particle B toward and particle C away from the central symmetry axis. The magnet therefore behaves as a focusing element in one plane and a defocusing element in the other plane. A combination of two quadrupole magnets focuses in both planes if the second magnet is rotated around the central axis by 90° with respect to the first one. Such *quadrupole doublets* form essential elements of all modern particle accelerators and also of the beam lines that lead from the machines to the experiments. With these focusing devices, a beam can be transported over distances of many km with small intensity loss.

2.6 Synchrotrons

Why do we need another accelerator type? The linac obviously can produce particles of arbitrary energy. However, consider the price: since the 50 GeV Stanford linac is already 3 km long, a 1-TeV accelerator would have to be about 60 km long with the same technology; construction and power costs would be enoermous. (Nevertheless a 1/2-1 TeV International Linear Collider with superconducting magnets is being planned.) It makes more sense to let the particles run around a smaller track repeatedly. The first circular accelerator, the *cyclotron*, was proposed by Lawrence in 1930.[8] Cyclotrons have been of enormous importance in the development of subatomic physics, and some very modern and sophisticated ones are currently in operation. We omit discussion of the cyclotron here because its cousin, the *synchrotron*, has many similar features and achieves higher energies.

[8] E. O. Lawrence and N. E. Edlefsen, *Science* **72**, 376 (1930); E. O. Lawrence and M. S. Livingston, *Phys. Rev.* **40**, 19 (1932).

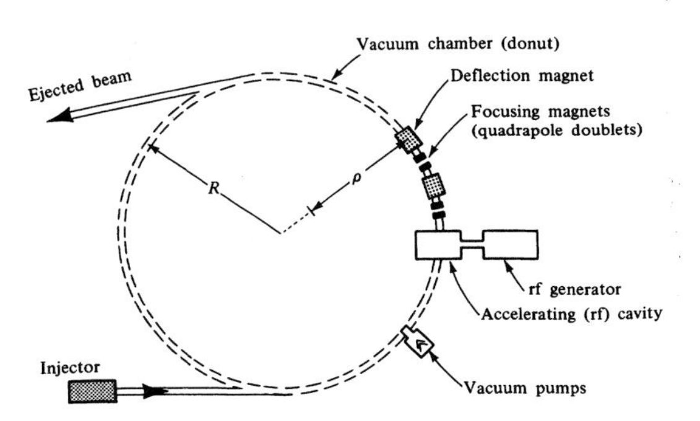

Figure 2.10: Essential elements of a synchrotron. Only a few of the repetitive elements are shown.

The synchrotron was proposed independently by McMillan and by Veksler in 1945.[(9)] Its essential elements are shown in Fig. 2.10. The injector sends particles of an initial energy E_i into the ring. Dipole magnets with a radius of magnetic curvature ρ bend the particles around the ring while quadrupole systems maintain the collimation. The particles are accelerated in a number of rf cavities which are supplied with a circular frequency ω. The actual path of the particles consists of straight segments in the accelerating cavities, the focusing elements, and some other elements and of circular segments in the bending magnets. The radius of the ring, R, is therefore larger than the radius of curvature, ρ.

Now consider the situation just after injection of the particles with energy E_i and momentum p_i, where energy and momentum are connected by Eq. (1.2). Assume that the rf power has not yet been turned on. The particles will then coast around the ring with a velocity v, and the time T for one full turn is given with Eq. (1.8) as

$$T = \frac{2\pi R}{v} = \frac{2\pi R E_i}{p_i c^2}. \tag{2.16}$$

The corresponding circular frequency, Ω, is

$$\Omega = \frac{2\pi}{T} = \frac{p_i c^2}{R E_i}, \tag{2.17}$$

and the magnetic field required to keep them on the track follows from Eq. (2.15) as

$$B = \frac{p_i c}{|q|\rho}. \tag{2.18}$$

Once the rf power is turned on, the situation changes. First, the radio frequency ω, must be an integer multiple, k, of Ω in order to always give the circulating

[9] E. M. McMillan, *Phys. Rev.* **68**, 143 (1945); V. Veksler, *J. Phys.* (*USSR*) **9**, 153 (1945).

particles the push at the right time. Equation (2.17) then shows that the applied rf must increase with increasing energy up to the point where the particles are fully relativistic so that $pc = E$. The magnetic field also must increase:

$$\omega = k\Omega = \frac{kc}{R}\frac{pc}{E} \longrightarrow \frac{kc}{R}; \quad B = \frac{pc}{|q|\rho}. \tag{2.19}$$

If these two conditions are satisfied, then the particles are properly accelerated. The procedure is as follows: A burst of particles of energy E_i is injected at the time $t = 0$. The magnetic field and the rf are then increased from their initial values B_i and ω_i to final values B_f and ω_f, always maintaining the relations (2.19). The energy of the bunch of particles is increased during this process from the injection energy E_i to the final energy E_f. The time required for bringing the particles up to the final energy depends on the size of the machine; for very big machines, a pulse per sec is about par.

Equation (2.19) shows another feature of these big accelerators: particles cannot be accelerated *from start* to the final energy in one ring. The range over which the rf and the magnetic field would have to vary is too big. The particles are therefore preaccelerated in smaller machines and then injected. Consider, for instance, the 1000 GeV synchrotron at FNAL: The enormous dimensions of the entire enterprise are evident from Photo 4.[10]

Synchrotrons can accelerate protons or electrons. Electron synchrotrons share one property with other circular electron accelerators: they are an intense source of short-wavelength light. The origin of *synchrotron radiation* can be explained on the basis of classical electrodynamics. Maxwell's equations predict that any accelerated charged particle radiates. A particle that is forced to remain in a circular orbit is continuously accelerated in the direction toward the center, and it emits electromagnetic radiation. The power radiated by a particle with charge e moving with velocity $v = \beta c$ on a circular path of radius R is given by[11]

$$P = \frac{2e^2c}{3R^2}\frac{\beta^4}{(1-\beta^2)^2}. \tag{2.20}$$

The velocity of a relativistic particle is close to c; with Eqs. (1.6) and (1.9) and with $\beta \approx 1$, Eq. (2.20) becomes

$$P \approx \frac{2e^2c}{3R^2}\gamma^4 = \frac{2e^2c}{3R^2}\left(\frac{E}{mc^2}\right)^4. \tag{2.21}$$

The time T for one revolution is given by Eq. (2.16), and the energy lost in one revolution is

[10] J. R. Sanford, *Annu. Rev. Nucl. Sci.* **26**, 151 (1976); H. T. Edwards, *Annu. Rev. Nucl. Part. Sci.* **35**; 605 (1985).

[11] Jackson, Eq. (14.31).

Photo 3: The photographs a-d show the essential parts of the 1 TeV proton synchrotron (Tevatron) at the Fermilab. Protons are accelerated to 750 keV in an electrostatic accelerator (Cockcroft–Walton, photo a); a linear accelerator (photo b) then brings the energy up to 400 MeV and injects the protons into a booster synchrotron. The booster synchrotron (photo c) raises the energy to about 8 GeV and the main ring (lower on photo d) to 150 GeV. The final energy of approximately 1 TeV is achieved in the Tevatron ring. [Courtesy Fermilab.]

Photo 4: Aerial photograph of Fermi National Accelerator Laboratory (FNAL), at Batavia, Illinois. The beam originates at the top left and is accelerated in the linac (visible as a straight line) to get into the main injector (the bottom ring) where it is brought to 150 GeV. The Tevatron is the top ring, approximately 2 kilometers in diameter, where the beam is accelerated to 1 TeV. (Courtesy Fermi National Accelerator Laboratory.)

$$-\delta E = PT \approx \frac{4\pi e^2}{3R}\left(\frac{E}{mc^2}\right)^4 . \tag{2.22}$$

The difference between the proton and electron synchrotron is obvious from Eq. (2.22). For equal radii and equal total energies E, the ratio of energy losses is

$$\frac{\delta E(e^-)}{\delta E(p)} = \left(\frac{m_p}{m_e}\right)^4 \approx 10^{13}. \tag{2.23}$$

The energy loss must be taken into account in the design of electron synchrotrons. Fortunately, the emitted radiation permits unique research in many other fields, from solid-state physics to surface science and biology.[(12)]

2.7 Laboratory and Center-of-Momentum Frames

Trying to achieve higher energies with ordinary accelerators is somewhat like trying to earn more money—you do not keep all you earn. In the second case, the tax collector takes an increasing bite, and in the first case, an increasing fraction of the total energy in a collision goes into center-of-mass motion and is not available for exciting internal degrees of freedom. To discuss this fact, we briefly describe the laboratory (lab) and center-of-momentum (c.m.) coordinates. Consider the following two-body reaction,

$$a + b \longrightarrow c + d, \tag{2.24}$$

and call a the projectile and b the target particle. In the *laboratory frame*, the target is at rest and the projectile strikes it with an energy E^{lab} and a momentum $\boldsymbol{p}^{\text{lab}}$. After the collision both particles in the final state, c and d, are usually moving. In the center-of-mass frame or, more correctly, the *center-of-momentum frame*, both particles approach each other with equal but opposite momenta. The two frames are defined by

$$\text{lab frame} : \boldsymbol{p}_b^{\text{lab}} = 0, \quad E_b^{\text{lab}} = m_b c^2 \tag{2.25}$$

$$\text{c.m. frame} : \boldsymbol{p}_a^{\text{c.m.}} + \boldsymbol{p}_b^{\text{c.m.}} = 0. \tag{2.26}$$

It is only the energy of one particle relative to the other one that is available for producing particles or for exciting internal degrees of freedom. The uniform motion of the center of momentum of the whole system is irrelevant. The energies and momenta in the c.m. system are thus the important ones.

[12] *Synchrotron Radiation Research*, H. Winick and S. Doniach, eds., Plenum, New York (1980); *Neutron and Synchrotron Radiation for Condensed Matter Studies*, Vols. I and II, ed. J. Baruchel et al., Springer Verlag, New York, 1993; H. Wiedemann, *Synchrotron Radiation*, Springer , New York, 2003.

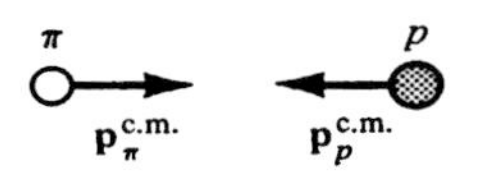

Before collision

After collision

Figure 2.11: Production of a new particle, N^*, in a collision $\pi p \longrightarrow \pi N^*$, seen in the c.m. frame.

A simple example can provide an understanding of how much one is robbed in the laboratory system. New particles can, for instance, be produced by bombarding protons with pions,

$$\pi p \longrightarrow \pi N^*,$$

where N^* is a particle of high mass ($m_{N^*} > m_p \gg m_\pi$). In the c.m. frame, the pion and proton collide with opposite momenta; the total momentum in the initial and hence also in the final state is zero. The highest mass can be reached if the pion and the N^* in the final state are produced at rest because then no energy is wasted to produce motion.

This collision in the c.m. frame is shown in Fig. 2.11. The total energy in the final state is

$$W^{\text{c.m.}} = (m_\pi + m_{N^*})c^2 \approx m_{N^*}c^2. \tag{2.27}$$

The total energy is conserved in the collision so that

$$W^{\text{c.m.}} = E_\pi^{\text{c.m.}} + E_p^{\text{c.m.}}. \tag{2.28}$$

The pion energy, E_π^{lab}, required in the laboratory system to produce the N^*, can be computed by using the Lorentz transformation. Here we make use of the relativistic invariance of $W^{c.m.}$ that was introduced in Chapter 1, Eq.(1.12). Consider a system of i particles with energies E_i and momenta $\boldsymbol{p}_i$. In a derivation similar to the one that leads to Eq. (1.12) it is possible to show that one can write

$$\left(\sum_i E_i\right)^2 - \left(\sum_i \boldsymbol{p}_i\right)^2 c^2 = M^2c^4. \tag{2.29}$$

where M is called the *total mass* or *invariant mass* of the system of i particles; it is equal to the sum of the rest masses of the i particles only if they are all at rest in their common c.m. frame. The right-hand side (RHS) is a constant and must therefore be the same in *all* coordinate systems. It then follows that the left-hand side (LHS) is also a relativistic invariant (sometimes called a relativistic scalar) that has the same value in all coordinate systems. We apply this invariance to the collision equation (2.24) as seen in the c.m. and the lab systems,

$$\begin{aligned}(E_a^{\text{c.m.}} + E_b^{\text{c.m.}})^2 &- (\boldsymbol{p}_a^{\text{c.m.}} + \boldsymbol{p}_b^{\text{c.m.}})^2c^2 \\ &= (E_a^{\text{lab}} + E_b^{\text{lab}})^2 - (\boldsymbol{p}_a^{\text{lab}} + \boldsymbol{p}_b^{\text{lab}})^2c^2,\end{aligned} \tag{2.30}$$

or with Eqs. (2.25) and (2.26),

$$\begin{aligned} W^2 &= (E_a^{\text{c.m.}} + E_b^{\text{c.m.}})^2 = (E_a^{\text{lab}} + m_b c^2)^2 - (\boldsymbol{p}_a^{\text{lab}} c)^2 \\ &= 2E_a^{\text{lab}} m_b c^2 + (m_a^2 + m_b^2)c^4. \end{aligned} \tag{2.31}$$

Equation (2.31) connects W^2, the square of the total c.m. energy, to the laboratory energy. With $E_a^{\text{lab}} \gg m_a c^2, m_b c^2$, the energy W becomes

$$W \approx (2E_a^{\text{lab}} m_b c^2)^{1/2}. \tag{2.32}$$

Only the energy available in the c.m. frame is useful for producing new particles or exploring internal structure. Equation (2.32) shows that this energy, W, increases only as the square root of the laboratory energy at *high* energies.

2.8 Colliding Beams

The price for working in the laboratory system is high, as is stated plainly by Eq. (2.32). If the machine energy is increased by a factor of 100, the effective gain is only a factor of 10. In 1956, Kerst and his colleagues and O'Neill therefore suggested the use of colliding beams to attain higher energies.(13)

Two proton beams of 21.6 GeV colliding head-on would be equivalent to one 1 TeV accelerator with a fixed target. The main technical obstacle is intensity; both beams must be much more intense than the ones available in normal accelerators in order to produce sufficient events in the regions where they collide.

The solution to this problem came in part from progress in vacuum technology, and in *beam storage and cooling*, techniques that are described further below. As an example, Fig. 2.12 shows the colliding beam arrangement at CERN, where an electron–positron collider (LEP) of 2×50 GeV was completed in 1989 and ran until 2000, and where the next Large Hadron Collider will soon start running. At DESY in Hamburg, the HERA electron-proton collider was constructed in the 1990's. Electrons are accelerated to 28 GeV, protons to 820 GeV in the same tunnel, with the proton accelerator on top of the electron one. The proton accelerator uses superconducting magnets with coils cooled to liquid helium temperatures, whereas the electron ring uses normal magnets.

2.9 Superconducting Linacs

A limiting factor in obtaining beams at the highest energies is the maximum attainable strength of the magnetic fields. Consider a circular accelerator. Equations (1.3) and (2.18) imply that, for a given radius of curvature in a magnet, the particle energy E is proportional to the magnetic field B. In an iron magnet, the field can be

[13] D. W. Kerst et al., *Phys. Rev.* **102**, 590 (1956); G. K. O'Neill, *Phys. Rev.* **102**, 1418 (1956).

of the order of 20 kG (or 2 Tesla ≡ 2 T), and it becomes expensive to exceed this value. (In recent years, power costs have largely determined the fraction of time during which large accelerators are used.)

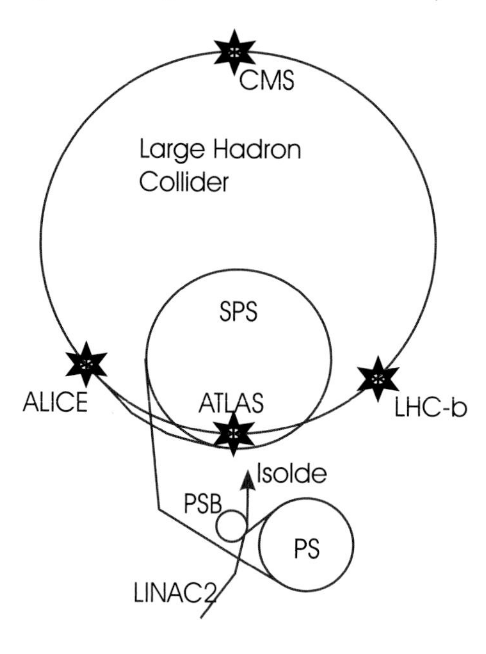

Figure 2.12: Sketch (not to scale) of parts of the complex accelerator system at CERN. The proton beams are initially accelerated by a LINAC, then futher accelerated in the proton synchrotron (SPS) and injected into the LHC loop. The stars indicate beam-collision points. Isolde is a fixed-target experiment used to produce radioactive beams for studies of nuclear astrophysics among other things.

Superconducting magnets can yield fields up to about 85 kG (8.5 T) and use less energy. Despite the enormous technical difficulties, "superconducting" accelerators can be built. Examples are the Main Injector and Tevatron at Fermilab in Batavia, the Large Electron-Positron collider and its successor, the Large Hadron Collider, at CERN in Switzerland, the electron linear accelerator at Thomas Jefferson Laboratory in Virginia, and the Relativistic Heavy Ion Collider in Long Island.

Fig. 2.13 shows that the available energy has grown exponentially in time and the progress in the last decades was been greatly helped by the use of superconducting magnets.

2.10 Beam Storage and Cooling

In addition to reaching higher energies accelerators for finding new physical processes need higher beam intensities as well.(14)

One technique that made possible for colliding beam experiments to reach the necessary intensities was *beam cooling.* This has been crucial for particle-anti-particle colliders, like the e^+e^- collider at CERN and the $\overline{p}p$ collider at Fermi

[14] N. Dikansky and D. Pestrikor, *The Physics of Intense Beams and Storage Rings*, Am. Inst. Phys. New York (1993); P.J. Bryant and K. Johnson, *The Principles of Circular Accelerators and Storage Rings*, Cambridge Univ. Press, Cambridge, 1993. An up-to-date and complete guide can be found in A.W. Chao and M. Tigner, *Handbook of Acceleartor Physics and Engineering*, World Sci., Singapore, 1999.

lab. The antiparticles (of which we will learn more in Chapters 5 and 7) are generated with initial collisions in an accelerator. For a $p\bar{p}$ collider,[15], for example, in order to get sufficient $p\bar{p}$ collisions, the total number of antiprotons circulating in the ring must be larger than 10^{11}. This is achieved at Fermilab by bombarding a Ni target with protons with energies of 120 GeV from the main injector. However, the number of antiprotons that can be produced and accelerated onto a beam per unit time is approximately 10^8 per sec. Antiprotons must consequently be accumulated and stored for approximately 10^3 sec. Since the antiprotons are produced by high energy collisions, they also have considerable random motions in various directions, or in other words, the antiproton beam has considerable temperature and entropy.

The beam can only be stored efficiently if it is focused, so as to have a small diameter and a small momentum spread. To reach such a state, the beam must be "cooled." In order to cool a "hot" system, it is brought into contact with a system of low temperature and entropy. For a hot antiproton beam, cooling can be achieved through contact with a colder electron beam.[16] The antiprotons are first confined in a storage ring of very large aperture. Electrons are passed through a straight section of the ring so that they move parallel to the average path of the antiprotons with the same average speed. The electrons have a much lower temperature and through collisions carry off the randomly directed momentum components of the antiprotons. The hot antiproton gas transfers heat and entropy to the cold electron gas. At the end of the straight section, antiprotons and electrons are separated by a magnet; the electrons are removed but the antiprotons continue and are recirculated through the cooling section. Electron cooling was first proposed by Budker in 1966 and demonstrated in Novosibirsk in 1974. Another method is stochastic cooling, first suggested by van der Meer in 1972[17] and used at CERN for the high energy $p\bar{p}$ colliders. In stochastic cooling the temperature of the beam is lowered through a feedback mechanism. At Fermilab a combination of electron and stochastic cooling is employed. Cooling is also helpful at lower energy accelerators and was used, for instance, at CERN for low energy antiprotons at LEAR.

Although the main trend observed in Fig. 2.13 is reaching for ever higher energies, a high-intensity e^+-e^- collider with energies on the TeV range, could be extremely useful in finding new physics. Because the energy radiated by these particles moving on a circle would be too large the only possibility is to build a linac. This machine would be ≈ 30 km long and would consist of approximately 21,000 RF cavities, each providing an acceleration of ≈ 50 MeV.[18]

---

[15] M.D. Shapiro and J.L. Siegrist, *Annu. Rev. Nucl. Part. Sci.* **41**, 97 (1991); N. Ellis and T.S. Vira, *ibid*, **44**, 413 (1994).

[16] G. I. Budker, *Atomnaya Energiya* **22,** 346 (1967); *Part. Accel.* **7,** 197 (1976).

[17] S. van der Meer, CERN/ISR, P.O./72-31 (1972).

[18] *Future Colliders* by I. Hinchliffe and M. Battaglia, *Physics Today*, **57**, 49 (2004).

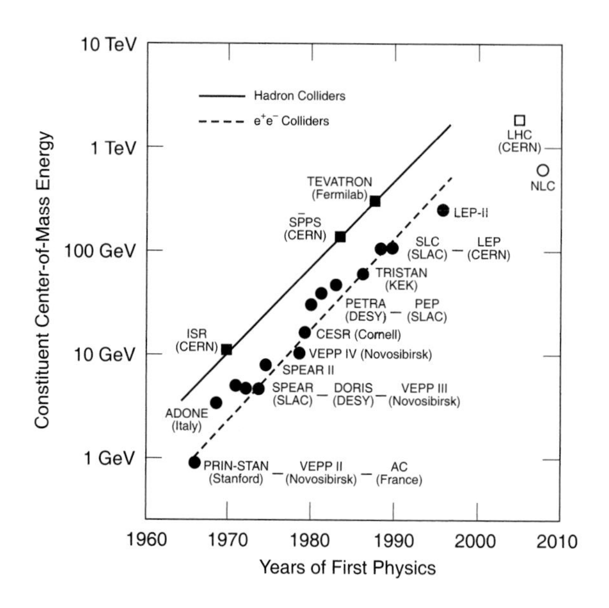

Figure 2.13: The energy in the center-of-mass frame of e^+-e^- and hadron colliders: filled circles and squares, constucted; open circle and square, planned. The energy in the hadron colliders has been reduced by factors of 6-10 because the incident proton energy is shared by its quark and gluon constituents. [From W. K. H. Panofsky and M. Breidenbach, *Rev. Mod. Phys.* **71**, S121 (1999).]

2.11 References

A resource letter with useful references is given in A.W. Chao, *Am. J. Phys.* **74**, 855 (2006).

There are a few introductions to modern particle accelerators; see for instance E.J.N. Wilson, *An Introduction to Particle Accelerators*, Oxford Univ. Press, Oxford, 2001. Other books at about the right level are M. Conte and W.M. MacKay, *An Introduction to the Physics of Particle Accelerators*, World Sci., Singapore, 1991; S.Y. Lee, *Accelerator Physics*, 2nd ed., World Sci., Singapore, 2004.

Beam cooling is reviewed in F.T. Cole and F.E. Mills, *Annu. Rev. Nucl. and Part. Sci.* **31,** 295 (1981).

Synchrotrons and accumulators for high-intensity proton beams and applications was recently reviewed by Jie Wei, *Rev. Mod. Phys.* **75**, 1383 (2003).

More details and concise stories of the development of accelerators can be found in M. S. Livingston and J. P. Blewett, *Particle Accelerators*, McGraw-Hill, New York (1962). Edwin McMillan wrote on "A History of the Synchrotron" in *Phys. Today* **37,** 31 (February 1984).

Relativistic kinematics, which we have only touched upon briefly, is treated in detail in R. Hagedorn, *Relativistic Kinematics.*, Benjamin, Reading, Mass. (1963), and in E. Byckling and K. Kajantie, *Particle Kinematics*, Wiley, New York (1973).

Beam optics is discussed in D.C. Carey, *The Optics of Charged Particle Beams*, Harwood, New York, 1987; J.D. Lawson, *The Physics of Charged Particle Beams*, 2nd. Edition, Clarendon Press, Oxford, 1988; M. Reiser, *Theory and Design of Charged Particle Beams*, J. Wiley, New York, 1994.

The RHIC is described in M. Harrison, S. Peggs, T. Roser, *Annu. Rev. Nucl. Part. Sci.* **52**, 425 (2002). A future possible linear collider is discussed in S. Dawson, M. Oreglia, *Annu. Rev. Nucl. Part. Sci.* **54**, 269 (2004).

Problems

2.1. An electron accelerator is to be designed to study properties of linear dimensions of 1 fm. What kinetic energy is required?

2.2. Estimate the capacity of a typical Van de Graaff terminal with respect to the ground (order of magnitude only). Assume that the terminal is to be charged to 1 MV. Compute the charge on the terminal. How long does it take to reach this voltage if the belt carries a current of 0.1 mA?

2.3. Consider a proton linac, working with a frequency of $f = 200$ MHz. How long must the drift tubes be at the point where the proton energy is

(a) 1 MeV?

(b) 100 MeV?

What is approximately the smallest energy with which the protons can be injected, and what determines the lower limit? Why does the frequency at the Los Alamos linac change from 200 to 800 MHz at a proton energy of about 200 MeV?

2.4. A proton beam of kinetic energy of 10 MeV enters a dipole magnet of 2 m length. It should be deflected by 10°. Compute the field that is necessary.

2.5. A proton beam of kinetic energy 200 GeV enters a 2 m long dipole magnet with a magnetic field of 20 kG. Compute the deflection of the beam.

2.6. The magnetic field that can be obtained in a superconducting magnet is about 50 kG. Assume an accelerator that follows the Earth's equator. What is the maximum energy to which protons can be accelerated in such a machine?

2.7. Use photo 4 and the data given in Section 2.6 to estimate over what range the frequency and the magnetic field must be changed in the main ring of the FNAL machine during one accelerating cycle.

2.8. Verify Eq. (2.29).

2.9. Assume collisions of protons from the accelerator described in Problem 2.6 with stationary protons. Compute the total energy, W, in GeV in the c.m. frame. Compare W with the corresponding quantity obtained in a colliding beam experiment, with each beam having a maximum energy E_0. How big must E_0 be in order to get the same W?

2.10. (a) Verify Eq. (2.20).

(b) Compute the energy loss per turn for a 10 GeV electron accelerator if the radius R is 100 m.

(c) Repeat part (b) for a radius of 1 km.

2.11. * Describe a typical ion source. What are the physical processes involved? How is one constructed?

2.12. In what way is a conventional cyclotron different from a synchrotron? What limits the maximum energy obtainable in a cyclotron? Why are high-energy accelerators predominantly synchrotrons?

2.13. What is meant by *phase stability*? Discuss this concept for linacs and for synchrotrons.

2.14. What is the duty cycle of an accelerator? Discuss the duty cycle for the Van de Graaff generator, the linac, and the synchrotron. Sketch the *beam structure*, i.e., the intensity of the ejected beam as a function of time for these three machines.

2.15. How is the beam ejected in a synchrotron?

2.16. How and why is superconductivity important in the field of accelerator physics?

2.17. Why is it expensive to build very-high-energy electron synchrotrons or very-high-energy proton linacs?

2.18. * Modern cyclotrons exist in various places, for instance, at the Paul Scherrer Institute (PSI) and at Michigan State University (superconducting cyclotron). Sketch the principles on which two of these cyclotrons are designed. In what way do they differ from the classic cyclotrons?

2.19. Discuss the direction of emission and the polarization of synchrotron radiation. Why is it useful in solid-state studies?

2.20. Compare the ratio of the appropriate (kinetic or total) c.m. energy to the laboratory energy for

(a) Nonrelativistic energies.

(b) Extreme relativistic energies.

2.21. Compare a typical colliding beam luminosity ($\sim 10^{34}$ particles per second) to that for a beam of protons of 1 μA colliding with a stationary liquid hydrogen target 30 cm long.

2.22. (a) Why is beam cooling important for $p\bar{p}$ colliders?

(b) Describe electron cooling.

(c) Describe stochastic cooling.

(d) * Describe the arrangement at Fermilab for beam cooling and $p\bar{p}$ collisions.

(e) Why can thin foils not be used for beam cooling?

2.23. Discuss heavy ion accelerators. What are the similarities and differences to proton accelerators? How are the heavy ions produced? List some of the ions that have been accelerated and give the maximum energies per nucleon.

2.24. Find the center-of-mass energy at HERA (see Section 2.7).

2.25. (a) An imaginary accelerator consists of colliding beams of electrons and protons, each of 2 TeV total energy. What laboratory energy would be required to achieve the same center-of-mass energy if electrons collide with stationary protons (hydrogen)?

(b) Repeat part (a) for an energy of 2 GeV instead of 2 TeV.

2.26. An electron beam of 10-GeV energy and a current of 10^{-8} A is focused onto an area of 0.5 cm^2. What is the flux F?

2.27. Assume that a beam pulse at a 100-GeV accelerator contains 10^{13} protons, is focused onto a 2 cm^2 area, and is extracted uniformly over a time of 0.5 sec. Compute the flux.

2.28. A copper target of thickness 0.1 cm intercepts a particle beam of 4 cm^2 area. Nuclear scattering is observed.

(a) Compute the number of scattering centers intercepted by the beam.

(b) Assume a total cross section of 10 mb for an interaction. What fraction of the incident beam is scattered?

2.29. Positive pions of kinetic energy of 190 MeV impinge on a 50 cm long liquid hydrogen target. What fraction of the pions undergoes pion–proton scattering? (See Fig. 5.35.)

2.30. Beams of electrons and protons, both traveling at almost the speed of light, collide. The electrons and protons are in bunches 2 cm in length in two rings of 300 m circumference, each of which contains one bunch. Each bunch contains 3×10^{11} particles, and the circulating frequency is $10^6/\text{sec}$ for each beam, so that 10^6 bunches collide with each other per second. Assume that the particles are distributed uniformly over a cross sectional area of 0.2 mm^2, and that this is also the area of the intersecting collision region.

(a) Determine the luminosity.

(b) If the cross section for collisions is 10 μb, determine the number of scattering events that would be observed in a counter totally surrounding the intersection region.

(c) Find the *average* flux of electrons.

(d) If the beam of electrons scatters from a stationary target of liquid hydrogen (density $\approx 0.1\ \text{g/cm}^3$) 2 cm long, rather than with the circulating proton beam, find the number of scattering events and compare to the answer of (b).

2.31. Experimenters A and B are trying to produce as much ^{47}Ca as possible using the $^{46}\text{Ca}(d,p)$ reaction. They have a limited amount of ^{46}Ca and a choice of two situations: a small-diameter beam and a thick, small-diameter target or a large-diameter beam with a thin, large-diameter target. The number of target atoms (the volume and the density) and beam current are identical, and the beam energy loss in the target is negligible for both situations. Experimenter A proposes to use the smaller-diameter beam because the number of incident particles per unit area and time (flux) is larger. Experimenter B argues that there should be no difference in the production of ^{47}Ca per unit time since the number of (^{46}Ca) target atoms exposed to the beam and the beam currents are identical. Who is correct and why?

Chapter 3

Passage of Radiation Through Matter

In everyday life we constantly use our understanding of the passage of matter through matter. We do not try to walk through a closed steel door, but we brush through if the passage is only barred by a curtain. We stroll through a meadow full of tall grass but carefully avoid a field of cacti. Difficulties arise if we do not realize the appropriate laws; for example, driving on the right-hand side of a road in England or Japan can lead to disaster. Similarly, a knowledge of the passage of radiation through matter is a crucial part in the design and the evaluation of experiments. The present understanding has not come without surprises and accidents. The early X-ray pioneers burned their hands and their bodies; many of the early cyclotron physicists had cataracts. It took many years before the exceedingly small interaction of the neutrino with matter was experimentally observed because it can pass through a light year of matter with only small attenuation. Then there was the old cosmotron beam at Brookhaven which was accidentally found a few km away from the accelerator, merrily traveling down Long Island.

The passage of charged particles and of photons through matter is governed primarily by *atomic* physics. True, some interactions with nuclei occur. However, the main energy loss and the main scattering effects come from the interaction with the atomic electrons. We shall therefore give few details and no theoretical derivations in the present chapter but shall summarize the important concepts and equations.

3.1 Concepts

Consider a well-collimated beam of monoenergetic particles passing through a slab of matter. The properties of the beam after passage depend on the nature of the particles and of the slab, and we first consider two extreme cases, both of great interest. In the first case, shown in Fig. 3.1(a), a particle undergoes many interactions. In each interaction, it loses a small amount of energy and suffers a small-angle scattering. In the second, shown in Fig. 3.1(b), the particle either passes unscathed through the slab or it is eliminated from the beam in one "deadly" encounter. The first case applies, for instance, to heavy charged particles, and the

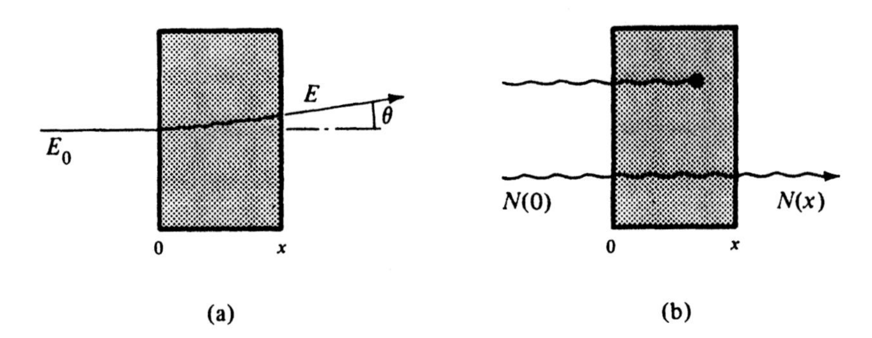

Figure 3.1: Passage of a well-collimated beam through a slab. In (a), each particle suffers many interactions; in (b), a particle is either unharmed or eliminated.

second one approximates the behavior of photons. (Electrons form an intermediate case.) We shall now discuss the two cases in more detail.

Many Small Interactions. Each interaction produces an energy loss and a deflection. Losses and deflections add up statistically. After passing through an absorber the beam will be degraded in energy, will no longer be monoenergetic, and will show an angular spread. Characteristics of the beam before and after passage are shown in Fig. 3.2. The number of particles left in the beam can be observed as a function of the absorber thickness x. Up to a certain thickness, essentially all particles will be transmitted. At some thickness, some of the particles will no longer emerge; at a thickness R_0, called the mean range, half of the particles will be stopped, and finally, at sufficiently large thickness, no particles will emerge. The behavior of the number of transmitted particles versus absorber thickness is shown in Fig. 3.3. The fluctuation in range is called range straggling.

"All-or-Nothing" Interactions. If an interaction eliminates the particle from the beam, the characteristics of the transmitted beam are different from the one just discussed. Since the transmitted particles have not undergone an interaction, the transmitted beam has the same energy and angular spread as the incident one. In each elementary slab of thickness dx the number of particles undergoing interactions is proportional to the number of incident particles, and the coefficient of proportionality is called the absorption coefficient μ:

$$dN = -N(x)\mu dx.$$

Integration gives

$$N(x) = N(0)e^{-\mu x}. \tag{3.1}$$

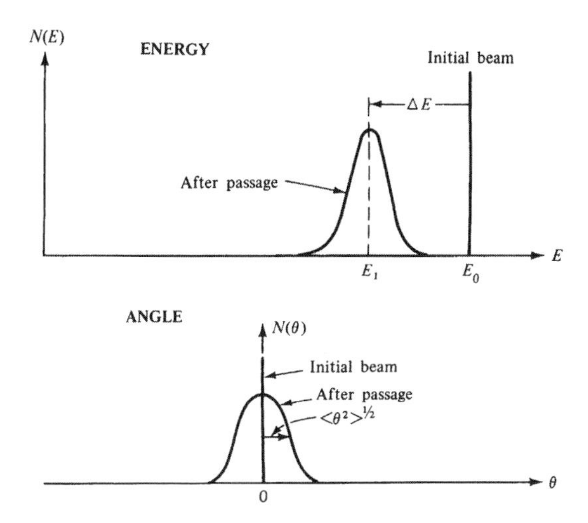

Figure 3.2: Energy and angular distribution of a beam of heavy charged particles before and after passing through an absorber.

The number of transmitted particles decreases exponentially, as indicated in Fig. 3.4. No range can be defined, but the *average distance* traveled by a particle before undergoing a collision is called the *mean free path*, and it is equal to $1/\mu$.

3.2 Heavy Charged Particles

Heavy charged particles lose energy mainly through collisions with bound electrons via Coulomb interactions. The electrons can be lifted to higher discrete energy levels (excitation), or they can be ejected from the atom (ionization). Ionization dominates if the particle has an energy large compared to atomic binding energies. The rate of energy loss due to collisions with electrons has been calculated classically by Bohr and quantum mechanically by Bethe and by Bloch.[1] The result, called the Bethe equation, is

$$-\frac{dE}{dx} = \frac{4\pi n z^2 Z^2 e^4}{m_e v^2}\left[\ln\frac{2m_e v^2}{I[1-(v/c)^2]} - \left(\frac{v}{c}\right)^2\right]. \tag{3.2}$$

Here $-dE$ is the energy lost in a distance dx, n the number of electrons per cm^3 in the stopping substance and Z its atomic number; m_e the electron mass; ze the charge and v the speed of the particle and I is the mean excitation potential of the atoms of the stopping substance. (Eq. (3.2) is an approximation, but it suffices for our purpose.)

[1] N. Bohr, *Phil. Mag.* **25**, 10 (1913); H. A. Bethe, *Ann. Physik* **5**, 325 (1930); F. Bloch, *Ann. Physik* **16**, 285 (1933).

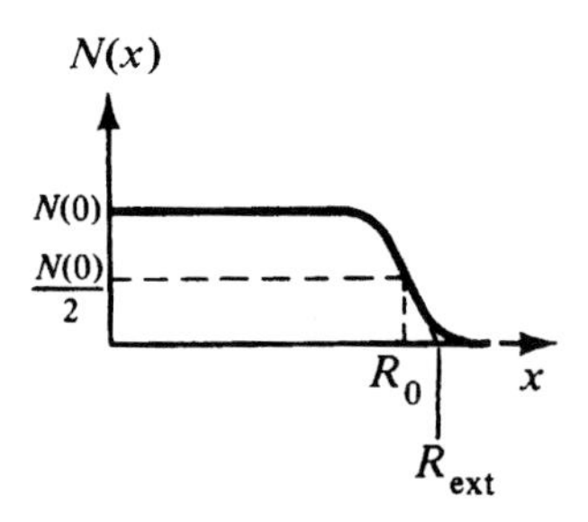

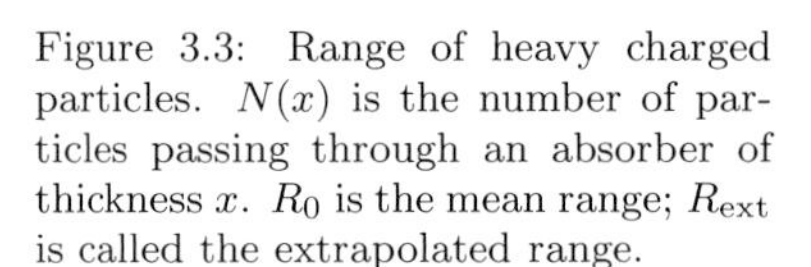
Figure 3.3: Range of heavy charged particles. $N(x)$ is the number of particles passing through an absorber of thickness x. R_0 is the mean range; R_{ext} is called the extrapolated range.

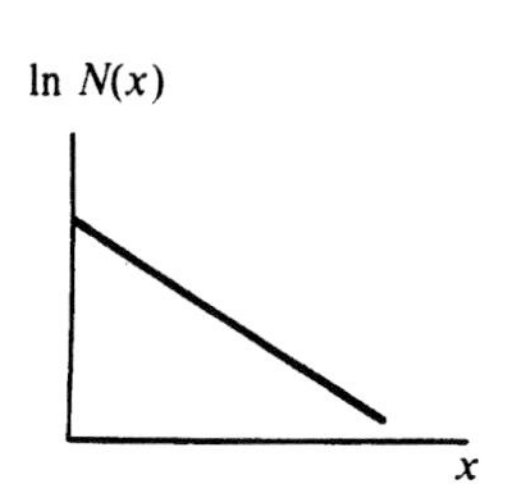

Figure 3.4: In all-or-nothing interactions, the number of transmitted particles, $N(x)$, decreases exponentially with the absorber thickness x.

In practical applications, the thickness of an absorber is not measured in length units but in terms of ρx, where ρ is the density of the absorber. ρx is usually given in g/cm^2, and it can be found experimentally by determining the mass and the area of the absorber and taking the ratio of the two. The specific energy loss tabulated or plotted is then

$$\frac{dE}{d(\rho x)} = \frac{1}{\rho}\frac{dE}{dx}.$$

Figure 3.5 gives the specific energy loss of protons, pions, and muons in several materials as a function of the momentum p. Figure 3.5 and Eq. (3.2) show the salient features of the energy loss of heavy particles in matter clearly. The specific energy loss is proportional to the number of electrons in the absorber and proportional to the *square* of the particle charge. At a certain energy, for protons about 1 GeV, an *ionization minimum* occurs. Below the minimum, $dE/d(\rho x)$ is proportional to $1/v^2$. Consequently, as a nonrelativistic particle slows down in matter, its energy loss increases. However, Eq. (3.2) breaks down when the particle speed becomes comparable to, or less than, the speed of the electrons in the atoms. The energy loss then decreases again, and the curves in Fig. 3.5 turn down below about 1 MeV. Above the ionization minimum, $dE/d(\rho x)$ increases slowly. It is often useful to remember that the energy loss at the minimum and for at least two decades above is about the same for all materials and that it is of the order

$$-\frac{dE}{d(\rho x)}(\text{at minimum}) \approx 1.6z^2 \text{ MeV/g cm}^{-2}. \tag{3.3}$$

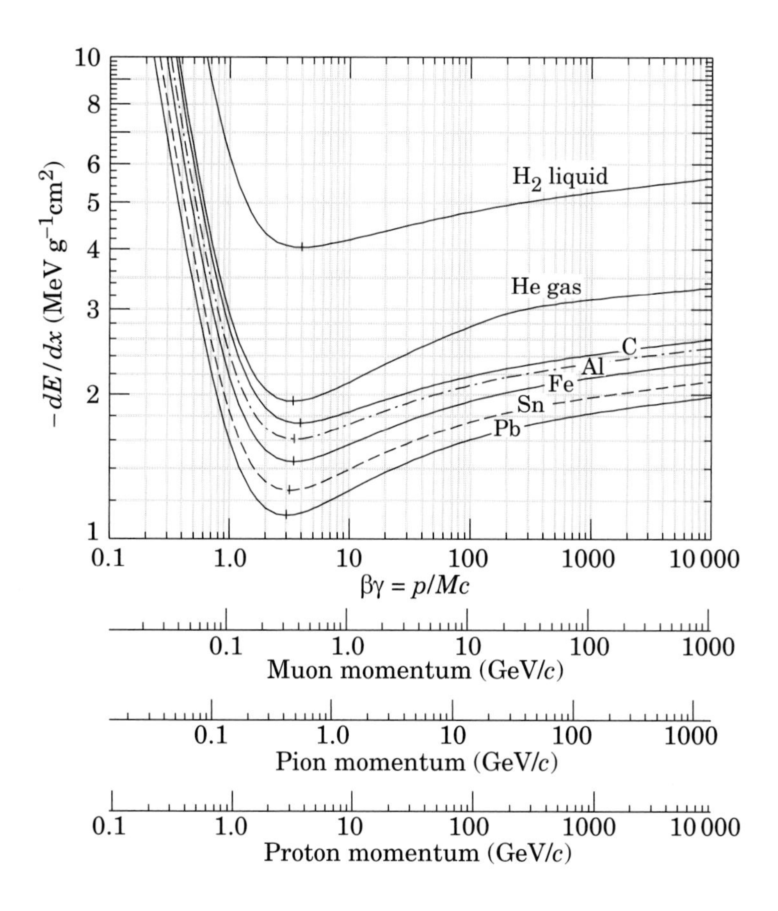

Figure 3.5: Specific energy loss, $dE/d(\rho x)$, for protons, pions, and muons in several materials.[From PDG.]

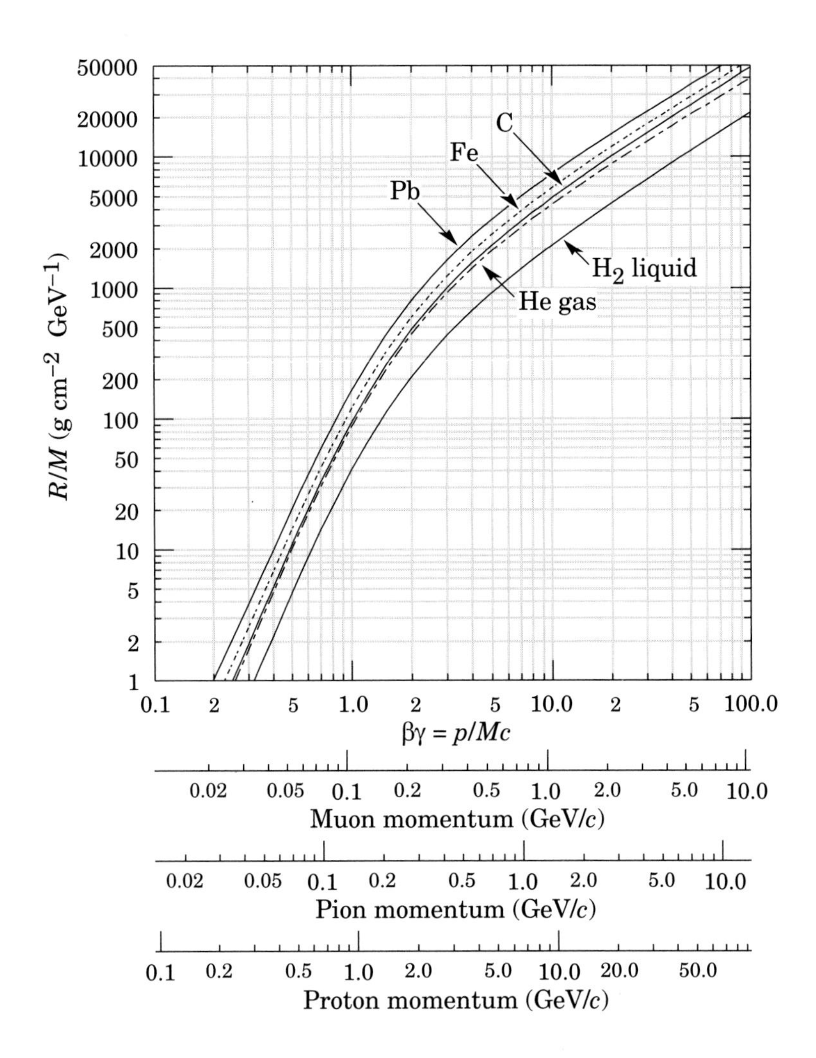

Figure 3.6: Range of particles in liquid hydrogen (bubble chamber), helium gas, carbon, iron, and lead. For example, for a pion of momentum 230 MeV/c, $\beta\gamma = 1.4$. For lead we read $R/M \approx$ 400 g cm^{-2} GeV^{-1}, and so the range is $\approx$ 56 g cm^{-2}. [From PDG.]

Equation (3.2) also shows that the specific energy loss does not depend on the mass of the particle (provided it is much heavier than the electron) but only on its charge and speed. The curves in Fig. 3.5 therefore are valid also for particles other than the protons if the energy scale is appropriately shifted.

The *range* of a particle in a given substance is obtained from Eq. (3.2) by integration:

$$R = \int_{T_0}^{0} \frac{dT}{(dT/dx)}. \tag{3.4}$$

Here T is the kinetic energy and the subscript 0 refers to the initial value. Some useful information concerning range and specific energy loss is summarized in Fig. 3.6.

Two more quantities shown in Fig. 3.2, the spread in energy and the spread in angle, are important in experiments, but they are not essential for a first view of the subatomic world. We shall therefore not discuss them here; the relevant information can be found in the references given in Section 3.6.

3.3 Photons

Photons interact with matter chiefly by three processes:

1. Photoelectric effect.
2. Compton effect.
3. Pair production.

A complete treatment of the three processes is rather complicated and requires the tools of quantum electrodynamics. The essential facts, however, are simple. In the photoelectric effect, the photon is absorbed by an atom, and an electron from one of the shells is ejected. In the Compton effect, the photon scatters from an atomic electron. In pair production, the photon is converted into an electron–positron pair. This process is impossible in free space because energy and momentum cannot be conserved simultaneously when a photon decays into two massive particles. It occurs in the Coulomb field of a nucleus which is needed to balance energy and momentum.

The energy dependences of processes 1–3 are very different. At low energies, below tens of keV, the photoelectric effect dominates (which accounts for the sharp edges), the Compton effect is small, and pair production is energetically impossible. At an energy of $2m_ec^2$, pair production becomes possible, and it soon dominates completely. Two of the three processes, photoelectric effect and pair production, eliminate the photons undergoing interaction. In Compton scattering, the scattered photon is degraded in energy. The all-or-nothing situation described in Section 3.1 and depicted in Fig. 3.1(b) is therefore a good approximation, and the transmitted beam should show an exponential behavior, as described by Eq. (3.1). The absorption coefficient μ is a sum of three terms,

$$\mu = \mu_{\text{photo}} + \mu_{\text{Compton}} + \mu_{\text{pair}} \tag{3.5}$$

and each term can be computed accurately.

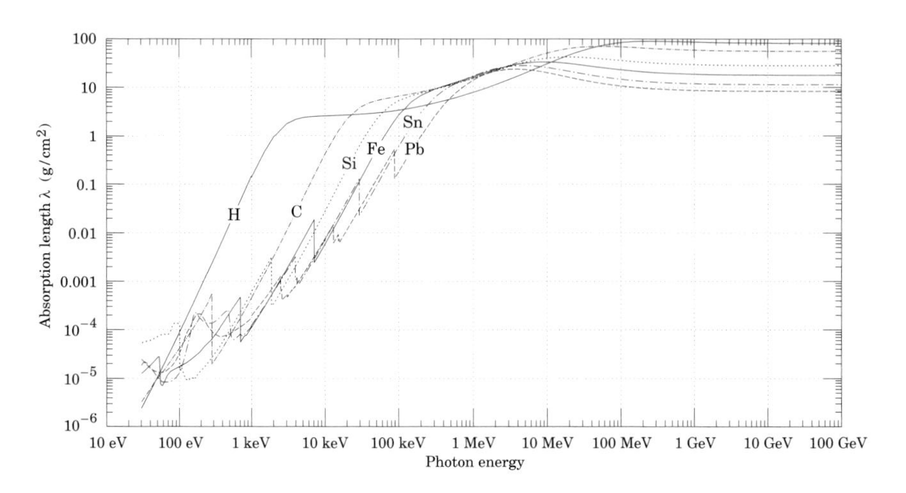

Figure 3.7: Mean free path ($\lambda = \rho/\mu$) versus photon energy. [From PDG.]

3.4 Electrons

The energy-loss mechanism of electrons differs from that of heavier charged particles for several reasons. The most important difference is energy loss by radiation; this mechanism is unimportant for heavy particles but dominant for high-energy electrons. Radiation makes it necessary to consider two energy regions separately. At energies well below the *critical energy* E_c, given approximately by

$$E_c \approx \frac{600 \text{ MeV}}{Z}, \tag{3.6}$$

excitation and ionization of the bound absorber electrons dominate. [In Eq. (3.6), Z is the charge number of the absorber's atoms.] Above the critical energy, radiation loss takes over. We shall treat the two regions separately.

Ionization Region ($E < E_c$) In this region, the energy loss of an electron and a proton of equal speed are nearly the same and Eq. (3.2) can be taken over with some small modifications. There is, however, one major difference, as sketched in Fig. 3.8. The path of the heavy particle is straight and the $N(x)$ against x curve is as given in Fig. 3.3. The electron, owing to its small mass, suffers many scatterings with considerable angles. The behavior of the number of transmitted electrons versus absorber thickness is sketched in Fig. 3.8. An extrapolated range R_p is defined as shown in Fig. 3.8. Between about 0.6 and 12 MeV the extrapolated range in aluminum is well represented by the linear relation

$$R_p(\text{in g/cm}^2) = 0.526 E_{\text{kin}}(\text{in MeV}) - 0.094. \tag{3.7}$$

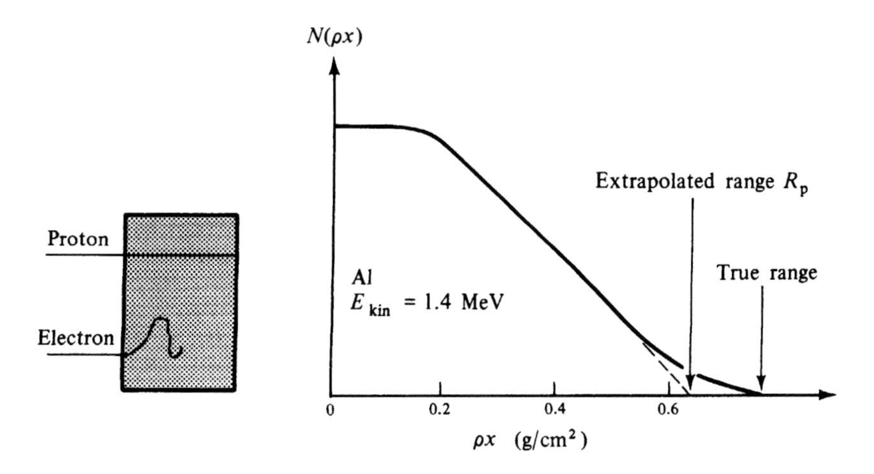

Figure 3.8: Passage of a proton and an electron with equal total pathlength through an absorber. The $N(x)$ against x behavior for electrons is given at right.

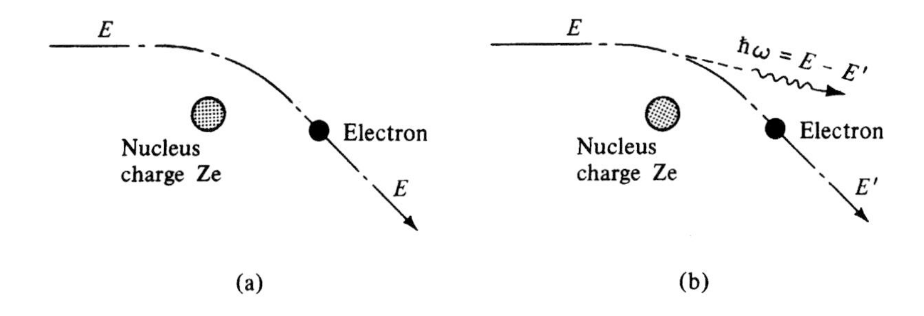

Figure 3.9: Coulomb scattering. (a) Elastic scattering. (b) The accelerated electron radiates and loses energy in the form of a photon (Bremsstrahlung).

Radiation Region ($E > E_c$) A charged particle passing by a nucleus of charge Ze experiences the Coulomb force and it is deflected (Fig. 3.9(a)). The process is called *Coulomb scattering.* The deflection accelerates (decelerates) the passing particle. As pointed out in Section 2.6, acceleration produces radiation. In the case of electrons in a synchrotron, it is called *synchrotron radiation;* in the case of charged particles scattered in the Coulomb field of nuclei, it is called *Bremsstrahlung* (braking radiation). Equations (2.21) and (2.22) show that, for equal acceleration, the energy carried away by photons will be proportional to $(E/mc^2)^4$. Bremsstrahlung is thus an important energy-loss mechanism for electrons, but it is very small for heavier particles, such as muons, pions, and protons.

Table 3.1: VALUES OF THE CRITICAL ENERGY E_c AND THE RADIATION LENGTH X_0 FOR VARIOUS SUBSTANCES.

Material	Z	Density (g/cm^3)	Critical Energy (MeV)	Radiation Length (g/cm^2)	Radiation Length (cm)
H_2 (liquid)	1	0.071	340	62.8	887
He (liquid)	2	0.125	220	93.1	745
C	6	1.5	103	43.3	28
Al	13	2.70	47	24.3	9.00
Fe	26	7.87	24	13.9	1.77
Pb	82	11.35	6.9	6.4	0.56
Air		0.0012	83	37.2	30870
Water		1	93	36.4	36.4

Actually, Eq. (2.21) has been calculated by using classical electrodynamics. Bremsstrahlung, however, must be treated quantum mechanically. Bethe and Heitler have done so, and the essential results are as follows.[2] The number of photons with energies between $\hbar\omega$ and $\hbar(\omega + d\omega)$ produced by an electron of energy E in the field of a nucleus with charge Ze is proportional to Z^2/ω:

$$N(\omega)d\omega \propto Z^2 \frac{d\omega}{\omega}. \tag{3.8}$$

Owing to the emission of these photons, the electron loses energy, and the distance over which its energy is reduced by a factor e is called the *radiation* or *attenuation length* and conventionally denoted by X_0. In terms of X_0, the radiative energy loss for large electron energies is

$$-\left(\frac{dE}{dx}\right)_{\text{rad}} \approx \frac{E}{X_0} \quad \text{or} \quad E = E_0 e^{-x/X_0}. \tag{3.9}$$

The radiation length is given either in g/cm^2 or in cm; a few values of X_0 and of the critical energy E_c are given in Table 3.1.

According to Eq. (3.9), a highly energetic electron loses its energy exponentially and after about seven radiation lengths has only 10^{-3} of its initial energy left. However, concentrating on the primary electron is misleading. Many of the Bremsstrahlung photons have energies greatly in excess of 1 MeV and can produce electron-positron pairs (Section 3.3). In fact, the mean free path, that is, the average distance, X_p, traveled by a photon before it produces a pair, is also related to the radiation length:

$$X_p = \frac{9}{7} X_0. \tag{3.10}$$

[2] H. A. Bethe and W. Heitler, *Proc. R. Soc. (London)* **A146**, 83 (1934).

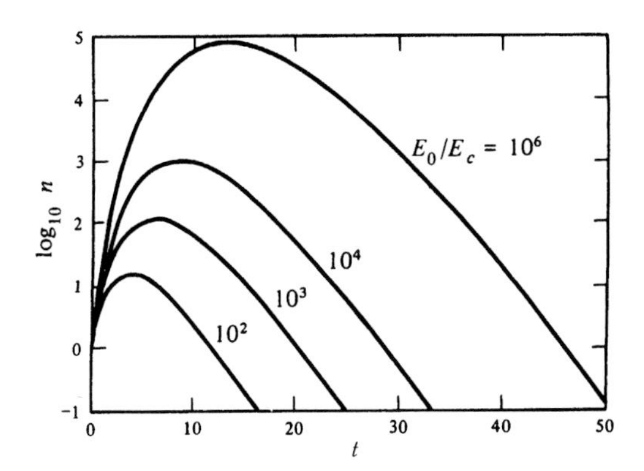

Figure 3.10: Number n of electrons in a shower as a function of the thickness traversed, t, in radiation lengths. [These curves were taken from the work of B. Rossi and K. Greisen, *Rev. Modern Phys.* **13**, 240 (1941).]

In successive steps, a high-energy electron creates a *shower*. (Of course a shower can also be initiated by a photon.) The detailed theory of such a shower is very complicated and in practice computer calculations are performed. Figure 3.10 shows the number n of electrons in a shower as a function of the thickness of the absorber. The energy E_0 of the incident electron is measured in units of the critical energy; the thickness is expressed in units of the radiation length X_0. Figure 3.10 expresses the development and death of a shower: The increase in the number of electrons is very rapid at the beginning. As the cascade progresses, the average energy per electron (or per photon) becomes smaller. At some point it becomes so small that the photons can no longer produce pairs, and the shower dies.

3.5 Nuclear Interactions

If the passage of particles through matter were governed entirely by the phenomena described in Sections 3.1–3.4, neutral particles would pass through matter without being affected, and muons and protons of the same energy would nearly have the same range. The facts, however, are different; the electrically neutral neutrons have a strong short-distance interaction with matter, and high-energy protons have a much shorter range than muons. The reason for this behavior, and for the discrepancy between naive expectation and reality, is the neglect of nuclear interactions. The treatment in Sections 3.1–3.4 is based entirely on the electromagnetic interaction, and nonelectromagnetic forces between the nucleus and the passing particle are neglected. These interactions, the hadronic and the weak ones, form the central topic of subatomic physics and they will be explored and described in the following parts.

3.6 References

The basic ideas underlying the computation of the energy loss of charged particles in matter are described lucidly in N. Bohr, "Penetration of Atomic Particles Through Matter," *Kgl. Danske Videnskab. Selskab Mat-fys Medd.* **XVIII**, No. 8 (1948), and in E. Fermi, *Nuclear Physics*, notes compiled by J. Orear, A. H. Rosenfeld, and R. A. Schluter, University of Chicago Press, Chicago, (1950); J.F. Ziegler, J.P. Biersack, and W. Littmark, *Stopping Powers and Ranges* Pergamon Press, New York, 1985; M.A. Kumak and E.F. Komarov, *Radiation from Charged Particles in Solids*, transl. G. Kurizki, Amer. Inst. Phys., New York, 1989; see also PDG for an up-to-date review and further references.

Problems

3.1. An accelerator produces a beam of protons with kinetic energy of 100 MeV. For a particular experiment, a proton energy of 50 MeV is required. Compute the thickness of

(a) a carbon and

(b) a lead absorber,

both in cm and in g/cm^2, necessary to reduce the beam energy from 100 to 50 MeV. Which absorber would be preferable? Why?

3.2. A counter has to be placed in a muon beam of 100-MeV kinetic energy. No muons should reach the counter. How much copper is needed to stop all muons?

3.3. We have stated that the transmission of charged particles through matter is dominated by atomic, and not nuclear, interactions. When is this statement no longer true; i.e., when do nuclear interactions become important?

3.4. A beam stop is required at the end of accelerators to prevent the particles from running wild. How many meters of solid dirt would be required at FNAL to completely stop the 200 GeV protons, assuming only electromagnetic interactions? Why is the actual beam stop length less?

3.5. Cosmic-ray muons are still observed in mines that are more than 1 km underground. What is the minimum initial energy of these muons? Why are no cosmic-ray protons or pions observed in these underground laboratories?

3.6. Discuss and understand the simplest derivation of Eq. (3.2).

3.7. Show that the mean free path of a particle undergoing exponential absorption as described by Eq. (3.1) is given by $1/\mu$.

3.8. A beam of 1-mA protons of kinetic energy of 800 MeV passes through a 1-cm^3 copper cube. Compute the maximum energy deposited per sec in the copper. Assume the cube to be thermally insulated, and compute the temperature rise per sec.

3.9. Compare the energy loss of nonrelativistic π^+, K^+, d, $^3\text{He}^{2+}$, $^4\text{He}^{2+} \equiv \alpha$ to that of protons of the same energy in the same material.

3.10. In an experiment, alpha particles of 200 MeV energy enter a scattering chamber through a copper foil that is 0.1 mm thick.

(a) Use the form of Eq. (3.2) to find approximately the energy of the proton beam that has the same energy loss as the α beam.

(b) Compute the energy loss.

3.11. Use Eq. (3.2) and Fig. 3.5 to sketch the ionization along the path of a heavy charged particle (Bragg curve).

3.12. Use Eq. (3.2) to calculate numerically the energy loss of a 20 MeV proton in aluminum ($I = 150$ eV).

3.13. A radioactive source emits gamma rays of 1.1 MeV energy. The intensity of these gamma rays must be reduced by a factor 10^4 by a lead container. How thick (in cm) must the container walls be?

3.14. ^{57}Fe has a gamma ray of 14 keV energy. A source is contained in a metal cylinder. It is desired that 99% of the gamma rays escape the cylinder. How thin must the walls be made if the cylinder is

(a) Aluminum?

(b) Lead?

3.15. A source emits gamma rays of 14 and 6 keV. The 6 keV gamma rays are 10 times more intense than the 14 keV rays. Select an absorber that cuts the intensity of the 6 keV rays by a factor of 10^3 but affects the 14 keV rays as little as possible. What is your choice? By what factor is the 14 keV intensity reduced?

3.16. The three processes discussed in Section 3.3 are not the only interactions of photons. List and briefly discuss other types of photon interactions.

3.17. A radioactive source contains two gamma rays of equal intensity with energies of 85 and 90 keV, respectively. Compute the intensity of the two gamma lines after passing through a 1 mm lead absorber. Explain your result.

3.18. Electrons of 1 MeV kinetic energy should be stopped in an aluminum absorber. How thick, in cm, must the absorber be?

3.19. What is the energy of an electron that has approximately the same total (true) pathlength as a 10 MeV proton?

3.20. An electron of 10^3GeV energy strikes the surface of the ocean. Describe the fate of the electron. What is the maximum number of electrons in the resulting shower? At which depth, in m, does the maximum occur?

3.21. A 10-GeV electron passes through a 1-cm aluminum plate. How much energy is lost?

3.22. Show that pair production is not possible without the presence of a nucleus to take up momentum.

3.23. Show that the maximum energy that can be transferred to an electron in a single collision by a non-relativistic particle of kinetic energy T and mass $M(M \gg m_e)$ is $(4m_e/M)T$.

Chapter 4

Detectors

What would a physicist do if he were asked to study ghosts and telepathy? We can guess. He would probably (1) perform a literature search and (2) try to design detectors to observe ghosts and to receive telepathy signals. The first step is of doubtful value because it could easily lead him away from the truth. The second step, however, would be essential. Without a detector that allows the physicist to *quantify* his observations, his announcement of the discovery of ghosts would be rejected by *Physical Review Letters*. In experimental subatomic physics, detectors are just as important and the history of progress is to a large extent the history of increasingly more sophisticated detectors. Even without accelerators and using only neutrinos or cosmic-ray particles, a great deal can be learned by making the detectors bigger and better. In the following sections, we shall discuss different types of detectors. Many beautiful and elegant tools are not treated here; however, once the ideas behind typical instruments are understood, it is easy to pick up more details concerning others. We also add a brief section about electronics because it is an integral part of any detection system.

4.1 Scintillation Counters

The first scintillation counter, called spinthariscope, was constructed in 1903 by Sir William Crookes. It consisted of a ZnS screen and a microscope; when alpha particles hit the screen, a light flash could be seen. In 1910, Geiger and Marsden performed the first coincidence experiment. As Fig. 4.1 shows, they used two screens, S_1 and S_2, and two observers with microscopes M_1 and M_2. If the radioactive gas between the two screens emitted two alpha particles within a "short" time and if each hit one screen, each observer would see a flash. They probably shouted to indicate the time of arrival. The human eye is slow and unreliable and the scintillation counter was abandoned for many years. It was reintroduced in 1944 with a photomultiplier replacing the eye. The basic arrangement for a modern scintillation counter is shown in Fig. 4.2.

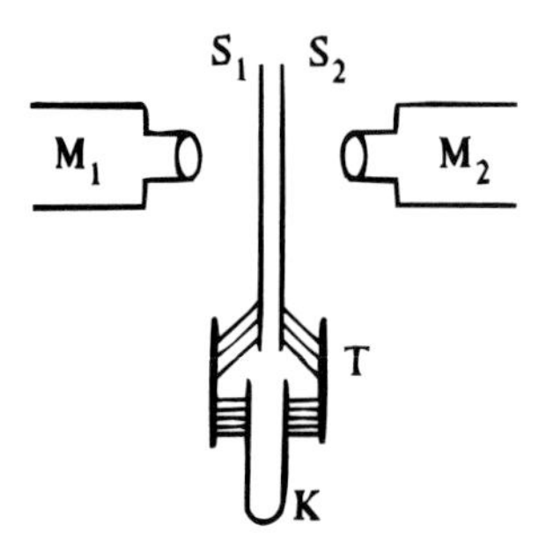

Figure 4.1: Coincidence observation “by eye”. (From E. Rutherford, *Handbuch der Radiologie*, Vol. II, Akademische Verlagsgesellschaft, Leipzig, 1913.)

A scintillator is joined to one (or more) photomultipliers through a light pipe. A particle passing through the scintillator produces excitations; deexcitation occurs through emission of photons. These photons are transmitted through a shaped light pipe to the photocathode of a photomultiplier. There, photons release electrons which are accelerated and focused onto the first dynode. For each primary electron hitting a dynode, two to five secondary electrons are released.

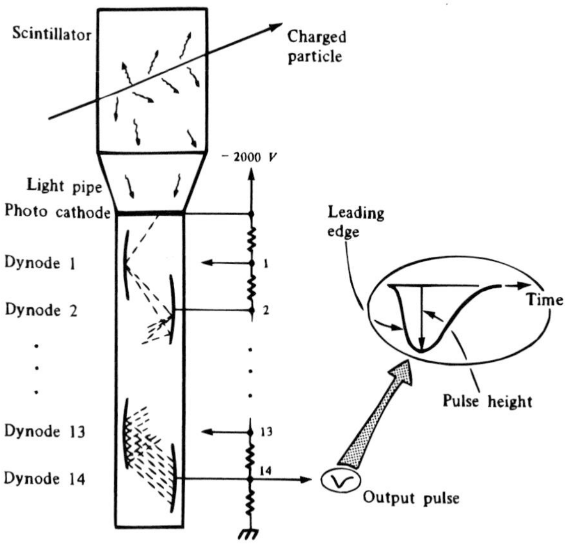

Figure 4.2: Scintillation counter. A particle passing through the scintillator produces light which is transmitted through a light pipe onto a photomultiplier.

Up to 14 multiplying stages are used, and overall multiplying factors of up to 10^9 can be achieved. The few incident photons therefore produce a measurable pulse at the output of the multiplier. The shape of the pulse is shown schematically in the insert of Fig. 4.2. The pulse height is proportional to the total energy deposited in the scintillator.

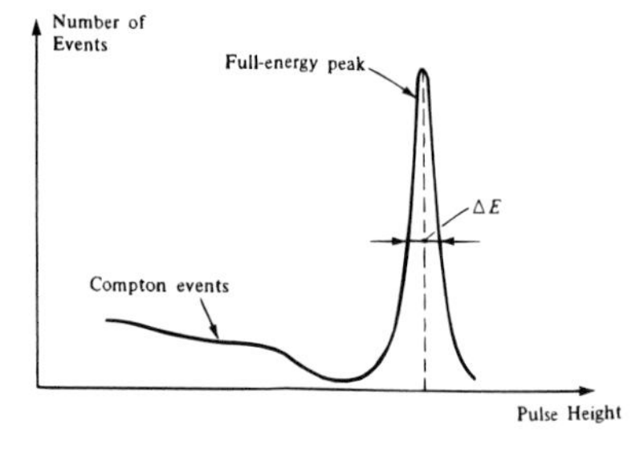

Figure 4.3: Scintillation spectrum, NaI(Tl) crystal.

Two types of scintillators are widely used, sodium iodide and plastics. *Sodium iodide* crystals are usually doped with a small amount of thallium and denoted by NaI(Tl). The Tl atoms act as luminescence centers.

The efficiency of these inorganic crystals for gamma rays is high, but the decay of each pulse is slow, about 0.25μs. Moreover, NaI(Tl) is hygroscopic and large crystals are very expensive. *Plastic scintillators*, for instance polystyrene with terphenyl added, are cheap; they can be bought in large sheets and can be machined in nearly any desired shape. The decay time is only a few ns, but the efficiency for photons is low. They are therefore mainly used for the detection of charged particles.

A few remarks are in order concerning the mechanism of observation of gamma rays in NaI(Tl) crystals. For a gamma ray of less than 1 MeV, only photoeffect and Compton effect have to be considered. Photoeffect results in an electron with an energy $E_e = E_\gamma - E_b$, where E_b is the binding energy of the electron before it was ejected by the photon. The electron will usually be completely absorbed in the crystal. The energy deposited in the crystal produces a number of light quanta that are then detected by the photomultiplier. In turn, these photons result in a pulse of electric charge proportional to E_e and with a certain width ΔE. This photo or full-energy peak is shown in Fig. 4.3. The energy of the electrons produced by the Compton effect depends on the angle at which the photons are scattered. The Compton effect therefore gives rise to a spectrum, as indicated in Fig. 4.3. The width of the full-energy peak, measured at half-height, depends on the number of light quanta produced by the incident gamma ray; typically $\Delta E/E_\gamma$ is of the order of 20% at $E_\gamma = 100$ keV and 6–8% at 1 MeV. At energies above 1 MeV, the incident gamma ray can produce an electron–positron pair; the electron is absorbed, and the positron annihilates into two 0.51 MeV photons. These two photons can escape from the crystal. The energy deposited is E_γ if no photon escapes, $E_\gamma - m_ec^2$ if one escapes, and $E_\gamma - 2m_ec^2$ if both annihilation photons escape.

The energy resolution $\Delta E/E$ deserves some additional consideration. Is a resolution of about 10% sufficient to study the gamma rays emitted by nuclei? In some cases, it is. In many instances, however, gamma rays have energies so close together that a scintillation counter cannot separate them. Before discussing a counter with better resolution, it is necessary to understand the sources contributing to the width. The chain of events in a scintillation counter is as follows: The incident gamma ray produces a photoelectron with energy $E_e \approx E_\gamma$. The photoelectron, via excitation and ionization, produces n_{lq} light quanta, each with an energy of $E_{\text{lq}} \approx 3$ eV($\lambda \approx 400$nm). (For clarity we call the incident photon *the gamma ray* and the optical photon *the light quantum.*) The number of light quanta is given by

$$n_{\text{lq}} \approx \frac{E_\gamma}{E_{\text{lq}}}\epsilon_{\text{light}},$$

where ϵ_{light} is the efficiency for the conversion of the excitation energy into light quanta. Of the n_{lq} light quanta, only a fraction ϵ_{coll} are collected at the cathode of the photomultiplier. Each light quantum hitting the cathode has a probability $\epsilon_{\text{cathode}}$ of ejecting an electron. The number n_e of electrons produced at the input of the photomultiplier is therefore

$$n_e = \frac{E_\gamma}{E_{1q}} \epsilon_{\rm light} \epsilon_{\rm coll} \epsilon_{\rm cathode}. \tag{4.1}$$

Typical values for the efficiencies are

$$\epsilon_{\rm light} \approx 0.1, \quad \epsilon_{\rm coll} \approx 0.4, \quad \epsilon_{\rm cathode} \approx 0.2,$$

so that the number of electrons released at the photocathode after absorption of a 1 MeV gamma ray is $n_e \approx 3 \times 10^4$. (The value $\epsilon_{\rm light} \approx 0.1$ is appropriate for a NaI crystal; for plastic scintillator $\epsilon_{\rm light} \approx 0.03$. The value $\epsilon_{\rm coll}$ is only a nominal value. The transmission of light through a scintillator decreases exponentially with its length, as seen in Chapter 3. Typical attenuation lengths are: $\sim 1-5$ m.) Since all processes in Eq. (4.1) are statistical, n_e will be subject to fluctuations, and these produce most of the observed line width. An additional broadening comes from the multiplication in the photomultiplier which is also statistical. To discuss the line width, we digress to present some of the fundamental statistical concepts.

4.2 Statistical Aspects

Random processes play an important part in subatomic physics. The standard example is a collection of radioactive atoms, each atom decaying independently of all the others. We shall consider here an equivalent problem that came up in the previous section, the production of electrons at the photocathode of a multiplier. The question to be answered is illustrated in Fig. 4.4.

Figure 4.4: Production of photoelectrons as a random process.

Each incident photon produces n photoelectrons as output. We can repeat the measurement of the number of output electrons N times, where N is very large. In each of these N identical measurements, we shall find a number $n_i, i = 1, \ldots, N$. The *average* number of output electrons is then given by

$$\overline{n} = \frac{1}{N} \sum_{i=1}^{N} n_i. \tag{4.2}$$

The question of interest can be stated: How are the various values n_i distributed around $\overline{n}$? Another way of phrasing the same question is: What is the probability $P(n)$ of finding a particular value n in a given measurement if the average number is $\overline{n}$?

Or, to make it more specific, consider a process where the average number of output electrons is small, say $\overline{n} = 3.5$. What is the probability of finding the value $n = 2$? This problem has occupied mathematicians for a long time, and the answer is well known[1]: The probability $P(n)$ of observing n events is given by the *Poisson distribution*,

$$P(n) = \frac{(\overline{n})^n}{n!} e^{-\overline{n}}, \tag{4.3}$$

where $\overline{n}$ is the average defined by Eq. (4.2). As behooves a probability, the sum over all possible values n is 1, $\sum_{n=0}^{\infty} P(n) = 1$. With Eq. (4.3), the previous questions can now be answered, and we first turn to the most specific one. With $\overline{n} = 3.5, n = 2$, Eq. (4.3) gives $P(2) = 0.185$. It is straightforward to compute the probabilities for all interesting values of n. The corresponding histogram is shown in Fig. 4.5. It shows that the distribution is very wide. There is a nonnegligible probability of measuring values as small as zero or as large as 9. If we perform only one measurement and find, for instance, a value of $n = 7$, we have no idea what the average value would be.

A glance at Fig. 4.5 shows that it is not enough to measure and record the average, $\overline{n}$. A measure of the *width* of the distribution is also needed. It is customary to characterize the width of a distribution by the *variance* σ^2:

$$\sigma^2 = \sum_{n=0}^{\infty} (\overline{n} - n)^2 P(n), \tag{4.4}$$

or by the square root of the variance, called the *standard deviation.* For the Poisson distribution, Eq. (4.3), variance and standard distribution are easy to compute, and they are given by

$$\sigma^2 = \overline{n}, \quad \sigma = \sqrt{\overline{n}}. \tag{4.5}$$

For small values of $\overline{n}$, the distribution is not symmetric about $\overline{n}$, as is evident from Fig. 4.5.

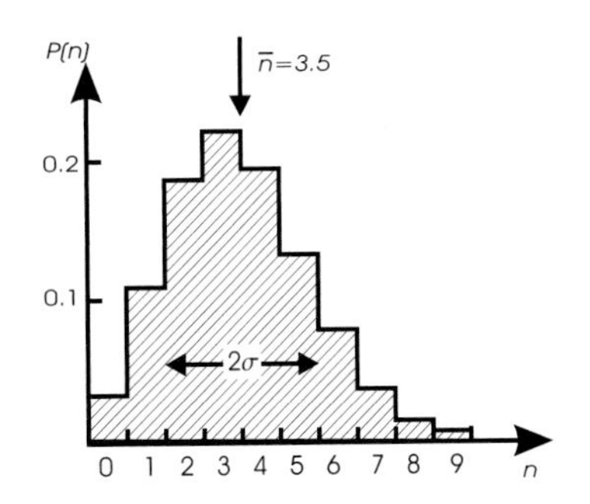

Figure 4.5: Histogram of the Poisson distribution for $\overline{n} = 3.5$. The distribution is not symmetric about $\overline{n}$.

So far we have discussed the Poisson distribution for *small* values of $\overline{n}$. Experimentally, such a situation arises, for instance, at the first dynode of a photomultiplier, where each incident electron produces two to five secondary electrons. Data are then given in the form of histograms, as in Fig. 4.5.

[1]A derivation can, for instance, be found in H. D. Young, *Statistical Treatment of Experimental Data*, McGraw-Hill, New York, 1962, Eq.(8.5); R.A. Fisher, *Statistical Methods, Experimental Design, and Scientific Inference*, Oxford Univ. Press, Oxford, 1990.

In many instances, $\overline{n}$ can be very *large*. In the case of the scintillation counter discussed in the previous section, the number of photoelectrons at the photomultiplier is on the average $\overline{n} = 3 \times 10^3$. For $\overline{n} \gg 1$, Eq. (4.3) is cumbersome to evaluate. However, for large n, $\overline{n}$ can be considered a continuous variable, and Eq. (4.3) can be approximated by

$$P(n) = \frac{1}{(2\pi n)^{1/2}} \exp\left[\frac{-(\overline{n}-n)^2}{2n}\right] \qquad (4.6)$$

which is easier to evaluate. Moreover, the behavior of $P(n)$ is now dominated by the factor $(\overline{n}-n)^2$ in the exponent. Particularly near the center of the distribution, n can be replaced by $\overline{n}$ except in the factor $(\overline{n}-n)^2$, and the result is

$$P(n) = \frac{1}{(2\pi \overline{n})^{1/2}} \exp\left[\frac{-(\overline{n}-n)^2}{2\overline{n}}\right]. \qquad (4.7)$$

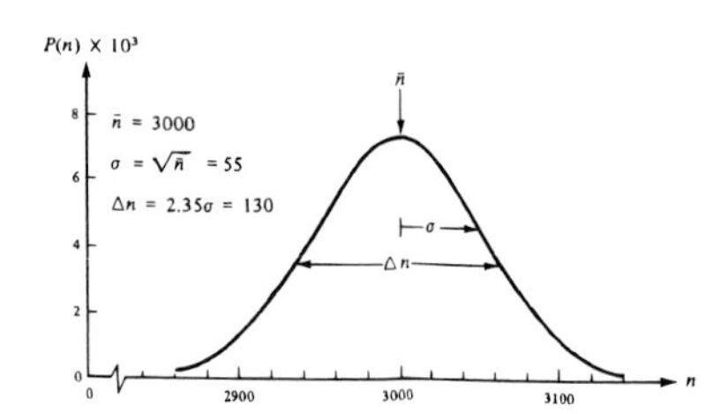

Figure 4.6: Poisson distribution for $\overline{n} \gg 1$ where it becomes a normal distribution.

This expression is symmetric about $\overline{n}$ and is called a *normal* or *Gaussian distribution*. The standard deviation and the variance are still given by Eq. (4.5).

As an example of the limiting case where the Poisson distribution can be represented by the normal one, we show in Fig. 4.6 $P(n)$ for $\overline{n} = 3 \times 10^3$, the number of photoelectrons of our example in the previous section. The standard deviation is equal to $(3 \times 10^3)^{1/2} = 55$, resulting in a fractional deviation $\sigma/\overline{n} \approx 2\%$. To compare this value to $\Delta E/E_\gamma$ we note that ΔE is the *full* width at half maximum (FWHM). With Eqs (4.5) and (4.7) it is straightforward to see that Δn, the full width at half maximum, is related to the standard deviation by

$$\Delta n = 2.35\sigma.$$

With $\Delta E/E_\gamma = \Delta n/\overline{n}$, the expected fractional energy resolution becomes about 5%. Since the value must still be corrected for additional fluctuations, for instance, in the multiplier, the agreement with the experimentally observed resolution of 6–8% is satisfactory.

As another example, we apply the statistical considerations to an experiment in which a quantity n is measured N times and where the distribution of n is Gaussian. The variance σ^2 is determined from the measured values n_i and the average $\overline{n}$ as

$$\sigma^2 = \overline{(\overline{n}-n_i)^2} = \frac{1}{N}\sum_{i=1}^{N}(\overline{n}-n_i)^2.$$

Note that σ^2 does not decrease with increasing number N; it describes the width of the distribution. Nevertheless, with increasing N, the value of $\overline{n}$ becomes better known. This fact is expressed through the *variance of the mean*, given by

$$\sigma_{\rm m}^2 = \frac{\sum_{i=1}^{N}(\overline{n}-n_i)^2}{N(N-1)} = \frac{\sigma^2}{(N-1)}. \tag{4.8}$$

The measured quantity with its standard deviation σ_m is usually quoted as

$$\text{result} = \overline{n} \pm \sigma_m. \tag{4.9}$$

4.3 Semiconductor Detectors

Scintillation counters started a revolution in the detection of nuclear radiations, and they reigned unchallenged from 1944 to the late 1950s. They are still essential for many experiments, but in many areas they have been replaced by semiconductor detectors. Before discussing these, we compare in Fig. 4.7 a complex gamma-ray spectrum as seen by a semiconductor and by a scintillation detector. The superior energy resolution of the solid-state counter is obvious. How is it achieved? In the scintillation counter, the efficiencies in Eq. (4.1) reduce the number of photoelectrons counted; it is difficult to imagine how each of the efficiency factors in Eq. (4.1) could be improved to about 1. A different approach is therefore needed and the solid-state (semiconductor) detector offers one. The idea underlying the semiconductor counter is old and it is used in ionization chambers: A charged particle with kinetic energy E_e moving through a gas or a solid produces ion pairs, and the number of

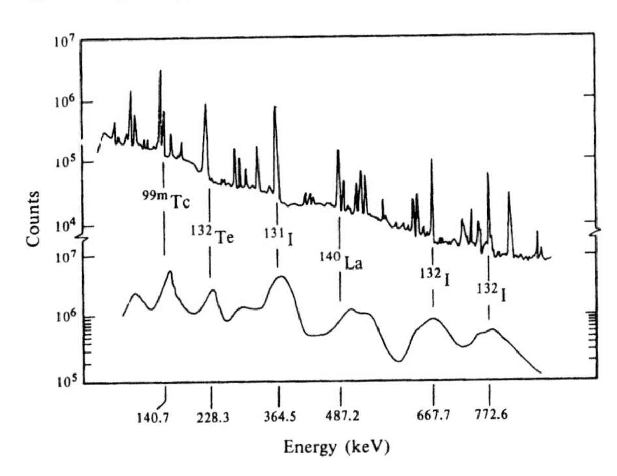

Figure 4.7: Complex gamma-ray spectrum, due to gross fission products, observed by a germanium detector (upper curve) and a scintillation detector (lower curve). [From F.S. Goulding and Y. Stone, *Science* **170**, 280 (1970). Copyright 1970 by the American Association for the Advancement of Science.]

these pairs is given by

$$n_{\text{ion}} = \frac{E_e}{W}, \tag{4.10}$$

where W is the energy needed to produce one ion pair. If the ion pairs are separated in an electric field and if the total charge is collected and measured, the energy of the electron can be found.

A gas-filled ionization chamber uses this principle, but it has two disadvantages: (1) The density of a gas is low so that the energy deposited by a particle is small. (2) The energy needed for the production of an ion pair is large ($W = 42$ eV for He, 22 eV for Xe, and 34 eV for air). Both disadvantages are avoided in a semiconductor detector, as sketched in Fig. 4.8. If a charged particle passes through a semiconductor, ion pairs will be created. The energy W is about 2.9 eV for germanium and 3.5 eV for silicon. The energies are so low because ionization does not occur from an atomic level to the continuum but from the valence band to the conduction band.(2) The electric field will sweep the negative charges toward the positive and the positive charges toward the negative surface. The resulting current pulse is fed to a low-noise amplifier. At room temperature, thermal excitation can produce an unwanted current, and many semiconductor detectors are therefore cooled to liquid nitrogen temperature. The low value of W and the collection of all ions explains the high energy resolution of semiconductor detectors shown in Fig. 4.7. Figure 4.9 presents the energy resolution as a function of particle energy for germanium and silicon detectors.

While semiconductor detectors have a much higher density than gas-filled ionization chambers, they are much more expensive for large volumes. Semiconductor counters can have volumes of $\sim 1000\ \text{cm}^3$. Scintillation counters can be made orders of magnitude larger, and they do not have to be cooled. For any given application one must therefore consider which type of counter will be more suitable and more convenient.

Typically Ge detectors are used for detection of gamma rays, while Si detectors are used to detect charged particles. Stripped Si detectors have become available which allow position resolution of $\lesssim 0.1$ mm.

Over the past three decades arrays of multiple Ge detectors have been produced and used mainly to measure gamma rays from fastly rotating nuclei. Figure 4.10 shows one example composed of 110 Ge detectors. A new generation of detectors, presently under developement, would track photons and allow for more efficient detection, determination of the polarization, better determination of multiplicity and better determination of original photon directions. Here instead of having an array of Ge detectors, one would use fewer highly-segmented detectors. Figure 4.11 shows an example.

2The band structure of semiconductors can be found in C.Hamaguchi, *Basic Semiconductor Physics*, Springer Verlag, New York, 2001, K.F. Brennan, *The Physics of Semiconductors*, Cambridge Univ. Press, Cambridge, 1999, or in the *Feynman Lectures*, Vol. III, Chapter 14.

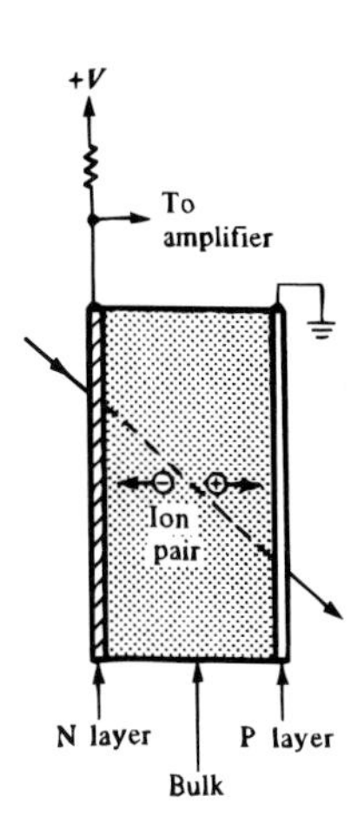

Figure 4.8: Ideal, fully depleted semiconductor detector with heavily doped surface layers of opposite types.

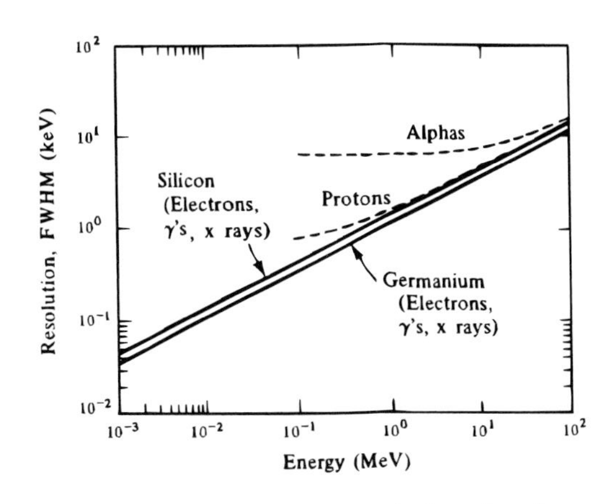

Figure 4.9: Optimal energy resolution of semiconductor counters as a function of energy. [From F.S. Goulding and Y. Stone, *Science* **170**, 280 (1970). Copyright 1970 by the American Association for the Advancement of Science.] FWHM means full width at half maximum.

Figure 4.10: Gammasphere: 110 high-purity Ge detectors were put together in a 4π array. This photo shows approximately half of the array around the chamber that holds the target where the beam impinges under vacuum. [Courtesy of A.O. Macchiavelli.]

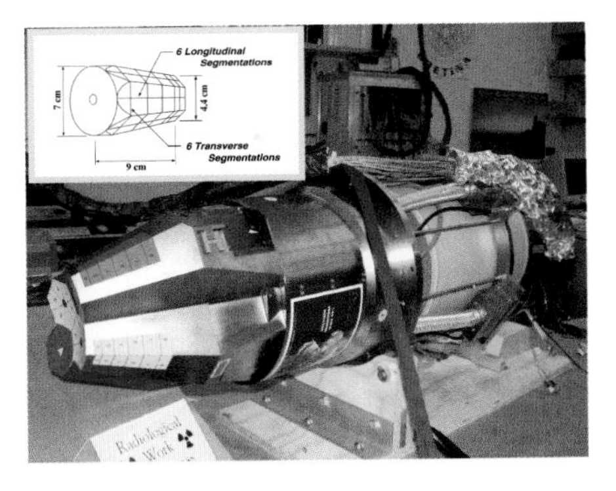

Figure 4.11: Developments for GRETA: Gamma-ray energy tracking array. This detector has 36 segments. The energy deposited in each segment can be read separately which allows for photon tracking. An array covering 4π would be built with several of these units. [Courtesy of A.O. Macchiavelli.]

4.4 Bubble Chambers

Bubble chambers became popular in the 1950–1980's as a tool to track particles through large volumes. Since its invention by Glaser in 1952, it played a crucial role in the elucidation of the properties of subatomic particles.[3]

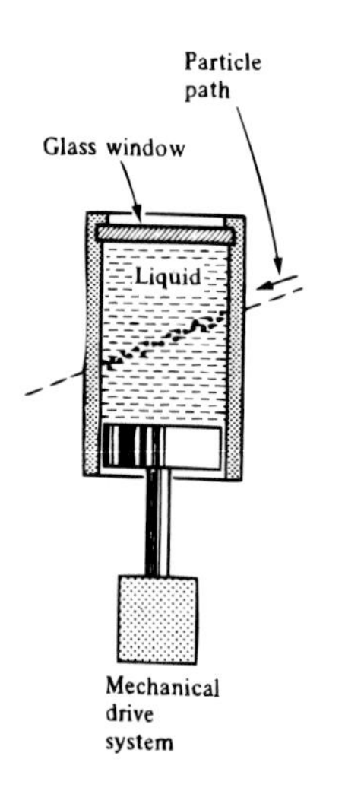

Figure 4.12: Bubble chamber—schematic diagram.

The physical phenomenon underlying the bubble chamber is best described in Glaser's own words[4]: "A bubble chamber is a vessel filled with a transparent liquid which is so highly superheated that an ionizing particle moving through it starts violent boiling by initiating the growth of a string of bubbles along its path." A superheated liquid is at a temperature and pressure such that the actual pressure is lower than the equilibrium vapor pressure. The condition is unstable, and the passage of a single charged particle initiates bubble formation. To achieve the superheated condition, the liquid in the chamber (Fig. 4.12) is first kept at the equilibrium pressure; the pressure is then rapidly dropped by moving a piston.

A few ms after the chamber becomes sensitive, the process is reversed and the chamber pressure is brought back to its equilibrium value. The bubbles are illuminated with an electronic photoflash and recorded.

In the times when bubble chambers were popular for high-energy experiments, the time during which the chamber was sensitive was synchronized with the arrival time of pulses of particles from an accelerator. Pictures were taken and later analyzed visually. Glaser's first chambers contained only a few cm^3 of liquid. Development was rapid, however, in less than twenty years, the volume increased by more than 10^6. Eventually bubble chambers became very large and costed millions of dollars. They required enormous magnets to curve the paths of the charged particles. The superheated liquid, often hydrogen, was explosive when in contact with oxygen, and accidents did occur. Bubble chambers could produce tens of millions of photographs/y, and data evaluation was complex.

Two examples demonstrate the beautiful and exciting events that were seen. Figure 1.4 shows the production and the decay of the omega minus, a most remarkable particle that we shall encounter later. Figure 4.13 represents the first neutrino interaction observed in pure hydrogen. It was found on November 13, 1970, in the

[3]L.W. Alvarez, *Science* **165**, 1071 (1969).

[4]D. A. Glaser and D. C. Rahm, *Phys. Rev.* **97**, 474 (1955).

Photo 5: Bubble chamber. Some versions become very large and sophisticated. [Courtesy Lawrence Berkeley National Laboratory.]

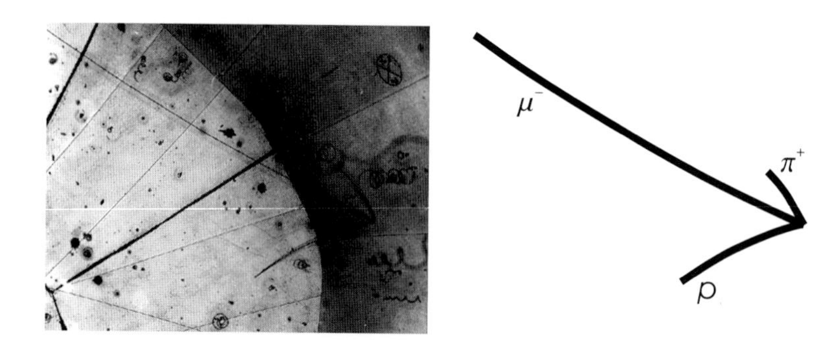

Figure 4.13: Neutrino interaction in a hydrogen bubble chamber. A neutrino enters from the right and interacts with the proton of a hydrogen atom to yield a muon (the long track that extends to the top left), a positive pion (the short top track), and a proton (the short bottom track). [Courtesy Argonne National Laboratory.]

3.6 m (12 ft) hydrogen bubble chamber of the Argonne National Laboratory which contained about 20,000 liters of hydrogen. A superconducting magnet produced a field of about 18 kG in the chamber volume of 25 m^3.

Because of their slow response, bubble chambers are seldomly used in present days for experiments with high counting rates. However, they are still being used for applications with low counting rates in combination with CCD cameras.(5) Some versions can be triggered using the spike in pressure when a pulse develops. In the next section we explain in more detail what *triggering* means.

4.5 Spark Chambers

Whereas spark chambers are no longer state-of-the-art, they illustrate the basic principles of a triggerable detector clearly. Spark chambers are based on a simple fact. If the voltage across two metal plates, spaced by a distance of the order of cm, is increased beyond a certain value, a breakdown occurs. If an ionizing particle passes through the volume between the plates, it produces ion pairs, and the breakdown takes the form of a spark that follows the track of the particle. Since the ions remain between the plates for a few μs, the voltage can be applied *after* passage of the particle: A spark chamber is a triggerable detector.

The elements of a spark chamber system are shown in Fig. 4.14. The problem to be studied in this simplified arrangement is the reaction of an incoming charged particle with a nucleus in the chamber, giving rise to at least two charged products. Thus the *signature* of the desired events is "one charged in, two charged out." Three scintillation counters, A, B, and C, detect the three charged particles. If the particles pass through the three counters, the LOGIC circuit activates the high-

[5] See, for example, W.J. Bolte et al., *Journal of Physics: Conference Series* **39**, 126 (2006); http://collargroup.uchicago.edu/

voltage supply, and a high-voltage pulse (10–20 kV) is applied to the plates within less than 50 ns. The resulting sparks are recorded on stereophotographs.

The standard spark chamber arrangement of the type just discussed has been used in many experiments, and chambers have been designed to solve many problems. Thin plates are employed if only the direction of charged particles is desired; thick lead plates are used if gamma rays are to be observed or if electrons have to be distinguished from muons. The electrons produce showers in the lead plates and can thus be recognized.

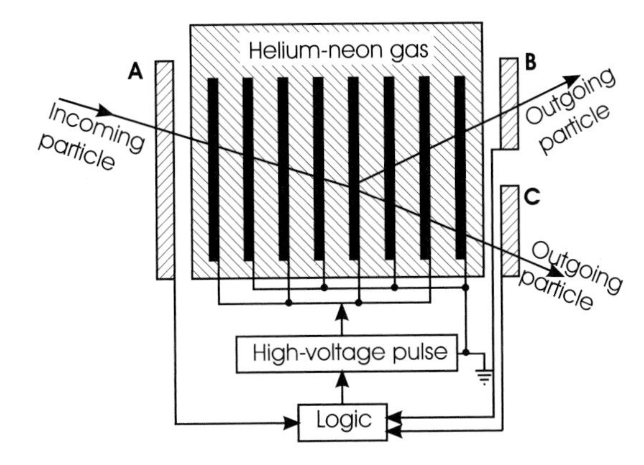

Figure 4.14: Spark chamber arrangement. The spark chamber consists of an array of metal plates in a helium–neon mixture. If the counter-and-logic system has decided that a wanted event has occurred, a high-voltage pulse is sent to alternate plates, and sparks are produced along the ionization trails.

Spark chambers have been replaced in high-energy experiments by silicon semiconductor detectors and by drift chambers, but they are still used in some experiments because they are simple and inexpensive.

4.6 Wire Chambers

Bubble and spark chambers share one disadvantage: Events must be photographed and then evaluated later. In experiments where a large amount of data is collected this approach is cumbersome.[6]

Wire chambers (multi-wire proportional counters), pioneered by Charpak, avoid this disadvantage. Wire chambers have very good time resolution, very good position accuracy, and are self-triggered. Their use has spread from high-energy physics to many other fields such as nuclear medicine, heavy ion astronomy, and protein crystallography. A cross section through a wire chamber is sketched in Fig. 4.15. A chamber may be a few m long and high. Tungsten wires of diameter $2a(\approx 20\mu\text{m})$ are stretched in one direction and a voltage of a few kV is applied between the anode wires and the cathode surfaces. The resulting field lines are indicated for two wires in Fig. 4.15. An ionizing particle passing through the chamber creates ion pairs. Electrons produced close to the wire are accelerated towards the wire with an energy sufficient to produce additional pairs and an avalanche results which leads to a negative pulse on the wire. In many wire chambers, each wire is connected to

[6] G. Charpak and F. Sauli, *Annu. Rev. Nucl. Part. Sci.* **34**, 285 (1984).

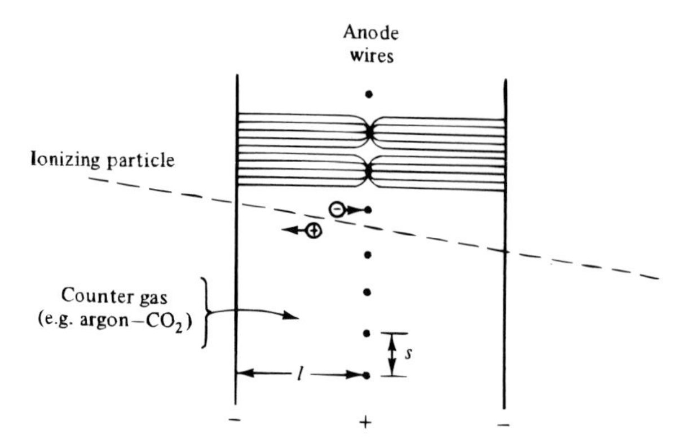

Figure 4.15: Cross section through a multi-wire proportional counter. Typical dimensions are $l = 8$ mm, $s = 2$ mm. Field lines are shown for two wires.

a separate amplifier and pulse shaper; the output pulse indicates position and time of the particle.

4.7 Drift Chambers

Drift chambers[7] are like wire detectors, but can provide much better spatial resolution ($\leq 200\mu$m) at lower cost because fewer wires are required. Drift chambers use a low electric field (~ 1 keV/cm) to make electrons drift to one or more anode wires. To produce a relatively constant electric field strength, potential wires are introduced between neighboring anode wires. Close to the anode wires, the electric field gets very large and an avalanche results. The drift time is used to define the position of the particle. The drift velocity is given by

$$v_D = \frac{e\tau E}{2m} , \tag{4.11}$$

where e is the charge of the particle, τ is the mean collision time, E is the electric field intensity, and m is the mass of the particle. The distance traversed to reach the avalanche region is

$$\Delta x = \int_{t_0}^{t_1} v_D dt , \tag{4.12}$$

where t_0 is the creation time and t_1 is the arrival time of the electron. For electrons, for which the chamber is most useful, the approximate drift speed is about 50 mm/μs and the drift distance of the order of 5-10 cm.

[7]W. Blum and L. Rolandi, *Particle Detection with Drift Chambers*, Springer Verlag, New York, 1993.

Figure 4.16: Drift chambers can be used to track charged particles. In this picture a scientist checks part of the TWIST apparatus,[8] which used 44 drift chamber planes to do precision measurements of muon decay. The drift chambers are circular planes whose edges can be seen in the photograph. [Courtesy Bob Tribble.]

Drift chambers are still very popular because they are reliable and not expensive. Figure 4.16 shows a picture of the TWIST apparatus which used drift chambers to track electrons from muon decay.[8]

A drift chamber can be planar or cylindrical. In the latter form it can be made into a time projection chamber.

4.8 Time Projection Chambers

Wire chambers have one major disadvantage: they only yield information about one spatial direction. To determine both coordinates, a second wire chamber must be used. This requirement makes the experimental arrangement complicated and reduces the solid angle subtended by the detector. Time projection chambers (TPCs), invented in 1974 by David Nygren, avoid this limitation and are nearly ideal detectors: TPCs have large solid angles, give excellent spatial resolution in three dimensions, yield charge and mass information, and allow good pattern recognition.[9,10]

TPCs can be as small as a grapefruit or weigh as much as 10 tons. The main features are illustrated in Fig. 4.17. The drift chamber is filled with a gas, usually a mixture of Ar and CH_4 because it is inexpensive and allows high electron mobility. Uniform electric ($\boldsymbol{E}$) and magnetic ($\boldsymbol{B}$) fields are applied parallel to the axis (beam pipe). A charged particle passing through the chamber produces ion pairs along its

[8] TWIST Collaboration, *Phys. Rev. Lett.* **94**, 101805 (2005).

[9] R. J. Madaras and P. J. Oddone, *Phys. Today* **37**, 38 (August 1984).

[10] "The Time Projection Chamber", ed. J.A. MacDonald, *AIP Conference Proceedings No. 108*, American Institute of Physics, New York, 1984.

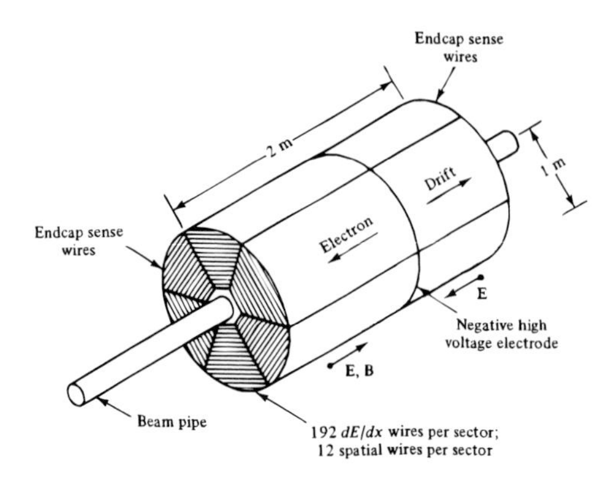

Figure 4.17: Schematic drawing of a Time Projection Chamber. Charged particles which traverse the chamber ionize the gas of the TPC; the electrons that result drift to the endcaps under the influence of the axial electric and magnetic fields. [Courtesy of Lawrence Berkeley Laboratory.]

trajectory. The applied electric field accelerates the electrons of these pairs towards one end of the chamber. The magnetic field causes the electron trajectories to be tiny spirals along the $\boldsymbol{B}$ field parallel to the beam axis. The point of impact of the electrons at the ionization (wire) chambers on the end caps consequently traces the projection of the particle trajectory, thereby yielding two coordinates. The third coordinate is determined by the arrival time of the electrons. The total charge deposited at the ends gives the total ionization and hence the total energy lost by the particle in passing through the chamber. Eq.(3.2) then permits calculation of the particle speed v. The curvature of the particle in the magnetic field $\boldsymbol{B}$ can be computed from the particle coordinates; Eq.(2.15) then yields the momentum. Momentum and velocity together determine the particle mass and thus identify the particle. Since the detectors can surround the beam pipe completely, the solid angle is very large. The large number of sensitive elements at each end permits the simultaneous observation of many particles and thus allows efficient pattern recognition. Because of their many advantages, TPCs are now used in many nuclear and high-energy laboratories.

4.9 Čerenkov Counters

Čerenkov counters use the light emitted by Čerenkov radiation to obtain the velocity of a particle; if the momentum is also measured, then the mass of the particle can be obtained and the particle can be identified.

If the speed, v, of a particle is faster than that of light in a medium with index of refraction n, then radiation is emitted at an angle θ, with $\cos\theta = c/(vn)$. Thus the angle can be used to determine the particle's speed. The maximum cone angle is $\theta_{max} = \cos^{-1} 1/n$. The energy loss per path length is small, of the order of 500 eV/cm in the visible region.

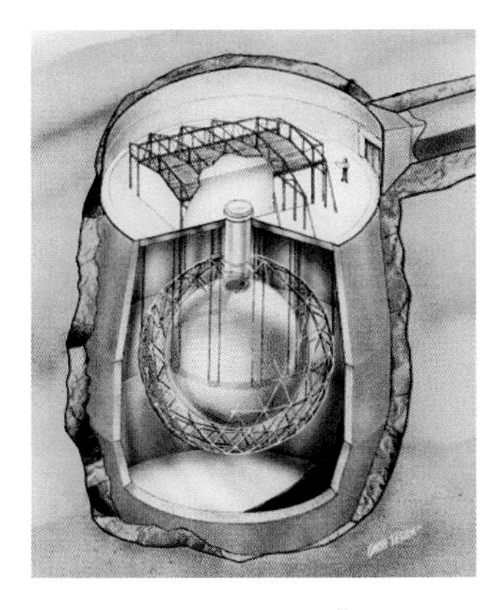

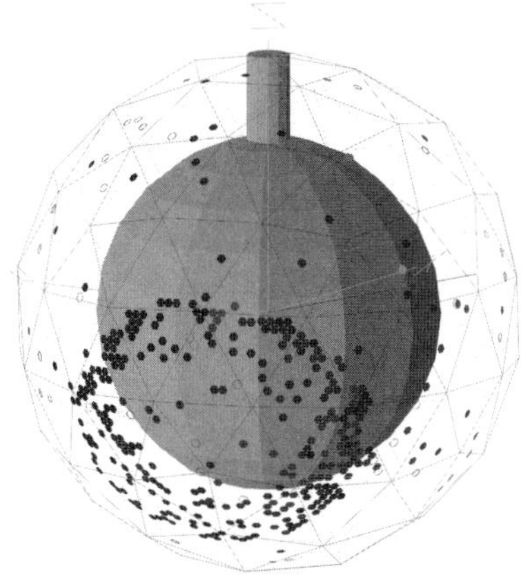

Figure 4.18: Example of the use of Čerenkov detectors: the SNO detector consited of 1000 Tons of heavy water and 9456 photomultiplier tubes in a spherical array. Left: artist's view of the detector, which was located deep underground. Right: the appearance of a neutrino event is shown. Each dot indicates a photomultiplier tube that detected Čerenkov light. [Courtesy J. Wilkerson]

The number of photons emitted per unit path length and energy is [11]

$$\frac{d^2N}{dxdE} = \frac{\alpha z^2}{\hbar c}\left(1 - \frac{c^2}{v^2 n^2}\right), \tag{4.13}$$

where ze is the charge of the particle and α is the fine structure constant $\alpha = e^2/(\hbar c) = 1/137$.

A recent spectacular application of Čerenkov detectors has been their use to detect neutrinos. In a typical situation a neutrino scatters from electrons in water and the electrons generate Čerenkov 'rings' that are detected with photomultiplier tubes. Figure 4.18 shows an example of a neutrino event from the SNO detector.

4.10 Calorimeters

Modern high energy accelerators and heavy ion accelerators both produce a multitude of events per collision. Calorimeters are used to measure the energies of particles by stopping them and thus having them deposit all of their energy inside the detector. Thus a large mass detector is required. There are two types of calorimeters, namely those for electrons and those for hadrons.

As discussed in Section 3.4, high energy electrons slow down primarily by bremsstrahlung, with the photons then producing electron-positron pairs; these pairs produce further photons and the process results in a "shower" of e^+e^- pairs. In one radiation length, X_0, the inital energy E_0 leads to two particles of energy $E_0/2$. After n iterations or in a distance of nX_0, there will be 2^n particles with an average

[11] See Jackson, Sect. 13.4.

energy $\bar{E} = E_0/2^n$. The shower stops when $\bar{E} = E_c$, when loss of energy by ionization becomes important.

The impact point of any particle can be obtained from the lateral spread of the shower. In the case of an incident electron, the shower is well defined and can be traced back.

High energy hadrons are generally not contained in an electron calorimeter, so that a hadron calorimater tends to surround or be placed behind an electron one. Hadrons slow through collisions with nuclei and give rise to secondary hadrons which produce more hadrons. The exception is a particle like a π^0, which decays primarily into two photons and thence produces an electron shower. The mean free path of a hadron depends on the cross section for collisions with nuclei and on the density of the material. A typical hadron will traverse about 135 g/cm^2 in Fe. A typical calorimater of Fe may be 2 m deep and 1/2 m in a transverse direction. For 95% containment of the particle in the calorimeter, its length $L \sim (9.4 \ln E(\text{GeV}) + 39)$ cm.

The shower development for electrons and for hadrons is a statistical process. Thus, the relative accuracy increases with energy, the error being proportional to $1/\sqrt{E_0}$, where E_0 is the incident energy.

Muons, tauons, and neutrinos do not produce showers. Muons leave an ionization trail which can be identified and then detected in a muon chamber (like a calorimeter, but the muons have a high probability of not being absorbed and reaching the layers of the chamber). In Fig. 4.19 we show examples of expected tracks of particles through a detector planned for the LHC showing the calorimeters. The size of the detector can be gauged by the scale on top.

4.11 Counter Electronics

The original scintillation counter, and even the original coincidence arrangement (Fig. 4.1), needed no electronics; the human eye and the human brain provided the necessary elements, and recording was achieved with paper and pen. Nearly all modern detectors, however, contain electronic components as integral elements. A typical example is the circuitry associated with the scintillation counter (Fig. 4.22).

A well-regulated *power supply* provides the voltage for the photomultiplier. The output pulse of the multiplier is shaped and amplified in the *analog* part. The height V of the final pulse is proportional to the height of the original pulse. In the ADC, the *analog-to-digital converter*, the information is transformed into digital form. The output is an integer number (usually expressed in binary units) that is proportional to the pulse height (or area) and can be recorded by a computer.

The example here is a simple one in which only one *parameter*, the height of the pulse, is digitized and stored. In most experiments, for every *event*, many parameters are recorded. In modern experiments events rates can be too large for all of them to be recorded, so an electronic system to decide which events are

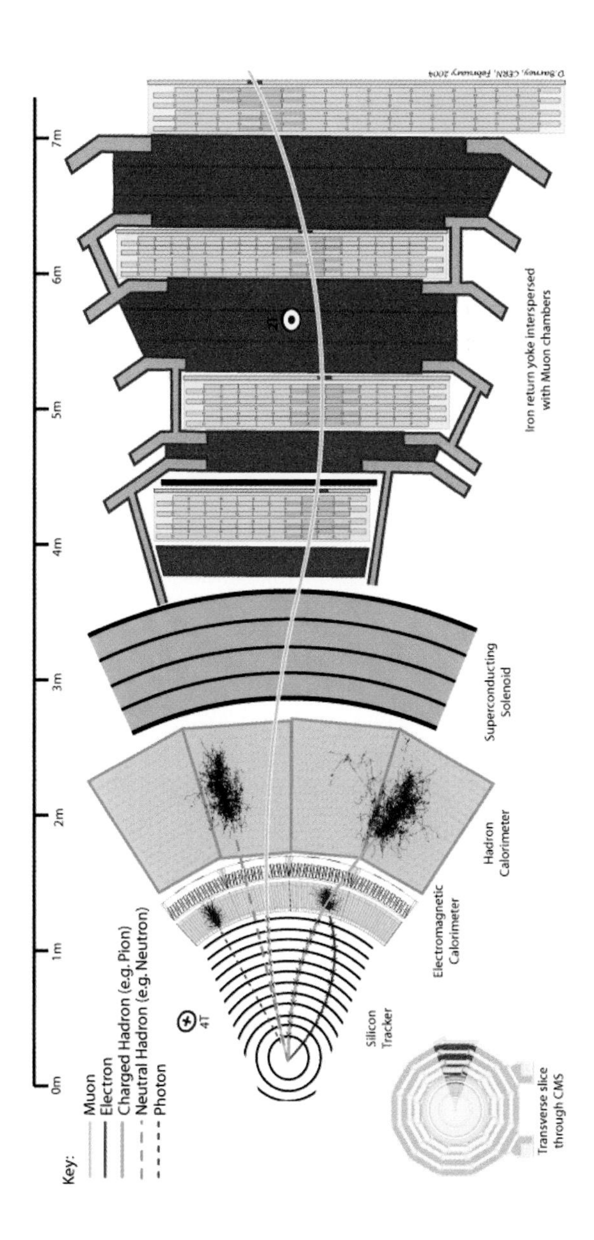

Figure 4.19: Tracks of particles through the CMS detector being constructed for the Large Hadron Collider. The longest (central) track corresponds to a muon, the shortest track that stops in the electromagnetic calorimeter is an electron, and hadrons stop in the hadron calorimeter. [Courtesy CMS collaboration.]

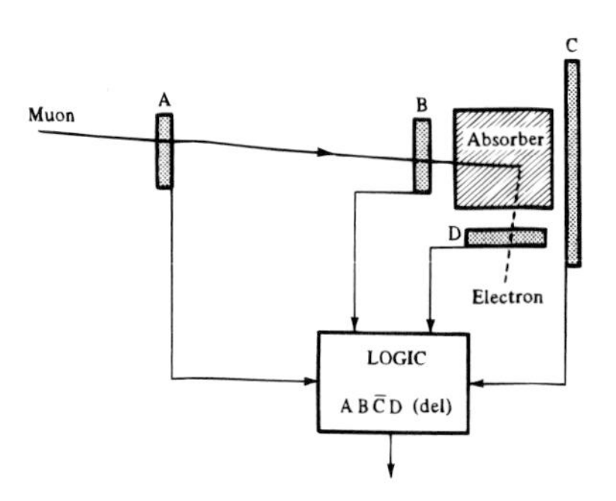

Figure 4.20: Logic elements in a counting system.

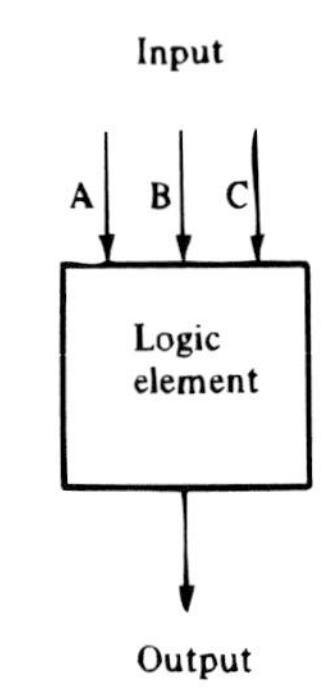

Figure 4.21: Logic element.

interesting enough to be recorded is used. This *electronic trigger* needs to be very fast and can be a very sophisticated electronic system.

Decades ago, nuclear and particle physicists assembled their electronics from components, resistors, capacitors, and vacuum tubes (yes). Later, transistors made the electronics smaller, faster, and more reliable. Now integrated circuits of continuously increasing complexity have become the building blocks. Moreover, much of the instrumentation has been standardized and can be bought; several international standard for modular instrumentation (CAMAC, VME) exist. Setting up a detector electronics system is usually straightforward because many standardized building blocks can be bought; the physicist selects and matches the proper components. We shall not discuss the building blocks here in detail.

4.12 Electronics: Logic

As mentioned before electronic units do considerably more than just process the data from one counter. A simple example, shown in Fig. 4.20, is the stopping of muons in matter. Muons from an accelerator pass through two counters and

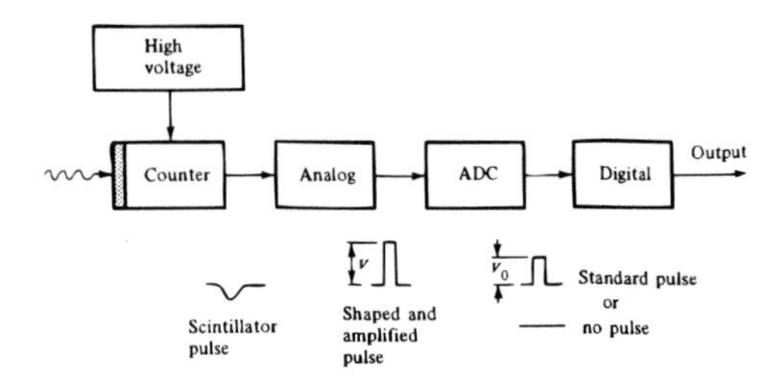

Figure 4.22: Schematic representation of the main components of counter electronics.

Table 4.1: FUNCTION OF THE FOUR LOGIC ELEMENTS AND, OR, NAND, AND NOR. 1 denotes a standard pulse, 0 no pulse. The elements are symmetric in A, B, and C. Only typical cases are shown.

Input			Output			
A	B	C	AND	NAND	OR	NOR
1	1	1	1	0	1	0
1	1	0	0	1	1	0
1	0	0	0	1	1	0
0	0	0	0	1	0	1

enter an absorber where they slow down and finally decay into an electron and two neutrinos:

$$\mu \longrightarrow e\nu\overline{\nu}.$$

We have already mentioned in Section1.3 that the mean life for the decay of a muon at rest is 2.2μs. The procedure in the experiment sketched in Fig. 4.20 is now as follows: the muon should pass through counters A and B but should stop in the absorber and therefore *not* traverse counter C. After a delay of about 1μs, an electron should be observed in counter D. The *logic* must record a muon only if these events happen as described. In shorthand, the requirement can be written as $AB\overline{C}D$(delayed), where the $AB\overline{C}D$ means a coincidence between ABD and an anticoincidence of this threefold coincidence with C. Furthermore, D must respond at least 1μs later than A and B. Such problems can be solved in a straightforward way with logic circuits.

Four *logic elements* are particularly important and useful: AND, OR, NAND, and NOR. The function of these four types can be explained with the aid of Fig. 4.21. The general logic element shown has three inputs and one output. Input and output pulses are of standard size (called 1); 0 denotes no pulse. An AND element produces no output (0) if only one or two pulses arrive. If, however, three pulses arrive within the resolving time (a few ns), a standard output pulse (1) results. OR produces an output pulse if one or more input pulses arrive. NAND (NOT AND) and NOR (NOT OR) are the logical complements; they produce pulses whenever AND, respectively OR, would *not* produce a pulse. The functions of the four elements are summarized in Table 4.1. The element NOR requires one remark. It puts out a steady signal as long as there is no input pulse present; the signal disappears if at least one pulse arrives.

4.13 References

Particle detectors are reviewed by PDG.

Some references appear throughout the chapter. Additional references are:

W. R. Leo, *Techniques for Nuclear and Particle Physics Experiments*, 2nd edition,

Springer, New York, NY, 1994; C. Leroy P.-G. Rancoita *Principles of Radiation Interaction in Matter and Detection*, World Sci., Singapore, 2004; A.C. Melissinos, J. Napolitano, *Experiments in Modern Physics*, 2nd edition, Academic Press, Elsevier, (2003); *Experimental Techniques in High Energy Physics*, (T. Ferbel, ed.), Addison-Wesley, Menlo Park, CA, 1987; C. Grupen, *Particle Detectors*, Cambridge Univ. Press, Cambridge 1996; D. Green, *The Physics of Particle Detectors*, Cambridge Univ.Press, Cambridge, 2000.

There exist many good books on the application of statistics to experiments: P. Bevington, D.K. Robinson, *Data Reduction and Error Analysis for the Physical Sciences*, McGraw-Hill, 2003; B.P. Roe, *Probability and Statistics in Experimental Physics*, Springer-Verlag, NY, 2001; J.R. Taylor, *An Introduction to Error Analysis: The Study of Uncertainties in Physical Measurements*, University Science Books, 1997. Detailed treatments of statistical methods are given in D. Drijard, W.T. Eadie, F.E. James, M.G.W. Roos, and B. Sadoulet, *Statistical Methods in Experimental Physics*, North-Holland, Amsterdam, 1971; T. Tanaka, *Methods of Statistical Physics*, Cambridge University Press, New York, 2002.

Various aspects of data gathering and evaluation are surveyed in *Data Acquisition in High-Energy Physics.* (G. Gologna and M. Vincelli, eds.), North-Holland, Amsterdam, 1982.

Electronics is treated in a number of texts, for instance: P. Horowitz, W. Hill, *The Art of Electronics*, Cambridge University Press, 1989; J.J. Brophy, *Basic electronics for scientists*, McGraw-Hill, 1983.

A recent review on all components of detectors for the LHC can be found in D. Froidevaux, P. Sphicas, *Annu. Rev. Nuc. Part. Sci.* **56**, 375 (2006).

Problems

4.1. * Find the circuit diagram for a photomultiplier. Discuss the importance and the choice of the components.

4.2. A proton with kinetic energy E_k impinges on a 5 cm thick plastic scintillator. Sketch the light output as a function of E_k.

4.3. Three-MeV photons are counted by a $7 \times 7\text{cm}^2$ NaI(Tl) counter.

   (a) Sketch the spectrum.

   (b) Find the probability of observing the photon in the full-energy peak.

4.4. The 14 keV gamma rays from ^{57}Fe must be counted with a NaI(Tl) counter. Higher-energy gamma rays are a nuisance. Find the optimum thickness of the NaI(Tl) crystal.

4.5. Compute and draw the Poisson distribution for $\overline{n} = 1$ and $\overline{n} = 100$.

4.6. Sketch the derivation of Eq. (4.3). Verify Eq. (4.5).

4.7. Compute the variance of $P(n)$ in Eq. (4.7).

4.8. Verify that Eq. (4.7) is the limiting case of a Poisson distribution.

4.9. For the Poisson distribution, compare

$$\frac{P(2\overline{n})}{P(\overline{n})} \quad \text{for } \overline{n} = 1, 3, 10, 100.$$

4.10. A scintillation counter used underground counts, on the average, eight muons/hr. An experiment is run for 10^3hr, and counts are recorded every hr. How often do you expect to find $n = 2$, 4, 7, 8, 16 counts in the records?

4.11. Consider a germanium counter. Discuss the processes in more detail than in the text. In particular, answer the questions

(a) Why does the major part of the counter have to be depleted?

(b) Why is it not possible to simply use metal foils on both sides to collect the charge?

(c) How big a current pulse can be expected for a 100 keV photon?

(d) What limits the low-energy range of such a counter?

4.12. Compute the efficiency of a 1 cm thick germanium counter for photons of

(a) 100 keV.

(b) 1.3 MeV.

4.13. Sketch the construction of a large bubble chamber.

4.14. Consider the 12 ft. Argonne bubble chamber. What is the highest-energy proton that will stop in the chamber? Assume that the same chamber is filled with propane. Compute the range of the proton in this chamber. What energy proton can now be stopped?

4.15. Estimate the magnetic energy stored in the Argonne 12 ft bubble chamber. From what height (in m) would an average car have to be dropped to equal this energy?

4.16. * Discuss the principle of a streamer chamber. How is the voltage produced that is necessary to cause streamers?

4.17. * What limits the speed with which a spark chamber can be triggered? Find typical delay times in the various components of the logical chain.

4.18. Use the elements listed in Table 4.1 to sketch the logic for the experiment of Fig. 4.20.

4.19. Sketch electronic circuits with which the four logic elements AND, OR, NAND, and NOR can be realized.

4.20. If the time resolution of a drift chamber is 1 ns and the drift speed is 5 cm/μs, what spatial resolution can be achieved?

4.21. If the index of refraction of a material is independent on frequency, what energy is lost in Čerenkov radiation between frequency f_1 and f_2 ?

4.22. A muon is created in the ocean by an upward going neutrino and continues to move vertically upwards

(a) What is the minimum energy of the muon to emit Čerenkov radiation? Take $n_{water} = 1.33$.

(b) If the muon has an energy of 200 GeV, will the Čerenkov light be totally (internally) reflected at the surface of the ocean? If not, what will be the angle of emission (refraction) of the light coming out of the ocean?

4.23. What is the number of generations that develop in a shower after n radiation lengths?

Part II

Particles and Nuclei

The situation is familiar. At a meeting we are introduced to some stranger. A few minutes later we realize with embarrassment that we have already forgotten his name. Only after being reintroduced a few times do we begin to fit the stranger into our catalog of people. The same phenomenon takes place when we encounter new concepts and new facts. At first they slip away rapidly, and only after grappling with them a number of times do we become familiar with them. The situation is particularly true with particles and nuclei. There are so many that at first they seem not to have sharp identities. So what is the difference between a muon and a pion?

In Part II we shall introduce many subatomic particles and describe some of their properties. Such a first introduction is not sufficient to give a clear picture, and we shall therefore return again to particle and nuclear characteristics in later chapters. They will lose their "look-alike" status, and it will become clear, for instance, that muons and pions have less in common than man and microbe. The first and most obvious questions are: What are particles? Can composite and elementary particles be distinguished? We shall try to explain why it is difficult to respond unambiguously to the apparently simple questions. Consider first the Franck–Hertz experiment[(1)] in which a gas, for instance helium or mercury, is studied by the passage of electrons through it. Below an energy of 4.9 eV in mercury vapor, the Hg atom behaves like an elementary particle. At an electron energy of 4.9 eV the first excited state of Hg is reached, and the mercury atom begins to reveal its structure. At 10.4 eV, an electron is knocked out; at 18.7 eV, a second electron is removed and it is apparent that electrons are atomic constitutents. A similar situation exists with nuclei. At low electron energies, the electron cannot excite the nuclear levels, and the nucleus appears as an elementary particle. At higher electron energies, the nuclear levels become apparent, and it is possible to knock out nuclear constituents, protons and neutrons. The question is now shifted to the new actors, proton and neutron. Are they elementary? Protons and neutrons can also be probed with electrons. At energies of a few hundred MeV it becomes apparent that the nucleons,

[1]R. Eisberg, Section 5.5; W. Kendall and W.K.H. Panofsky, *Sci. Amer.* **224**, 60 (June 1971).

neutron and proton, are not point particles but have a "size" of the order of 1 fm. It also turns out that the nucleons have excited states, just as atoms and nuclei do. These excited states decay very rapidly, usually with the emission of a particle, the pion. At still higher energies, more particles are created; finally, above 10 GeV, it becomes clear that proton, neutron, and all the created particles are not elementary, but are composed of quarks.(2) At present we believe that quarks, like electrons, are point particles; electron scattering reveals no structure at the level of 10^{-18} m. Thus, the conceptually simple experiment of hitting a target with electrons of ever increasing energy reveals that the notion of "elementary particle" has no simple meaning and depends on the energy and means of observation. It also shows, however, that the very large number of observed particles can be explained in terms of a relatively small number of "elementary constituents", the quarks. Thus leptons and quarks are the building blocks of the present particle zoo. It is not known if these building blocks are, in turn, composed of even more fundamental entities,(3) possibly "superstrings".(4) A second set of particles, called gauge bosons, appear when we consider the forces between leptons and/or quarks. It is now accepted that the forces between particles are carried by fields and their quanta.(5) In subatomic physics, these quanta, the gauge bosons, all have spin $= 1\hbar$; the best known one is the photon which transmits the electro-magnetic force between charged particles. The hadronic force is mediated by gluons and the weak force by the exchange of "intermediate bosons", of which there are three.(6) In the next two chapters we describe some of the salient experimental facts concerning subatomic particles.

[2] H. Fritzsch, *Quarks*, Basic Books, New York, 1983.

[3] H. Harari, *Sci. Amer.* **248**, 56 (April 1983).

[4] M.B. Green, *Sci. Amer.* **255**, 48 (September 1986), B. Greene, *The Elegant Universe: Superstrings, Hidden Dimensions, and the Quest for the Ultimate Theory*, W.W. Norton, New York, 1999.

[5] C. Quigg, *Sci. Amer.* **252**, 84 (April 1985).

[6] C. Rubbia, *Rev. Mod. Phys.* **57**, 699 (1985), P. Watkins, *Story of the W and Z*, Cambridge Univ. Press, Cambridge, 1986.

Chapter 5

The Subatomic Zoo

A conventional zoo is a collection of various animals, some familiar and some strange. The subatomic zoo also contains a great variety of inhabitants, and a number of questions concerning the catching, care, and feeding of these come to mind: (1) How can the particles be produced? (2) How can they be characterized and identified? (3) Can they be grouped in families? In the present chapter, we concentrate on the second question. In the first two sections, the properties that are essential for the characterization of the particles are introduced. Some members of the zoo already appear in these two sections as examples. In the later sections, the various families are described in more detail. Since there are so many animals in the subatomic zoo, some initial confusion in the mind of the reader is unavoidable. We hope, however, that the confusion will give way to order as the same particles appear again and again.

5.1 Mass and Spin. Fermions and Bosons

A first identification of a particle is usually made by measuring its *mass*, m. In principle, the mass can be found from Newton's law by observing the acceleration, $\boldsymbol{a}$, in a force field, $\boldsymbol{F}$:

$$m = \frac{|\boldsymbol{F}|}{|\boldsymbol{a}|}. \tag{5.1}$$

Equation (5.1) is not valid relativistically, but the correct generalization poses no problems. We only note that with mass we always mean *rest mass*. The actual determination of masses will be discussed in Section 5.3. The rest masses of subatomic particles vary over a wide range. The photon has zero rest mass. The lightest massive particles are the neutrinos with rest masses less than $1eV/c^2$; the electron is the next lightest particle with a mass, m_e, of about $10^{-27}\text{g} \approx 0.51MeV/c^2$. Then comes the muon with a mass of about $200m_e$. From there on, the situation gets more complex, and many particles with strange and wonderful properties have masses that lie between about 270 times the electron mass to several orders of magnitude higher. Nuclei, which of course are also subatomic particles, start with the proton, the nucleus of the hydrogen atom, with a mass of about $2000m_e$. The heaviest known nucleus is about 260 times more massive than the proton. The masses (not counting zero) consequently vary by a factor of over a billion. We shall return to the masses a few more times, and details will become clearer as more specific examples appear. However, just as it is impossible to understand chemistry without a thorough knowledge of the periodic table, it is difficult to obtain a clear picture of the subatomic world without an acquaintance with the main occupants of the subatomic zoo.

A second property that is essential in classifying particles is the *spin* or *intrinsic angular momentum.* Spin is a purely quantum mechanical property, and it is not easy to grasp this concept at first. As an introduction we therefore begin to discuss the *orbital* angular momentum which has a classical meaning. Classically, the orbital angular momentum of a particle with momentum $\boldsymbol{p}$ is defined by

$$\boldsymbol{L} = \boldsymbol{r} \times \boldsymbol{p}, \tag{5.2}$$

where $\boldsymbol{r}$ is the radius vector connecting the center of mass of the particle to the point to which the angular momentum is referred. Classically, orbital angular momentum can take any value. Quantum mechanically, the magnitude of $\boldsymbol{L}$ is restricted to certain values. Moreover, the angular momentum vector can assume only certain orientations with respect to a given direction. The fact that such a *spatial quantization* exists appears to violate intuition. However, the existence of spatial quantization is beautifully demonstrated in the Stern–Gerlach experiment,[(1)] and it follows logically from the postulates of quantum mechanics. In quantum mechanics, $\boldsymbol{p}$ is replaced by the operator $-i\hbar(\partial/\partial x, \partial/\partial y, \partial/\partial z) \equiv -i\hbar\boldsymbol{\nabla}$ and the orbital angular momentum consequently also becomes an operator[(2)] whose z component, for instance, is given by

$$L_z = -i\hbar\left(x\frac{\partial}{\partial y} - y\frac{\partial}{\partial x}\right) = -i\hbar\frac{\partial}{\partial\varphi}, \tag{5.3}$$

where φ is the azimuthal angle in polar coordinates. The wave function of a particle with definite angular momentum can then be chosen to be an eigenfunction of $\boldsymbol{L}^2$ and L_z:[(3)]

$$\begin{aligned} \boldsymbol{L}^2\psi_{lm} &= l(l+1)\hbar^2\psi_{lm} \\ L_z\psi_{lm} &= m\hbar\psi_{lm}. \end{aligned} \tag{5.4}$$

[1]Tipler and Llewellyn, Chapter 7; Feynman Lectures, II-35-3.

[2]Tipler and Llewellyn, Chapter 7; Merzbacher, Chapter 9.

[3]Some confusion can arise from the usual convention that classical quantities (e.g., $\boldsymbol{L}$) and the corresponding quantum mechanical operators (e.g., $\boldsymbol{L}$) are denoted by the same symbol. Moreover, the quantum numbers are often also denoted by similar symbols (l or L). We follow this convention because most books and papers use it. After some initial bewilderment, the meaning of all symbols should become clear from the context. Occasionally we use the subscript *op* for quantum mechanical operators.

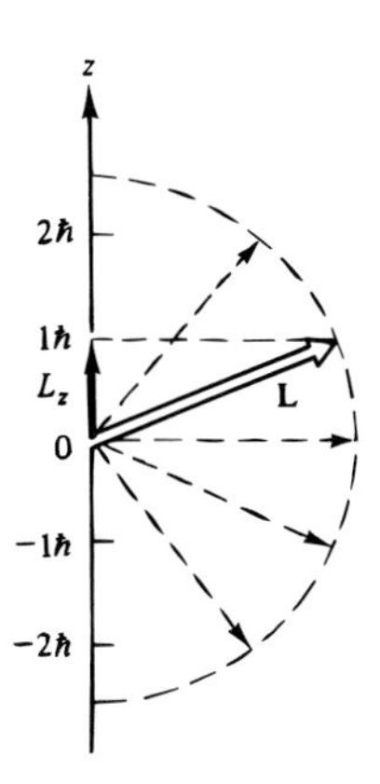

Figure 5.1: Vector diagram for an angular momentum with quantum number $l = 2, m = 1$. The other possible orientations are indicated by dashed lines.

The first equation states that the magnitude of the angular momentum is quantized and restricted to values $[l(l+1)]^{1/2}\hbar$. The second equation states that the component of the angular momentum in a given direction, called z by general agreement, can assume only values $m\hbar$. The quantum numbers l and m must be *integers*, and for a given value of l, m can assume the $2l + 1$ values from $-l$ to $+l$. The spatial quantization is expressed in a *vector diagram*, shown in Fig. 5.1 for $l = 2$. The component along the arbitrarily chosen z direction can assume only the values shown.

We repeat again that the quantization of the orbital angular momentum Eq. (5.2) leads to integral values of l and hence to *odd* values of $2l + 1$, the number of possible orientations. It was therefore a surprise when the alkali spectra showed unmistakable doublets. Two orientations demand $2l + 1 = 2$ or $l = \frac{1}{2}$. Many attempts were made before 1924 to explain this half-integer number. The first half of the correct solution was found by Pauli in 1924; he suggested that the electron possesses a classically nondescribable two-valuedness, but he did not associate a physical picture with this property. The second half of the solution was provided by Uhlenbeck and Goudsmit, who postulated a spinning electron. The two-valuedness then arises from the two different directions of rotation.

Of course, a way has to be found to incorporate the value $\frac{1}{2}$ into quantum mechanics. It is easy to see that the quantum mechanical operators that correspond to L, Eq. (5.2), satisfy the *commutation relations*

$$\begin{aligned} L_xL_y - L_yL_x &= i\hbar L_z \\ L_yL_z - L_zL_y &= i\hbar L_x \\ L_zL_x - L_xL_z &= i\hbar L_y. \end{aligned} \tag{5.5}$$

It is postulated that the commutation relations, Eq. (5.5), are more fundamental than the classical definition, Eq. (5.2). To express this fact, the symbol $\boldsymbol{L}$ is reserved for the orbital angular momentum, and a symbol $\boldsymbol{J}$ is introduced that stands for any angular momentum. $\boldsymbol{J}$ is assumed to satisfy the commutation relations

$$\begin{aligned} J_xJ_y - J_yJ_x &= i\hbar J_z \\ J_yJ_z - J_zJ_y &= i\hbar J_x \\ J_zJ_x - J_xJ_z &= i\hbar J_y. \end{aligned} \tag{5.6}$$

The consequences of Eq. (5.6) can be explored by using algebraic techniques.(4) The result is a vindication of Pauli's and of Goudsmit and Uhlenbeck's proposals. The operator $\boldsymbol{J}$ satisfies eigenvalue equations analogous to the ones for the orbital operator, Eq. (5.4):

$$J^2\psi_{JM} = J(J+1)\hbar^2\psi_{JM} \tag{5.7}$$

$$J_z\psi_{JM} = M\hbar\psi_{JM}. \tag{5.8}$$

However, the allowed values of J are not only integers but also half-integers:

$$J = 0, \tfrac{1}{2}, 1, \tfrac{3}{2}, 2, \ldots. \tag{5.9}$$

For each value of J, M can assume the $2J+1$ values from $-J$ to $+J$.

Equations (5.7)–(5.9) are valid for any quantum mechanical system. As for any angular momentum, the particular value of J depends not only on the system but also on the reference point to which the angular momentum is referred. Now we return to *particles.* It turns out that each particle has an *intrinsic angular momentum,* usually called *spin.* Spin cannot be expressed in terms of the classical position and momentum coordinates, as in Eq. (5.2), and it has no analog in classical mechanics. Spin is often pictured by assuming the particle to be a small fast-spinning top (see Fig. 5.2.) However, for any acceptable radius of the particle the velocity at the surface of the particle then exceeds the velocity of light, and the picture therefore is not really tenable. In addition, even particles with zero rest mass, such as the photon and the neutrino, possess a spin. The existence of spin has to be accepted as a fact. In the rest frame of the particle, any orbital contribution to the total angular momentum disappears, and the spin is the angular momentum in the rest frame. It is an immutable characteristic of a particle. The spin operator is denoted by $\boldsymbol{J}$ or by $\boldsymbol{S}$;(5) it satisfies the eigenvalue equations (5.7) and (5.8). The quantum number J is a constant and characterizes the particle, while the quantum number M describes the orientation of the particle in space and depends on the choice of the reference axis.

---

[4] A clear and concise derivation is given in Messiah, Chapter XIII.

[5] S will later also be used for *strangeness,* and therefore S does not always denote the spin quantum number.

How can J be determined experimentally? For a macroscopic system, the classical angular momentum can be measured. For a particle such a measurement is not feasible. However, if we succeed in determining the number of possible orientations in space, the spin quantum number J, usually just called *the spin*, follows because there are $2J + 1$ possible orientations.

We have noted above that integer J values occur in connection with orbital angular momentum, which has a classical limit, but that half-integral values have no classical counterpart. As we shall see soon, particles with integer and half-integer spins exist. Examples for the integer class are the photon and the pion, whereas electrons, neutrinos, muons, and nucleons have spin $\frac{1}{2}$. Does the difference between integer and half-integer values express itself in some profound way? It indeed does, and the two classes of particles behave very differently. The difference becomes apparent when the properties of wave functions are studied. Consider a system of two *identical* particles, denoted by 1 and 2. The particles have the same spin J, but their orientation, given by $J_z^{(i)}$, can be different. The wave function of the system is written as

$$\psi(\boldsymbol{x}^{(1)}, J_z^{(1)}; \boldsymbol{x}^{(2)}, J_z^{(2)}) \equiv \psi(1,2).$$

If the two particles are interchanged, the wave function becomes $\psi(2,1)$. It is a remarkable fact of nature that all wave functions for identical particles are either symmetric or antisymmetric under the interchange $1 \rightleftharpoons 2$:

$$\begin{aligned} \psi(1,2) &= +\psi(2,1), \quad \text{symmetric} \\ \psi(1,2) &= -\psi(2,1), \quad \text{antisymmetric.} \end{aligned} \tag{5.10}$$

Complete symmetry or antisymmetry under interchange of any two particles is easily extended to n identical particles.[6]

There exists a profound connection between *spin and symmetry* that was first noted by Pauli and that was proved by him using relativistic quantum field theory: The wave function of a system of n identical particles with half-integer spin, called *fermions*, changes sign if any two particles are interchanged. The wave function of a system of n identical particles with integer spin, called *bosons*, remains unchanged under the interchange of any two particles. The spin-symmetry relation is summarized in Table 5.1.

The connection between spin and symmetry leads to the *Pauli exclusion principle.* Assume that two particles have exactly the same quantum numbers. The two particles are then said to be in the same *state.* An interchange $1 \rightleftharpoons 2$ will leave the wave function unchanged. However, if the two particles are fermions, the wave function changes sign, and it consequently must vanish. The exclusion principle hence states that one quantum mechanical state can be occupied by only one *fermion.*[7] The principle is extremely important in all of subatomic physics.

---

[6] Park, Chapter 11.

[7] Pauli describes the situation in the following words:

Table 5.1: BOSONS AND FERMIONS.

Spin J	Particles	Behavior of Wave Function Under Interchange of Any Two Identical Particles
Integer	Bosons	Symmetric
Half-integer	Fermions	Antisymmetric

5.2 Electric Charge and Magnetic Dipole Moment

Many particles possess *electric charges.* In an external electromagnetic field, the force on a particle of charge q will be given by Eq. (2.21),

$$\boldsymbol{F} = q\left(\boldsymbol{E} + \frac{1}{c}\boldsymbol{v} \times \boldsymbol{B}\right). \tag{5.11}$$

The deflection of the particle in a purely electric field $\boldsymbol{E}$ determines q/m. If m is known, q can be determined. Historically, progress went the inverse way: The electron charge was determined by Millikan in his oil drop experiment. With q and q/m known, the electron mass was found.

The *total charge* of a subatomic particle determines its interaction with $\boldsymbol{E}$ and $\boldsymbol{B}$, as expressed by the Lorentz equation (5.11). It is a remarkable and not understood observation that, for all observed particles, the charge always appears in integer multiples of the elementary quantum e. Because of this fact, the total charge gives little information about the structure of a subatomic system. Other electromagnetic properties, however, do so, and the most prominent is the *magnetic dipole moment.* A classical particle with charge and spin contains currents and consequently presents a magnetic dipole moment (Fig. 5.2).

If electric charges are distributed throughout the particle, they will spin also and give rise to current loops, which produce a magnetic dipole moment, $\boldsymbol{\mu}$. How does such a current distribution interact with an external magnetic field $\boldsymbol{B}$? Classical electrodynamics shows that a current loop as in Fig. 5.3. leads to an energy

$$E_{\text{mag}} = -\boldsymbol{\mu} \cdot \boldsymbol{B}, \tag{5.12}$$

"If one pictures by boxes the nondegenerate states of an electron in an atom, the exclusion principle maintains that a box can contain no more than one electron. This, for example, makes the atoms much larger than if many electrons could be contained in the innermost shell. Quantum theory maintains that other particles such as photons or light particles show opposite behavior; that is, as many as possible fill the same box. One can call particles obeying the exclusion principle the 'antisocial' particles, while photons are 'social.' However, in both cases sociologists will envy the physicists on account of the simplifying assumption that all particles of the same type are exactly alike."

From W. Pauli, *Science* **103**, 213 (1946). Reprinted in *Collected Scientific Papers by Wolfgang Pauli* (R. Kronig and V. F. Weisskopf, eds), Wiley-Interscience, New York, 1964.

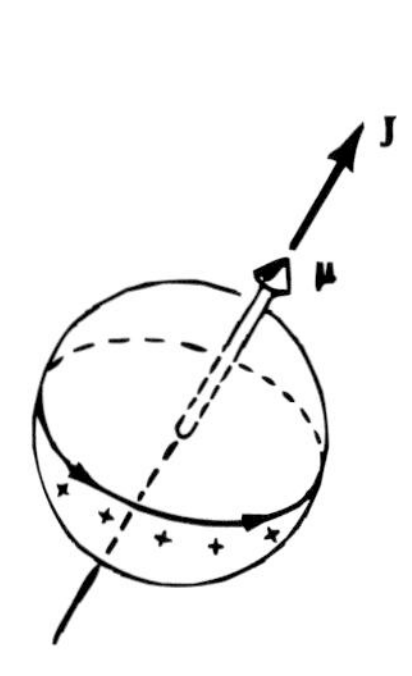

Figure 5.2: Magnetic dipole moment. In a classical picture the spinning particle gives rise to electric current loops, which, in turn, produce a magnetic dipole moment.

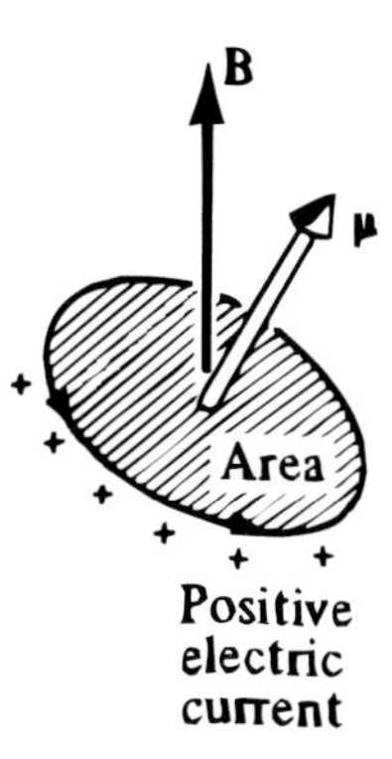

Figure 5.3: A current loop gives rise to a magnetic moment $\boldsymbol{\mu}$. The direction of the magnetic moment is perpendicular to the plane bounded by the current.

where the magnitude of the magnetic dipole moment $\boldsymbol{\mu}$ is, in Gaussian units, given by

$$\mu = \frac{1}{c}\text{current} \times \text{area}. \tag{5.13}$$

The direction of $\boldsymbol{\mu}$ is perpendicular to the plane of the current loop; positive current and $\boldsymbol{\mu}$ form a right-handed screw.[8] A connection between magnetic moment and angular momentum is established by considering a particle of charge q moving with velocity $\boldsymbol{v}$ in a circular orbit of radius r (Fig. 5.4).

The particle revolves $v/(2\pi r)$ times/sec and hence produces a current $qv/2\pi r$. With Eqs. (5.2) and (5.13), $\boldsymbol{\mu}$ and $\boldsymbol{L}$ are related by

$$\boldsymbol{\mu} = \frac{q}{2mc}\boldsymbol{L}. \tag{5.14}$$

This result suffers from two defects. It has been derived by using classical physics, while the subatomic particles we are interested in here are not *classical*, and it applies to a point particle moving in a circular orbit.

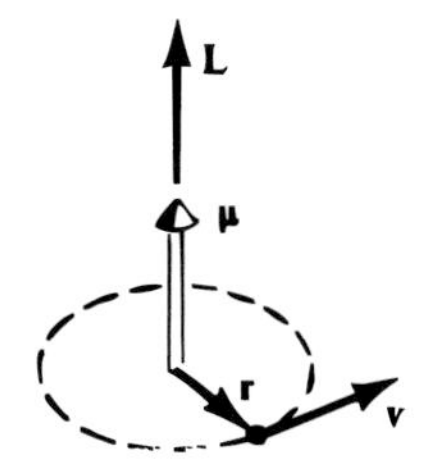

Figure 5.4: A particle of mass m and charge q on a circular orbit produces a magnetic moment $\boldsymbol{\mu}$ and an orbital angular moment $\boldsymbol{L}$.

Nevertheless, Eq. (5.14) exhibits two significant facts: $\boldsymbol{\mu}$ points in the direction of $\boldsymbol{L}$, and the ratio μ/L is given by $q/2mc$. These two facts indicate a way to define a quantum mechanical operator $\boldsymbol{\mu}$ for a particle with mass m and spin J. Even in

[8]Jackson, Eqs. (5.57) and (5.59).

this case, $\boldsymbol{\mu}$ should be parallel to $\boldsymbol{J}$ because there is no other preferred direction; the operators $\boldsymbol{\mu}$ and $\boldsymbol{J}$ are consequently related by

$$\boldsymbol{\mu} = \text{const.}\boldsymbol{J}.$$

According to Eq. (5.14), the constant has the dimension e/mc, and it is convenient to write const. $= g(e/2mc)$. The new constant g is then dimensionless, and the relation between $\boldsymbol{\mu}$ and $\boldsymbol{J}$ becomes

$$\boldsymbol{\mu} = g\frac{e}{2mc}\boldsymbol{J}. \tag{5.15}$$

The constant g measures the deviation of the actual magnetic moment from the simple value $e/2mc$. Note that e and not q is used in Eq. (5.15). While q can be positive or negative, e is defined to be positive, and the sign of $\boldsymbol{\mu}$ is given by the sign of the g factor. $\boldsymbol{J}$ has the same units as $\hbar$ so that $\boldsymbol{J}/\hbar$ is dimensionless. Equation (5.15) is therefore rewritten as

$$\boldsymbol{\mu} = g\mu_0\frac{\boldsymbol{J}}{\hbar} \tag{5.16}$$

$$\mu_0 = \frac{e\hbar}{2mc}. \tag{5.17}$$

The constant μ_0 is called a *magneton*, and it is the unit in which magnetic moments are measured. Its value depends on the mass that is used. In atomic physics and in all problems involving electrons, m in Eq. (5.17) is taken to be the electron mass, and the unit is called the *Bohr magneton* (μ_B):

$$\mu_B = \frac{e\hbar}{2m_ec} = 5.7884 \times 10^{-15}\ \text{MeV/G}. \tag{5.18}$$

In subatomic physics, magnetic moments are expressed in terms of *nuclear magnetons*, obtained from Eq. (5.17) with $m = m_p$:

$$\mu_N = \frac{e\hbar}{2m_pc} = 3.1525 \times 10^{-18}\ \text{MeV/G}. \tag{5.19}$$

The nuclear magneton is about 2000 times smaller than the Bohr magneton.

Information about the structure of a particle is contained in the g factor. For a large number of nuclear states and for a small number of particles, the g factor has been measured. It is the problem of theory to account for the observed values.

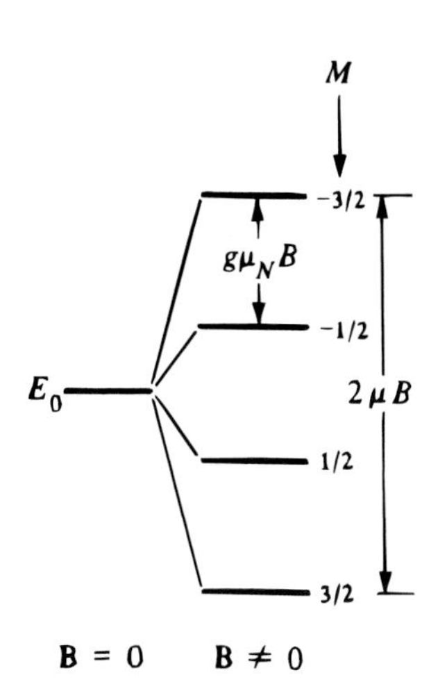

Figure 5.5: Zeeman splitting of the energy levels of a subatomic particle with spin J and g factor g in an external magnetic field $\boldsymbol{B}$. $\boldsymbol{B}$ is along the z axis, $g > 0$.

The energy levels of a particle with magnetic moment $\boldsymbol{\mu}$ in a magnetic field $\boldsymbol{B}$ are obtained from the Schrödinger equation,

$$H\psi = E\psi,$$

where the Hamiltonian H is assumed to have the form

$$H = H_0 + H_{\text{mag}} = H_0 - \boldsymbol{\mu} \cdot \boldsymbol{B},$$

or, with Eq. (5.16),

$$H = H_0 - \frac{g\mu_0}{\hbar}\, \boldsymbol{J} \cdot B. \tag{5.20}$$

The spin-independent Hamiltonian H_0 gives rise to an energy $E_0 : H_0\psi = E_0\psi$. To find the energy values corresponding to the complete Hamiltonian, the z axis is conveniently chosen along the magnetic field so that $\boldsymbol{J} \cdot B = J_z B_z \equiv J_z B$.

With Eq. (5.8), the eigenvalues E of the Hamiltonian H are

$$E = E_0 - g\mu_0 M B. \tag{5.21}$$

where M assumes the $2J + 1$ values from $-J$ to $+J$. The corresponding *Zeeman* splitting is shown in Fig. 5.5 for a spin $J = \frac{3}{2}$.

Experimentally the splitting $\Delta E = g\mu_0 B$ between two Zeeman levels is determined. If B is known, g follows. Nevertheless the value quoted in the literature is usually not g but a quantity μ, defined by

$$\mu = g\mu_0 J, \tag{5.22}$$

where J is the quantum number defined in Eq. (5.7). As can be seen from Fig. 5.5, $2\mu B$ is the total splitting of the Zeeman levels. (Quantum mechanically, μ is the expectation value of the operator Eq. (5.16) in the state $M = J$). To determine μ, g and J have to be known. J can in principle be found from the Zeeman effect because the total number of levels is equal to $2J + 1$.

5.3 Mass Measurements

The mass is the home address of a particle or nucleus, and it is therefore no surprise that there exist many methods for its measurement. We shall discuss only three here, and we have selected three that are different in character and apply to very different situations.

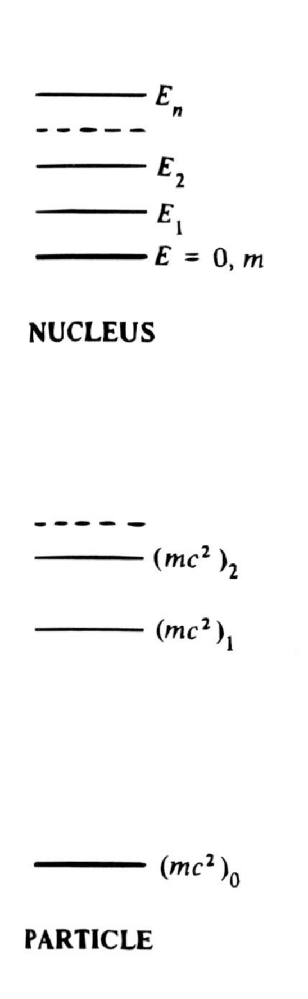

Figure 5.6: Level diagrams of nuclei and particles. The notation is explained in the text.

Subatomic particles are quantum systems, and nearly all of the ones that are not elementary particles possess excited states. Schematically the level diagrams appear as shown in Fig. 5.6. Even though the basic aspects are similar for nuclei and particles, units and notation differ. In the case of *nuclei*, the mass of the ground state is quoted not for the nucleus alone but for the neutral atom, including all electrons. The international unit for the *atomic mass* is one twelfth of the atomic mass of ^{12}C. This unit is called the *atomic mass unit* and is abbreviated u. In terms of grams and MeV, it is

$$\begin{aligned} 1\ u &\approx 1.66054 \times 10^{-24}\ \text{g (mass)} \\ &\approx 931.494\ \text{MeV}/c^2. \end{aligned} \tag{5.23}$$

The masses of nuclear ground states are given in u. The excited nuclear states are not characterized by their masses but by their excitation energies (MeV above ground state). In the case of *particles*, rest energies are given, and they are quoted in MeV or GeV. This procedure is arbitrary but makes sense because in the nuclear case excitation energies are small compared to the rest energy of the ground state, whereas in the particle case excitation energies and ground-state energies are comparable.

After these preliminary remarks we turn to *mass spectroscopy*, the determination of nuclear masses. The first mass spectrometer was built in 1910 by J. J. Thomson, advanced by F. W. Aston. The components of Aston's mass spectrometer are shown in Fig. 5.7. Atoms are ionized in an ion source. The ions are accelerated by a voltage of 20–50 kV. The beam is collimated by slits and passes through an electric and a magnetic field. These fields are so chosen that ions of different velocity but with the same charge-to-mass ratio are focused on the photographic plate. The positions of the various ions on the photographic plate permit a determination of the relative masses with accuracy. However, the most accurate determination of nuclear masses have been performed with ion traps (see Section 6.5 for a description of Penning traps) where instead of measuring the *deflection* of charged particles in a field one determines the *frequency* of oscillations in a field. In recent years there has been great progress in using these techniques to accurately determine the masses of

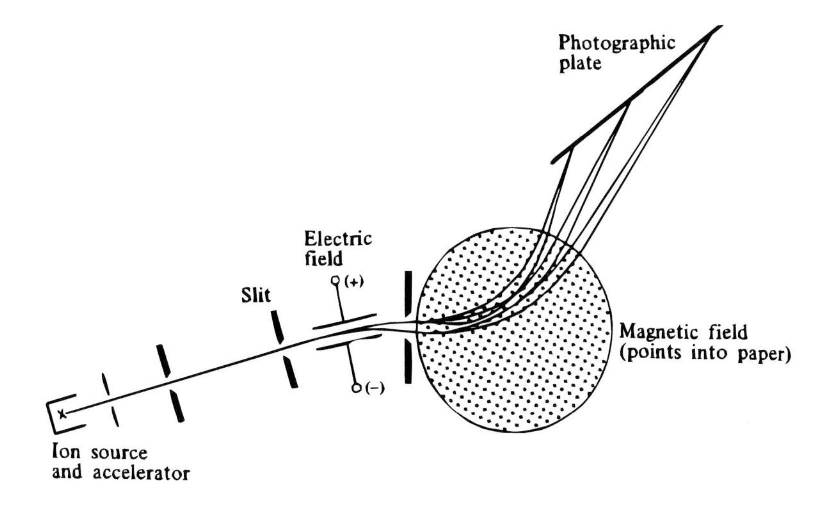

Figure 5.7: Aston's mass spectrometer.

short-lived isotopes,[9] an issue of great importance to understand the production of elements in stars (see Chapter 19.)

Mass spectroscopy works well for nuclei, but it is difficult (or impossible) to apply to most particles. In the mass spectrometer, all ions start with a very small (thermal) velocity and are accelerated in the same field. Their relative masses can therefore be determined very accurately. However, particles are produced in reactions, and their initial velocities are not accurately known. Moreover, some of the particles are neutral and cannot be deflected. Different approaches are necessary, and they are based on Eqs. (1.2) and (1.7):

$$E^2 = p^2c^2 + m^2c^4 \tag{1.2}$$

$$\boldsymbol{p} = m\gamma\boldsymbol{v} \tag{1.7}$$

$$\gamma = \frac{1}{(1-(v/c)^2)^{1/2}}. \tag{1.6}$$

These relations show that the mass of a particle can be computed if momentum and energy or momentum and velocity are known. Many techniques are based on this fact, and the arrangement shown in Fig. 5.8 provides an example. A magnet selects particles with momentum $\boldsymbol{p}$. Two scintillation counters, S_1 and S_2, record the passage of a particle. The time delay between pulses S_2 and S_1 can be measured and, with the distance between S_1 and S_2 known, the velocity can be computed. Together, momentum and velocity give the mass.

The method just discussed fails if the particle is neutral or if its life-time is so

[9] These techniques have been brought to a fine point by H.-J. Kluge and collaborators, see K. Blaum, *Phys. Rep.* **425**, 1 (2006).

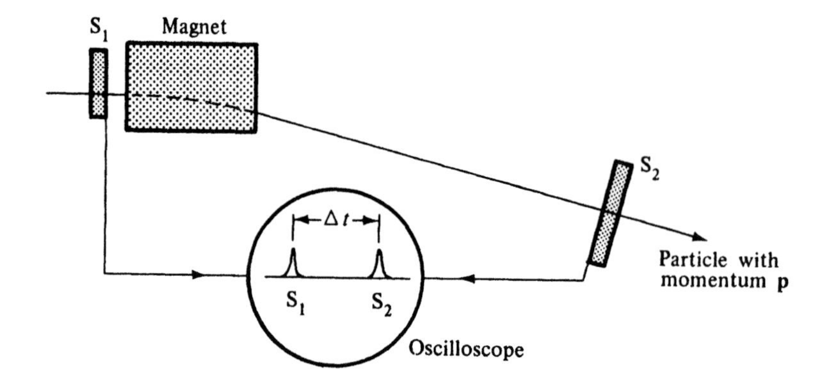

Figure 5.8: Determination of the mass of a particle by selecting its momentum $\boldsymbol{p}$ and measuring its velocity v.

short that neither momentum nor velocity can be measured. As an example of how it is even then possible to obtain a mass, we discuss the *invariant mass plot*. Consider the reaction

$$p\pi^- \longrightarrow n\pi^+\pi^-, \tag{5.24}$$

taking place in a hydrogen bubble chamber.

The reaction can proceed in two different ways, shown in Fig. 5.9. If it proceeds as in Fig. 5.9(a), the three particles in the final state will be created incoherently. It is, however, also possible that a neutron and a new particle, called a *neutral rho*, will be produced (Fig. 5.9(b)). The neutral rho then decays into two pions. Is it possible to distinguish between the two cases? Yes, as we see now. If the rho lives for a sufficiently long time, there will be a gap between the proton and the pion tracks. We shall see in Section 5.7 that the lifetime of the ρ^0 is about 6×10^{-24} sec. Even if the ρ^0 moves with the velocity of light, it will travel only about 1.5 fm during one mean life, about a factor 10^{10} less than needed for observation. How can the ρ^0 be detected and its mass be determined? To see how the trick is done, consider the energies and momenta involved (Fig. 5.10).

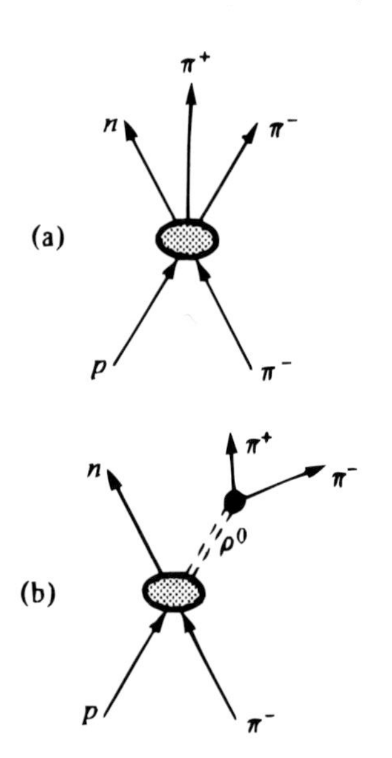

Figure 5.9: The reaction $p\pi^- \rightarrow n\pi^+\pi^-$ can proceed in two different ways: (a) The three particles in the final state can all be produced in one step, or (b) in the first step, two particles, n and ρ^0, are created. ρ^0 then decays into two pions.

Earlier, in Eq. (2.29), we defined the total or invariant mass of a system of particles. Applying this definition to the two pions and using the notation defined in Fig. 5.10, the invariant mass m_{12} of the two pions is

$$m_{12} = \frac{1}{c^2}[(E_1 + E_2)^2 - (\boldsymbol{p}_1 + \boldsymbol{p}_2)^2 c^2]^{1/2}. \quad (5.25)$$

If a magnetic field is applied to the bubble chamber, the momenta of the two charged pions can be determined. The energy can be found from their range (Fig. 3.6) or their ionization. For every observed pion pair, the invariant mass m_{12} can then be computed from Eq. (5.25). If the reaction proceeds according to Fig. 5.9(a), with no correlation between the two pions and the neutron, they will share energy and momentum statistically. The number of pion pairs with a certain invariant mass, $N(m_{12})$, can be calculated in a straightforward way, and the result is called a *phase-space spectrum*. (Phase space will be discussed in Section 10.2.) It is sketched in Fig. 5.11.

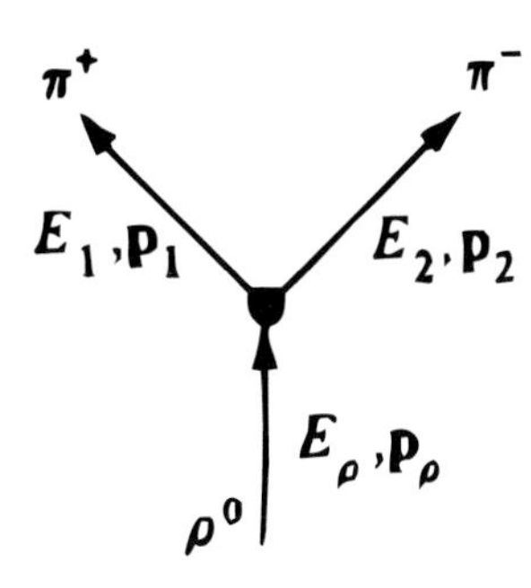

Figure 5.10: Energies and momenta involved in the decay of the ρ^0.

If, on the other hand, the reaction proceeds via the production of a ρ, energy and momentum conservation demand

$$E_\rho = E_1 + E_2, \qquad \boldsymbol{p}_\rho = \boldsymbol{p}_1 + \boldsymbol{p}_2. \quad (5.26)$$

The mass of the rho is given by Eq. (1.2) as

$$m_\rho = \frac{1}{c^2}(E_\rho^2 - \boldsymbol{p}_\rho^2 c^2)^{1/2};$$

or, with Eqs. (5.25) and (5.26), as

$$m_\rho = m_{12}. \quad (5.27)$$

If the pions result from the decay of a particle, their invariant mass will be a constant and will be equal to the mass of the decaying particle. Figure 5.12 shows an early result, the invariant mass spectrum of pion pairs produced in the reaction Eq. (5.24) with pions of momentum 1.89 GeV/c. A broad peak at an invariant mass of 765 MeV/c^2 is unmistakable. The particle giving rise to this peak is called the rho. Even though it lives only about 6×10^{-24} sec, its existence is well established and its mass known.

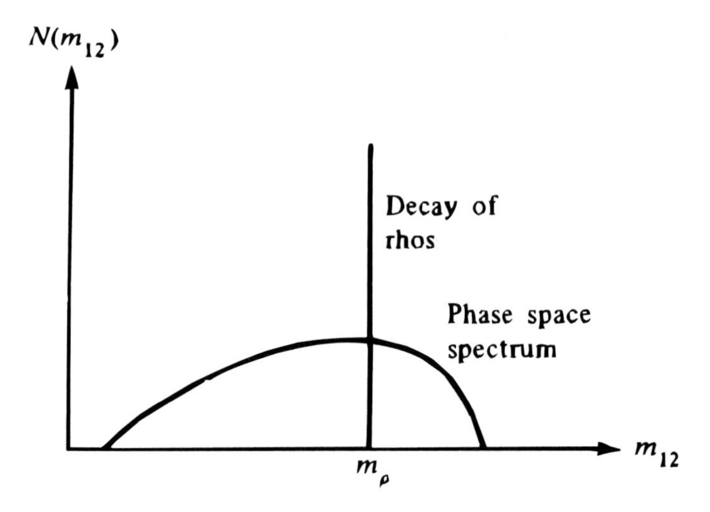

Figure 5.11: Invariant mass spectrum if pion pairs are produced independently (phase space) or if they result from the decay of a rho of small decay width.

The invariant mass spectrum is not restricted to particle physics; it has also been used in nuclear physics. Consider, for instance, the reaction

$$p +^{11} \mathrm{B} \rightarrow \left\{ \begin{array}{l} 3\alpha \\ {}^{8}\mathrm{Be} + \alpha. \end{array} \right. \tag{5.28}$$

Since ^{8}Be lives only for 2×10^{-16} sec before decaying into two alpha particles, three alphas are observed in either case. Nevertheless, the formation of ^{8}Be can be studied with the invariant mass spectrum.

5.4 A First Glance at the Subatomic Zoo

The techniques discussed so far have led to the discovery of well over 100 particles and a much larger number of nuclei. How can these be ordered in a meaningful way? A first separation is achieved by considering the interactions that act on each particle. Four interactions are known to exist, as pointed out in Section 1.1. In order of increasing strength they are the gravitational, the weak, the electromagnetic, and the hadronic interaction.[10] In principle, then, the four interactions can be used to classify subatomic particles. However, the gravitational interaction is so weak that it plays no role in present-day subatomic physics. For this reason we shall restrict

[10] We shall see later that all but the gravitational interaction are connected within the standard model.

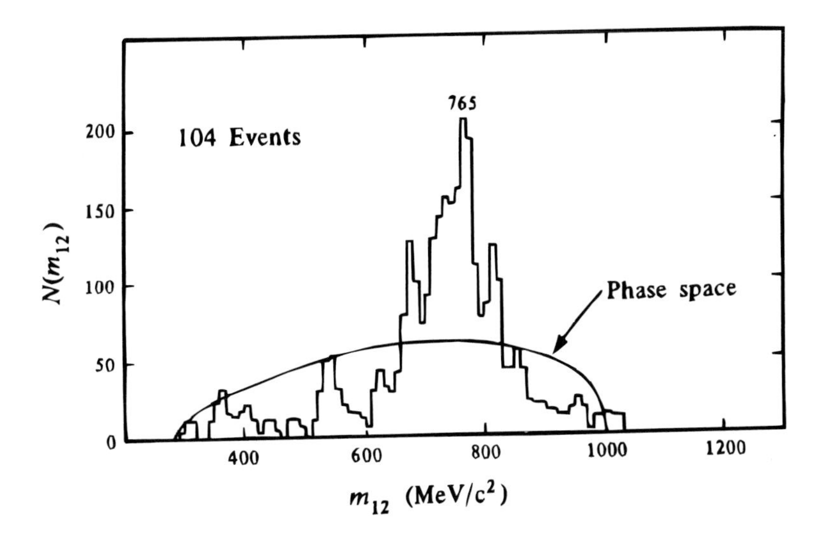

Figure 5.12: Invariant mass spectrum of the two pions produced in the reaction $p\pi^- \to n\pi^+\pi^-$. [After A. R. Erwin, R. March, W. D. Walker, and E. West, *Phys. Rev. Lett.* **6**, 628 (1961).]

our attention to the three other interactions. We sall see later that the standard model connects the weak and electromagnetic interactions into an electroweak one.

How can we discover which interactions govern the behavior of a particular particle? First consider the electron. It clearly is subject to the electromagnetic interaction because it carries an electric charge and is deflected in electromagnetic fields. Does it participate in the weak interaction? The prototype of a weak process is the neutron decay,

$$n \longrightarrow pe^-\overline{\nu}.$$

This decay is very slow; the neutron lives on average for about 15 min before decaying into a proton, an electron, and a neutrino. If we call the neutron decay a weak decay, then the electron participates in it. Does the electron interact hadronically? To find out, nuclei are bombarded with electrons, and the behavior of the scattered electrons is investigated. It turns out that the scattering can be explained by invoking the electromagnetic force alone; the electron does *not* interact hadronically. Decay and collision processes are also used to investigate the interactions of all other particles. The result is summarized in Table 5.2.

Subatomic particles can be divided into three groups, the gauge bosons, leptons, and hadrons. Among the gauge bosons, the best known is the photon which takes part in the electromagnetic interaction, despite the fact that it has no electric charge. This fact follows, for instance, from the emission of photons by accelerated charges [Eq. (2.20)]. The massive gauge bosons, $W^\pm$ and Z^0 take part in the weak interaction and the gluon mediates the strong interaction. Neutrinos, electron, muon, and tau are grouped together under the name *leptons*. All leptons have

Table 5.2: INTERACTIONS AND SUBATOMIC PARTICLES. Entries not in parentheses are for particles that exist free in nature. The particles in parentheses are permanently confined.

Particle	Type	Weak	Electromagnetic	Hadronic
Photon	Gauge boson	No	Yes	No
$W^{\pm}, Z^0$	Gauge bosons	Yes	Yes	No
(Gluon)	Gauge boson	No	No	Yes
Leptons				
Neutrino	Fermion	Yes	No	No
Electron	Fermion	Yes	Yes	No
Muon	Fermion	Yes	Yes	No
Tau	Fermion	Yes	Yes	No
Hadrons				
Mesons	Bosons	Yes	Yes	Yes
Baryons	Fermions	Yes	Yes	Yes
(Quarks)	Fermions	Yes	Yes	Yes

a weak interaction. The charged leptons, in addition, are also subject to the electromagnetic force. All other particles, including nuclei, are hadrons; their behavior is governed by the strong, the electromagnetic, and the weak interactions. In the following sections we describe the particles listed in Table 5.2 in more detail. We include quarks and gluons; they cannot be observed directly but their existence is based on firm arguments.

5.5 Gauge Bosons

The first group of particles in Table 5.2 lists three types of quanta, called gauge bosons, the photon, the W^+, W^- and Z^0, and the gluons. We are all familiar with the photon, but the other quanta and the name "gauge boson" require some introductory remarks. These particles are the carriers of forces as will be discussed in Section 5.8. Three types of forces are important in subatomic physics, the hadronic, the electromagnetic, and the weak. We therefore expect three types of particles to be responsible for the three forces between the leptons and quarks. Indeed, the photon mediates the electromagnetic force, the massive bosons, $W^{\pm}$ and Z^0 carry the weak force, and the gluons are the field quanta of the hadronic force. As we will show later, the form of the interaction is determined by a symmetry principle called gauge invariance; hence the name gauge bosons. We begin the discussion of the gauge bosons with the photon, the quantum of light. The particle properties of light invariably lead to some confusion. It is not possible to eliminate all confusion at an elementary level because a satisfactory treatment of photons requires quantum electrodynamics. However, a few remarks may at least make some of the important physical properties clearer.

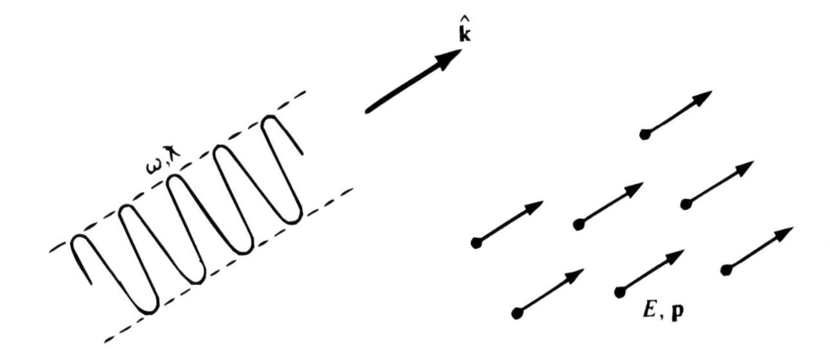

Figure 5.13: An electromagnetic wave can be said to be composed of photons with energy E and momentum $\boldsymbol{p}$.

Consider an electromagnetic wave with circular frequency ω and with a reduced wavelength $\lambda\!\!\!^{-} = \lambda/2\pi$ moving in a direction given by the unit vector $\hat{\boldsymbol{k}}$ (Fig. 5.13). Instead of giving $\hat{\boldsymbol{k}}$ and $\lambda\!\!\!^{-}$ separately, a wave vector $\boldsymbol{k} = \hat{\boldsymbol{k}}/\lambda\!\!\!^{-}$ is introduced. It points in the direction $\hat{\boldsymbol{k}}$ and has a magnitude $1/\lambda\!\!\!^{-}$. According to Einstein, a monochromatic electromagnetic wave is composed of N monoenergetic photons, each with energy E and momentum $\boldsymbol{p}$, where

$$E = \hbar\omega, \qquad \boldsymbol{p} = \hbar\boldsymbol{k}. \tag{5.29}$$

The number of photons in the wave is such that the total energy $W = NE = N\hbar\omega$ is equal to the total energy in the electromagnetic wave. Equation (5.29) shows that photons are endowed with energy and momentum. How about angular momentum? In 1909, Poynting predicted that a circularly polarized electromagnetic wave carries angular momentum, and he proposed an experiment to verify this prediction: If a circularly polarized wave is absorbed, the angular momentum contained in the electromagnetic field is transferred to the absorber, which should then rotate. The first successful experiment was performed by Beth in 1935.[11]

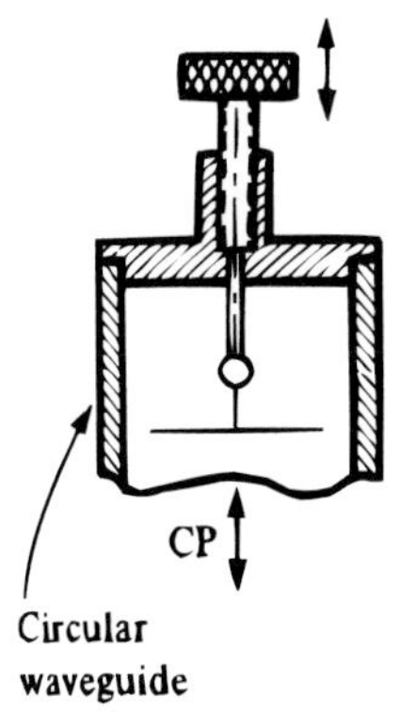

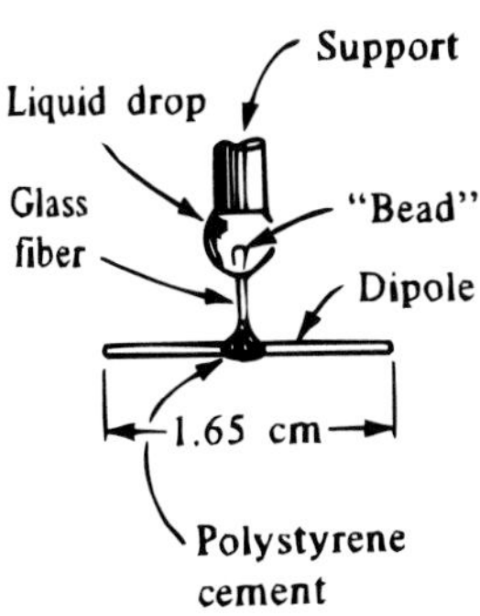

Figure 5.14: A drop-suspended dipole exposed to a circularly polarized microwave rotates because the angular momentum of the electromagnetic field exerts a torque. [From P. J. Allen, *Am. J. Phys.* **34**, 1185 (1964).]

A modern variant, a *microwave motor*, is shown in Fig. 5.14. A circularly polarized microwave impinges on a suspended dipole at the end of a circular wave guide. Some energy and some angular momentum are absorbed by the dipole and it begins to rotate. The ratio of absorbed energy to absorbed angular momentum can easily be calculated, and it is(11)

$$\frac{\Delta E}{\Delta J_z} = \omega. \tag{5.30}$$

This relation shows that the torque experiment is easier with microwaves than with optical light because the angular momentum transfer for a given energy transfer increases as $1/\omega$. Equation (5.30) has been computed on the basis of classical electro-magnetism. It can be translated into quantum mechanics by assuming that n photons, moving along the z axis with energy $\Delta E = n\hbar\omega$ and with angular momentum $\Delta J_z = nJ_z$, are absorbed. Equation (5.30) then yields

$$J_z = \hbar. \tag{5.31}$$

The angular momentum carried by one photon is $\hbar$. This result can be restated by saying that the photon has *spin 1*.

Spin 1 for the photon is not surprising. Remember that a spin-1 particle has three independent orientations. To describe the three orientations, a quantity with three independent components is needed. A vector fills the bill, since it has three independent components. The electromagnetic field is a vector field: It is described by vectors $\boldsymbol{E}$ and $\boldsymbol{B}$ and corresponds to a vector particle—a particle with spin 1.(12,13)

There is, however, a fly in the ointment. It is well known from classical optics that an electromagnetic wave has only *two* independent polarization states. Could it be that the photon has spin $\frac{1}{2}$? This possibility can be ruled out quickly. The connection between spin and symmetry, discussed in Section 5.1, would make a spin-$\frac{1}{2}$ photon a fermion, and it would obey the exclusion principle. Not more than one photon could be in one state; classical electromagnetic waves and television would be impossible. The solution to the apparent paradox comes not from quantum theory but from relativity. The photon has zero mass; it is light and moves with the velocity of light. There is no coordinate system in which the photon is at rest. The argument leading to Eq. (5.8) and to the $2J+1$ possible orientations is, however, made in the rest system, and it breaks down for the photon. In fact, any massless particle can at most have two spin orientations, parallel or antiparallel

[11] R. A. Beth, *Phys. Rev.* **50**, 115 (1936). Reprinted in *Quantum and Statistical Aspects of Light*, American Institute of Physics, New York, 1963.

[12] See, for instance, R. T. Weidner and R. L. Sells, *Elementary Classical Physics*, Allyn and Bacon, Boston, 1965, Eq. (47.5).

[13] The situation is actually somewhat more complicated. The correct description of the electro-magnetic field is through the potential; the scalar and the vector potential together form a four vector, $(A^0, \boldsymbol{A})$. It therefore appears at first as if this four vector corresponded to four degrees of freedom. However, the Lorentz subsidiary condition removes one degree and we are back to three.

to its momentum, regardless of its spin.[14] We can summarize the result of the previous arguments by saying that the free photon is a spin-1 particle that can have its spin either parallel or antiparallel to the direction of motion.[15] The two states are called right- and left-circularly polarized or states of positive and negative helicity, respectively.

The carriers of the weak force, the gauge bosons $W^{\pm}$ and Z^0, were found after a long search;[16] their masses are 81 GeV/c^2 for the $W^{\pm}$ and 91 GeV/c^2 for the Z^0. Their spin is also $1\hbar$; since they are massive, the spin can have three orientations. The evidence for gluons is indirect, because gluons cannot exist freely. They are "confined" and only occur inside hadrons. They are massless and have spin $1\hbar$.[17]

5.6 Leptons

Electrons, muons, taus, and neutrinos are all called leptons. Originally the name indicated that these particles were much lighter than nucleons. With the discovery of the tau,[18,19] with a mass of 1.78 GeV/c^2, the name "lepton" has become a misnomer, but it has been retained. The properties of the electron and muon are extremely well measured and the theoretical description of some of their properties, in particular the g-factor, is incredibly successful. Until recently, however, the "*raison d'être*" of the muon was a mystery and it appeared as an unwelcome intruder. With the discovery of the tau, a reason for the number of leptons has emerged as we will sketch in Section 5.11. Half of all leptons are listed in Table 5.3. The word *half* requires preliminary explanation. One of the best-documented facts of subatomic physics is that each particle has an antiparticle, with opposite charge, but otherwise very similar properties. Each of the leptons in Table 5.3 has an antilepton, e^+, μ^+, and τ^+ (and $\overline{\nu_i}$ for each neutrino.) A more careful explanation of the idea of antiparticles will follow in Section 5.10.

The manner in which we have introduced the neutrino and the muon here is really terrible. It can be compared to introducing a master criminal, such as Professor Moriarty,[20] by listing his weight, height, and hair color rather than by telling of

[14] E. P. Wigner, *Rev. Mod Phys.* **29**, 255 (1957).

[15] Two words of warning are in order here. Single photons do not have to be eigenstates of momentum and angular momentum. It is possible to form linear combinations of eigenstates that correspond to single photons but do not have well-defined momentum and angular momentum. The second remark concerns the term *polarization vector.* In electromagnetism it is conventional to call the direction of the electric vector the polarization direction. *A photon with its spin along the momentum has its electric vector perpendicular to the momentum.*

[16] C. Rubbia, *Rev. Mod. Phys.* **57**, 699 (1985).

[17] PLUTO collaboration, *Phys. Lett.* **99B**, 292 (1981).

[18] M.L. Perl et al., *Phys. Rev. Lett.* **35**, 1489 (1975); reprinted in *New Particles.* Selected Reprints. (J. L. Rosner, ed.), American Association Physics Teachers, Stony Brook, NY 1981.

[19] M.L. Perl, *Ann. Rev. Nucl. Part. Sci.* **30**, 299 (1980); B.C. Barish and R. Stroynowski, *Phys. Rep.* **157**, 1 (1987).

[20] A. C. Doyle, *The Complete Sherlock Holmes*, Doubleday, New York, 1953.

Table 5.3: CHARGED LEPTONS.*

Lepton	Spin	(Mass)c^2	Magnetic Moment Unit $(eh/2mc)$	Lifetime
e^-	1/2	0.5109989 MeV	−1.001 159 652 1859	Stable
μ^-	1/2	105.6584 MeV	−1.001 165 9208	2.197 14 μsec
τ^-	1/2	1777 MeV	−1.0	2.91×10^{-13} sec

*The neutral leptons are called neutrinos, with mass eigenstates ν_1, ν_2, ν_3, are known to have masses below $\sim$ 2 eV and are stable. Upper limits on their magnetic moments are given by PDG. The neutrinos are produced by the weak interaction and appear as linear combinations (called ν_e, ν_μ, ν_τ) of the mass eigenstates.

his feats. In reality, the neutrino behaved like a master criminal, and it escaped suspicion at first and then detection for a long time. The muon arrived disguised as a hadron and managed to confuse physicists for a considerable period before it was unmasked as an imposter. The introduction, as we have performed it, can be excused only by noting that excellent accounts of the histories of the neutrino and muon exist.[(21)]

5.7 Decays

Two facts compel us to digress and talk about decays before attacking the hadrons. The first is the comparison of muon and electron. The electron is stable, whereas the muon decays with a lifetime of 2.2 μsec. Does this fact indicate that the electron is more fundamental than the muon? The second fact emerges from comparing Figs. 5.11 and 5.12. In Fig. 5.11, the rho is indicated as a sharp line with mass m_ρ; the actually observed rho displays a wide *resonance* with a width of over 100 MeV/c^2. Is this width of experimental origin, or does it have fundamental significance? To answer the questions raised by the two observations we turn to a discussion of *decays*.

Consider an assembly of independent particles, each having a probability λ of decaying per unit time. The number decaying in a time dt is given by

$$dN = -\lambda N(t)\, dt, \tag{5.32}$$

where $N(t)$ is the number of particles present at time t. Integration yields the exponential decay law,

$$N(t) = N(0)e^{-\lambda t}. \tag{5.33}$$

[21]W.C. Haxton and B.R. Holstein, *Am. Jour. Phys.* **72**, 18 (2004).

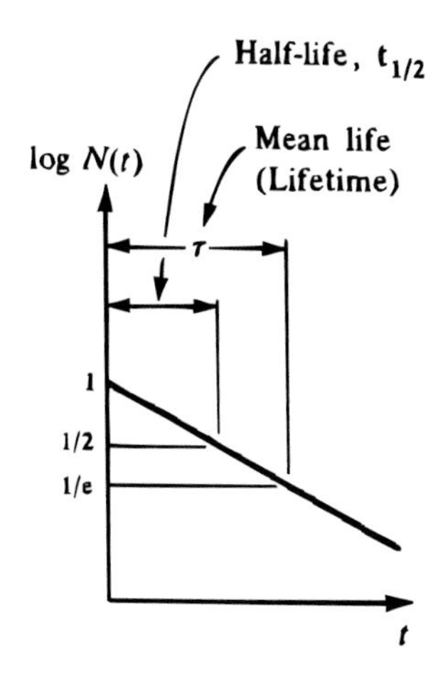

Figure 5.15: Exponential decay.

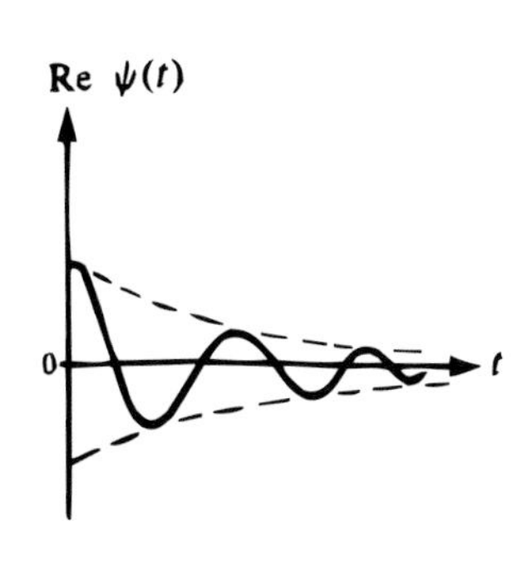

Figure 5.16: Real part of the wave function of a decaying state. It is assumed that the decaying state is formed at $t = 0$.

Figure 5.15 shows log $N(t)$ plotted against t. Half life and mean life are indicated. In one half life, one half of all atoms present decay. The mean life is the average time a particle exists before it decays; it is connected to λ and $t_{1/2}$ by

$$\tau = \frac{1}{\lambda} = \frac{t_{1/2}}{\ln 2} \cong 1.44 t_{1/2}. \tag{5.34}$$

To relate the exponential decay to properties of the decaying state, the time dependence of the wave function of a particle at rest ($\boldsymbol{p} = 0$) is shown explicitly as

$$\psi(t) = \psi(0) \exp\left(-\frac{iEt}{\hbar}\right). \tag{5.35}$$

If the energy E of this state is *real*, the probability of finding the particle is *not* a function of time because

$$|\psi(t)|^2 = |\psi(0)|^2.$$

A particle described as a wave function of the type of Eq. (5.35) with real E does not decay. To introduce an exponential decay of a state described by $\psi(t)$, a small imaginary part is added to the energy,

$$E = E_0 - \tfrac{1}{2} i\Gamma, \tag{5.36}$$

where E_0 and Γ are real and where the factor $\frac{1}{2}$ is chosen for convenience. With Eq. (5.36), the probability becomes

$$|\psi(t)|^2 = |\psi(0)|^2 \exp\left(\frac{-\Gamma t}{\hbar}\right). \tag{5.37}$$

It agrees with the decay law (Eq. (5.33)) if

$$\Gamma = \lambda\hbar. \tag{5.38}$$

With Eqs. (5.35) and (5.36) the wave function of a decaying state is

$$\psi(t) = \psi(0)\exp\left(\frac{-iE_0 t}{\hbar}\right)\exp\left(\frac{-\Gamma t}{2\hbar}\right). \tag{5.39}$$

The real part of $\psi(t)$ is shown in Fig. 5.16 for positive times. The addition of a small imaginary part to the energy permits a description of an exponentially decaying state, but what does it mean? The energy is an observable; does an imaginary component make sense? To find out we note that $\psi(t)$ in Eq. (5.39) is a function of time. What is the probability that the emitted particle has an energy E? In other words, we would like to have the wave function as a function of energy rather than time. A change from $\psi(t)$ to $\psi(E)$ is effected by a Fourier transformation, a generalization of the ordinary Fourier expansion. A short and readable introduction is given by Mathews and Walker;[22] here we present only the essential equations. Consider a function $f(t)$. Under rather general conditions it can be expressed as an integral,

$$f(t) = (2\pi)^{-1/2}\int_{-\infty}^{+\infty} d\omega\; g(\omega)\;\exp(-i\omega t). \tag{5.40}$$

The expansion coefficient in the ordinary Fourier series has become a function $g(\omega)$. Inversion of Eq. (5.40) gives

$$g(\omega) = (2\pi)^{-1/2}\int_{-\infty}^{+\infty} dt f(t)\exp(+i\omega t). \tag{5.41}$$

The variables t and ω are chosen so that the product ωt is dimensionless; otherwise $\exp(i\omega t)$ does not make sense. Thus t and ω can be time and frequency or coordinate and wave number. We now set $f(t)$ in Eq. (5.41) equal to $\psi(t)$, Eq. (5.39). If the decay starts at the time $t = 0$, the lower limit on the integral can be set equal to zero, and $g(\omega)$ becomes

$$g(\omega) = (2\pi)^{-1/2}\psi(0)\int_0^{\infty} dt\exp\left[+i\left(\omega - \frac{E_0}{\hbar}\right)t\right]\exp\left(-\frac{\Gamma t}{2\hbar}\right) \tag{5.42}$$

or

$$g(\omega) = \frac{\psi(0)}{(2\pi)^{1/2}}\frac{i\hbar}{(\hbar\omega - E_0) + i\Gamma/2}. \tag{5.43}$$

[22]Mathews and Walker, Chapter 4. Short tables of Fourier transforms are given in the *Standard Mathematical Tables*, Chemical Rubber Co., Cleveland, Ohio. Extensive tables can be found in A. Erdelyi, W. Magnus, F. Oberhettinger, and F. G. Tricomi, *Tables of Integral Transforms*, McGraw-Hill, New York, 1954.

The function $g(\omega)$ is proportional to the probability amplitude that the frequency ω occurs in the Fourier expansion of $\psi(t)$. Since $E = \hbar\omega$, the probability density $P(E)$ of finding an energy E is also proportional to $|g(\omega)|^2 = g^*(\omega)g(\omega)$[23]:

$$P(E) = \text{const.}\ g^*(\omega)g(\omega) = \text{const.}\frac{\hbar^2}{2\pi}\frac{|\psi(0)|^2}{(E-E_0)^2+\Gamma^2/4}.$$

The condition

$$\int_{-\infty}^{+\infty} P(E)dE = 1 \tag{5.44}$$

yields

$$\text{const.} = \frac{\Gamma}{\hbar^2|\psi(0)|^2},$$

and $P(E)$ finally becomes

$$P(E) = \frac{\Gamma}{2\pi}\frac{1}{(E-E_0)^2+(\Gamma/2)^2}. \tag{5.45}$$

The energy of a decaying state is not sharp. The small imaginary part in Eq. (5.36) leads to a decay *and* it introduces a broadening of the state. The width acquired by the state because of its decay is called *natural line width.* The shape is called a Lorentzian or Breit–Wigner curve; it is sketched in Fig. 5.17. Γ turns out to be the full width at half maximum. With Eqs. (5.34) and (5.38), the product of lifetime and width becomes

$$\tau\Gamma = \hbar. \tag{5.46}$$

This relation can be interpreted as a Heisenberg uncertainty relation, $\Delta t \Delta E \geq \hbar$. To measure the energy of the state or particle to within an uncertainty $\Delta E = \Gamma$, a time $\Delta t = \tau$ is needed. Even if a longer time is used, the energy cannot be measured more accurately.

We can now answer the second question posed at the beginning of this section: The width observed in the decay of the rho is caused by decay; the instrumental width is much smaller. Since $\Gamma_\rho \approx 150$ MeV, the lifetime becomes

$$\tau_\rho = \frac{\hbar}{\Gamma_\rho} \approx 4.4 \times 10^{-24}\,\text{sec}.$$

We still have not answered the first question: Are decaying particles less fundamental than stable ones? To answer it, a few examples of unstable particles are listed in Table 5.4. A number of facts emerge from this Table:

[23]For photons, the relation $E = \hbar\omega$ connects the energy to the frequency of the electromagnetic wave. For massive particles, it *defines* the frequency ω; the derivation leading to Eq. (5.45) remains correct because it is independent of the actual form of ω.

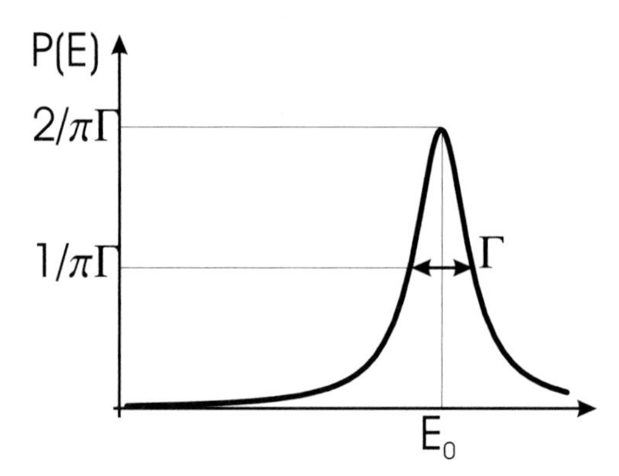

Figure 5.17: Natural line shape of a decaying state. Γ is the full width at half maximum.

1. No connection between simplicity and decay appears. Electron and muon differ only in mass, yet the muon decays. The deuteron, a composite of neutron and proton, is not listed because it is stable, but the free neutron decays. The charged pions decay slowly, but the neutral one decays rapidly. The data suggest that a particle decays if it can and that it is stable only if there is no state of lower energy (mass) to which it is allowed to decay. Stability does not appear to be a criterion for *elementarity*.

2. Comparison of particles with about the same decay energy shows that classes occur. We know that hadronic, electromagnetic, and weak forces exist and thus expect corresponding decays. Indeed, all three types show up. Detailed calculations are required to justify that the three interactions can give rise to decays with the listed lifetimes. Nevertheless, a very crude idea of typical lifetimes can be gained by comparing the delta (Δ), the neutral pion, and the lambda. These have decay energies between 40 and 160 MeV and decay into two particles. Approximate values for the corresponding lifetimes are

$$\begin{array}{ll} \text{hadronic decay}(\Delta) & 10^{-23}\,\text{sec} \\ \text{electromagnetic decay}(\pi^0) & 10^{-18}\,\text{sec} \\ \text{weak decay}(\Lambda) & 10^{-10}\,\text{sec}\,. \end{array} \tag{5.47}$$

   The ratios of these lifetimes give approximately the ratios of strengths of the three forces. To obtain better measures of the relative strengths, the interactions must be studied in more detail, as will be done in Part IV.

3. The type of particle or quantum emitted is not always an indication of the interaction at work. Lambda and delta both decay into proton and pion, yet the delta decays about 10^{14} times faster. *Selection rules* must be involved, and it will be one of the tasks of later chapters to find these rules.

Table 5.4: SELECTED DECAYS. The entry under Class indicates the type of decay. W means weak, EM electromagnetic, and H hadronic.

Particle	Mass (MeV/c²)	Main Decays	Decay Energy (MeV)	Lifetime (sec)	Class
μ	106	$e\nu\bar{\nu}$	105	2.2×10^{-6}	W
$\pi^{\pm}$	140	$\mu\nu$	34	2.6×10^{-8}	W
π^0	135	$\gamma\gamma$	135	8.7×10^{-17}	EM
η	549	$\gamma\gamma, \pi\pi\pi$	549	6.3×10^{-19}	EM
ρ	769	$\pi\pi$	489	4.3×10^{-24}	H
n	940	$pe^-\bar{\nu}$	0.8	0.90×10^3	W
Λ	1116	$p\pi^-, n\pi^0$	39	2.6×10^{-10}	W
Δ	1232	$N\pi$	159	6×10^{-24}	H
$D^{\pm}$	1869	$\overline{K^0} + \cdots$		9.2×10^{-13}	W
D^0	1865	$K^{\pm} + \cdots$		4.3×10^{-13}	W
^{8}Be*	3726	2α	3	6×10^{-22}	H

5.8 Mesons

In Table 5.2, hadrons are separated into mesons and baryons. We shall explain the difference between these two types of hadrons in more detail in Chapter 7, where a new quantum number, the *baryon number*, will be introduced. It is similar to the electric charge: Particles can have baryon numbers $0, \pm 1, \pm 2, \ldots$. The prototype of a baryon-number-1 particle is the nucleon. Like the electric charge, baryon number is "conserved," and a state with baryon number 1 can decay only to another state with baryon number 1. Mesons are hadrons with baryon number 0. All mesons have a transient existence and decay through one of the three interactions discussed in the previous section.

The first meson to appear in the zoo was the *pion*. Since its existence was predicted more than 10 years before it was found experimentally, it is worth explaining the basis of the prophecy. To do so, it is necessary to return to the photon and the electromagnetic interaction. Because of relativity, it is generally assumed that no interactions at a distance exist.[(24)] The electromagnetic force between two electrons, for instance, is assumed to be mediated by photons.

[24] In Newton's theory of gravitation it is assumed that the interaction between two bodies is instantaneous. A rapid acceleration of the Sun, for instance, would affect the Earth immediately and not after 8 min. This basic tenet is in conflict with the special theory of relativity which assumes that no signal can travel faster than the speed of light. This inconsistency led Einstein to his general theory of relativity. [S. Chandrasekhar, *Am. J. Phys.* **40**, 224 (1972).] In quantum theory a force that is transmitted with at most the speed of light is pictured as being caused by the exchange of quanta. Even the possible existence of particles with speed exceeding that of light (tachyons) does not change the argument. [O. M. Bilaniuk and E. C. G. Sudarshan, *Phys. Today*, **22**, 43 (May 1969); G. Feinberg, *Phys. Rev.* **159** 1089 (1967), L. M. Feldman, *Am. J. Phys.* **42**, 179 (1974).]

Figure 5.18 explains the idea. One electron emits a photon which is absorbed by the other electron. The exchange of photons or *field quanta* gives rise to the electromagnetic interaction between the two charged particles, whether it occurs in a collision or in a bound state, such as positronium (e^-e^+ atom). The exchange process is best considered in the c.m. of the two colliding electrons. Since the collision is elastic, the energies of the electrons are unchanged so that $E_1' = E_1, E_2' = E_2$. Before the emission of the photon, the total energy is $E = E_1 + E_2$.

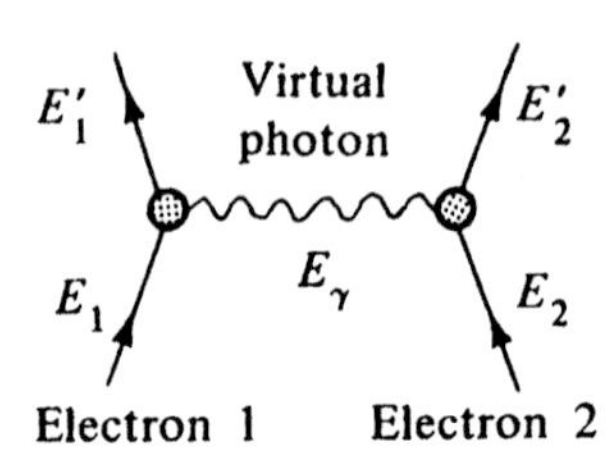

Figure 5.18: Exchange of a photon between two electrons, 1 and 2. The virtual photon is emitted by one and absorbed by the other electron.

After emission but before reabsorption of the quantum the total energy is given by $E = E_1 + E_2 + E_\gamma$, and energy is not conserved. Is such a violation allowed? Energy conservation can indeed be broken for a time Δt because of the Heisenberg uncertainty relation

$$\Delta E \, \Delta t \geq \hbar. \tag{5.48}$$

Equation (5.48) states that the time Δt required to observe an energy to within the uncertainty ΔE must be greater than $\hbar/\Delta E$. Nonconservation of energy within an amount ΔE is therefore unobservable if it occurs within a time T given by

$$T \leq \frac{\hbar}{\Delta E}. \tag{5.49}$$

A photon of energy $\Delta E = \hbar\omega$ consequently cannot be observed if it exists for less than a time

$$T = \frac{\hbar}{\hbar\omega} = \frac{1}{\omega}. \tag{5.50}$$

Since the unobserved photon exists for less than the time T, it can travel at most a distance

$$r = cT = \frac{c}{\omega}. \tag{5.51}$$

The frequency ω can be arbitrarily small, and the distance over which a photon can transmit the electromagnetic interaction is arbitrarily large. Indeed, the Coulomb force has a distance dependence $1/r^2$ and presumably extends to infinity. Since the exchanged photon is not observed, it is called a *virtual photon.*

By 1934, it was known that the strong force is very strong and that it has a range of about 2 fm, but there was total ignorance as to what caused it. Yukawa, a

Japanese theoretical physicist, then suggested in a brilliant paper that a "new sort of quantum" could be responsible.[25]

Yukawa's arguments are more mathematical than we can present here, but the analogy to the virtual photon exchange permits an estimate of the mass m of the "new quantum," the pion. In Yukawa's approach, the force between two hadrons, for instance two neutrons, is mediated by an unobserved pion, as sketched in Fig. 5.19.

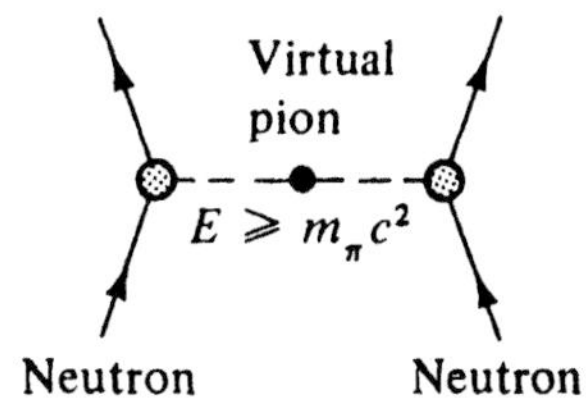

Figure 5.19: Exchange of a virtual pion between two neutrons.

The minimum energy of the virtual pion is given by $E = m_\pi c^2$ and its maximum velocity by c. With Eq. (5.49), the maximum distance that the virtual pion is allowed to travel by the uncertainty relation is given by

$$R \leq cT = \frac{\hbar}{m_\pi c} \approx 1.4 \text{ fm}. \tag{5.52}$$

The range is therefore at most equal to the Compton wavelength of the pion. Originally, of course, the argument was turned around, and the mass of the postulated hadronic quantum was estimated by Yukawa as 100 MeV/c^2.

Physicists were delighted when a particle with a mass of about 100 MeV/c^2 was found in 1938. Delight turned to dismay when it was realized that the newcomer, the muon, did not interact strongly with matter and hence could not be held responsible for the hadronic force. In 1947, the true Yukawa particle, the pion, was finally discovered in nuclear emulsions.[26] After 1947, more mesons kept turning up, and at present the list is long. Some of these new mesons live long enough to be studied by conventional techniques. Some decay so rapidly that the invariant-mass-spectra method, discussed in Section 5.3, had to be invented. A list of the known mesons can be found in PDG.

The fact that the idea of virtual quanta led to the prediction of the existence of a new particle is important. Even more important, however, is the powerful concept that forces between elementary particles are caused by the exchange of virtual particles and we will return to this concept again later.

5.9 Baryon Ground States

The spectrum of baryons is even richer than that of mesons. We begin the survey by considering *nuclear ground states*. By about 1920 it was well established that the

[25] H. Yukawa, *Proc. Math. Soc. Japan* **17**, 48 (1935). Reprinted in D.M. Brink, *Nuclear Forces*, Pergamon, Elmsford, N. Y., 1965. This book also contains a reprint of the articles by G.C. Wick on which our discussion of the connection between force range and quantum mass is based.

[26] C. M. G. Lattes, H. Muirhead, G. P. S. Occhialini, and C. F. Powell, *Nature* **159**, 694 (1947).

Table 5.5: HADRONICALLY STABLE MESONS. The mesons listed here decay either by weak or by electromagnetic processes.

Particle	Mass (MeV/c^2)	Charge (e)	Mean Life (sec)
π^0	135.0	0	0.84×10^{-16}
$\pi^\pm$	139.6	$+, -$	2.60×10^{-8}
$K^\pm$	493.7	$+, -$	1.24×10^{-8}
K^0	497.7	0	Complicated
η	547.8	0	5.1×10^{-19}
$D^\pm$	1869	$+, -$	1.0×10^{-12}
D^0	1865	0	4.1×10^{-13}
$B^\pm$	5279	$+, -$	$\sim 1.7 \times 10^{-12}$
B^0	5279	0	$\sim 1.5 \times 10^{-12}$

electric charge Q and the mass M of a particular nuclear species are characterized by two integers, Z and A:

$$Q = Ze \tag{5.53}$$

$$M \approx Am_p. \tag{5.54}$$

The first relation was found to hold accurately, and the second one approximately. The nuclear charge number Z was determined by Rutherford's alpha-particle scattering, by X-ray scattering, and by the measurement of the energy of characteristic X rays. It was also found that Z is identical to the chemically determined *atomic number* of the corresponding element. The *mass number* A was extracted from mass spectroscopy, where it turned out that a given element can have nuclei with different values of A. The ground state of any nuclear species can, according to Eqs. (5.53) and (5.54), be characterized by two integers, A and Z. Before the discovery of the neutron, the interpretation of these facts was rather unclear. When the neutron was finally found by Chadwick in 1932,[27] everything fell into place: A nucleus (A, Z) is composed of Z protons and $N = A - Z$ neutrons; since neutrons and protons are about equally heavy, the total mass is approximately given by Eq. (5.54). The mass number, A, is thus the sum of the number of neutrons and protons and is also called the baryon number. The charge is entirely due to the protons so that Eq. (5.53) is also satisfied.

At this point, we can get some definitions out of the way: A *nuclide* is a particular nuclear species with a given number of protons and neutrons: *isotopes* are nuclides with the same number of protons, Z; *isotones* are nuclides with the same

[27] J. Chadwick, *Nature* **129**, 312 (1932); *Proc. R. Soc. (London)* **A136**, 692 (1932).

neutron number, N; *isobars* are nuclides with the same total number of nucleons, A. A particular nuclide is written as (A, Z) or ${}^{A}_{Z}$element. The alpha particle, for instance, is characterized by (4, 2) or ${}^{4}_{2}$He or simply ^{4}He.

Stable nuclides, characterized by $N = A - Z$ and Z, are represented as small squares in an $N - Z$ plot in Fig. 5.20. The plot indicates that stable nuclides exist only in a small band in the $N - Z$ plane. The band starts off at 45° (equal proton and neutron numbers) and slowly veers toward neutron-rich nuclides. This behavior will provide a clue to an understanding of properties of the nuclear force. Figure 5.20 contains only stable nuclides. In Section 5.7 we have pointed out that stability is not an essential criterion in considering hadrons. Unstable nuclear ground states therefore can also be added to the $N - Z$ plot. We shall explore some properties of such an extended plot in Chapter 16.

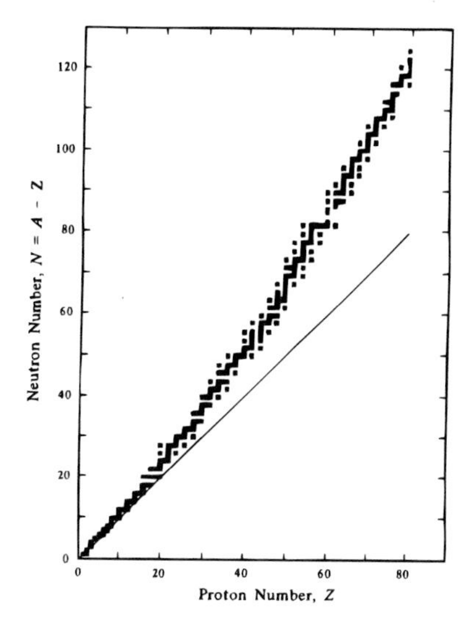

Figure 5.20: Plot of the stable nuclides. Each stable nuclide is indicated as a square in this $N - Z$ plot. The solid line would correspond to nuclides with equal proton and neutron numbers. (After D.L. Livesey, *Atomic and Nuclear Physics*, Blaisdell, Waltham, MA, 1966.)

At the mass number $A = 1$, nuclear and particle physics meet. The proton and the neutron, the two building blocks of all heavier nuclides, can either be considered the simplest nuclei or they can be called particles. It is a surprising fact that the two nucleons are not the only $A = 1$ hadrons. Other baryons with the mass number $A = 1$ exist; they are called hyperons.

As an example of the investigation of *hyperons*, we consider the production of the lambda. If negative pions of a few GeV of energy pass through a hydrogen bubble chamber, events such as the one shown in Fig. 5.21 are observed: The negative pion "disappears," and further downstream two V-like events appear. At first, the two Vs seem to be very similar.

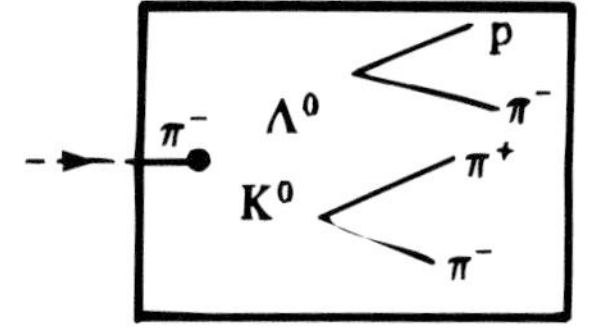

Figure 5.21: Observation of the process $p\pi^- \to \Lambda^0 K^0$ in a hydrogen bubble chamber.

However, when the energies and momenta of the four particles are determined (Section 5.3) it turns out that one V consists of two pions, and the other of a pion and a proton. Invariant mass plots, such as explained in Section 5.3, show that the particle giving rise to the two pions has a mass of about 500 MeV/c^2, while

Table 5.6: HADRONICALLY STABLE BARYONS.

Particle	Charge (e)	Mass (MeV/c^2)	Mean Life (sec)
N	+	938.3	$\gtrsim 3 \times 10^{37} \approx 10^{30}$ y
	0	939.6	0.89×10^3
Λ	0	1115.7	2.63×10^{-10}
Σ	+	1189.4	8.02×10^{-11}
	0	1192.6	7.4×10^{-20}
	−	1197.4	1.48×10^{-10}
Ξ	0	1314.8	2.90×10^{-10}
	−	1321.3	1.64×10^{-10}
Ω	−	1672.5	0.82×10^{-10}
Λ_c	+	2284.9	2.0×10^{-13}

the particle decaying into proton and pion has a mass of 1116 MeV/c^2. The first particle is the neutral kaon, and the second particle is called *lambda*. (The name, of course, refers to the characteristic appearance of the tracks of the proton and the pion.) The lifetime of each particle can be computed from the distance traveled in the bubble chamber and from its momentum. A complete reaction reads

$$p\pi^- \rightarrow \Lambda^0 K^0 \quad , \tag{5.55}$$

$$\Lambda^0 K^0 \rightarrow (p\pi^-)\,(\pi^+\pi^-). \tag{5.56}$$

The lambda is not the only hyperon; a number of other hadronically stable particles of similar character have been found. These earn the designation *hadronically stable* because their lifetimes are much longer than 10^{-22} sec, and they are called baryons because they all ultimately decay to *one* proton or neutron. The hadronically stable baryons are listed in Table 5.6.

5.10 Particles and Antiparticles

We have mentioned antiparticles many times, but have not yet explained the concept. The particle–antiparticle concept is actually one of the most fascinating ones in physics. The present section is brief and restricted and will leave many problems unsolved. At the same time some of the aspects that are needed in later sections and chapters should become somewhat clearer.

The story begins about 1927 with Eq. (1.2):

$$E^2 = (pc)^2 + (mc^2)^2. \tag{1.2}$$

Consider a particle with momentum $\boldsymbol{p}$ and mass m. What is its energy? All of us were taught early in our life to write a square root with two signs,

$$E^{\pm} = \pm[(pc)^2 + (mc^2)^2]^{1/2}. \tag{5.57}$$

Two solutions appear, a positive and a negative one. What does the negative energy solution mean? In classical physics, it did not cause havoc. When the classical gods created the world, they chose the initial conditions without negative energies. Continuity then guaranteed that none would appear later. In quantum mechanics, the situation is far more serious.

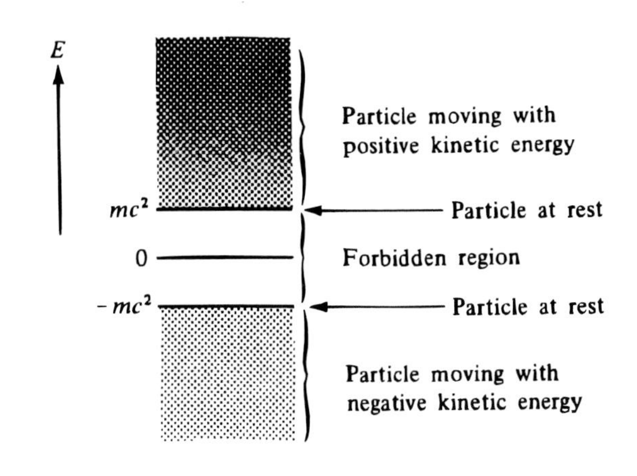

Figure 5.22: Positive and negative energy states of a particle with mass m.

Consider the energy levels of a particle with mass m. Equation (5.57) states that positive and negative energy levels are possible, and these levels are shown in Fig. 5.22. The smallest possible positive energy is $E = mc^2$; the largest negative energy is $-mc^2$. According to Eq. (5.57), the particle can have any energy from mc^2 to $+\infty$ and from $-mc^2$ to $-\infty$. Do the negative energy states lead to observable consequences? We shall see that they do and that there is an enormous amount of experimental evidence to back up this claim. Before doing so, we mention a mathematical argument that also calls for their existence: One of the most fundamental theorems in quantum mechanics states that any observable has a complete set of eigenfunctions.(28) It can be shown in relativistic quantum mechanics that eigenfunctions do *not* form a complete set without the negative energy states.

If the negative energy states exist, what do they mean? They cannot be normal energy states as indicated in Fig. 5.22; otherwise, ordinary particles could make transitions to the negative energy states with emission of energy, and matter would rapidly disappear. The first workable interpretation of the negative energy states is due to Dirac,(29) who identified particles missing from the negative energy states (holes) with antiparticles. We shall not discuss his *hole theory* but proceed immediately to a more modern interpretation, first proposed by Stueckelberg and later

---

[28] Merzbacher, Section 8.3.

[29] P. A. M. Dirac, *Proc. R. Soc.* (*London*) **A126**, 360 (1930).

again in more powerful form by Feynman.[30]

We present this approach in a pedestrian version and first consider a particle moving along the positive x axis with positive momentum p and positive energy E^+. The trajectory of this particle is shown in an xt plot in Fig. 5.23. Its wave function is of the form

$$\psi(x,t) = \exp\left[\frac{i(px - E^+t)}{\hbar}\right]. \tag{5.58}$$

The fact that it moves to the right can be seen most easily by noting that the phase of the wave function is constant if

$$px - E^+t = \text{const.}$$

or if

$$x = \frac{E^+}{p}t. \tag{5.59}$$

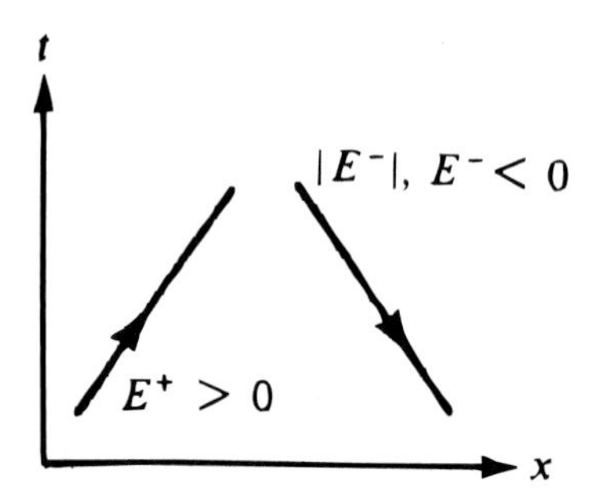

Figure 5.23: The particle with positive energy, E^+, moves like any ordinary particle. The particle with negative energy, E^-, is represented as a particle with positive energy $|E^-|$, but moving backwards in time. Both travel to the right.

The point x moves to the right. (This argument can be made more rigorous by using a wave packet.) For the negative energy solution,

$$\psi(x,t) = \exp\left[\frac{i(px - E^-t)}{\hbar}\right] \quad E^- < 0, \tag{5.60}$$

the relation (5.59) becomes

$$x = \frac{E^-}{p}t = -\frac{|E^-|}{p}t = \frac{|E^-|}{p}(-t), \tag{5.61}$$

and it can be interpreted as a particle moving backward in time but having a positive energy, $|E^-|$.

What is a particle moving backward in time? The classical equation of motion of a particle of charge $-q$ in a magnetic field becomes, with the Lorentz force [Eq.(2.21)],

$$m\frac{d^2\boldsymbol{x}}{dt^2} = \frac{-q}{c}\frac{d\boldsymbol{x}}{dt} \times \boldsymbol{B} = \frac{q}{c}\frac{d\boldsymbol{x}}{d(-t)} \times \boldsymbol{B}. \tag{5.62}$$

A particle with charge q moving backward in time satisfies the same equation of motion as a particle with charge $-q$ moving forward in time.[31]

[30] E. C. G. Stueckelberg, *Helv. Phys. Acta*, **14**, 588 (1941); R. P. Feynman, *Phys. Rev.* **74**, 939 (1948).

[31] The argument becomes more convincing in the covariant formulation, given, for instance, in Jackson, Chapter 12.

The content of Eqs. (5.61) and (5.62) can be combined: Eq. (5.61) suggests that a negative energy solution can be looked at as a particle moving backward in time but having a positive energy. Equation (5.62) demonstrates that a particle moving backward in time satisfies the same equation of motion as a particle with opposite charge moving forward in time. Taken together, the two relations imply that a particle with charge q and *negative* energy behaves like a particle with charge $-q$ and *positive* energy. The negative energy states thus behave like antiparticles.

With this interpretation the processes shown in Fig. 5.24 can be described in two different but equivalent ways: in the conventional language, a particle–antiparticle pair is produced at time t_1 and position x_1. The antiparticle meets another particle at time t_2 and position x_2, giving rise to two gamma quanta that propagate forward in time. In Stueckelberg–Feynman language, the particle is the primary object and it weaves through space and time, backward and forward: at time t_2, the particle emits two photons and turns back in time to reach the spot (x_1, t_1). There it is scattered by a photon and again moves forward in time. What is the advantage of this way of looking at negative energy states? Negative energy states have disappeared from the discussion, and they are replaced by antiparticles with positive energy. The description makes it obvious that the antiparticle concept applies just as well to bosons as to fermions.

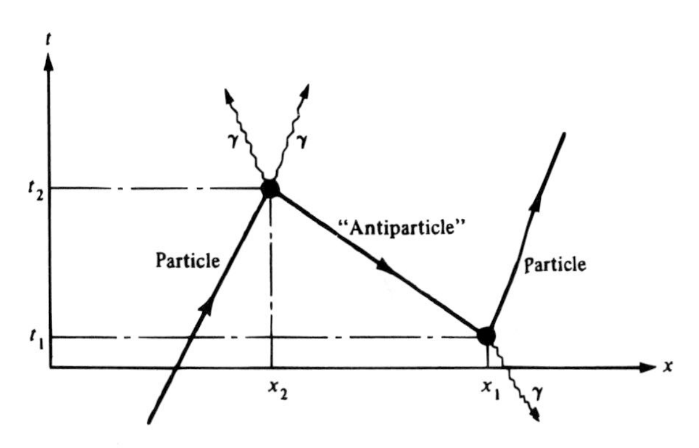

Figure 5.24: Pair production at (x_1, t_1) and particle–antiparticle annihilation at (x_2, t_2). As noted in Chapter 3, pair production can occur only in the field of a nucleus that takes up momentum. A nucleus is implied near point (x_1, t_1).

Assuming an antiparticle to be a particle moving backward in time, a number of conclusions can be drawn immediately. A particle and its antiparticle must have the same mass and the same spin because they are the same particle, just moving in a different direction in time:

$$\begin{aligned} m(\text{particle}) &= m(\text{antiparticle}) \\ J(\text{particle}) &= J(\text{antiparticle}). \end{aligned} \tag{5.63}$$

However, particle and antiparticle are expected to have opposite additive internal (not connected to space–time) quantum numbers. Consider the pair production at the time t_1 in Fig. 5.24. For times $t < t_1$, only a photon is present in the region around x_1, and its additive quantum numbers q, A, L, and L_μ are zero. If these quantum numbers are conserved, the sum of the corresponding quantum numbers for the particle–antiparticle pair must also add up to zero so that

$$N(\text{particle}) = -N(\text{antiparticle}). \tag{5.64}$$

Here N stands for any additive quantum number whose value for the photon is zero.

A final remark about a technical point in labeling Feynman diagrams may help prevent some confusion. A pair production process is usually drawn as shown in Fig. 5.25(a). The outgoing particle has its arrow along its momentum. The antiparticle, however, is shown with the arrow reversed. This convention makes reading diagrams unambiguous, and the example in Fig. 5.25(b) should be clear. Are the Stueckelberg–Feynman concepts of particles and antiparticles correct? Only experiment can tell, and experiment has indeed provided impressive support. Dirac predicted the antielectron in 1931, and it was found in 1933.[32] After this major success, the question arose whether an antiproton existed, but even persistent search in cosmic rays failed to turn it up. It was finally discovered in 1955 when the Bevatron in Berkeley began working.[33] Since then, antiparticles to essentially all particles have been found.

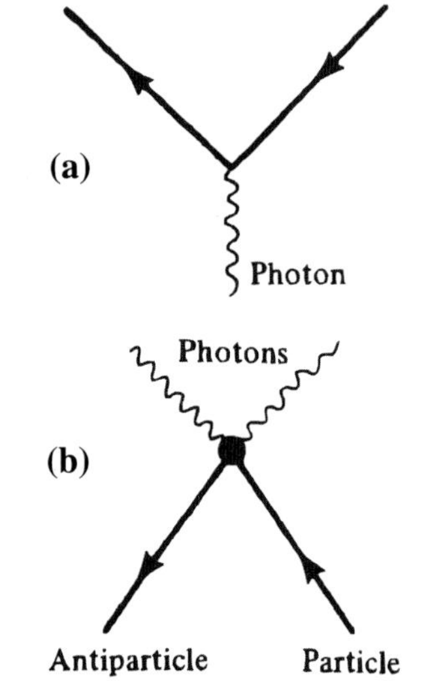

Figure 5.25: *Arrow convention* for particles and antiparticles.

A spectacular example is the observation of the antiomega.[34] This hyperon was produced in the reaction

$$dK^+ \longrightarrow \overline{\Omega}\Lambda\Lambda p\pi^+\pi^- \tag{5.65}$$

An example of detection of the production and decay of the antiomega are shown in Figs. 5.26 and 5.27.

Finally we note that a neutral particle can be its own antiparticle, e.g., π^0, or it might be different, e.g. K^0 and $\bar{K}^0$. We will come back to this issue in Chapter 11.

5.11 Quarks, Gluons, and Intermediate Bosons

When is a particle officially admitted to the zoo? This question has no simple answer, as we learn from history. The photon, introduced by Einstein in 1905, was

[32] C. D. Anderson, *Phys. Rev.* **43**, 491 (1933); *Am. J. Phys.* **29**, 825 (1961).
[33] O. Chamberlain, E. Segrè, C. Wiegand, and T. Ypsilantis, *Phys. Rev.* **100**, 947 (1955).
[34] A. Firestone, G. Goldhaber, D. Lissauer, B.M. Sheldon, and G.H. Trilling, *Phys. Rev. Lett.* **26**, 410 (1971).

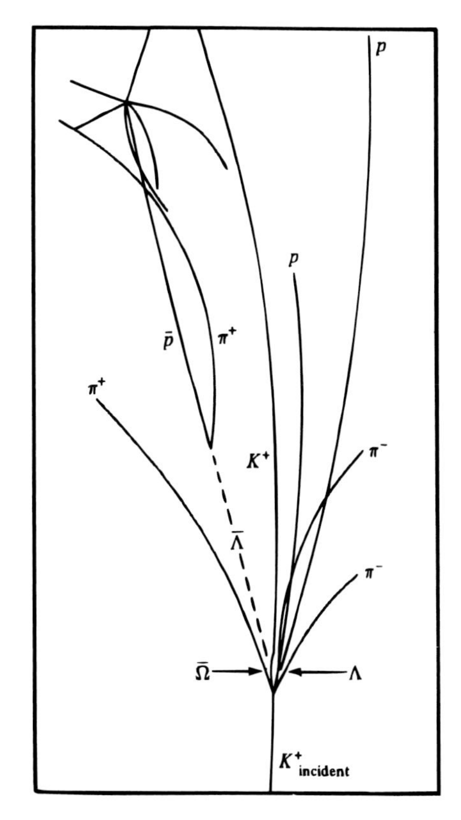

Figure 5.26: Drawing of the reaction $dK^+ \rightarrow \overline{\Omega}\Lambda\Lambda p\pi^+\pi^-$ and the resulting decays. [A. Firestone et al., *Phys. Rev. Lett.* **26**, 410 (1971).]

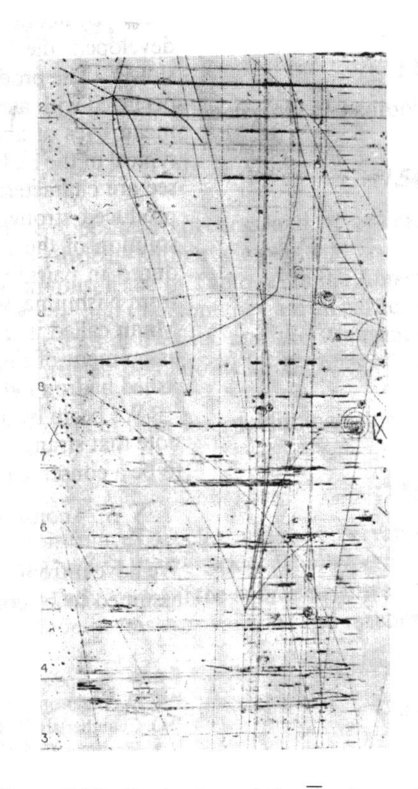

Figure 5.27: Production of the $\overline{\Omega}$, observed in a study of K^+d interactions at a momentum of 12 GeV/c, in the 2 m SLAC (Stanford Linear Accelerator Center) bubble chamber. (Courtesy Gerson Goldhaber, Lawrence Berkeley Laboratory.)

not accepted by Planck for at least 15 years. The neutrino, postulated by Pauli in 1930, was considered to be speculative for many years even by Bohr. In the case of the photon, the observation of the Compton effect dispelled the doubt; in the case of the neutrino, detection in absorption by Reines and coworkers in 1956 convinced the last disbelievers. As we have said in Section 5.3, we will never see the rho "directly". Can we still consider it a particle? We will not establish a firm criterion here, but instead introduce particles for which either the experimental evidence is very strong, or for which the theoretical arguments are convincing. In either case, the introduction of these particles makes the discussion of experiments and results much more elegant.

We have already stated in the introduction to Part II that experiments at energies above 10 GeV reveal that the proton, for instance, is not elementary but composed of subunits. These experiments, discussed in Section 6.7, and many additional data provide unambiguous evidence for the existence of *quarks*.(35) We will treat quarks in detail in Chapter 15. Here we only describe the properties that we will need for a preliminary understanding. Baryons are fermions built primarily from three quarks, and mesons are bosons built from a quark (q) and an antiquark ($\bar{q}$):

$$\text{baryon } (qqq)$$
$$\text{meson } (q\bar{q}).$$

In order to describe the presently known baryons and mesons, six quarks and the corresponding antiquarks are needed. In Table 5.7, we give the most important properties of quarks. At the same time we list the leptons again, in order to point out a striking similarity in the grouping of the two otherwise very different sets of particles.

Leptons and quarks are fermions; all particles in Table 5.7 have spin 1/2 and possess antiparticles. The particles divide into three generations or families, light, intermediate, and heavy. Recent evidence from the decay of the Z^0 shows conclusively that there are only three generations of neutrinos of small mass. (36) Within each family, there are two different "flavors", and the table contains six flavors of leptons and six of quarks.

The quark property that immediately catches the eye is the *electric charge*: quarks have charges $(2/3)e$ and $-(1/3)e$! These charges, of course, permit the assignment Eq. (5.66). With the charges given in Table 5.7 it is easy to see that the combination (uud) has the correct charge to be a proton, (udd) a neutron. Despite great efforts to catch a free quark, none has been seen (see Chapter 15); strong theoretical arguments imply that quarks must remain confined within hadrons.(37)

35 S. L. Glashow, *Sci. Amer.* **33**, 38 (October 1975).
36 The present limit is $N_\nu = 2.984 \pm 0.008$ See PDG.
37 Y. Nambu, *Sci. Amer.* **235**, 48 (November 1976); K. A. Johnson, *Sci. Amer.* **241**, 112 (July 1979); C. Rebbi, *Sci. Amer.* **248**, 54 (February 1982).

Table 5.7: LEPTONS AND QUARKS. From PDG†.

Leptons			Quarks		
	Charge (e)	Mass eV/c^2	Flavor	Charge (e)	Mass‡ MeV/c^2
ν_1	0	< 2	u up	2/3	3
e	-1	5.1×10^5	d down	$-1/3$	6
ν_2	0	< 2	c charmed	2/3	1.3×10^3
μ	-1	1.1×10^8	s strange	$-1/3$	110
ν_3	0	< 2	t top	2/3	1.8×10^5
τ	-1	1.8×10^9	b bottom	$-1/3$	4.2×10^3

†We show the mass eigenstates. As will be shown in Chapter 11, the weak eigenstates are linear combinations of the latter.
‡The masses for the quarks are only approximate because they are deduced from composite states in which their strong interactions have to be taken into account. All quarks come in three colors.

Since no free quarks are available, their masses cannot be measured and the mass estimates in Table 5.7 are based on theoretical arguments.[38]

Quarks have another remarkable property, color! Each quark comes in three colors, red, green, and blue. Of course, flavor and color have nothing to do with taste or vision; they are names chosen to describe previously unknown but well-defined physical properties. While flavor denotes the type of quark $(u, d, s, \ldots)$, color charge refers to a hadronic "charge." Just as the electric charge characterizes the strength of a particle's interaction with an electromagnetic field [Eq. (5.11)], color charge represents its interaction with the hadronic field of force. Antiquarks, like quarks, also have three colors, antired, antigreen, and antiblue. Since no colored particle has ever been observed, the combinations in (5.55) must be colorless or white. Consequently, a proton can, for instance, contain a red and a green up quark and a blue down quark, but not two red u quarks. If you, the reader, at this point feel you have inadvertently picked up a science fiction story, you are forgiven. Nature, however, is strange (and charmed) and the concepts introduced here without justification do make sense. We will justify the concepts later in more detail. Table 5.8 lists the principal quark composition of some mesons and baryons.

More particles or quanta emerge when we consider the forces that rule subatomic physics. In Section 5.8 we told the story of the prediction of the pion as the quantum mediating the interaction between nucleons. The conviction that no action at a distance exists and that all forces are transmitted by quanta[24] leads to the quanta listed in Table 5.9.

[38] J. Gasser and H. Leutwyler, *Phys. Rep.* **87**, 77 (1982).

Table 5.8: PRINCIPAL QUARK COMPOSITION OF SOME MESONS AND BARYONS†.

Mesons		Baryons	
π^+, π^0, π^-	$u\overline{d}$, $u\overline{u} + d\overline{d}$, $d\overline{u}$	p, n	uud, ddu
ρ^+, ρ^0, ρ^-	$u\overline{d}$, $u\overline{u} - d\overline{d}$, $d\overline{u}$		
ω^0	$u\overline{u} + d\overline{d}$		
η^0	$u\overline{u} + d\overline{d} + \epsilon\, s\overline{s}$	Λ^0	uds
ϕ^0	$s\overline{s}$	Σ^+, Σ^0, Σ^-	uus, uds, dds
K^+, K^-	$u\overline{s}$, $s\overline{u}$	Ξ^0, Ξ^-	uss, dss
K^0, $\overline{K^0}$	$d\overline{s}$, $s\overline{d}$	Ω^-	sss
D^+, D^-	$c\overline{d}$, $d\overline{c}$	Λ_c^+	udc
D^0, $\overline{D^0}$	$c\overline{u}$, $u\overline{c}$	Σ_c^{++}, Σ_c^+, Σ_c^0,	uuc, udc, ddc
B^+, B^-	$u\overline{b}$, $b\overline{u}$		
B^0, $\overline{B^0}$	$d\overline{b}$, $b\overline{d}$		

†The numbers are only approximate and not normalized.

Table 5.9: FIELDS AND QUANTA.

Field	Quanta	Mass	Spin	"Charge"
Electromagnetic	Photon	0	1	0
Hadronic	Gluon	0	1	8 colors
Weak	$W^\pm$	81 GeV/c^2	1	$\pm e$
	Z^0	91 GeV/c^2	1	0
Gravitational	Graviton	0	2	?

Table 5.10: THE BASIC PARTICLES AND FORCES OF THE STANDARD MODEL OF SUBATOMIC PHYSICS.

Constituents†	Forces	Gauge boson
Quarks	Hadronic	Gluon
u c t		
d s b		
	Electromagnetic	Photon
Leptons		
ν_1 ν_2 ν_3		
e μ τ		
	Weak	$W^\pm$, Z^0

†We show the mass eigenstates. As will be shown in Chapter 11, the weak eigenstates are linear combinations of the latter.

We have already encountered the photon, the $W^{\pm}$ and Z^0 gauge bosons, and have sketched in Fig. 5.18 how the force between two electrically charged particles is transmitted by a virtual photon. Similarly the *gluons* are the quanta that transmit the force between two quarks. They are the gauge bosons of the strong force, akin to the photon in the electromagnetic force. The electromagnetic interaction between two particles with electric charges q_1 and q_2 is proportional to the product q_1q_2. Similarly the hadronic charge on a quark, called the color charge, is introduced and the hadronic force between the two quarks is proportional to the product of the two color charges. There are, however, major differences between the photon and the gluon. The photon is electrically neutral and leaves the electric charges of the two interacting particles unchanged. Moreover, two photons cannot interact directly with each other. Gluons, however, carry color and consequently can change the color of the interacting quarks. Gluons also can interact directly with each other; the theory predicts that they can form bound states, called glueballs.

The weak interaction is transmitted by three quanta, W^+, W^-, and Z^0.[39] In Chapters 11 and 13, we will discuss weak processes in detail. One well-known example of a weak process is the decay of the neutron, $n \rightarrow p\ e^-\overline{\nu}_e$. In 1938, Klein[40] suggested that this decay was, in reality, a two-step process,

$$\begin{aligned} n \rightarrow p \quad & W^- \quad , \\ & W^- \quad \rightarrow e^-\overline{\nu}. \end{aligned}$$

In the quark model, depicted in Fig. 5.28, protons and neutrons consist of quarks, and the weak interaction occurs between the quarks. One quark, a d for instance, may emit a W, and as a result, the neutron changes into a proton:

$$\begin{aligned} d \rightarrow u \quad & W^- \quad , \\ & W^- \quad \rightarrow e^-\overline{\nu}_e. \end{aligned}$$

or

$$n(udd) \rightarrow p(uud)\ e^-\,\overline{\nu}_e.$$

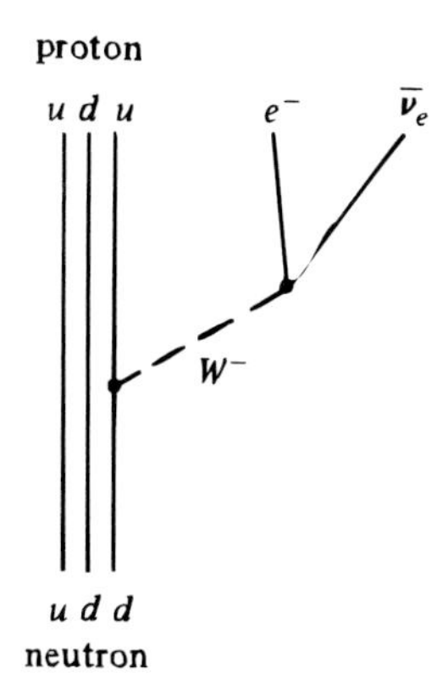

Figure 5.28: Quark model description of the beta decay of a neutron.

The $W^{\pm}$ and the corresponding neutral Z^0 are gauge bosons and sometimes are called "intermediate bosons". Theory predicted the masses of the $W^{\pm}$ and Z^0 before they were discovered; the predictions are given in Table 5.9. The large masses

[39] P. Q. Hung and C. Quigg, *Science* **210**, 1205 (1980).

[40] O. Klein in *Les Nouvelles Théories de la Physique*, Institut International de Coopération Intellectuelles, Paris, 1939.

of the $W(\sim 80\ \mathrm{GeV}/c^2)$ and of the $Z(\sim 90\ \mathrm{GeV}/c^2)$ imply that their production requires extremely high energies. The long search for the W finally came to an end in 1983 when five clear cases of W production and decay were observed in $p\bar{p}$ collisions at 2×270 GeV at the CERN SPS (Fig. 2.12).[(41)] The Z^0 was found shortly thereafter.[(42)]

Why have we not listed the pion as a field quantum in Table 5.9? In the picture we have presented, the pion itself is viewed as a quark–antiquark state and the long-range force between nucleons, mediated by the pion, is not elementary. At the more basic level, all three forces—strong, electromagnetic, and weak—are mediated by gauge bosons of spin one.

Together with the basic constituents of matter, the three subatomic forces make up the so-called "standard model." Its basic features have been introduced in this and previous sections and will be discussed in more detail in later chapters. We summarize its main features in Table 5.10.

The standard model is believed to be a rather accurate description of nature: The basic constituents of matter are three families of point quarks and three of point leptons. There are also three basic non-gravitational gauge-type forces. The quarks interact through all three forces and the (charged) leptons interact only through the electromagnetic and weak forces. All three forces are carried by gauge bosons.

5.12 Excited States and Resonances

In atomic physics, the development of concepts and theories is intimately linked with the exploration of excited states, in particular those of the hydrogen atom. The Balmer series, the Ritz combination principle, the Bohr theory, the Schrödinger equation, the Dirac equation, and the Lamb shift are all connected with the hydrogen spectrum. Without the simplicity *and* the richness of the hydrogen spectrum, progress would have been slower. In subatomic physics, the situation is more complex. The nuclear system that most closely resembles the hydrogen atom is the deuteron, a bound system consisting of a proton and a neutron. This system has only one bound state and consequently does not provide the richness of information that the hydrogen atom yielded. It is necessary to consider the excited states of more complicated systems, such as heavier nuclides. Moreover, excited states of baryons and mesons exist, and they must be studied in detail in the hope that they will provide clues to an understanding of hadronic physics.

An understanding of the features of excited hadronic states requires a knowledge of some results of quantum mechanics, and these can be discussed most easily by

[41] G. Arnison et al., *Phys. Lett.* **122B**, 103 (1983); M. Banner et al., *Phys. Lett.* **122B**, 476 (1983).

[42] G. Arnison et al., *Phys. Lett.* **126B**, 398 (1983); P. Bagnaia et al., *Phys. Lett.* **129B**, 130 (1983); for a summary, see E. Rademacher in *Progress Particle Nuclear Physics*, Vol 14 (A. Faessler, ed) (Pergamon, New York), p. 231 (1985) and P. Watkins, *Story of the W and Z* (Cambridge University Press, Cambridge, 1986).

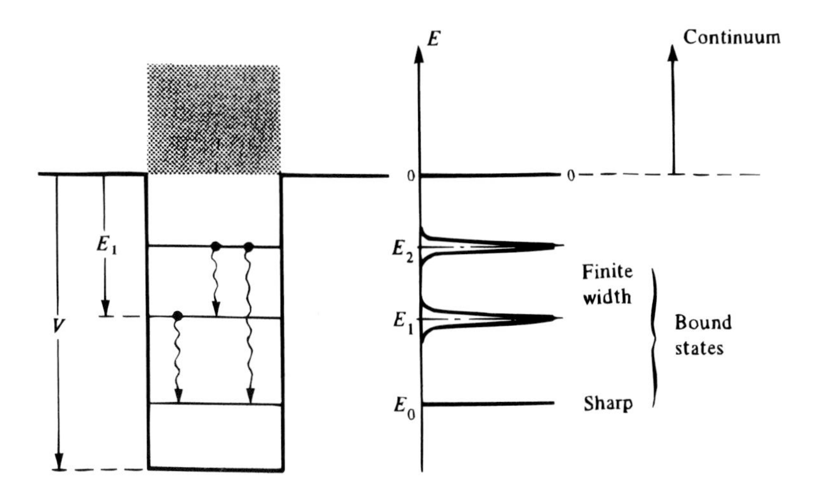

Figure 5.29: Energy levels in a square well. The ground state is sharp. The excited states can decay to the ground state by photon emission, and they display a natural line width. States with positive energy form a continuum.

treating the square well. Consider a particle with mass m in a square well as shown in Fig. 5.29. It is straightforward to solve the Schrödinger equation for this problem and to find the allowed energy levels. First consider the case $E < 0$, where the numerical or graphical solution of the Schrödinger equation produces a number of bound states. *Bound* indicates that a particle in one of these levels will remain attached to the force center.

The Schrödinger equation for the square well is an eigenvalue equation, $H\psi = E_i\psi$, and the eigenvalues E_i represent sharp energy states. In reality, however, all states but the lowest one usually decay, for instance by photon emission. We have seen in Section 5.7 that decaying states possess a finite width and that the energy is composed of a large real and a small imaginary part, as in Eq. (5.36). For a bound state, the large real component is negative if the zero point of the energy is taken to be the value of the potential at infinity, as in Fig. 5.29.

For positive energies, E can have any value. In other words, the spectrum forms a *continuum*. One would therefore guess that nothing interesting can happen in this region. This guess is false. To study the situation, scattering events have to be considered. In the one-dimensional case, as in Fig. 5.30, scattering is simple: A particle beam is assumed to impinge on the potential well from the left (Fig. 5.30). Classically, such a particle will pass unhindered over the well. In quantum mechanics the situation is more interesting. The Schrödinger equation can easily be solved, and it turns out that only a fraction of the incident beam is transmitted; another fraction is reflected at the barrier. The transmitted fraction, T, is given by[43]

[43] Tipler and Llewellyn, Chapter 6; Park, Eq. (4.38).

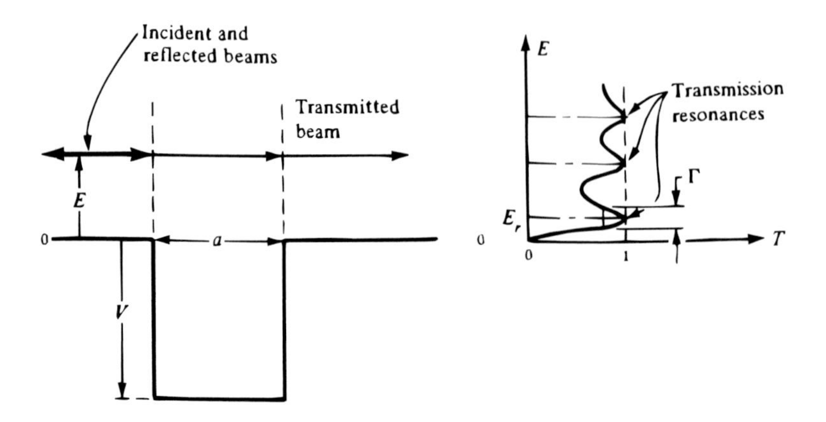

Figure 5.30: Scattering of a particle with energy E from a one-dimensional potential well. Classically, all incident particles will be transmitted. Quantum mechanically, at small energies, the transmission coefficient T is unity only at certain energies. The appearance of *transmission resonances* in the behavior of the transmission as a function of particle energy E is shown at the right.

$$\frac{1}{T} = 1 + \frac{V^2}{4E(E + |V|)} \sin^2 ka, \tag{5.66}$$

where E is the kinetic energy of the incident particles, $V(< 0)$ the depth, and a the width of the potential well. The wave number k is given by

$$k^2 = \frac{2m}{\hbar^2}(E + |V|). \tag{5.67}$$

Equations (5.66) and (5.67) demonstrate that the transmission coefficient T is unity only at certain energies. The behavior of T as a function of E is sketched in Fig. 5.30, where the appearance of *transmission resonances* is evident. The behavior of a particle with an energy E_r corresponding to maximum transmission can be investigated by using wave packets rather than plane waves to describe the incident beam. It turns out that the incident particle remains in the well region for a time that is much longer than that expected from classical mechanics.[44] The mean time spent in the well region, τ, and the width of the corresponding resonance, Γ, satisfy Eq. (5.46). Mathematically, the existence of a resonance at the energy E_r can again be described, in analogy to Eq. (5.36), by introducing a complex energy,

$$E = E_r - \tfrac{1}{2}i\Gamma.$$

Here E_r is positive, and Γ can be comparable to E_r.

The appearance of a resonance in the continuum is not restricted to the simple one-dimensional case just discussed but is a more general phenomenon. To treat

[44] Detailed discussions can be found in Merzbacher, Chapter 6, and in D. Bohm, *Quantum Theory*, Prentice Hall, Englewood Cliffs, N. J., 1951, Chapters 11 and 12.

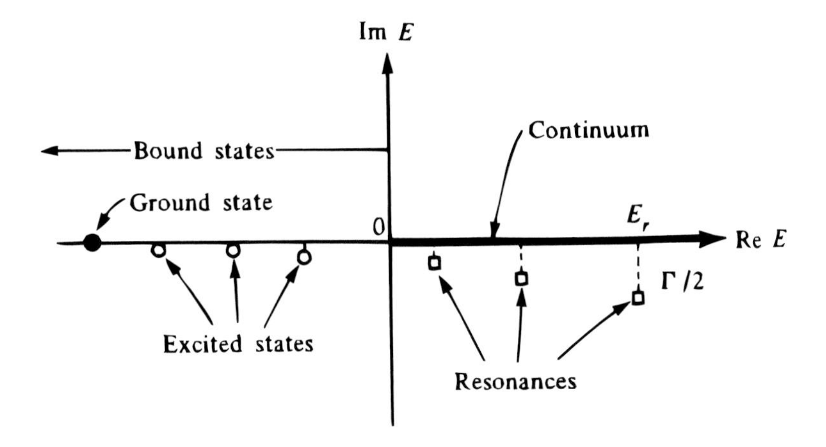

Figure 5.31: Classification of the energy levels of a quantum system in the complex energy plane. Re $E = 0$ is determined by the potential at infinity. The widths Γ in actual resonances are usually much smaller than indicated here.

the problem with more relevance to actual situations, scattering of particles from a three-dimensional potential has to be studied. The basic ideas, however, are already contained in our simple example: Resonances can appear in the continuous energy spectrum, and they are characterized by the energy of their maximum, E_r, and by their width, Γ. Width and position together can be described by introducing a complex energy, $E = E_r - \frac{1}{2}i\Gamma$.

The use of a complex energy allows a classification of the energy levels of a quantum system. The classification is illustrated in Fig. 5.31. A point in the complex energy plane represents energy and width of a particular state. In addition to resonances, every positive energy corresponds to a permissible solution of the scattering problem. This fact is expressed in Fig. 5.31 by drawing the continuum along the positive energy axis.(45)

Resonances are characterized by unique quantum numbers; energy, width, and quantum numbers of the states appearing in a particular system depend on the constituents of the system and on the forces acting among them. It is the task of experimental subatomic physics to find the levels and determine their quantum numbers, and it is the goal of theoretical subatomic physics to explain and predict the properties of the observed bound states and resonances in terms of models and forces.

---

[45] In a more advanced treatment of scattering, the bound states and the resonances appear as poles, and the continuum as a cut of the scattering matrix in the complex energy plane.

5.13 Excited States of Baryons

The problem of finding all excited states of the baryons is probably hopeless. It is crucial, however, to find enough states to be able to discover regularities, get clues to the construction of theories, and test the theories. Even this more restricted requirement is very difficult to fulfill in subatomic physics. A great deal of ingenuity and effort has been expended on *nuclear and particle spectroscopy*, the study of nuclear and particle states. In the present section we shall give some examples of how excited states and resonances are found.

As a first example, we consider the nuclide ^{58}Fe, with a natural abundance of 0.31%. Two ways in which the energy levels of ^{58}Fe have been investigated are sketched in Fig. 5.32. An accelerator, for instance, a Van de Graaff, produces a proton beam of well-defined energy. The beam is momentum-analyzed and transported to a scattering chamber where it hits a thin target. The target consists of an iron foil that has been enriched in ^{58}Fe. The transmission through the foil can be studied as a function of the energy of the incident proton, or the scattered protons can be momentum-analyzed. Consider the second case, denoted by (p, p'). The notation (p, p') indicates that incoming and scattered particles are protons but that the scattered particle has a different energy in the c.m. The momentum and hence the energy of the scattered proton p' are determined in a magnetic spectrometer, i.e., a combination of bending magnet, slits, and detectors. If the kinetic energy of the incident proton is E_p and that of the scattered one is E'_p, the nucleus received an energy $E_p - E'_p$, and a level at this energy was excited. The experiment constitutes a nuclear Franck–Hertz effect. (A correction has to be applied because the ^{58}Fe* nucleus recoils, and the recoil energy must be subtracted from $E_p - E'_p$ in order to find the correct excitation energy.) A typical result of such an experiment is shown in Fig. 5.33. The appearance of many excited levels is unmistakable. The reaction (p, p') is only one of many that are used to excite and study nuclear levels. Other possibilities are $(e, e'), (\gamma, \gamma'), (\gamma, n)$, $(p,\ n)$, (p, γ), $(p,\ 2p)$, $(d,\ p)$, $(d,\ n)$, and so forth. Decays are also sources of information, and Fig. 4.7 gives an example of a partial gamma-ray spectrum. Data from a large variety of experiments are used to piece together a level diagram of a particular nuclide. For ^{58}Fe, the level diagram is shown in Fig. 5.37.

As the excitation energy is increased, the situation becomes more complex. In a simplified picture it can be discussed by referring to Fig. 5.30 with the essential aspects shown in Fig. 5.34. At an excitation energy of about 8 MeV, the top of the well is reached, and it becomes possible to eject a nucleon from the nucleus, for instance, by a reaction $(\gamma, n), (\gamma, p)$, $(e,\ ep)$, or $(e,\ en)$. Just above the well, such processes are still not very likely, and most excited states will return to the nuclear ground state by the emission of one or more photons, because particle emission is inhibited by reflections from the nuclear surface (Fig. 5.30), angular momentum effects, and the small number of states available per unit energy (small phase space).

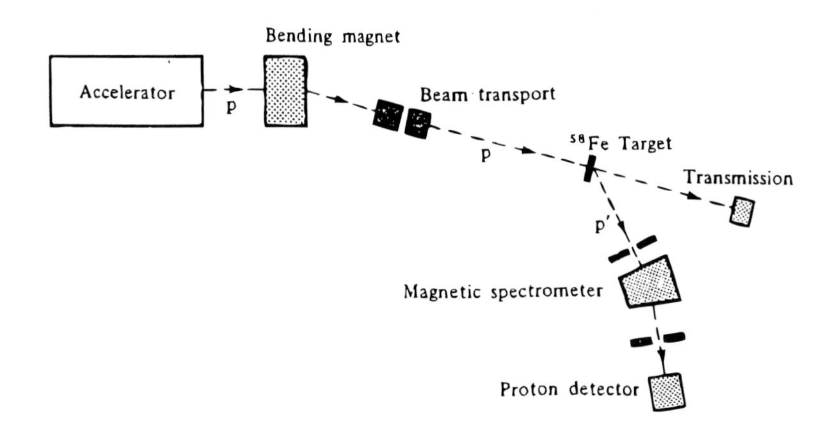

Figure 5.32: Investigation of energy levels by transmission and by inelastic scattering.

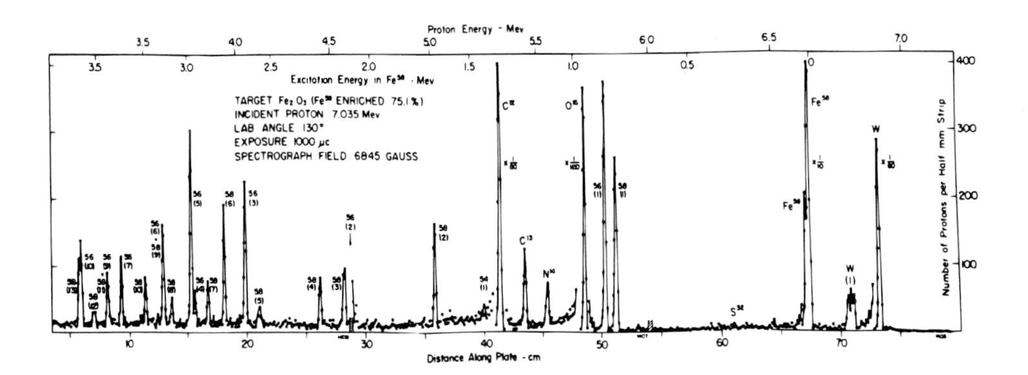

Figure 5.33: Spectrum of protons scattered from enriched ^{58}Fe (75.1%) target. The detector consists of photographic plates so that many lines can be observed simultaneously. [From A. Sperduto and W. W. Buechner, *Phys. Rev.* **134**, B142 (1964).] Since the target still contains some isotopes other than ^{58}Fe, additional lines appear. The iron lines are labeled by the mass number A.

Table 5.11: NUCLEAR ENERGY LEVEL CHARACTERISTICS FOR THE THREE REGIONS SHOWN IN FIG. 5.34. E is the excitation energy, $\overline{\Gamma}$ the average level width, and $\overline{D}$ the average level spacing.

Region	Characteristics	E (MeV)	Typical Values	
			$\overline{\Gamma}$ (eV)	$\overline{D}$ (eV)
I. Bound states	$\overline{\Gamma} \ll \overline{D} \approx E$	1	10^{-3}	10^{5}
II. Resonance region	$\overline{\Gamma} < \overline{D} \ll E$	8	1	10^{2}
III. Statistical region	$\overline{D} \ll \overline{\Gamma} \ll E$	20	10^{4}	1

Nevertheless, the states are no longer bound but are now classed as resonances. In the idealized cross-section curve in Fig. 5.34, the individual resonances are shown in region II. As the energy is further increased, the resonances become more numerous and their widths increase. They begin to overlap, and the individual structure averages out. In region III, called the statistical region, the envelope of the overlapping individual resonances is measured, and it displays a prominent feature, called the giant resonance: At around 20 MeV excitation energy, the total cross section goes through a pronounced maximum. At much higher energies, the continuum loses all features.

The three regions shown in Fig. 5.34 are characterized by three numbers, the average level width, $\overline{\Gamma}$; the average distance between levels, $\overline{D}$; and the excitation energy, E. Typical values of these three quantities for the three regions are given in Table 5.11. Details vary widely from nuclide to nuclide, but the gross features remain. Exploration of the excited states of baryons with $A = 1$ is more difficult for three reasons: (1) No bound states exist and resonances are harder to study than bound states. (2) Most of the resonances decay by hadronic processes, their widths are large, and it is difficult to separate individual levels. (3) The only stable baryon that can be used as a target is the proton; liquid hydrogen targets are standard equipment in all high-energy laboratories. No isolated neutron targets exist. All other baryons (Table 5.6) have such a short lifetime that experiments of the type shown in Fig. 5.32 are not possible, and indirect methods must be used.

The first excited proton state was discovered by Fermi and collaborators in 1951. They measured the scattering of pions from protons and found that the cross section increased rapidly with energy up to about 200 MeV pion kinetic energy and then leveled off or decreased again.(46) Brueckner suggested that this behavior could be interpreted as being due to a nucleon isobar (excited nucleon state) with spin 3/2.(47) It took some more time and many more experiments before it became clear that the *Fermi resonance* is only the first of many excited states of the nucleon.

The investigation of excited proton states proceeds similarly to the study of

[46] H. L. Anderson, E. Fermi, E. A. Long, and D. E. Nagle, *Phys. Rev.* **85**, 936 (1952).
[47] K. A. Brueckner, *Phys. Rev.* **86**, 106 (1952).

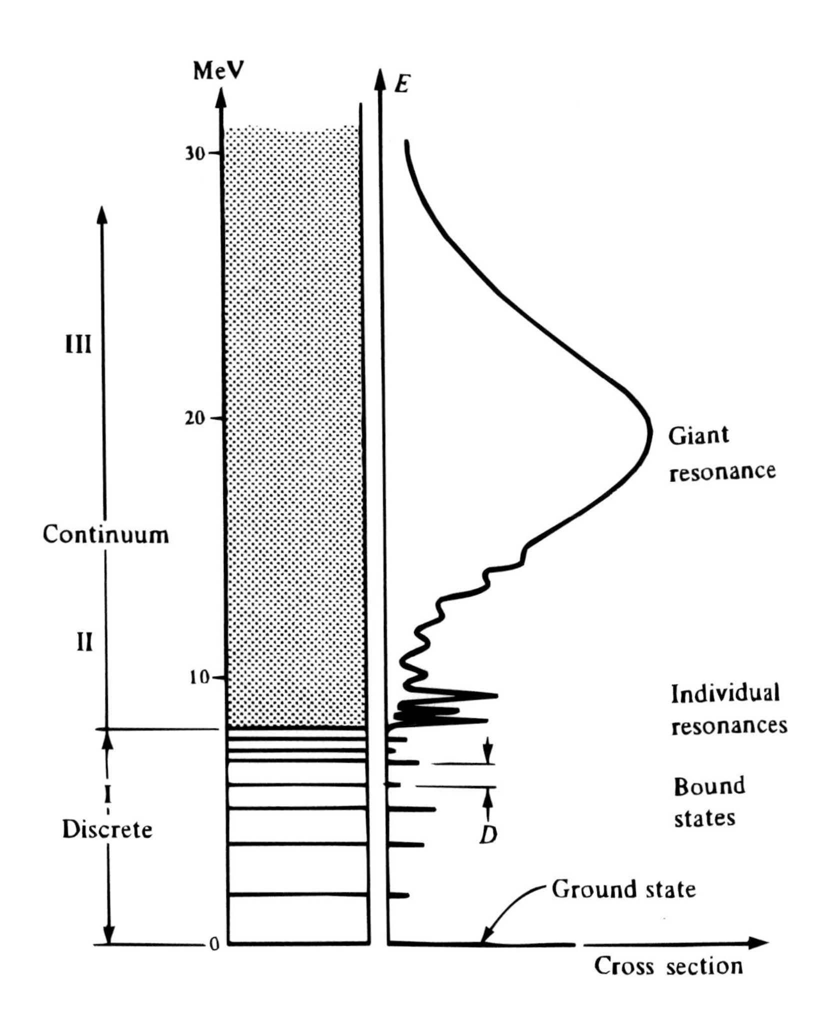

Figure 5.34: Typical features of the excited states of a nucleus. The cross-section curve is idealized; it can be investigated by inelastic electron scattering or by studying the absorption of gamma rays as a function of gamma-ray energy. Three regions are distinguished: I, bound (discrete) states; II, individual resonances; and III, statistical region (overlapping resonances).

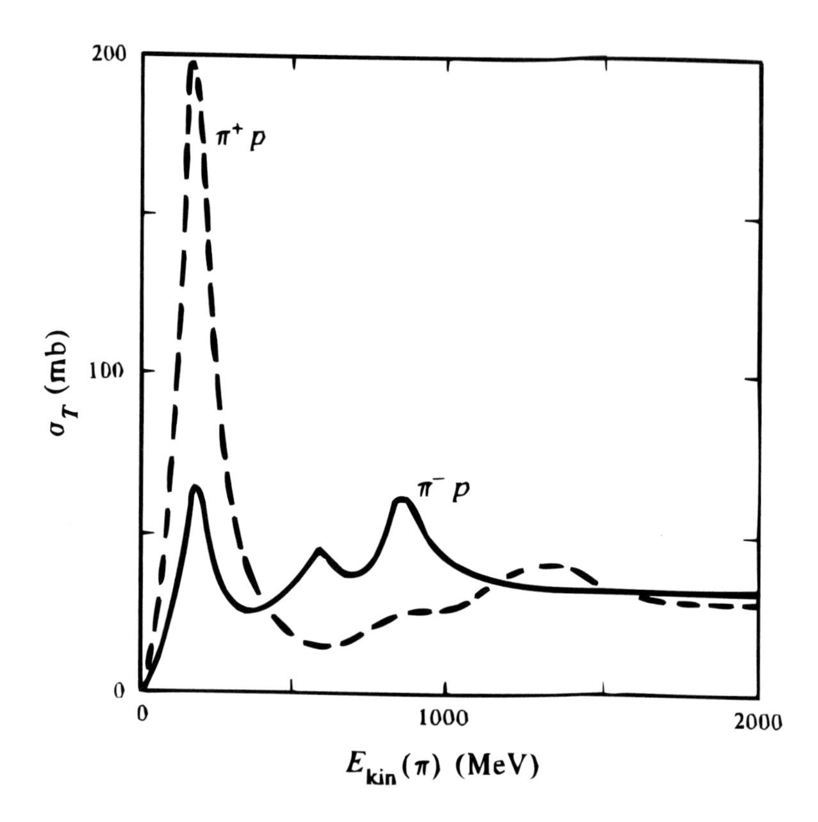

Figure 5.35: Total cross section as a function of pion kinetic energy for the scattering of positive and negative pions from protons. (1 mb = 1 millibarn = 10^{-27} cm^2.)

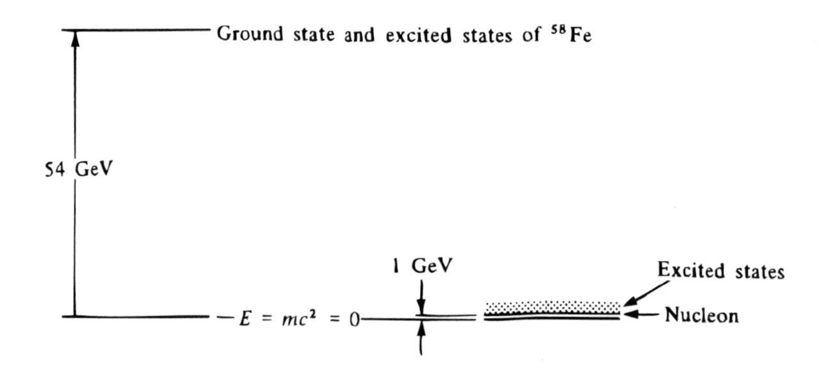

Figure 5.36: Total rest energies of the states in ^{58}Fe and of the nucleon and its excited states. On the scale shown here, the excited states of the nuclide ^{58}Fe are so close to the ground state that they cannot be distinguished without magnification. A magnified spectrum is provided in Fig. 5.37.

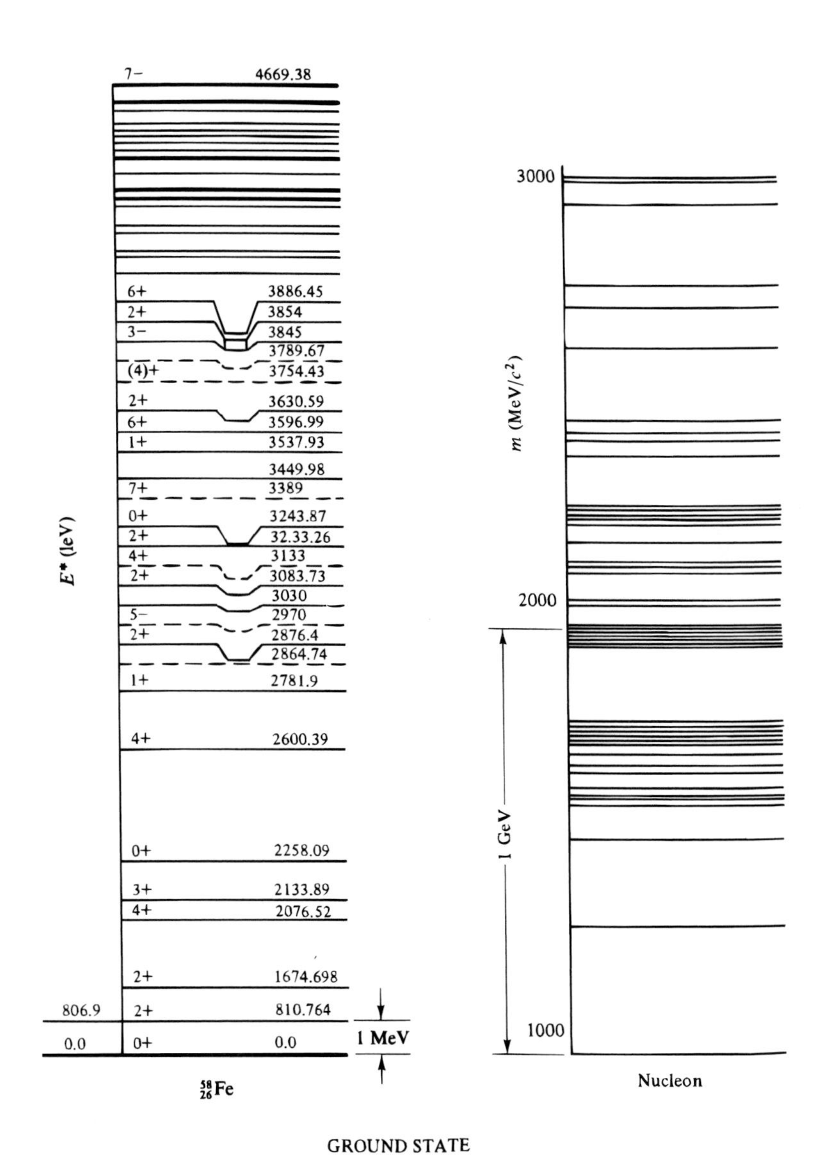

Figure 5.37: Ground state and excited states of the nuclide ^{58}Fe and of the nucleon (neutron and proton). The region above the nuclear ground state in Fig. 5.36 has been enlarged by a factor of about 5000. The spectrum of the nucleon in Fig. 5.36 has been magnified about 40 times. The nuclear states have widths of the order of eV or less and consequently can be observed separately. The excited particle states or resonances, on the other hand, have widths of the order of a few hundred MeV; they overlap and are often very difficult to find. It is likely that many additional levels exist.

excited nuclear states. High-energy particles, mainly electrons or pions, impinge on a hydrogen target, and the transmitted and the scattered beams are detected and analyzed. The behavior of the total cross section for pions on protons is given in Fig. 5.35. The appearance of resonances is evident. Since 1951, a great deal of effort has been expended to find such resonances and determine their quantum numbers. The Fermi resonance discussed above and shown as the first peak in Fig. 5.35 is called $\Delta(1232)$, where the number denotes the rest energy of the resonance in MeV.

In Figs. 5.36 and 5.37, we compare the energy spectra of the nuclide ^{58}Fe and of the nucleon. Figure 5.36 depicts the total masses (rest energies), while Fig. 5.37 presents the excitation spectra, namely the energies above the ground states. The figures make it clear that the nuclear excitation energies are very small compared to the rest energy of the ground state, whereas the particle excitation energies can be large compared to the rest energy of the ground state. The particle excitation energies are 2 to 3 orders of magnitude larger than nuclear excitation energies. Another difference exists between nuclear and particle excited states: Nuclei possess bound states *and* resonances, as indicated in Fig. 5.34. The excited particle states, on the other hand, are all resonances.

Finally, we note that we have treated nuclear and particle spectroscopy here extremely briefly; we have sketched only one way of finding the excited states. Many other ones exist. Moreover, the determination of the various quantum numbers of a state (spin, parity, charge, isospin, magnetic moment, quadrupole moment) can be an exceedingly difficult business. In fact, some of these quantum numbers can be measured only for very few states. The references in Section 5.14 describe most of the techniques and ideas of subatomic spectroscopy, but we shall not treat this topic further.

5.14 References

The properties of elementary particles are reviewed in PDG. The properties of nuclear levels are summarized in *Table of Isotopes*, 8th Ed. (R.B. Firestone, V.S. Shirley, eds.) John Wiley & Sons, New York, 1996. The information can be found online at http://www.nndc.bnl.gov/. Current information can be found in the journals *Nuclear Data Tables and Nuclear Data Sheets* published by Academic Press, as well as in special issues of *Nucl. Phys. A*.

Nuclear spectroscopy is reviewed in many places, and the following books provide additional information on most of the problems treated in the present chapter: F. Ajzenberg-Selove, ed., *Nuclear Spectroscopy*, Academic Press, New York, 1960 (two volumes); K. Siegbahn, ed. *Alpha-, Beta-, and Gamma-Ray Spectroscopy*, North-Holland, Amsterdam, 1965 (two volumes); J. Cerny, *Nuclear Spectroscopy and Reactions*, Academic, New York, 1974.

A nice introduction to particles can be found in S. Weinberg, *The Discovery of Subatomic Particles*, Cambridge, 2003. The *photon concept*, treated very briefly in Section 5.5, often leads to long and heated arguments. An interesting brief discussion is given in M.O. Scully and M. Sargent III, "The Concept of the Photon," *Phys. Today* **25**, 38 (March 1972). A more complete exposition can be found in M. Sargent III, M.O. Scully, and W.E. Lamb, Jr., *Laser Physics*, Addison-Wesley, Reading, 1974. A recent review of the photon's history and on upper limits on its mass and charge are given by L.B. Okun *Acta Phys. Polon.* **B37**, 565 (2006); also at hep-ph/0602036.

Charged leptons are discussed by M.L. Perl, *Phys. Tod.* **50**, 34 (Oct. 1997), and neutrinos by W.C. Haxton and B.R. Holstein, *Am. Jour. Phys.* **72**, 18 (2004).

Examples of recent findings of new particles can be found in T.M. Liss, P.L. Tipton *Sci. Am.* **277**, 54 (1997) (on the top quark) and in R.M. Thurman-Keup, A.V. Kotwal, M. Tecchio, A. Byon-Wagner, *Rev. Mod. Phys.* **73**, 267 (2001) (on the W boson).

Problems

5.1. * Does a vanishing mass indicate that the corresponding particle has no gravitational interaction? If not, how can the force in a gravitational field be defined?

5.2. * Discuss the Mössbauer experiment that indicates that photons falling in the earth's gravitational field gain energy. Why can such an experiment not be performed with optical photons? [R.V. Pound and J.L. Snider, *Phys. Rev.* **140B**, 788 (1965).]

5.3. Use Eq. (5.4) and the corresponding complete expressions for the operators L^2 and L_z to find the eigenvalues l and m for the functions

$$Y_0^0(\theta,\varphi) = (4\pi)^{-1/2}$$
$$Y_1^0(\theta,\varphi) = \frac{1}{2}\left(\frac{3}{\pi}\right)^{1/2} \cos\theta$$
$$Y_1^{\pm 1}(\theta,\varphi) = \pm\frac{1}{2}\left(\frac{3}{2\pi}\right)^{1/2} \sin\theta \exp(\pm\, i\varphi).$$

Here θ and φ are the angles defining spherical coordinates.

5.4. Verify Eq. (5.5).

5.5. Assume that electron and muon are uniform spheres with a radius of 0.1 fm. Compute the velocity at the surface caused by the rotation with spin $(\frac{3}{4})^{1/2}\hbar$.

5.6. Consider a system consisting of two identical particles and assume that the total wave function is of the form

$$\psi(\boldsymbol{x}_1,\boldsymbol{x}_2) = A\psi(\boldsymbol{x}_1)\varphi(\boldsymbol{x}_2) + B\psi(\boldsymbol{x}_2)\varphi(\boldsymbol{x}_1).$$

If ψ and ϕ are orthonormal, find the values of A and B that make the total wave function normalized to unity and (a) symmetric, (b) antisymmetric, or (c) neither under interchange $1 \rightleftharpoons 2$.

5.7. Does a particle with zero electric charge necessarily have no interaction with an external electromagnetic field? Give an example of a neutral particle that does interact with an external electromagnetic field. Find an example for a particle that does not. Does a particle with electric charge necessarily interact with an external electromagnetic field?

5.8. A nucleus with a spin $J = 2$ and a g factor of $g = -2$ is placed in a magnetic field of 1 MG.

(a) Where can such a field be found?

(b) Sketch the corresponding splitting of the energy levels. Label the levels with magnetic quantum numbers M. Find the value of the splitting between two adjacent levels in eV and in K.

5.9. Show that the magnetic dipole moment of a particle with spin $J = 0$ must vanish.

5.10. *Discuss the setup and basic features of the experiment to determine masses of short-lived isotopes using Penning traps. (See K. Blaum, *Phys. Rep.* **425**, 1 (2006).)

5.11. The determination of the mass of a particle often requires knowledge of its velocity. Discuss the principle of the Cerenkov counter. Show that the Cerenkov counter is a velocity-dependent detector.

5.12. How were the masses of the following particles determined:

(a) Muon

(b) Charged pion

(c) Neutral pion

(d) Charged kaon

(e) Charged sigma

(f) Cascade particle (Ξ).

5.13. Use wave packets to justify the interpretation of a particle with negative energy being a particle with positive energy but moving backward in time.

5.14. Use the covariant formulation of the equation of motion of a charged particle in an electromagnetic field to show that a particle with charge $-q$ moving backward in time behaves like an antiparticle of charge q moving forward in time.

5.15. In Eq. (5.24), $\pi^- p \to n\pi^+\pi^-$, the neutron in the final state escapes unobserved. The fact that the "missing" particle is a neutron is verified by using a *missing mass plot*: Assume a reaction of the form $a + b \to 1 + 2 + 3 + \cdots$. Denote the total energy by $E_\alpha = E_a + E_b$ and the total momentum of the two colliding particles by $p_\alpha = p_a + p_b$. Similarly, denote the corresponding sums for all *observed* particles in the final state by E_β and p_β. The unobserved (neutral) particles then carry away the "missing" energy $E_m = E_\alpha - E_\beta$ and the "missing" momentum $p_m = p_\alpha - p_\beta$. The "missing mass" is defined by

$$m_m^2 c^4 = E_m^2 - p_m^2 c^2.$$

(a) Sketch a missing mass plot, i.e., a plot of the number of events expected with mass m_m against m_m, if the only unobserved particle is a neutron.

(b) Repeat part (a) for the case where a neutron and a neutral pion escape.

(c) Find a missing mass plot in the literature.

5.16. * Discuss the reaction $d\pi^+ \to pp\pi^+\pi^-\pi^0$. The invariant mass spectrum of the three pions in the final state provides evidence for two short-lived mesons. Read the relevant literature and discuss how these mesons have been found.

5.17. Consider Eq. (5.24). Assume that the two pions do not form a resonant state (rho) but are emitted independently. Compute the upper and lower limit on the phase-space spectrum in Fig. 5.11.

5.18. Verify Eq. (5.30).

5.19. * Discuss the determination of the present limit on the mass of

(a) The electron neutrino and

(b) The muon neutrino.

(c) How can the limit on the mass of the muon neutrino be improved?

5.20. How can the stability of electrons be measured? Try to design a simple experiment and estimate the limit on the lifetime that you expect to get from your experiment.

5.21. What was Professor Moriarty's profession? Where did he finally disappear?

5.22. Describe the experimental facts that led Pauli to postulate the existence of the neutrino.

5.23. ^{64}Cu decays with a branching ratio of 62% to ^{64}Ni and with a branching ratio of 38% by electron emission to ^{64}Zn. The overall half life of ^{64}Cu is 12.8 hr. A spectrometer (magnet and scintillation counter) is adjusted so that only the electron decay to ^{64}Zn is observed. How long does it take until the intensity of this decay mode is reduced by a factor of 2?

5.24. Verify Eq. (5.34).

5.25. Find the Fourier transform of the function

$$f(x) = \begin{cases} 1, & |x| < a, \\ 0, & |x| > a. \end{cases}$$

5.26. Find the Fourier transform of

$$f(x) = \begin{cases} 0, & x < -1, \\ \frac{1}{2}, & -1 < x < 1, \\ 0, & x > 1. \end{cases}$$

5.27. Verify Eq. (5.43).

5.28. The level giving rise to the 14.4 keV gamma ray in ^{57}Fe decays with a half life of 98 nsec. Compute Γ, the full width at half-height, in eV.

5.29. Verify Eq. (5.45).

5.30. * Discuss methods to measure lifetimes of the order of

(a) 10^6 y

(b) 1 sec

(c) 10^{-8} sec

(d) 10^{-12} sec

(e) 10^{-20} sec.

5.31. The rho is believed to contribute to the hadronic force between hadrons. Compute the range of this force.

5.32. * What experiments would you perform to check if the muon is the quantum predicted by Yukawa? Compare your proposal to the actual evidence that led to the conclusion that the muon is not the Yukawa particle. [M. Conversi, E. Pancini, and O. Piccioni, *Phys. Rev.* **71**, 209 (1947); E. Fermi, E. Teller, and V.F. Weisskopf, *Phys. Rev.* **71**, 314 (1947).]

5.33. Does an electron bound in an atom satisfy Eq. (1.2)?

5.34. Discuss the following methods for determining the nuclear charge Z:

(a) X-ray scattering.

(b) Observation of characteristic X rays.

5.35. Before the discovery of the neutron, the nucleus was pictured as consisting of A protons and $A - Z$ electrons. Discuss arguments against this hypothesis.

5.36. At which pion kinetic energy does the process $p\pi^- \rightarrow \Lambda^\circ K^\circ$ begin to occur? (i.e., determine the threshold for the reaction).

5.37. List two reactions that lead to the production of the Ξ^-; compute the corresponding threshold energies.

5.38. (a) Derive Eq. (5.66).

(b) Sketch the transmission T as a function of E/V_0 for a one-dimensional square well with the parameters $(2mV_0)^{1/2}a/\hbar = 100$.

5.39. Consider a one-dimensional potential well with a half-width $a = 1$ fm and a depth $V_0 = 100$ MeV. Find (numerically or graphically) the lowest two energy levels of a proton in this well.

5.40. Consider a well as shown in Fig. 5.38.

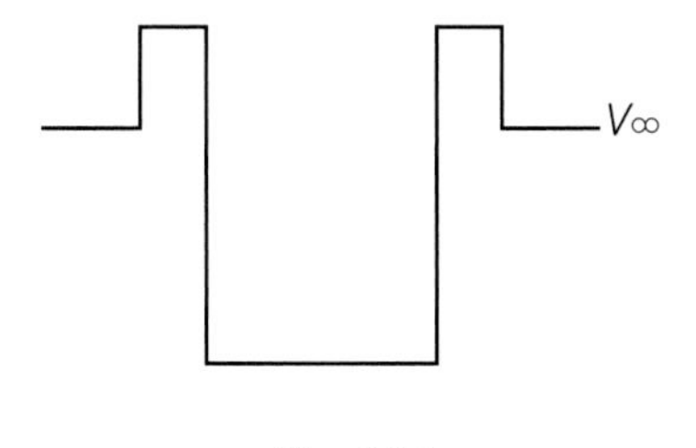

Fig. 5.38

(a) Indicate the energy region where bound states exist.

(b) How will particles behave in the region above V_∞?

5.41. * The experiment discussed in Section 5.12 demands the use of enriched ^{58}Fe.

(a) How is enriched iron prepared?

(b) What is the price of 1 mg of enriched ^{58}Fe?

5.42. In elastic and in inelastic scattering, some energy is given to the target particle in the form of recoil.

(a) Consider the reaction $^{58}\text{Fe}(p,p')^{58}\text{Fe}^*$. Assume that the incident protons have an energy of 7 MeV, that the scattered proton is observed at 130° in the laboratory, and that excitation to the first excited state of ^{58}Fe is studied. What is the energy of the scattered proton?

(b) Assume that you try to excite the first nucleon resonance, $N^*(1232)$, by inelastic proton-proton scattering and that the primary proton kinetic energy is 1 GeV. What is the maximum scattering angle at which the scattered proton can be observed? At which energy will the peak in the inelastically scattered protons occur at this angle?

5.43. * Discuss resonance fluorescence:

(a) What is the process?

(b) How can resonance fluorescence be observed in nuclei?

(c) What information can be obtained from it?

5.44. * Describe the discovery of the $W^\pm$.

5.45. (a) How are taus produced in e^+e^- collisions?

(b) If the e^+ and e^- beams have equal energy, what is the minimum beam energy required for τ production?

(c) Protons striking a stationary hydrogen target can produce τ's through the reaction $pp \to \tau^+\tau^-X$, where X is any set of hadron(s). What is the minimum proton energy for this reaction to occur?

5.46. (a) Can the Z^0 be produced in e^+e^- collisions? What is the minimum energy required?

(b) How can you determine that the Z^0 has been produced in the reaction (a)?

5.47. Based on the masses of the heavy gauge bosons ($W^\pm, Z^0$), what is the range of the weak force?

Chapter 6

Structure of Subatomic Particles

In Chapter 5 the members of the subatomic zoo have been classified according to interaction, symmetry, and mass. In the present chapter, we shall investigate some particles in more detail; in particular, we shall study the charged leptons, some hadrons and the ground-state structure of some nuclides. What do we mean by *ground-state structure*? For *atoms*, the answer is familiar: Structure denotes the spatial distribution of the electrons, and it is described by the ground-state wave function. For the hydrogen atom, neglecting spin, the probability density $\rho(\boldsymbol{x})$ at point $\boldsymbol{x}$ is given by

$$\rho(\boldsymbol{x}) = \psi^*(\boldsymbol{x})\psi(\boldsymbol{x}), \tag{6.1}$$

where $\psi(\boldsymbol{x})$ is the electron wave function at $\boldsymbol{x}$. The electric charge density is given by $e\rho(\boldsymbol{x})$; the charge and the electron probability density are proportional to each other. Actually, the structure includes the excited states, and only if the wave functions of all possible atomic states are known is the structure completely determined. We shall, however, restrict the discussion to the ground state.

For *nuclei*, the concept of a charge distribution still makes sense, but charge and matter distribution are not identical. For *nucleons*, a new problem arises. The momenta needed to investigate the structure are so high that the nucleons, which are initially at rest, recoil with velocities that are close to the velocity of light. It is then very difficult to compute the nucleon charge distribution from the observed cross section. To avoid this problem, the nucleon structure is described in terms of *form factors*. While it takes some time to get used to this concept, it is closer to the experimental information than the charge distribution. For *leptons*, no structure is found at all, even at the smallest distances studied, less than 10^{-18} m. They appear to be true pointlike Dirac particles.

6.1 The Approach: Elastic Scattering

Elastic scattering experiments have provided a great deal of insight into the structure of subatomic particles. How do such studies differ from the spectroscopic experiments discussed in Chapter 5? There is no sharp boundary, but the essential aspects can be described as follows. Both kinds of studies use an arrangement of the type shown in Fig. 5.32. In spectroscopy one angle is selected, and the spectrum of the scattered particles is explored at this angle. The energy levels of the nuclide under investigation can be taken from data similar to the ones given in Fig. 5.34. In structure (elastic form factor) experiments the detector looks only at the elastic peak. The intensity of the elastic peak is then determined as a function of the scattering angle. (Note that the energy at the elastic peak changes with scattering angle because of the recoil of the target particle; the detector must be adjusted correspondingly at each angle.) The observed intensity is translated into a differential cross section, a quantity that we shall define in Section 2.2. From the cross section, the information concerning the structure of the target particle can be obtained.

In 1911, Rutherford observed the elastic scattering of alpha particles from nuclei; he found a small deviation from the scattering law derived for point nuclei and therefrom got a good idea concerning the size of the nucleus.(1) Many of the later investigations were also done with hadrons, mainly alpha particles or protons. These experiments, however, have one serious drawback: Nuclear size effects are intertwined with nuclear force effects, and the two must be disentangled. *Leptonic* probes do not suffer from this handicap, and the most detailed information concerning the nuclear charge distribution has been obtained with electrons and muons.

6.2 Rutherford and Mott Scattering

The classical picture of elastic scattering of an alpha particle by the Coulomb field of a nucleus of charge Ze is shown in Fig. 6.1. This event is called *Rutherford scattering* if the nucleus is spinless; the alpha particle also has spin 0. The cross section for scattering of a spin-0 particle by a spinless nucleus can be computed classically or quantum mechanically, with the same result. The *Rutherford scattering formula* is one of the few equations that can be taken over into quantum mechanics without change, and this fact was a source of great pride to Rutherford.(2)

A fast way to derive the differential cross section for Rutherford scattering is based on the first Born approximation. In general, the differential cross section is written as

$$\frac{d\sigma}{d\Omega} = |f(\boldsymbol{q})|^2, \tag{6.2}$$

where $f(\boldsymbol{q})$ is called the scattering amplitude and $\boldsymbol{q}$ is the momentum transfer,

$$\boldsymbol{q} = \boldsymbol{p} - \boldsymbol{p}'. \tag{6.3}$$

$\boldsymbol{p}$ is the momentum of the incident and $\boldsymbol{p}'$ that of the scattered particle. For elastic scattering, Fig. 6.1(b) shows that the magnitude of the momentum transfer is connected to the scattering angle θ by

---

[1] E. Rutherford, *Phil. Mag.* **21**, 669 (1911).

[2] Rutherford scorned complicated theories and used to say that a theory is good only if it could be understood by a barmaid. (G. Gamow, *My World Line*, Viking, New York, 1970.)

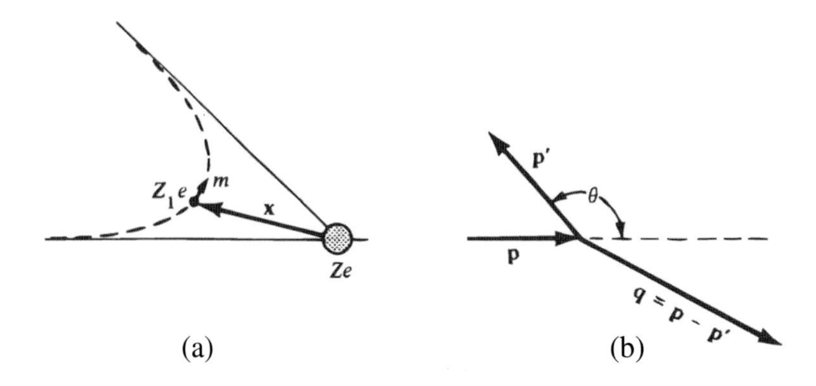

Figure 6.1: Rutherford scattering. (a) Classical trajectory of a particle with charge Z_1e in the field of a heavy nucleus with charge Ze. (b) Representation of the collision in momentum space.

$$q = 2p \sin \tfrac{1}{2}\theta. \tag{6.4}$$

In the first Born approximation it is assumed that the incident and the scattered particle can be described by plane waves. The scattering amplitude can then be written as[3]

$$f(\boldsymbol{q}) = -\frac{m}{2\pi\hbar^2}\int V(\boldsymbol{x}) \exp\left(\frac{i\boldsymbol{q}\cdot\boldsymbol{x}}{\hbar}\right) d^3x. \tag{6.5}$$

$V(\boldsymbol{x})$ is the scattering potential. If it is spherically symmetric, integration over angles can be performed, and the scattering amplitude becomes, with $x = |\boldsymbol{x}|$,

$$f(\boldsymbol{q}^2) = -\frac{2m}{\hbar q}\int_0^\infty dx x \sin\left(\frac{qx}{\hbar}\right) V(x). \tag{6.6}$$

Since f no longer depends on the direction of $\boldsymbol{q}$ but only on its magnitude, it is now written as $f(\boldsymbol{q}^2)$.

For Rutherford scattering, the potential $V(x)$ is the Coulomb potential.[4] Ordinarily, the Coulomb interaction between two charges q_1q_2 at a distance x is written as

$$V(x) = \frac{q_1q_2}{x}.$$

[3]We introduce Eq. (6.2) and the Born approximation here without derivation. This omission will be rectified later, in Section 6.11 and, with a different approach, in Problem 10.3. The student who has not yet encountered Eqs. (6.2) and (6.5) should simply use them as a tool here and then study their derivation later. Derivations are also given in Merzbacher, Section 13.4; and Park, Section 9.3.

[4]In the original Rutherford experiments, the probing particles were α particles. These are hadrons, and if they get close to the nucleus, the hadronic force must also be taken into account. The experiments discussed here are performed with electrons, and no problems from hadronic forces arise.

In the scattering experiment shown in Fig. 6.1, the nucleus is surrounded by its electron cloud, and the nuclear charge Ze is shielded. Shielding is taken into account by writing

$$V(x) = \frac{Z_1 Z e^2}{x} \exp\left(\frac{-x}{a}\right) \tag{6.7}$$

where a is a length characteristic of *atomic* dimension. Eq. (6.7) enables the integral in Eq. (6.6) to be done, and the scattering amplitude becomes

$$f(\boldsymbol{q}^2) = -\frac{2mZ_1 Z e^2}{q^2 + (\hbar/a)^2}. \tag{6.8}$$

In all collisions exploring the structure of nuclei, the momentum transfer q is at least of the order of a few MeV/c, and the term $(\hbar/a)^2$ can be neglected completely. With Eqs. (6.8) and (6.2) the Rutherford differential cross section becomes

$$\left(\frac{d\sigma}{d\Omega}\right)_R = \frac{4m^2(Z_1 Z e^2)^2}{q^4}. \tag{6.9}$$

The Rutherford scattering formula, Eq. (6.9), is based on a number of assumptions. The four most important ones are

1. The Born approximation.
2. The target particle is very heavy and does not take up energy (no recoil).
3. The incident and target particle have spin 0.
4. The incident and target particle have no structure; they are assumed to be point particles.

These four restrictions have to be justified or removed. We shall retain and justify the first two and partially remove the second two.

1. The Born approximation assumes that the incident and the outgoing particle can be described by plane waves. Such an assumption is allowed as long as

$$\frac{Z_1 Z e^2}{\hbar c} \ll 1. \tag{6.10}$$

If condition (6.10) is not satisfied, a more detailed calculation is necessary (phase-shift analysis or higher Born approximations).[5] The essential physical aspects can, however, be understood by using the first Born approximation, and we shall not go beyond it.

[5] D.R. Yennie, D.G. Ravenhall, and R.N. Wilson, *Phys. Rev.* **95**, 500 (1954).

2. Only elastic scattering is considered here. The target particle remains in its ground state, and it does not accept excitation energy. Moreover, it is assumed to be so heavy that its recoil energy can be neglected. However, as Fig. 6.1(b) shows, a very large momentum can be transferred to the target particle. At first the idea of a collision with large momentum transfer but with negligible energy transfer seems unrealistic. A simple experiment will convince an unbeliever that such a process is possible: take a car or motorcycle and race straight into a concrete wall. If well constructed, the wall will take up the entire momentum but will accept very little energy. Most of the later discussion will be concerned with the scattering of electrons from nuclei and nucleons. In this case, restriction 2 is satisfied as long as the ratio of incident electron energy to target rest energy is small. At higher energy, the cross section can be corrected for nucleon or nuclear recoil in a straightforward manner. Essential results remain unaffected, and we shall therefore not treat the recoil corrections.

3. As just pointed out, most experiments to be discussed concern the scattering of electrons. In this case, the spin has to be taken into account. Scattering of spin-$\frac{1}{2}$ particles with charge $|Z_1| = 1$ from spinless target particles has been treated by Mott, and the cross section for Mott scattering is[(6)]

$$\left(\frac{d\sigma}{d\Omega}\right)_{\text{Mott}} = 4(Ze^2)^2\frac{E^2}{(qc)^4}\left(1 - \beta^2\sin^2\frac{\theta}{2}\right). \tag{6.11}$$

E is the energy of the incident electron and $v = \beta c$ its velocity. The term $\beta^2 \sin^2 \theta/2$ comes from the interaction of the electron's magnetic moment with the magnetic field of the target. In the rest frame of the target, this field vanishes, but in the electron's rest frame, it is present. The term is peculiar to spin $\frac{1}{2}$, it disappears as $\beta \to 0$, and it is as important as the ordinary electric interaction as $\beta \to 1$ since the magnetic and electric forces are then of equal strength. In the limit $\beta \to 0 (E \to mc^2)$, the Mott cross section reduces to the Rutherford formula, Eq. (6.9).

4. The aim of the present chapter is the exploration of the structure of subatomic particles, and restriction 4 must consequently be removed. This task will be performed in the following section.

[6] A relatively easy-to-read derivation of Eq. (6.11) can be found in R. Hofstadter, *Annu. Rev. Nucl. Sci.* **7**, 231 (1958). A more sophisticated proof is given in J.D. Bjorken and S.D. Drell, *Relativistic Quantum Mechanics*, McGraw-Hill, New York, 1964, p. 106, or in J. J. Sakurai, *Advanced Quantum Mechanics*, Addison-Wesley, Reading, Mass., 1967, p. 193.

6.3 Form Factors

How is the cross section modified if the colliding particles possess extended structures? We shall treat leptons in Section 6.5 and find that they behave like point particles. This fact renders them ideal as probes, and the modification of Eq. (6.11) must take only the spatial distribution of the target particle into account. For simplicity, we shall assume here that the target particle possesses a spherically symmetric density distribution. It will then be shown below that the cross section for scattering of electrons from such a target is of the form

$$\frac{d\sigma}{d\Omega} = \left(\frac{d\sigma}{d\Omega}\right)_{\text{Mott}} |F(\boldsymbol{q}^2)|^2. \tag{6.12}$$

The multiplicative factor $F(\boldsymbol{q}^2)$ is called the *form factor*, and

$$\boldsymbol{q}^2 = (\boldsymbol{p} - \boldsymbol{p}')^2 \tag{6.13}$$

is the square of the momentum transfer.

Form factors play an important role in subatomic physics because they are the most convenient link between experimental observation and theoretical analysis. Equation (6.12) expresses the fact that the form factor is the direct result of a measurement. To discuss the theoretical side, consider a system that can be described by a wave function $\psi(\boldsymbol{r})$, which in turn can be found as the solution of a Schrödinger equation. For an object of charge Q, the charge density can be written as $Q\rho(\boldsymbol{r})$, where $\rho(\boldsymbol{r})$ is a normalized probability density, $\int d^3r\rho(\boldsymbol{r}) = 1$. It will be shown below that the form factor can be written as the Fourier transform of the probability density

$$F(\boldsymbol{q}^2) = \int d^3r\rho(\boldsymbol{r}) \exp(i\boldsymbol{q} \cdot \boldsymbol{r}/\hbar). \tag{6.14}$$

The form factor at zero momentum transfer, $F(0)$, is usually normalized to be 1 for a charged particle; however for a neutral one, $F(0) = 0$. The chain linking the experimentally observed cross section to the theoretical point of departure can thus be sketched as follows:

Experiment Comparison Theory

$$\frac{d\sigma}{d\Omega} \longrightarrow |F(q^2)| \Leftrightarrow F(q^2) \longleftarrow \rho(\boldsymbol{r}) \longleftarrow \psi(\boldsymbol{r}) \longleftarrow \text{Schrödinger equation}$$

In reality, individual steps can be more complicated than shown here, but the essential aspects of the chain remain.

We verify these introductory remarks by computing the scattering of a spinless electron from a finite spherically symmetric nucleus in the first Born approximation (Fig. 6.2).

The scattering potential $V(x)$ in Eq. (6.5) at the position of the electron consists of contributions from the entire nucleus. Each volume element d^3r contains a charge $Ze\rho(r)d^3r$ and gives a contribution (Eq. 6.7)

$$dV(x) = -\frac{Ze^2}{z}\exp\left(-\frac{z}{a}\right)\rho(r)\,d^3r,$$

so that

$$V(x) = -Ze^2\int d^3r\frac{\rho(r)}{z}\exp\left(-\frac{z}{a}\right) \tag{6.15}$$

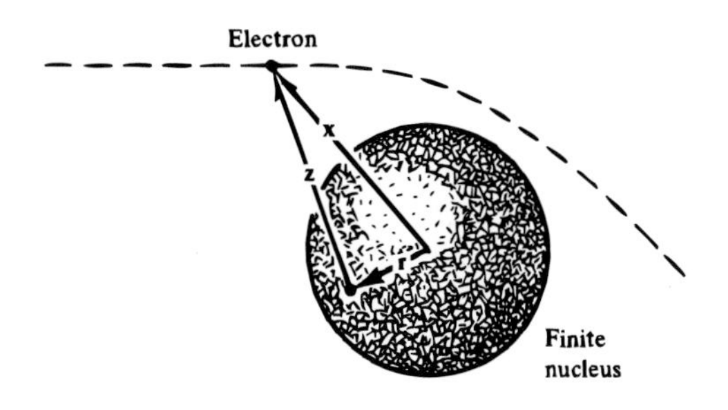

Figure 6.2: Scattering of a spinless electron by a spinless nucleus with extended charge distribution.

where $z = |\boldsymbol{z}|$ and the vector $\boldsymbol{z}$ is shown in Fig. 6.2. Introducing $V(x)$ into Eq. (6.5) and using $\boldsymbol{x} = \boldsymbol{r} + \boldsymbol{z}$ yields

$$f(\boldsymbol{q}^2) = \frac{mZe^2}{2\pi\hbar^2}\int d^3r\exp\left(\frac{i\boldsymbol{q}\cdot\boldsymbol{r}}{\hbar}\right)\rho(r)\int d^3x\frac{\exp(-z/a)}{z}\exp\left(\frac{i\boldsymbol{q}\cdot\boldsymbol{z}}{\hbar}\right).$$

For fixed $\boldsymbol{r}$, d^3x can be replaced by d^3z. The integral over d^3z is then the same as encountered in the evaluation of Eq. (6.8), and it gives

$$\int d^3z\frac{\exp(-z/a)}{z}\exp\left(\frac{i\boldsymbol{q}\cdot\boldsymbol{z}}{\hbar}\right) = \frac{4\pi\hbar^2}{\boldsymbol{q}^2 + (\hbar/a)^2} \longrightarrow \frac{4\pi\hbar^2}{\boldsymbol{q}^2}. \tag{6.16}$$

The integral over d^3r is the form factor, defined in Eq. (6.14), and the cross section $d\sigma/d\Omega = |f|^2$ becomes

$$\frac{d\sigma}{d\Omega} = \left(\frac{d\sigma}{d\Omega}\right)_R |F(\boldsymbol{q}^2)|^2. \tag{6.17}$$

The computation for electrons with spin follows the same lines; Eq. (6.12) is the correct generalization of Eq. (6.17). One remark is in order concerning the density $\rho(r)$. By Eq. (6.14), the density $\rho(r)$ has been defined in such a way that

$$\int \rho(r)d^3r = 1. \tag{6.18}$$

Equation (6.12) indicates how the form factor $|F(\boldsymbol{q}^2)|$ can be determined experimentally: The differential cross section is measured at a number of angles, the Mott cross section is computed, and the ratio gives $|F(\boldsymbol{q}^2)|$. The step from $F(\boldsymbol{q}^2)$ to $\rho(r)$ is less easy. In principle, Eq. (6.14) can be inverted and then reads

$$\rho(r) = \frac{1}{(2\pi)^3}\int d^3q\;F(\boldsymbol{q}^2)\exp\left(-\frac{i\boldsymbol{q}\cdot\boldsymbol{r}}{\hbar}\right). \tag{6.19}$$

Equations (6.14) and (6.19) are the three-dimensional generalization of Eqs. (5.40) and (5.41). The expression for $\rho(r)$ shows that the probability distribution is determined completely if $F(q^2)$ is known for all values of q^2. Experimentally, however, the maximum momentum transfer is limited by the available particle momentum. Moreover, as we shall see soon, the cross section becomes very small at large values of q^2, and it is then extremely difficult to determine $F(q^2)$. The practical approach is therefore different: Forms for $\rho(\boldsymbol{r})$ with a number of free parameters are assumed. The parameters are determined by computing $F(\boldsymbol{q}^2)$ with Eq. (6.14) and fitting the expression to the measured form factors.[7]

To provide some insight into the meaning of form factors and probability distributions, we shall connect $F(\boldsymbol{q}^2)$ to the nuclear radius and give examples of the relation between form factor and probability distribution. For $qR \ll \hbar$, where R is approximately the nuclear radius, the exponential in Eq. (6.14) can be expanded, and $F(q^2)$ becomes

$$F(\boldsymbol{q}^2) = 1 - \frac{1}{6\hbar^2}\boldsymbol{q}^2\langle r^2\rangle + \cdots \tag{6.20}$$

where $\langle r^2\rangle$ is defined by

$$\langle r^2\rangle = \int d^3r\, r^2\rho(r) \tag{6.21}$$

and is called the mean-square radius. For small values of the momentum transfer, only the zeroth and second moments of the charge distribution are measured, and further details cannot be obtained.

If the probability density is Gaussian,

$$\rho(r) = \rho_0 \exp\left[-\left(\frac{r}{b}\right)^2\right] \tag{6.22}$$

then the form can be computed easily, and it becomes

$$F(\boldsymbol{q}^2) = \exp\left(-\frac{q^2b^2}{4\hbar^2}\right), \qquad \langle r^2\rangle = \frac{3}{2}b^2. \tag{6.23}$$

If b becomes very small, the distribution approaches a point charge and the form factor tends toward unity. This limiting case is the point from which we started. A few probability densities and form factors are given in Table 6.1.

A final word concerns the dependence of the form factor on experimental quantities. Equation (6.14) shows that $F(\boldsymbol{q}^2)$ depends only on the square of the momentum transferred to the target particle and not on the energy of the incident

[7] One famous problem is apparent from the chain shown after Eq. (6.14). Experimentally, the absolute square of the form factor is obtained and not the form factor. The same problem appears in X-ray structure determinations. To get more information on the form factor, interference effects must be studied. In X-ray investigations of large molecules, interference is produced by substituting a heavy atom, for instance, gold, into the large molecule, and the resultant change of the X-ray pattern is observed. What can be used in subatomic physics?

Table 6.1: PROBABILITY DENSITIES AND FORM FACTORS FOR SOME ONE-PARAMETER CHARGE DISTRIBUTIONS. [After R. Herman and R. Hofstadter, *High-Energy Electron Scattering Tables*, Stanford University Press, Stanford, CA, 1960.]

Probability Density, $\rho(r)$	Form Factor, $F(\boldsymbol{q}^2)$
$\delta(r)$	1
$\rho_0 \exp(-r/a)$	$(1+\boldsymbol{q}^2a^2/\hbar^2)^{-2}$
$\rho_0 \exp[-(r/b)^2]$	$\exp(-\boldsymbol{q}^2b^2/4\hbar^2)$
$\left.\begin{matrix}\rho_0, r \le R\\ 0, r \ge R\end{matrix}\right\}$	$\frac{3[\sin(\lvert\boldsymbol{q}\rvert R/\hbar)-(\lvert\boldsymbol{q}\rvert R/\hbar)\cos(\lvert\boldsymbol{q}\rvert R/\hbar)]}{(\lvert\boldsymbol{q}\rvert R/\hbar)^3}$

particle. $F(\boldsymbol{q}^2)$, for a specific value of $\boldsymbol{q}^2$, can therefore be determined with projectiles of different energies. Equation (6.4) indicates that it is only necessary to change the scattering angle correspondingly, and the same value of $F(\boldsymbol{q}^2)$ should result. Incidentally, the fact that $F(\boldsymbol{q}^2)$ depends only on $\boldsymbol{q}^2$ is true only in the first Born approximation; it is not valid in higher order. It can therefore be used to test the validity of the first Born approximation.

6.4 The Charge Distribution of Spherical Nuclei

The investigation of nuclear structure by electron scattering has been pioneered by Hofstadter and his collaborators.[(8)] The basic arrangement is similar to the one shown in Fig. 5.32: An electron accelerator produces an intense beam of electrons with energies between 250 MeV and a few GeV. The electrons are transported to a scattering chamber where they strike the target. The intensity of the elastically scattered electrons is determined as a function of the scattering angle. Many improvements have occurred since the early experiments by Hofstadter. In addition to higher energies and higher intensity electron beams, which allow higher momentum transfers to be studied, much higher resolution ($\sim$ 100 keV or $\lesssim 10^{-3}$ of the beam energy) has been achieved. The high resolution allows one to separate elastic from inelastic scattering and to study inelastic scattering to individual levels in addition to elastic scattering. The differential cross section for the scattering of 500 MeV electrons from ^{40}Ca is shown in Fig. 6.3. The data can be seen to extend over 12 orders of magnitude; they yield values of $|F(\boldsymbol{q}^2)|$ and from these values information about the charge distribution is obtained.[(9)]

The crudest approximation to the nuclear charge distribution is a one-parameter

[8] R. Hofstadter, H.R. Fechter, and J.A. McIntyre, *Phys. Rev.* **92**, 978 (1953); for a review, see C.J. Batty, E. Friedman, H.J. Gils, and H. Rebel, *Adv. Nucl., Phys.*, ed. J.W. Negele and E. Vogt, Plenum Press, New York, **19**, 1 (1989).

[9] A nice review with data tables can be found at H. De Vries, C.W. De Jager and C. De Vries, *Atom. Data Nucl. Data Tabl.* **36**, 495 (1987).

function, for instance, a uniform or a Gaussian distribution. Such distributions give poor fits, and the simplest useful approximation is the two-parameter *Fermi distribution*

$$\rho(r) = \frac{N}{1 + \exp[(r - c)/a]}. \tag{6.24}$$

N is a normalization constant and c and a are the parameters describing the nucleus. The Fermi distribution is shown in Fig. 6.4; c is called the half-density radius and t the surface thickness. The parameter a in Eq. (6.24) and t are related by

$$t = (4 \ln 3)a. \tag{6.25}$$

The results of many experiments can be summarized in terms of the parameters defined in Eqs. (6.21) and (6.24):

1. For medium- and heavyweight nuclei the root-mean-square charge radius can be approximated by the relation

$$\langle r^2 \rangle^{1/2} = r_0 A^{1/3}, \qquad r_0 = 0.94 \text{ fm}, \tag{6.26}$$

   where A is the mass number (number of nucleons). The nuclear volume consequently is proportional to the number of nucleons. The nuclear density is approximately constant; nuclei behave more like solids or liquids than atoms.

2. The half-density radius and the skin thickness satisfy approximately

$$c(\text{in fm}) = 1.18\ A^{1/3} - 0.48, \qquad t \approx 2.4 \text{ fm}. \tag{6.27}$$

   From these values, the density of nucleons at the center follows as

$$\rho_n \approx 0.17 \text{ nucleon/fm}^3. \tag{6.28}$$

   This value approaches the density of nuclear matter, namely the density that an infinitely large nucleus, without surface effects, is presumed to have.

3. In the older literature, written at a time when the shape of nuclei was not yet well known, it was customary to describe the nuclear radius differently. A nucleus of uniform density and radius R was assumed. From Eq. (6.21) it follows that R^2 and $\langle r^2 \rangle$ are connected by

$$\langle r^2 \rangle = 4\pi \int_0^R \frac{3r^4 dr}{4\pi R^3} = \frac{3}{5} R^2. \tag{6.29}$$

   R approximately satisfies the relation

$$R = R_0 A^{1/3}, \qquad R_0 = 1.2 \text{ fm}. \tag{6.30}$$

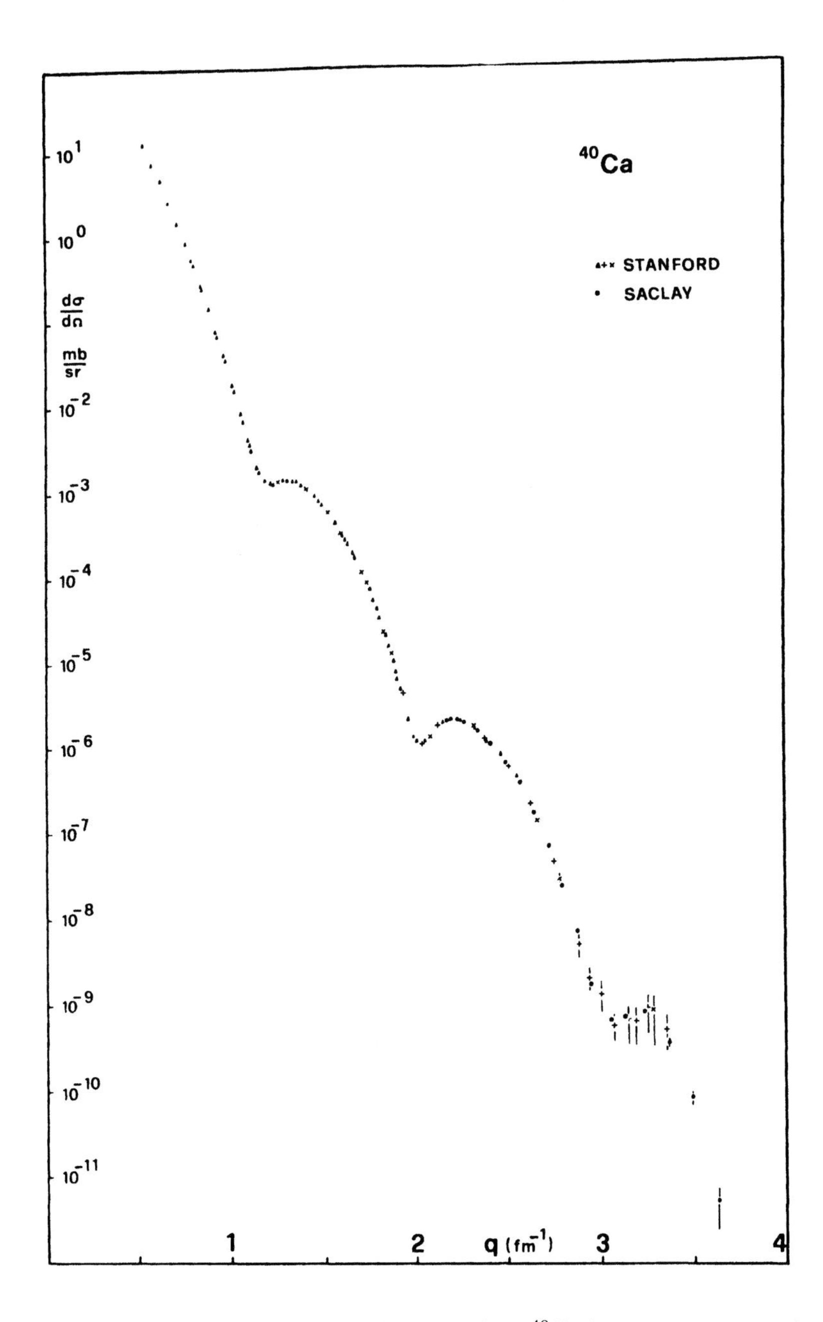

Figure 6.3: Elastic scattering cross section of electrons from ^{40}Ca from experiments performed at Stanford and Saclay, France. [Courtesy I. Sick, *Phys. Lett.* **88B**, 245 (1979).]

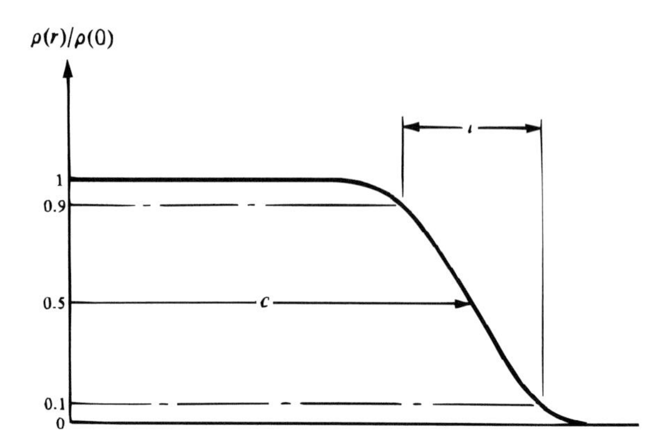

Figure 6.4: Fermi distribution for the nuclear charge density. c is the approximate half-density radius and t the surface thickness.

4. The actual charge distribution is more complex than the two-parameter Fermi distribution. In particular the density in the interior of nuclei is not constant as assumed in Eq. (6.24); it can decrease or increase toward the center, as shown in Fig. 6.5 for ^{40}Ca and ^{208}Pb.(9) These variations arise, primarily, from shell structure effects; see Chapter 17. It is possible to extract the charge distribution from the measured electron scattering cross section in an almost model-independent manner(10) by writing the charge distribution as a superposition of Gaussians,

$$\rho \propto \sum_{i=1}^{N} A_i \exp\left[-\frac{(r-R_i)^2}{\delta^2}\right].$$

The charge distributions shown in Fig. 6.5 were obtained in this manner.(9)

5. Nuclei that have nonzero spins also possess magnetic moments; the distribution of the magnetization can also be described by a form factor. Experimental information about the magnetization density is obtained from large angle (backward)(11,12) electron scattering.

The information given so far in this section provides a glimpse into the structure of nuclei. Considerably more is known—finer details have been investigated,(9,13) still higher momentum transfers have been studied with 4 and 10–20 GeV electrons, particularly in the lightest nuclei, ^{2}H,^{3}He,^{3}H.(12) In addition, inelastic scattering

[10] I. Sick, *Nucl. Phys.* **A218**, 509 (1974).

[11] S.K. Platchkov et al., *Phys. Rev.* **C25**, 2318 (1982); S. Auffret, *Phys. Rev. Lett.* **54**, 649 (1985); T.W. Donnelly and I. Sick, *Rev. Mod. Phys.* **56**, 461 (1984).

[12] R.G. Arnold et al., *Phys. Rev. Lett.* **35**, 776 (1975); B.T. Chertok, *Prog. Part. Nucl. Phys.*, (D.H. Wilkinson, ed.), **8**, 367 (1982); P.S. Justen, *Phys. Rev. Lett.* **55**, 2261 (1985); R.G. Arnold et al., *Phys. Rev. Lett.* **58**, 1723 (1987).

[13] J.M. Cavedon et al., *Phys. Rev. Lett.* **58**, 1723 (1987).

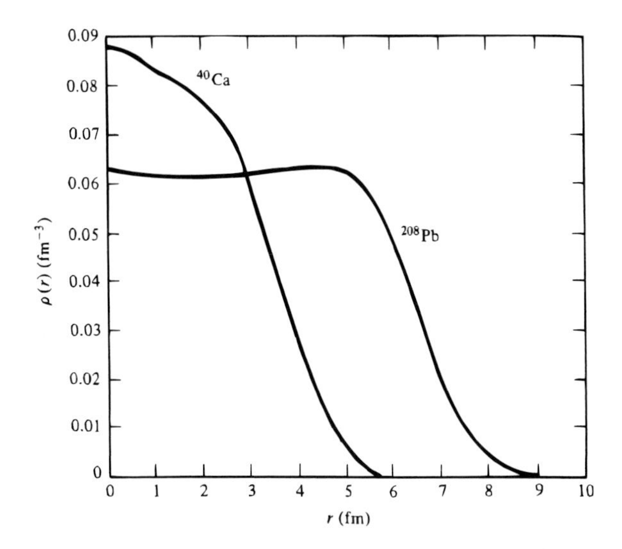

Figure 6.5: Probability distribution for ^{40}Ca and ^{208}Pb, as obtained from electron scattering. [From I. Sick. *Phys. Lett.* **88B**, 245 (1979).]

to many excited nuclear states have been examined.(14) We must, however, remember that the information provided by charged lepton scattering concerns the nuclear charge and current distributions and that corresponding data on hadronic structure (matter distribution) require a different probe, such as hadrons(15) or the weak interaction of electrons.(16)

6.5 Leptons Are Point Particles

We return now to the g factor of the electron. By 1926, the idea of the spinning electron and its magnetic moment was generally accepted,(17) but the value of the g factor (Eq. (5.16)),

$$g(1926) = -2,$$

had to be taken from experiment. (The minus sign indicates that the magnetic moment points in the direction opposite to the spin for a negative electron.) It was exactly twice as large as the g factor for orbital motion, Eq. (5.14). In other words, even though the electron has spin $\frac{1}{2}$, it carries *one* Bohr magneton. In 1928, Dirac

[14] J. Heisenberg and H. P. Blok, *Annu. Rev. Nucl. Part. Sci.* **33**, 569 (1983).

[15] A.W. Thomas, *Nucl. Phys.* **A354**, 51c (1981); R. Campi, *Nucl. Phys.* **A374**, 435c (1982).

[16] C.J. Horowitz, S.J. Pollock, P.A. Souder, and R. Michael, *Phys. Rev. C* **63**, 025501 (2001).

[17] A fascinating description of the history of the spin is presented by B.L. Van der Waerden, in *Theoretical Physics of the Twentieth Century* (M. Fierz and V.F. Weisskopf, eds.), Wiley-Interscience, New York, 1960. See also S.A. Goudsmit, *Phys. Today* **14**, 18 (June 1961) and P. Kusch, *Phys. Today* **19**, 23 (February 1966).

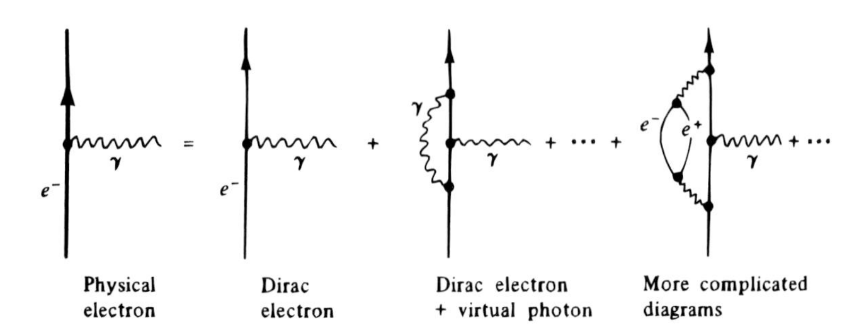

Figure 6.6: A physical electron is not just a pure Dirac electron. The presence of virtual photons affects the properties of the electron; in particular it changes the g factor by an amount that can be calculated and measured.

introduced his famous equation; the existence of a magnetic moment and the value $g = -2$ turned out to be natural consequences.[18]

In 1947, Kusch and Foley measured the g factor carefully by using the then-new microwave technique and discovered that it showed a small deviation from -2.[19] Within a very short time, Schwinger could explain the deviation. The experiment was accurate to about 5 parts in 10^5, and the theory was somewhat better. Since then, theoretical and experimental physicists have been in a race to improve the numbers. The winner has consistently been physics, because everybody has learned more. Since the comparison between theory and experiment is very important, a few words on both are in order here.

The theoretical explanation invokes virtual photons, a concept already discussed in Section 5.8. A physical electron does not always exist as a Dirac electron. Part of the time it emits a virtual photon which it then reabsorbs. (Classically, this process corresponds to the electron's interaction with its own electromagnetic field.) The measurement of the g factor involves the interaction of the electron with photons; the presence of virtual photons changes the interaction and consequently also the g factor. Figure 6.6 shows how the simple interaction of a photon with a Dirac electron is altered and complicated by the electron's own electromagnetic field. The net effect is to add an *anomalous* magnetic moment. An enormous amount of labor has been put into calculating the magnetic moment of a Dirac particle taking into account corrections of the type shown in Fig. 6.6. The result is expressed in terms of the number

$$a = \frac{|g| - 2}{2}. \tag{6.31}$$

[18]For a derivation of the magnetic moment of the electron in Dirac theory, see, for instance, Merzbacher, Section (24.7), or Messiah, Section XX, 29. Actually, the magnetic moment can already be derived as a nonrelativistic phenomenon, as, for instance, in A. Galindo and C. Sanchez del Rio, *Am. J. Phys.* **29**, 582 (1961), or R.P. Feynman, *Quantum Electrodynamics*, Benjamin, Reading, Mass., 1961, p. 37.

[19]P. Kusch and H.M. Foley, *Phys. Rev.* **72**, 1256 (1947); **74**, 250 (1948).

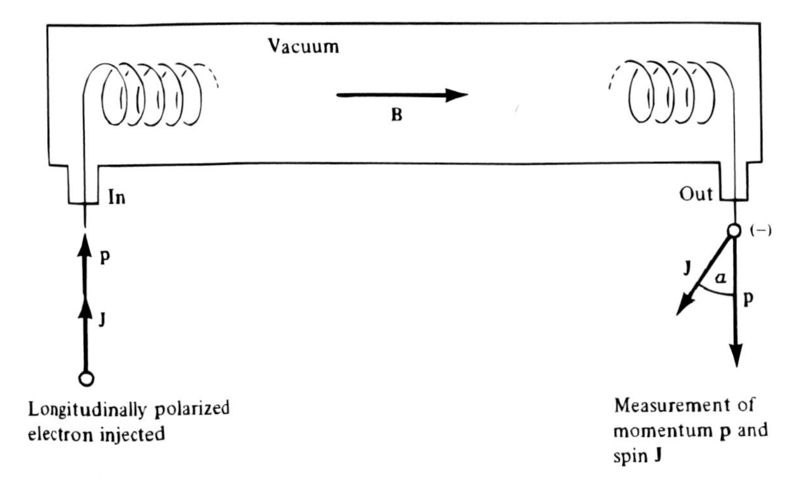

Figure 6.7: Basic approach underlying the direct determination of $a = (|g| - 2)/2$. For details see the text.

A pure *Dirac particle*, that is, a particle with properties as predicted by the Dirac equation alone, would have a value $a = 0$. The value of a for a *physical electron* has been computed by many people, and the present best theoretical value is[20]

$$a_e^{\text{th}} = \frac{1}{2}\left(\frac{\alpha}{\pi}\right) - 0.328478965\left(\frac{\alpha}{\pi}\right)^2 + 1.181241456\left(\frac{\alpha}{\pi}\right)^3 - 1.7366160\left(\frac{\alpha}{\pi}\right)^4 + \cdots, \tag{6.32}$$

where α is the fine structure constant, $\alpha = e^2/\hbar c$.

The early experimental results for a_e were based on an approach that can be explained with Fig. 5.5: if an electron is placed in an external magnetic field, Zeeman splitting results. A precise determination of the energy difference between levels and of the externally applied field yields g. Indeed, the discovery of a nonvanishing parameter a_e occurred with such a technique. Present experiments determine $|g| - 2$, and not g.[21] Two different approaches exist and because they are of such importance to subatomic physics, we will sketch both.

The first approach, pioneered by Crane,[22] is based on the following idea. In a uniform magnetic field, the spin and the momentum of a particle with spin $\frac{1}{2}$ and $|g| = 2$ retain a constant angle between them. Now consider an experimental arrangement as in Fig. 6.7. Longitudinally polarized electrons, i.e., electrons with

---

[20] T. Kinoshita *An Isolated Atomic Particle at Rest in Free Space in A Tribute to Hans Dehmelt, Nobel Laureate; E. Henley, N. Fortson, W. Nagourney, eds., Alpha Science Limited International, Pangbourne, UK, (2005)*; V.W. Hughes and T. Kinoshita, Rev. Mod. Phys. **71**, S133 (1999).

[21] A more detailed description of the ideas underlying the $|g| - 2$ experiments is given in R. D. Sard, *Relativistic Mechanics*, Benjamin, Reading, Mass., 1971.

[22] H. R. Crane, *Sci. Amer.* **218**, 72 (January 1968); A. Rich and J. C. Wesley, *Rev. Mod. Phys.* **44**, 250 (1972).

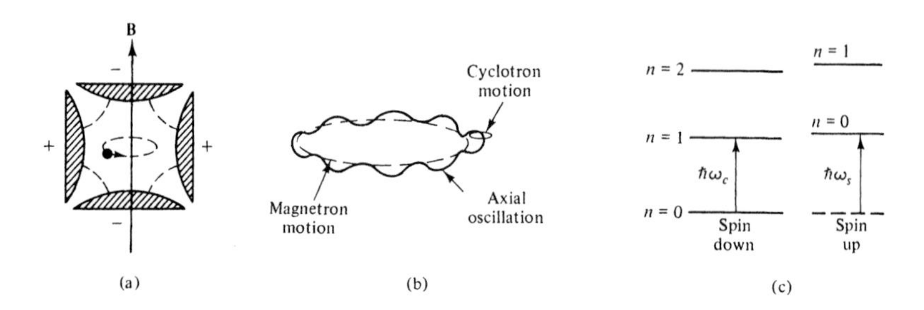

Figure 6.8: (a) Penning trap—a combination of a magnetic field $\boldsymbol{B}$ and a cylindrical electric quadrupole field. (b) Motion of an electron in the combined fields of the Penning trap. (c) Magnetic energy levels of the electron in the trap.

spin and momentum pointing in the same or opposite direction, are injected into a solenoidal magnetic field. In this field, the electrons move in circular orbits, and their spins and momenta are observed after a large number of revolutions. If the g factor were exactly 2, spin and magnetic moment of the outcoming electrons would still be parallel, regardless of the time spent in the field $\boldsymbol{B}$. The small anomalous part a, however, causes a slightly different rotation for spin and magnetic moment. After a time t in the field $\boldsymbol{B}$, the angle α between $\boldsymbol{p}$ and $\boldsymbol{J}$ becomes

$$\alpha = a\omega_c t, \tag{6.33}$$

where

$$\omega_c = \frac{eB}{mc} \tag{6.34}$$

is the cyclotron frequency. If the product Bt is very large, α also becomes very large and a can be measured very accurately. This method has been applied to electrons and muons of both signs.

The linear field arrangement shown in Fig. 6.7 works well for electrons because they are stable and reach the end of the coil after many turns even if they have a small velocity. Muons, however, decay and it is desirable to use muons with large velocity in order to gain flight time and distance [Eq. (1.9)]. The number of turns of high-energy muons in a linear field is too small to achieve the desired accuracy. The problem was overcome at CERN by replacing the linear by a circular field.

Pions of 3.1 GeV/c momentum were injected into a storage ring of 14 m diameter; their decay in flight into muons produced polarized muons in the storage ring. With such an arrangement, $|g| - 2$ could be determined with great accuracy for muons of both signs.[23] A more recent experiment at the Brookhaven National Laboraory AGS uses the same energy pions, but the muons from their decays are injected

[23] F.J.M. Farley and E. Picasso, *Annu. Rev. Nucl. Part. Sci.* **29**, 243 (1979).

directly into the ring. This method provides a gain in the number of stored muons by a factor of about 10. The magnetic field in the storage ring is optimized for uniformity. The results are sumarized in Table 6.2.

The second approach to the measurement of $|g| - 2$, pioneered by Dehmelt and his collaborators,[24] is based on a sophisticated form of a Zeeman experiment and constitutes a triumph of experimental ingenuity. A single electron is confined for weeks in a "trap" formed by a combination of a magnetic and an electric quadrupole field (Penning trap). Electron and apparatus constitute an atom with macroscopic dimensions that is called geonium, the earth atom. In the trap, sketched in Fig. 6.8(a), the electron performs a motion that consists of three components illustrated in Fig. 6.8(b): a cyclotron motion in the uniform magnetic field, an axial motion in the electric field, and a magnetron motion in the combined fields. Consider first an electron with spin down. The motion of this electron in the magnetic field is quantized. The orbits shown in Fig. 6.8(a) and (b) can have only energies allowed by quantization; the higher the energy the larger the radius. The energy difference between any two Zeeman levels [Fig. 6.8(c)] is given by the cyclotron frequency ω_c, Eq. (6.34), as

$$\hbar\omega_c = 2\ \mu_B B. \tag{6.35}$$

The energy can, however, also be changed by flipping the spin. If the spin is reversed from down to up, the corresponding energy change, indicated in Fig. 6.8(c) is

$$\hbar\omega_s = g\ \mu_B B. \tag{6.36}$$

By applying the proper rf field, transitions can be induced in which only the orbit is changed, or in which spin and orbit both change. The resonance frequency is given by ω_c in the first case and by

$$\omega_a = \omega_s - \omega_c = \frac{(|g| - 2)\mu_B B}{\hbar}$$

in the second case. The ratio of the two frequencies yields

$$\frac{\omega_a}{\omega_c} = \frac{(|g| - 2)}{2}. \tag{6.37}$$

By measuring these frequencies accurately, the values of $|g| - 2$ for the electron and the positron were measured with extreme accuracy.[25,26] In Table 6.2, we list

[24] R.S. VanDyck, Jr., P.B. Schwinberg and H.G. Dehmelt in *New Frontiers in High Energy Physics*, (B. Kursunoglu, A. Perlmutter, and L. Scott, eds) Plenum, New York, 1978, p. 159; P. Ekstrom and D. Wineland, *Sci. Amer.* **243**, 105 (August 1980); H. Dehmelt, in *Atomic Physics*, Vol 7. (D. Kleppner and F. Pipkin, eds) Plenum, New York, 1981.

[25] R.S. VanDyck, Jr., P.B. Schwinberg, and H.G. Dehmelt, *Phys. Rev. Lett.* **38**, 310 (1977); P. B. Schwinberg, R. S. VanDyck, Jr., and H. G. Dehmelt, *Phys. Rev. Lett.* **47**, 1679 (1981); R.S. VanDyck Jr., P.B. Schwinberg and H.G. Dehmelt, *Phys. Rev. Lett.* **59**, 26 (1987); B. Odom et al., *Phys. Rev. Lett.* **97**, 030801 (2006).

[26] G. Gabrielse, D. Hanneke, T. Kinoshita, M. Nio, and B. Odom, *Phys. Rev. Lett.* **97**, 030802 (2006).

Table 6.2: COMPARISON OF THEORETICAL AND EXPERIMENTAL VALUES OF $a = (|g| - 2)/2$.

Particle	Exp./Th.	a
e^-	Exp.	$1\ 159\ 652\ 180.85(76) \times 10^{-12\dagger}$
e^+	Exp.	$1\ 159\ 652\ 187.9(43) \times 10^{-12\dagger}$
$e^\pm$	Th.	$1\ 159\ 652\ 180.85 \times 10^{-12*}$
μ^-	Exp.	$1\ 165\ 921\ 4(9) \times 10^{-10}$ ††
μ^+	Exp.	$1\ 165\ 920\ 3(8) \times 10^{-10}$ ††
$\mu^\pm$	Th.	$1\ 165\ 918\ 8(8) \times 10^{-10}$ ††

†See Ref. (25) and references therein.
* The uncertainty in the theory is about 1/3 of the experimental uncertainty. The number quoted here is identical to the measurement because presently these values are used to extract the value of the fine structure constant. See Ref. (26).
††G.W. Bennet et al., *Phys. Rev. Lett.* **92**, 161802 (2004), and references therein. The uncertainties in the calculation are dominated by uncertainties on the contribution from virtual loops that can be better estimated by using data from e^+e^- collisions and from τ decays. We use an uncertainty that encompasses both.

values of $a = (|g| - 2)/2$.

For the case of the electron the experimental and theoretical uncertainties are small enough that one can use the magnetic moment measurements to get the fine structure constant with better precision than any other experiment.(26) The values from different experiments agree to within experimental uncertainties. Quantum electrodynamics (QED), the quantum theory of the interactions of charged leptons and photons, is a superbly successful theory.

The theoretical calculations for the electron are performed under the assumption that the leptons are point particles with only electromagnetic interactions. For the more massive leptons, the muon and the tau, strong and weak interactions also become important at the level of accuracy obtained experimentally. In addition to the diagrams of the kind shown in Fig. 6.6, strong and weak vacuum polarization terms, illustrated in Fig. 6.9 must be taken into account. For the muon, strong corrections are of the order of 7×10^{-9}, weak ones are estimated as 1×10^{-9}. These corrections are much less important for the electron because they scale as the square of the mass of the lepton.(27)

The agreement between experiment and theory expressed in Table 6.2 not only confirms the strong interaction correction for the muon, but can also be used to set an upper limit on the size of the leptons. Both the muon and electron must be smaller than 10^{-18} m.

Experiments performed with high energy charged leptons also demonstrate that

[27] K. Hagiwara et al, Phys. Lett. B **557**, 69 (2003).

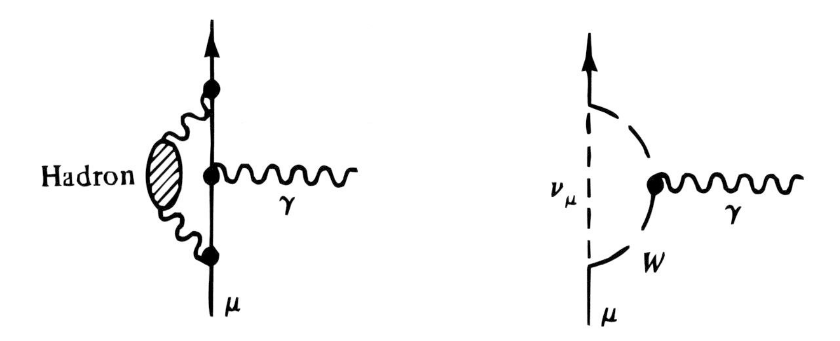

Figure 6.9: Strong and weak correction terms that appear in the interaction of a charged lepton with photons.

quantum electrodynamics predicts all observed phenomena correctly if proper theoretical corrections, such as the strong vacuum polarization shown in Fig. 6.9, are carried out. Measurements in colliding beam experiments, in particular,

$$e^-e^+ \longrightarrow e^-e^+, \quad e^-e^+ \longrightarrow \mu^-\mu^+, \quad \text{and} \quad e^-e^+ \longrightarrow \tau^-\tau^+$$

show that QED holds to distances smaller than about 10^{-18} m.(28) We consequently cannot yet answer the question raised by the incredible success of QED: Will the theory break down, and if so, at what scale?

6.6 Nucleon Elastic Form Factors

By 1932 it was well known that electrons have spin $\frac{1}{2}$ and a magnetic moment of $1\mu_B$, (Bohr magneton), as predicted by the Dirac equation. Two other spin-$\frac{1}{2}$ particles were also known to exist, the proton and the neutron. It was firmly believed that these would also have magnetic moments as predicted by the Dirac equation, one nuclear magneton for the proton and zero moment for the neutron. Enter Otto Stern. Stern had principles in selecting his experiments: "Try only crucial experiments. Crucial experiments are those that test universally accepted principles." When he started setting up equipment to measure the magnetic moment of the proton, his friends teased him and told him that he should not waste his time on an experiment whose outcome was foreordained. The surprise was great when Stern and his collaborators found a magnetic moment of about 2.5 μ_N for the proton and about -2 μ_N for the neutron.(29)

How can the departure of the magnetic moments of the proton and the neutron from the "Dirac values" be understood? Before quarks were introduced, the explanation of the anomalous magnetic moments of the nucleons was based on virtual

[28] K.G. Gan and M.L. Perl, *Int. J. Mod. Phys.* **A3**, 531 (1988).

[29] I. Estermann, R. Frisch, and O. Stern, *Nature* **132**, 169 (1933); R. Frisch and O. Stern, *Z. Physik* **85**, 4 (1933).

mesons that are present in their structures. The virtual mesons surround ("clothe") the Dirac ("bare") nucleon. It is now clear that nucleons are composed primarily of three quarks, the proton has the composition (uud), the neutron (udd), where u stands for an up quark and d for a down one. Nucleons contain not just one point particle and a meson cloud; three point particles reside there. The interaction among the quarks is transmitted by gluons; the force is weak at short distances ($\lesssim$0.1 fm) and strong at large ones ($\gtrsim$0.5 fm). The corresponding theory is called "QCD," quantum chromodynamics. As the interaction is a strong one, it is difficult to calculate detailed structure effects from first principles. The mesons are an effective means of describing "large" distance hadronic structure. Pions are the lightest mesons, thus they account for the outermost part of the structure and are therefore the most important ones to consider in addition to the quarks. However, the quark composition given above is sufficient to give the correct ratio of the magnetic moments of the neutron to proton;[(30)] this result was considered one of the early successes of the use of quarks. In addition, a number of "bag" models have been constructed; some of the more successful ones include a pion cloud in addition to quarks to explain the structure of the nucleon.[(31)] In such a picture, illustrated in Fig. 6.10, a photon interacts not only with the core (bare proton or quarks), but also with the surrounding meson cloud. Since the pions do not leave the nucleon and have to return, they can only go to about half the pion Compton wavelength [Eq. (5.52)]. The radius of the nucleons consequently is expected to be about $\hbar/2\ m_\pi c$ or about 0.7 fm. In this model, which can account for the static properties of both the proton and neutron, the quarks and the pion cloud contribute to the magnetic moment. The anomalous magnetic moments of the nucleons are due to hadronic effects, thus they cannot be computed to anywhere near the accuracy of the anomalous g factors for the leptons.

The best way to explore the charge and current distributions of nucleons is again electron scattering. Experimentally, the problem is straightforward for protons. A liquid hydrogen target is placed in an electron beam, and the differential cross section of the elastically scattered electrons is determined. For neutrons, the situation is not so easy. No neutron targets exist, and it is necessary to use deuteron targets and subtract the effect of the proton. The subtraction procedure introduces uncertainties. The e^-n elastic scattering cross section is consequently less well known than the e^-p cross section.[(32)]

For spinless target particles, the form factor can be extracted from the cross section by using Eq. (6.12). Nucleons have spin $\frac{1}{2}$, and Eq. (6.12) must be generalized. Without calculation, we can guess some features of the result. $F(q^2)$ in Eq. (6.12) describes the distribution of the electric charge, and it can be called an *electric* form

---

[30] F. E. Close, *An Introduction to Quarks and Partons*, Academic Press, New York, 1979; Chs. 4 and 7.

[31] A.W. Thomas and G.A. Miller, *Phys. Rev* **D24**, 216 (1981). See also the review by D.O. Riska, *Adv. Nucl. Phys.* **22**, 1 (1996), ed. J.W. Negele and E. Vogt, Plenum Press, New York.

[32] G. Warren et al, Phys. Rev. Lett. **92**, 042301 (2004).

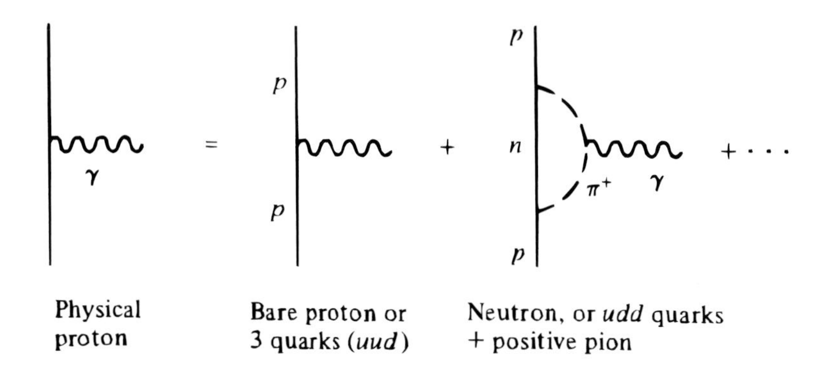

Figure 6.10: A physical proton is pictured as a superposition of many states, for instance a bare proton or three quarks, a bare neutron plus a pion, and so forth.

factor. The proton also possesses, in addition to its charge, a magnetic moment. It is unlikely that it behaves like a point moment and sits at the center of the proton. It is to be expected that the magnetization is also distributed over the volume of the nucleon and this distribution will be described by a magnetic form factor.[33] The detailed computation indeed proves that elastic electromagnetic scattering from a spin-$\frac{1}{2}$ particle with structure must be described by two form factors; the laboratory cross section can be written as

$$\frac{d\sigma}{d\Omega} = \left(\frac{d\sigma}{d\Omega}\right)_{\text{Mott}} \left[\frac{G_E^2 + bG_M^2}{1+b} + 2bG_M^2 \tan^2\left(\frac{\theta}{2}\right)\right], \tag{6.38}$$

where

$$b = \frac{-q^2}{4m^2c^2}. \tag{6.39}$$

Equation (6.38) is called the Rosenbluth formula;[34] m is the mass of the nucleon, θ the scattering angle, and q the four-momentum transferred to the nucleon.[35] The Mott cross section is given by Eq. (6.11). G_E and G_M are the electric and magnetic

[33]Nuclei with spin $J \geq 1/2$ also possess magnetic moments, and the magnetization is also distributed over the volume of the nucleus. For such nuclei, the discussion given in Section 6.4 must be generalized.

[34]M.N. Rosenbluth, *Phys. Rev.* **79**, 615 (1950).

[35]Here a word of explanation is in order: The variable q is the four-momentum transfer. It is defined as

$$q = \left\{\frac{E}{c} - \frac{E'}{c}, \boldsymbol{p} - \boldsymbol{p}'\right\}.$$

Its square,

$$q^2 = \frac{1}{c^2}(E - E')^2 - (\boldsymbol{p} - \boldsymbol{p}')^2 = \frac{1}{c^2}(E - E')^2 - \boldsymbol{q}^2,$$

is a Lorentz-invariant quantity. Since q^2 is a Lorentz scalar, its use is preferred in high-energy physics. For elastic scattering in the c.m. or at low energies, $q^2 = -\boldsymbol{q}^2$.

form factors, respectively, and they are both functions of q^2. The designations *electric* and *magnetic* stem from the fact that for $q^2 = 0$, the static limit, they are given by

$$G_E(q^2 = 0) = \frac{Q}{e}, \qquad G_M(q^2 = 0) = \frac{\mu}{\mu_N}, \tag{6.40}$$

where Q and μ are the charge and magnetic moment, respectively, of the nucleon. Specifically, $G_E(0)$ and $G_M(0)$ for the proton and the neutron are

$$\begin{aligned} G_E^p(0) &= 1, \qquad & G_E^n(0) &= 0, \\ G_M^p(0) &= 2.79, \qquad & G_M^n(0) &= -1.91. \end{aligned} \tag{6.41}$$

Early electron–proton scattering experiments,[36] performed with an electron energy of 188 MeV, were analyzed by fitting the observed differential cross section with an expression of the form of Eq. (6.38) with fixed values of the parameters G. An example is shown in Fig. 6.11. Comparison of the various theoretical curves with the experimental one indicates that the proton is not a point particle. The conclusion based on the discussion of the anomalous magnetic moment is consequently verified by a direct measurement. However, an electron energy of about 200 MeV is too small to permit studies at significant values of the momentum transfer and to get information on the q^2 dependence of G_E and G_M. Since 1956, many experiments have been performed at accelerators with much higher electron energies. To extract the form factors from the measured elastic scattering cross sections, the cross section for a fixed value of q^2 is normalized by division by the Mott cross section and plotted against $\tan^2 \theta/2$, as shown in Fig. 6.12. Such a plot should yield a straight line; from the slope, the value of G_M^2 is obtained. The intersection with the y axis then yields G_E^2.

Figure 6.13 gives the magnetic form factor of the proton. For convenience, $G_M/(\mu/\mu_N)$ is plotted, where μ is the proton magnetic moment. For comparison we show also a plot of the function:

$$G_D(q^2) = \frac{1}{(1 + |q|^2/q_0^2)^2}, \tag{6.42}$$

with $q_0^2 = 0.71(\text{GeV}/c)^2$. This function in conjunction with Table 6.1 can help the reader picture the distribution of magnetism in the proton. Although it is clear that at values of $|q|^2 > 10$ $(\text{GeV/c})^2$ the dipole function does not reproduce the data very well, it has become customary to compare the form factors to G_D[37]. Initially both the electric and magnetic form factors were determined by the procedure sketched in Fig. 6.12. This method has the disadvantage that, as $|q|^2$ gets larger it becomes more difficult to extract G_E as is apparent from Eq. 6.38. Recently there has

[36] R.W. McAllister and R. Hofstadter, *Phys. Rev.* **102**, 851 (1956).

[37] G_D goes down as $|q|^{-4}$ as $|q|^2 \to \infty$, a behavior that is predicted by QCD.

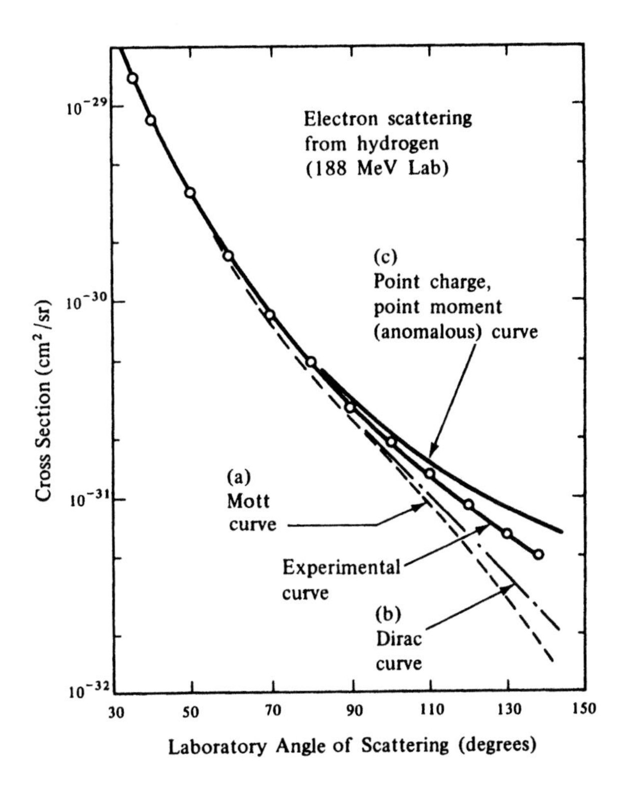

Figure 6.11: Electron–proton scattering with 188 MeV electrons. [R. W. McAllister and R. Hofstadter, *Phys. Rev.* **102**, 851 (1956).] The theoretical curves correspond to the following values of G_E and G_M: Mott (1;0), Dirac (1;1), anomalous (1;2.79).

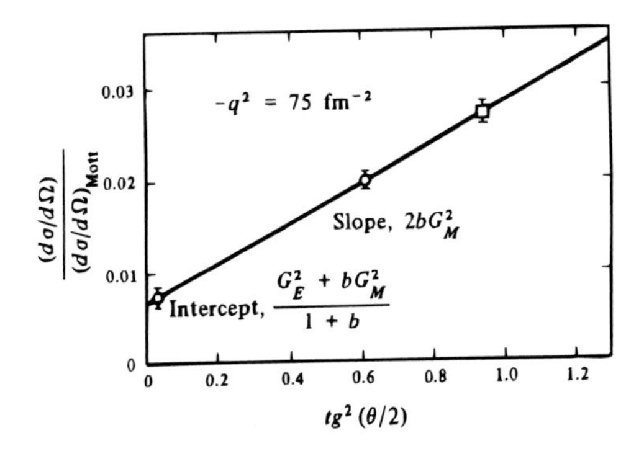

Figure 6.12: Rosenbluth plot. See the text for description.

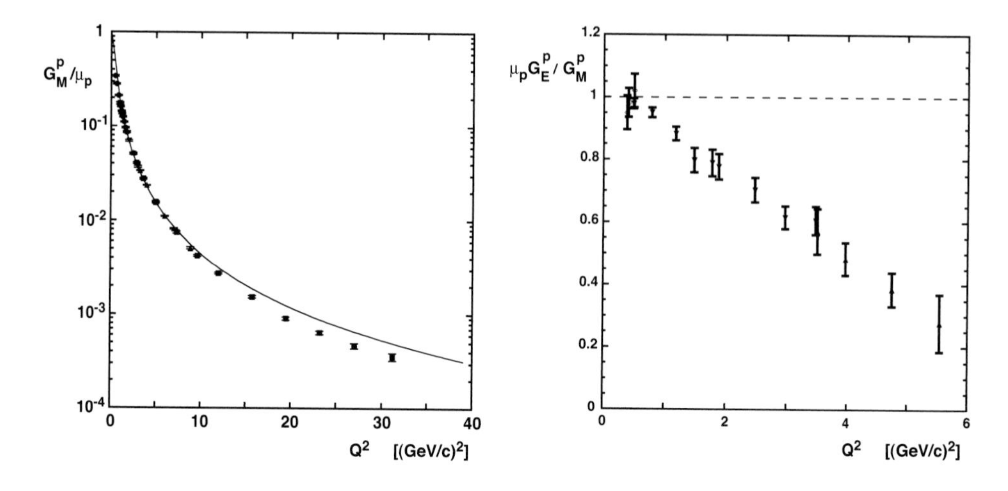

Figure 6.13: Left: Magnetic form factor for the proton plotted against the squared momentum transfer $|q|^2$. The different symbols correspond to different experiments. The 'dipole' function –described in the text and shown as a continuous line– describes the G_M data quite accurately below $|q|^2 \approx 10$ (GeV/c)^2. Right: G_E/G_M. The distributions of charge and magnetism in the proton are quite different. [See C.Hyde-Wright and K. de Jager, *Annu. Rev. Nucl. Part. Sci.* **54**, 217 (2004).]

been significant progress using polarized electron scattering on polarized targets to extract directly the ratio of the electric and magnetic form factors. The conclusions are summarized in Fig. 6.13.

Some features of the nucleon structure emerge from these relations:

1. Nucleons are not point particles. For point particles, the form factors are constant.

2. The proton charge distribution, although not acurately described by the dipole formula, shows that nucleons are extended systems but do not have well-defined surfaces.

3. The charge distribution is small within the neutron:

$$G_E^n \approx 0. \tag{6.43}$$

4. The proton and neutron magnetic form factors are roughly described by the dipole formula, Eq. (6.42), so the radial distribution follows from Table 6.1 as

$$\rho(r) = \rho(0)\exp\left(-\frac{r}{a}\right) \qquad a = \frac{\hbar}{q_0} = 0.23 \text{ fm}. \tag{6.44}$$

   One remark must be added: The Fourier transform used here is valid only for small values of $|q|^2$. For large values of $|q|^2$, the proton that was initially at rest recoils with a velocity approaching that of light, and G_E (G_M) no longer

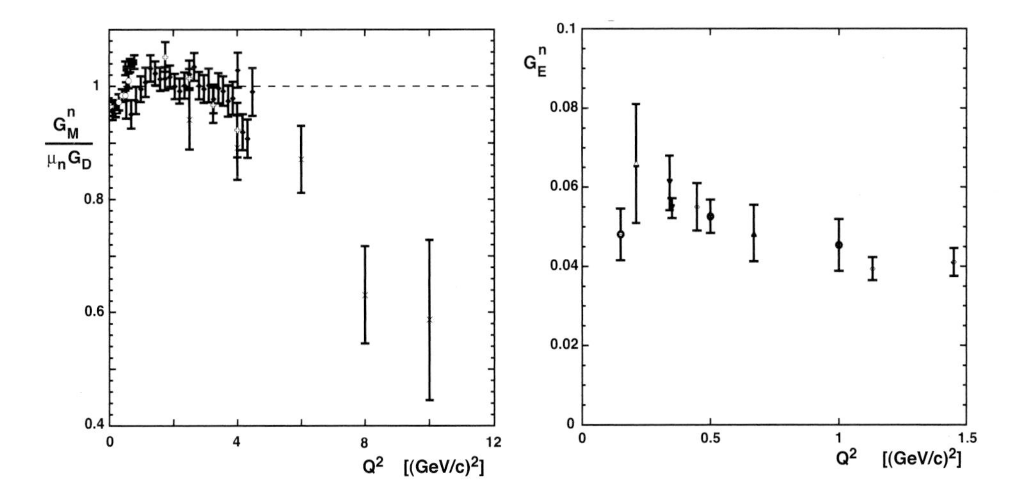

Figure 6.14: Magnetic (left) and electric (right) form factors for the neutron. Here we show the magnetic form factor divided by the dipole formula. The magnetic form factor shows rough agreement with the dipole formula for $|q|^2 < 5$ (GeV/c)^2. [See C.Hyde-Wright and K. de Jager, *Annu. Rev. Nucl. Part. Sci.* **54**, 217 (2004).]

represents the charge (magnetic) distribution. The contributions of charge and magnetism are mixed in both G_E and G_M.

5. The distribution of electric and magnetic charges within the proton are significantly different. Since G_E falls faster with q^2 than G_M the electric charge is spread out more than the magnetic one.

6. If a certain property, for instance the charge, is described by a form factor G, with $G(0) = 1$, then Eq. (6.20) shows that the mean-square radius for this property can be found from the slope of $G(q^2)$ at the origin:

$$\langle r^2 \rangle = -6\hbar^2 \left(\frac{dG(q^2)}{dq^2} \right)_{q^2=0} . \tag{6.45}$$

From the dipole fit, Eq. (6.42), one obtains $\langle r_E^2(\text{proton})\rangle \approx 0.7$ fm^2. However, a more accurate estimation[38] yields $\langle r_E^2(\text{proton})\rangle \approx 0.8$ fm^2. Nevertheless, the mean-square radii are in the range

$$\begin{aligned} \langle r_E^2(\text{proton})\rangle &\approx \langle r_M^2(\text{proton})\rangle \\ &\approx \langle r_M^2(\text{neutron})\rangle \approx 0.7 - 0.8 \text{ fm}^2. \end{aligned} \tag{6.46}$$

The estimate for the proton radius, given earlier in this section, by considering virtual pions, qualitatively agrees with this value. The assumption that the

[38] I. Sick, *Phys. Lett.* **576**, 62 (2003).

deviation of the electromagnetic moments from the Dirac values is caused by the hadronic structure is therefore verified.

7. Determination of the mean square charge radius of the neutron is made difficult by uncertainties that arise from the use of a deuterium target. Fortunately, there is another way to determine $\langle r_E^2(\text{neutron})\rangle$, namely by scattering low-energy neutrons from electrons bound in atoms. Much ingenuity goes into disentangling the different components to the scattering to extract the charge radius.(39,40) The result is:

$$\langle r_E^2(\text{neutron})\rangle = -0.116 \pm 0.002.\ \text{fm}^2 \tag{6.47}$$

The negative sign (keeping in mind Eqs. (6.20), (6.14) and the fact that the charge distribution is given by $Q\rho(r)$) implies that the neutron, although of net zero charge, has negatively-charged consituents further from the center than positive ones.(41) This can be understood by considering the neutron partially as a virtual negative pion around a proton or in terms of quarks.

6.7 The Charge Radii of the Pion and Kaon

So far we have learned that the lepton radius is extremely small or vanishes altogether, while the radius of the proton charge distribution is given by Eq. (6.46) as $r_p \approx 0.8$ fm. The intense pion and kaon beams available at accelerators have made it possible to determine also the charge radii of the charged pion(42) and charged kaon.(43) Pions and kaons have spin 0, and scattering of electrons and pions or electrons and kaons is described by Eq. (6.12), with just one form factor. The experiments are performed by observing the elastic scattering of high energy pion or kaon beams from the electrons in a liquid hydrogen target. Evaluation of the scattering cross section with Eq. (6.12) gives the form factor as a function of q^2; the slope of the form factor at the origin determines the radius as shown in Eq. (6.45). The root mean square radii are

$$\sqrt{\langle r_\pi^2\rangle} = 0.67 \pm 0.01\ \text{fm}, \quad \sqrt{\langle r_K^2\rangle} = 0.56 \pm 0.03\ \text{fm}. \tag{6.48}$$

The pion radius is smaller than the proton radius, but larger than that of the kaon. These differences are not fully understood.

(39) S. Kopecky et al., *Phys. Rev. Lett* **74**, 2427 (1995); PDG.

(40) For recent proposals to improve on this determination, see J.-M. Sparenberg, H. Leeb *Phys. Rev. C* **66**, 055210 (2002) and F. Wietfeldt et al., *Physica B* **385**, 1374 (2006).

(41) An additional relativistic correction to Eq. (6.47) associated with the magnetic moment of the neutron, called the Foldy term, which used to be considered dominant, has been shown to cancel in constituent quark models; see N. Isgur *Phys. Rev. Lett.* **83**, 272 (1999).

(42) G.T. Adylov et al., *Phys. Lett.* **51B**, 402 (1974); E.B. Dally et al., *Phys. Rev. Lett.* **48**, 375 (1982); T.F. Hoang et al., *Z. Physik* **C12**, 345 (1982); S. Amendolia et al., *Nucl. Phys.* **B277**, 168 (1986).

(43) E.B. Dally et al., *Phys. Rev. Lett.* **45**, 232 (1980); S.R. Amendolia et al., *Phys. Lett.* **178B**, 435 (1986).

6.8 Inelastic Electron and Muon Scattering

In inelastic scattering, the differential cross section is measured for electrons that have lost a certain amount of energy to the target. The diagrams for elastic and inelastic electron scattering from a proton are shown in Fig. 6.15. The interaction between the electron and proton, or nucleus, is mediated by a photon, as in Fig. 5.18. In elastic scattering, the final state is the same as the initial one and no new particles are created. In inelastic scattering, excited nuclear states are reached or additional particles are produced. For a nuclear target, a typical scattering spectrum is sketched in Fig. 6.16. Several features stand out, an elastic peak, relatively narrow resonances, a broad shoulder or resonance, and a continuum. The narrow resonances correspond to excited states of the nucleus, which can be studied in detail;[14,44] for example *transition* form factors can be obtained. The shoulder or broad resonance is called a *quasi-elastic* peak; the name stems from its explanation as elastic scattering from a single nucleon rather than the whole nucleus. In the laboratory system, the recoil energy of the nucleus in elastic scattering is also the energy loss, ν, of the electron

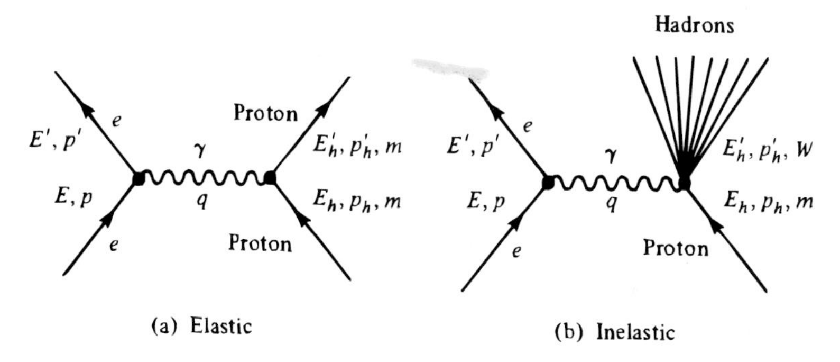

Figure 6.15: Elastic and inelastic electron scattering.

$$\nu = E - E'. \tag{6.49}$$

It is given by

$$\nu = \frac{|q^2|}{2m_A}, \tag{6.50}$$

where m_A is the mass of the nucleus and q^2 is the square of the four-momentum transferred from the electron to the nucleus,[35]

$$q^2 = \frac{\nu^2}{c^2} - (\boldsymbol{p} - \boldsymbol{p}')^2 = \frac{\nu^2}{c^2} - \boldsymbol{p}_h'^2, \tag{6.51}$$

where $\boldsymbol{p}$ and $\boldsymbol{p}'$ are the electron momenta before and after the collision, respectively, and $\boldsymbol{p}_h'$ the momentum of the hadron after the collision, as shown in Fig. 6.15(a),

[44] B. Frois, *Annu. Rev. Nucl. Part. Sci.* **37**, 133 (1987).

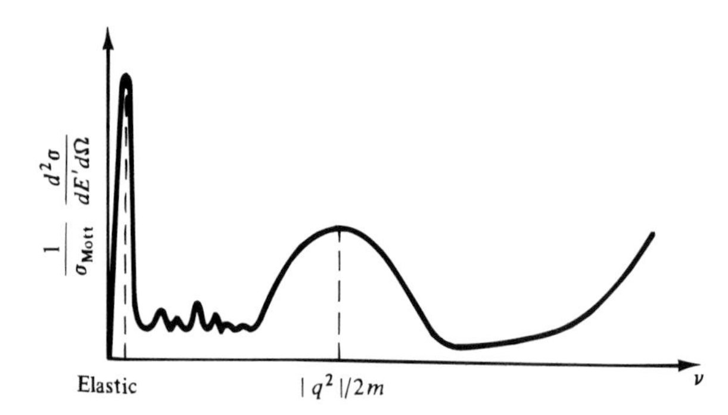

Figure 6.16: Typical double differential cross section, normalized by dividing through by the Mott cross section, for inelastic electron scattering from a nucleus. The final rise shown is due to the onset of pion production.

but in the laboratory system where $\boldsymbol{p}_h = 0$. For quasi-elastic scattering, on the other hand, the energy loss is taken up by a single nucleon that is usually ejected from the nucleus; ν is

$$\nu = \frac{|q^2|}{2m}, \tag{6.52}$$

where m is the mass of a nucleon. The peak is not sharp because the nucleon is bound in the nucleus and therefore has a momentum spread of order of magnitude given by the uncertainty principle, namely $\hbar/R \sim 100$ MeV/c, where R is the nuclear radius. Finally, one reaches a characterless continuum region where many broad states are excited. For the measurement of the differential cross section in this continuum region and for broad resonances it is necessary to determine the *double differential cross section* $d^2\sigma/dE'd\Omega$, which is proportional to the probability of a scattering occurring in a given solid angle $d\Omega$ and into an energy interval between E' and $E' + dE'$. At still higher energies, barely shown in Fig. 6.16 pion production occurs and new features appear.

A scattering spectrum on a proton target is sketched in Fig. 6.17. Its appearance resembles that of Fig. 6.16 except that it is plotted as a function of E' rather than ν and there is no quasi-elastic peak. The reason for this absence is that quarks are permanently confined inside the proton and cannot be ejected. The elastic cross section, already discussed in Section 6.6, is shown in Fig. 6.18 normalized by division through the Mott cross section, Eq. (6.11). The differential cross sections for the production of particular resonances can also be studied; their angular distributions have features similar to the elastic case. Like the nucleus, the nucleon in its excited states has a spatial extension similar to that in its ground state.

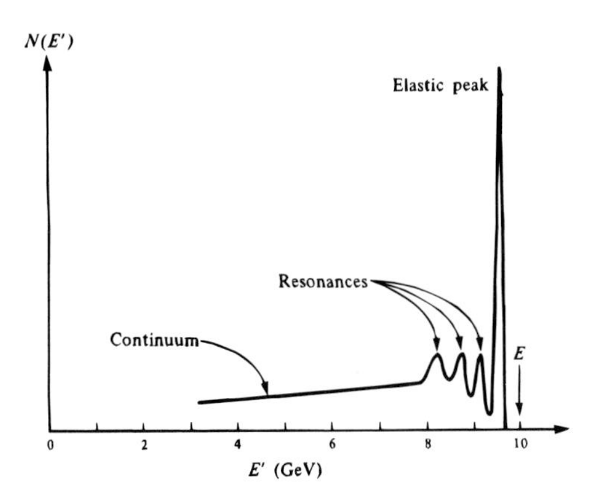

Figure 6.17: Inelastic electron scattering from protons. $N(E')$ gives the number of scattered electrons with energy E'. Note that this figure is backwards relative to Fig. 6.16.

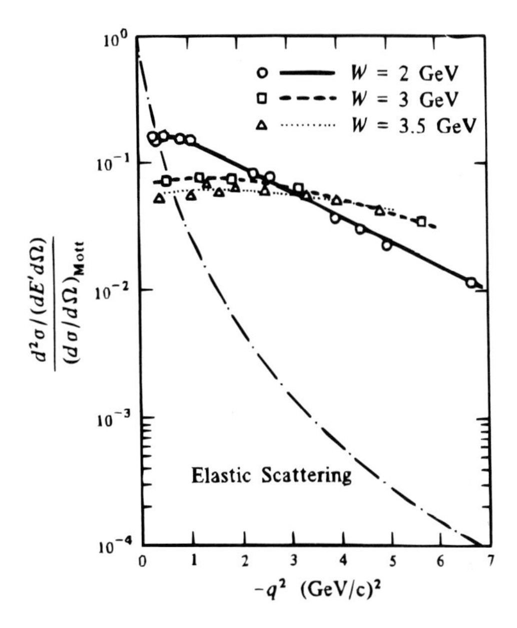

Figure 6.18: Elastic and double differential cross sections, normalized by division with $\sigma_{\text{Mott}} \equiv (d\sigma/d\Omega)_{\text{Mott}}$. $(d^2\sigma/dE'd\Omega)/\sigma_{\text{Mott}}$, in GeV^{-1}, is given for $W = 2$, 3, and 3.5 GeV. [After M. Breidenbach et al., *Phys. Rev. Lett.* **23**, 935 (1969).] Better data now exist, but we show these results because they demonstrate the salient features clearly.

6.9 Deep Inelastic Electron Scattering

The Thomson model of the atom, in vogue before 1911, assumed that the positive and negative charges were distributed uniformly throughout the atom. Rutherford's scattering experiment[(1)] proved that one charge is concentrated in the nucleus; this discovery profoundly affected atomic physics and founded nuclear physics. Highly inelastic electron scattering has had a similar impact on particle physics and we consequently discuss the most surprising results of these experiments here.

In *deep inelastic scattering*, usually only the energies and momenta of the initial and final electron are observed, but not the particles produced from the target. These measurements result in what is often called *inclusive* cross sections. Nevertheless some kinematical information about the final hadronic state can be gleaned. Energy and momentum conservation give for the energy E'_h and momentum $\boldsymbol{p}'_h$ of the final hadrons in the laboratory system (see Fig. 6.15)

$$E'_h = \nu + mc^2, \quad \boldsymbol{p}'_h = \boldsymbol{p} - \boldsymbol{p}', \tag{6.53}$$

where m is the mass of the struck particle. In terms of E'_h and $\boldsymbol{p}'_h$, or q and ν one can define the relativistically invariant effective mass, W, of all the hadrons in the final state

$$W^2 = E'^2_h - (\boldsymbol{p}'_h c)^2 = m^2c^4 + q^2c^2 + 2\nu mc^2. \tag{6.54}$$

Since q^2 and W^2 are relativistic scalars or invariants, Eq. (6.54) makes it clear that ν is also a Lorentz invariant, and therefore has the same value in any frame of reference. Indeed, we can write ν in terms of the target particle's energy E_h and momentum $\boldsymbol{p}_h$[(32)]

$$\nu = \frac{p_h \cdot q}{m} = \left(\frac{E_h q_0}{mc^2} - \frac{\boldsymbol{p}_h \cdot \boldsymbol{q}}{m}\right), \tag{6.55}$$

which makes its Lorentz invariance manifest.

At different scattering angles, what energies E' should be selected? The answer can be obtained from elastic scattering and inelastic scattering to resonances: elastic scattering corresponds to looking at a final state with $W = mc^2$; observation of a resonance means selecting a final state with $W = m_{\text{res}}c^2$, where m_{res} is the mass of the resonance. W characterizes the total mass of the hadrons in the final state here also, and the cross section $d^2\sigma/dE'd\Omega$ for the continuum is consequently determined as a function of q^2 for a fixed value of W.

Inelastic electron–proton scattering into the continuum has been studied both at medium energies ($E \sim 0.5 - 4$ GeV) on nuclei and with high energy electrons and positrons.[(45)] At SLAC the primary electron energy was varied between about 4.5 and 24 GeV; ν reached values as high as 15 GeV and $|q^2|$ over $20(\text{GeV}/c)^2$. At the

[45] A. Abramowicz and A.C. Caldwell, *Rev Mod. Phys.* **71**, 1275 (1999).

HERA collider with 27.5 GeV $e^{\pm}$ on 820 GeV protons $|q|^2$ can be varied between 0.1 $(\mathrm{GeV/c})^2$ and 5000 $(\mathrm{GeV/c})^2$. Since the late 1970s muon beams at Fermilab and CERN have also been used for deep inelastic scattering from hydrogen, deuterium, and heavier nuclei.(46,47) The ratios

$$\frac{d^2\sigma}{dE'd\Omega} : \left(\frac{d\sigma}{d\Omega}\right)_{\mathrm{Mott}}$$

for three values of W from early measurements are shown in Fig. 6.18. The difference between the elastic and the inelastic continuum scattering is dramatic: The ratio for the elastic cross section decreases rapidly with increasing $|q^2|$, whereas it is nearly independent of $|q^2|$ for the inelastic case. The ratio represents a form factor, and Table 6.1 states that a constant form factor implies a point scatterer. This conclusion is reinforced by looking at the magnitude of the cross section ratio. The cross section $d^2\sigma/dE'd\Omega$ displayed in Fig. 6.18 represents the cross section for scattering into the energy interval between E' and $E' + dE'$, where dE' is 1 GeV. To get the total inelastic cross section from the continuum, $d^2\sigma/dE'd\Omega$ must be integrated over all values of E'. To do this integration crudely, we note that the cross section ratio shown in Fig. 6.18 is nearly independent of q^2 and W over a wide range. Equation (6.53) implies that it is then also independent of E'. Integration over dE' can hence be replaced by multiplication with the total range of E'. E' ranges over nearly 10 GeV. Thus the *total* cross section for inelastic scattering into the continuum is nearly 10 times bigger than $d^2\sigma/dE'd\Omega$ in Fig. 6.18, or

$$\left(\frac{d\sigma}{d\Omega}\right)_{\mathrm{cont}} \approx \frac{1}{2}\left(\frac{d\sigma}{d\Omega}\right)_{\mathrm{Mott}}.$$

Shades of Rutherford. The Mott cross section applies to a point scatterer, and the deep inelastic scattering thus behaves nearly as if it were produced by point scatterers inside the proton.

Further evidence for the existence of point constituents inside the nucleon has come from other experiments. The cross section for the production of muon pairs by 10 GeV photons, for instance, is much larger than expected on the basis of a smooth charge distribution.(48) Initially, the nature of these point scatterers was not clear. Feynman coined the word "partons" to describe them.(49) By now, it is generally acknowledged that the charged subunits are quarks and in the context of

[46] B. Adeva et al., *Phys. Lett* **B420**, 180 (1998); M.R. Adams et al. (Fermilab E665 Collaboration) *Phys. Rev. D* **54**, 3006 (1996).

[47] J.J. Aubert et al., *Phys. Lett.* **123B**, 123 (1983); D. Bollikni et al., *Phys. Lett.* **104B**, 403 (1981); J. Ashman et al., Phys. Lett. 202B, 603 (1988).

[48] J.F. Davis, S. Hayes, R. Imlay, P.C. Stein, and P.J. Wanderer, *Phys. Rev. Lett.* **29**, 1356 (1972).

[49] R.P. Feynman, in *High Energy Collisions*, Third International Conference, State University of New York, Stony Brook, 1969 (C.N. Yang, J.A. Cole, M. Good, R. Hwa, and J. Lee-Franzini, eds.), Gordon and Breach, New York, 1969; R.P. Feynman, *Photon-Hadron Interactions*, W.A. Benjamin, Reading, MA, 1972, Lectures 25–35.

deep inelastic scattering are often called quark–partons.[50] Indeed, deep inelastic scattering provided some of the first evidence for quarks. Some conclusions concerning the subunits can be obtained with simple arguments from the pioneering experiments (Fig. 6.18).

The wavelength corresponding to the momentum transferred is sufficiently small that the interaction of the electron is with individual quarks, as shown in Fig. 6.19. The collision with each quark is elastic, and that with different quarks incoherent. The charge Ze in Eq. (6.9) then is the charge of a quark, and the observed scattering should be obtained by summing the square of the charges of the three quarks in a proton, and dividing by the number of quarks, so that we can talk of an "average quark":

$$uud: \quad \langle (Ze)^2 \rangle = (\tfrac{1}{3})[(\tfrac{2}{3})^2 + (\tfrac{2}{3})^2 + (\tfrac{1}{3})^2]e^2 = (\tfrac{1}{3})e^2.$$

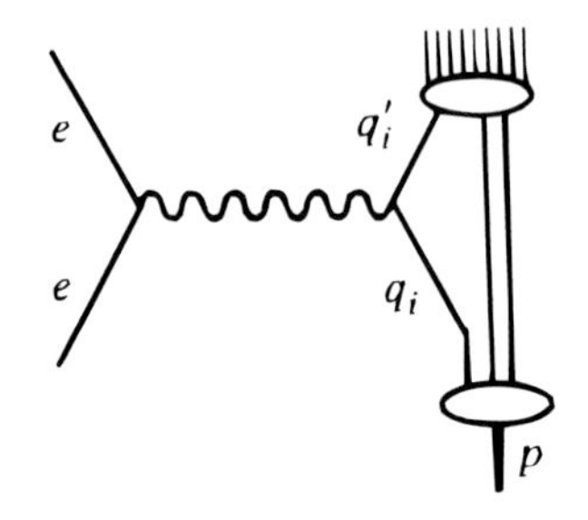

Figure 6.19: Deep inelastic scattering of electrons from the quarks of a proton.

The cross section for deep inelastic scattering from an average quark in the proton should consequently be about 1/3 of that for a point scatterer of charge e. this estimate is in good agreement with experiment. (The argument is unaltered by the fact that each quark comes in three colors because the electromagnetic interaction is color-blind.)

As for elastic scattering, two form factors are required to describe deep inelastic scattering from protons. These two functions are related to each other if $\nu \gg mc^2$. To a good approximation scaling holds for these two functions in that they are independent of q^2 and depend only on $q^2/2m\nu$.[35] These features are explained in more detail in the next section.

6.10 Quark–Parton Model for Deep Inelastic Scattering

We can gain further insight by examining deep inelastic scattering more quantitatively. First we note that the masses of the leptons can be neglected at the energies being considered. The momentum transfer to the target is so large that the interaction of the electron with the quarks is almost instantaneous and certainly very fast relative to the period of the quark motion in the nucleon. These conditions suggest that an *impulse* approximation can be used. In this approximation the binding (confinement) of the quarks can be neglected during the collision. The quarks can

[50] J. D. Bjorken, *Phys. Rev.* **163**, 1767 (1967); J. D. Bjorken and E. A. Paschos, *Phys. Rev.* **185**, 1975 (1969); J. Kuti and V. F. Weisskopf, *Phys. Rev.* **D4**, 3418 (1971).

be visualized as being free, but with a momentum distribution determined by their wavefunctions (see Fig. 6.19). The impulse approximation is well known from nuclear physics,[(51)] where it has been used successfully for studying the collision of fast particles with nuclei. The nucleons are considered to be free during the short collision time, but with a momentum distribution that is determined by their bound state wave function. A simple picture is to consider a collision with a particle attached to the end of a spring. If the collision time is short compared to the spring oscillation period, the spring can be neglected at the time of collision except for giving the particle a momentum determined by the spring constant and the particle's position. Thus, in deep inelastic scattering from a hydrogen target, we can measure the momentum distribution of the quarks in a proton. With a deuterium target, the momentum distribution of the quarks in a neutron can also be found. What happens to the particles after the very fast collision is on such a relatively long time scale that it does not affect the cross section, so that "final state" interactions among the particles can be neglected. Since the collision with each quark is elastic, the cross section is given by Eq. (6.11) if the quarks have spin zero and are very heavy. Since experiments provide clear evidence that the quark–partons have spin 1/2 and are very light, the formula must be generalized. For two spin 1/2 point particles of charge e and of negligible mass compared to their energies, the differential cross section in the laboratory system is given by Eq.(6.46) with $G_E = G_M = 1$ For the application that follows, it is more useful to have the cross section in terms of the four-momentum transfer, q, rather than the solid angle,

$$\frac{d\sigma}{d|q|^2} = \frac{2\pi\alpha^2\hbar^2}{q^4}\left[1 + \left(\frac{E'}{E}\right)^2\right]. \tag{6.56a}$$

In an arbitrary frame of reference the differential cross section is given by

$$\frac{d\sigma}{d|q^2|} = \frac{2\pi\alpha^2\hbar^2}{q^4}\left[1 + \left(\frac{p_h \cdot p'}{p_h \cdot p}\right)^2\right]. \tag{6.56b}$$

where $p_i \cdot p_j = E_i E_j/c^2 - \boldsymbol{p}_i \cdot \boldsymbol{p}_j$. In Eq. 6.56b, p and p' are the four-momenta of the electron before and after the collision, respectively, and p_h and p'_h are those of the target particle, as in Fig. 6.15.

The deep inelastic cross section can be described by an equation similar to Eq. (6.38) with two different form factors,

$$\frac{d^2\sigma}{d|q|^2 d\nu} = \frac{4\pi\alpha^2\hbar^2 E'}{q^4 mc^2 E} \times \{W_2(q^2,\nu) + [2W_1(q^2,\nu) - W_2(q^2,\nu)]\sin^2\frac{1}{2}\theta\}, \tag{6.57}$$

[51] See e.g., L. S. Rodberg and R. M. Thaler, *Introduction to the Quantum Theory of Scattering*, Academic Press, New York, NY, 1967, Ch. 12.

where the momentum transfer q^2 is

$$q^2c^2 = -4EE' \sin^2 \frac{1}{2}\theta, \tag{6.58}$$

with E and E' the electron energies before and after collision; in the energy region considered here $E = |\boldsymbol{p}|c$ and $E' = |\boldsymbol{p}'|c$. For *inelastic* scattering, W_1 and W_2 are functions of both the momentum transfer *and* the energy loss; they are referred to as *structure functions.* For elastic scattering, in the laboratory system, ν is given by Eq. (6.52), and W_1 and W_2 can be related to G_E and G_M by (see Eq. (6.38)]

$$W_2 = \frac{G_E^2 + bG_M^2}{1+b}, \qquad W_1 = bG_M^2. \tag{6.59}$$

In the region of deep inelastic scattering, Bjorken[50,52] conjectured that, in the limit $q^2 \to \infty$ and $\nu \to \infty$, but q^2c^2/ν finite, the structure functions depend only on a single dimensionless parameter, x,

$$x = \frac{-q^2}{2m\nu}. \tag{6.60}$$

This conjecture is based on the absence of a dimension to set the scale in this limit; the conjecture is called a *scaling* property. Instead of W_1 and W_2 one introduces in this limit

$$F_1 = W_1 \quad \text{and} \quad F_2 = \frac{\nu}{mc^2}W_2, \tag{6.61}$$

and these structure functions are most closely connected with the quark momentum distributions, as we shall now show. We will also see that W_1 and W_2 are related to each other in this limit. Of course, if infinite momentum transfers or energy losses really had to be reached, the conjecture of Bjorken would not be useful. As shown in Fig. 6.20,[47,52] however, scaling sets in at quite low values of q^2 and ν (e.g., a few GeV^2).

To build a picture of deep inelastic collisions, we consider quark i to carry a fraction x_i of the longitudinal (along the direction of motion) momentum of the proton of momentum p_h.[53] Because p_h is large in the frame of reference being considered, it is unlikely that any quark moves with a velocity opposite to p_h, so that we have

$$0 \leqslant x_i \leqslant 1, \quad \text{and} \quad \sum_i x_i = 1 \tag{6.62}$$

where the sum on i is over all quarks. The dimensionless fraction of momentum, x, is equal to the kinematical variable x introduced in Eq. (6.60). Thus, for an elastic

[52] J. T. Friedman and W. H. Kendall, *Annu. Rev. Nucl. Sci.* **22**, 203 (1972).

[53] The analysis is actually carried out in a momentum frame in which a proton moves with a speed almost equal to that of light both before and after the collision. In this frame, the momentum perpendicular to the motion can be neglected and will not be mentioned in our derivation.

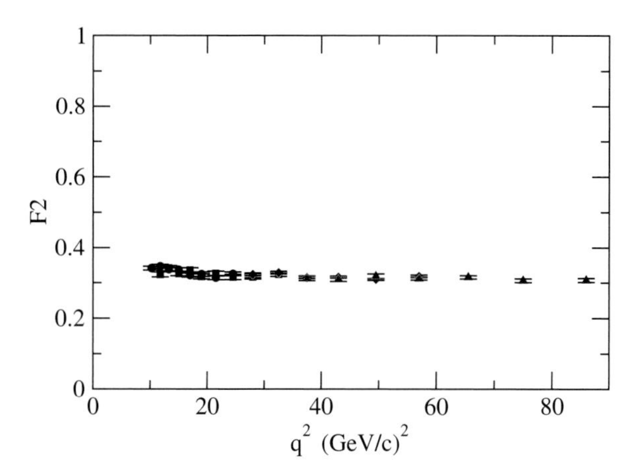

Figure 6.20: F_2 for the proton as a function of $|q|^2$ for $x = 0.225$. [From HEPDATA.]

collision of an electron with a quark of momentum xp_h, we have with the use of energy and momentum conservation

$$\begin{aligned}(xp_h')^2 = m_q^2c^2 &= (xp_h + q)^2, \\ x &= \frac{-q^2}{2p_h \cdot q}.\end{aligned} \tag{6.63}$$

But Eq. (6.55) gives $\nu = p_h \cdot q/m$, so that with Eq. (6.63) we obtain $\nu = -q^2/2mx$ and thus $x = -q^2/2m\nu$.

Let $\mathcal{P}(x_i)$ be the probability of finding quark i with momentum $x_i p_h$. The cross section for elastic scattering from the quark is then given by Eqs. (6.56a) and (6.56b) and for the proton we have in the laboratory system

$$\begin{aligned}\frac{d^2\sigma}{dx\, d|q|^2} &= \frac{2\pi\alpha^2\hbar^2}{q^4}\left[1 + \left(\frac{E'}{E}\right)^2\right]\mathcal{P}(x), \\ &= \frac{4\pi\alpha^2\hbar^2}{q^4}\frac{E'}{E}\left(1 - \frac{\nu}{mx}\frac{q^2}{4EE'}\right)\mathcal{P}(x)\end{aligned} \tag{6.64}$$

since $E^2 + E'^2 = \nu^2 + 2EE'$ and $x = -q^2/2m\nu$. We have defined $\mathcal{P}(x)$ by

$$\mathcal{P}(x) \equiv \sum_i \frac{e_i^2}{e^2}\mathcal{P}(x_i). \tag{6.65}$$

We see that the deep inelastic scattering can be described by a single structure function related to the probability of finding a quark with momentum fraction x. Equation (6.64), of course, resembles Eq. (6.57). We see the correspondence more clearly if we note that

$$dx = (q^2/2m\nu^2)\, d\nu = -(x/\nu)\, d\nu, \tag{6.66}$$

so that Eq. (6.64) can be rewritten as

$$\frac{d^2\sigma}{d|q|^2 d\nu} = \frac{4\pi\alpha^2\hbar^2}{q^4}\frac{E'}{E}\left(\frac{x}{\nu} + \frac{1}{mc^2}\sin^2\frac{1}{2}\theta\right)\mathcal{P}\left(x\right) \tag{6.67}$$

By comparing Eq. (6.67) with Eqs. (6.57) and (6.61), we obtain

$$\begin{aligned} F_2(x) &= x\mathcal{P}\left(x\right), \\ 2F_1(x) - \frac{mc^2}{\nu}F_2(x) &= \mathcal{P}\left(x\right). \end{aligned} \tag{6.68}$$

Since $x_i \leq 1$ and $\nu/mc^2 \gg 1$, we obtain the Callan–Gross relation[54]

$$F_2(x) = 2xF_1(x), \tag{6.69}$$

and thus note that W_1 and W_2 are related. The Callan–Gross relationship is specific to spin-1/2 particles; for spin-zero quarks $F_1 = 0$. In Fig. 6.21 we show an experimental comparison of F_2 and xF_1. This shows that quarks have spin 1/2.

Let us, for a moment, return to the probability $\mathcal{P}$. If we call the probability of finding an up quark in the proton u^p and a down quark d^p, then we can write[55]

$$\mathcal{P}(x) = \frac{4}{9}u^p + \frac{1}{9}d^p, \tag{6.70}$$

since the charges of the up and down quarks are $\frac{2}{3}$ and $-\frac{1}{3}$, respectively. However, we know the total probability, namely

$$\int_0^1 u^p(x)dx = 2 \quad \text{and} \quad \int_0^1 d^p(x)dx = 1, \tag{6.71}$$

since there are two up quarks and one down quark in a proton. The average momentum carried by the quarks can be written as

$$\langle \boldsymbol{p}_q \rangle = \int_0^1 x\boldsymbol{p}_h(u^p + d^p)dx. \tag{6.72}$$

The same analysis can, of course, be repeated for a neutron. Experimentally, it is found that $\langle \boldsymbol{p}_q \rangle \approx 0.5\boldsymbol{p}_h$, so that the quarks carry only about 50% of the nucleon's momentum. Therefore other, neutral, particles must carry the remaining 50% of the momentum; these particles are assumed to be the gluons.

If we are more careful we must include a correction to Eq. 6.70. In addition to the valence quarks, the nucleons contain *sea* quarks which provide a non-negligible background. These sea quarks are assumed to arise from gluons and vacuum fluctuations splitting into quark-antiquark pairs and are particularly important for $x \leq 0.2$.

---

[54] C.G. Callan and D.G. Gross, *Phys. Rev. Lett.* **21**, 311 (1968); *Phys. Rev.* **D22**, 156 (1969).

[55] For simplicity, we neglect all but "valence" quarks; there is a small contribution from other "sea quarks."

[56] For recent data, see PDG and HEPDATA.

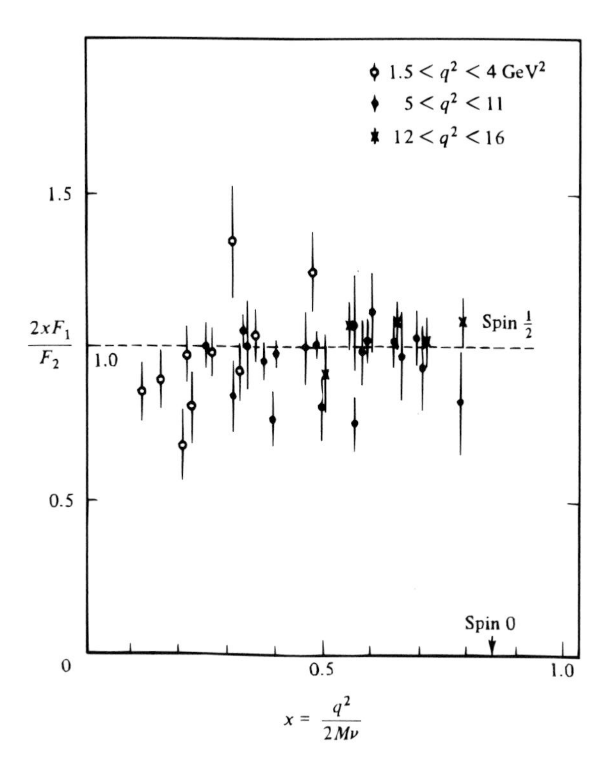

Figure 6.21: The ratio $2xF_1/F_2$ from SLAC electron–nucleon scattering experiments. The Callan–Gross relation predicts unity for this ratio. [From D. H. Perkins, *Introduction to High Energy Physics*, 3rd ed, Addison Wesley, Menlo Park, CA, 1987.]

A plot of the parton distribution functions of the proton, multiplied by x, is shown in Fig. 6.22.(56) The corrected formula is:

$$\mathcal{P}(x) = \frac{4}{9}(u^p + \bar{u}^p) + \frac{1}{9}(d^p + \bar{d}^p + s^p + \bar{s}^p), \tag{6.73}$$

where $\bar{u}^p$, $\bar{d}^p, \bar{s}^p$ and s^p represent pure sea quarks whereas u^p and d^p include both valence and sea quarks.

Further surprises were in store. Experiments at CERN by the European Muon Collaboration (EMC) revealed that the structure functions deduced from deep inelastic scattering in iron and copper differed from those in deuterium. In Fig. 6.23 we show the ratio of $F_2(\mathrm{Fe})/F_2(\mathrm{d})$ and $F_2(\mathrm{Cu})/F_2(\mathrm{d})$. Since deuterium is bound by a very small energy, these results appear to indicate that a nucleon in a nucleus is different from a free one. The difference at very small x is thought to be due to "shadowing" of the struck nucleon by other ones in the nucleus, (57) a concept

[57] F. E. Close and R. G. Roberts, *Phys. Lett.* **213B**, 91 (1988); P.R. Norton, *Rept. Prog. Phys.* **66**, 1253 (2003).

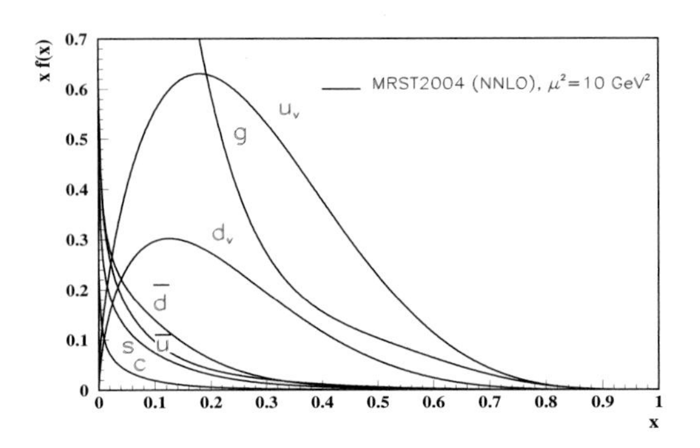

Figure 6.22: Plot of parton distribution functions, $f(x)$, times x, as a function of x for the proton. From PDG.

we will discuss in more detail in Chapter 10. The decrease in the ratio of F_2 for $0.2 \lesssim x \lesssim 0.7$ is now known to be, at least in part, due to the binding of the nucleon in the nucleus and the increase beyond $x \approx 0.7$ is caused by the motion of these bound nucleons (see Chapter 16).[58] Is this the complete explanation, or are there subtle differences between a bound and free nucleon? Is a nucleon somewhat larger (say $\sim 5\%$) in a nucleus than when free? Such questions have been raised and the so-called EMC effect remains of keen interest, because it has not yet been fully explained.

6.11 More Details on Scattering and Structure

The material in Sections 6.3–6.10 demonstrates that much information concerning subatomic structure can be obtained from scattering experiments. Even a glance at a differential cross section, without detailed computation, can reveal gross features. As an example, the information contained in Figs. 6.3, 6.5, 6.11, and 6.13 is reproduced schematically in Fig. 6.24. It highlights one difference between heavy nuclei and nucleons: Typical heavy nuclei have well-defined surfaces; as in optics, interference effects then produce diffraction minima and maxima in the differential cross section. Nucleons, in contrast, do not have such surfaces; their density decreases smoothly, and they do not show prominent diffraction effects.

The Scattering Amplitude In the present section, we shall treat scattering in somewhat more detail than we have done before. A glance at any current book on scattering[59] will show that the material presented here constitutes only a minute

[58] D.F. Geesaman, K. Saito, and A.W. Thomas, *Annu. Rev. Nucl. Part. Sci.* **45**, 337 (1995).

[59] M. L. Goldberger and K. M. Watson, *Collision Theory*, Wiley, New York, 1964; R. G. Newton, *Scattering Theory of Waves and Particles*, McGraw-Hill, New York, 1966; L. S. Rodberg and R. M.

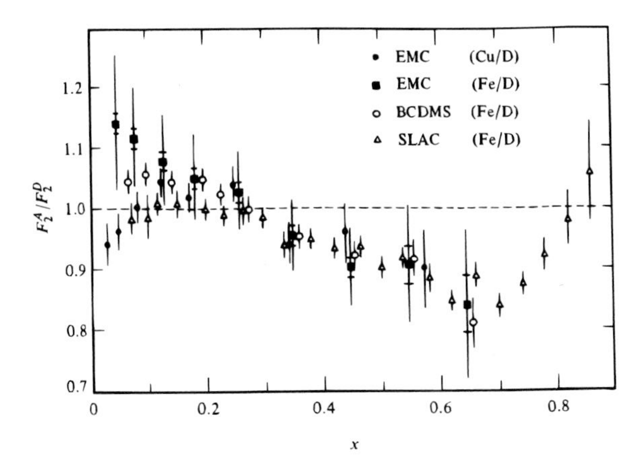

Figure 6.23: Ratios of the nucleon structure functions deduced from $F_2(\text{Cu})/F_2(d)$ and $F_2(\text{Fe})/F_2(d)$. [From J. Ashman et. al., European Muon Collaboration, *Phys. Lett.* **202B**, 603 (1988).] Later data looks similar.

fraction of what is actually used in research. Even so, it should provide some insight into the connection between scattering and structure.

We begin the discussion with a simple case, nonrelativistic scattering by a fixed potential, $V(\boldsymbol{x})$, and we approximate the incoming particle by a plane wave moving along the z axis, $\psi = \exp(ikz)$.

The solution to the scattering problem is a solution of the time-independent Schrödinger equation,

$$-\frac{\hbar^2}{2m}\nabla^2\psi + V\psi = E\psi \quad \text{or} \qquad (6.74)$$
$$(\nabla^2 + k^2)\psi = \frac{2m}{\hbar^2}V\psi,$$

where the wave number k is related to the energy E by

$$k = \frac{p}{\hbar} = \frac{1}{\hbar}\sqrt{2mE}. \qquad (6.75)$$

Far away from the scattering center, the scattered wave will be spherical, and it will originate at the scattering center, which is assumed to be at the origin of the coordinate system. The total asymptotic wave function, shown in Fig. 6.25, consequently will be of the form

$$\psi = e^{ikz} + \psi_s, \qquad \psi_s = f(\theta, \varphi)\frac{e^{ikr}}{r}. \qquad (6.76)$$

Thaler, *Introduction to the Quantum Theory of Scattering*, Academic Press, New York, 1967; W.O. Amrein, J.M. Jauch, K.B. Sinha, *Scattering theory in quantum mechanics : physical principles and mathematical methods*, Reading, Mass. : W. A. Benjamin, Advanced Book Program, 1977.

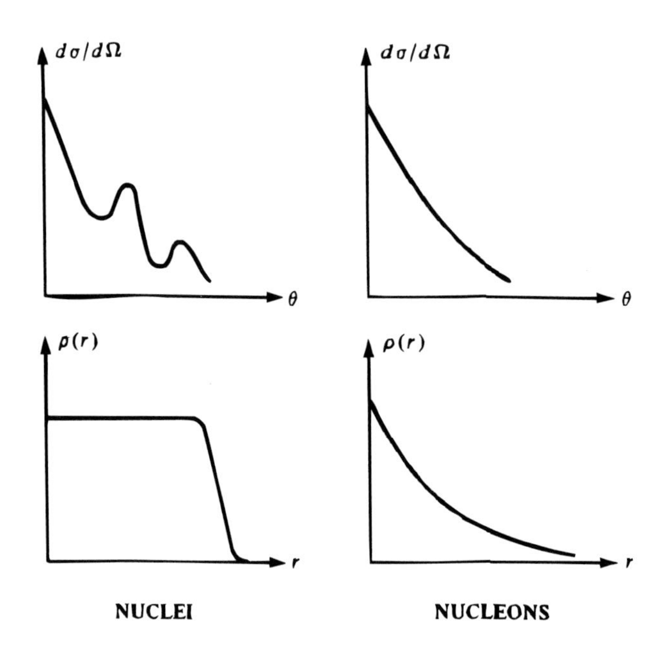

Figure 6.24: Cross section and charge distribution: The appearance of diffraction minima in the cross section for heavy nuclei implies the existence of a well-defined nuclear surface. Nucleons, in contrast, possess a charge density that decreases smoothly.

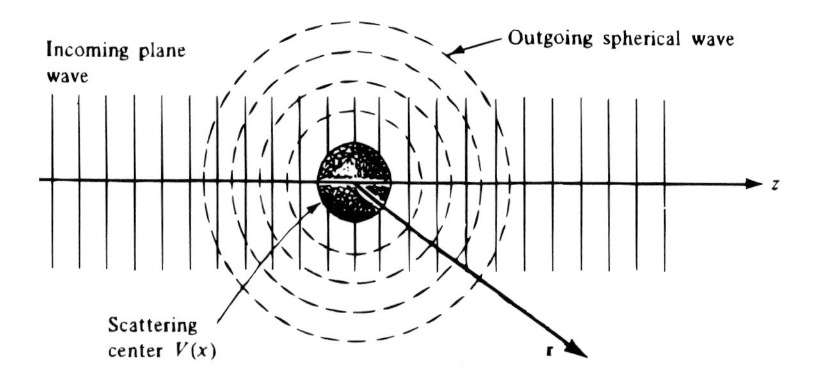

Figure 6.25: The asymptotic wave function consists of an incoming plane wave and an outgoing spherical wave.

The scattering amplitude f describes the angular dependence of the outgoing spherical wave; its determination is the goal of the scattering experiment.

The connection between differential cross section and scattering amplitude is given by Eq. (6.2). To verify the relation, we note that for the present case of one scattering center ($N = 1$), Eqs. (2.12) and (2.13) give for the differential cross section

$$\frac{d\sigma}{d\Omega} = \frac{(d\mathcal{N}/d\Omega)}{F_{\text{in}}}.$$

The outgoing flux, the number of particles crossing a unit area a at distance r per unit time, is connected to $d\mathcal{N}/d\Omega$ by

$$F_{\text{out}} = \frac{d\mathcal{N}}{da} = \frac{d\mathcal{N}}{r^2\, d\Omega}$$

so that

$$\frac{d\sigma}{d\Omega} = \frac{r^2 F_{\text{out}}}{F_{\text{in}}}. \tag{6.77}$$

Since the flux is given by the probability density current, the computation of $d\sigma/d\Omega$ is now easy. For the incident wave, $\psi = \exp(ikz)$, we find

$$F_{\text{in}} = \frac{\hbar}{2mi}|\psi^*\nabla\psi - \psi\nabla\psi^*| = \frac{\hbar k}{m}.$$

In all directions except forward (0°), the scattered wave is given by the second term in Eq. (6.76) so that

$$F_{\text{out}} = \frac{\hbar k}{mr^2}|f(\theta, \phi)|^2.$$

With Eq. (6.77), the relation (6.2) between scattering amplitude and cross section is verified.(60)

In the *forward direction*, the interference between the incident and the scattered wave can no longer be neglected. It is necessary for the conservation of flux: The scattered particles deplete the incident beam, and the scattering in the forward direction and the total cross section must be related. The relation is called the *optical theorem:* The total cross section and the imaginary part of the forward scattering amplitude are connected by(61)

$$\sigma_{\text{tot}} = \frac{4\pi}{k}\text{Im} f(0^\circ). \tag{6.78}$$

---

60 The derivation given here is superficial. A careful treatment can be found in K. Gottfried, *Quantum Mechanics*, Benjamin, Reading, Mass., 1966, Subsection 12.2.

61 For derivations of the optical theorem, see Park, p. 376; Merzbacher, p. 532; and Messiah, p. 867.

The Scattering Integral Equation • To find the general solution of the Schrödinger equation, Eq. (6.74), we recall that it can be written as the sum of a special solution and of the appropriate solution of the corresponding homogeneous equation, where $V = 0$. To find a special solution of Eq. (6.74), it is convenient to consider the term $(2m/\hbar^2)V\psi$ on the right-hand side as the given inhomogeneity, even though it contains the unknown wave function ψ. As a first step, then, we solve the scattering problem for a point source for which the inhomogeneity becomes a three-dimensional Dirac delta function and Eq. (6.74) takes on the form

$$(\nabla^2 + k^2)G(\boldsymbol{r}, \boldsymbol{r}') = \delta(\boldsymbol{r} - \boldsymbol{r}'). \tag{6.79}$$

The solution of this equation that corresponds to an outgoing wave is

$$G(\boldsymbol{r}, r') = \frac{-1}{4\pi} \frac{e^{ik|\boldsymbol{r}-\boldsymbol{r}'|}}{|\boldsymbol{r} - \boldsymbol{r}'|}. \tag{6.80}$$

To verify that this *Green's function* indeed satisfies Eq. (6.79), we set, for simplicity, $\boldsymbol{r}' = 0, |\boldsymbol{r}| = r$, and use the relations[62]

$$\nabla^2 \left(\frac{1}{r}\right) = -4\pi\delta(\boldsymbol{r}) \tag{6.81}$$

$$\begin{aligned} \nabla^2(FG) = (\nabla^2 F)G \\ + 2(\boldsymbol{\nabla} F) \cdot (\boldsymbol{\nabla} G) + F\nabla^2 G \end{aligned} \tag{6.82}$$

$$\begin{aligned} \nabla^2(\text{polar coord.}) = & \frac{1}{r^2}\frac{\partial}{\partial r}\left(r^2 \frac{\partial}{\partial r}\right) \\ & + \frac{1}{r^2 \sin\theta}\frac{\partial}{\partial \theta}\left(\sin\theta \frac{\partial}{\partial\theta}\right) + \frac{1}{r^2\sin^2\theta}\frac{\partial^2}{\partial\phi^2}. \end{aligned} \tag{6.83}$$

After some calculations we obtain

$$\begin{aligned} (\nabla^2 + k^2)\frac{e^{ikr}}{r} & = -4\pi\delta(\boldsymbol{r})e^{ikr} \\ & = -4\pi\delta(\boldsymbol{r}). \end{aligned} \tag{6.84}$$

The second step in this identity follows from the fact that

$$\int d^3r\delta(\boldsymbol{r})f(r) \quad \text{and} \quad \int d^3r\delta(\boldsymbol{r})\exp(ikr)f(\boldsymbol{r})$$

give the same result, $f(0)$, for any continuous function f. The solution of Eq. (6.55) for a potential $V(\boldsymbol{r})$ is found by assuming that the inhomogeneity $(2m/\hbar^2)V(\boldsymbol{r})\psi(\boldsymbol{r})$

[62] For a derivation of Eq. (6.81) see, for instance, Jackson, Section 1.7.

is built up from delta functions, $\delta(\boldsymbol{r}')$, each with a weight $(2m/\hbar^2)V(\boldsymbol{r}')\psi(\boldsymbol{r}')$ so that

$$\psi_s(\boldsymbol{r}) = \frac{2m}{\hbar^2}\int d^3r' G(\boldsymbol{r}, r')V(\boldsymbol{r}')\psi(\boldsymbol{r}'), \tag{6.85}$$

where $G(\boldsymbol{r}, r')$ is the Green's function for a delta function potential, Eq. (6.80). The appropriate solution of the homogeneous Schrödinger equation describes a particle that impinges on the target along the z axis; the general solution is therefore

$$\psi(\boldsymbol{r}) = e^{ikz} + \frac{2m}{\hbar^2}\int d^3r' G(\boldsymbol{r}, r')V(\boldsymbol{r}')\psi(\boldsymbol{r}'). \tag{6.86}$$

The original Schrödinger differential equation for the wave function ψ has been transformed into an integral equation, called the *scattering integral equation.* For many problems, it is more convenient to start from such an integral equation rather than from the differential equation.

In scattering experiments, the incident beam is prepared far outside the scattering potential, and the scattered particles are also analyzed and detected far away. The detailed form of the wave function inside the scattering region is consequently not investigated, and what is needed is the *asymptotic* form of the scattered wave, $\psi_s(\boldsymbol{x})$. With $\hat{\boldsymbol{r}} = \boldsymbol{r}/r$ and $\boldsymbol{k} = k\hat{\boldsymbol{r}}$, as indicated in Fig. 6.26, $|\boldsymbol{r} - \boldsymbol{r}'|$ becomes

$$|\boldsymbol{r} - \boldsymbol{r}'| = r\left\{1 - \frac{2\boldsymbol{r}\cdot\boldsymbol{r}'}{r^2} + \frac{r'^2}{r^2}\right\}^{1/2} \underset{r\to\infty}{\longrightarrow} r - \hat{\boldsymbol{r}}\cdot\boldsymbol{r}' \tag{6.87}$$

and the Green's function takes on the asymptotic value

$$G(\boldsymbol{r}, r') \underset{r\to\infty}{\sim} \frac{-1}{4\pi}\frac{\exp(ikr)}{r}\exp(-i\boldsymbol{k}\cdot\boldsymbol{r}'). \tag{6.88}$$

Inserting $G(\boldsymbol{r}, r')$ into Eq. (6.85) and comparing with Eq. (6.76) yields the expression for the scattering amplitude,

$$f(\theta, \varphi) = \frac{-m}{2\pi\hbar^2}\int d^3r' e^{i\boldsymbol{k}\cdot\boldsymbol{r}'}V(\boldsymbol{r}')\psi(\boldsymbol{r}').\bullet \tag{6.89}$$

The First Born Approximation The first Born approximation corresponds to the case of a weak interaction. If the interaction were negligible, the scattering amplitude would vanish and $\psi(\boldsymbol{r}')$ would be given by $\exp(ikz') \equiv \exp(i\boldsymbol{k}_0\cdot\boldsymbol{r}')$. As a first approximation, this value of the wave function is inserted in Eq. (6.89), with the result

$$f(\theta, \varphi) = \frac{-m}{2\pi\hbar^2}\int d^3r' V(\boldsymbol{r}')\exp(i\boldsymbol{q}\cdot\boldsymbol{r}'/\hbar), \tag{6.90}$$

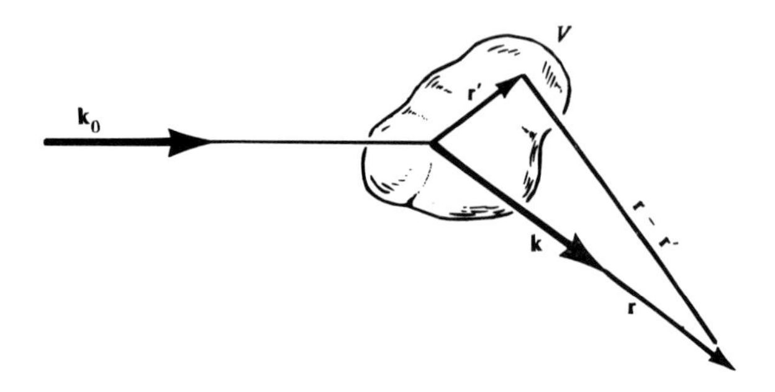

Figure 6.26: Vectors involved in the description of scattering.

where $\boldsymbol{q} = \hbar(\boldsymbol{k}_o - \boldsymbol{k})$ is the momentum that the scattered particle imparts to the scattering center, as already defined in Eq. (6.3). Equation (6.90) is called the first Born approximation; we quoted this expression in Eq. (6.5) without proof. The scattering of high-energy electrons by nucleons and light nuclei and weak processes can be described adequately by the Born approximation. In Section 6.2, we used it to derive the Rutherford cross section. Next we shall turn to an approximation that is valid under certain conditions even if the force is strong.

Diffraction Scattering—Fraunhofer Approximation When the wavelength of the incident particle is short compared to the size of the interaction region, a semiclassical approach can be used, even if the force is strong. Such an approximation is justified because the average trajectory followed by the particle approaches the classical one. The approximation used for elastic scattering is well known from optics, namely Fraunhofer diffraction. In the scattering of electromagnetic waves, optical or microwaves, the appearance of diffraction patterns has been known for a long time, and their description is well understood.[63] A characteristic example, diffraction from a black disk, is shown in Fig. 6.27. *Black* means that any photon hitting the disk is absorbed. Optical diffraction displays a number of characteristic features of which we stress three:

1. A large forward peak, called diffraction peak.

2. The appearance of minima and maxima, with the first minimum approximately at an angle

$$\theta_{\text{min}} \approx \frac{\lambda}{2R_0}, \tag{6.91}$$

   where R_0 is the radius of the disk.

[63] E. Hecht. *Optics*, 4th. Ed., Addison-Wesley, Reading, MA 2002; Jackson, Chapter 10.

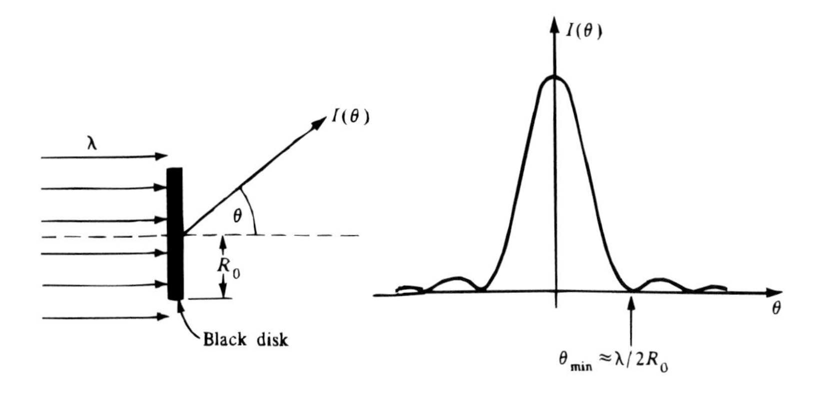

Figure 6.27: Optical diffraction pattern produced by a black disk.

3. At very small wavelengths (corresponding to the energy going to infinity) the total cross section for the scattering of light by the disk tends to a constant value,

$$\sigma \longrightarrow \text{const. for } E \longrightarrow \infty. \tag{6.92}$$

A detailed examination of the diffraction pattern for a number of wavelengths permits conclusions to be drawn concerning the shape of the scattering object. Diffraction scattering occurs not only in optics but also in subatomic physics, where it is a useful tool for structure investigations. Diffraction phenomena appear because the wavelength of the incident particles can be chosen to be smaller than the dimension of the target particle. The Fraunhofer approximation applies because the incident and the outgoing wave can be taken to be plane waves. To illustrate Fraunhofer diffraction we will present some examples in nuclear and particle physics. Consider first *nuclei*. Figure 6.28 shows the differential cross section for elastic scattering of 42 MeV alpha particles from ^{24}Mg.[(64)] A sharp forward peak and pronounced diffraction minima and maxima stand out clearly. A simple model that considers the nucleus as a dark disk reproduces the position of the minima and maxima well, but with increasing scattering angle, the observed maxima are increasingly smaller than the predicted ones.

The reason for the disagreement is that nuclei are not exactly 'black disks'. First, Figure 6.5 indicates that they have a skin of considerable thickness rather than sharp edges, and, further, nuclei are not always spherical but may have a permanent deformation, as will be discussed in Section 18.1. Finally, nuclei are partially transparent for low- and medium-energy hadrons. The simple theory can

[64] I. M. Naqib and J. S. Blair, *Phys. Rev.* **165**, 1250 (1968); S. Fernbach, R. Serber, and T. B. Taylor, *Phys. Rev.* **75**, 1352 (1949).

[65] E. Gadioli and P. E. Hodgson, *Rep. Prog. Phys.* **49**, 951 (1986); P. E. Hodgson, *Growth Points in Nuclear Physics*, Vol. 1, Pergamon, Elmsford, NY, 1984.

be modified to take these complications into account, and the resulting theory fits the experimental data reasonably well.[64,65]

Diffraction phenomena appear also in *high-energy* physics.[66,67] We restrict the discussion to elastic proton–proton scattering because it already displays characteristic diffraction features. Differential cross sections, $d\sigma/d|t|$, with $|t| = |q|^2$, for elastic pp scattering at various momenta are shown in Fig. 6.29.[68] The spectacular forward peak stands out clearly, and some other diffraction traits are also evident. In particular, the value of $d\sigma/d|t|$ at $|t| = 0$ is approximately independent of the incident momentum, and this turns out to be a prediction of the simple dark-disk model mentioned above. The total cross section can be extracted from these measurements via the optical theorem, Eq. (6.78) and it is shown in Fig. 6.30.

Fig. 6.30 shows also the $\bar{p}p$ cross section and confirms a prediction of high energy physics, namely, that particle and antiparticle cross sections on a given target should approach each other at very high energies because there are so many possible reactions that the difference becomes blurred.

In nuclear physics, the most outstanding diffraction structure is the occurrence of maxima and minima as shown in Fig. 6.28. In particle physics, the smooth distribution of the electric charge and presumably also of nuclear matter washes out the diffraction structure up to momenta of at least 20 GeV/c. At higher momenta, however, the first minimum and the following maximum appear as shown in the lowest curve in Fig. 6.29.

The Profile Function[69] The black-disk approximation reproduces the coarse features, but not the finer details, of diffraction scattering. It can be improved by assuming the scatterer to be *gray*. The shadow of a gray scatterer is not uniformly black; its grayness (transmission) is a function of $\boldsymbol{\rho}$, where $\boldsymbol{\rho}$ is the radius vector in the shadow plane (Fig. 6.31). Knowing the shadow allows calculation of the scattering amplitude, $f(\theta)$. In the black-disk approximation the total wave, $\psi(\boldsymbol{r}') \equiv \psi(\boldsymbol{\rho})$, in the shadow plane is zero behind the scatterer. For a gray scatterer it is assumed that the total wave behind the scatterer in the shadow plane is given by

$$\psi(\boldsymbol{\rho}) = e^{i\boldsymbol{k}_0\cdot\boldsymbol{\rho}} e^{i\chi(\boldsymbol{\rho})}. \tag{6.93}$$

[66] F. Zachariasen, *Phys. Rep.* **C2**, 1 (1971); B. T. Feld, *Models of Elementary Particles*, Ginn/Blaisdell, Waltham, Mass., 1969, Chapter 11. M. Kawasaki et al, Phys. Rev. D **70**, 114024 (2004).

[67] M. M. Islam, *Phys. Today* **25**, 23 (May 1972); for details see *Diffraction 2000*, R. Fiore et al. eds, North-Holland, Elsevier (2001), Nucl. Phys. **B** Proceedings, suplements; **99A** (2001).

[68] J. V. Allaby et al., *Nucl. Phys.* **B52**, 316 (1973); G. Barbiellini et al., *Phys. Lett.* **39B**, 663 (1972); A. Böhm et al., *Phys. Lett.* **49B**, 491 (1974).

[69] R.J. Glauber, in *Lectures in Theoretical Physics*, Vol. 1 (W. E. Brittin et al., eds.), Wiley-Interscience, New York, 1959, p. 315; R.J. Glauber, in *High Energy Physics and Nuclear Structure* (G. Alexander, ed.), North-Holland, Amsterdam, 1967, p. 311; W. Czyz, in *The Growth Points of Physics*, Rivista Nuovo Cimento **1**, Special No., 42 (1969) (From Conf. European Physical Society).

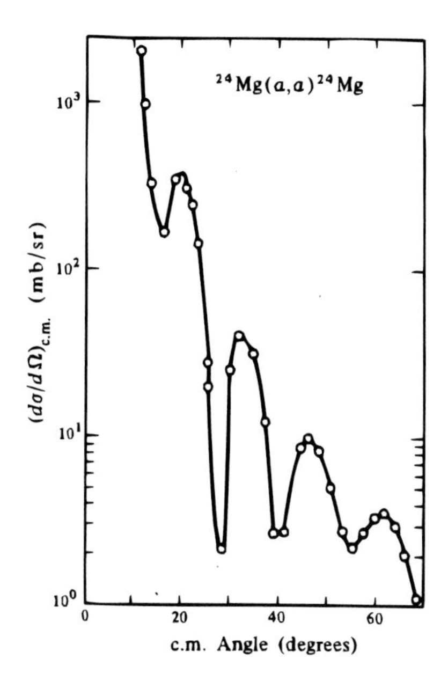

Figure 6.28: Differential cross section for the elastic scattering of alpha particles from ^{24}Mg. [I. M. Naqib and J. S. Blair, *Phys. Rev.* **165**, 1250 (1968).]

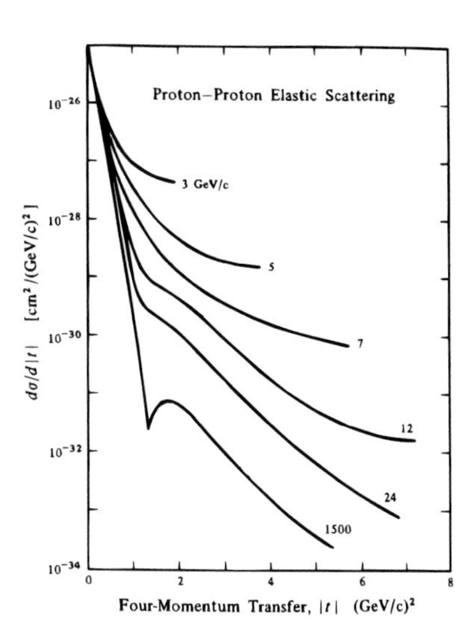

Figure 6.29: Differential cross section for elastic pp scattering. The parameter assigned to the curves gives the laboratory momentum of the incident protons. The cross sections up to $p_{\text{lab}} = 19.3$ GeV/c have been measured at the CERN proton synchrotron; the one for $p_{\text{lab}} = 1500$ GeV/c has been obtained with the CERN Intersecting Storage Rings (ISR).

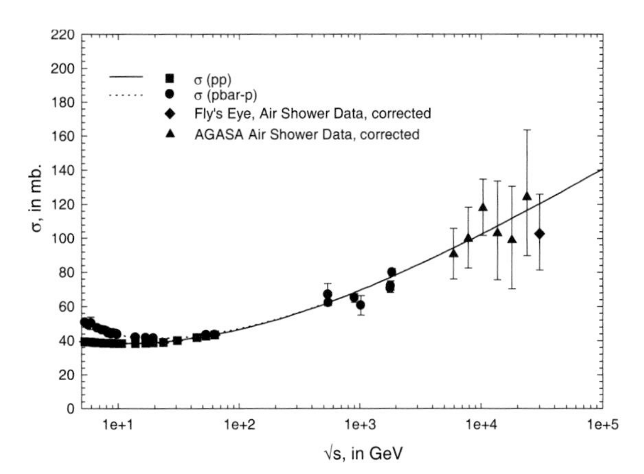

Figure 6.30: Total proton-proton and antiproton-proton cross sections as a function of laboratory momentum and the equivalent square of the c.m. energy. The cross section is roughly constant around the region of the relatively wide minimum. The lines show calculations [From M.M. Block and F. Halzen, Phys. Rev. D **63**, 114004 (2001).]

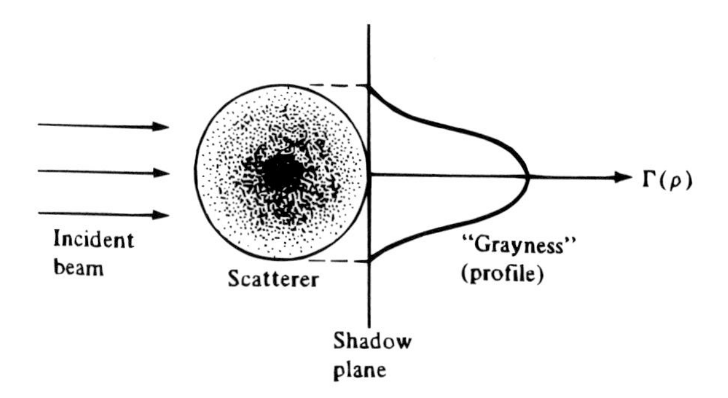

Figure 6.31: Gray scatterer and profile of its shadow. $\Gamma(\rho)$ and ρ are discussed in the text.

The total wave is modified by a multiplicative factor. For a black disk, the phase χ is purely imaginary and large. The factor $\exp(i\boldsymbol{k}_0 \cdot \boldsymbol{\rho})$ is equal to 1, but we keep it because it will turn out to be convenient. Since

$$\psi(\boldsymbol{\rho}) = e^{ikz} + \psi_s(\boldsymbol{\rho}) \tag{6.94}$$

and $kz = \boldsymbol{k}_0 \cdot \boldsymbol{\rho}$ in the shadow plane, the scattered wave is:

$$\psi_s(\boldsymbol{\rho}) = -\exp(i\boldsymbol{k}_0 \cdot \boldsymbol{\rho})\Gamma(\boldsymbol{\rho}), \tag{6.95}$$

where

$$\Gamma(\boldsymbol{\rho}) = 1 - e^{i\chi(\boldsymbol{\rho})} \tag{6.96}$$

is called the profile function.[69]

For small scattering angles, $\cos\theta \approx 1$, the scattering amplitude can be shown to be:

$$f(\boldsymbol{q}) = \frac{ik}{2\pi}\int d^2\rho \exp\left(\frac{i\boldsymbol{q}\cdot\boldsymbol{\rho}}{\hbar}\right)\Gamma(\boldsymbol{\rho}). \tag{6.97}$$

where $\boldsymbol{q} = \hbar(\boldsymbol{k}_0 - \boldsymbol{k})$ is the momentum transfer. The scattering amplitude is the Fourier transform of the profile function. If the scatterer possesses azimuthal symmetry, integration over the azimuthal angle yields

$$f(\theta) = ik\int d\rho\, \rho\Gamma(\rho)J_0(k\rho\theta). \tag{6.98}$$

This expression coincides with $f(\theta)$ for a black scatterer if $\Gamma(\rho) = 1$ (see Problem 6.31.) The relation connecting $\Gamma(\rho)$ and $f(\theta)$ in Eq. (6.98) is called a Fourier-Bessel (or Hankel) transform.[70] Given a profile function, the scattering amplitude can be calculated. As an example, assume a Gaussian profile function,

[70] W. Magnus, F. Oberhettinger and R.P. Soni, *Formulas and Theorems for the Functions of Mathematical Physics*, 3d. Ed. (English), Springer Verlag, New York, 1966, p. 397; see also P.M. Morse and H. Feshbach, *Methods of Thoretical Physics*, McGraw-Hill, New York, 1953, p. 944-962.

$$\Gamma(\rho) = \Gamma(0) \exp\left[-\left(\frac{\rho}{\rho_0}\right)^2\right]. \tag{6.99}$$

The Fourier–Bessel transform then becomes[70]

$$f(\theta) = \frac{1}{2} ik\Gamma(0)\rho_0^2 \exp\left[-\left(\frac{k\theta\rho_0}{2}\right)^2\right].$$

With $-t = |q^2| \approx (\hbar k\theta)^2$, the corresponding differential cross section is

$$-\frac{d\sigma}{dt} = \frac{\pi}{4\hbar^2}\Gamma^2(0)\rho_0^4 \exp\left[-\left(\frac{\rho_0^2}{2\hbar^2}\right)|t|\right]. \tag{6.100}$$

A Gaussian profile function leads to an exponentially decreasing cross section $d\sigma/dt$.

The physical interpretation of the profile function becomes clear by considering the total cross section. The optical theorem, Eq. (6.78), with Eq. (6.97) for $\theta = 0°$, yields

$$\sigma_{\text{tot}} = 2 \int d^2\rho Re\Gamma(\rho). \tag{6.101}$$

For a black scatterer, $\Gamma(\rho) = 1$ is real, and $f(\theta)$ is purely imaginary. If we assume that in the limit of very high energy the amplitude is imaginary,[71] then Γ is real, and Eq. (6.101) becomes

$$\sigma_{\text{tot}} = 2 \int d^2\rho\Gamma(\rho). \tag{6.102}$$

$2\Gamma(\rho)$ can consequently be interpreted as the probability that scattering occurs in the element $d^2\rho$ at the distance ρ from the center (see Fig. 6.31.) $\Gamma(\rho)$ is the scattering probability density distribution in the shadow plane; hence the name profile function.

As an application of these considerations, we return to elastic *pp* scattering.[69] Figure 6.29 shows that the diffraction peak drops exponentially for many orders of magnitude. This behavior suggests that the cross section in the region of the forward peak can be approximated by

$$\frac{d\sigma}{dt}(s,t) = \frac{d\sigma}{dt}(s,t=0)\ e^{-b(s)|t|}, \tag{6.103}$$

where s is the conventional symbol for the square of the total energy of the colliding protons in their c.m. and $b(s)$ is called the slope parameter. It is remarkable that the experimental data over a wide range of s and t can indeed be fitted by such a simple expression. The slope parameter turns out to be a slowly varying logarithmic function of the total energy s, as shown in Fig. 6.32. The exponential drop of $d\sigma/dt$

[71] The ratio between the real and the imaginary part of the proton–proton forward scattering amplitude is expected to become small at high incident momenta.

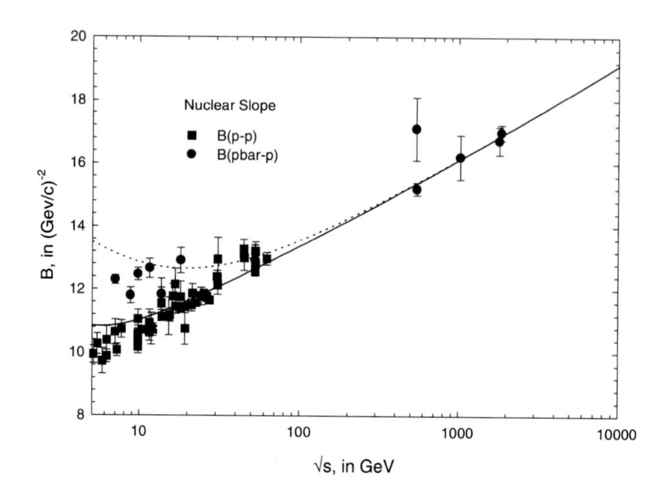

Figure 6.32: Slope parameters, b, corresponding to the cross sections shown in Fig. 6.30 [From M.M. Block and F. Halzen, Phys. Rev. D **63**, 114004 (2001).] It is seen from the figure that $b(\bar{p}p)$ approaches $b(pp)$ asymptotically.

can be interpreted in terms of a Gaussian profile function, as given in Eq. (6.99). Identification of Eqs. (6.100) and (6.103) leads to the relation

$$\rho_o = \hbar(2b)^{1/2}. \tag{6.104}$$

ρ_o characterizes the width of the Gaussian profile function describing the scattering of two extended protons by hadronic forces. It is therefore not legitimate to compare ρ_o^2, or a corresponding mean-square radius, directly with the mean-square radius of the proton as determined with electromagnetic probes. Nevertheless, it is reassuring that the two measures of the proton size are comparable: The electromagnetic radius is given by Eq. (6.46) as $\langle r^2 \rangle \approx 0.7$ fm, whereas a value of $b = 10(\text{ GeV}/c)^{-2}$, taken from Fig. 6.32, leads to $\rho_0 \approx 0.9$ fm.

The "size" of the proton and slope parameter $b(s)$ are related through Eq. (6.104); a constant ρ_0 implies a constant $b(s)$. Fig. (6.32) shows, however, that at the highest energies $b(s)$ increases logarithmically with the square of the c.m. energy, s. Since $b(s)$ describes the width of the diffraction peak, an increase of $b(s)$ means a shrinking diffraction peak, and it suggests an increase in the size, ρ_0, of the interaction region. This behavior can be understood with a geometric picture in which the area of the interaction region is related to the total cross section.[72] We saw in Fig. (6.30) that the total cross section increases with s or laboratory momenta at very high energies. Indeed, the ratio $b/\sigma_{\text{tot}} \approx$ constant,[73] as can be noted from a comparison of Figs. (6.32) and (6.30).

[72] M. Kamran, *Phys. Rep.* **108**, 275 (1984); K. Goulianos, *Phys. Rep.* **101**, 169 (1983).
[73] M.M. Block and F. Halzen, *Phys. Rev. D* **63**, 114004 (2001).

The Glauber Approximation(69,74) So far we have treated diffraction scattering from a single object. We shall now turn to the coherent scattering of a projectile from a target made up of several subunits, for instance, a nucleus built from nucleons. An incoming high-energy particle can collide with a single nucleon, with many in succession, or it can interact strongly with several at once. The treatment of such a multiscattering process is difficult, but diffraction theory makes the problem manageable; it leads to the Glauber approximation.(74)

To arrive at the Glauber approximation, we consider first the optical analog, the passage of a light wave with momentum $p = \hbar k$ through a medium with index of refraction n and thickness d. The electric vector, $\boldsymbol{E}_1$, after passage of the wave through the absorber is related to the electric vector of the incident wave, $\boldsymbol{E}_0$, by(75)

$$\boldsymbol{E}_1 = \boldsymbol{E}_0 \exp(i\chi_1), \quad \chi_1 = k(1-n)\,d. \tag{6.105}$$

If the index of refraction is complex, then its imaginary part describes the absorption of the wave. If the wave traverses successive absorbers, each characterized by a phase χ_i, the end result is

$$\begin{aligned} \boldsymbol{E}_n &= \boldsymbol{E}_0 \exp(i\chi_1)\exp(i\chi_2)\cdots\exp(i\chi_n) \\ &= \boldsymbol{E}_0 \exp[i(\chi_1 + \cdots + \chi_n)] \end{aligned} \tag{6.106}$$

The phases of the various absorbers add. The same technique can be applied to the scattering of high-energy particles. Equation (6.93) shows that the wave behind a single scatterer is related to the incident wave as the electric waves are related in Eq. (6.105). In the Glauber approximation it is assumed that the phases from the individual scatterers in a compound system, such as a nucleus, also add. To formulate the approximation, we assume that the individual scatterers are arranged as shown in Fig. 6.33. The distance of the center of each scatterer to the axis perpendicular to the shadow plane is denoted by s_i. The distance that determines the profile function for each nucleon is no longer ρ but $\rho - s_i$, and the phase factor for the ith nucleon is given by Eq. (6.96) as

$$e^{i\chi_i} = 1 - \Gamma_i(\rho - s_i).$$

For the total phase factor, additivity of the individual phases gives

$$\begin{aligned} \exp(i\chi) &= \exp(i\chi_1)\exp(i\chi_2)\cdots\exp(i\chi_A) \\ &= \prod_{i=1}^{A}[1 - \Gamma_i(\rho - s_i)], \end{aligned}$$

[74] R.J. Glauber, *Phys. Rev.* **100**, 242 (1955).
[75] *The Feynman Lectures* 1-31-3.

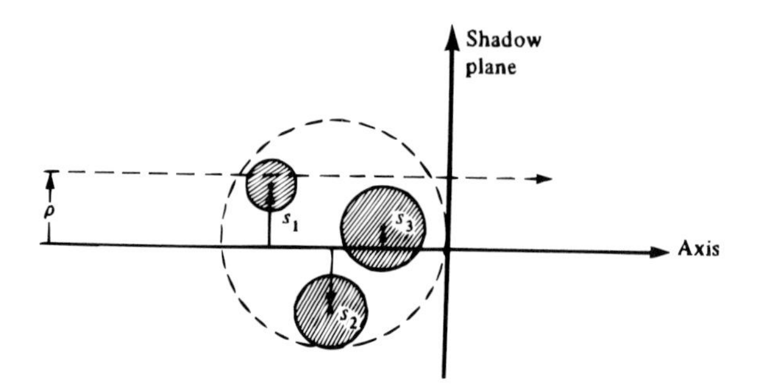

Figure 6.33: Arrangement of the individual scatterers in a nucleus.

and for the complete profile function

$$\Gamma(\rho) = 1 - \prod_{i=1}^{A}[1 - \Gamma_i(\rho - s_i)]. \tag{6.107}$$

This relation describes the Glauber approximation. If the profile functions for the individual nucleons are known, the profile function for the entire nucleus can be calculated. One more step is needed to arrive at the Glauber expression for the scattering amplitude. Nucleons are not fixed, as shown in Fig. 6.33; they move around and their probability distribution is given by the relevant wave function. For *elastic scattering*, initial and final wave functions are identical, and $\Gamma(\rho)$ in Eq. (6.97) must be replaced by

$$\int d^3x_1 \cdots d^3x_A \psi^*(\boldsymbol{x}_1, \ldots, \boldsymbol{x}_A)\Gamma(\rho)\psi(\boldsymbol{x}_1, \ldots, \boldsymbol{x}_A)$$
$$\equiv \langle i|\Gamma(\rho)|i\rangle.$$

The scattering amplitude equation (6.97) thus becomes

$$f(\boldsymbol{q}) = \frac{ik}{2\pi}\int d^2\rho \exp\left(\frac{i\boldsymbol{q}\cdot\boldsymbol{\rho}}{\hbar}\right)\langle i|\Gamma(\rho)|i\rangle, \tag{6.108}$$

with an inverse which is

$$\langle i|\Gamma(\rho)|i\rangle = \frac{1}{2\pi ik}\int \exp\left(-\frac{i\boldsymbol{q}\cdot\boldsymbol{\rho}}{\hbar}\right) f(\boldsymbol{q})\, d^2q.$$

As an example, we consider the elastic scattering of a high-energy projectile from the simplest nucleus, the deuteron (Fig. 6.34). When the energy of the incident particle is so high that its wavelength is much smaller than the deuteron radius ($R \approx 4$ fm), one could at first assume that neutron and proton scatter independently and that the total cross section is simply the sum of the individual ones. Use of the

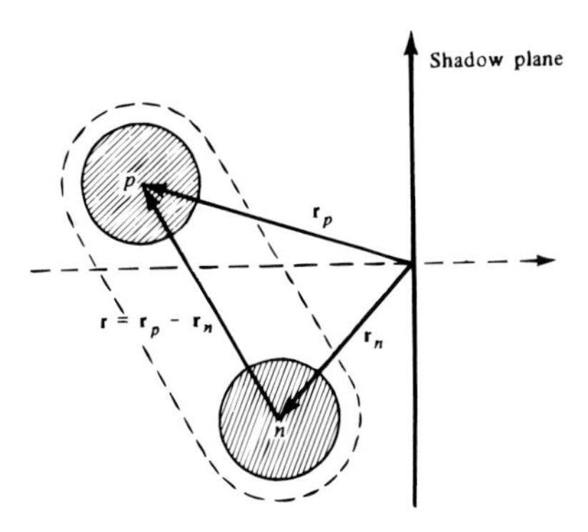

Figure 6.34: Coordinates used in the description of the scattering from deuterons.

Glauber approximation shows that this assumption is wrong, and experiment bears out the calculations. For the deuteron, with $\boldsymbol{r} = \boldsymbol{r}_p - \boldsymbol{r}_n$, Eq. (6.107) becomes

$$\begin{aligned}\Gamma_d(\boldsymbol{\rho}) &= \Gamma_p\left(\boldsymbol{\rho} + \frac{1}{2}\boldsymbol{r}\right) + \Gamma_n\left(\boldsymbol{\rho} - \frac{1}{2}\boldsymbol{r}\right) \\ &\quad - \Gamma_p\left(\boldsymbol{\rho} + \frac{1}{2}\boldsymbol{r}\right)\Gamma_n\left(\boldsymbol{\rho} - \frac{1}{2}\boldsymbol{r}\right). \end{aligned} \tag{6.109}$$

Inserting $\Gamma_d(\boldsymbol{\rho})$ into Eq. (6.108), and using the fact that the deuteron wave function, $\psi_d(\boldsymbol{r})$, is only a function of the relative coordinate $\boldsymbol{r}$, gives, for the scattering function of the deuteron,

$$\begin{aligned} f_d(\boldsymbol{q}) &= f_p(\boldsymbol{q})F\left(\frac{1}{2}\boldsymbol{q}\right) + f_n(\boldsymbol{q})F\left(\frac{1}{2}\boldsymbol{q}\right) + \frac{i}{2\pi k} \\ &\quad \times \int F(\boldsymbol{q}')f_p\left(\frac{1}{2}\boldsymbol{q} - \boldsymbol{q}'\right) f_n\left(\frac{1}{2}\boldsymbol{q} + \boldsymbol{q}'\right) d^2q', \end{aligned} \tag{6.110}$$

where $F(\boldsymbol{q})$ is the form factor for the deuteron ground state,

$$F(\boldsymbol{q}) = \int d^3r \exp\left(\frac{i\boldsymbol{q}\cdot\boldsymbol{r}}{\hbar}\right) |\psi_d(\boldsymbol{r})|^2. \tag{6.111}$$

Note that because of the symmetry of the deuteron wave function $F(\boldsymbol{q}) = F(-\boldsymbol{q})$. The first two terms in Eq. (6.110) describe the individual scatterings; the last one represents the double scattering correction. For the total cross section, the optical theorem Eq. (6.78) yields

$$\begin{aligned} \sigma_d &= \sigma_p + \sigma_n \\ &\quad + \frac{2}{k^2}\int d^2q\, F(\boldsymbol{q})\mathrm{Re}[f_p(-\boldsymbol{q})f_n(\boldsymbol{q})]. \end{aligned} \tag{6.112}$$

The deuteron radius is considerably larger than the range of the hadronic interaction; the form factor $F(\boldsymbol{q})$ hence is sharply peaked in the forward direction, and the total cross section becomes

$$\sigma_d \approx \sigma_p + \sigma_n + \frac{2}{k^2}\text{Re}[f_p(0)f_n(0)]\langle r^{-2}\rangle_d,$$

where $\langle r^{-2}\rangle_d$ is the expectation value of r^{-2} in the deuteron ground state. If the scattering is again assumed to be entirely absorptive so that the forward scattering amplitudes are imaginary, then

$$\sigma_d \approx \sigma_p + \sigma_n - \frac{1}{4\pi}\sigma_p\sigma_n\langle r^{-2}\rangle_d. \qquad (6.113)$$

The last term here shows the shadow effect of one nucleon on the other one. The shadow or double scattering term has a negative sign: the total cross section is smaller than the sum of that from the individual nucleons. This feature follows already from Eq. (6.109), where the double scattering contribution has the opposite sign from the single scattering one. More generally, expansion of Eq. (6.107) shows that the signs of successive terms alternate. This behavior has been verified experimentally.

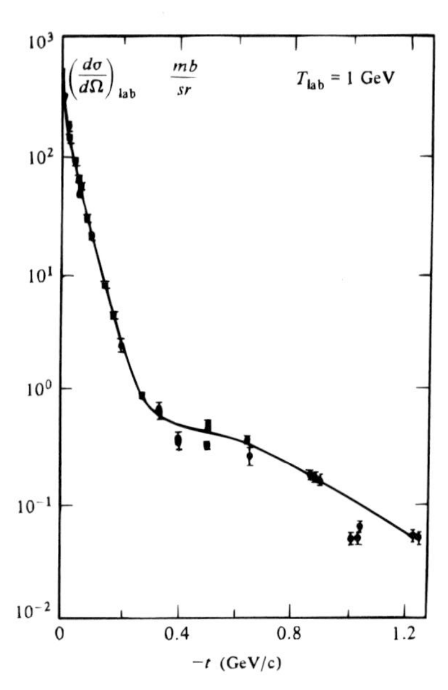

Figure 6.35: Measured and calculated pd elastic scattering cross section versus $-t = q^2$. [After M. Bleszynsky et al., *Phys. Lett.* **87B**, 198 (1979).]

The angular distribution of the scattering from deuterons provides considerably more information than the total cross section. Using Eq. (6.2) and $t = -\boldsymbol{q}^2 = (2\hbar k \sin\theta)^2$, $d\sigma/dt$ is

$$\frac{d\sigma}{dt} = \frac{-\pi}{\hbar^2 k^2}|f(\boldsymbol{q})|^2. \qquad (6.114)$$

To compute $d\sigma/dt, f_d(\boldsymbol{q})$ from Eq. (6.110) is inserted into Eq. (6.114). Consider specifically proton–deuteron scattering. The scattering amplitudes f_n and f_p can then be obtained from electron scattering on the proton and neutron; the corresponding ideas have already been treated in Sect. 6.7. To find the form factor $F(\boldsymbol{q})$, a specific form of the deuteron wave function must be assumed; for a given ψ_d, $f_d(\boldsymbol{q})$ and hence $d\sigma/dt$ can be calculated. Figure 6.35 shows $d\sigma/dt$ for scattering of 1 and 2 GeV protons from deuterons.

Some characteristic features stand out: an initial rapid drop, a shallow minimum, and then a slower decrease in $d\sigma/dt$. These features can be understood with Eq. (6.110). The first two terms, corresponding to single scattering, possess diffraction peaks of widths $\propto 1/k$, as expected from diffraction from a dark disk (see Eq. 6.91.) In double scattering, each nucleon absorbs half the momentum transfer; the corresponding diffraction width is larger. The first rapid drop-off is due to single scattering; the double scattering dominates at larger values of t. The explicit calculation of $d\sigma/dt$ shows that scattering indeed explores the structure of a nucleus.[76] As we shall discuss in more detail in Section 14.5, the two nucleons in the deuteron are predominantly in a state with relative orbital angular momentum $L = 0$ (s state), but there is a small admixture of angular momentum $L = 2$ (d state) (Fig. 14.8). To obtain the good agreement exhibited by the solid lines, this small d-state admixture (4–6%) is required; it washes out the deep interference minimum between single and double scattering.

The technique described here for the deuteron has been used to explore the structure of other nuclides.[69,77] It can also be applied if particles other than the proton, for instance, pion or antiproton, are employed as probes. •

6.12 References

Results on elastic and inelastic electron scattering from nucleons and nuclei can be found in a number of references: *Electromagnetic Form Factors of the Nucleon and Compton Scattering* C.E. Hyde-Wright and K. de Jager, *Annu. Rev. Nucl. Part. Sci.* **54**, 217 (2005); *Nuclear charge-density-distribution parameters from elastic electron scattering*, H. De Vries, C.W. De Jager and C. De Vries, *Atom. Data Nucl. Data Tabl.* **36**, 495 (1987); B. Frois in *Nuclear Structure 1985*, (R. Broglia, G. Hageman, and B. Herskind, eds) North-Holland, Amsterdam, 1985, p. 25; J. Heisenberg and H. P. Blok, *Annu. Rev. Nucl. Part. Sci.* **33**, 569 (1983); D. Drechsel and M. M. Giannini, *Rep. Prog. Phys.* **52**, 1089 (1989).

The theory behind nuclear structure studies by electron scattering can be found in T.W. Donnelly and J.D. Walecka, *Annu. Rev. Nucl. Sci.* **25**, 329 (1975); and in J.D. Walecka, *Theoretical Nuclear and Subnuclear Physics*, World Scientific, 2004.

In the present chapter, only one technique for determining the nuclear charge distribution has been treated, namely elastic electron scattering. However, many other approaches exist. Of particular importance is the observation of muonic X rays. This topic is reviewed in the following publications: F. Scheck, *Leptons, Hadrons and Nuclei*, North-Holland, Amsterdam, 1983; J. Hufner and C.S. Wu, in *Muonic Physics*, (V.W. Hughes and C.S. Wu, eds) Vol. I, Academic Press, N.Y., 1975, Ch. 3; also R.C. Barrett in the Appendix; *Exotic Atoms.* (K. Crowe, G. Fiorentini, and G. Torelli, eds.), Plenum, New York, 1980.

---

[76] M. Bleszynski et al., *Phys. Lett.* **87B**, 198 (1979).
[77] W. Czyz, *Adv. Nucl. Phys.* **4**, 61 (1971).

Helpful introductions to the modern aspects of nucleon structure and deep inelastic scattering can be found in: R.P. Feynman, *Photon-Hadron Interactions*, W.A. Benjamin, Inc., Reading, MA, 1972; F.E. Close, *An Introduction to Quarks and Partons*, Academic Press, New York, 1979; *Pointlike Structures Inside and Outside Hadrons*, (A. Zichichi, ed.), Plenum, New York, 1982; K. Gottfried and V.F. Weisskopf, *Concepts of Particle Physics*, Vol. 2, Oxford University Press, New York, NY, 1986; R. Jaffe, *Comm. Nucl. Part. Phys.*, **13**, 39 (1984); D.H. Perkins, *Introduction to High Energy Physics*, 4th. ed., Addison-Wesley, Reading, MA, 2000. See also A.W. Thomas and W. Weise, *The Structure of the Nucleon*, Wiley-VCH, New York, 2001.

Descriptions on the origin of the spin of the nucleon can be found in B. Filippone and X. Ji, *Adv. Nucl. Phys.*, (J.W. Negele and E.W. Vogt eds.) **26**, 1 (2001) and D. Drechsel, L. Tiator *Annu. Rev. Nucl. Part. Sci.* **54**, 69 (2004).

A detailed description of the recent progress in determinations of the magnetic moment of the muon is given in D.W. Hertzog, W.M. Morse, *Annu. Rev. Nucl. Part. Sci.* **54**, 141 (2004).

Problems

6.1. Consider the collision of an alpha particle with an electron. Show that the maximum energy loss and the maximum momentum transfer in one collision are small. Compute the maximum energy loss that a 10-MeV alpha particle can suffer by striking an electron at rest.

6.2. Sketch the derivation of the Rutherford scattering formula.

6.3. Show that Eq. (6.6) follows from Eq. (6.5) for a spherically symmetric potential.

6.4. Verify Eq. (6.8).

6.5. (a) Show that in all experiments that can give information concerning the structure of subatomic particles the term $(\hbar/a)^2$ in Eq. (6.8) can be neglected.

(b) For what scattering angles is the correction term $(\hbar/a)^2$ important?

6.6. Rewrite Eq. (6.9) in terms of the kinetic energy of the incident particle and of the scattering angle. Verify that the resulting expression agrees with the standard Rutherford formula.

6.7. An electron of 100 MeV energy strikes a lead nucleus.

(a) Compute the maximum possible momentum transfer.

(b) Compute the recoil energy given to the lead nucleus under the conditions of part (a).

(c) Show that the electron can be treated as a massless particle for this problem.

6.8. Verify Eq. (6.20) and find the next term in the expansion.

6.9. Assume that the probability distribution is given by ($x = |\boldsymbol{x}|$)

$$\rho(x) = \rho_0 \quad x \le R$$
$$\rho(x) = 0 \quad \text{for } x > R.$$

(a) Compute the form factor for this "uniform charge distribution."

(b) Calculate $\langle x^2 \rangle^{1/2}$.

6.10. 250 MeV electrons are scattered from ^{40}Ca.

(a) Use equations given in the text to compute numerically values of the cross section as a function of the scattering angle for the following assumptions:

(a1) Spinless electrons, point nucleus.

(a2) Electrons with spin, point nucleus.

(a3) Electrons with spin, "Gaussian" nucleus [Eq. (6.23)].

(b) Find experimental values for the cross section and compare with your computations. Determine a value for b in Eq. (6.23).

6.11. (a) What are muonic atoms?

(b) Why can muonic atoms be used to study nuclear structure?

(c) Compute the energy of the $2p - 1s$ muonic transition in ^{208}Pb under the assumption that Pb is a point nucleus. Compare with the observed value of 5.8 MeV.

(d) Use the values computed and given in part (c) to give an order-of-magnitude estimate of the nuclear radius of Pb (whose actual nuclear charge radius is ≈ 6 fm).

6.12. Use Eq. (6.18) to determine the normalization constant N in Eq. (6.24).

6.13. Use the values given in Eq. (6.27) to find an average value for the internucleon distance in a nucleus.

6.14. Discuss the $g - 2$ experiments for the electron and the muon.

(a) Derive Eq. (6.33) for the nonrelativistic case.

(b) Sketch the experimental arrangement for the $g - 2$ experiment for negative electrons. How were the electrons polarized? How was the polarization at the end measured?

(c) Repeat part (b) for muons.

6.15. * How did Stern, Estermann, and Frisch determine the magnetic moment of the proton?

6.16. * (a) How was the magnetic moment of the neutron first determined (indirect method)?

(b) Discuss a direct method to determine the magnetic moment of the free neutron.

(c) Can storage rings for neutrons be designed? If yes, sketch a possible arrangement and describe the physical idea.

6.17. Assume that a neutron consists part of the time of a Dirac neutron with 0 magnetic moment and part of the time of a Dirac proton (1 nuclear magneton) plus a negative pion. Assume that the negative pion and the Dirac proton form a system with an orbital angular momentum of 1. Estimate the fraction of time during which the physical neutron has to be in the proton–pion state in order to get the observed magnetic moment.

6.18. Verify Eq. (6.45).

6.19. * Discuss one of the methods used to determine the mean-square electric charge radius of the neutron from the scattering of slow neutrons from matter.

6.20. In the determination of the elastic form factor of the proton by electron scattering, q^2 values higher than $20(\text{ GeV}/c)^2$ are reached. In pion-electron scattering, the highest q^2 values are of the order of $1(\text{ GeV}/c)^2$. Why?

6.21. * Describe the Penning trap (Section 6.5) in detail. Could you trap a $\overline{p}$? Could the Dehmelt technique be used to measure $|g| - 2$ for the $\overline{p}$?

6.22. What squared momentum transfer t is required to observe the structure of the electron if its radius is 1 am (10^{-18} m). What beam energy is required for the experiment in e^-e^+ collisions? In collisions of energetic e^- with a stationary heavy atom target?

6.23. Show that the argument for the cross section in deep inelastic scattering of electrons with the three quarks of charges $\frac{2}{3}$ and $-\frac{1}{3}$ in a proton, i.e., $\langle Ze^2 \rangle = \frac{1}{3}e^2$, are unaltered by the property that each quark comes in three colors as long as all three colors are present in equal proportion.

6.24. The order of magnitude of a cross section is very roughly related to the strength of an interaction. Use ideas similar to those which led to Eq. (5.47) to derive approximate total cross sections for hadronic, electromagnetic, and weak interactions.

6.25. Estimate the width of the quasi-elastic peak, centered at $|q^2|/2m$, found in the scattering of electrons from nuclei, Fig. 6.16.

6.26. (a) Show the correctness of Eq. (6.50).

(b) Prove Eq. (6.55) and show that it corresponds to Eq. (6.49).

6.27. What are the maximum values of W, Eq. (6.54), which could be reached at Fermilab with muons scattering on hydrogen?

6.28. (a) Show that Eq. (6.58) is correct.

(b) Obtain the relation between dq^2 and $d\Omega$.

(c) Use parts (a) and (b) to show the equality of the two equations (6.57).

6.29. Show that $q^2 = -2p_h \cdot q$ for elastic scattering. Here p_h and q are 4-vectors with $p_h \cdot q = E_h q_0/c^2 - \boldsymbol{p}_h \cdot \boldsymbol{q}$ and p_h is the initial momentum of the hadron. (See Section 6.10).

6.30. (a) Determine the ratio for the deep inelastic cross section of electrons on neutrons to that on protons.

(b) Determine the ratio of the deep inelastic cross section of electrons on an isospin zero target (i.e., with an equal number of u and d quarks) to that on protons.

6.31. Use Eq. 6.97 to calculate the scattering amplitude from a black disk and show that the elastic cross section is πR_0^2, where R_0 is the radius of the disk. Use the optical theorem to calculate the *total* scattering cross section.

Part III

Symmetries and Conservation Laws

If the laws of the subatomic world were fully known, there would no longer be a need for investigating symmetries and conservation laws. The state of any part of the world could be calculated from a master equation that would contain all symmetries and conservation laws. In classical electrodynamics, for example, the Maxwell equations already contain the symmetries and the conservation laws. In subatomic physics, however, the fundamental equations are not yet established, as we shall see in Part IV. The exploration of the various symmetries and conservation laws, and of their consequences, therefore provides essential clues for the construction of the missing equations. One particular consequence of a symmetry is of the utmost importance: *Whenever a law is invariant under a certain symmetry operation there usually exists a corresponding conservation principle.* Invariance under translation in time, for instance, leads to conservation of energy; invariance under spatial rotation leads to conservation of angular momentum. This profound connection is used both ways: If a symmetry is found or suspected, the corresponding conserved quantity is searched for until it is discovered. If a conserved quantity turns up, the search is on for the corresponding symmetry principle. One word of warning is in place here: Intuitive feelings can be misleading. Often a certain symmetry principle looks attractive but turns out to be partially or completely wrong. Experiment is the only judge as to whether a symmetry principle holds.

Conserved quantities can be used to label states. A particle can be characterized by its mass or rest energy because energy is conserved. Or consider the electric charge, q. It is conserved and comes only in units of the elementary quantum e. The value of q/e can thus be used to distinguish particles of the same mass. Positive, neutral, and negative pions can be christened; pion is the family and positive the first name.

In the next three chapters we shall discuss a number of symmetries and conservation laws. Additional symmetries exist, and we shall encounter some later on. Some of the symmetries are perfect even under closest scrutiny, and no breakdown in the corresponding conservation law has ever been found. Rotational symmetry and conservation of angular momentum are one example of this "perfect" class.

Other symmetries are "broken," and the corresponding conservation law holds only approximately. There are two kinds of symmetry breaking; one is a symmetry broken by small effects.Invariance under mirroring (parity) provides one example of such a broken symmetry. A second kind of symmetry breaking is called "spontaneous". Here the forces have the symmetry, but the ground state does not. We shall encounter both types of symmetry breaking, the first kind in Chapter 7 and the second kind in Chapter 12. At the present time it is not understood why some symmetries are broken and others are not. It is not even clear whether the question should be phrased "Why are symmetries broken?" or "Why are some symmetries perfect?" We must continue to explore symmetries and their consequences and hope that a more complete understanding will be reached at some point.(1)

---

[1]The meaning of symmetries in physics, and more generally, in human endeavor are beautifully described in the following references: R. P. Feynman, R. B. Leighton, and M. L. Sands, *The Feynman Lectures on Physics*, Vol. I, Addison-Wesley, Reading, Mass., 1963, Chapter 52; H. Weyl, *Symmetry*, Princeton University Press, Princeton, N.J., 1952; E. P. Wigner, *Symmetries and Reflections*, Indiana University Press, Bloomington, 1967; C. N. Yang, *Elementary Particles*, Princeton University Press, Princeton, N.J., 1962; R. P. Feynman, *The Character of Physical Law*, MIT Press, Cambridge, MA, 1965; A. V. Shubnikov and V. A. Kopstik, *Symmetry in Science and Art*, Plenum, New York, 1974; J. P. Elliott and P. G. Dawber, *Symmetry in Physics*, Oxford University Press, New York, 1979; F. Close, *Lucifer's Legacy, the Meaning of Asymmetry*, Oxford University Press, New York, 2000.

Chapter 7

Additive Conservation Laws

In this chapter we shall first discuss the connection between conserved quantities and symmetries in a general way. Such a discussion is somewhat formal, but it paves the way for an understanding of the connection between symmetries and invariances.[1] We shall then treat some additive conservation laws, beginning with the electric charge. The electric charge is the prototype of a quantity that satisfies an additive conservation law: The charge of an assembly of particles is the algebraic sum of the charges of the individual particles. Moreover it is quantized and has only been found in multiples of the elementary quantum e. Other additive conserved and quantized observables exist, and in the present chapter we shall discuss the ones that are established beyond doubt.

7.1 Conserved Quantities and Symmetries

When Is a Physical Quantity Conserved? To answer this question, we consider a system described by a time-independent Hamiltonian H. The wave function of this system satisfies the Schrödinger equation,

$$i\hbar\frac{d\psi}{dt} = H\psi. \tag{7.1}$$

The value of an observable[2] F in the state $\psi(t)$ is given by the expectation value, $\langle F\rangle$. When is $\langle F\rangle$ independent of time? To find out, we assume that the operator F does not depend on t, and we compute $(d/dt)\langle F\rangle$:

$$\frac{d}{dt}\langle F\rangle = \frac{d}{dt}\int d^3x\psi^* F\psi = \int d^3x\frac{d\psi^*}{dt}F\psi + \int d^3x\psi^* F\frac{d\psi}{dt}.$$

[1]The connection between symmetries and invariants was first discovered by E. Noether; See Emmy Noether, *Collected Papers*, Springer-Verlag 1983.

[2]It is a well-known fact that the concepts of *observable* and *matrix element* are at first foreign to most students. Continuous exposure and occasional rereading of a quantum mechanics text—

To evaluate the last expression, the complex conjugate Schrödinger equation is needed:

$$-i\hbar\frac{d\psi^*}{dt} = (H\psi)^* = \psi^* H. \tag{7.2}$$

Here the reality of H has been used. With Eqs (7.1) and (7.2), $(d/dt)\langle F\rangle$ becomes

$$\frac{d}{dt}\langle F\rangle = \frac{i}{\hbar}\int d^3x\psi^*(HF - FH)\psi. \tag{7.3}$$

The term $HF - FH$ is called the *commutator* of H and F and it is denoted by brackets:

$$HF - FH \equiv [H, F]. \tag{7.4}$$

Equation (7.3) shows that $\langle F\rangle$ is conserved (i.e., is a constant of the motion) if the commutator of H and F vanishes:

$$[H, F] = 0 \to \frac{d}{dt}\langle F\rangle = 0. \tag{7.5}$$

If H and F commute, the eigenfunctions of H can be chosen so that they are also eigenfunctions of F,

$$\begin{aligned} H\psi &= E\psi \\ F\psi &= f\psi. \end{aligned} \tag{7.6}$$

Here, E is the energy eigenvalue and f the eigenvalue of the operator F in the state ψ.

---

for instance, Chapter 8 of Merzbacher—will remove the problem. We only remark that an observable is represented by a quantum mechanical operator F whose expectation value corresponds to a measurement. The expectation value of F in the state ψ_a is defined as

$$\langle F\rangle = \int d^3x\psi_a^*(\boldsymbol{x})F\psi_a(\boldsymbol{x}).$$

Since the expectation value of F can be measured, it must be real, and F therefore must be Hermitian. If two states are considered, a quantity similar to $\langle F\rangle$ can be formed by writing

$$F_{ba} = \int d^3x\psi_b^*(\boldsymbol{x})F\psi_a(\boldsymbol{x}).$$

F_{ba} is called the matrix element of F between states a and b. The expectation value of F in state a is the diagonal element of F_{ba} for $b = a$:

$$\langle F\rangle = F_{aa}.$$

The off-diagonal elements do not correspond directly to classical quantities. However, transitions between states a and b are related to F_{ba} (Merzbacher, Section 5.4).

How Can Conserved Quantities Be Found? After resolving the question as to when an observable is conserved, we attack the more physical problem: *How can conserved quantities be found*? The direct approach, writing down H and inserting all observables into the commutator, is usually not feasible because H is not fully known. Fortunately, H does not have to be known explicitly; a conserved observable can be found if the invariance of H under a symmetry operation is established. To define *symmetry operation*, we introduce a transformation operator U. U changes a wave function $\psi(\boldsymbol{x}, t)$ into another wave function $\psi'(\boldsymbol{x}, t)$:

$$\psi'(\boldsymbol{x}, t) = U\psi(\boldsymbol{x}, t). \tag{7.7}$$

Such a transformation is admissible only if the normalization of the wave function is not changed:

$$\int d^3x \psi^* \psi = \int d^3x (U\psi)^* U\psi = \int d^3x \psi^* U^\dagger U \psi.$$

The transformation operator U consequently must be *unitary*,[(3)]

$$U^\dagger U = UU^\dagger = I. \tag{7.8}$$

U is a *symmetry* operator if $U\psi$ satisfies the same Schrödinger equation as ψ. From

$$i\hbar \frac{d(U\psi)}{dt} = HU\psi \quad \text{it follows that} \quad i\hbar \frac{d\psi}{dt} = U^{-1}HU\psi,$$

where U is assumed to be time independent and where U^{-1} is the inverse operator. Comparison with Eq. (7.1) gives

$$H = U^{-1}HU = U^\dagger HU \quad \text{or} \quad HU - UH \equiv [H, U] = 0. \tag{7.9}$$

The symmetry operator U commutes with the Hamiltonian.

Comparison of Eqs. (7.5) and (7.9) shows the way to find conserved observables. If U is Hermitian, it will be an observable. If U is not Hermitian, a Hermitian operator can be found that is related to U and satisfies Eq. (7.5). Before giving an example of such a related operator, we recapitulate the essential facts about the operators F and U.

---

[3] *Notation and definitions*: If A is an operator, the Hermitian adjoint operator $A^\dagger$ is *defined* by

$$\int d^3x (A\psi)^* \phi = \int d^3x \psi^* A^\dagger \phi.$$

The operator A is Hermitian if $A^\dagger = A$; it is unitary if $A^\dagger = A^{-1}$ or $A^\dagger A = 1$. Unitary operators are generalizations of $e^{i\alpha}$, the complex numbers of absolute value 1 (Merzbacher, Chapter 14). *Notation*: If A is a matrix with elements a_{ik}, A^* with elements a^*_{ik} is the complex conjugate matrix. $\tilde{A}$ with elements a_{ki} is the transposed matrix. $A^\dagger$ with elements a^*_{ki} is the Hermitian conjugate (H.C.) matrix. $(AB)^\dagger = B^\dagger A^\dagger$. I is the unit matrix. The matrix F is called Hermitian if $F^\dagger = F$. The matrix U is unitary if $U^\dagger U = UU^\dagger = I$.

The operator F is an *observable*; it represents a physical quantity. Its expectation values must be real in order to correspond to measured values, and F consequently must be *Hermitian*,

$$F^\dagger = F. \tag{7.10}$$

Note the difference between F and U which is a *transformation operator*. The latter is unitary and changes one wave function into another one, as in Eq. (7.7).

In general, transformation operators are not Hermitian and consequently do not correspond to observables. However, there exist exceptions, and to discuss these we note that nature contains two types of transformations, *continuous* and *noncontinuous* ones. The continuous ones connect smoothly to the unit operator; the noncontinuous ones do not. Among the latter category we find the operators that are simultaneously unitary and Hermitian. Consider, for instance, the parity operation (space inversion) which changes $\boldsymbol{x}$ into $-\boldsymbol{x}$ and represents a mirroring at the origin. Such an operation is obviously not continuous; it is impossible to mirror "just a little bit." Mirroring is either done or not done. If space inversion is performed twice, the original situation is regained; noncontinuous operators often have this property:

$$U_h^2 = 1. \tag{7.11}$$

As can be seen from Eqs. (7.8) and (7.10), U_h then is unitary *and* Hermitian and it is an observable.

A well-known example of a continuous transformation is the ordinary rotation. A rotation about a given axis can occur through any arbitrary angle, α, and α can be made as small as desired. In general, a continuous transformation can always be made so small that its operator approaches the unit operator. The operator U for a continuous transformation can be written in the form

$$U = e^{i\epsilon F} \tag{7.12}$$

where ϵ is a real parameter and where F is called the generator of U. The action of such an exponential operator on a wave function ψ is defined by

$$U\psi = e^{i\epsilon F}\psi \equiv \left(1 + i\epsilon F + \frac{(i\epsilon F)^2}{2!} + \cdots\right)\psi.$$

As a rule $\exp(i\epsilon F) \neq \exp(-i\epsilon F^\dagger)$ and U is not Hermitian. However, the unitarity condition, Eq. (7.8), yields (if $[F, F^\dagger] = 0$)

$$\exp(-i\epsilon F^\dagger)\exp(i\epsilon F) = \exp[i\epsilon(F - F^\dagger)] = 1$$

or

$$F^\dagger = F. \tag{7.13}$$

The generator F of the transformation operator U is a Hermitian operator, and it is the observable connected to U if U is not Hermitian. To find F, it is usually most advantageous to consider only infinitesimally small transformations:

$$U = e^{i\epsilon F} \longrightarrow U = 1 + i\epsilon F, \quad \epsilon F \ll 1. \tag{7.14}$$

If a system is invariant under the finite transformation, it surely is invariant under the infinitesimal transformation, and investigation of infinitesimal transformations is much less cumbersome than that of finite transformations. In particular, if U is a symmetry operator, it commutes with H, as shown by Eq. (7.9). Inserting the expansion (7.14) into Eq. (7.9) gives

$$H(1 + i\epsilon F) - (1 + i\epsilon F)H = 0$$

or

$$[H, F] = 0. \tag{7.15}$$

The generator F is a Hermitian operator that is conserved if U is conserved.

The arguments in the present section have been quite formal and abstract. The applications will show, however, that the rather dry considerations have far-reaching consequences. Continuous and noncontinuous transformations play important roles in subatomic physics. Invariance under a continuous transformation leads to an additive conservation law, and relevant examples will be discussed in the present and the following chapters. Invariance under a noncontinuous transformation can lead to a multiplicative conservation law, and specific examples will be given in Chapter 9.

An Example. The treatment in the following sections and chapters is concentrated, and we therefore present first one simple example in considerable detail, in order to make the following cases easier to digest.

We consider the behavior of a particle (or system) moving in one dimension, x. Two positions of the particle, together with the corresponding wave functions, are shown in Fig. 7.1. $\psi(x)$ is the wave function of the particle centered at position x_0 and $\psi^\Delta(x)$ is the wave function of the particle that has been displaced by the distance Δ. According to Eq. (7.7), ψ and ψ^Δ *at the same point* x are connected by a transformation operator U,

$$\psi^\Delta(x) = U(\Delta)\psi(x). \tag{7.7a}$$

So far, no invariance arguments have been used, and the wave functions ψ and ψ^Δ can have completely different shapes. If the system is *invariant under translation*, ψ and ψ^Δ satisfy the same Schrödinger equation, and H and U commute. The

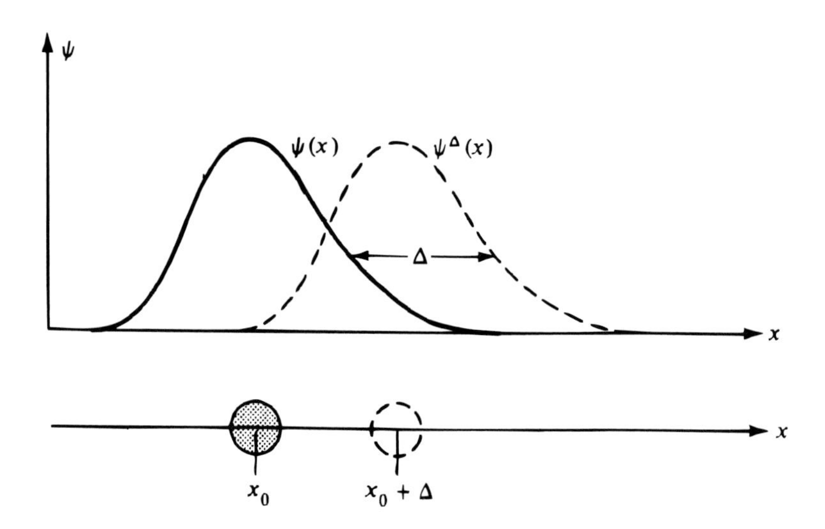

Figure 7.1: Particle in one dimension. Two different positions and the corresponding wave functions are shown. The two positions are displaced by a distance Δ.

invariance implies that the wave function does not change shape as it is displaced with the particle along x, and hence, as is apparent from Fig. 7.1,

$$\psi(x) = \psi^{\Delta}(x + \Delta).$$

The goal is now to find an explicit expression for the symmetry operator U and for the corresponding generator F. For infinitesimally small displacements Δ, expansion of the last equation gives

$$\psi(x) \approx \psi^{\Delta}(x) + \frac{d\psi^{\Delta}(x)}{dx}\Delta = \left(1 + \Delta\frac{d}{dx}\right)\psi^{\Delta}(x).$$

Multiplication from the left with $(1 - \Delta d/dx)$ and neglecting the term proportional to Δ^2 yields

$$\psi^{\Delta}(x) \approx \left(1 - \Delta\frac{d}{dx}\right)\psi(x).$$

Comparison with Eq. (7.7a) shows that

$$U(\Delta) \approx 1 - \Delta\frac{d}{dx}.$$

The general infinitesimal operator U is shown in Eq. (7.14); identifying the real parameter ϵ with the displacement Δ demonstrates that the generator F is proportional to the momentum operator p_x:

$$F = i\frac{d}{dx} = -\frac{1}{\hbar}p_x.$$

Since U commutes with H, so does F, as shown in Eq. (7.15). Invariance under translation along x leads to conservation of the corresponding momentum p_x.

7.2 The Electric Charge

As a further example of a conserved quantity we consider the electric charge. We are so used to the fact that electricity does not appear or disappear spontaneously that we often forget to ask: How well is electric charge conservation known? A good way to look for a possible violation of charge conservation is to search for a decay of the electron. If charge were not conserved, the decay of the electron into a neutrino and a photon,

$$e \longrightarrow \nu\gamma,$$

would be allowed by all known conservation laws. How could such a process be observed? If an electron bound in an atom decays, it will leave a hole in the shell. The hole will be filled by an electron from a higher state, and an X ray will be emitted. No such X rays have ever been seen, and the mean life of an electron is longer than 4.6×10^{26} y.[4] The result is generalized by saying that the total charge in any reaction is conserved; the electric charge in the initial and final state of any reaction must be the same:

$$\sum q_{\text{initial}} = \sum q_{\text{final}}. \tag{7.16}$$

The conservation law is in agreement with all observations.

Quantization of the electric charge permits us to express charge conservation in a somewhat different form. Quantization follows from Millikan's oil droplet experiment; all investigations are in agreement with the observation that the electric charge of a particle is always an integral multiple of the elementary quantum e:

$$q = Ne. \tag{7.17}$$

N is called the electric charge number, or sometimes, loosely, the electric charge. If free quarks were to exist, charges could occur in multiples of $e/3$. Relation (7.17) implies that the neutron charge must be exactly zero and that the charges of electron and proton must be equal in magnitude. Indeed, observation of the behavior of neutron and neutral-atom beams in electric fields indicates that the neutron charge is less than $2 \times 10^{-21}e$ and that the electron–proton charge sum is less than $1 \times 10^{-21}e$.[4] An electric charge number N is therefore assigned to all particles. Conservation of the electric charge, Eq. (7.16), demands that N satisfies an *additive conservation law:* In any reaction

$$a + b \longrightarrow c + d + e$$

the sum of the charge numbers remains constant,

$$N_a + N_b = N_c + N_d + N_e. \tag{7.18}$$

[4] PDG.

Equation (7.16) is an example of a conservation law. We have stated in the introduction that each conservation law is related to a corresponding symmetry principle. What is the symmetry principle that gives rise to the conservation of the electric charge? To answer this question, we repeat the arguments of Section 7.1 specifically for electric charge conservation. While reading the following derivation, it is a good idea to follow the more general steps in Section 7.1 in parallel. Assume that ψ describes a state with charge q and that it satisfies a Schrödinger equation, Eq. (7.1):

$$i\hbar\frac{d\psi}{dt} = H\psi. \tag{7.19}$$

If Q is the charge operator, we know from Eqs. (7.5) and (7.6) that $\langle Q\rangle$ is conserved if H and Q commute. ψ then can also be chosen to be an eigenfunction of Q,

$$Q\psi = q\psi, \tag{7.20}$$

and the eigenvalue q is also conserved. What symmetry guarantees that H and Q commute? The answer to this question was given by Weyl[5] who considered a transformation of the type of Eq. (7.12):

$$\psi' = e^{i\epsilon Q}\psi \tag{7.21}$$

where ϵ is an arbitrary real parameter and Q the charge operator. The transformation is called a "global" gauge transformation,[6] since it is independent of space and time coordinates. *Gauge invariance* means that ψ' satisfies the same Schrödinger equation as does ψ:

$$i\hbar\frac{d\psi'}{dt} = H\psi'$$

or

$$i\hbar\frac{d}{dt}(e^{i\epsilon Q}\psi) = He^{i\epsilon Q}\psi.$$

Multiplying from the left with $\exp(-i\epsilon Q)$, noting that Q is a time-independent and Hermitian operator, and comparing with Eq. (7.19) give

$$e^{-i\epsilon Q}He^{i\epsilon Q} = H. \tag{7.22}$$

Since ϵ is an arbitrary parameter, it can be taken to be so small that $\epsilon Q \ll 1$. Expanding the exponential yields

$$(1 - i\epsilon Q)H(1 + i\epsilon Q) = H$$

[5] H. Weyl, *The Theory of Groups and Quantum Mechanics*, Dover, New York, 1950, pp. 100, 214.

[6] The word "gauge" stems from a translation of Hermann Weyl's first introduction of the subject in 1919 as a scale invariance; H. Weyl, *Ann. Physik* **59**, 101 (1919). The idea lay dormant for about forty years because Weyl's use of it was shown to be incorrect.

or

$$[Q, H] = 0. \tag{7.23}$$

Invariance under the gauge transformation (7.21) guarantees conservation of the charge q. It is an additive conservation because when products of wavefunctions are transformed by the operator in Eq. (7.21), the Hermitian operator Q occurs in the exponent, so that Eq. (7.18) is obtained for the charges.

In addition to a global gauge transformation, we can define a "local" gauge transformation, where the parameter ϵ in Eq. (7.21) becomes an arbitrary function $\epsilon(\boldsymbol{x}, t)$ of space and time. In that case, the phases at two different space–time points are no longer related. This local gauge transformation and the associated symmetry is the crucial underpinning of all modern subatomic physical forces, the hadronic, electromagnetic, and weak. Here we only illustrate the usefulness of the local gauge symmetry by a simple example. We will return to local gauge transformations in more detail in Chapter 12.

We have proven that a global gauge invariance leads to charge conservation, but we have not identified the charge as an electric one. To do so requires a local gauge invariance, as we shall now show. We assume that q is an *electric* charge and place the system in a static electric field, $\boldsymbol{E}$, defined in terms of the scalar potential A_0,

$$\boldsymbol{E} = -\boldsymbol{\nabla} A_0. \tag{7.24}$$

The Hamiltonian H in the Schrödinger equation (7.1) can then be written as

$$H = H_0 + qA_0 \tag{7.25}$$

where H_0 describes the system in the absence of the field A_0; for a free particle of mass m,

$$H_0 = \frac{p^2}{2m} = \frac{-\hbar^2\nabla^2}{2m}.$$

It is well known from classical electricity and magnetism that the electric and magnetic field vectors $\boldsymbol{E}$ and $\boldsymbol{B}$ are unchanged by a gauge transformation $A_0 \to A_0'$, $\boldsymbol{A} \to \boldsymbol{A}'$,

$$A_0' = A_0 - \frac{1}{c}\frac{\partial\Lambda(\boldsymbol{x}, t)}{\partial t}, \quad \boldsymbol{A}' = \boldsymbol{A} + \boldsymbol{\nabla}\Lambda(\boldsymbol{x}, t) \tag{7.26}$$

where $\Lambda(\boldsymbol{x}, t)$ is an arbitrary function of $\boldsymbol{x}$ and t.[7] We replace the global gauge transformation of Eq. (7.21) by a local gauge transformation

$$\psi' = e^{i\epsilon(\boldsymbol{x},t)Q}\psi. \tag{7.27}$$

Although in general, the phase $\epsilon(\boldsymbol{x}, t)$ is an arbitrary function of space and time, it is sufficient for our purpose here to take Λ and ϵ to be constant in space and only

[7] Jackson, Section 6.3.

functions of time, i.e., $\Lambda(t)$ and $\epsilon(t)$. This restriction simplifies the arithmetic and will be removed in Chapter 12. Invariance under the local gauge transformation requires that the Schrödinger equation for ψ and ψ' have the same form,

$$i\hbar\frac{\partial\psi'}{\partial t} = (H_0 + qA_0')\psi'. \tag{7.28a}$$

Under the simultaneous gauge transformations of ψ and A_0, Eqs. (7.26) and (7.27), and with Eq. (7.24), the Schrödinger equation (7.28) becomes

$$\begin{aligned} i\hbar\frac{\partial}{\partial t}e^{i\epsilon(t)Q}\psi &= \left(\frac{-\hbar^2\nabla^2}{2m} + qA_0 - \frac{q}{c}\frac{\partial\Lambda}{\partial t}\right)e^{i\epsilon(t)Q}\psi, \\ e^{i\epsilon(t)Q}\left(\frac{i\hbar\partial\psi}{\partial t} - \hbar Q\psi\frac{\partial\epsilon}{\partial t}\right) &= e^{i\epsilon(t)Q}\left(-\frac{\hbar^2\nabla^2}{2m} + qA_0 - \frac{q}{c}\frac{\partial\Lambda}{\partial t}\right)\psi. \end{aligned} \tag{7.28b}$$

Comparison of Eqs. (7.1) with (7.25) and (7.28) shows that the invariance condition implies

$$\hbar Q\frac{\partial\epsilon(t)}{\partial t} = \frac{q}{c}\frac{\partial\Lambda(t)}{\partial t}. \tag{7.29}$$

Since $\epsilon(t)$ and $\Lambda(t)$ are arbitrary functions of space and time, we set

$$\Lambda(t) = \hbar c\epsilon(t) \tag{7.30}$$

so that Eq. (7.29) becomes identical with the eigenvalue equation (7.20). Eq. (7.25) means that q is the electric charge and Q, therefore, is the electric charge operator. The global gauge transformation leads to the introduction of a conserved quantum number, the local gauge transformation (7.27) together with the gauge transformation of the electromagnetic field, Eqs. (7.26), identifies the charge. The phase of the wavefunction varies in space and time as described by $\epsilon(\boldsymbol{x}, t)$; the variation is counteracted by corresponding changes in the electromagnetic potential as given by

$$\Lambda(\boldsymbol{x}, t) = \hbar c\epsilon(\boldsymbol{x}, t)$$

so that no net effect is observable.

7.3 The Baryon Number

Conservation of the electric charge alone does not guarantee stability against decay. The proton, for instance, could decay into a positron and a gamma ray without violating either charge or angular momentum conservation. What prevents such a decay? Stueckelberg first suggested that the total number of nucleons should be conserved.[8] This law can be formulated compactly by assigning a *baryon number* $A = 1$ to the proton and the neutron and $A = -1$ to the antiproton and the

[8] E. C. G. Stueckelberg, *Helv. Phys. Acta* **11**, 225, 299 (1938); E. P. Wigner, *Proc. Am. Phil. Soc.* **93**, 521 (1949).

antineutron. (See Section 5.10 for a discussion of antiparticles.) Leptons, photons, and mesons are assigned $A = 0$. (Particle physicists use B for baryon number, but we follow the convention of the nuclear physicists here.) The additive conservation law for the baryon number then reads

$$\sum A_i = \text{const.} \tag{7.31}$$

The extent to which Eq. (7.31) holds can be described by a limit on the lifetime of the nucleons. A geochemical method examining decays of nucleons in ^{130}Te gives a lower limit of $1.6 \times 10^{25} y$.(9) A better limit is found by measuring possible decays in a large quantity of water, which contains many protons, and with very large counters that are shielded from cosmic rays by being deep underground.(10) The limit then becomes about 10^{30}y; for the specific decay $p \to e^+\pi^0$, the lower limit is 1.6×10^{33}y.(11) We do not have to live in fear of wasting away through the decay of nucleons.

The discovery of strange particles led to a generalization of the law of nucleon conservation. Consider, for instance, the decays

$$\begin{aligned} &\Lambda^\circ \longrightarrow n\pi^0 \\ &\Sigma^+ \begin{cases} \longrightarrow p\pi^0 \\ \longrightarrow \Lambda e^+ v \end{cases} \\ &\Sigma^- \longrightarrow n\pi^-. \end{aligned}$$

In each of these decays, the baryon number is conserved if it is generalized to read

$$A = 1 \quad \text{for } p, n, \Lambda, \Sigma, \Xi, \Omega$$

and $A = -1$ for the corresponding antiparticles. Similarly, *resonances* and *nuclei* can be characterized by their baryon number A. Since nuclei are built up from protons and neutrons, the baryon number A is identical to the mass number, introduced in Section 5.9. *Hypernuclei* are similar to nuclei, but one or two nucleons are replaced by a hyperon.

As in the case of the electric charge, the question of the symmetry responsible for baryon conservation arises. Again, a global gauge transformation

$$\psi' = \psi e^{i\epsilon A} \tag{7.32}$$

leads formally to the conservation law, Eq. (7.31). If the gauge invariance were a local one then there should be a long range field, similar to the electromagnetic

[9] J.C. Evans and R.J. Steinberg, *Science* **197**, 989 (1977).

[10] S. Weinberg, *Sci. Amer.* **231**, 50 (July 1974); J. M. Lo Secco, F. Reines, and D. Sinclair, *Sci. Amer.* **252**, 54 (June 1985).

[11] PDG. Also J. Bartelt et al., *Phys. Rev. Lett.* **50**, 651 (1983); M. Goldhaber in *Interactions and Structures in Nuclei*, (R. J. Blin-Stoyle and W. D. Hamilton, eds.) Adam Hilger, Philadelphia, 1988, p. 99.

one, associated with it. No such field has been found. This is one reason that it is believed that the symmetry is not an exact one and that the proton decays.

The data given so far appear to indicate that further searches for a violation of baryon conservation are unnecessary since the limits of 10^{30}y and 1.6×10^{33y} are very long compared to the age of the universe, which is only about 10^{10}y. Theoretical arguments, however, suggest that the proton lifetime, although long, is finite. It is important to realize that there is a profound difference between the conservation laws for electric charge and baryon number. The conservation of electric charge is related to, or obtained from, the continuity equation for the electric current and to gauge invariance, which in turn are connected to the Maxwell equations. No such sound theoretical basis has been found for baryon conservation, which thus is an empirical rule based on precise experimental measurements. Furthermore, the success of the unification of the weak and electromagnetic interactions, which we will discuss in Chapter13, has led theorists to speculate about a (grand unified) theory that also encompasses gravity and the strong interactions.(12,13) All of these theories and connected arguments for the excess of matter over antimatter in our universe contain a very small violation of baryon conservation.(14) The predicted lifetime of the proton depends on the particular theory, but many models place it somewhere between 10^{33} and 10^{38}y.

7.4 Lepton and Lepton Flavor Number

In Section 5.6 the basic characteristics of six leptons (electron, muon, tau and the three neutrinos) were sketched, and we pointed out that six antileptons also exist. To explain the absence of some decay modes allowed by all other conservation laws, Konopinski and Mahmoud introduced a lepton number, L, and lepton number conservation.(15) They assigned $L = 1$ to e^-, μ^-, ν_e and ν_μ, $L = -1$ to the antileptons $e^+, \mu^+, \overline{\nu}_e$, and $\overline{\nu}_\mu$; and $L = 0$ to all other particles.

$$\sum L_i = \text{const.} \tag{7.33}$$

If lepton conservation indeed holds, leptons can be destroyed or created only in particle–antiparticle pairs. High-energy photons can produce pairs such as

$$\gamma \longrightarrow e^-e^+, \quad \gamma \longrightarrow p\overline{p},$$

[12] G. Oycho, *Grand Unified Theorem*, Nova Science Publ, Commack, New York, 1999.

[13] P. Ramond, *Ann. Rev. Nucl. Part. Sci.* **33**, 31 (1984); H. P. Niles, *Phys. Rep.* **110**, 1 (1984); P. Davies, *Superforce*, W. Heinemann Ltd, London, 1984; J. Griblin, *The Search for Superstrings, Symmetry, and the Theory of Everything*, Little Brown, and Co, Boston, 1988; B. Greene, *The Fabric of the Cosmos*, A. Knopf, New York, 2004.

[14] R.S. Chivukula et al, *Ann. Rev. Nucl. Part. Sci.*, **45**, 255 (1995); .Lepton and Baryon Number Violation in Particle Physics, Astrophysics and Cosmology, ed. H. V. Klapdor-Kleingrothaus and I.V. Krivosheina, Bristol, Philadelphia, 1999.

[15] E. J. Konopinski and H. M. Mahmoud, *Phys. Rev.* **92**, 1045 (1953).

but not $\gamma \to e^- p$. (Remember that these processes can happen only in the field of a nucleus that takes up momentum; see Problem 3.22 .)

Evidence for lepton conservation comes partially from neutrino reactions. Consider first antineutrino capture,

$$\overline{\nu}_e p \longrightarrow e^+ n. \tag{7.34}$$

This process is allowed by lepton conservation because the lepton number on both sides of the equation is -1. Antineutrino capture has been observed by Reines, Cowan, and collaborators with antineutrinos from a nuclear reactor.[16] A reactor produces predominantly antineutrinos because fission yields neutron-rich nuclides. These decay through processes involving the mode

$$n \longrightarrow p e^- \overline{\nu}_e. \tag{7.35}$$

Since the neutron has $L = 0$, the right-hand side must also have $L = 0$, and the particle emitted together with the negative electron must be an antineutrino. The observation of the reaction Eq. (7.34) is in agreement with Eq. (7.35). However, reactions of the type $\overline{\nu}_e n \to e^- p$ and $\nu_e p \to e^+ n$ are forbidden by lepton conservation. Davis has searched for a reaction of this type,

$$\overline{\nu}_e {}^{37}\text{Cl} \longrightarrow e^- {}^{37}\text{Ar}, \tag{7.36}$$

again using antineutrinos from reactors. Here, $L = -1$ on the left-hand side and $L = +1$ on the right-hand side, and lepton conservation would be violated if the reaction were observed. Davis did not see reaction (7.36) and thus was able to set a limit[17] ($2 \times 10^{-42}\text{cm}^2/\text{atom}$) on the cross section of the reaction caused by antineutrinos. Note, however, that the reaction

$$\nu_e {}^{37}\text{Cl} \longrightarrow e^- {}^{37}\text{Ar} \tag{7.37}$$

should occur and was observed by Davis. This result shows that antineutrinos and neutrinos have different characteristics.

---

[16] F. Reines, C. L. Cowan, F. B. Harrison, A. D. McGuire, and H. W. Kruse, *Phys. Rev.* **117**, 159 (1960).

[17] R. Davis, *Phys. Rev.* **97**, 766 (1955); J. K. Rowley et al., in *Solar Neutrinos and Neutrino Astronomy*, (M. L. Cherry, K. Lande and W. A. Fowler, eds) American Institute of Physics, New York, 1985) p. 1.

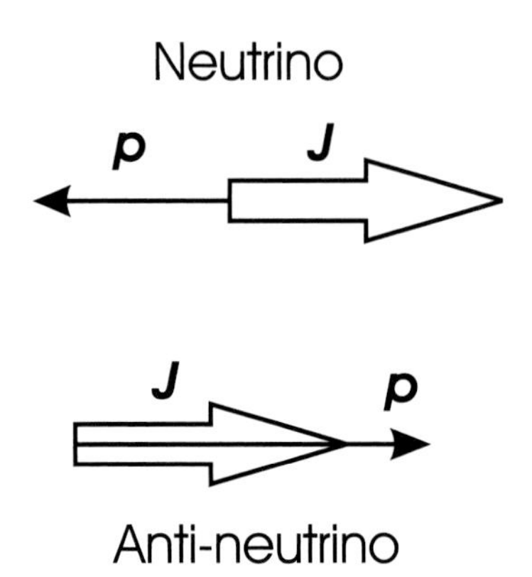

Figure 7.2: Neutrino and antineutrino are always *polarized* if we neglect their very small masses. The neutrino has its spin always opposite to its momentum; the antineutrino has parallel spin and momentum.

In Chapter 5 we pointed out that photons present two states of *helicity*, according as to whether their spin points along or opposite the direction of their momentum. Experiments on beta decay have shown that antineutrinos are produced mainly with right-handed helicity[19] and neutrinos mainly with left-handed helicity (see Fig. 7.2.) Under the assumption that neutrinos are different than antineutrinos[18] (Dirac neutrinos) we find reactions like Eq. 7.36 to put severe constraints on lepton number non-conservation.

Why have we also distinguished a muon (or tau) and an electron neutrino? Both have $L = 1$. In what way are they different? To attack this question, another of the puzzles that surround neutrinos must be told.

The muon, for example, decays through the mode

$$\mu \longrightarrow e\overline{\nu}\nu, \tag{7.38}$$

but the possibility

$$\mu \longrightarrow e\gamma \tag{7.39}$$

is allowed by all the conservation laws discussed so far. Over the years many groups have searched for the gamma decay of the muon, without any success, and the limit[20] on the branching ratio is less than 1.2×10^{-11}. The simplest way to explain the absence of the muon gamma decay is a new conservation law, conservation of flavor $(e, \mu, \text{or } \tau)$ number, e.g., L_μ. $L_\mu = +1$ is assigned to the negative and $L_\mu = -1$ to the positive muon. The lepton number of the neutrinos associated with muons can then be found from the pion decays:

$$L_\mu : \begin{array}{cccccc} \pi^- & \longrightarrow & \mu^- \overline{\nu}_\mu, & \pi^+ & \longrightarrow & \mu^+ \nu_\mu \\ 0 & & 1 - 1 & 0 & & -1 \;\; 1 \end{array} \tag{7.40}$$

$\overline{\nu}_\mu$ is labeled an antineutrino because it is right-handed. The muon neutrino has a flavor number $L_\mu = 1$, and the muon antineutrino $L_\mu = -1$. These particles

[18] Because the neutrino is neutral it can be its own antiparticle (Majorana νs.) This interesting scenario implies lepton number nonconservation and will be discussed in Chapter 11. Here we assume that the neutrino and the antineutrino are not identical particles (Dirac νs.)

[19] See M. Goldhaber, L. Grodzins, and A.W. Sunyar, Phys. Rev. **109**, 1015 (1958) for the first determination of antineutrino helicity.

[20] M.L. Brooks et al.(MEGA collaboration), *Phys. Rev. Lett.* **83**, 1521 (1999); see also PDG.

all belong to the same "family". All other particles are assigned $L_\mu = 0$. Helicity considerations would not forbid the reaction in Eq. (7.39), but we shall see in Ch. 11 that neutrino flavor (whether ν_e, ν_μ, ν_τ) is not strictly respected; during their propagation neutrinos can and do transform (oscillate) from one flavor to another.

Conservation of the flavor number accounts for the absence of the decay $\mu \to e\gamma$. However, if the introduction of the muon number does nothing else, it is not meaningful. Actually, it does lead to new predictions, as can be seen by considering the two reactions

$$\nu_\mu n \longrightarrow \mu^- p, \quad \nu_\mu n \longrightarrow e^- p. \tag{7.41}$$

If the muon number is conserved, only the first one is allowed; the second one is forbidden. The reactions can be tested because the pion decay, Eq. (7.40), produces almost only muon neutrinos. The experimental observation is difficult because neutrinos have an extremely small cross section and the detector for the reaction equation (7.41) must be guarded against all other particles. In 1962, a Columbia group performed a successful experiment at the Brookhaven accelerator and indeed found that no electrons were produced by muon neutrinos.[21] Since this first experiment, the fact has been verified many times, but muon number is not conserved exactly, since, for example, $\nu_e \leftrightarrow \nu_\mu$.

The discovery of the tau lepton has led to the introduction of yet a new lepton quantum number, the tau flavor number. Allowed decays of the tau are numerous and include

$$\begin{aligned}\tau^- &\longrightarrow \mu^- \overline{\nu}_\mu \nu_\tau \\ &\longrightarrow \pi^- \nu_\tau \\ &\longrightarrow e^- \overline{\nu}_e \nu_\tau .\end{aligned}$$

These modes and others have been seen.[22]

7.5 Strangeness Flavor

In 1947, Rochester and Butler observed the first V particles[23] (Fig. 5.21). By about 1952, many V events had been seen, and a mystery had developed: the V particles were produced copiously but decayed very slowly. The production, for instance, through Eq. (5.56), $p\pi^- \to \Lambda^0 K^0$, occurred with a cross section of the order of mb, whereas the decays had mean lives of about 10^{-10} sec. Cross sections of the order of mb are typical of the hadronic interactions, whereas decays of the order of 10^{-10} sec are characteristic of the weak interaction: kaons and hyperons

[21] G. Danby, J.M. Gaillard, K. Goulianos, L.M. Lederman, N. Mistry, M. Schwartz, and J. Steinberger, *Phys. Rev. Lett.* **9**, 36 (1962). See also *Adventures in Experimental Physics.*, Vol. α, World Sci. Communic., Princeton, NJ, 1972.

[22] PDG; A.J. Weinstein, R. Stroynowski, *Ann. Rev. Nucl. Part. Sci.* **43**, 457 (1993).

[23] G.D. Rochester and C.C. Butler, *Nature* **160**, 855 (1947).

are produced strongly but decay weakly. Pais made the first step to the solution of the paradox by suggesting that V particles are always produced in pairs.[24] The complete solution came from Gell-Mann and from Nishijima, who both introduced a new quantum number.[25] Gell-Mann called it *strangeness* and the name stuck. We shall describe the assignment of this new additive quantum number by using well-established hadronic reactions.[26]

We begin by assigning strangeness $S = 0$ to nucleons and pions, and note that strangeness is not defined for leptons. Strangeness is assumed to be a conserved quantity in all interactions that are not weak:

$$\sum_i S_i = \text{const. in hadronic and electromagnetic interactions.} \tag{7.42}$$

We have introduced here the first example of a "broken" symmetry: S is assumed to be conserved in hadronic and electromagnetic interactions but violated in weak ones. With such a quantum number, the mystery of copious production and slow decay can be explained easily. Consider the production reaction $p\pi^- \to \Lambda^0 K^0$ and assign a strangeness $S = 1$ to K^0. The total strangeness on both sides of the reaction must be zero, since only nonstrange particles are present initially. The Λ^0 consequently must have strangeness -1 and Pais' rule is explained: In reactions involving only nonstrange particles in the initial state, strange particles must be produced in pairs. Moreover, a single strange particle cannot decay hadronically or electromagnetically to a state involving only nonstrange particles; such decays must proceed by the weak interaction, and they are therefore slow. Thus the observed long lifetime of the strange particles is also explained.

The assignment of strangeness flavor to the various hadrons is based on reactions that are observed to proceed hadronically. *By definition*, the strangeness of the positive kaon is set equal to 1:

$$S(K^+) = 1. \tag{7.43}$$

The reaction

$$p\pi^- \longrightarrow nK^+K^- \tag{7.44}$$

is observed to proceed with a cross section characteristic of hadronic interactions, and it therefore yields

$$S(K^-) = -1. \tag{7.45}$$

[24] A. Pais, *Phys. Rev.* **86**, 663 (1952).

[25] M. Gell-Mann, *Phys. Rev.* **92**, 833 (1953); T. Nakano and K. Nishijima, *Prog. Theor. Phys.* **10**, 581 (1953).

[26] The assignment is much easier now than in 1952 or 1953. An enormous number of reactions are known now, whereas Pais, Gell-Mann, and Nishijima had to work with very few clues and had to make imaginative guesses.

Positive and negative kaons have opposite strangeness, and we assume, with Eq. (5.64), that they form a particle–antiparticle pair.

Next we turn to the stable baryons, like the proton and neutron[27]. We first see that all have $A = 1$, and therefore they are all particles. The corresponding set of antiparticles also exists, and the strangeness quantum numbers for the antiparticles are opposite to the ones of the particles that we are about to find.

The two charged kaons are excellent tools for establishing values of S. Consider first the reaction

$$p\pi^- \longrightarrow XK. \tag{7.46}$$

The initial state contains only nonstrange particles, and the observation of reaction 7.46 consequently gives $S(X) = -S(K)$. The hyperon X has $S = -1$ if the kaon is positive and $S = +1$ if the kaon is negative. At modern accelerators, separated kaon beams are available, and reactions of the type

$$pK^- \begin{array}{l} \nearrow X\pi \\ \searrow X'K^+ \end{array} \tag{7.47}$$

or the corresponding ones with positive kaons can also readily be observed. In the first of the reactions (Eq. (7.47)), $S(X) = S(K^-) = -1$ and in the second $S(X') = -2$. Reactions 7.46 and 7.47 are only two prototypes; far more involved processes occur and serve to find S.

As an example of reaction 7.46, the process

$$p\pi^- \longrightarrow \Sigma^- K^+$$

assigns $S = -1$ to the negative sigma. An example of Eq. (7.47) is

$$pK^- \longrightarrow \Sigma^+ \pi^-,$$

which gives $S(\Sigma^+) = -1$. Σ^- and Σ^+ are both baryons with $A = 1$; they have the same strangeness but opposite charge. This fact does not contradict Eq. (5.64), which demands only that antiparticles have opposite charge but does not state that a pair with opposite charges has to be a particle–antiparticle pair.

The reactions

$$pp \longrightarrow p\Sigma^0 K^+ \quad \text{and} \quad pK^- \longrightarrow \Lambda^0 \pi^0$$

assign strangeness -1 to Λ^0 and Σ^0. The reaction

$$pK^- \longrightarrow \Xi^- K^+$$

yields $S = -2$ for Ξ^-. Similarly, the strangeness of Ω^- is found to be -3, and the strangeness of $\overline{\Omega}^-$ follows from Eq. (5.64) as $+3$.

Now we return to the *kaons*. Reaction (5.54),

$$p\pi^- \longrightarrow \Lambda^0 K^0,$$

[27] See PDG for a complete list.

determines the strangeness of the K^0 as positive. This assignment raises a question. We have

$$\begin{array}{ll} S(K^+) = 1, & S(K^-) = -1 \\ S(K^0) = 1, & ? \end{array}$$

Something is missing: We have two kaons with $S = 1$ and only one with $S = -1$. Gell-Mann therefore suggested that K^0 should also have an antiparticle, $\overline{K^0}$, with $S = -1$. This antiparticle was found; it can, for instance, be produced in the reaction.

$$p\pi^+ \longrightarrow pK^+\overline{K^0}.$$

The existence of the two neutral kaons, different only in their strangeness but in no other quantum number, gives rise to truly beautiful quantum mechanical interference effects; they will be discussed in Chapter 9. These effects are the subatomic analog to the inversion spectrum of ammonia.

For some discussions it has become customary to use the hypercharge Y rather than strangeness for ordinary and strange particles; the *hypercharge* Y is defined by

$$Y = A + S. \tag{7.48}$$

In Table 7.1 we list the values of baryon number, strangeness, and hypercharge for some hadrons. In the last column we give the average value of the charge number of the particles listed in the relevant row. This quantity will be used later.

Table 7.1 provides considerable food for thought, and a few remarkable facts stand out. Some of these we shall be able to explain later. First we note that the number of particles in each row varies. There are three pions, two kaons, two nucleons, one lambda, and so forth. Why? We shall give an explanation in Chapter 8. Second, we remark that all antiparticles exist and have been found. In some cases the set of antiparticles is identical to the set of particles. When can this happen? Equation (5.64) states that a particle can be identical to its antiparticle only if all additive quantum numbers vanish. The only particles in Table 7.1 satisfying this condition are the photon and the neutral pion. The pion set is identical to its own antiset, and the positive pion is the antiparticle of the negative one. All other entries in Table 7.1 are different from their antiparticles. Third, we note that for physical particles

$$Y = 2\langle N_q \rangle = 2\left\langle \frac{q}{e} \right\rangle, \tag{7.49}$$

and this relation will be used later.

7.6 Additive Quantum Numbers of Quarks

The additive quantum numbers listed in Table 7.1 are not complete; additional ones have been discovered. Before discussing the newer ones, we change the basic style of assignments. Up to now we have discussed the quantum numbers of the observed

Table 7.1: BARYON NUMBER A, STRANGENESS S, HYPERCHARGE Y, AND AVERAGE VALUE OF THE CHARGE NUMBER $N_q = q/e$.

Particle		A	S	Y	$\langle N_q \rangle$
Photon	γ	0	0	0	0
Pion	$\pi^+\pi^0\pi^-$	0	0	0	0
Kaon	K^+K^0	0	1	1	$\frac{1}{2}$
Nucleon	pn	1	0	1	$\frac{1}{2}$
Lambda	Λ^0	1	-1	0	0
Sigma	$\Sigma^+\Sigma^0\Sigma^-$	1	-1	0	0
Cascade	$\Xi^-\Xi^0$	1	-2	-1	$-\frac{1}{2}$
Omega	Ω^-	1	-3	-2	-1

particles, baryons and mesons. The principles become much more transparent, however, if we assign additive quantum numbers to the quarks, which are the counterparts to the leptons. Recall that a baryon is composed of three quarks, (qqq), a meson of a quark and antiquark, $(q\overline{q})$. Each quark has a specific individual additive flavor quantum number, which distinguishes it from the others and is conserved in hadronic and electromagnetic interactions. By assigning additive quantum numbers to each quark, we easily find the quantum numbers of any hadron as the sum of those of its component quarks. In order to agree with the values assigned by early experiments, it is necessary to assign strangeness -1 to the s quark. Then the K^+, composed of $(u\overline{s})$, has the assigned strangeness of $+1$; the Λ^0, composed of (uds), has the desired strangeness -1; values of S for other hadrons are readily obtained. These assignments also explain why baryons can have strangeness S ranging from 0 to -3, with the Ω^- being composed of all s quarks (sss), whereas mesons only can have strangeness $S = 0$, and ± 1. The additive quantum number S, connected to the quark s and the antiquark $\overline{s}$, can appear in a covert or overt way: $(s\overline{s})$ contains two strange objects, a strange quark and strange antiquark, but appears to the outside as nonstrange. On the other hand, $(u\overline{s})$ contains one strange object, and exhibits strangeness explicitly.

By 1964, three quarks had been introduced, but four leptons were known. Suggestions for the existence of a fourth quark were made, for instance, by Bjorken and Glashow,[28] who described the hypothetical quark by the additive quantum number "charm." In 1970, Glashow, Iliopoulos, and Maiani[29] introduced a model that included the fourth quark, charm, showed quark–lepton symmetry, and explained one unsolved problem, the strong suppression or absence of decays like $K^0 \to \mu^+\mu^-$ and $K^\pm \to \pi^\pm e^+ e^-$ (see Section11.4). The major breakthrough occurred with the "November revolution" in 1974. Ting and his group at Brookhaven[30] and Richter

[28] J. D. Bjorken and S. L. Glashow, *Phys. Lett.* **11**, 255 (1964).
[29] S. L. Glashow, J. Iliopoulos, and L. Maiani, *Phys. Rev.* **D2**, 1285 (1970).
[30] J. J. Aubert et al., *Phys. Rev. Lett.* **33**, 1404 (1974).

Table 7.2: QUANTUM NUMBER ASSIGNMENTS FOR THE SIX QUARKS.

	Quantum Number					
Quark	A	S	C	B	T	Y_{gen}
d	1/3	0	0	0	0	1/3
u	1/3	0	0	0	0	1/3
s	1/3	−1	0	0	0	−2/3
c	1/3	0	1	0	0	4/3
b	1/3	0	0	−1	0	−2/3
t	1/3	0	0	0	1	4/3

and his collaborators at SLAC[31] simultaneously discovered a new particle, J/ψ. The long lifetime, the decay characteristics, and the excited states of this particle proved that it was the bound state $(c\bar{c})$. We will return to the J/ψ in Section10.9.

Here we use only one result of these experiments, namely the existence of the new additive quantum number C. With four leptons and four quarks, lepton–quark symmetry is satisfied, and nature might have stopped here. However, more particles with new additive quantum numbers were discovered. In Section 5.6, we briefly described the heaviest known lepton, the tau. If lepton–quark symmetry holds, and there are sound theoretical reasons for this symmetry, the tau and its neutrino call for two more quarks called bottom and top with associated quantum numbers B and T. Indeed, in 1977, Lederman and his collaborators found a new particle which they called upsilon (Υ).[32] The experimental evidence implies that the upsilon is a $(b\bar{b})$ bound state; we will return to it in Section 10.9. The particle $(t\bar{t})$ has also been found, and we list some of the quantum numbers of all six quarks in Table 7.2.

With the new additive quantum numbers C, B, and T, a generalized hypercharge can be introduced and Eqs. (7.48) and (7.49) become

$$Y_{\text{gen}} = A + S + C + B + T = 2\langle q/e\rangle. \tag{7.50}$$

7.7 References

A guide to the literature on new particles and reprints of many papers quoted in the present chapter can be found in J.L. Rosner, *New Particles*, A.A.P.T., Stony Brook, New York, 1981. It is based on "Resource Letter NP-1", *Am. J. Phys.* **48**, 290 (1980). A further guide is *Quarks*, (O.W. Greenberg, ed.) A.A.P.T., Stony Brook, New York, 1986 based on "Resource Letter Q-1" *Am. J. Phys.* **50**, 1074 (1982).

[31] J. E. Augustin et al., *Phys. Rev. Lett.* **33**, 1406 (1974).
[32] S.W. Herb et al., *Phys. Rev. Lett.* **39**, 252 (1977); L.M. Lederman, *Sci. Amer.* **239**, 72 (October 1978).

There are also beginners' books: H. Fritzsch, *Quarks*, Penguin Books, London, 1983;; F Close,M. Martin, and C. Sutton, *The Particle Odyssey: A Journey to the Heart of Matter*, Oxford University Press, New York, 2002; S Weinberg *The Discovery of Subatomic Particle*, Scientific Amer. Books, New York, 1983. See also, J.A. Appelquist, *Ann. Rev. Part. Nucl. Sci*,**42**, 367 (1992) and R. Cester and P.A. Rapidis, loc.cit. **44**, 329 (1994).

Charm and related aspects are discussed in S. D. Drell, *Sci. Amer.* **232**, 50 (June 1975); S. L. Glashow, *Sci. Amer.* **233**, 38 (October 1975); S. C. Ting, *Science* **196**, 1167 (1977), *Rev. Mod. Phys.* **49**, 235 (1977); B. Richter, *Science* **196** , 1286 (1977); see also R.M. Barnett, H. Müehry, and H.R. Quin, *The Charm of Strange Quarks: Mysteries and Revolutions of Particle Physics*, AIP Press, Springer, New York, 2000.

Symmetries and invariance principles are the subject of the following books: J.J. Sakurai, *Invariance Principles and Elementary Particles*, Princeton University Press, Princeton, N.J., 1964. Easier books are H.R.T. Pagels, *Perfect Symmetry*, Simon and Schuster, New York, 1985; A. Zee, *Fearful Symmetry*, Macmillan, New York, 1986; F. Close, *Lucifer's Legacy, The Meaning of Asymmetry*, Oxford University Press, New York, 2000.; despite its title the author also discusses symmetry; L.M. Lederman and C. Hill, *Symmetry and the Beautiful Universe*, Prometheus Books, Amherst, NY, 2004. The limits set on the various conservation laws as of 1959 are treated in G. Feinberg and M. Goldhaber, *Proc. Natl. Acad. Sci. U.S.* **45**, 1301 (1959). Although old this article is clear and interesting to read.

Problems

7.1. Show that the reality of the expectation value $\langle F \rangle$ demands that the operator F be Hermitian.

7.2. Discuss more carefully and in more detail than in the text

(a) Quantum mechanical operators and matrices associated with these operators. How is a matrix associated with an observable F and a transformation operator U?

(b) How is Hermiticity defined for operators and for the corresponding matrices?

(c) How is unitarity defined for operators and for matrices?

7.3. Discuss the evidence for conservation of the electric charge and the electric current in macroscopic systems (classical electrodynamics).

7.4. Devise an experiment that would measure a possible neutron charge. Use realistic values of neutron flux, neutron velocity, electric field strength, and spatial resolution of neutron counters to obtain an estimate on the limit that could be obtained.

7.5. Assume that nucleons decay with a lifetime of 10^{15}y and that all the energy of the nucleons decaying in the earth is transformed into heat. Compute the heat flow at the surface of the earth. Compare the energy produced with the energy that the earth receives from the sun during the same time.

7.6. * Sketch an experimental arrangement for measuring the lifetime of protons and explain its basic functioning. [See e.g., C. McGrew et al., *Phys. Rev. D* **59**, 052004 (1999) or K. Kobayashi et al., *Phys. Rev. D* **72**, 052007 (2005).]

7.7. The cross section for the absorption of antineutrinos with energies as emitted by nuclear reactors is about 10^{-43}cm^2.

(a) Compute the thickness of a water absorber needed to reduce the intensity of an antineutrino beam by a factor of 2.

(b) Consider a liquid scintillator with a volume of 10^3 liters and an antineutrino beam with an intensity of $10^{13}\overline{\nu}/\text{cm}^2\text{sec}$. How many capture events [Eq. (7.34)] are expected per day?

(c) How can the antineutrino capture be distinguished from other reactions?

7.8. * How can the reaction of Eq. (7.37) be observed? [See e.g., R. Davis, Jr. *Rev. Mod. Phys.* **75**, 985(2003).]

7.9. Suppose we assign an additive quantum number to a pion: +1 for π^+, 0 for π^0, and −1 for π^-. What are the simplest reactions which allow the production of pions by photons on protons? What are they for π^- on protons? For π^+ on protons?

7.10. Can the following reactions occur? If so, do they proceed via strong, electromagnetic, or weak interactions? Give reasons.

(a) $\Sigma^+ p \to pp\pi^0$

(b) $p\overline{p} \to \lambda^0\overline{\Sigma}^0$

(c) $n\overline{p} \to \Sigma^-\Sigma^0$

(d) $p\overline{p} \to \Xi^- p$

(e) $e\overline{p} \to \Sigma^-\overline{p}n\pi^0\nu$

7.11. Can strange particles be produced singly by reactions that involve only non-strange particles? If yes, give a possible reaction.

7.12. Follow the production and decay of $\overline{\Omega}$ in Figs. 5.26 and 5.27 and verify that the additive quantum numbers A and q are conserved in every interaction. Where is S conserved and where not?

7.13. *Discuss the reaction(s) that allows the assignment $S = -3$ to Ω^- and $S = +3$ to $\overline{\Omega^-}$.

7.14. Which of the following reactions can take place? If forbidden, state by what selection rule. If allowed, indicate through which interaction the reaction will proceed.

(a) $p\overline{p} \to \pi^+\pi^-\pi^0\pi^+\pi^-$.

(b) $pK^- \to \Sigma^+\pi^-\pi^+\pi^-\pi^0$.

(c) $p\pi^- \to pK^-$.

(d) $p\pi^- \to \Lambda^0\overline{\Sigma^0}$.

(e) $\overline{\nu}_\mu p \to \mu^+ n$.

(f) $\overline{\nu}_\mu p \to e^+ n$.

(g) $\nu_e p \to e^+\Lambda^0 K^0$.

(h) $\nu_e p \to e^-\Sigma^+ K^+$.

7.15. Estimate the lifetime of the proton if it decayed through gravitational forces.

7.16. * Sketch the experiment of Ting and collaborators that led to the discovery of the J/ψ.

7.17. (a) Assume fermion number conservation, but not separate lepton and baryon number conservation. List some of the possible decay modes of a proton into a lepton and other particles. What is the minimum number of other particles required? Why?

(b) List some decays of the proton that do not conserve B and L separately but conserve B+L; repeat for B-L.

(c) Repeat a) for decays into antileptons plus other particles.

Chapter 8

Angular Momentum and Isospin

In this chapter we shall show that invariance under rotation in space leads to conservation of angular momentum. We shall then introduce isospin, a quantity that has many properties similar to ordinary spin, and discuss the "breaking" of isospin invariance.

8.1 Invariance Under Spatial Rotation

Invariance under spatial rotation provides an important application of the general considerations presented in Section 7.1.

Consider an idealized experimental arrangement, shown in Fig. 8.1. We assume for simplicity that the equipment is in the xy plane; its orientation is described by the angle φ. We further assume that the result of the experiment is described by a wave function $\psi(\boldsymbol{x})$. Next, the equipment is rotated by an angle α about the z axis. This rotation is denoted by $R_z(\alpha)$, and it carries a point $\boldsymbol{x}$ into a point $\boldsymbol{x}^R$:

$$\boldsymbol{x}^R = R_z(\alpha)\boldsymbol{x}. \tag{8.1}$$

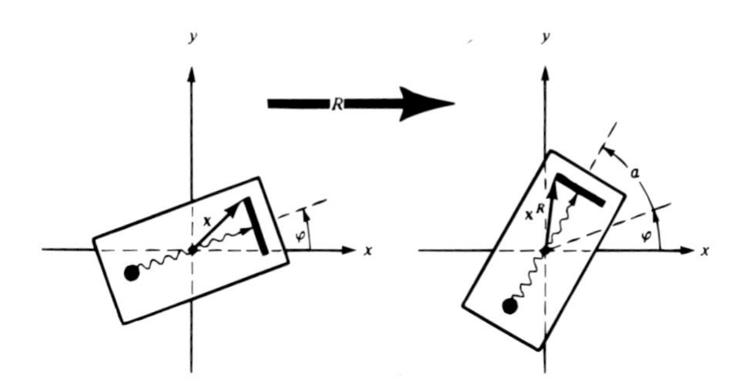

Figure 8.1: Rotation around the z axis. The angle φ fixes the position of the original equipment axis; it does not denote a rotation. The equipment is rotated about the z axis by an angle α. Invariance under rotation means that the outcome of the experiment is not affected by the rotation.

The rotation changes the wave function; the relation between the rotated and unrotated wave function at point $\boldsymbol{x}$ is given by Eq. (7.7) as

$$\psi^R(\boldsymbol{x}) = U_z(\alpha)\psi(\boldsymbol{x}). \tag{8.2}$$

The notation indicates that the rotation is by an angle α about the z axis. So far, no invariance properties have been used and Eqs. (8.1) and (8.2) are valid even if the system changes during rotation.

Invariance arguments can now be used to find U. If the state of the system is unaffected by rotation, the wave function at point $\boldsymbol{x}$ in the original system is identical to the rotated wave function at the rotated point $\boldsymbol{x}^R$,

$$\psi(\boldsymbol{x}) = \psi^R(\boldsymbol{x}^R). \tag{8.3}$$

Note the difference between Eqs. (8.2) and (8.3). The first connects $\psi(\boldsymbol{x})$ to ψ^R at the same point, and the second to ψ^R at the rotated point $\boldsymbol{x}^R$. U can be found if $\psi^R(\boldsymbol{x}^R)$ can be expressed in terms of $\psi^R(\boldsymbol{x})$. Because the rotation is continuous, any rotation by a finite angle can be built up from rotations by infinitely small angles. An *infinitesimal rotation* suffices to find U. If the system is rotated by an infinitesimal angle $\delta\alpha$ about the z axis, $\psi^R(\boldsymbol{x}^R)$ becomes

$$\psi^R(\boldsymbol{x}^R) = \psi^R(\boldsymbol{x}) + \frac{\partial \psi^R(\boldsymbol{x})}{\partial \varphi}\delta\alpha = \left(1 + \delta\alpha\frac{\partial}{\partial\varphi}\right)\psi^R(\boldsymbol{x}).$$

This relation can be inverted by multiplication by $[1 - \delta\alpha(\partial/\partial\varphi)]$. Neglecting terms in $\delta\alpha^2$ and using Eq. (8.3) then yields

$$\psi^R(\boldsymbol{x}) = \left(1 - \delta\alpha\frac{\partial}{\partial\varphi}\right)\psi(\boldsymbol{x}). \tag{8.4}$$

Comparison with Eq. (8.2) shows that the operator in front of $\psi(\boldsymbol{x})$ is $U_z(\delta\alpha)$. The general expression for the operator for an infinitesimal unitary transformation is given by Eq. (7.14). Identifying ϵ with $\delta\alpha$ and comparing the two expressions for U yields the desired Hermitian operator F,[1]

$$F = i\frac{\partial}{\partial\varphi}. \tag{8.5}$$

If U commutes with H, so will F, according to Eq. (7.15), and we have found the desired conserved observable. We could start exploring the physical consequences

[1]Some confusion can arise because formally $F^\dagger = -i\partial/\partial\varphi$ looks different from F. However, Hermiticity is not a property of an operator alone but also of the wave functions and the region of integration. For a Hermitian operator, with $F^\dagger = F$, the equation in footnote 3 in Chapter 7 reads

$$\int d^3x (F\psi)^*\phi = \int d^3x \psi^* F\phi.$$

$F = i\partial/\partial\varphi$ satisfies this relation:

$$\int d^3x \left(i\frac{\partial\psi}{\partial\varphi}\right)^*\phi = \int d^3x \left(-i\frac{\partial}{\partial\varphi}\right)\psi^*\phi = \int d^3x \psi^* i\frac{\partial\phi}{\partial\varphi}.$$

In the last step, a partial integration has brought the operator to the right of ψ^*. The explicit form of a Hermitian operator depends on its position with respect to the wave functions.

of F and find the eigenfunctions and eigenvalues. This procedure is not necessary because F is an old friend. Equation (5.3) shows that

$$F = -\frac{L_z}{\hbar}. \tag{8.6}$$

Not unexpectedly, F is proportional to the z component of the orbital angular momentum. Invariance of a system under rotation around the z axis leads to conservation of F and thus also of L_z.

Two generalizations are physically reasonable, and we give them without proof: (1) If the system has a total angular moment $\boldsymbol{J}$ (spin plus orbital), then L_z is replaced by J_z. (2) U for a rotation by an angle δ around the arbitrary direction $\hat{\boldsymbol{n}}$ (where $\hat{\boldsymbol{n}}$ is a unit vector) is

$$U_{\boldsymbol{n}}(\delta) = \exp\left(\frac{-i\delta\hat{\boldsymbol{n}}\cdot\boldsymbol{J}}{\hbar}\right). \tag{8.7}$$

If the system is invariant under rotation about $\hat{\boldsymbol{n}}$, the Hamiltonian will commute with $U_{\boldsymbol{n}}$ and consequently also with $\hat{\boldsymbol{n}}\cdot\boldsymbol{J}$:

$$[H, U_{\boldsymbol{n}}] = 0 \longrightarrow [H, \hat{\boldsymbol{n}}\cdot\boldsymbol{J}] = 0. \tag{8.8}$$

The component of the angular momentum along $\hat{\boldsymbol{n}}$ is conserved. If $\hat{\boldsymbol{n}}$ can be taken to be any direction, all components of $\boldsymbol{J}$ are conserved, and $\boldsymbol{J}$ is a constant of the motion.

With Eq. (8.7) it is straightforward to find the commutation relations for the components of $\boldsymbol{J}$:

$$\begin{aligned}&[J_x, J_y] = i\hbar J_z,\\ &\text{cyclic.}\end{aligned} \tag{5.6}$$

The steps in the derivation are outlined in Problem 8.1 The commutation relations (Eq. (5.6)) are a consequence of the unitary transformation (Eq. (8.7)), which in turn is a consequence of the invariance of H under rotation.

8.2 Symmetry Breaking by a Magnetic Field

A particle with spin $\boldsymbol{J}$ and magnetic moment μ can be described by a Hamiltonian

$$H = H_0 + H_{\text{mag}}, \tag{8.9}$$

where H_{mag} is given in Eq. (5.20). Usually, H_0 is isotropic, and the system described by H_0 is invariant under rotations about any direction. This fact is expressed by

$$[H_0, \boldsymbol{J}] = 0. \tag{8.10}$$

The energy of the particle is independent of its orientation in space. If a magnetic field is switched on, the symmetry is broken, and Eq. (8.10) no longer holds:

$$[H, \boldsymbol{J}] = [H_0 + H_{\text{mag}}, \boldsymbol{J}] \neq 0. \tag{8.11}$$

(If needed, the commutator can be calculated with Eqs. (5.20) and (5.6).) The component of the angular momentum along the field, however, still remains conserved. It is customary to select the quantization axis z along the magnetic field. Equations (5.6) and (5.20) then give

$$[H_0 + H_{\text{mag}}, J_z] = 0. \tag{8.12}$$

The system is still invariant under rotations about the direction of the externally applied field, namely the z axis. However, the introduction of a preferred direction through the application of the magnetic field has broken the overall symmetry, and $\boldsymbol{J}$ is no longer conserved. Before the application of the field, the energy levels of the system were $(2J+1)$-fold degenerate, as shown on the left-hand side of Fig. 5.5. The introduction of the field results in a removal of the degeneracy, and the corresponding Zeeman splitting is shown in Fig. 5.5.

8.3 Charge Independence of Hadronic Forces

In 1932, when the neutron was discovered, the nature of the forces holding nuclei together was still mysterious. By about 1936, crucial features of the nuclear force had emerged.[2] Particularly revealing was the analysis of *pp* and *np* scattering data. Of course, at that time, such scattering experiments could be performed only at very low energies, but the outcome was still surprising: After subtracting the effect of the Coulomb force in *pp* scattering, it was found that the *pp* and the *np* hadronic force were of about equal strength and had about equal range.[3] This result was corroborated by studies of the masses of ^{3}H and ^{3}He which gave approximately equal values for the *pp*, *np*, and *nn* interactions. Strong evidence for a *charge independence* of the nuclear forces was also found by Feenberg and Wigner.[4] Charge independence for nuclear forces can be formulated by stating that the forces between any two nucleons in the same state are the same, apart from electromagnetic effects. Today, the experimental evidence for charge independence is very strong, and it is known that all hadronic forces, not just the one between nucleons, are charge-independent.[5] We shall not discuss the experimental evidence for charge independence here but only point out that the concept of isospin, which will be discussed in the following sections, is a direct consequence of the charge independence of hadronic forces.

---

[2] In 1936 and 1937, Bethe and collaborators surveyed the state of the art in a series of three articles, later known as the *Bethe bible*. These admirable reviews in *Rev. Mod Phys.* **8**, 82 (1936), **9**, 69 (1937), and **9**, 245 (1937), reprinted in *Basic Bethe, Am. Inst. Phys., New York*, 1986, can still be read with profit.

[3] G. Breit, E.U. Condon, and R.D. Present, *Phys. Rev.* **50**, 825 (1936).

[4] E. Feenberg and E.P. Wigner, *Phys. Rev.* **51**, 95 (1937).

[5] The evidence for charge independence of the hadronic forces is discussed by G.A. Miller and W.T.H. van Oers in *Symmetries and Fundamental Interactions*, ed. W.C. Haxton and E.M. Henley, World Sci., Singapore (1995), p. 127.

8.4 The Nucleon Isospin

Charge independence of nuclear forces leads to the introduction of a new conserved quantum number, isospin. As early as 1932, Heisenberg treated the neutron and the proton as two states of one particle, the nucleon N.[6] The two states presumably have the same mass, but the electromagnetic interaction makes the masses slightly different. (The mass difference of the u and d quarks also contributes, but we neglect this effect here and throughout this chapter.)

To describe the two states of the nucleon, an isospin space (internal charge space) is introduced, and the following analogy to the two spin states of a spin-$\frac{1}{2}$ particle is made:

	Spin-$\frac{1}{2}$ Particle in Ordinary Space	Nucleon in Isospin Space
Orientation	Up	Up, proton
	Down	Down, neutron

The two states of an ordinary spin-$\frac{1}{2}$ particle are not treated as two particles but as two states of one particle. Similarly, the proton and the neutron are considered as the *up* and the *down* state of the nucleon. Formally, the situation is described by introducing a new quantity, *isospin* $\vec{I}$.[7] The nucleon with isospin $\frac{1}{2}$ has $2I+1=2$ possible orientations in isospin space. The three components of the isospin vector $\vec{I}$ are denoted by I_1, I_2, and I_3. The value of I_3 distinguishes, *by definition*, between the proton and the neutron. $I_3 = +\frac{1}{2}$ is the proton and $I_3 = -\frac{1}{2}$ is the neutron.[8] The most convenient way to write the value of I and I_3 for a given state is by using a Dirac ket:

$$|I, I_3\rangle.$$

Then proton and neutron are

$$\text{proton } |\tfrac{1}{2}, \tfrac{1}{2}\rangle, \qquad \text{neutron } |\tfrac{1}{2}, -\tfrac{1}{2}\rangle. \tag{8.13}$$

The charge for the particle $|I, I_3\rangle$ is given by

$$q = e(I_3 + \tfrac{1}{2}). \tag{8.14}$$

With the values of the third component of I_3 given in Eq. (8.13), the proton has charge e, and the neutron charge 0.

[6] W. Heisenberg, *Z. Physik* **77**, 1 (1932). [Translated in D. M. Brink, *Nuclear Forces*, Pergamon, Elmsford, N.Y., 1965].

[7] To distinguish spin and isospin, we write isospin vectors with an arrow.

[8] In nuclear physics, isospin is sometimes called isobaric spin; it is often denoted by T, and the neutron is taken to have $I_3 = \frac{1}{2}$ and the proton $I_3 = -\frac{1}{2}$, because there are more neutrons than protons in stable nuclei and $I_3(T_3)$ is then positive for these cases.

8.5 Isospin Invariance

What have we gained with the introduction of isospin? So far, very little. Formally, the neutron and the proton can be described as two states of one particle. New aspects and new results appear when charge independence is introduced and when isospin is generalized to all hadrons.

Charge independence states that the hadronic forces do not distinguish between the proton and the neutron. As long as only the hadronic interaction is present, the isospin vector $\vec{I}$ can point in any direction. In other words, there exists rotational invariance in isospin space; the system is invariant under rotations about any direction. As in Eq. (8.10), this fact is expressed by

$$[H_h, \vec{I}] = 0. \tag{8.15}$$

With only H_h present, the $2I + 1$ states with different values of I_3 are degenerate; they have the same energy (mass). Said simply, with only the hadronic interaction present, neutron and proton would have the same mass. The electromagnetic interaction (and the up–down quark mass difference) destroy the isotropy of isospin space; it breaks the symmetry, and, as in Eq. (8.11), it gives

$$[H_h + H_{em}, \vec{I}] \neq 0. \tag{8.16}$$

However, we know from Section 7.1 that the electric charge is always conserved, even in the presence of H_{em}:

$$[H_h + H_{em}, Q] = 0. \tag{8.17}$$

Q is the operator corresponding to the electric charge q; it is connected to I_3 by Eq. (8.14): $Q = e(I_3 + \frac{1}{2})$. Introducing Q into the commutator, Eq. (8.17), gives

$$[H_h + H_{em}, I_3] = 0. \tag{8.18}$$

The third component of isospin is conserved even in the presence of the electromagnetic interaction. The analogy to the magnetic field case is evident; Eq. (8.18) is the isospin equivalent of Eq. (8.12).

It was pointed out in Section 8.4 that charge independence holds not only for nucleons but for all hadrons. Before generalizing the isospin concept to all hadrons and exploring the consequences of such an assumption, a few preliminary remarks are in order concerning isospin space. We stress that $\vec{I}$ is a vector in isospin space, not in ordinary space. The direction in isospin space has nothing to do with any direction in ordinary space, and the value of the operator $\vec{I}$ or I_3 in isospin space has nothing to do with ordinary space. So far, we have related only the third component of $\vec{I}$ to an observable, the electric charge q (Eq. (8.14)). What is the physical significance of I_1 and I_2? These two quantities cannot be connected directly to a physically measurable quantity. The reason is nature: in the laboratory, two magnetic fields can be set up. The first can point in the z direction, and the second in the x direction. The effect of such a combination on the spin of the particle can be computed, and the measurement along any direction is meaningful (within the limits of the uncertainty relations). The electromagnetic field in the isospin space, however, cannot be switched on and off. The charge is always related to one component of $\vec{I}$, and this component is traditionally taken to be I_3. Renaming the components and connecting the charge, for instance, to I_2 does not change the situation.

We now assume the general existence of an isospin space, with its third component connected to the charge of the particle by a linear relation of the form

$$q = aI_3 + b. \tag{8.19}$$

With such a relationship, conservation of the electric charge implies conservation of I_3. I_3 is therefore a good quantum number, even in the presence of the electromagnetic interaction. The unitary operator for a rotation in isospin space by an angle ω about the direction $\hat{\alpha}$ is

$$U_{\hat{\alpha}}(\omega) = \exp(-i\omega\hat{\alpha} \cdot \vec{I}\,), \tag{8.20}$$

where $\vec{I}$ is the Hermitian generator associated with the unitary operator U, and we expect $\vec{I}$ to be an observable. As in the case of the angular momentum operator $\boldsymbol{J}$, the arguments follow the general steps outlined in Section 7.1. To study the physical properties of $\vec{I}$, we assume first that only the hadronic interaction is present. Then the electric charge is zero for all systems, and Eq. (8.19) does not determine the direction of I_3. Charge independence thus implies that a hadronic system without electromagnetic interaction is invariant under any rotation in isospin space. We know from Section 7.1, Eq. (7.9), that U then commutes with H_h:

$$[H_h, U_{\hat{\alpha}}(\omega)] = 0. \tag{8.21}$$

As in Eq. (7.15), conservation of isospin follows immediately,

$$[H_h, \vec{I}\,] = 0.$$

Charge independence of the hadronic forces leads to conservation of isospin.

In the case of the ordinary angular momentum, the commutation relations for $\boldsymbol{J}$ follow from the unitary operator (8.7) by straightforward algebraic steps. No further assumptions are involved. The same argument can be applied to $U_{\hat{\alpha}}(\omega)$, and the three components of the isospin vector must satisfy the commutation relations

$$[I_1, I_2] = iI_3, \quad [I_2, I_3] = iI_1, \quad [I_3, I_1] = iI_2. \tag{8.22}$$

The eigenvalues and eigenfunctions of the isospin operators do not have to be computed because they are analogous to the corresponding quantities for ordinary spin.

The steps from Eq. (5.6) to Eqs. (5.7) and (5.8) are independent of the physical interpretation of the operators. All results for ordinary angular momentum can be taken over. In particular, I^2 and I_3 obey the eigenvalue equations

$$I_{\text{op}}^2|I, I_3\rangle = I(I+1)|I, I_3\rangle \tag{8.23}$$

$$I_{3,\text{op}}|I, I_3\rangle = I_3|I, I_3\rangle. \tag{8.24}$$

Here I_{op}^2 and $I_{3,\text{op}}$ on the left-hand side are operators, and I and I_3 on the right-hand side are quantum numbers. The symbol $|I, I_3\rangle$ denotes the eigenfunction ψ_{I,I_3}. (In a situation where no confusion can arise, the subscripts "op" will be omitted.) The allowed values of I are the same as for J, Eq. (5.9), and they are

$$I = 0, \tfrac{1}{2}, 1, \tfrac{3}{2}, 2, \ldots. \tag{8.25}$$

For each value of I, I_3 can assume the $2I+1$ values from $-I$ to I.

In the following sections, the results expressed by Eqs. (8.22)–(8.25) will be applied to nuclei and to particles. It will turn out that isospin is essential for understanding and classifying subatomic particles.

• We have noted above that the components I_1 and I_2 are not directly connected to observables. However, the linear combinations

$$I_\pm = I_1 \pm iI_2 \tag{8.26}$$

have a physical meaning. Applied to a state $|I, I_3\rangle, I_+$ raises and I_- lowers the value of I_3 by one unit:

$$I_\pm|I, I_3\rangle = [(I \mp I_3)(I \pm I_3 + 1)]^{1/2}|I, I_3 \pm 1\rangle. \tag{8.27}$$

Equation (8.27) can be derived with the help of Eqs. (8.22) to (8.24).[9] •

8.6 Isospin of Particles

The isospin concept was first applied to nuclei, but it is easier to see its salient features in connection with particles. As stated in the previous section, isospin is presumably a good quantum number as long as only the hadronic interaction is present. The electromagnetic interaction destroys the isotropy of isospin space, just as a magnetic field destroys the isotropy of ordinary space. Isospin and its manifestations should consequently appear most clearly in situations where the electromagnetic interaction is small. For nuclei, the total electric charge number Z can be as high as 100, whereas for particles it is usually 0 or 1. Isospin should therefore be a better and more easily recognized quantum number in particle physics.

If isospin is an observable that is realized in nature, then Eqs. (8.15) and (8.23)–(8.25) predict the following characteristics: The quantum number I can take on

[9] Merzbacher, Section 16.2; Messiah, Section XIII.I.

the values $0, \frac{1}{2}, 1, \frac{3}{2}, \ldots$. For a given particle, I is an immutable property. In the absence of the electromagnetic interaction, a particle with isospin I is $(2I+1)$-fold degenerate, and the $2I+1$ *subparticles* all have the same mass.

Since H_h and $\vec{I}$ commute, all subparticles have the same hadronic properties and are differentiated only by the value of I_3. The electromagnetic interaction partially or completely lifts the degeneracy, as shown in Fig. 8.2, and it thus gives rise to the isospin analog of the Zeeman effect. The $2I+1$ subparticles belonging to a given state with isospin I are said to form an *isospin multiplet*. The electric charge of each member is related to I_3 by Eq. (8.19). Quantum numbers that are conserved by the electromagnetic interaction are unaffected by the switching on of H_{em}. Since most quantum numbers have this property, the members of an isospin multiplet have very nearly identical properties; they have, for instance, the same spin, baryon number, hypercharge, and intrinsic parity. (Intrinsic parity will be discussed in Section 9.2.)

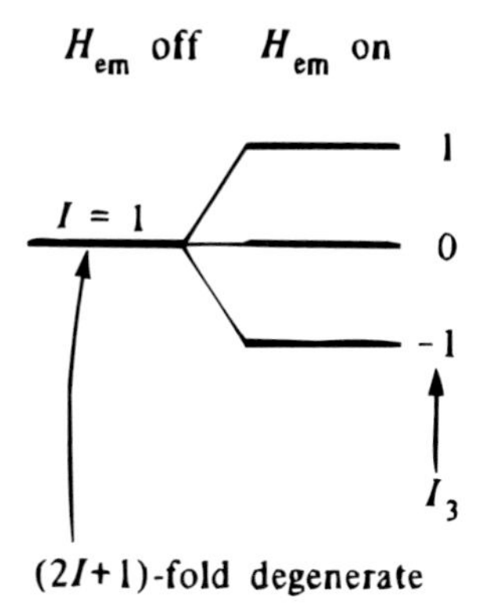

Figure 8.2: A particle with isospin I is $(2I+1)$-fold degenerate in the absence of the electromagnetic interaction. H_{em} lifts the degeneracy, and the resulting subparticles are labeled by I_3.

The different members of an isospin multiplet are in essence the same particle appearing with different orientations in isospin space, just as the various Zeeman levels are states of the same particle with different orientations of its spin with respect to the applied magnetic field. The determination of the quantum number I for a given state is straightforward if all subparticles belonging to the multiplet can be found: Their number is $2I+1$ and thus yields I. Sometimes counting is not possible, and it is then necessary to resort to other approaches, such as the use of selection rules.

The arguments given so far can be applied most easily to the pion. The possible values of the isospin of the pion can be found by looking at Fig. 5.19: If virtual pions are exchanged between nucleons, the basic Yukawa reaction

$$N \longrightarrow N' + \pi$$

should conserve isospin. Nucleons have isospin $\frac{1}{2}$; isospins add vectorially like angular momenta, and the pion consequently must have isospin 0 or 1. If I were 0, only one pion would exist. The assignment $I = 1$, on the other hand, implies the existence of three pions.[10] Indeed, three and only three hadrons with mass of

[10] N. Kemmer, *Proc. Cambridge Phil. Soc.* **34**, 354 (1938).

about 140 MeV/c² are known, and the three form an *isovector* with the assignment

$$I_3 = \begin{cases} +1 & \pi^+, \quad m = 139.569 \text{ MeV}/c^2, \\ 0 & \pi^0, \quad m = 134.964 \text{ MeV}/c^2, \\ -1 & \pi^-, \quad m = 139.569 \text{ MeV}/c^2. \end{cases}$$

The charge is connected to I_3 by the relation

$$q = eI_3, \tag{8.28}$$

which is a special case of Eq. (8.19). The pion shows particularly clearly that the properties in ordinary and in isospin space are not related because it is a vector in isospin space but a scalar (spin 0) in ordinary space.

In the ordinary Zeeman effect, it is easy to demonstrate that the various sublevels are members of one Zeeman multiplet: if the applied magnetic field is reduced to zero, they coalesce into one degenerate level. This method cannot be applied to an isospin multiplet because the electromagnetic interaction cannot be switched off. It is necessary to resort to calculations to show that the observed splitting can be blamed solely on H_{em}. Comparison of the pion and the nucleon shows that the problem is not straight forward: the proton is lighter than the neutron, whereas the charged pions are heavier than the neutral one. Nevertheless, the computations performed up to the present time account for the mass splitting by the electromagnetic interaction and the mass difference between the *up* and *down* quarks.[11]

After having spent considerable time on the isospin of the pion, the other hadrons can be discussed more concisely.

The *kaon* appears in two particle and two antiparticle states. The assignment $I = \frac{1}{2}$ is in agreement with all known facts.

The assignment of I to *hyperons* is also straightforward. It is assumed that hyperons with approximately equal masses form isospin multiplets. The lambda occurs alone, and it is a singlet. The sigma shows three charge states, and it is an isovector. The cascade particle is a doublet, and the omega is a singlet.

The hadrons encountered so far can all be characterized by a set of additive quantum numbers, A, q, Y, and I_3. For pions, charge and I_3 are connected by Eq. (8.28). Gell-Mann and Nishijima showed how this relation can be generalized to apply also to strange particles. They assumed charge and I_3 to be connected by a linear relation as in Eq. (8.19). The constant a in Eq. (8.19) is determined from Eq. (8.28) as e. To find the constant b, we note that I_3 ranges from $-I$ to $+I$. The average charge of a multiplet is therefore equal to b:

$$\langle q \rangle = b.$$

[11] See e.g., A. De Rújula, H. Georgi, and S. L. Glashow, *Phys. Rev.* **D12**, 147 (1975); N. Isgur and G. Karl, *Phys. Rev. D* **20**, 1191 (1979); J. Gaisser and H. Leutwyler, *Phys. Repts.* **87**, 77 (1982); E.M. Henley and G.A. Miller, *Nucl. Phys.* **A518**, 207 (1990).

The average charge of a multiplet has already been determined in Eq. (7.49):

$$\langle q \rangle = \tfrac{1}{2}eY. \tag{8.29}$$

Only particles with zero hypercharge have the center of charge of the multiplet at $q = 0$; for all others, it is displaced. Consequently the generalization of Eqs. (8.14) and (8.28) is

$$q = e(I_3 + \tfrac{1}{2}Y) = e(I_3 + \tfrac{1}{2}A + \tfrac{1}{2}S). \tag{8.30}$$

This equation is called the Gell-Mann–Nishijima relation. If q is considered to be an operator, it can be said that the electric charge operator is composed of an isoscalar $(\frac{1}{2}eY)$ and the third component of an isovector (eI_3). For particles with charm, bottom, or top quantum numbers, Y in Eq. (8.30) is replaced by Y_{gen}, Eq. (7.50).

The Gell-Mann–Nishijima relation can be visualized in a Y versus q/e diagram, shown in Fig. 8.3. A few isospin multiplets are plotted. The multiplets with $Y \neq 0$ are *displaced*: Their center of charge is not at zero but, as expressed by Eq. (8.29), at $\frac{1}{2}eY$.

The considerations in the present section have shown that isospin is a useful quantum number in particle physics. The value of I for a given particle determines the number of subparticles belonging to this particular isospin multiplet. The third component, I_3, is conserved in hadronic and electromagnetic interactions, whereas $\vec{I}$ is conserved only by the hadronic force.

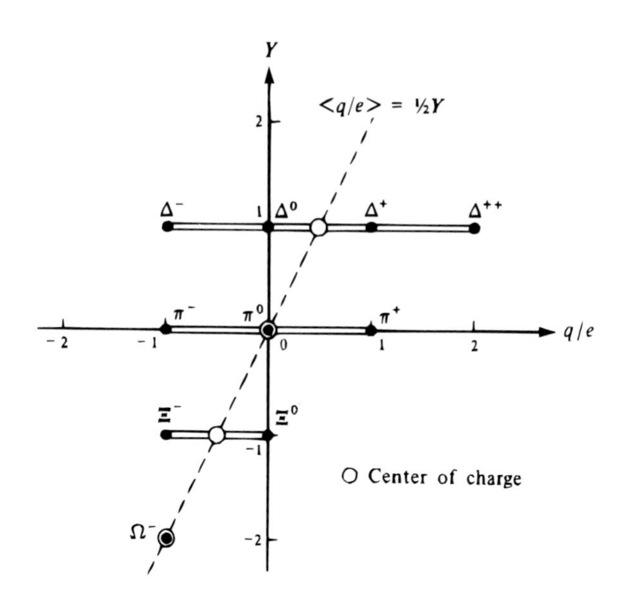

Figure 8.3: Isospin multiplets with $Y \neq 0$ are displaced: Their center of charge (average charge) is at $\frac{1}{2}eY$. A few representative multiplets are shown, but many more exist.

In the following section we shall demonstrate that isospin is also a valuable concept in nuclear physics.

8.7 Isospin in Nuclei

A nucleus with A nucleons, Z protons, and N neutrons, has a total charge Ze. The total charge can be written[12] as a sum over all A nucleons with the help of Eq. (8.14):

$$Ze = \sum_{i=1}^{A} q_i = e(I_3 + \tfrac{1}{2}A), \tag{8.31}$$

where the third component of the total isospin is obtained by summing over all nucleons,

$$I_3 = \sum_{i=1}^{A} I_{3,i}. \tag{8.32}$$

The isospin $\vec{I}$ behaves algebraically like the ordinary spin $\boldsymbol{J}$, and the total isospin of the nucleus A is the sum over the isospins from all nucleons:

$$\vec{I} = \sum_{i=1}^{A} \vec{I}_i. \tag{8.33}$$

Do these equations mean something? All states of a given nuclide are characterized by the same values of A and Z. What are the values of I and I_3? According to Eq. (8.31), all states of a nuclide have the same value of I_3, namely

$$I_3 = Z - \tfrac{1}{2}A = \tfrac{1}{2}(Z - N). \tag{8.34}$$

The assignment of the total isospin quantum number I is not so simple. There are A isospin vectors with $I = \frac{1}{2}$, and, since they add vectorially, they can add up to many different values of I. The maximum value of I is $\frac{1}{2}A$, and it occurs if the contributions from all nucleons are parallel. The minimum value is $|I_3|$, because a vector cannot be smaller than one of its components. I therefore satisfies

$$\tfrac{1}{2}|Z - N| \le I \le \tfrac{1}{2}A. \tag{8.35}$$

Can a value of I be assigned to a given nuclear level, and can it be determined experimentally? To answer these questions, we return to a world where all but the hadronic interactions are switched off, and we consider a nucleus formed from A nucleons. I is a good quantum number in a purely hadronic world, and each state of the nucleus can be characterized by a value of I. Equation (8.35) shows that I is integer if A is even and half-integer if A is odd. The state is $(2I+1)$-fold degenerate.

[12] E. P. Wigner, *Phys. Rev.* **51**, 106, 947 (1937); *Proc. Robert A. Welch Confer. Chem. Res.* **1**, 67 (1958).

If the electromagnetic interaction is switched on, the degeneracy is broken, as indicated in Fig. 8.4. Each of the substates is characterized by a unique value of I_3 and, as shown by Eq. (8.31), appears in a different isobar. As long as the electromagnetic interaction is reasonably small $[(Ze^2/\hbar c) \ll 1]$ it is expected that real nuclear states will behave as described and consequently can be labeled by I.

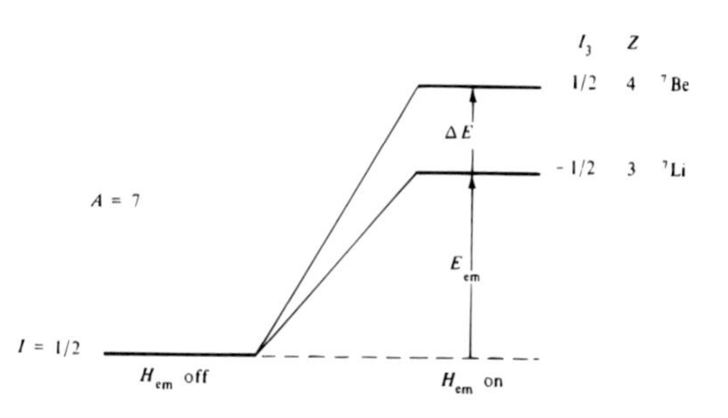

Figure 8.4: Isospin doublet. Without the electromagnetic interaction, the two substates are degenerate. With H_{em} switched on, the degeneracy is lifted, and each sublevel appears in a different isobar. The levels in the real nuclides are said to form an isospin multiplet.

It turns out that I can even be assigned to states in heavy nuclei where this condition is not fulfilled. Such states are called *isobaric analog states*; they were discovered in 1961.[13] Figure 8.4 is the nuclear analog to Fig. 8.2. Both are the isospin analogs of the Zeeman effect shown in Fig. 5.5. In the magnetic (spin) case, the levels are labeled by J and J_z, and in the isospin case by I and I_3. In the magnetic case, the splitting is caused by the magnetic field, and in the isospin case by the Coulomb interaction.

The way to find the value of I is similar to the one used for particles: If all members of an isospin multiplet can be found, their number can be counted; it is $2I + 1$, and I is determined. As pointed out in Section 8.6, all members of an isospin multiplet are expected to have the same quantum numbers, apart from I_3 and q. Properties other than discrete quantum numbers can be affected by the electromagnetic force but should still be approximately alike. The search is started in a given isobar, and levels with similar properties are looked for in neighboring isobars. In contrast to particle physics, where the effect of the electromagnetic interaction is difficult to compute, the positions of the levels can be predicted with confidence: The electromagnetic force produces two effects, a repulsion between the protons in the nucleus and a mass difference between neutron and proton. The Coulomb repulsion can be calculated, and the mass difference is taken from experiment. The energy difference between members of an isospin multiplet in isobars $(A, Z+1)$ and (A, Z) is

$$\Delta E = E(A, Z+1) - E(A, Z) \approx \Delta E_{\text{Coul}} - (m_n - m_H)c^2. \tag{8.36}$$

The energies refer to the neutral atoms and include the electrons; $(m_n - m_H)c^2 = 0.782$ MeV is the neutron–hydrogen atomic mass difference. The simplest estimate of the Coulomb energy is obtained by assuming that the charge Ze is distributed

[13] J. D. Anderson and C. Wong, *Phys. Rev. Lett.* **7**, 250 (1961). Isobaric analog states are discussed in Section 17.6.

uniformly through a sphere of radius R. The classical electrostatic energy is then given by

$$E_{\text{Coul}} = \frac{3}{5}\frac{(Ze)^2}{R}, \tag{8.37}$$

and it gives rise to the shift shown in Fig. 8.4. The energy difference between isobars with charges $Z+1$ and Z becomes approximately

$$\Delta E_{\text{Coul}} \approx \frac{6}{5}\frac{e^2}{R}Z \tag{8.38}$$

if both nuclides have equal radii. (They should have equal radii since their hadronic structures are alike.) Values for R can be taken from Eq. (6.30), and the Coulomb energy difference can then be calculated.

The values of nuclear spins vary all the way from 0 to more than 10. Does a similar richness exist in the values of isospin? It does, many isospin values occur, and we shall discuss a few in order to show the importance of the isospin concept. All examples will show one regularity: The isospin of the nuclear ground state always assumes the smallest value allowed by Eq. (8.35), $I_{\text{min}} = |Z - N|/2$.

Isospin *singlets*, $I = 0$, can appear only in nuclides with $N = Z$, as is evident from Eq. (8.35). Such nuclides are called self-conjugate. The ground states of ^{2}H, ^{4}He, ^{6}Li, ^{8}Be, ^{12}C, ^{14}N, and ^{16}O have $I = 0$. ^{14}N is a good example, and the lowest levels of the $A = 14$ isobars are shown in Fig. 8.5. Since A is even, only integer isospin values are allowed. If the ^{14}N ground state had a value of $I \neq 0$, similar levels would have to appear in ^{14}C and ^{14}O, with $I_3 = \pm 1$. These levels should have the same spin and parity as the ^{14}N ground state, namely 1^+.

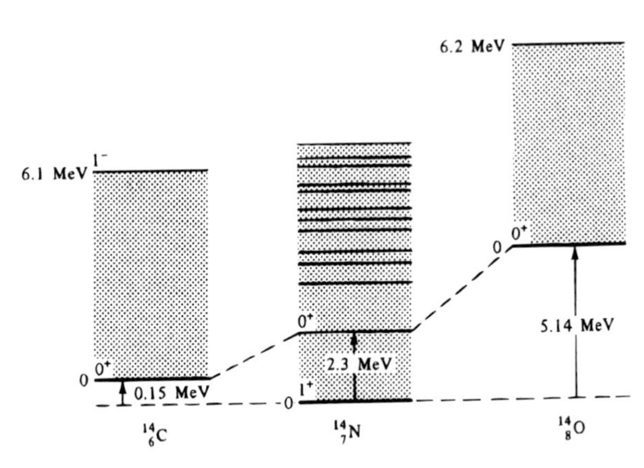

Figure 8.5: $A = 14$ isobars. The labels denote spin and parity, for instance, 0^+. The ground state of ^{14}N is an isospin singlet; the first excited state is a member of an isospin triplet.

Equation (8.36) permits a calculation of the approximate position: The level in ^{14}O should be about 3.0 MeV higher, and the level in ^{14}C should be about 2.5 MeV lower than the ^{14}N ground state. No such states exist. On the oxygen side, the first level appears at 5.14 MeV and it has spin 1 and negative parity. On the ^{14}C side, the first level is higher and not lower, and it also has spin 1 and negative parity. All evidence indicates that the ^{14}N ground state has isospin 0.

Isospin *doublets* occur in mirror nuclides for which $Z = (A \pm 1)/2$. An example is shown in Fig. 8.6. The ground state and the first five excited states have isospin $\frac{1}{2}$. Equation (8.36) predicts an energy shift of 1.3 MeV, which is in reasonable agreement with the observed shift of 0.86 MeV.

An example of an isospin *triplet* is shown in Fig. 8.5. The ground states of ^{14}C and ^{14}O form an $I = 1$ triplet with the first excited state of ^{14}N. All three states have spin 0 and positive parity. The energies agree reasonably well with the prediction of Eq. (8.36). *Quartets* and *quintets* have also been found,[14] and the existence of isospin multiplets in isobars is well established.

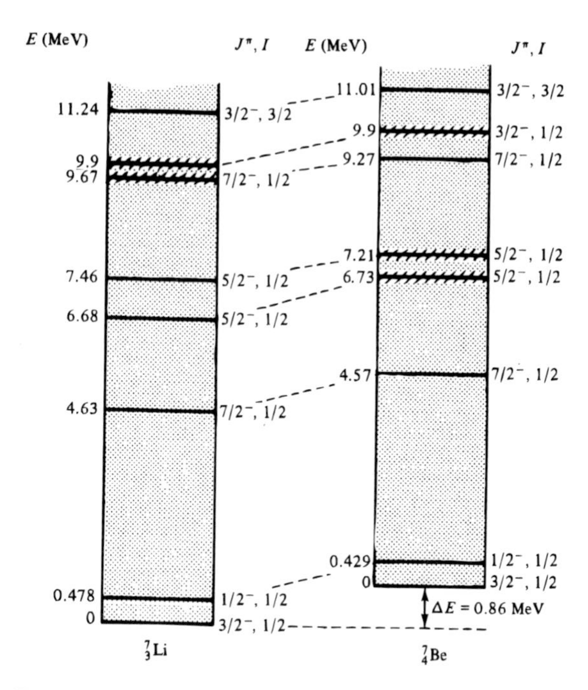

Figure 8.6: Level structure in the two isobars ^{7}Li and ^{7}Be. These two nuclides contain the same number of nucleons; apart from electromagnetic effect, their level schemes should be identical. J^x denotes spin and parity of a level, I its isospin. Parity will be discussed in Chapter 9. [For reference see F. Ajzenberg-Selove, *Nucl. Phys.* **A490**, 1 (1988).]

8.8 References

General references to invariance properties are given in Section 7.6. In addition to these, the following books and articles are recommended.

Rotations in ordinary space and the ensuing quantum mechanics of angular momentum are important in all parts of subatomic physics. We have only scratched the surface. For further details, the texts by Messiah and Merzbacher are useful. The subject is treated in more detail in D.M. Brink and G.R. Satchler, *Angular Momentum*, Oxford University Press, London, 1968.

The early ideas concerning isospin are lucidly described in E. Feenberg and E. P. Wigner, *Rep. Prog. Phys.* **8**, 274 (1941), and in W. E. Burcham, *Prog. Nucl. Phys.* **4**, 171 (1955). A later review is D. Robson, *Annu. Rev. Nucl. Sci.* **16**, 119 (1966). The book *Isospin in Nuclear Physics* (D. H. Wilkinson, ed.), North-Holland, Amsterdam, 1969, provides a review of the entire field. Even though many

[14] J. Cerny, *Annu. Rev. Nucl. Sci.* **18**, 27 (1968); W. Benenson and E. Kashy, *Rev. Mod. Phys*, **51**, 527 (1979); F. Ajzenberg-Selove, *Nucl. Phys.* **A449**, 1 (1986).

contributions in this volume are far above the level of the present course, the book can be consulted if questions arise. A further review is that of E.M. Henley and G.A. Miller in *Mesons in Nuclei*, (M. Rho and D. Wilkinson, eds.), North-Holland, Amsterdam, 1979 p. 406. An up-to-date account of the present status is given by E.M. Henley in *Prog. Part. Nucl. Phys.* (A. Faessler, ed.) **20**, 387 (1987); G.A. Miller, B.M.K. Nefkens, and I Slaus, *Phys. Rept.* **194**, 1 (1990) and G.A. Miller and W.T.H. van Oers in *Symmetries and Fundamental Interactions*, ed. W.C. Haxton and E.M. Henley, World Sci., Singapore (1995), p. 127.

Equation (8.37) for the Coulomb energy is good enough for estimates. For detailed arguments, it must be improved. A thorough discussion of Coulomb energies is given in a review by J.A. Nolan, Jr., and J.P. Schiffer, *Annu. Rev. Nucl. Sci.* **19**, 471 (1969).

Problems

8.1. Derive the commutation relation between J_x and J_y:

(a) Equation (8.2) gives the relation between a wave function before and after rotation, $\psi^R = U\psi$. Matrix elements of an operator F can be taken between the original and the rotated states. It is, however, also possible to consider rotation of the operator F and leave the states unchanged. Justify that the relation between the rotated and the original operator is given by

$$F^R = U^\dagger F U.$$

(b) Assume $\boldsymbol{J} \equiv (J_x, J_y, J_z)$ to be a vector. Consider an infinitesimal rotation of $\boldsymbol{J}$ by the angle ϵ about the y axis. Express $\boldsymbol{J}^R \equiv (J_x^R, J_y^R, J_z^R)$ in terms of $\boldsymbol{J}$ and ϵ.

(c) Assume $\boldsymbol{J}$ to be the generator of the rotation U, Eq. (8.7). Use infinitesimal rotations to derive the commutation relation between J_x and J_y by setting $F = J_x$ in part (a) and using the result of part (b).

8.2. Consider the operator $U = \exp(-i\boldsymbol{a} \cdot \boldsymbol{p}/\hbar)$, where $\boldsymbol{a}$ is a displacement in real space and $\boldsymbol{p}$ is a momentum vector.

(a) What operation is described by U?

(b) Assume that H is invariant under translation in space. Find the conserved quantity corresponding to this symmetry operation and discuss its eigenfunctions and eigenvalues.

8.3. Discuss some evidence for charge independence in the pion-nucleon interaction.

8.4. Verify the steps in footnote 1.

8.5. Calculate the commutator (8.11).

8.6. Justify that the isospin of the deuteron is zero

(a) By using experimental information.

(b) By considering the generalized Pauli principle stating that the total wave function, assumed to be a product of space, spin, and isospin parts, must be antisymmetric under the exchange of the two nucleons.

8.7. The reaction

$$dd \longrightarrow \alpha\pi^0$$

has been observed (see E.J. Stephenson et al., *Phys. Rev. Lett.* **91**, 142302 (2003)), but with a very small cross section. The isospin of the deuteron and the alpha particle are known to be zero. What does the abnormally small cross section of the reaction tell us?

8.8. Verify Eq. (8.37), Eq. (8.38).

8.9. * Study the energy levels of the $A = 12$ isobars.

(a) Sketch the energy level diagrams.

(b) Justify that the ground state and the first few excited states of ^{12}C have isospin zero.

(c) Find the first $I = 1$ state in ^{12}C and justify that it forms an isospin triplet with the ground states of ^{12}B and ^{12}N.

8.10. Consider the reactions

$$d^{16}\mathrm{O} \longrightarrow \alpha^{14}\mathrm{N}$$
$$d^{12}\mathrm{C} \longrightarrow p^{13}\mathrm{C}.$$

Assume isospin invariance. What are the values of I of the states in ^{14}N and ^{13}C that can be reached by these reactions? (^{16}O, ^{12}C, α, and d denote ground states; ^{14}N and ^{13}C can be excited.)

8.11. * Consider the beta decay of ^{14}O to the first and second excited states in ^{14}N. Normally, a beta decay will have a lifetime that is approximately proportional to E^{-5}, where E is the maximum energy of the beta particles. Use isospin invariance to explain the observed branching ratio.

8.12. * Compare ΔE_{Coul} for $A = 10, 80$, and 200. Why is it more difficult (or impossible) to find all the members of an isospin multiplet in heavy nuclei than in light nuclei?

8.13. Consider the reactions

$$\begin{aligned} \gamma A &\longrightarrow nA' \\ dA &\longrightarrow pA' \\ dA &\longrightarrow \alpha A' \\ {}^3\text{He}A'' &\longrightarrow {}^3\,\text{H}A'. \end{aligned}$$

If A is a self-conjugate ($N = Z$) nuclide, what are the isospin states in A' that can be reached by these reactions? The photon "carries" isospin 0 *and* 1. If A'' has isospin 0, or $\frac{1}{2}$, or $\frac{3}{2}$, what are the possible values of the isospin states in A'?

8.14. (a) Prove the commutation relations

$$[I_\pm, I^2] = 0, \quad [I_3, I_\pm] = \pm I_\pm, \quad [I_+, I_-] = 2I_3.$$

(b) Use these commutation relations and Eq. (8.24) to prove Eq. (8.27).

8.15. (a) Use the generalization of Eq. (8.30) to deduce the strangeness content of the D^0 meson of isospin 1/2, the η_c meson of isospin 0, the Λ_c^+ baryon of isospin 0. Assume that $B = T = 0$.

(b) Repeat part (a) for bottomness content for the B^- of isospin $\frac{1}{2}$, if $C = T = 0$.

8.16. The angular distribution of neutral pions produced in the reaction $np \to d\pi^0$ is found to be (almost) symmetrical about 90° in the c.m. (see A.K. Opper et al., *Phys. Rev. Lett.* **91**, 212302 (2003)) Show that this follows from isospin conservation.

8.17. Projection operators have the properties that $P|a\rangle$ and $P|b\rangle = 0$ if $\langle a|b\rangle = 0$ and $P^2 = P$. In terms of the isospin operator $I_{3,op}$ determine a projection operator P_p for the proton and P_n for the neutron, such that $P_p|p\rangle = 1$, $P_p|n\rangle = 0$, $P_n|n\rangle = 1$, $P_n|p\rangle = 0$. (*For isospin 1/2* $I_{3,op}^2 = 1/4$).

Chapter 9

P, C, CP, and T

In the previous chapter we have discussed two continuous symmetry operations: rotations in ordinary space and in isospin space. These rotations can be made as small as desired and consequently can be studied by employing infinitesimal transformations. Invariance under these rotations leads to conservation of spin and isospin, respectively. In this chapter we shall discuss examples of discontinuous transformations, which can lead to operators of the type already given in Eq. (7.11), namely

$$U_h^2 = 1.$$

Such operators are Hermitian *and* unitary. Invariance under U_h leads to a multiplicative conservation law in which the product of quantum numbers is an invariant.

9.1 The Parity Operation

Parity invariance, loosely stated, means invariance under an interchange left $\rightleftharpoons$ right, or symmetry of mirror image and object. For many years, physicists were convinced that all natural laws should be invariant under such mirror reflections. Clearly this belief has little to do with everyday observations because our world is not left–right-invariant. Keys, screws, and DNA have a handedness. Why, then, the belief in invariance under space reflection? The history of the parity operation shows how a concept is found, how a concept is understood, how a concept becomes a dogma, and how finally the dogma falls: In 1924, Laporte discovered that atoms have two different classes of levels;[1] he established selection rules for transitions between the two classes, but he could not explain their existence. Wigner then showed that the two classes follow from invariance of the wave function under space reflection.[2] This symmetry was so appealing that it was elevated to a dogma. The observed left–right asymmetries in nature were all blamed on initial conditions. It came, therefore, as a rude shock when Lee and Yang, in 1956, showed that no evidence for parity conservation in the weak interaction existed[3] and parity

[1]O. Laporte, *Z. Physik* **23**, 135 (1924).
[2]E. P. Wigner, *Z. Physik* **43**, 624 (1927).
[3]T. D. Lee and C. N. Yang, *Phys. Rev.* **104**, 254 (1956).

nonconservation was subsequently found by Wu and collaborators in beta decay.[4] The fall of parity, however, was only partial. Parity is conserved in hadronic and electromagnetic processes.

The *parity operation* (space inversion), P, changes the sign of any true (polar) vector:

$$\boldsymbol{x} \xrightarrow{P} -\boldsymbol{x}, \quad \boldsymbol{p} \xrightarrow{P} -\boldsymbol{p}. \tag{9.1}$$

Axial vectors, however, remain unchanged under P. An example is the orbital angular momentum, $\boldsymbol{L} = \boldsymbol{r} \times \boldsymbol{p}$. Under P, both $\boldsymbol{r}$ and $\boldsymbol{p}$ change sign, and $\boldsymbol{L}$ consequently remains unchanged. A general angular momentum vector, $\boldsymbol{J}$, behaves the same way:

$$\boldsymbol{J} \xrightarrow{P} \boldsymbol{J}. \tag{9.2}$$

This behavior follows from the observation that P commutes with an infinitesimal rotation and hence also with $\boldsymbol{J}$. Moreover, the transformation (9.2) leaves the commutation relations for angular momentum, Eq. (5.6), invariant. The effect of the parity operation on momentum and on angular momentum is shown in Fig. 9.1.

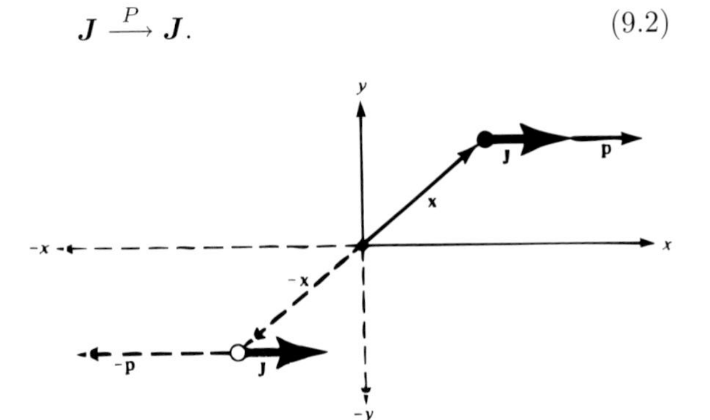

Figure 9.1: The parity operation changes $\boldsymbol{x}$ into $-\boldsymbol{x}$, $\boldsymbol{p}$ into $-\boldsymbol{p}$, but leaves the angular momentum $\boldsymbol{J}$ unchanged. For clarity, only two dimensions are shown.

The parity operator is a special case of the transformation operator U discussed in Section 7.1; P changes a wave function into another wave function:

$$P\psi(\boldsymbol{x}) = \psi(-\boldsymbol{x}). \tag{9.3}$$

If P is applied a second time to Eq. (9.3), the original state is regained,[5]

$$P^2\psi(\boldsymbol{x}) = P\psi(-\boldsymbol{x}) = \psi(\boldsymbol{x}), \tag{9.4}$$

and P consequently satisfies the operator equation

$$P^2 = I. \tag{9.5}$$

[4] C. S. Wu, E. Ambler, R. W. Hayward, D. D. Hoppes, and R. P. Hudson, *Phys. Rev.* **105**, 1413 (1957).

[5] For relativistic wave functions, Eq. (9.4) must be generalized.

P is an example of the operator (7.11), which was denoted by U_h and which is Hermitian and unitary at the same time. Equation (9.5) shows that the eigenvalues of P are $+1$ and -1.

Up to this point, no invariance arguments have been introduced. The discussion was restricted to the parity operation, and it dealt only with what happens under P. The wave functions $\psi(\boldsymbol{x})$ and $\psi(-\boldsymbol{x})$ can be wildly different. The situation becomes orderly when invariance under parity is introduced. Assume a system to be described by a Hamiltonian H that commutes with P:

$$[H, P] = 0. \tag{9.6}$$

In this case, the wave function $\psi(\boldsymbol{x})$ can be chosen to be an eigenfunction of the parity operator, as can be seen as follows. $\psi(\boldsymbol{x})$ is an eigenfunction of H,

$$H\psi(\boldsymbol{x}) = E\psi(\boldsymbol{x}).$$

Operating with P and using Eq. (9.6) give

$$HP\psi(\boldsymbol{x}) = PH\psi(\boldsymbol{x}) = PE\psi(\boldsymbol{x}) \quad \text{or} \quad H\psi'(\boldsymbol{x}) = E\psi'(\boldsymbol{x}),$$

where

$$\psi'(\boldsymbol{x}) \equiv P\psi(\boldsymbol{x}).$$

The wave functions $\psi(\boldsymbol{x})$ and $P\psi(\boldsymbol{x})$ satisfy the same Schrödinger equation with the same energy eigenvalue E, and two possibilities now exist. The state with energy E can be degenerate so that two different physical states, described by the wave functions $\psi(\boldsymbol{x})$ and $\psi'(\boldsymbol{x}) \equiv P\psi(\boldsymbol{x})$, have the same energy. If the state is *not* degenerate, then $\psi(\boldsymbol{x})$ and $P\psi(\boldsymbol{x})$ must describe the same physical situation, and they must be proportional to each other:

$$P\psi(\boldsymbol{x}) = \eta_P\psi(\boldsymbol{x}). \tag{9.7}$$

This relation has the form of an eigenvalue equation, and the eigenvalue η_P is called the parity of the wave function $\psi(\boldsymbol{x})$. The argument following Eq. (9.5) implies that the eigenvalue must be $+1$ or -1:

$$\eta_P = \pm 1. \tag{9.8}$$

The corresponding wave functions are said to have even $(+)$ or odd $(-)$ parity. Since P commutes with H, according to Eq. (9.6), parity is conserved, and η_P is the observable eigenvalue associated with the Hermitian operator P.

A particularly useful example of a parity eigenfunction is $Y_l^m(\theta, \varphi)$, the eigenfunction of the orbital angular momentum operator. In Eq. (5.4), we wrote this eigenfunction as $\psi_{l,m}$ and defined it as the eigenfunction of the operators L^2 and

L_z. The function Y_l^m is called a spherical harmonic. [6] In polar coordinates, the parity operation $\boldsymbol{x} \rightarrow -\boldsymbol{x}$ is given by

$$r \longrightarrow r, \qquad \theta \longrightarrow \pi - \theta, \qquad \varphi \longrightarrow \pi + \varphi, \tag{9.9}$$

and under such a transformation, Y_l^m changes sign if l is odd and remains unchanged if l is even:

$$PY_l^m = (-1)^l Y_l^m. \tag{9.10}$$

Conservation of parity leads to a *multiplicative* conservation law, as can be seen by considering a reaction

$$a + b \longrightarrow c + d.$$

Symbolically, the initial state can be described as

$$|\text{initial}\rangle = |a\rangle|b\rangle|\text{relative motion}\rangle,$$

where $|a\rangle$ and $|b\rangle$ describe the internal state of the two subatomic particles and $|\text{relative motion}\rangle$ is the part of the wave function characteristic of the relative motion of a and b. Space inversion affects each factor so that

$$P|\text{initial}\rangle = P|a\rangle P|b\rangle P|\text{relative motion}\rangle. \tag{9.11}$$

Equation (9.9) shows that the radial part of the relative-motion wave function is unaffected by P and the orbital part gives the contribution $(-1)^l$, where l is the relative orbital angular momentum of the two particles a and b. The expressions $P|a\rangle$ and $P|b\rangle$ refer to the internal wave functions of the two particles. We can assign intrinsic parities to particles so that, for instance,

$$P|a\rangle = \eta_P(a)|a\rangle.$$

Equation (9.11) then becomes

$$\eta_P(\text{initial}) = \eta_P(a)\eta_P(b)(-1)^l. \tag{9.12}$$

A similar equation holds for the final state, and parity conservation in the reaction demands that

$$\eta_P(a)\eta_P(b)(-1)^l = \eta_P(c)\eta_P(d)(-1)^{l'}, \tag{9.13}$$

where l' is the relative orbital angular momentum of the particles c and d in the final state. Equation (9.13) implies that parity is a conserved multiplicative quantum number.

---

[6] Properties of the Y_l^m and their explicit form can be found for example in Morse and Feshbach.

Why does a gauge transformation lead to an additive quantum number while P leads to a multiplicative one? P is a Hermitian operator in itself, while in a gauge transformation, the Hermitian operator appears in the exponent. A product of exponentials leads to a sum of exponents and hence to an additive law.

9.2 The Intrinsic Parities of Subatomic Particles

Can intrinsic parities be assigned to subatomic particles as assumed in Section 9.1? We shall show that such assignments are feasible, but we shall also encounter a fine example of an unsuspected trap.

As in all cases where a sign is involved, the starting point must be defined. In electricity, the charge on cat fur is defined to be positive, whence the proton acquires a positive charge. The intrinsic parity of the proton is also defined to be positive,

$$\eta_P(\text{proton}) = +. \tag{9.14}$$

The determination of the parity of other particles is based on relations of the type of Eq. (9.13). As an example, we consider the capture of negative pions by deuterium.[7] Low-energy negative pions impinge on a deuterium target, and the reaction products are observed. Of the three reactions,

$$d\pi^- \longrightarrow nn \tag{9.15}$$

$$d\pi^- \longrightarrow nn\gamma \tag{9.16}$$

$$d\pi^- \longrightarrow nn\pi^0 \tag{9.17}$$

only the first two are observed; the third one is absent. Parity conservation for the first reaction leads to the relation

$$\eta_P(d)\eta_P(\pi^-)(-1)^l = \eta_P(n)\eta_P(n)(-1)^{l'} = (-1)^{l'}.$$

First consider spin and parity of the initial state. The deuteron is the bound state of a proton and a neutron. The nucleon spins are parallel and add up to a deuteron spin 1. The relative orbital angular momentum of the two nucleons is predominantly zero. (We shall discuss the deuteron in more detail in Chapter 14.) Consequently the deuteron parity is $\eta_P(d) = \eta_P(p)\eta_P(n)$. The negative pion slows down in the target and is finally captured around a deuteron, forming a pionic atom. With emission of photons, the pion rapidly falls to an orbit with zero orbital angular momentum from where reactions (9.15) and (9.16) occur. Consequently the orbital angular momentum l is zero, and the parity of the initial state is given by $\eta_P(\pi^-)\eta_P(p)\eta_P(n)$. The angular momentum l' in the final state can also be obtained easily: The total wave function in the final state must be antisymmetric

[7] W. K. H. Panofsky, R. L. Aamodt, and J. Hadley, *Phys. Rev.* **81**, 565 (1951).

(two identical fermions). If the spins of the two neutrons are antiparallel, the spin state is antisymmetric, and the space state must be symmetric; consequently l' must be even, and the possible total angular momenta are 0, 2, ... The total angular momentum in the initial state is 1; angular momentum conservation therefore rules out the antisymmetric spin state. For the symmetric spin state, where the two spins are parallel, the orbital angular momentum l' must be odd, $l' = 1, 3, \ldots$. Only in the state $l' = 1$ can the total angular momentum be 1, and the final state therefore is $^3P_1$. With $l' = 1$ the parity relation becomes

$$\eta_P(p)\eta_P(n)\eta_P(\pi^-) = -1. \tag{9.18}$$

Two solutions exist, and with the standardization (9.14) they are

$$\eta_P(p) = \eta_P(n) = 1, \qquad \eta_P(\pi^-) = -1, \tag{9.19}$$

and

$$\eta_P(p) = \eta_P(\pi^-) = 1, \qquad \eta_P(n) = -1. \tag{9.19a}$$

The two solutions are equivalent, experimentally. No experiment can be devised that gets around the ambiguity and measures the relative parity between proton and neutron. The choice is made on theoretical grounds: proton and neutron form an isodoublet. According to Eq. (8.15), the members of an isospin multiplet should have the same hadronic properties, and it is assumed that they do have the same intrinsic parity. By setting

$$\eta_P(\text{neutron}) = + \tag{9.20}$$

the parity of the pion becomes negative; the pion is a *pseudoscalar* particle. The absence of the reaction (9.17) indicates that the neutral pion is also a pseudoscalar.

• Why can the relative parity of the proton and the neutron, or of the positive and the neutral pion, not be measured? The reason is connected with the existence of additive conservation laws. Consider the parity equations for the proton and the neutron.

$$P|p\rangle = |p\rangle$$
$$P|n\rangle = |n\rangle.$$

A modified parity operator, P', is introduced through the definition

$$P' = Pe^{i\pi Q} \tag{9.21}$$

where Q is the electric charge operator. Physically, the new operator P' is indistinguishable from P. It performs the same function (for instance, changes $\boldsymbol{x}$ into $-\boldsymbol{x}$), and, according to Eq. (7.22), it commutes with H. P and P' therefore are equally good parity operators. Applied to $|p\rangle$ and $|n\rangle, P'$ gives

$$P'|p\rangle = Pe^{i\pi Q}|p\rangle = -P|p\rangle = -|p\rangle,$$
$$P'|n\rangle = |n\rangle.$$

The modified parity operator assigns negative intrinsic parity to the proton and leaves the neutron parity unchanged. Since P and P' are equally good parity operators and we have no reason to prefer one over the other, we conclude that the relative parity between systems of different electric charge is not a measurable concept. Then, there is no way to determine experimentally which of the two solutions given in Eq. (9.19) is correct; the assignment of equal parities to the proton and neutron cannot be verified by a measurement, but it rests on firm theoretical grounds.

Instead of the modification (9.21), parity operators of the form

$$P'' = Pe^{i\pi A}$$

can be introduced, where A is the baryonic (or another conserved additive) number operator. The arguments proceed as above, and it becomes clear that the relative parity is observable only for systems that have equal additive quantum numbers.

We have just shown that the relative parity of two systems is measurable only if the two systems have equal additive quantum numbers. This restriction limits the usefulness of the parity concept, but not as much as could be suspected. It is only necessary to fix the intrinsic parities of as many hadrons as there are additive quantum numbers; the parities of all other hadrons can be found by building composite systems of the *standard particles* and measuring the relative parities of all other states with respect to these. The parities of the proton and the neutron have already been set positive; next it is customary to add the lambda as the third standard particle so that

$$\eta_P(\text{proton}) = \eta_P(\text{neutron}) = \eta_P(\text{lambda}) = +. \tag{9.22}$$

With this definition, the parities of all nonstrange and strange hadrons, including all nuclear states, can in principle be determined experimentally. To include particles with other additive quantum numbers, for instance charm, the parities of a corresponding number of particles with these quantum numbers must be defined. The gauge bosons γ, gluon, $W^{\pm}, Z^0$ all have negative intrinsic parities; that of the photon has been determined from experiments. Leptons have been omitted here for reasons that will become clear in Section 9.3.

We have restricted the above paragraph to particle systems; the intrinsic parities of antiparticles is also needed and is not arbitrary. For bosons, the parity of an antiparticle is the same as that of the particle. The π^0 is its own antiparticle and the antiparticle of the π^+ is the π^-. The parity of an antiboson is thus seen to be the same as that of the boson. This no longer holds for fermions. As predicted by the Dirac theory, the intrinsic parity of an antiparticle is opposite to that of the particle. The parities of e^+, μ^+, and $\bar{p}$ are opposite to those of the e^-, μ^-, p, respectively. These assignments can be checked experimentally, for instance, in the annihilation of $\bar{p}p$ into two pions (see problems 9.44 and 9.45) and was first shown experimentally by Wu and Shaknov [8] by means of the decay of positronium (a bound state of e^+, e^-) in the $^1S_0$ state to two photons, $e^+e^- \rightarrow \gamma\gamma$. For an angular momentum 0 state the decay amplitude must be a scalar under rotation. For two photons of polarization ϵ_1 and ϵ_2, the two such scalars under rotation which can be formed are

$$\begin{aligned} A_s &= \boldsymbol{\epsilon}_1 \cdot \boldsymbol{\epsilon}_2 , \\ A_{ps} &= \boldsymbol{\epsilon}_1 \cdot \boldsymbol{\epsilon}_2 \times \boldsymbol{k} , \end{aligned}$$

where $\boldsymbol{k}$ is the relative momentum of the two photons. A_s is even under a parity transformation, but because the momentum is odd under parity, A_{ps} is a pseudoscalar, odd under parity. Wu and Shaknov measured the polarization of the two emitted photons and showed that they tended to be perpendicular to each other as predicted by A_{ps}, rather than parallel, as predicted by A_s. Since the electromagnetic interaction conserved parity, this implies that the $^1S_0$ state of positronium is a pseudoscalar of negative parity. For an S-state the orbital angular momentum has positive parity (see Eq. (9.10)); thus the intrinsic parity of the e^+ must be opposite to that of the electron, e^-. •

A first example of the determination of the parity of a particle has already been given above where it was shown that the reaction (9.15) leads to the assignment of negative parity to the pion. As a second example, consider the following reactions:

$$dd \longrightarrow p\,^3\text{H} \tag{9.23}$$

$$dd \longrightarrow n\,^3\text{He} \tag{9.24}$$

$$d\,^3\text{H} \longrightarrow n\,^4\text{He}. \tag{9.25}$$

Spin and parity of the deuteron, d, have already been discussed above where it was found that the assignment is 1^+. The spins of ^{3}H, ^{3}He, and ^{4}He can be measured with standard techniques; studies of reactions (9.23)–(9.25) yield values of l and l' and the assignments J^π becomes $\frac{1}{2}^+$ for ^{3}H and ^{3}He and 0^+ for ^{4}He.

In principle, parities of other states can be investigated with similar reactions. One more example is shown in Fig. 9.2. Assume that the assignment 0^+ for ^{228}Th is known and that the spins of the various states in ^{224}Ra have also been determined.

[8] C.S. Wu and I. Shaknov, *Phys. Rev.* **77**, 136 (1950).

As stated above, the alpha particle has spin 0 and positive parity. If it is emitted with orbital angular momentum L, it carries a parity $(-1)^L$. Since the initial state of the decay has spin 0, an alpha emitted with angular momentum L can only reach states with spin $J = L$. The parities of these states then must be $(-1)^L = (-1)^J$ or $0^+, 1^-, 2^+, 3^-, 4^+, \ldots$. Such states indeed are seen to be populated by the alpha decay in Fig. 9.2.

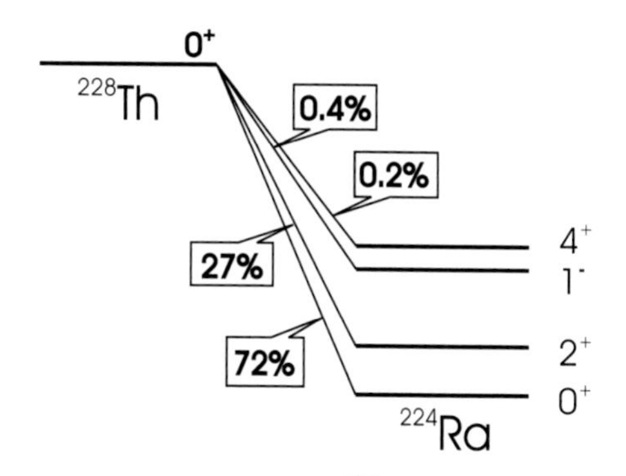

Figure 9.2: Alpha decay of ^{228}Th. The intensities of the various alpha branches are given in %.

The examples given so far are simple. In the actual assignment of parities to particles and excited nuclear states, more complex methods are often necessary, but the basic ideas remain the same. The various methods used in nuclear and in particle physics are described in the references listed in Section 5.14.

9.3 Conservation and Breakdown of Parity

In the previous section we have discussed the experimental determination of the intrinsic parities of some subatomic particles. Implied in all arguments was conservation of parity in the processes used to find η_P. How good is the evidence for parity conservation in the various interactions? To answer this question in a quantitative way, a measure for the degree of parity conservation must be introduced. If $|\alpha\rangle$ is a nondegenerate state of a system with, for instance, even parity, it is written as

$$|\alpha\rangle = |\text{even}\rangle.$$

If parity is not conserved, $|\alpha\rangle$ can be written as a superposition of an even and an odd part,

$$|\alpha\rangle = c|\text{even}\rangle + d|\text{odd}\rangle, \quad |c|^2 + |d|^2 = 1. \tag{9.26}$$

A state of this form, with $c \neq 0$ and $d \neq 0$, is no longer an eigenstate of the parity operator P because

$$P|\alpha\rangle = c|\text{even}\rangle - d|\text{odd}\rangle \neq \eta_P|\alpha\rangle.$$

$\mathcal{F}_P = d/c$ is a measure for the degree of parity nonconservation ($d \leq c$). Parity violation is maximal if the state contains equal amplitudes of $|\text{even}\rangle$ and $|\text{odd}\rangle$, or if $|\mathcal{F}_P| = 1$.

A sensitive test for parity conservation in the *hadronic* and the *electromagnetic* interaction is based on selection rules for alpha decay. In Fig. 9.2 it was shown how the occurrence of an alpha decay can be used to determine the parity of a state to which a transition occurs. The approach can be inverted: Since an alpha particle with orbital angular momentum L carries a parity $(-1)^L$, decays such as $1^+ \xrightarrow{\alpha} 0^+$ or $2^- \xrightarrow{\alpha} 0^+$ are parity-forbidden. They can occur only if one or both of the states involved contain an admixture of the opposite parity. Figure 9.3 shows the levels used for an experiment[9]: A 1^- state in ^{16}O at an excitation energy of about 9.6 MeV is populated by the decay of ^{16}N, and it can decay by alpha emission leaving ^{12}C in its ground state. This transition is parity-allowed, because vector addition of angular momenta permits emission of an alpha particle with $L = 1$ in a transition $1^- \xrightarrow{\alpha} 0^+$.

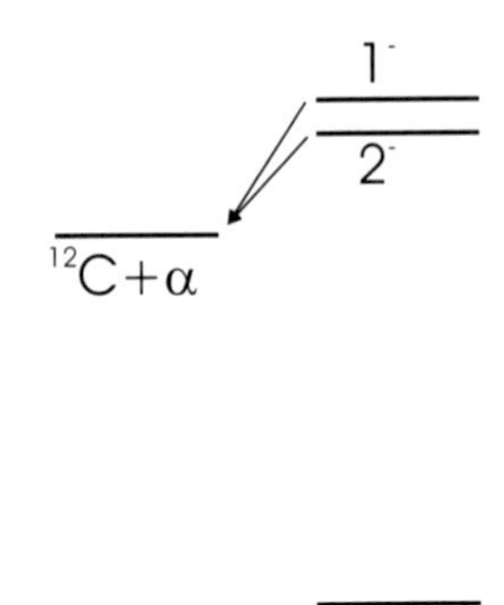

Figure 9.3: Alpha decays from 1^- and 2^- states in ^{16}O. The decays from the latter violate parity. Only levels of interest are shown.

However, ^{16}N decays also to a 2^- state in ^{16}O with excitation energy of 8.9 MeV, which can only go with $L = 2$; the corresponding parity is positive, and the decay $2^- \xrightarrow{\alpha} 0^+$ is parity-forbidden. Seeking such a parity-forbidden branch consequently constitutes a search for $|\mathcal{F}_P|^2$. Analysis of the data show that the decay occurs with a width of $\Gamma = (1.03 \pm 0.28) \times 10^{-10} eV$; when it is compared to the typical alpha decay width of 2^+ states in ^{16}O, we deduce that for the strong interaction:

$$|\mathcal{F}_P|^2 \lesssim 10^{-15}. \tag{9.27}$$

This tiny parity violation is due to the weak interaction. Such a small number provides very good evidence for parity conservation in the hadronic interaction. At the same time, it shows that parity is also conserved in the electromagnetic interaction. If parity were violated electromagnetically, the nuclear wave functions would also be of the form of Eq. (9.26), and parity-forbidden alpha decays would become possible. Since the electromagnetic force is weaker than the hadronic one by about a factor of 100, the limit on the corresponding violation is less stringent than Eq. (9.27) by about $10^4 (|\mathcal{F}_P|^2 \lesssim 10^{-11})$, which is still very low.

Before 1957, the limits were much less convincing. However, since parity conservation had already become a dogma, very few physicists were willing to spend their time improving a number that was considered to be safe anyway. The astonishment was therefore great when it was found early in 1957 that parity was not conserved in the *weak interaction.*[10] The puzzle that motivated the crucial thinking developed

[9] N. Neubeck et al., *Phys. Rev.* **C10**, 320 (1974).

[10] The discovery of parity nonconservation in the weak interaction came as a great shock to most physicists. The background and the story is described in a number of books and reviews. We recommend R. Novick, ed., *Thirty Years Since Parity Nonconservation—A Symposium for T.*

before 1956. By 1956, it had become clear that two strange particles with remarkable properties existed. They were called the tau and the theta, and they appeared to be identical in every respect (mass, production cross section, spin, charge) except in their decay. One decayed to a state of negative parity, and the other to a state of positive parity. The dilemma thus was as follows: either two practically identical particles with opposite parities existed or parity conservation had to be given up. Lee and Yang studied the problem in depth(3) and found, much to their surprise, an overlooked fact: Evidence for parity conservation existed, but only for the hadronic and the electromagnetic interactions, and not for the weak one. The decays of the tau and the theta were so slow that they were known to be weak; Lee and Yang suggested experiments to test parity conservation specifically in the weak interaction. The first experiment was performed by Wu and collaborators, and it brilliantly showed the correctness of Lee and Yang's conjecture.(4) The tau and theta are now known to be one and the same particle, the kaon.

The concept underlying the Wu et al. experiment is explained in Fig. 9.4. ^{60}Co nuclei are polarized so that their spins $\boldsymbol{J}$ point along the positive z axis. When the nuclei decay through the intensity of the emitted electrons is measured in the two directions 1 and 2. The electron momenta are denoted by $\boldsymbol{p}_1$ and $\boldsymbol{p}_2$, and the corresponding intensities by I_1 and I_2. Under the parity transformation, the spins remain unchanged, but the momenta $\boldsymbol{p}_1$ and $\boldsymbol{p}_2$, and the intensities I_1 and I_2, are interchanged. Invariance under the parity operation means that the original and the parity-transformed situations cannot be distinguished. Figure 9.4 shows that the two situations give identical intensities if $I_1 = I_2$. Parity conservation demands that the intensity of electrons emitted parallel to $\boldsymbol{J}$ is the same as for electrons emitted anti-parallel to $\boldsymbol{J}$.

$$^{60}\text{Co} \longrightarrow ^{60}\text{Ni} + e^- + \overline{v},$$

In a more formal way, the essential aspect of the experiment is the observation of the expectation value of the operator

$$\mathcal{P} = \boldsymbol{J} \cdot \boldsymbol{p}, \tag{9.28}$$

where $\boldsymbol{J}$ is the spin of the nucleus and $\boldsymbol{p}$ is the momentum of the emitted electron.

D. Lee, Birkhäuser, Boston, 1988. A letter from Pauli to Weisskopf (German but with English translation) is reprinted in W. Pauli, *Collected Scientific Papers*, Vol. 1 (R. Kronig and V. F. Weisskopf, eds.), Wiley-Interscience, New York, 1964, p. xii. The letter shows how much the fall of parity affected physicists.

$\mathcal{P}$ is a pseudoscalar; under the parity operation it transforms as

$$\boldsymbol{J} \cdot p \xrightarrow{P} -\boldsymbol{J} \cdot p. \qquad (9.29)$$

Invariance under the parity opera-
tion means that the transition rates in the two situations, $\boldsymbol{J} \cdot p$ and $-\boldsymbol{J} \cdot p$, are identical. Equation (9.29) instructs the experimental physicist how to test parity invariance: Measure the transition rate for a fixed orientation of $\boldsymbol{J}$ and $\boldsymbol{p}$ and compare the result to the transition rate for the state $-\boldsymbol{J} \cdot p$.

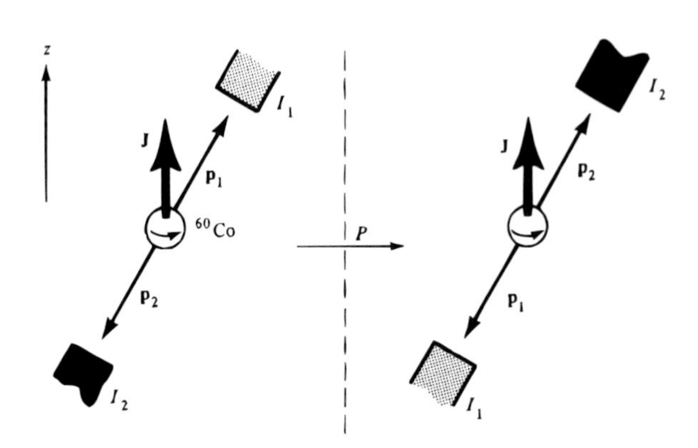

Figure 9.4: Concept of the Wu et al. experiment. A polarized nucleus emits electrons with momenta $\boldsymbol{p}_1$ and $\boldsymbol{p}_2$. The original situation is shown at the left, and the parity-transformed one at the right. Invariance under parity means that the two situations cannot be distinguished.

The state $-\boldsymbol{J} \cdot p$ can be reached by inverting $\boldsymbol{J}$ or $\boldsymbol{p}$. The experiment of Wu and collaborators consisted of comparing the transition rates for $\boldsymbol{J} \cdot p$ and $-\boldsymbol{J} \cdot p$ by inverting $\boldsymbol{J}$ through inverting the polarization of the ^{60}Co nuclei.

In a radioactive source at room temperature, the nuclear spins are randomly oriented. It is necessary to polarize the nuclei so that all spins $\boldsymbol{J}$ point in the same direction. The transition rate for electron emission parallel and antiparallel to $\boldsymbol{J}$ can then be compared. To describe the experimental approach, we use a hypothetical decay, shown in Fig. 9.5(a). A nuclide with spin 1 and g factor $g > 0$ decays by emission of an electron and an antineutrino to a state with spin 0. To polarize the nuclei, the sample is placed in a strong magnetic field $\boldsymbol{B}$ and cooled to a very low temperature T. The magnetic sublevels of the initial state split as in Fig. 9.5; the energy of a state with magnetic quantum number M is given by Eq. (5.21) as $E(M) = E_0 - g\mu_N BM$. The ratio of populations, $N(M')/N(M)$, of two states, M' and M, is determined by the Boltzmann factor,

$$\frac{N(M')}{N(M)} = \exp\{-[E(M') - E(M)]/kT\}, \qquad (9.30)$$

or, with Eq. (5.21),

$$\frac{N(M')}{N(M)} = \exp\left[\frac{(M' - M)g\mu_N B}{kT}\right]. \qquad (9.31)$$

If the condition

$$kT \ll g\mu_N B \qquad (9.32)$$

is satisfied, only the lowest Zeeman level is populated, the nucleus is fully polarized, and its spin points in the direction of the magnetic field [Fig. 9.5(b)]. The change $\boldsymbol{J} \cdot p \rightarrow -\boldsymbol{J} \cdot p$ is obtained by reversing the direction of the external field, $\boldsymbol{B}$. The experimental arrangement requires mastery of many techniques.

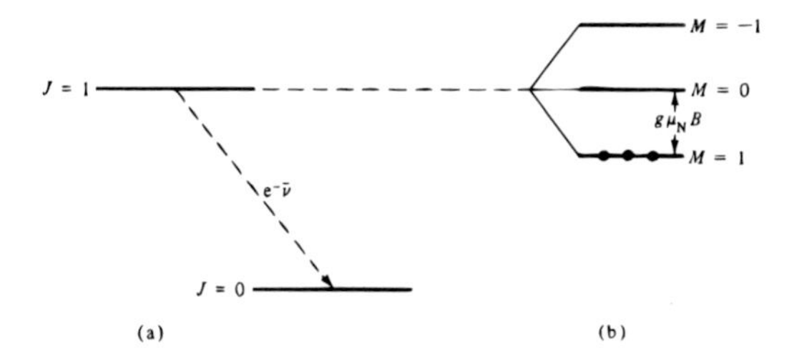

Figure 9.5: (a) Beta decay from a state with spin 1 to a state with spin 0. (b) At very low temperatures in a high magnetic field, only the lowest Zeeman level is populated, and the nucleus (with $g > 0$) is fully polarized and points in the direction of $\boldsymbol{B}$.

The radioactive nuclei are introduced into a cerium–magnesium-nitrate crystal and cooled to a temperature of 0.01 K by adiabatic demagnetization. The magnetic field required to satisfy Eq. (9.32) is very high. To obtain such a high field, paramagnetic atoms are chosen, and the field at the nucleus is then predominantly produced by its own electronic shell. The radioactive source must be thin so that the electrons can escape and be counted in a detector placed in the cryogenic system [Fig. 9.6(a)]. Data are reproduced in Fig. 9.6(b). The result is striking. The expectation value of $\mathcal{P} = \boldsymbol{J} \cdot p$ does not vanish, and parity is not conserved in beta decay. Many additional experiments have borne out the remarkable result that parity is violated in weak interactions. We can now return to an earlier figure and understand it better. In Fig. 7.2, neutrino and antineutrino are shown to be fully polarized. Full polarization means that neutrino and antineutrino have a nonvanishing value of $\boldsymbol{J} \cdot p$ and therefore are a permanent expression of parity nonconservation in the weak interaction.

It is customary to describe the polarization of a spin-$\frac{1}{2}$ particle not by $\boldsymbol{J} \cdot p$, (particularly for massless particles or for particles with energy $\gg mc^2$) but by the helicity operator

$$\mathcal{H} = 2\frac{\boldsymbol{J} \cdot \hat{p}}{\hbar}, \tag{9.33}$$

where $\hat{\boldsymbol{p}}$ is a unit vector in the direction of the momentum. The expectation value of $\mathcal{H}$ for a particle that has its spin along its momentum is $+1$ and such a particle is said to be right-handed; $\langle|\mathcal{H}|\rangle = -1$ characterizes a particle with spin opposite to $\hat{\boldsymbol{p}}$, a left-handed particle. Particles with nonvanishing helicity can be produced in many experiments; common to all these is the existence of a preferred direction, for instance, given by a magnetic field. If no preferred direction exists, a nonvanishing value of $\langle|\boldsymbol{J} \cdot \hat{p}|\rangle$ and hence also of $\langle|\mathcal{H}|\rangle$ is a sign of parity nonconservation. An example is the helicity of leptons emitted from isotropic weak sources, such as beta or muon decay. The helicity of both neutral and charged leptons in such weak decays has been measured.[11]

[11] H. Frauenfelder and R. M. Steffen, in *Alpha-, Beta- and Gamma-Ray Spectroscopy*, Vol. 2, (K. Siegbahn, ed.), North-Holland, Amsterdam, 1965; M. Goldhaber, L. Grodzins and A. W. Sunyar,

The result,

$$\begin{aligned} \langle\mathcal{H}(e^-)\rangle &= -\tfrac{v}{c}, \\ \langle\mathcal{H}(e^+)\rangle &= +\tfrac{v}{c}, \end{aligned} \qquad (9.34)$$

where v is the lepton velocity, confirms parity nonconservation in the weak interaction.

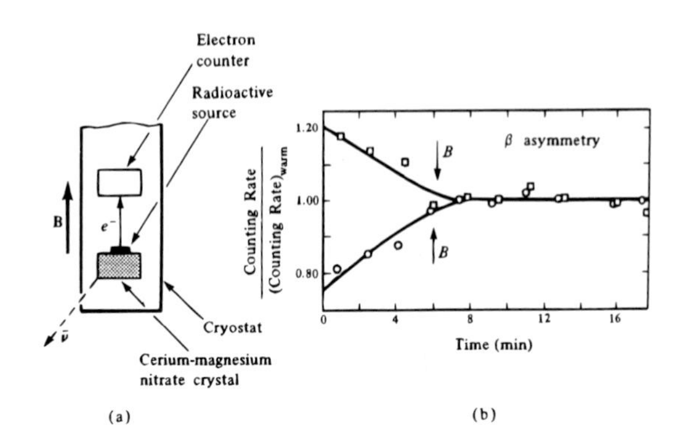

Figure 9.6: (a) Arrangement to measure beta emission from polarized nuclei. (b) Result of the earliest experiment showing parity nonconservation [C. S. Wu, E. Ambler, R. W. Hayward, D. D. Hoppes, and R. P. Hudson, *Phys. Rev.* **105**, 1413 (1957).] A normalized counting rate in the beta detector is shown for two directions of the external magnetic field. After adiabatic demagnetization, the source warms up, the polarization decreases, and the effect disappears.

We have stated above that parity is conserved in the electromagnetic and the hadronic interaction. This statement requires some explanations. Neglecting the gravitational interaction, the total Hamiltonian can be written as

$$H = H_h + H_{\text{em}} + H_w.$$

Cross sections or transition probabilities are always proportional to $|H|^2$; consequently interference terms between the weak and the other two interactions will occur. Since H_w does not conserve parity, the interference terms should also show parity violation. Experiments to detect these interference terms are extremely difficult, but parity violating asymmetries of the expected order of magnitude have indeed been seen in many experiments.(12) The interference of the weak and hadronic interactions has been observed in nuclear reactions, radiative transitions and in nucleon–nucleon scattering. The effect for the electromagnetic interaction has been verified in atomic physics,(13) in electron-electron scattering,(14) and in polarized electron- proton and -nucleus scattering(15) experiments.

9.4 Charge Conjugation

In Section 5.10, the concept of antiparticles was introduced. This concept gives rise to long and mainly philosophical discussions centered around questions such as

Phys. Rev. **109**, 1015 (1958).

[12] E. G. Adelberger and W. Haxton, *Annu. Rev. Nucl. Part. Sci.* **35**, 501 (1985); E. M. Henley in *Prog. Part. Nucl. Phys.*, (A. Faessler, ed.) **20**, 387 (1987); W. Haeberli and B.R. Holstein in *Symmetries and Fundamental Interactions*, ed. W.C. Haxton and E.M. Henley, World Scientific Singapore, 1995, p. 17.

[13] E. A. Hinds, *Amer. Sci.* **69**, 430 (1981); E. N. Fortson and L. L. Lewis, *Phys. Rept* **113**, 289 (1984); M. C. Noecker, B. P. Masterson, and C. E. Wieman, *Phys. Rev. Lett.* **61**, 310 (1988).

[14] P.L. Anthony et al., SLAC E158 Collaboration, *Phys. Rev. Lett.* **92**, 181602 (2004).

[15] C.Y. Prescott et al., *Phys. Lett.* **77B**, 347 (1978), **84B**, 524 (1979); T. M. Ito et al. (SAMPLE Collaboration), *Phys. Rev. Lett.* **92**, 102003 (2004); K.A. Aniol et al. (HAPPEX Collaboration), *Phys. Rev. C* **69**, 065501 (2004); D. S. Armstrong et al. (G0 Collaboration), *Phys. Rev. Lett.* **95**, 092001 (2005).

"Is there really a sea of negative energy states?" or "Can a particle really move backward in time?" The important features, however, are not connected with such vague aspects but concern the undeniable fact that antiparticles exist. In the present section, the particle–antiparticle connection will be put into a more formal frame than in Section 5.10. Many of the ideas are similar to the ones already introduced in connection with parity in Section 9.1 so that the discussion can be brief.

We describe a particle by the ket $|q_{\text{gen}}\rangle$, where q_{gen} stands for all internal additive quantum numbers such as A, q, S, L, and L_μ. The operation of charge conjugation, C, is then defined by

$$C|q_{\text{gen}}\rangle = |-q_{\text{gen}}\rangle. \quad (9.35)$$

Charge conjugation reverses the sign of the additive quantum numbers but leaves momentum and spin unchanged.

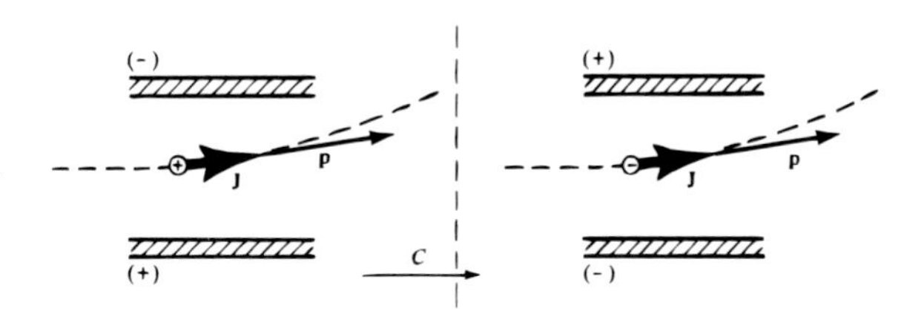

Figure 9.7: Charged particle traversing an electric field. Charge conjugation, acting on the whole system, reverses the additive quantum numbers of a particle but leaves space–time properties ($\boldsymbol{p}, J$) unchanged. If the charges of the external field are also reversed the trajectories of particle and antiparticle are the same.

C is sometimes also called particle–antiparticle conjugation to express the fact that not only the electric charge but all internal additive quantum numbers change sign. The situation is depicted in Fig. 9.7. Charge conjugation invariance means that to every particle there exists an antiparticle with the same mass, spin, and other space–time properties (e.g. decay lifetime), but with opposite internal additive quantum numbers. If C is applied twice the original charges are regained so that

$$C^2 = 1. \quad (9.36)$$

C, like P, is a discontinuous operator of the type of Eq. (7.11), and it is unitary and Hermitian.

Equation (9.36) indicates that, if $[C, H] = 0$, the eigenvalues of the charge conjugation operator are $+1$ and -1. However, as we shall see now, there is a considerable difference between P and C because C does *not* always have eigenstates. To explore this new feature, we write tentatively

$$C|q_{\text{gen}}\rangle \stackrel{?}{=} \eta_c|q_{\text{gen}}\rangle \quad (9.37)$$

and ask when such a relation is meaningful. As an example, the state $|q_{\text{gen}}\rangle$ is taken to be an eigenstate of the charge operator, Q. For a particle with charge q, described by $|q\rangle$, the eigenvalue equation

$$Q|q\rangle = q|q\rangle \quad (9.38)$$

holds. By Eq. (9.35) however, C applied to $|q\rangle$ gives

$$C|q\rangle = |-q\rangle.$$

The commutator of the two operators Q and C, when operating on $|q\rangle$, can now be obtained in a straightforward way:

$$\begin{aligned} CQ|q\rangle &= qC|q\rangle = q|-q\rangle \\ QC|q\rangle &= Q|-q\rangle = -q|-q\rangle \end{aligned}$$

or

$$(CQ - QC)|q\rangle = 2q|-q\rangle = 2CQ|q\rangle. \tag{9.39}$$

The operators C and Q do not commute; this result can be expressed as an operator equation,

$$[C, Q] = 2CQ. \tag{9.40}$$

Since the two operators C and Q do not commute, it is, in general, not possible to find states that are simultaneous eigenstates. A charged particle cannot satisfy an eigenvalue equation of the form of Eq. (9.37) since nature has chosen particles to be eigenstates of Q. The argument just given applies to all quantum numbers q_{gen}. Particles appear in nature as eigenstates of operators corresponding to q_{gen} and these operators also do not commute with C. There is one loophole, however. Fully neutral particles, that is particles for which all quantum numbers q_{gen} vanish, can be in an eigenstate of C. For such systems, Eq. (9.37) applies:

$$C|q_{\text{gen}} = 0\rangle = \eta_c|q_{\text{gen}} = 0\rangle, \qquad \eta_c = \pm 1, \tag{9.41}$$

and η_c is called the *charge parity* (or charge conjugation quantum number). It satisfies a multiplicative conservation law.

What is the charge parity of the fully neutral particles, such as the photon, the gluon, the neutral pion, and η^0? A satisfactory answer requires quantum field theory, but the correct values can be obtained with some hand waving. The photon is described by its vector potential $\boldsymbol{A}$. The potential is produced by charges and currents and consequently changes sign under C:

$$\boldsymbol{A} \xrightarrow{C} -\boldsymbol{A}. \tag{9.42}$$

An example of this sign change has already been shown in Fig. 9.7. Equation (9.42) suggests the assignment

$$\eta_c(\gamma) = -1. \tag{9.43}$$

The π^0 and η^0 decay electromagnetically into two photons.

$$\pi^0 \longrightarrow 2\gamma \qquad \text{and} \qquad \eta^0 \longrightarrow 2\gamma,$$

and therefore must have positive C parity if C is conserved in the decay:

$$\eta_c(\pi^0) = 1, \qquad \eta_c(\eta^0) = 1. \tag{9.44}$$

If C parity were applicable only to the photon, π^0 and η^0, it would not be very useful. However, there exist many particle–antiparticle systems that are fully neutral. Examples are positronium $(e^+e^-), \pi^+\pi^-, p\overline{p}, n\overline{n}$. The C parity of these systems depends on angular momentum and spin, and it is a useful quantity for discussing the possible decay modes.

Use of charge parity for discussion of a decay requires η_c to be a good quantum number. It is conserved if C commutes with the Hamiltonian H. It is easy to see that C is not conserved in the weak interaction,

$$[H_w, C] \neq 0. \tag{9.45}$$

Fig. 7.2 shows that neutrino and antineutrino have opposite polarization (helicity). If charge conjugation were conserved in the weak interaction, the two particles would have to have the same helicity.

C conservation in the hadronic interactions has been tested in numerous reactions, such as

$$p\overline{p} \longrightarrow \pi^+\pi^-\pi^0. \tag{9.46a}$$

C acting on the reaction gives

$$\overline{p}p \longrightarrow \pi^-\pi^+\pi^0. \tag{9.46b}$$

If the proton produces the π^+ forward and the $\overline{p}$ the π^- backwards in the reaction (9.46a), then the reaction (9.46b) would give rise to π^- forward and π^+ backward. Thus, if the hadronic Hamiltonian commutes with C, the angular distribution and energy spectra of the positive and negative pion must be identical. Comparison of the two distributions and similar tests in other reactions show the expected symmetry. The result can be stated as[16]

$$\left|\frac{C\text{-nonconserving amplitude}}{C\text{-conserving amplitude}}\right| \lesssim 0.01. \tag{9.47}$$

To test conservation of C in the electromagnetic interaction, charge–parity-forbidden decays are looked for. Consider the decays

$$\pi^0 \longrightarrow 3\gamma \qquad \text{and} \qquad \eta^0 \longrightarrow 3\gamma.$$

[16] C. Baltay, N. Barash, P. Franzini, N. Gelfand, L. Kirsch, G. Lütjens, J. C. Severiens, J. Steinberger, D. Tycko, and D. Zanello, *Phys. Rev. Lett.* **15**, 591 (1965).

π^0 and η^0 have positive charge parity; the three photons in the final states have negative charge parity, and the decay is forbidden. The decays have not been found. Perhaps the best limit comes from the reaction

$$e^+e^- \longrightarrow \mu^+\mu^-.$$

Charge conjugation invariance requires the angular distribution of the positive or negative muon to be symmetric about 90°. Experimentally, a small asymmetry is found (see Chapter 10), of a magnitude that is consistent with it being caused by the weak interaction. This experiment shows that C is conserved in the electromagnetic interaction. Thus, the present evidence indicates that charge conjugation is a valid symmetry for both the hadronic and electromagnetic Hamiltonians.

9.5 Time Reversal

In the two previous sections, the discrete transformations P and C were introduced. Both operations are unitary and Hermitian and give rise to multiplicative quantum numbers. In the present section, a third discrete transformation is introduced, time reversal, T. It will turn out that T is not unitary, and a complication is thus introduced; no conserved quantity such as parity or charge parity is associated with it. Nevertheless, time-reversal *invariance* is a very useful symmetry in subatomic physics.

Formally, the time-reversal operation is defined by

$$t \xrightarrow{T} -t, \qquad \boldsymbol{x} \xrightarrow{T} \boldsymbol{x}. \tag{9.48}$$

Since classically $\boldsymbol{p} = d\boldsymbol{x}/dt$, momentum and angular momentum change sign under T:

$$\boldsymbol{p} \xrightarrow{T} -\boldsymbol{p}, \qquad \boldsymbol{J} \xrightarrow{T} -\boldsymbol{J}. \tag{9.49}$$

In classical mechanics and electrodynamics, the basic equations are invariant under T: Newton's law of motion and Maxwell's equations are second-order differential equations in t and are therefore unaffected by the replacement of t by $-t$.

The essential aspects of time-reversal invariance appear already in the treatment of a nonrelativistic spinless particle, described by the Schrödinger equation,

$$i\hbar\frac{d\psi(t)}{dt} = H\psi(t). \tag{9.50}$$

This equation is formally similar to the diffusion equation which is *not* invariant under $t \to -t$. The feature that distinguishes T from P and C turns up when the connection between ψ and $T\psi$ is explored. According to the arguments given in Section 7.1, T is a symmetry operator and satisfies

$$[H, T] = 0 \tag{9.51}$$

if $T\psi(t)$ and $\psi(t)$ obey the same Schrödinger equation. The Schrödinger equation for $T\psi(t)$ is

$$i\hbar\frac{dT\psi(t)}{dt} = HT\psi(t). \tag{9.52}$$

The simplest attempt to satisfy this equation,

$$T\psi(t) = \psi(-t), \tag{9.53}$$

is incorrect: inserting Eq. (9.53) into Eq. (9.52) and writing $-t = t'$ gives

$$-i\hbar\frac{d\psi(t')}{dt'} = H\psi(t'). \tag{9.54}$$

This equation is not the same as Eq. (9.50). The fact that Eq. (9.54) is written in terms of t' rather than t is immaterial because t is only a parameter. What counts is *form invariance*: $\psi(t)$ and $T\psi(t)$ must satisfy equations that have the same form.

The correct time-reversal transformation was found by Wigner, who set[17]

$$T\psi(t) = \psi^*(-t). \tag{9.55}$$

Inserting $\psi^*(-t)$ into Eq. (9.52) and taking the complex conjugate of the entire equation produces a relation that has the same form as the original Schrödinger equation if H is real.

The simplest application of the time-reversal transformation (9.55) is to a free particle with momentum $\boldsymbol{p}$, described by the wave function

$$\psi(\boldsymbol{x}, t) = \exp\left[\frac{i(\boldsymbol{p}\cdot x - Et)}{\hbar}\right].$$

The time-reversed wave function is

$$\begin{aligned} T\psi(\boldsymbol{x}, t) &= \psi^*(\boldsymbol{x}, -t) \\ &= \exp\left[\frac{-i(\boldsymbol{p}\cdot x + Et)}{\hbar}\right] = \exp\left[\frac{i(-\boldsymbol{p}\cdot x - Et)}{\hbar}\right]. \end{aligned} \tag{9.56}$$

The time-reversed wave function describes a particle with momentum $-\boldsymbol{p}$, in accord with Eq. (9.49). It is not necessary to interpret the function $T\psi(\boldsymbol{x}, t)$ as describing a particle going backward in time. The more physical interpretation of T is *motion reversal:* T reverses momentum and angular momentum,

$$T|\boldsymbol{p}, \boldsymbol{J}\rangle = |-\boldsymbol{p}, -\boldsymbol{J}\rangle. \tag{9.57}$$

[17]E. Wigner, *Nachr. Akad. Wiss. Goettingen, Math. Physik. Kl. IIa*, **31**, 546 (1932).

When we played the game with P and C, at this point we asked for conserved eigenvalues. The answers were parity η_P and charge parity η_c. Does T have observable and conserved eigenvalues? Such eigenvalues would be solutions of the equation

$$T\psi(t) = \eta_T\psi(t).$$

Equation (9.55) shows, however, that T changes ψ into its complex conjugate, and the eigenvalue equation makes no sense. This fact is connected with the *antiunitarity* of T. P and C are unitary operators; unitary operators are linear and satisfy the relation

$$U(c_1\psi_1 + c_2\psi_2) = c_1U\psi_1 + c_2U\psi_2. \tag{9.58}$$

Antiunitary operators, however, obey the relation

$$T(c_1\psi_1 + c_2\psi_2) = c_1^*T\psi_1 + c_2^*T\psi_2. \tag{9.59}$$

The time-reversal transformation is antiunitary. Why are P and C unitary but not T? In Sections 9.1 and 9.4 we justified the choice of P and C as unitary operators by saying that they must leave the norm $\mathcal{N}$ invariant, where $\mathcal{N}$ is

$$\mathcal{N} = \int d^3x\psi^*(\boldsymbol{x})\psi(\boldsymbol{x}).$$

An antiunitary operator also leaves $\mathcal{N}$ invariant, as can be seen by inserting Eq. (9.55) into $\mathcal{N}$. The choice between the two possibilities is dictated by the physical nature of the transformation. For P and C, the transformed wave functions satisfy the original equations if the transformation is unitary. For T, form invariance demands that it be antiunitary.

We have just seen that T does not have observable eigenvalues; states can therefore not be labeled with such eigenvalues, and invariance under T cannot be tested by searching for *time–parity-forbidden decays*. Fortunately there are other approaches. Time-reversal invariance predicts, for instance, equality of transition probabilities for a reaction and its inverse (principle of detailed balance) and it demands that the electric dipole moments of particles vanish. A great deal of effort has gone into testing time-reversal invariance, but no evidence for a violation in the strong, electromagnetic, or in the flavor-conserving part of the weak interaction has been found.[18] All findings are in flavor-changing systems and can be accounted for as a phase in the CKM matrix, as will be discussed in Chapter 11. Among very sensitive tests are searches for the electric dipole moments of electrons,[19] ultracold bottled neutrons,[20] and of atoms.[21] The electric dipole moment of the electron is

[18] L. Wolfenstein, *Annu. Rev. Nucl. Part. Sci.* **36**, 137 (1986); E. M. Henley in *Progr. Part. Nucl. Phys.*, (A. Faessler, ed.) **20**, 387 (1987).

[19] B.C. Regan et al., *Phys. Rev. Lett.* **88**, 071805 (2002).

[20] C.A. Baker et al., *Phys. Rev. Lett.*,**97**, 131801 (2006); see also R. Golub and S.K. Lamoreaux, *Phys. Rep.* **237**, 1 (1994) for a proposal that is now being persued.

[21] M.V. Romalis et al., *Phys. Rev. Lett.* **86**, 2505 (2001).

found to be $\leq 7 \times 10^{-26}$ e-cm in magnitude; that of the neutron $\leq 6.3 \times 10^{-26}$ e-cm, and that of the mercury atom $\leq 2.1 \times 10^{-28}$ e-cm. Since the size of the neutron is roughly 1 fm, the upper limit on the size of the neutron electric dipole moment means that that the T-odd effect, $\mathcal{F}_T$, is less than about 10^{-12}. The electric dipole moment of the Hg atom improves this limit by about a factor of 5. To-date, no electric dipole moments have been found. These experiments probe for physics beyond the standard model, which predicts even smaller electric dipole moments. However, after more than thirty years of effort, a time reversal violation was observed in 1998 in a strangeness-changing reaction, an indirect comparison of the reaction rates $\bar{K}^0 \leftrightarrow K^0$ and in a correlation experiment in the final state of a particular decay of the neutral kaon.[22]

It is important to note that an electric dipole moment requires that both parity *and* time reversal invariance are violated; this can be illustrated by a simple picture, shown in Figure 9.8. Consider a particle with spin represented by a sphere. The spin defines a direction in space, which we here take to be upwards. We assume the particle to have a net positive charge distributed as shown in Fig. 9.8 so that it has a classical electric dipole moment, $\boldsymbol{d}_E$. Since the particle is rotating it also has a classical magnetic dipole moment, $\boldsymbol{d}_M$.

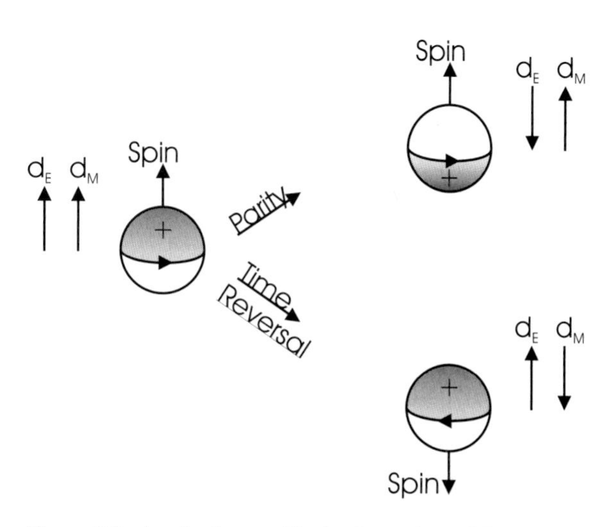

Figure 9.8: A spinning positively charged particle is represented here as a spherical object. Its mirror (located in the horizontal midplane) image and its time reversed image are shown. The magnetic d_M and electric d_E dipole moments are also shown.

By performing a parity inversion about the midplane, we see that the parity inverted particle has an electric dipole moment oppositely directed relative to the spin of the particle, whereas the magnetic dipole moment remains parallel to the spin. Thus, if parity is conserved, the particle cannot have an electric dipole moment since you can tell the mirror picture from the original one. If we perform a time reversal transformation, as shown, the particle will spin in the opposite direction; the magnetic dipole moment changes its direction as well and remains parallel to the spin, but the electric dipole moment fails to do so. Thus, you can tell the difference

[22] A. Angelopoulos et al., CPLEAR Collaboration, *Phys. Lett.* **B 444**, 43 (1998); A. Halavi-Harati et al., the KTEV Collaboration *Phys. Rev. Lett.* **84**, 48 (2000); see also L. Wolfenstein, *Int. J. Mod. Phys.* **E8**, 501 (1999) and E. M. Henley, *Fizika* **B10**, 161 (2002).

between the direct and time reversed pictures if the particle has an electric dipole moment, and this is not allowed if time reversal is a valid symmetry.

9.6 The Two-State Problem

As an introduction to the discussion of neutral kaons, we consider two identical unconnected potential wells L and R shown in Fig. 9.9(a). The energies of the stationary states $|L\rangle$ and $|R\rangle$ are given by the Schrödinger equations,

$$H_0|L\rangle = E_0|L\rangle, \qquad H_0|R\rangle = E_0|R\rangle.$$

Since H_0 does not connect the two wells, we write

$$\langle L|H_0|R\rangle = \langle R|H_0|L\rangle = 0.$$

For simplicity it is assumed that only the states $|L\rangle$ and $|R\rangle$ play a role. All other states are assumed to have so much higher energies that they can be neglected. If we

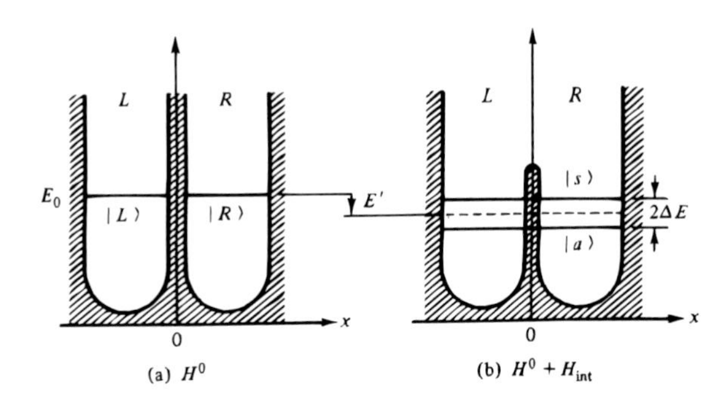

Figure 9.9: Eigenvalues and eigenfunctions of a particle in two identical potential wells, without and with transmission through the barrier.

switch on a perturbing interaction, H_{int}, that lowers the barrier between the wells and induced transitions $L \rightleftharpoons R$, the stationary states of the system are determined by

$$H|\psi\rangle \equiv (H_0 + H_{\text{int}})|\psi\rangle = E|\psi\rangle. \tag{9.60}$$

The problem consists of finding the eigenvalues and eigenfunctions of the total Hamiltonian $H \equiv H_0 + H_{\text{int}}$. Since the two unperturbed states $|L\rangle$ and $|R\rangle$ are degenerate, the solution requires use of the correct linear combinations of the unperturbed eigenfunctions.[23] These combinations can be found by symmetry considerations. Since the potentials are placed symmetrically about the origin, the

[23] Merzbacher, Section 17.5; Park, Section 8.4.

Hamiltonian is invariant under reflections through the origin, and H and the parity operator P commute,

$$[H, P] = [H_0 + H_{\text{int}}, P] = 0. \tag{9.61}$$

With the choice of coordinates shown in Fig. 9.9, the parity operator gives

$$P|L\rangle, = |R\rangle \qquad P|R\rangle = |L\rangle. \tag{9.62}$$

The simultaneous eigenfunctions of H_0 and P are easy to find; they are the symmetric and antisymmetric combinations of the unperturbed states $|L\rangle$ and $|R\rangle$:

$$|s\rangle = \sqrt{\tfrac{1}{2}}\{|L\rangle + |R\rangle\}, \qquad |a\rangle = \sqrt{\tfrac{1}{2}}\{|L\rangle - |R\rangle\}. \tag{9.63}$$

These combinations indeed are eigenstates of P,

$$P|s\rangle = +|s\rangle, \qquad P|a\rangle = -|a\rangle. \tag{9.64}$$

Eqs. (9.61) and (9.64) together prove that H does not connect $|a\rangle$ and $|s\rangle$:

$$\langle a|H|s\rangle = \langle a|HP|s\rangle = \langle a|PH|s\rangle = \langle a|P^\dagger H|s\rangle = -\langle a|H|s\rangle,$$

or

$$\langle a|H|s\rangle = 0. \tag{9.65}$$

Ordinary perturbation theory can consequently be applied to the states $|a\rangle$ and $|s\rangle$. The energy shift caused by the perturbation, H_{int}, is given by the expectation value of H_{int}, or

$$\begin{aligned} \langle s|H_{\text{int}}|s\rangle &= E' + \Delta E \\ \langle a|H_{\text{int}}|a\rangle &= E' - \Delta E, \end{aligned} \tag{9.66}$$

where

$$\begin{aligned} \langle L|H_{\text{int}}|L\rangle &= \langle R|H_{\text{int}}|R\rangle = E' \\ \langle L|H_{\text{int}}|R\rangle &= \langle R|H_{\text{int}}|L\rangle = \Delta E. \end{aligned} \tag{9.67}$$

The interaction lowers the center of the energy levels by E' and splits the degenerate levels by an amount $2\Delta E$, as indicated in Fig. 9.9(b). The splitting shows up in the hydrogen molecule ion and particularly clearly in the inversion spectrum of ammonia.[(24)]

---

[24] Two-state systems and the ammonia MASER are beautifully treated in R. P. Feynman, R. B. Leighton, and M. Sands, *The Feynman Lectures on Physics*, Vol. III, Addison-Wesley, Reading, Mass., 1965, Chapters 8–11.

What happens to a particle that is dropped into *one* potential well, say L, at time $t = 0$? Equation (9.63) gives its state at $t = 0$ as

$$|\psi(0)\rangle = |L\rangle = \sqrt{\tfrac{1}{2}}\{|s\rangle + |a\rangle\}; \tag{9.68}$$

the state does not have definite parity and is not an eigenstate of H. To investigate the behavior of the particle at later times, we use the time-dependent Schrödinger equation

$$i\hbar\frac{d}{dt}|\psi(t)\rangle = (H_0 + H_{\text{int}})|\psi(t)\rangle \tag{9.69}$$

and the expansion

$$\begin{aligned} &|\psi(t)\rangle = \alpha(t)|L\rangle + \beta(t)|R\rangle \\ &|\alpha(t)|^2 + |\beta(t)|^2 = 1. \end{aligned} \tag{9.70}$$

Inserting the expansion (9.70) into the Schrödinger equation (9.69) and multiplying in turn from the left by $\langle L|$ and $\langle R|$, yields a system of two coupled differential equations for $\alpha(t)$ and $\beta(t)$:

$$\begin{aligned} i\hbar\dot{\alpha}(t) &= (E_0 + E')\alpha(t) + \Delta E\beta(t) \\ i\hbar\dot{\beta}(t) &= \Delta E\alpha(t) + (E_0 + E')\beta(t). \end{aligned} \tag{9.71}$$

The solution of these equations with the initial conditions $\alpha(0) = 1$ and $\beta(0) = 0$ gives

$$|\psi(t)\rangle = \exp\left[\frac{-i(E_0 + E')t}{\hbar}\right]\left[\cos\left(\frac{\Delta Et}{\hbar}\right)|L\rangle - i\sin\left(\frac{\Delta Et}{\hbar}\right)|R\rangle\right]. \tag{9.72}$$

The probability of finding the particle, dropped into well L at $t = 0$, in well R at a time t is given by the absolute square of the expansion coefficient of $|R\rangle$, or

$$\text{prob}(R) = \sin^2\left(\frac{\Delta Et}{\hbar}\right). \tag{9.73}$$

The particle hence oscillates between the two wells with a circular frequency 2ω, where

$$\omega = \frac{\Delta E}{\hbar} = \langle L|H_{\text{int}}|R\rangle\frac{1}{\hbar}. \tag{9.74}$$

9.7 The Neutral Kaons

Hypercharge is the only quantum number that distinguishes the neutral kaon from its antiparticle: $Y(K^0) = 1, Y(\overline{K^0}) = -1$. Since the hadronic and the electromagnetic interactions conserve hypercharge, K^0 and $\overline{K^0}$ appear as two distinctly different particles in all experiments involving these two forces. However, the weak interaction does not conserve hypercharge, and virtual weak transitions between the two particles can occur. Both particles decay, for instance, into two pions, $K^0 \to 2\pi$ and $\overline{K^0} \to 2\pi$. They are therefore connected by virtual second-order weak transitions,

$$K^0 \rightleftharpoons 2\pi \rightleftharpoons \overline{K^0}, \tag{9.75}$$

shown in Fig. 9.10. The existence of these virtual transitions leads to remarkable effects, as first pointed out by Gell-Mann and Pais.[25]

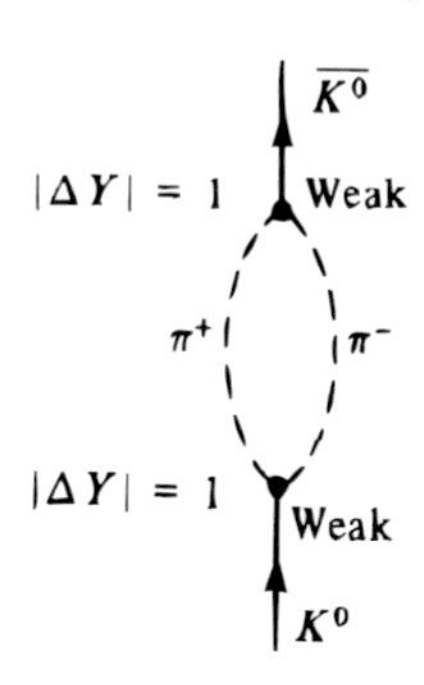

Figure 9.10: Example of a virtual second-order weak transition $K^0 \to \overline{K^0}$.

The effects are easy to understand if the analogy to the two-well problem is recognized: In the absence of the weak interaction, $|K^0\rangle$ and $|\overline{K^0}\rangle$ are two unconnected degenerate states just like $|L\rangle$ and $|R\rangle$ before switching on H_{int}. The weak interaction, H_w, then plays the same role as H_{int} and connects the two states $|K^0\rangle$ and $|\overline{K^0}\rangle$. With minor changes, the equations and results of the previous section can be applied to the neutral kaon system by setting

$$H_0 = H_h + H_{em} \equiv H_s, \qquad H_{\text{int}} = H_w. \tag{9.76}$$

To find the transformation that corresponds to Eq. (9.62), we note that charge conjugation changes K^0 into $\overline{K^0}$ and vice versa,

$$C|K^0\rangle = |\overline{K^0}\rangle, \quad C|\overline{K^0}\rangle = |K^0\rangle. \tag{9.77}$$

Gell-Mann and Pais used these relations in their original work in place of Eq. (9.62) in order to find the proper linear combinations of the unperturbed eigenstates $|K^0\rangle$ and $|\overline{K^0}\rangle$. When the breakdown of parity was discovered it became clear that C does not commute with the total Hamiltonian, and this fact is expressed in Eq. (9.45). The combined parity, CP, is a better choice, as can be seen as follows. C applied to a neutrino with negative helicity changes it into an antineutrino with negative helicity, in disagreement with experiment. CP, however, changes a negative helicity neutrino into an antineutrino with positive helicity, in agreement with observation. To find the effect of CP on states $|K^0\rangle$ and $|\overline{K^0}\rangle$, we note that the intrinsic parity of the kaons is negative,

[25] M. Gell-Mann and A. Pais, *Phys. Rev.* **97**, 1387 (1955).

$$P|K^0\rangle = -|K^0\rangle, \quad P|\overline{K^0}\rangle = -|\overline{K^0}\rangle, \tag{9.78}$$

so that the effect of the combined parity is given by

$$CP|K^0\rangle = -|\overline{K^0}\rangle, \quad CP|\overline{K^0}\rangle = -|K^0\rangle. \tag{9.79}$$

If the total Hamiltonian conserves CP,

$$[H, CP] = [H_s + H_w, CP] = 0, \tag{9.80}$$

then the eigenstates of H can be chosen to also be eigenstates of CP. (We shall return to the question of CP conservation in Section 9.8.) Just as in Eq. (9.63), we write these eigenstates as[26]

$$\begin{aligned} |K_1^0\rangle &= \sqrt{\tfrac{1}{2}}\{|K^0\rangle - |\overline{K^0}\rangle\} \\ |K_2^0\rangle &= \sqrt{\tfrac{1}{2}}\{|K^0\rangle + |\overline{K^0}\rangle\}, \end{aligned} \tag{9.81}$$

with

$$CP|K_1^0\rangle = +|K_1^0\rangle, \quad CP|K_2^0\rangle = -|K_2^0\rangle. \tag{9.82}$$

K_1^0 has a combined parity η_{CP} of $+1$, and K_2^0 one of -1.

The analogy with the two-well problem in Section 9.6 is obvious: The states $|K^0\rangle$ and $|\overline{K^0}\rangle$, just as the states $|L\rangle$ and $|R\rangle$ are eigenstates of the unperturbed Hamiltonian. The states $|K_1^0\rangle$ and $|K_2^0\rangle$, just as $|s\rangle$ and $|a\rangle$, are simultaneous eigenstates of the total Hamiltonian and of the relevant symmetry operator. The results of Section 9.6 can be applied to the neutral kaons and remarkable predictions ensue:

1. K^0 is the antiparticle of $\overline{K^0}$. The two should therefore have the same mass and the same lifetime. K_1^0, however, is not the antiparticle of K_2^0, and the two particles can have very different properties.

2. The thought experiment of "dropping the particle at $t = 0$ into one well," discussed in Section 9.6, can be realized with kaons. Kaons are produced by hadronic interactions, for instance by $\pi^- p \to K^0\Lambda^0$. Such a production in a state of well-defined hypercharge corresponds to dropping the particle into one well. Equations (9.72) and (9.73) predict that the particle will tunnel into the other well. The other well corresponds to the opposite hypercharge: A neutral kaon, produced in a state of $Y = 1$, should partially transform to a state with $Y = -1$ after a certain time.

[26]The freedom allowed by the arbitrary phases in the definitions of C and P has led to different ways of writing the linear combinations (9.81). The usual choice is to introduce a phase of 180° in either C or P so that $CP \mid K^0\rangle = + \mid K^0\rangle, CP \mid \bar{K}^0\rangle = + \mid K^0\rangle$. The observable consequences are unchanged by the phase choice.

3. The states $|s\rangle$ and $|a\rangle$ have slightly different energies, as is shown by Eq. (9.66) and Fig. 9.9. The corresponding kaon states, $|K_1^0\rangle$ and $|K_2^0\rangle$, should therefore have slightly different rest energies.

In the following we shall describe the verification of these three predictions.

1. *K_1^0 and K_2^0 Decay Differently.* Energetically, kaons can decay into two or three pions. Since the kaon spin is zero, the total angular momentum of the pions in the final state must also be zero. Consider first the two-pion system, $\pi^+\pi^-$. In the c.m. of the two pions, the parity operation exchanges π^+ and π^-. Charge conjugation exchanges π^- and π^+ again so that the combined operation CP leads back to the original state. The same argument holds for two neutral pions so that

$$CP|\pi\pi\rangle = +|\pi\pi\rangle \quad \text{in all states with } J = 0. \tag{9.83}$$

Two pions with total angular momentum zero have a combined parity $\eta_{CP} = +1$. If the total Hamiltonian conserves CP, as assumed by Eq. (9.80), CP must be conserved in the decays of the neutral kaons. K_1^0, with $\eta_{CP} = 1$, then can decay into two pions. K_2^0, with $\eta_{CP} = -1$, *cannot* decay into two pions; it must decay into at least three:

$$K_2^0 \not\rightarrow 2\pi \quad \text{if } CP \text{ conserved.} \tag{9.84}$$

The decay energy available for the two-pion mode is about 220 MeV, and for the three-pion mode about 90 MeV. The phase space available for decay into three pions is therefore considerably smaller than for that into two pions (Chapter 10), and the mean life τ_1 of K_1^0 is expected to be much smaller than the mean life τ_2 of K_2^0.

The decay of K^0 (or of $\overline{K^0}$) is more complicated. Consider, for instance, K^0 produced by a reaction such as $\pi^- p \to K^0\Lambda^0$. At $t = 0$, the state has hypercharge $Y = 1$; with Eq. (9.81) the initial state is

$$|t = 0\rangle \equiv |K^0\rangle = \sqrt{\tfrac{1}{2}}\{|K_1^0\rangle + |K_2^0\rangle\}. \tag{9.85}$$

If the particle is allowed to decay freely, it will do so through the weak interactions. We have observed above that K_1^0 and K_2^0 are expected to decay with different lifetimes τ_1 and τ_2. K^0 will therefore *not* decay with a single lifetime. Gell-Mann and Pais expressed their prediction in these words[25]: "To sum up, our picture of the K^0 implies that it is a particle mixture exhibiting two distinct lifetimes, that each lifetime is associated with a different set of decay modes, and that *not more than half of all K^0's* can undergo the familiar decay into two pions." They also stated "Since we should properly reserve the word 'particle' for an object with a unique lifetime, it is the K_1^0 and the K_2^0 quanta that are the true 'particles.' The K^0 and the $\overline{K^0}$ must, strictly speaking, be considered 'particle mixtures.' "

The unequivocal predictions of Gell-Mann and Pais concerning the decay properties of K^0 posed a challenge to the experimental physicists: Does K^0 possess a long-lived component that decays into three pions? At the time of Gell-Mann and Pais' paper, neutral kaons were known to decay with a lifetime of about 10^{-10} sec. A longer-lived component was found by a Columbia–Brookhaven group using a cloud chamber.[27] The experimental arrangement is sketched in Fig. 9.11. A 90 cm cloud chamber was exposed to the neutral beam emitted from a copper target hit by 3 GeV protons. Charged particles were eliminated by a sweeping magnet. The 6 m flight path from target to chamber corresponded to about 100 mean lives for the known decay component; the K_1^0 component hence was absent in the chamber. The observation of many V events that could not be fitted kinematically by two-pion decays established the existence of a long-lived three-pion decay of K_2^0 and constituted a clear verification of the brilliant proposal by Gell-Mann and Pais. Later experiments substantiated this conclusion, and the mean lives of the two components were found to be $\tau(K_2^0) = 0.517 \times 10^{-7}$ sec and $\tau(K_1^0) = 0.894 \times 10^{-10}$ sec.

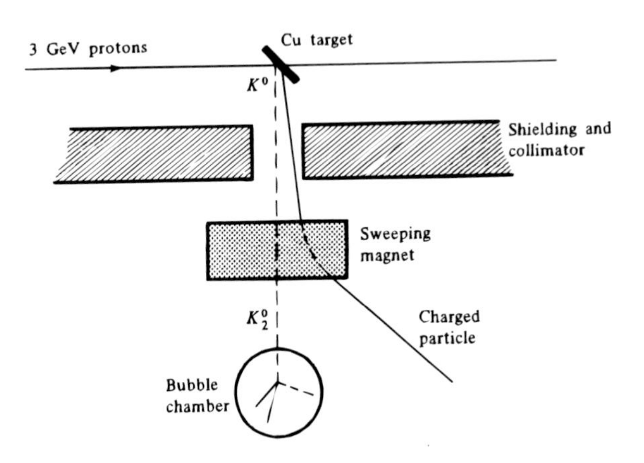

Figure 9.11: Observation of the long-lived neutral kaon component, K_2^0, by a Columbia–Brookhaven group in a cloud chamber. [K. Lande et al., *Phys. Rev.* **103**, 1901 (1956); **105**, 1925 (1957).] The charged particles are swept out of the beam by a magnet; the neutral particles in the beam are observed after a flight of about 3×10^{-8} sec. The observed V events cannot be explained by two-particle decays.

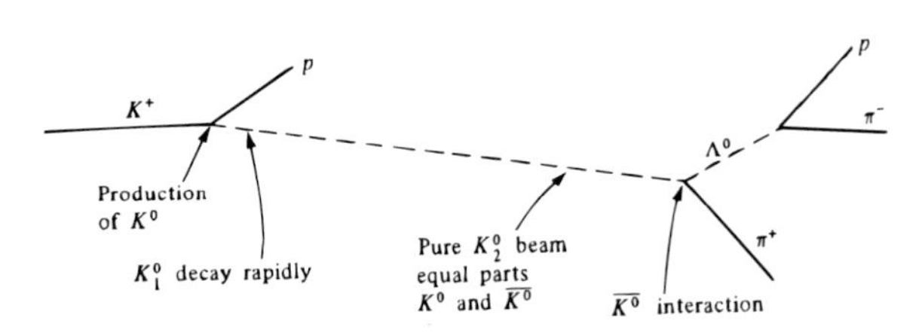

Figure 9.12: Observation of the $\overline{K^0}$ component of an initially pure K^0 beam.

2. *Hypercharge Oscillations.*[28] Equation (9.72) predicts that a particle that was dropped into one well at time $t = 0$ will continuously oscillate between the two wells, with a circular frequency given by Eq. (9.74). If neutral kaons were stable, they would do the same. However, they decay, and the oscillations are damped. Consider a situation where at time $t = 0$ a K^0 was produced, as described by Eq. (9.85). After a time that is long compared to $\tau(K_1^0)$, all $K_1^0 s$ will have decayed,

[27]K. Lande, E. T. Booth, J. Impeduglia, L. M. Lederman, and W. Chinowsky, *Phys. Rev.* **103**, 1901 (1956); **105**, 1925 (1957).

[28]A. Pais and O. Piccioni, *Phys. Rev.* **100**, 1487 (1955).

and only K_2^0s are left, as shown in Fig. 9.11. Equation (9.81) expresses K_2^0 in terms of the eigenstates of hypercharge as

$$|K_2^0\rangle = \sqrt{\tfrac{1}{2}}\{|K^0\rangle + |\overline{K^0}\rangle\}.$$

The kaon beam will consist of equal parts K^0 and $\overline{K^0}$. A kaon beam that has been produced in a pure $Y = 1$ state has changed to one containing equal parts $Y = 1$ and $Y = -1$. Experimentally, the appearance of the $\overline{K^0}$ component can be verified through the observation of hadronic interactions such as $\overline{K^0}p \to \pi^+\Lambda^0$. Since nucleons have $Y = 1$ and the Λ^0 has $Y = 0$, a state $\pi^+\Lambda^0$ can be produced only by $\overline{K^0}$, not by K^0. The features of the observation of the $\overline{K^0}$ component are shown in Fig. 9.12.

3. *Regeneration and Mass Splitting.* If the pure K_2^0 beam shown in Fig. 9.12 passes through matter, the short-lived component K_1^0 will reappear; this process is called regeneration and is sketched in Fig. 9.13.

Since the experiment involves the hadronic interaction of the kaons with matter, we return to the description in terms of K^0 and $\overline{K^0}$,

$$K_2^0 = \sqrt{\tfrac{1}{2}}\{|K^0\rangle + |\overline{K^0}\rangle\}.$$

K^0 and $\overline{K^0}$ interact differently with matter; the $\overline{K^0}$ can participate in reactions such as $\overline{K^0}p \to \pi^+\Lambda^0$ and $\overline{K^0}n \to \pi^0\Lambda^0$ that are forbidden to the K^0 because of strangeness conservation. We describe the effects of the regenerator by two complex numbers, f and $\bar{f}$.

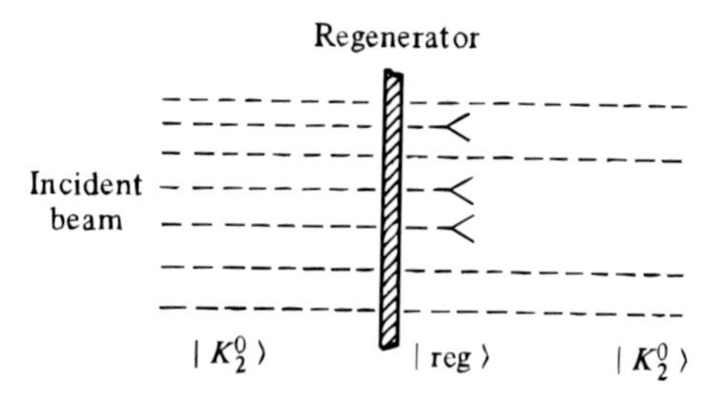

Figure 9.13: Regeneration of a K_1^0. A pure K_2^0 beam that passes through matter transforms into a beam that again contains a K_1^0 component. The K_1^0 s decay close to the regenerator into two pions and are thus unambiguously identifiable.

Neglecting decay effects, the amplitude of the regenerated beam immediately after the regenerator becomes

$$|\text{reg}\rangle = \sqrt{\tfrac{1}{2}}\{f|K^0\rangle + \bar{f}|\overline{K^0}\rangle\} = \tfrac{1}{2}(f - \bar{f})|K_1^0\rangle + \tfrac{1}{2}(f + \bar{f})|K_2^0\rangle. \qquad (9.86)$$

Because K^0 and $\overline{K^0}$ interact differently, f and $\bar{f}$ are different and the regenerated beam contains again a K_1^0 component. Experimentally, this component can be recognized by the emergence of two-prong events close to the regenerator.[29]

Regeneration is one of the methods by which the mass difference between the K_1^0 and the K_2^0 can be determined.[30] Consider the simplest case, coherent forward

[29] R.H. Good et al., *Phys. Rev.* **124**, 1223 (1961).

[30] The various methods for mass determination are described in T.D. Lee and C.S. Wu, *Annu. Rev. Nucl. Sci.* **16**, 511 (1966); for recent measurements see the CPLEAR (http://cplear.web.cern.ch/cplear) and KTEV (http://kpasa.fnal.gov:8080/public/) results.

regeneration. The wavefunction of the K_2^0 moving through the regenerator will be proportional to $\exp(ip_2x/\hbar)$. Thus, at each point x along the path, regeneration of the K_1^0 also will be proportional to $\exp(ip_2x/\hbar)$, but the regenerated wave will move through the absorber with a wavefunction proportional to $\exp(ip_1x/\hbar)$. The interference between the two waves at the end of the regenerator of length L will thus contain a term proportional to $\exp[i(p_2-p_1)L/\hbar]$. In the forward direction no energy is lost so that $p^2c^2 + m^2c^4 =$ constant, or

$$\Delta pc = \left(\frac{mc}{p}\right)\Delta mc^2,$$

where $\Delta p \equiv p_2 - p_1$, and $\Delta m = m_1 - m_2$ is the mass difference between the K_1^0 and the K_2^0. Measurements of the probability of finding a K_1^0 after a regenerator of length L as a function of L yields the mass difference.[31] Additional experiments also give the sign of the mass difference, with the result

$$\Delta m = m_1 - m_2 = -3.489 \times 10^{-6}\ \mathrm{eV}/c^2. \tag{9.87}$$

The mass splitting is incredibly small; it is of second order in the weak interaction strength. The ratio $\Delta m/m_K \approx 10^{-14}$ proves that the weak interaction is responsible for H_{int}, Eq. (9.76), as shown in Fig. 9.10.

All predictions of the Gell-Mann–Pais theory thus have been verified experimentally. In addition to yielding deep insight into the kaon system, the experiments also show that particles have wave-like properties and behave as demanded by quantum mechanics.

9.8 The Fall of CP Invariance

Kaons are a wonderful source of surprises. In Section 9.3 we described how the observation of two different decay modes of the charged kaons led to the fall of parity invariance. In the previous section, we showed that the coherence properties of the neutral kaons give rise to two different decay mean lives, to hypercharge oscillations, and to regeneration. The coherence properties were predicted theoretically, and the subsequent experimental verification was exciting but not unexpected. The breakdown of parity was unexpected, but it was taken in stride and was quickly incorporated into the theoretical framework. In this section we shall treat the next major surprise, the fall of CP invariance.

Three features that were discussed in the previous section underlie the experiments demonstrating CP violation:

1. A neutral kaon beam far away from the point of production is in a pure $|K_2^0\rangle$ state.

[31] T. Fujii et al., *Phys. Rev. Lett.* **13**, 253, 324 (1964); J. H. Christenson et al., *Phys. Rev.* **140B**, 74 (1965).

2. The state $|K_2^0\rangle$ is an eigenstate of the total Hamiltonian. In vacuum, no transitions from $|K_2^0\rangle$ to $|K_1^0\rangle$ can occur. For the two wells, the absence of such transitions is expressed by Eq. (9.65). The corresponding relation for kaons follows from Eqs. (9.80) and (9.81) as

$$\langle K_1^0|H|K_2^0\rangle = 0. \tag{9.88}$$

3. As stated by Eq. (9.84), K_2^0 cannot decay into two pions if CP is conserved.

In 1964, a Princeton group performed an experiment to set a lower limit on the two-pion decay of K_2^0.[32] Another experiment was simultaneously done by an Illinois group.[33] Both gave the astounding result that decays into two pions do occur; the branching ratio was found to be approximately

$$\frac{\text{Int}(K_L^0 \rightarrow \pi^+\pi^-)}{\text{Int}(K_L^0 \rightarrow \text{all charged modes})} \approx 2 \times 10^{-3}. \tag{9.89}$$

We have switched notation here and denote the long-lived neutral kaon with K_L^0 and the short-lived one with K_S^0. The reason for the switch is Eq. (9.82), which *defines* K_1^0 and K_2^0 to be eigenstates of CP. Equation (9.89) indicates, however, that the long-lived kaon is *not* an eigenstate of CP. It is customary to retain the notation K_1^0 and K_2^0 for the eigenstates of CP and to denote the real particles with K_S^0 and K_L^0.

The news of violation of CP traveled through the world of physics with nearly the speed of light, just as, seven years earlier, had the news of parity breakdown. It was greeted with even more scepticism. To describe the reason for the disbelief, we digress to describe the celebrated *CPT theorem*. The *CPT* theorem is easy to understand but difficult to prove. In a somewhat sloppy way, it can be stated as follows: the product of the three operations T, C, and P commutes with practically every conceivable Hamiltonian, or

$$[CPT, H] = 0. \tag{9.90}$$

In other words, our world and a time-reversed parity-reflected antiworld must behave identically. The order of the three operators T, C, and P is irrelevant.[34] The operation CPT is thus very different from the individual operations T, C, and P. It is easy to construct a Lorentz-invariant Hamiltonian that violates, for instance,

[32] J. H. Christenson, J. W. Cronin, V. L. Fitch, and R. Turlay, *Phys. Rev. Lett.* **13**, 138 (1964). V. L. Fitch, *Rev. Mod. Phys.* **53**, 367 (1981); *Science* **212**, 939 (1981); J. W. Cronin, *Rev. Mod. Phys.* **53**, 373 (1981); *Science* **212**, 1221 (1981).

[33] A. Abashian, R. J. Abrams, D. W. Carpenter, G. P. Fisher, B. M. K. Nefkens, and J. H. Smith, *Phys. Rev. Lett.* **13**, 243 (1964).

[34] Since the order of the operations T, C, and P does not matter, there exist 3! possibilities of naming the theorem. Lüders and Zumino checked that their choice,TCP, agreed with the name of a well-known gasoline additive. Despite this, we use the more standard order, namely CPT. [G. Lüders, *Physikalische Blätter* **22**, 421 (1966).]

P and C, and we shall discuss one in Chapter 11. However, it is extremely difficult to construct a Lorentz-invariant Hamiltonian that violates CPT. (These statements are somewhat oversimplified, but the essential features are correct.)

The CPT theorem was something of a sleeper. In preliminary form, it was discovered independently by Schwinger and by Lüders.(35) Pauli then generalized the theorem.(36) Up to 1956, however, it was considered to be rather esoteric. Dogma held that the three operations T, C, and P were separately conserved, and the CPT theorem was assumed to give little experimentally usable information. When violation of parity became a possibility, the CPT theorem suddenly acquired more meaning(37): Equation (9.90) states that if P is violated, some other operation must also be violated. Indeed, we have mentioned in Section 9.4 that C is also not conserved in the weak interaction.

The CPT theorem can be tested. For instance, it predicts that the masses and lifetimes of weakly decaying particles and antiparticles, such as the negative and positive muon, should be identical, even though charge conjugation invariance does not hold in the weak interactions. No violation of the CPT theorem has been found, despite a resurgence of interest caused by some (string) theories which try to unify gravity with the other interactions. Tests that are as good or better than the equality of the masses of the neutral kaons , $\bar{K}^0$ and K^0 to about 1 part in 10^{14} have been performed.(38)

After this digression, we return to the situation in 1964. The observed CP violation in the decay of the neutral kaons together with the CPT theorem leads nearly inescapably to one of two conclusions: either T is not conserved or the CPT theorem is wrong. Theorists had in the meantime found even stronger proofs for it(39) and were rather reluctant to give it up. On the other hand, time reversal is also a cherished symmetry. Certainly the easiest way out would have been capitulation of the experimentalists with an admission that the experiments were wrong. Additional data, however, strengthened the earliest conclusions. Detailed analysis of all the information from the decays of the neutral kaons at least provides some further insight. The analysis implies that the CPT theorem holds but that not only CP but also T invariance is violated.(40)

---

35 J. Schwinger, *Phys. Rev.* **82**, 914 (1951); **91**, 713 (1953); G. Lüders, *Kgl. Danske Videnskab Selskab, Mat.fys. Medd.* **28**, No. 5 (1954).

36 W. Pauli, in *Niels Bohr and the Development of Physics*, (W. Pauli, ed.) McGraw-Hill, New York, 1955.

37 T. D. Lee, R. Oehme, and C. N. Yang, *Phys. Rev.* **106**, 340 (1957).

38 *CPT and Lorentz Symmetry III*, (Alan Kostelecky, ed.), World Sci., Singapore, 2005.

39 Proofs of the *CPT* theorem require relativistic field theory and are never easy. For the reader who wants to convince himself of this fact, we list here a few references, approximately in order of increasing difficulty: J. J. Sakurai, *Invariance Principles and Elementary Particles*, Princeton University Press, Princeton, N.J., 1964; G. Lüders, *Ann. Phys.* (New York) **2**, 1 (1957); R. F. Streater and A. S. Wightman, *PCT, Spin, and Statistics, and All That*, Benjamin, Reading, Mass., 1964.

40 R. C. Casella, *Phys. Rev. Lett.* **21**, 1128 (1968); **22**, 554 (1969); K. R. Schubert, B. Wolff, J. C. Chollet, J. M. Gaillard, M. R. Jane, T. J. Ratcliffe, and J.-P. Repellin, *Phys. Lett.* **31B**, 662 (1970); G. V. Dass, *Fortsch. Phys.* **20**, 77 (1972).

• There are three possible causes of CP Violation in kaon decays: The first one is in the K_1 and K_2 mixing (mass mixing):

$$K_L = \frac{K_2 + \epsilon K_1}{\sqrt{1 + |\epsilon|^2}}, \qquad K_S = \frac{K_1 - \epsilon K_2}{\sqrt{1 + |\epsilon|^2}}, , \tag{9.91}$$

where ϵ is a measure of the CP-violation. It is the K_L and K_S which have definite lifetimes. Experimentally, it is found that

$$\epsilon = (2.284 \pm 0.014) \times 10^{-3}. \tag{9.92}$$

The second possibile cause of CP Violation is in the decay matrix element itself. A third one is an interference beween these two causes. The CP violation in the decay matrix element, measured by a parameter called ϵ', occurs due to an admixture of isospin 2 to isospin 0 (or change of isospin by 3/2 vs. 1/2 in the decay), see Problem 9.46. The ratio of ϵ' to ϵ is of the order of 10^{-3}. •

9.9 References

General references concerning invariance properties are given in Section 7.7.

Discontinuous transformations and unitary and antiunitary operators are treated in detail in Messiah, Vol. II, Chapter XV. Some of the theoretical aspects of parity and time reversal are discussed in E. M. Henley, "Parity and Time Reversal Invariance in Nuclear Physics," *Annu. Rev. Nucl. Sci.* **19**, 367 (1969), and in *Symetries and Fundamental Interactions in Nuclei*, ed. W.C. Haxton and E.M. Henley, World Sci., Singapore, 1995. A detailed treatise is Robert G. Sachs, *The Physics of Time Reversal Invariance*, University of Chicago Press, Chicago, 1987. The theory and status of these symmetries are discussed in a number of reviews: E.N. Fortson and L.L. Lewis, "Atomic Parity Nonconservation Experiments," *Phys. Rep.* **113**, 289 (1984); E.G. Adelberger and W. Haxton, "Parity Violation in the Nucleon–Nucleon Interaction," *Annu. Rev. Nucl. Part. Sci.* **35**, 501 (1985); E.M. Henley, "Status of Some Symmetries", *Prog. Part. Nucl. Phys.*, (A. Faessler, ed.) **20**, 387 (1987); and *Tests of Time Reversal Invariance*, (N.R. Roberson, C.R. Gould, and J.D. Bowman, eds.) World Scientific, Teaneck, NJ, 1988.

A popular account of CP and T violation is given by N. Fortson, P. Sandars, and S. Barr, *Phys. Tod.* **56**, 33 (June 2003); see also R. G. Sachs, *Science* **176**, 587 (1972). CP violation in the K and B meson systems can be found on R. Kleinknecht, *Experimental Clarification in the Neutral K Meson and B Meson Systems*, Springer, NY, 2003; and in P. Bloch and L. Tauscher, *Annu. Rev. Nucl. Part. Sci.* **53**, 123 (2003); see also *CP Violation in Particle, Nuclear and Astrophysics*, ed. M. Beyer, Springer, New York, 2002; and I.I. Bigi and A.I. Sanda, *CP Violation in Nature—A Status Report*, *Comm. Nucl. Part. Phys.* **14**, 149 (1985); I.B. Khriplovich, S.K. Lamoreaux, *CP violation without strangeness: electric dipole moments of particles, atoms, and molecules*, Berlin ; New York : Springer-Verlag, 1997.

The theory of CP invariance is treated in the books by G. Branco, L.Lavoura, and T. Silva, *CP Violation*, Clarendon Press, Oxford (1999) and I.I. Bigi and A.I. Sanda, *CP Violation*, Cambridge University Press, Cambridge (2000). While most of these reviews are written at a higher level, much useful information can be extracted even at the level of the present book.

Problems

9.1. (a) Show that an infinitesimal rotation, R, and space inversion (parity), P, commute by showing in a sketch that PR and RP transform an arbitrary vector $\boldsymbol{x}$ into the same vector $\boldsymbol{x}'$.

(b) Use part (a) to show that P and $\boldsymbol{J}$ commute, where $\boldsymbol{J}$ is the generator of the infinitesimal rotation R.

9.2. Show that the commutation relations for angular momentum remain invariant under the parity operation.

9.3. Use the Schrödinger equation with a Hamiltonian $H = (p^2/2m) + V(\boldsymbol{x})$. Show that $\psi(-\boldsymbol{x})$ satisfies the Schrödinger equation if $\psi(\boldsymbol{x})$ does, provided that $V(\boldsymbol{x}) = V(-\boldsymbol{x})$.

9.4. Show that the eigenfunctions ψ_{lm} given in Problem 5.3 are eigenfunctions of P. Compute the eigenvalues and compare the result with Eq. (9.10).

9.5. Use a gauge transformation of the form of Eq. (7.32), with a properly chosen value of ϵ, to show that the relative parity of the proton and the positive pion is not a measurable quantity.

9.6. Would it be possible to assign meaningful intrinsic parities to all hadrons if in Eq. (9.22) instead of the parity of the lambda the parity of

(a) π^0 or

(b) K^+

had been chosen? Justify your answers.

9.7. *Discuss the reaction

$$np \longrightarrow d\gamma$$

and use information in the literature (e.g., nuclear physics texts) to determine the intrinsic parity of the deuteron.

9.8. * Find information on the reactions

$$dd \longrightarrow p^3\text{H}$$
$$dd \longrightarrow n^3\text{He}$$

and discuss the parities of the ^{3}H and ^{3}He.

9.9. * Discuss the determination of the parity of a hyperon (not the lambda).

9.10. * How would you determine the parity of the kaon? Compare your proposal with actual experiments.

9.11. The operator for the emission of electric dipole gamma radiation is of the form $q\boldsymbol{x}$, where q is a charge. The matrix element for a transition $i \to f$ is of the form

$$F_{fi} = \int d^3x \psi_f^*(\boldsymbol{x}) q\boldsymbol{x} \psi_i(\boldsymbol{x}).$$

Use this expression to find the parity selection rule for electric dipole radiation.

9.12. Discuss the arguments and facts that assign spin 0 and positive parity to the alpha particle (ground state of ^{4}He).

9.13. Electrons and positrons emitted in weak interactions can be characterized by their momenta and their spins.

(a) Show that a nonvanishing value of the expectation value $\langle \boldsymbol{J} \cdot \boldsymbol{p} \rangle$ implies parity nonconservation.

(b) Discuss an experiment that can be used to measure the helicity of electrons.

9.14. Assume a nucleus with a magnetic moment g factor of $g = 1$ to be in a magnetic field of 1 MG. Compute the temperature at which at least 99% of the nuclei are polarized.

9.15. Use the information given in Figs. 7.2 and 9.6 to answer the following question. Are electron and antineutrino emitted predominantly in the same direction or in opposite directions? (For simplicity assume the ^{60}Co state to be 1^+ and that of ^{60}Ni to be 0^+.)

9.16. Discuss the evidence for parity nonconservation in the decay $\pi^+ \to \mu^+ \, \nu_\mu$:

(a) What polarization of the muon is expected?

(b) How can the muon polarization be observed?

9.17. Electrons emitted in nuclear beta decay are found to have negative helicity, whereas positrons show positive helicity. What can be deduced from this observation?

9.18. Consider a system consisting of a positive and a negative pion, with orbital angular momentum l in their c.m.

(a) Determine the C parity of this $(\pi^+\pi^-)$ system.

(b) If $l = 1$, can the system decay into two photons? Justify your answer.

9.19. Show that Maxwell's equations are invariant under time reversal.

9.20. Assume

$$\psi = \begin{pmatrix} \psi_1 \\ \psi_2 \end{pmatrix}$$

to be a two-component Pauli spinor, satisfying the Pauli equation. Find the wave function $T\psi$ that satisfies the Pauli equation.

9.21. Discuss one test of time-reversal invariance in the hadronic and one in the electromagnetic interaction.

9.22. Show that the helicity $\boldsymbol{J} \cdot \hat{\boldsymbol{p}}$ is invariant under the time-reversal operation.

9.23. A very small violation of parity invariance has been observed in nuclear decays $(\mathcal{F}_P \approx 10^{-7})$. How can this violation be explained without giving up parity conservation in the hadronic interaction?

9.24. Sketch the application of the two-well model to ammonia. How big is the total splitting $2\Delta E$ between states $|a\rangle$ and $|s\rangle$? Which state lies higher? Are transitions between states $|a\rangle$ and $|s\rangle$ observed? If yes, where are these transitions important?

9.25. (a) Find the general solution of Eqs. (9.71).

(b) Verify that Eq. (9.72) is the special solution of Eq. (9.71) with the initial conditions $\alpha(0) = 1$ and $\beta(0) = 0$.

9.26. Neutron and antineutron are neutral antiparticles, just as K^0 and $\overline{K^0}$ are. Why is it not meaningful to introduce linear combinations N_1 and N_2, similar to K_1^0 and K_2^0?

9.27. Assume that K^0 is produced at $t = 0$.

(a) Justify that the wave function of K^0 at rest at time t can be written as

$$|t\rangle = \sqrt{\frac{1}{2}}\left\{|K_1^0\rangle \exp\left(\frac{-im_1c^2t}{\hbar} - \frac{t}{2\tau_1}\right)\right.$$
$$\left. +|K_2^0\rangle \exp\left(\frac{-im_2c^2t}{\hbar} - \frac{t}{2\tau_2}\right)\right\},$$

where m_i and τ_i are mass and lifetime of K_i.

(b) Express $|t\rangle$ as a function of $|K^0\rangle$ and $|\overline{K^0}\rangle$.

(c) Compute the probability of finding $\overline{K^0}$ at time t as a function of $\Delta m = m_1 - m_2$.

(d) Sketch the probability for

$$\Delta m = 0, \qquad \Delta m = \frac{\hbar}{c^2\tau_1}, \qquad \Delta m = \frac{2\hbar}{c^2\tau_1}.$$

9.28. K_1^0 and K_2^0 have slightly different rest masses.

(a) Estimate the magnitude of the mass difference by assuming that the splitting is due to a second-order weak effect and that the weak interaction is about a factor of 10^7 weaker than the hadronic one.

(b) Describe how the magnitude of the mass difference can be determined.

(c) Compare the actually observed value with your estimate.

9.29. (a) Assume that K^0 and $\overline{K^0}$ beams, of equal energy, pass through a slab of matter. Will the beams be attenuated equally? If not, why not?

(b) A pure K_2^0 beam passes through a slab of matter. Will the emerging beam still be a pure K_2^0 beam? Explain your answer.

(c) How can it be experimentally decided if the K_2^0 beam is still pure after passage through the slab?

9.30. * Describe the experimental arrangements that were used to detect the two-pion decay of the long-lived neutral kaon.

9.31. Assume that you are in contact with physicists on another galaxy. The contact is restricted to exchange of information. Can you find out if the other physicists are built from matter or antimatter? Discuss the following three possibilities:

(a) C, P, and T are conserved in all interactions.

(b) C and P are violated but CP is conserved in the weak interaction.

(c) C, P, and CP are violated, as discussed in Section 9.8.

9.32. Show that CPT invariance guarantees that a particle and its antiparticle have equal mass.

9.33. Show that the decay of the K^0

(a) to $\pi^0\pi^0$ is forbidden if the spin of the K^0 is odd,

(b) to $\pi^0\gamma$ is allowed if the spin of the K^0 is not zero.

9.34. How can one determine the parity of the photon? Describe a possible experiment.

9.35. (a) For the cross section, determine the order of magnitude of the ratio of the interference term of the amplitudes of the parity-violating weak interaction to that of the electromagnetic interaction in elastic electron scattering on hydrogen at an energy of 20 GeV and a momentum transfer of 1 GeV/c. Compare to experiment.

(b) Repeat part (a) for the total cross section of proton–proton scattering at a laboratory energy of about 50 MeV.

9.36. Show that the rate of the parity-forbidden alpha decay of the 2^- level in ^{20}Ne or ^{16}O is proportional to the square of the weak interaction, i.e., to $|\mathcal{F}|^2$, and does not depend on an interference term between the weak and strong amplitudes.

9.37. The decay of the η is useful for testing C-invariance. Which of the following decays are allowed and which are forbidden by C-invariance?

$$\eta \longrightarrow \gamma\gamma$$
$$\eta \longrightarrow \pi^0\gamma$$
$$\eta \longrightarrow \pi^0\pi^0\pi^0$$
$$\eta \longrightarrow 3\gamma$$
$$\eta \longrightarrow \pi^+\pi^-\pi^0$$

9.38. Show that the neutron or any other non-degenerate system cannot have an electric dipole moment unless both P and T conservation are violated.

9.39. Compare the expected orders of magnitude of the electric dipole moments of a neutron and a heavy neutral atom. Explain or show reasoning.

9.40. A B^0 meson consists of a bottom antiquark and a down quark. Consider the system of a B^0 and $\overline{B^0}$ and compare it to that of a K^0 and $\overline{K^0}$. Should there be a B_1^0 and B_2^0? Do you expect CP to be violated in the decays? If the system is produced in e^+e^- collisions, can CP be tested? If so, suggest some possible experiments to test CP in this system?

9.41. Estimate the energy difference between a neutron and proton, if they are made up of up and down quarks, with the average mass of the quarks being 330 MeV/c^2, but the down quark being 5 MeV/c^2 heavier than the up quark.

9.42. The ρ^0 meson decays hadronically to two pions. Its spin is $1\hbar, \pi_\rho = -1, \eta_c = -1$, and its isospin $I = 1$. Can the ρ^0 decay to $\pi^0\pi^0$? to $\pi^+\pi^-$? Can it decay electromagnetically to $\pi^0\gamma$?

9.43. (a) If we define a spherical harmonic by $\mathcal{Y}_l^m = i^l Y_l^m(\theta, \phi)$, where Y_l^m is the usual spherical harmonic, show that under a time reversal transformation

$$T\mathcal{Y}_l^m(\theta, \phi) = (-1)^{l-m}\mathcal{Y}_l^{-m}(\theta, \phi).$$

(b) With the use of part (a), we can write

$$T|a, s, m\rangle = (-1)^{s-m}|a_T, s, -m\rangle,$$

where a stands for other quantum numbers than the spin s and its magnetic quantum number m, and a_T are the time reversed quantum numbers corresponding to a. Make use of this equation to show that the Hermitian operator T^2 has eigenvalues $+1$ for bosons and -1 for Fermions.

9.44. Consider the annihilation of an antiproton by a proton at rest (or in a $\bar{p}p$ atom in an S-state) into two pions.

(a) Show that this decay is forbidden from a $^1S_0$ state.

(b) Show that the decay can occur from a $^3S_1$ state into $\pi^+\pi^-$.

(c) Show that the decay is forbidden into $\pi^0\pi^0$.

9.45. Consider the annihilation of an antiproton by a proton in a P-state into two pions.

(a) Is the decay $\pi^+\pi^-$ allowed from both $^1P_1$ and $^3P_0$ states, from only one of these, or from neither? If allowed from only one of the states, which one is it?

(b) Is the decay $\pi^0\pi^0$ allowed from either of the two states listed in part (a), from only one of them or from neither? if allowed from only one of them, which one is it?

9.46. With respect to CP violation originating in the matrix element: it was mentioned that an admixture of isospin 2 into isospin 0 in the neutral Kaon system would give raise to CP-violating decays. Show that isospin 1 is not allowed.

Part IV

Interactions

In the previous nine chapters, we have used the concept of *interaction* without discussing it in detail. In the present part, we shall rectify this omission, and we shall outline the important aspects of the interactions that rule subatomic physics.

It is useful in the treatment of interactions to distinguish between bosons and fermions. Bosons can be created and destroyed singly. Lepton and baryon conservation guarantee that fermions are always emitted or absorbed in pairs. The simplest interaction is thus one in which a boson is emitted or absorbed. Two examples are shown in Fig. IV.1. The interactions occur at the vertices where three particle lines are joined. The fermion does not disappear, but the boson either is created or destroyed. In both cases, the strength of the interaction can be characterized by a coupling constant. This coupling constant is written next to the vertex. A boson can also transform into another boson, as shown in Fig. IV.2. There a photon disappears, and a vector meson, for instance, a rho, takes its place. Again the coupling constant is indicated near the vertex.

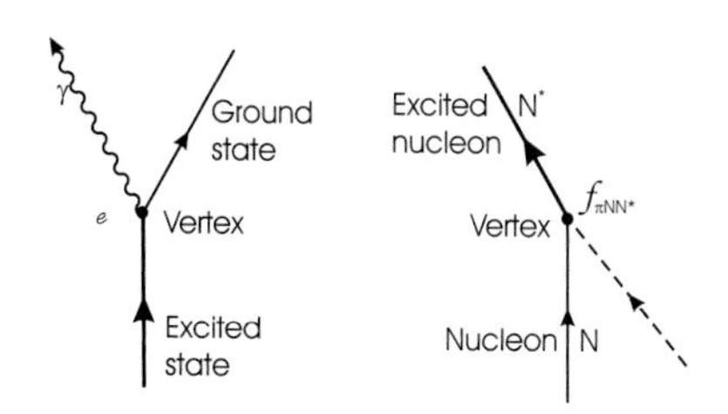

Figure IV.1: Emission and absorption of a boson by a fermion. The coupling constants are denoted by e and $f_{\pi NN^*}$.

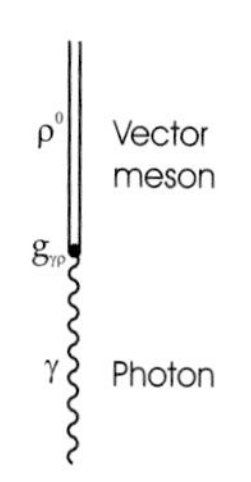

Figure IV.2: Transformation of one boson into another.

The force between two particles is usually assumed to be mediated by particles, as discussed in Section 5.8. The exchange of a pion between two nucleons, shown in Fig. 5.19, is again represented in Fig. IV.3. The forces represented by Figs. IV.1 and IV.3 are, however, no longer considered to be elementary. As discussed in Section 5.11, baryons and mesons are composed of quarks and the more fundamental

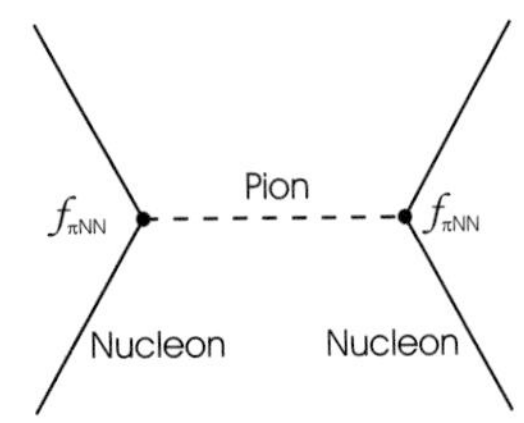

Figure IV.3: The force between two nucleons is mediated by the exchange of mesons, for instance, pions, as shown here.

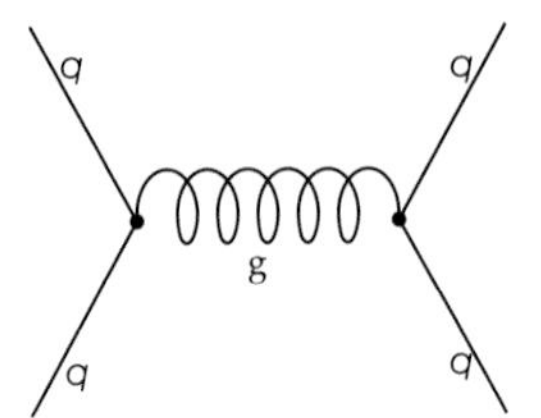

Figure IV.4: The force between two quarks, q, is produced through the exchange of gluons.

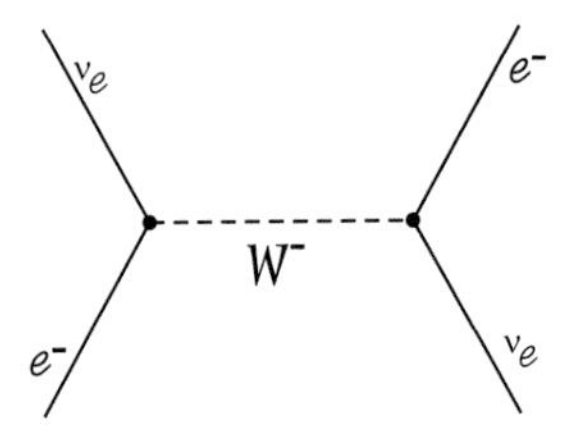

Figure IV.5: The weak force is mediated through the exchange of W's and Z's.

interactions occur between quarks and between leptons. Fig. IV.4 represents the hadronic force between quarks, mediated by a gluon; Fig. IV.5 shows the weak force between two leptons, mediated by a W boson. The examples given here provide some glimpses of the forces acting between particles in the standard model. In the following chapters we shall study interactions in more detail.

Chapter 10

The Electromagnetic Interaction

In this chapter we will examine the electroweak interaction of the standard model, and, in particular, the electromagnetic part of it. We relegate the weak part to the next chapter. The electromagnetic interaction is important in subatomic physics for two reasons. First, it enters whenever a charged particle is used as a probe. Second, it is the only interaction whose form can be studied in classical physics, and it provides a model after which other interactions can be patterned.

Without at least some approximate computations, interactions cannot be understood. In the simplest form, such computations are based on quantum mechanical perturbation theory and, in particular, on the expression for the transition rate from an initial state α to a final state β:

$$w_{\beta\alpha} = \frac{2\pi}{\hbar}|\langle\beta|H_{\text{int}}|\alpha\rangle|^2\rho(E). \tag{10.1}$$

Fermi called this expression *the golden rule*, because of its usefulness and importance. In Section 10.1 we shall derive this relation; in Section 10.2, we shall discuss the density-of-states factor $\rho(E)$. Readers who are familiar with these topics can omit these two sections.

10.1 The Golden Rule

Consider a system that is described by a time-independent Hamiltonian H_0; its Schrödinger equation is

$$i\hbar\frac{\partial\varphi}{\partial t} = H_0\varphi. \tag{10.2}$$

The stationary states of this system are found by inserting the ansatz,

$$\varphi = u_n(\boldsymbol{x})\exp\left(\frac{-iE_nt}{\hbar}\right) \tag{10.3}$$

into Eq. (10.2). The result is the time-independent Schrödinger equation

$$H_0u_n = E_nu_n. \tag{10.4}$$

For the further discussion it is assumed that this equation has been solved, that the eigenvalues E_n and the eigenfunctions u_n are known, and that the eigenfunctions form a complete orthonormal set, with

$$\int d^3x u_N^*(\boldsymbol{x}) u_n(\boldsymbol{x}) = \delta_{Nn}. \tag{10.5}$$

If the system is produced in one of the eigenstates u_n, it will remain in that state forever and no transitions to other states will occur.

We next consider a system that is similar to the one just discussed, but its Hamiltonian, H, differs from H_0 by a small term, the interaction Hamiltonian, H_{int},

$$H = H_0 + H_{\text{int}}.$$

The state of this system can, in zeroth approximation, still be characterized by the energies E_n and the eigenfunctions u_n. It is still possible to form the system in a state described by one of the eigenfunctions u_n, and we shall call a particular initial state $|\alpha\rangle$.

However, such a state will in general no longer be stationary; the perturbing Hamiltonian H_{int} will cause transitions to other states, for instance, $|\beta\rangle$. In the following we shall derive an expression for the transition rate $|\alpha\rangle \to |\beta\rangle$. Two examples of such transitions are shown in Fig. 10.1. In Fig. 10.1(a), the interaction is responsible for the decay of the state via the emission of a photon. In Fig. 10.1(b), an incident particle in state $|\alpha\rangle$ is scattered into the state $|\beta\rangle$.

To compute the rate for a transition, we use the Schrödinger equation,

$$i\hbar \frac{\partial \psi}{\partial t} = (H_0 + H_{\text{int}})\psi. \tag{10.6}$$

To solve this equation, ψ is expanded in terms of the complete set of unperturbed eigenfunctions, Eq. (10.3):

$$\psi = \sum_n a_n(t) u_n \exp\left(\frac{-iE_n t}{\hbar}\right). \tag{10.7}$$

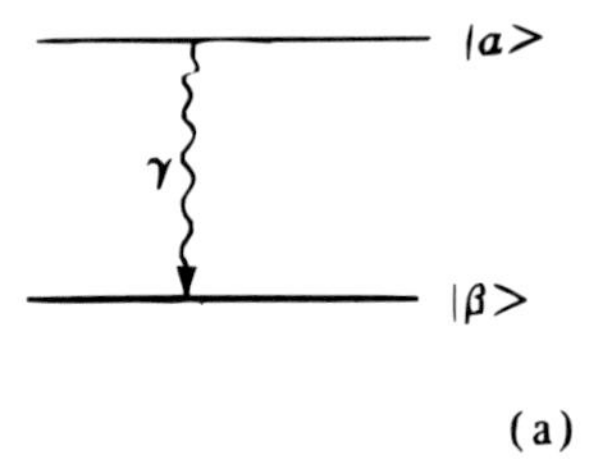

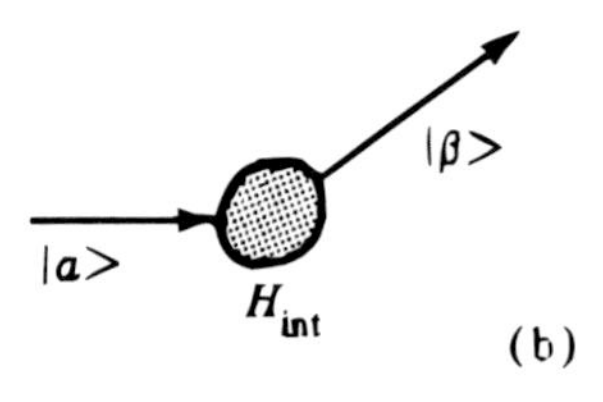

Figure 10.1: The interaction Hamiltonian H_{int} is responsible for transitions from the unperturbed eigenstate $|\alpha\rangle$ to the unperturbed eigenstate $|\beta\rangle$.

The coefficients $a_n(t)$ generally depend on time and $|a_n(t)|^2$ is the probability of finding the system at time t in state n with energy E_n. Inserting ψ into the

Schrödinger equation gives ($\dot{a}_n \equiv da_n/dt$)

$$i\hbar \sum_n \dot{a}_n u_n \exp\left(\frac{-iE_nt}{\hbar}\right) + \sum_n E_n a_n u_n \exp\left(\frac{-iE_nt}{\hbar}\right)$$
$$= \sum_n a_n (H_0 + H_{\text{int}}) u_n \exp\left(\frac{-iE_nt}{\hbar}\right).$$

With equation (10.4), the second term on the left-hand side and the first term on the right-hand side cancel. Multiplying by u_N^* from the left, integrating over all space, and using the orthonormality relation, produce the result

$$i\hbar\dot{a}_N = \sum_n \langle N|H_{\text{int}}|n\rangle a_n \exp\left[\frac{i(E_N - E_n)t}{\hbar}\right]. \tag{10.8}$$

Here, a convenient abbreviation for the matrix element of H_{int} has been introduced:

$$\langle N|H_{\text{int}}|n\rangle = \int d^3x u_N^*(\boldsymbol{x}) H_{\text{int}} u_n(\boldsymbol{x}). \tag{10.9}$$

The set of relations (10.8) for all N is equivalent to the Schrödinger equation (10.6) and no approximation is involved.

A useful approximate solution of Eq. (10.8) is obtained if it is assumed that the interacting system is initially in one particular state of the unperturbed system and if the perturbation H_{int} is weak. In Fig. 10.1, the initial state is $|\alpha\rangle$; it can, for instance, be a well-defined excited level. In terms of the expansion (10.7), the situation is described by

$$a_\alpha(t) = 1, \quad \text{all other } a_n(t) = 0, \quad \text{for } t < t_0. \tag{10.10}$$

Only one of the expansion coefficients is different from zero; all others vanish. The assumption that the perturbation is weak means that, during the time of observation, so few transitions have occurred that the initial state is not appreciably depleted, and other states are not appreciably populated. In lowest order it is then possible to set

$$a_\alpha(t) \approx 1, \quad a_n(t) \ll 1, \quad n \neq \alpha, \quad \text{all } t. \tag{10.11}$$

Equation (10.8) then simplifies to

$$\dot{a}_N = (i\hbar)^{-1} \langle N|H_{\text{int}}|\alpha\rangle \exp\left[\frac{i(E_N - E_\alpha)t}{\hbar}\right].$$

If H_{int} is switched on at the time $t_0 = 0$ and is time-independent thereafter, integration, for $N \neq \alpha$, gives

$$a_N(T) = (i\hbar)^{-1} \langle N|H_{\text{int}}|\alpha\rangle \int_0^T dt \exp\left[\frac{i(E_N - E_\alpha)t}{\hbar}\right]$$

or

$$a_N(T) = \frac{\langle N|H_{\text{int}}|\alpha\rangle}{E_N - E_\alpha}\left\{1 - \exp\left[\frac{i(E_N - E_x)T}{\hbar}\right]\right\}. \tag{10.12}$$

The probability of finding the system in the particular state N after time T is given by the absolute square of $a_N(T)$, or

$$P_{N\alpha}(T) = |a_N(T)|^2 = 4|\langle N|H_{\text{int}}|\alpha\rangle|^2 \frac{\sin^2[(E_N - E_\alpha)T/2\hbar]}{(E_N - E_\alpha)^2}. \tag{10.13}$$

If the energy E_N is different from E_α, then the factor $(E_N - E_\alpha)^{-2}$ depresses the transition probability so much that transitions to the corresponding states can be neglected for large times T. However, there may be a group of states with energies $E_N \approx E_\alpha$, such as shown in Fig. 10.2(a), for which the matrix element $\langle N|H_{\text{int}}|\alpha\rangle$ is almost independent of N. This case occurs, for instance, if the states N lie in the continuum.

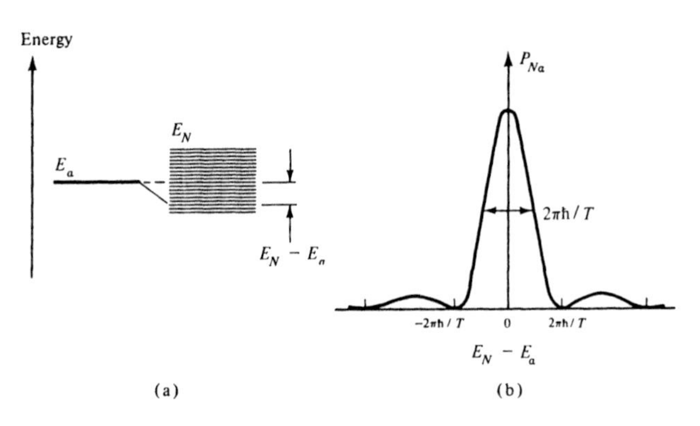

Figure 10.2: (a) Transitions occur mainly to states with energies E_N that are close to the initial energy E_α. (b) Transition probability as a function of the energy difference $E_N - E_\alpha$.

To express the fact that the matrix element is assumed to be independent of N, it is written as $\langle\beta|H_{\text{int}}|\alpha\rangle$. The transition probability is then determined by the factor $\sin^2[(E_N - E_\alpha)T/2\hbar](E_N - E_\alpha)^{-2}$, and it is shown in Fig. 10.2(b). The transition probability is appreciable only within the energy region

$$E_\alpha - \Delta E \text{ to } E_\alpha + \Delta E, \quad \Delta E = \frac{2\pi\hbar}{T}. \tag{10.14}$$

As time increases, the spread becomes smaller: within the limits given by the uncertainty relation, energy conservation is a consequence of the calculation and does not have to be added as a separate assumption.

Equation (10.13) gives the transition probability from one initial state to one final state. The total transition probability to all states E_N within the interval (10.14) is the sum over all individual transitions.

$$P = \sum_N P_{N\alpha} = 4|\langle\beta|H_{\text{int}}|\alpha\rangle|^2 \sum_N \frac{\sin^2[(E_N - E_\alpha)T/2\hbar]}{(E_N - E_\alpha)^2}, \tag{10.15}$$

where it has been assumed that the matrix element is independent of N. This assumption is good as long as $\Delta E/E_\alpha$ is small compared to 1. With Eq. (10.14), the condition becomes

$$T \gg \frac{2\pi\hbar}{E_\alpha} \approx \frac{4 \times 10^{-21}\text{ MeV-sec}}{E_\alpha(\text{in MeV})}, \tag{10.16}$$

where T is the time of observation. In most experiments, this condition is satisfied.

Now we return to the original problem, shown, for instance, in Fig. 10.1(a). Here, the energy in the initial state is well defined, but in the final state, the emitted photon is free and can have an arbitrary energy (Fig. 10.3). The discrete energy levels E_N of Fig. 10.2(a) consequently are replaced by a continuum. This fact is expressed by writing the energy as $E(N)$. N now labels the energy levels of the photon in the continuum, and it is a continuous variable. The total transition probability follows from Eq. (10.15) if the sum is replaced by an integral, $\sum_N \to \int dN$:

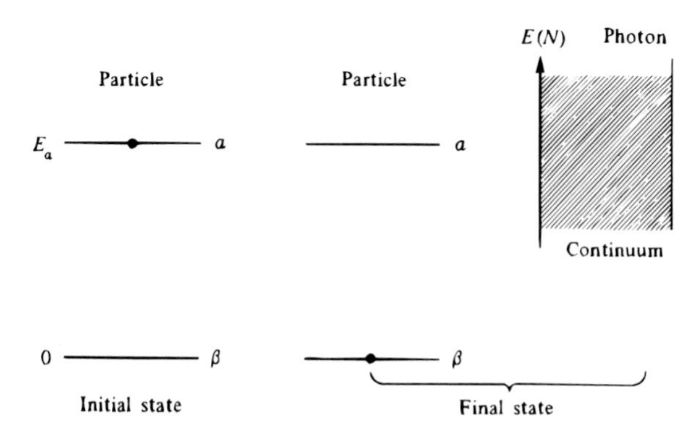

Figure 10.3: In the initial state the subatomic particle is in the excited state α, and no photon is present. In the final state, the subatomic system is in state β, and a photon with energy $E(N)$ has been emitted. The energy of the photon "is in the continuum."

$$P(T) = 4|\langle\beta|H_{\text{int}}|\alpha\rangle|^2 \int \frac{\sin^2[(E(N)-E_\alpha)T/2\hbar]}{(E(N)-E_\alpha)^2} dN. \tag{10.17}$$

The integral extends over the states to which the transitions can occur. Since the integral converges very rapidly, the limits can be extended to $\pm\infty$. With

$$x = \frac{(E(N)-E_\alpha)T}{2\hbar}, \quad dN = \frac{dN}{dE}dE = \frac{2\hbar}{T}\frac{dN}{dE}dx,$$

the transition probability becomes

$$P(T) = 4|\langle\beta|H_{\text{int}}|\alpha\rangle|^2 \frac{dN}{dE}\frac{T}{2\hbar}\int_{-\infty}^{+\infty} dx \frac{\sin^2 x}{x^2}.$$

The integral has the value π, so that the transition probability finally becomes

$$P(T) = \frac{2\pi T}{\hbar} |\langle\beta|H_{\text{int}}|\alpha\rangle|^2 \frac{dN}{dE}. \tag{10.18}$$

The notation $\langle\beta|H_{\text{int}}|\alpha\rangle$ indicates that the transition occurs from states $|\alpha\rangle$ to states $|\beta\rangle$. Since H_{int} is assumed to be time-independent, the transition probability is proportional to the time T. The transition *rate* is the transition probability per unit time, and it is

$$w_{\beta\alpha} = \dot{P}(T) = \frac{2\pi}{\hbar} |\langle\beta|H_{\text{int}}|\alpha\rangle|^2 \frac{dN}{dE}. \tag{10.19}$$

We have thus derived the golden rule. (Actually Fermi called it the *golden rule No. 2*.) It is extremely useful in all discussions of transition processes and we shall refer to it frequently. The factor

$$\frac{dN}{dE} \equiv \rho(E) \tag{10.20}$$

is called the *density-of-states factor*; it gives the number of available states per unit energy, and it will be discussed in Section 10.2.

• In some applications it happens that the matrix element $\langle\beta|H_{\text{int}}|\alpha\rangle$, connecting states of equal energy, vanishes. The approximation that leads to Eq. (10.18) can then be taken one step further. Fermi called this result the *golden rule No. 1*, and it can be stated simply: Replace the matrix element $\langle\beta|H_{\text{int}}|\alpha\rangle$ in Eq. (10.19) by

$$\langle\beta|H_{\text{int}}|\alpha\rangle \longrightarrow -\sum_n \frac{\langle\beta|H_{\text{int}}|n\rangle\langle n|H_{\text{int}}|\alpha\rangle}{E_n - E_\alpha}. \tag{10.21}$$

The one-step transition $|\alpha\rangle \longrightarrow |\beta\rangle$ from the initial to the final state is replaced by a sum over two-step transitions. These proceed from the initial state $|\alpha\rangle$ to all accessible *intermediate* states $|n\rangle$ and from there to the final state $|\beta\rangle$. •

10.2 Phase Space

In the present section, we shall derive an expression for the density-of-states factor $\rho(E) \equiv dN/dE$. We consider first a one-dimensional problem, where a particle moves along the x direction with momentum p_x. Position and momentum of the particles are described simultaneously in an $x - p_x$ plot (phase space). The representation is different in classical and in quantum mechanics. In classical mechanics, position and momentum can be measured simultaneously to arbitrary accuracy, and the state of a particle can be represented by a point (Fig. 10.4(a)). Quantum mechanics, however, limits the description in phase space. The uncertainty relation

$$\Delta x \Delta p_x \geq \hbar$$

states that position and momentum cannot be simultaneously measured to unlimited accuracy. The product of uncertainties must be bigger than $\hbar$, and a particle

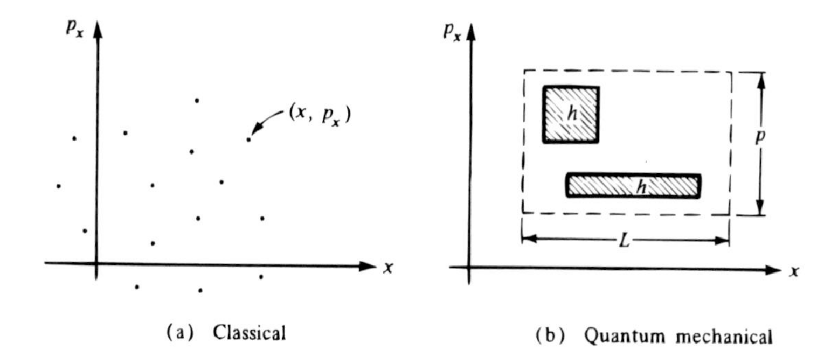

Figure 10.4: Classical and quantum mechanical one-dimensional phase space. In the classical case, the state of a particle can be described by a point. In the quantum case, a state must be described by a cell of *volume* $h = 2\pi\hbar$.

consequently must be represented by a cell rather than a point in phase space. The shape of the cell depends on the measurements that have been made, but the *volume* is always equal to $h = 2\pi\hbar$. In Fig. 10.4(b), a volume Lp is shown. The maximum number of cells that can be crammed into this volume is given by the total volume divided by the cell volume,

$$N = \frac{Lp}{2\pi\hbar}. \tag{10.22}$$

N is the *number of states* in the volume Lp.[1]

The *density of states factor* $\rho(E)$ in one dimension is obtained from Eq. (10.22), with $E = p^2/2m$, as

$$\begin{aligned} \rho(E) = \frac{dN}{dE} &= 2\frac{dN}{dp}\frac{dp}{dE} = \frac{L}{2\pi\hbar}\frac{2m}{p} \\ &= \frac{L}{2\pi\hbar}\sqrt{\frac{2m}{E}}. \end{aligned} \tag{10.23}$$

The factor 2 in Eq. (10.23) is introduced because for each energy E there are two degenerate states of momentum p and $-p$.

Equation (10.22) can be verified by considering a free wave in a one-dimensional "box" of length L. The normalized solution for the Schrödinger equation in the box,

$$\frac{d^2\psi}{dx^2} + \frac{2m}{\hbar^2}E\psi = 0 \quad \text{is} \quad \psi = \frac{1}{\sqrt{L}}e^{ikx}.$$

Periodic boundary conditions, $\psi(x) = \psi(x + L)$, give

$$\psi(0) = \psi(L), \quad \text{and} \quad k = \pm\frac{2\pi n}{L}, \quad n = 0, 1, 2, \ldots \tag{10.24}$$

[1]Note that N is the number of states, not particles. One state can accommodate one fermion but an arbitrary number of bosons.

The number of states per unit momentum interval for $n \gg 1$ is given by

$$\frac{\Delta n}{\Delta p} \approx \frac{dn}{dp} = \frac{1}{\hbar}\frac{dn}{dk} = \frac{L}{2\pi\hbar},$$

in agreement with Eq. (10.22).

Equation (10.22) is valid for a particle with one degree of freedom. For a particle in three dimensions, the volume of a cell is given by $h^3 = (2\pi\hbar)^3$, and the number of states in a volume $\int d^3x d^3p$ in the six-dimensional phase space is

$$N_1 = \frac{1}{(2\pi\hbar)^3}\int d^3x d^3p. \tag{10.25}$$

The subscript 1 indicates that N_1 is the number of states for one particle. If the particle is confined to a spatial volume V, integration over d^3x gives

$$N_1 = \frac{V}{(2\pi\hbar)^3}\int d^3p. \tag{10.26}$$

The density-of-states factor, Eq. (10.20), can now be computed easily:

$$\rho_1 = \frac{dN_1}{dE} = \frac{V}{(2\pi\hbar)^3}\frac{d}{dE}\int d^3p = \frac{V}{(2\pi\hbar)^3}\frac{d}{dE}\int p^2\,dp\,d\Omega, \tag{10.27}$$

where $d\Omega$ is the solid-angle element. With $E^2 = (pc)^2 + (mc^2)^2$, d/dE becomes

$$\frac{d}{dE} = \frac{E}{pc^2}\frac{d}{dp}$$

and consequently (with $(d/dp)\int dp \to 1$)

$$\rho_1 = \frac{V}{(2\pi\hbar)^3}\frac{pE}{c^2}\int d\Omega. \tag{10.28}$$

For transitions to all final states, regardless of the direction of the momentum $\boldsymbol{p}$, the density-of-states factor for one particle is

$$\rho_1 = \frac{VpE}{2\pi^2c^2\hbar^3}. \tag{10.29}$$

Next we consider the density of states for *two particles*, 1 and 2. If the total momentum of the two particles is fixed, the momentum of one determines the momentum of the other and the extra degrees of freedom are not really there. The total number of states in momentum space is the same as for one particle, namely N_1, as in Eq. (10.26). However, the density-of-states factor, ρ_2, is different from Eq. (10.28) because E is now the total energy of the *two* particles:

$$\rho_2 = \frac{V}{(2\pi\hbar)^3}\frac{d}{dE}\int d^3p_1 = \frac{V}{(2\pi\hbar)^3}\frac{d}{dE}\int p_1^2\,dp_1\,d\Omega_1, \tag{10.30}$$

where

$$dE = dE_1 + dE_2 = \frac{p_1 c^2}{E_1} dp_1 + \frac{p_2 c^2}{E_2} dp_2.$$

The evaluation is easiest in the c.m. where $\boldsymbol{p}_1 + \boldsymbol{p}_2 = 0$, or

$$p_1^2 = p_2^2 \longrightarrow p_1 dp_1 = p_2 dp_2, \quad \text{and} \quad dE = p_1 dp_1 \frac{(E_1 + E_2)}{E_1 E_2} c^2.$$

The density-of-states factor is then given by

$$\rho_2 = \frac{V}{(2\pi\hbar)^3 c^2} \frac{E_1 E_2}{(E_1 + E_2) p_1} \frac{d}{dp_1} \int p_1^2 dp_1 \, d\Omega_1$$

or

$$\rho_2 = \frac{V}{(2\pi\hbar)^3 c^2} \frac{E_1 E_2 p_1}{(E_1 + E_2)} \int d\Omega_1. \tag{10.31}$$

The extension of Eq. (10.30) to three or more particles is straightforward. Consider three particles; in their c.m. the momenta are constrained by

$$\boldsymbol{p}_1 + \boldsymbol{p}_2 + \boldsymbol{p}_3 = 0. \tag{10.32}$$

The momenta of two particles can vary independently, but the third one is determined. The number of states therefore is

$$N_3 = \frac{V^2}{(2\pi\hbar)^6} \int d^3p_1 \int d^3p_2, \tag{10.33}$$

and the density-of-states factor becomes

$$\rho_3 = \frac{V^2}{(2\pi\hbar)^6} \frac{d}{dE} \int d^3p_1 \int d^3p_2. \tag{10.34}$$

For n particles, the generalization of Eq. (10.34) is

$$\rho_n = \frac{V^{n-1}}{(2\pi\hbar)^{3(n-1)}} \frac{d}{dE} \int d^3p_1 \cdots \int d^3p_{n-1}. \tag{10.35}$$

We shall encounter an application of Eq. (10.34) in Chapter 11, and we shall discuss the further evaluation there.

10.3 The Classical Electromagnetic Interaction

The energy (Hamiltonian) of a free nonrelativistic particle with mass m and momentum $\boldsymbol{p}_{\text{free}}$ is given by

$$H_{\text{free}} = \frac{\boldsymbol{p}_{\text{free}}^2}{2m}. \tag{10.36}$$

How does the Hamiltonian change if the particle is subject to an electric field $\boldsymbol{E}$ and a magnetic field $\boldsymbol{B}$? The resulting modification can best be expressed in terms of potentials rather than the fields $\boldsymbol{E}$ and $\boldsymbol{B}$. A scalar potential A_0 and a vector potential $\boldsymbol{A}$ are introduced and the fields are related to the potentials through the vector relations[2]

$$\boldsymbol{B} = \boldsymbol{\nabla} \times \boldsymbol{A} \tag{10.37}$$

$$\boldsymbol{E} = -\boldsymbol{\nabla} A_0 - \frac{1}{c}\frac{\partial \boldsymbol{A}}{\partial t}. \tag{10.38}$$

The Hamiltonian of a point particle with charge q in the presence of the external fields is obtained from the free Hamiltonian by a procedure introduced by Larmor.[3] Energy and momentum of the free particle are replaced by

$$H_{\text{free}} \longrightarrow H - qA_0, \quad \boldsymbol{p}_{\text{free}} \longrightarrow \boldsymbol{p} - \frac{q}{c}\boldsymbol{A}, \tag{10.39}$$

or, in four-vector notation,

$$c(p_\mu)_{\text{free}} \longrightarrow (cp_\mu - qA_\mu). \tag{10.40}$$

Here p_0 is the Hamiltonian H. The resulting interaction is called *minimal electromagnetic interaction*. Eq. (10.40) satisfies local gauge invariance; that is, it is unchanged under a local gauge transformation (see Sec. 7.2). The term was coined by Gell-Mann to express the fact that only the charge q is introduced as a fundamental quantity. All currents are produced by the motion of particles. In particular, the current of a point particle is given by $q\boldsymbol{v}$. All higher moments (dipole moment, quadrupole moment, etc.) are assumed to be due to the particle's *structure*; they are not introduced as fundamental constants.

With the substitution (10.39), the Hamiltonian (10.36) changes to

$$H = \frac{1}{2m}\left(\boldsymbol{p} - \frac{q}{c}\boldsymbol{A}\right)^2 + qA_0 \tag{10.41}$$

or

$$H = H_{\text{free}} + H_{\text{int}} + \frac{q^2\boldsymbol{A}^2}{2mc^2}, \tag{10.42}$$

where H_{free} is given by Eq. (10.36) and H_{int} is

$$H_{\text{int}}(\boldsymbol{x}) = -\frac{q}{mc}\boldsymbol{p} \cdot \boldsymbol{A} + qA_0. \tag{10.43}$$

---

[2] Jackson, Section 6.2.

[3] J. Larmor, *Aether and Matter*, Cambridge University Press, Cambridge, 1900. See also Messiah, Sections 20.4 and 20.5; Jackson, Section 12.1; and Park, Section 7.6. Note that q can be positive or negative, whereas e is always positive.

For all practical field strengths, the last term in Eq. (10.42) is so small that it can be neglected. If no external charges are present, the scalar potential vanishes, and the interaction energy becomes

$$H_{\text{int}}(\boldsymbol{x}) = -\frac{q}{mc}\boldsymbol{p}\cdot\boldsymbol{A} = -\frac{q}{c}\boldsymbol{v}\cdot\boldsymbol{A}\,. \tag{10.44}$$

$H_{\text{int}}(\boldsymbol{x})$ in Eq. (10.43) is the interaction energy of the nonrelativistic point particle at the position $\boldsymbol{x}$ with the fields characterized by the potentials $\boldsymbol{A}$ and A_0. For many applications, this form is already sufficient. In particular, it allows a description of the emission and absorption of photons. For some other applications, for instance, the electromagnetic interaction between two particles, the equations must be rewritten by expressing the potentials in terms of the currents and charges producing them. Rather than deriving the general expression, we shall treat specific examples that are useful later.

The simplest situation arises if the electromagnetic field is produced by a point charge, q', at rest at $\boldsymbol{x}'$. The potential is then given by

$$A_0(\boldsymbol{x}) = \frac{q'}{|\boldsymbol{x}-\boldsymbol{x}'|}, \tag{10.45}$$

and the interaction is the ordinary *Coulomb energy*, already encountered in Eq. (6.7). If the charge q' is distributed over a volume, for instance the volume of a nucleus, the scalar potential is given by

$$A_0(\boldsymbol{x}) = q'\int d^3x'\,\frac{\rho'(\boldsymbol{x}')}{|\boldsymbol{x}-\boldsymbol{x}'|}, \tag{10.46}$$

and the interaction is of the form found in Eq. (6.15). The charge contained in the volume d^3x' at point $\boldsymbol{x}'$ is given by $q'\rho'(\boldsymbol{x}')d^3x'$, and the probability density $\rho'(\boldsymbol{x}')$ is normalized by Eq. (6.18).

The interaction of a point particle with a vector potential is given by Eq. (10.44). For a particle with an extended structure described by the charge distribution $q\rho(\boldsymbol{x})$, the factor $q\boldsymbol{p}/m = q\boldsymbol{v}$ in Eq. (10.44) must be replaced by

$$q\int d^3x\rho(\boldsymbol{x})\boldsymbol{v}(\boldsymbol{x}).$$

It is straightforward to see that

$$q\rho(\boldsymbol{x})\boldsymbol{v}(\boldsymbol{x}) = q\boldsymbol{j}(\boldsymbol{x}), \tag{10.47}$$

where $q\boldsymbol{j}(\boldsymbol{x})$ is the *charge current density*, namely the charge flowing through unit area per unit time. With Eq. (10.47), the interaction with an external potential $\boldsymbol{A}(\boldsymbol{x})$ becomes

$$H_{\text{int}} = -\frac{q}{c}\int d^3x\boldsymbol{j}(\boldsymbol{x})\cdot\boldsymbol{A}(\boldsymbol{x}). \tag{10.48}$$

Here the famous "jay-dot-A" has turned up. Equation (10.48) is one of the fundamental equations on which many calculations are based.

The vector potential $\boldsymbol{A}(\boldsymbol{x})$ produced by a current density $q'\boldsymbol{j}'(\boldsymbol{x}')$ is given by

$$\boldsymbol{A}(\boldsymbol{x}) = \frac{q'}{c}\int d^3x' \frac{\boldsymbol{j}'(\boldsymbol{x}')}{|\boldsymbol{x}-\boldsymbol{x}'|}. \tag{10.49}$$

Inserting this expression into Eq. (10.48) yields

$$H_{\text{int}} = -\frac{qq'}{c^2}\int d^3x d^3x' \frac{\boldsymbol{j}(\boldsymbol{x})\cdot\boldsymbol{j}'(\boldsymbol{x}')}{|\boldsymbol{x}-\boldsymbol{x}'|}. \tag{10.50}$$

Such *a current–current interaction* was first written down by Ampère, and it will be a helpful guide in elucidating the weak interaction.

One additional classical relation is a useful guide in subatomic physics, namely the *continuity equation.* Maxwell's equations show that the density ρ and the current density $\boldsymbol{j}$ satisfy

$$\frac{\partial\rho}{\partial t} + \boldsymbol{\nabla}\cdot\boldsymbol{j} = 0, \tag{10.51}$$

or in four-vector notation

$$g_{\mu\nu}\nabla_\mu j_\nu = 0\,. \tag{10.52}$$

A connection between the continuity equation and the conservation of the electric charge is established by integrating Eq. (10.51) over a volume V:

$$\int_V d^3x \frac{\partial\rho(\boldsymbol{x})}{\partial t} = -\int_V d^3x \boldsymbol{\nabla}\cdot\boldsymbol{j} = -\int_S d\boldsymbol{S}\cdot\boldsymbol{j}.$$

Here, S is the surface bounding the volume V. If the surface is far away from the system under consideration, the current through it will vanish. Interchanging integration and differentiation on the left-hand side and multiplication by the constant q give

$$\frac{\partial}{\partial t}\int_V d^3x q\rho(\boldsymbol{x}) = \frac{\partial}{\partial t}Q_{\text{total}} = 0. \tag{10.53}$$

The continuity equation implies conservation of the total electric charge.

10.4 Photon Emission

The relations in the previous section are classical and consequently cannot be applied to the elementary processes in quantum mechanics.[4] The task facing us then

[4]The problems inherent in any treatment of radiation theory make it difficult to write a really easy introduction. Probably the easiest-to-read first article is the beautiful review by E. Fermi, *Rev. Mod. Phys.* **4**, 87 (1932). A more modern readable introduction is R. P. Feynman, *Quantum Electrodynamics*, Benjamin, Reading, Mass., 1962. The basic ideas are explained lucidly in R. P. Feynman, *QED*, Princeton University Press, Princeton, 1985 and V.N. Gribov and J. Nyiri, *Quantum Electrodynamics*, Cambridge University Press, Cambridge, 2001. The present section is somewhat more difficult than the others, and parts of it can be omitted without losing information that is essential for later chapters.

is a twofold one. First, the interaction energy must be translated into quantum mechanics where it becomes an operator, the interaction Hamiltonian. Second, once H_{int} is found, the transition rate or the cross section for a particular process must be computed so that it can be compared with experiment. We cannot proceed very far with the solution of these tasks without hand waving. A major part of the problem lies with the photon. It always moves with the velocity of light, and a nonrelativistic description of the photon makes no sense. In addition, in most of the processes of interest, the particles involved have energies large compared to their rest energies, and they also must be treated relativistically. A proper discussion of quantum electrodynamics is far above our level. We shall only treat one process here in some detail, namely the emission of a photon by a quantum mechanical system.

Many of the ideas that are important in quantum electrodynamics will show up in this simple problem. The elementary radiation process, the emission or absorption of a quantum, is shown in Fig. 10.5. Two types of questions can be asked about such a process, kinematical and dynamical ones. The *kinematical* ones are of the type "What is the energy and momentum of the photon if it is emitted at a certain angle?" They can be answered by using energy and momentum conservation. The *dynamical* ones concern, for instance, the probability of decay or the polarization of the emitted radiation; they can be answered only if the form of the interaction is known.

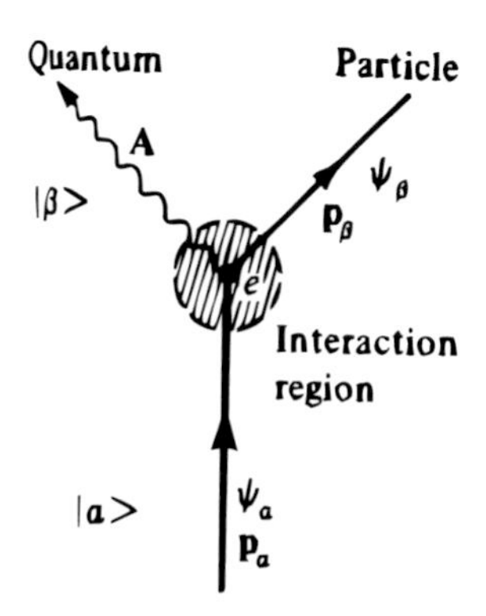

Figure 10.5: Emission of a photon by an atomic or subatomic system in a transition $|\alpha\rangle \to |\beta\rangle$.

In the present section we shall solve the simplest dynamical problem, the computation of the lifetime of an electromagnetic decay, by using the golden rule, Eq. (10.1). The first step is the choice of the proper interaction Hamiltonian, H_{int}. An appealing candidate is Eq. (10.44) in Section 10.3.[5] For an electron, with charge $q = -e, e > 0$, the interaction Hamiltonian, now denoted as H_{em}, is

$$H_{em} = e\frac{\boldsymbol{p} \cdot \boldsymbol{A}}{mc}. \tag{10.54}$$

The three factors in this expression can be associated with the elements of the diagram in Fig. 10.5: The vector potential $\boldsymbol{A}$ describes the emitted photon, $(\boldsymbol{p}/mc)$ characterizes the particle, and the constant e gives the strength of the interaction.

[5]Many students claim that the best way to solve physics problems in undergraduate courses is the following: list the physical quantities that appear in the problem. Find the equation in the text that contains the same symbols. Insert. Hand in. We are apt to laugh at such a naive approach but do the same when confronted with a new phenomenon. We see what observables nature has given us and then form the combination that has the properties expected from invariance laws.

The classical quantity H_{em} becomes an operator by translating $\boldsymbol{p}$ and $\boldsymbol{A}$ into quantum mechanics. The momentum $\boldsymbol{p}$ is straightforward; it becomes the momentum operator

$$\boldsymbol{p} \longrightarrow -i\hbar\boldsymbol{\nabla}. \tag{10.55}$$

This substitution is well known from nonrelativistic quantum mechanics. The corresponding substitution for $\boldsymbol{A}$ depends on the process under consideration. Two kinds of emission events occur from the state $|\boldsymbol{\alpha}\rangle$. The first takes place in the presence of an external electromagnetic field, produced, for instance, by photons incident on the system. $\boldsymbol{A}$ is the field due to these photons, and it gives rise to *stimulated* or *induced* emission of photons. Stimulated photon emission is the basic physical process involved in lasers. Here we are interested in the second kind of emission, called *spontaneous*. The state $|\boldsymbol{\alpha}\rangle$ can decay even in the absence of an external electromagnetic field. The expression for $\boldsymbol{A}$ for spontaneous emission cannot be obtained from nonrelativistic quantum mechanics, because photons are always relativistic. We circumvent quantum electrodynamics by *postulating* that $\boldsymbol{A}$ is the wave function of the created photon.[6] The form of $\boldsymbol{A}$ can be found by considering the vector potential of a classical electromagnetic plane wave,

$$\boldsymbol{A} = a_0\hat{\boldsymbol{\epsilon}}\cos(\boldsymbol{k}\cdot\boldsymbol{x} - \omega t). \tag{10.56}$$

Here $\hat{\boldsymbol{\epsilon}}$ is the polarization vector and a_0 the amplitude. If this wave is contained in a volume V, the average energy is given by

$$W = \frac{V}{4\pi}\overline{|\boldsymbol{E}|^2},$$

or with Eq. (10.38).

$$W = \frac{V\omega^2a_0^2}{4\pi c^2}\overline{\sin^2(\boldsymbol{k}\cdot\boldsymbol{x} - \omega t)} = \frac{V\omega^2a_0^2}{8\pi c^2}. \tag{10.57}$$

If $\boldsymbol{A}$ is to describe one photon in the volume V, W must be equal to the energy $E_\gamma = \hbar\omega$ of this photon. This condition fixes the constant a_0 as

$$a_0 = \left(\frac{8\pi\hbar c^2}{\omega V}\right)^{1/2}. \tag{10.58}$$

With $E_\gamma = \hbar\omega$ and $\boldsymbol{p}_\gamma = \hbar\boldsymbol{k}$, the wave function of the photon, Eq. (10.56), is determined. $\boldsymbol{A}$ is real because classically it is connected to the observable, and therefore real, fields $\boldsymbol{E}$ and $\boldsymbol{B}$ by Eqs. (10.37) and (10.38). For the application to emission and absorption it will turn out to be convenient to write Eq. (10.56) into

[6] This step can be justified by using quantum electrodynamics. Here we have no choice but to postulate it without further explanation. See Merzbacher, Chapter 23; Messiah, Section 21.27.

the form

$$\boldsymbol{A}(\text{one photon}) = \left(\frac{2\pi\hbar^2c^2}{E_\gamma V}\right)^{1/2} \hat{\boldsymbol{\epsilon}} \left\{ \exp\left[\frac{i(\boldsymbol{p}_\gamma \cdot \boldsymbol{x} - E_\gamma t)}{\hbar}\right] + \exp\left[\frac{-i(\boldsymbol{p}_\gamma \cdot \boldsymbol{x} - E_\gamma t)}{\hbar}\right] \right\}. \tag{10.59}$$

Here, $\boldsymbol{A}$ is no longer a classical vector potential, but it is postulated to be the wave function of the emitted photon. $\boldsymbol{A}$ is a vector, as is appropriate for photons which are spin-1 particles (Section 5.5). The next step is the construction of the matrix element of H_{em},

$$\begin{aligned} \langle\beta|H_{em}|\alpha\rangle &\equiv \int d^3x \psi_\beta^* H_{em} \psi_\alpha \\ &= \frac{e}{mc} \int d^3x \psi_\beta^* \boldsymbol{p} \psi_\alpha \cdot \boldsymbol{A} = -i\frac{e\hbar}{mc} \int d^3x \psi_\beta^* \boldsymbol{\nabla} \psi_\alpha \cdot \boldsymbol{A}. \end{aligned} \tag{10.60}$$

To evaluate $\langle\beta|H_{em}|\alpha\rangle$, we make approximations. The first is the *electric dipole approximation*. The momentum part of the exponent in $\boldsymbol{A}$ can be expanded,

$$\exp\left(\frac{\pm i\boldsymbol{p}_\gamma \cdot x}{\hbar}\right) = 1 \pm i\frac{\boldsymbol{p}_\gamma \cdot \boldsymbol{x}}{\hbar} + \cdots . \tag{10.61}$$

The exponential can be replaced by unity if $\boldsymbol{p}_\gamma \cdot \boldsymbol{x} \ll \hbar$. To obtain an approximate idea of what this condition implies, we assume that x has roughly the size of the system that emits the photon, and we denote this dimension by R. The condition imposed on the gamma-ray energy then is

$$E_\gamma = p_\gamma c \ll \frac{\hbar c}{R} \simeq \frac{197 \text{ MeV-fm}}{R(\text{in fm})}. \tag{10.62}$$

The second approximation applies to the decaying system. We assume it to be spinless and so heavy that it is at rest before and after the emission of the photon. The wave functions ψ_α and ψ_β can then be written as

$$\begin{aligned} \psi_\alpha(\boldsymbol{x}, t) &= \Phi_\alpha(\boldsymbol{x}) \exp\left(\frac{-iE_\alpha t}{\hbar}\right) \\ \psi_\beta(\boldsymbol{x}, t) &= \Phi_\beta(\boldsymbol{x}) \exp\left(\frac{-iE_\beta t}{\hbar}\right), \end{aligned} \tag{10.63}$$

where $\Phi_\alpha(\boldsymbol{x})$ and $\Phi_\beta(\boldsymbol{x})$ describe the spatial extension of the system before and after the photon emission (Chapter 6). E_α and E_β are the rest energies of the initial and final states. Energy conservation demands that

$$E_\alpha = E_\beta + E_\gamma. \tag{10.64}$$

With Eqs. (10.59), (10.61), and (10.63), the matrix element, Eq. (10.60), becomes

$$\langle\beta|H_{em}|\alpha\rangle = \frac{-i\hbar^2 e}{m}\left(\frac{2\pi}{E_\gamma V}\right)^{1/2}\left\{\exp\left[\frac{i(E_\beta - E_\gamma - E_\alpha)t}{\hbar}\right]\right.$$
$$\left. + \exp\left[\frac{i(E_\beta + E_\gamma - E_\alpha)t}{\hbar}\right]\right\}\hat{\boldsymbol{\epsilon}}\cdot\int d^3x\Phi_\beta^*\boldsymbol{\nabla}\Phi_\alpha. \tag{10.65}$$

The two exponential factors that appear in the matrix element behave very differently. With Eq. (10.64), the first one becomes $\exp(-2iE_\gamma t/\hbar)$. Perturbation theory in the form derived in Section 10.1 is valid only if, according to Eq. (10.16), the time t is large compared to $2\pi\hbar/E_\gamma$. For such times, the exponential factor is a very rapidly oscillating function of time. Any observation involves an averaging over times satisfying Eq. (10.16), and the rapid oscillation wipes out any contribution to the matrix element from the first term. The second exponential factor is unity because of energy conservation, Eq. (10.64), and the emission matrix element becomes

$$\langle\beta|H_{em}|\alpha\rangle = -i\frac{\hbar^2 e}{m}\left(\frac{2\pi}{E_\gamma V}\right)^{1/2}\hat{\boldsymbol{\epsilon}}\cdot\int d^3x\Phi_\beta^*\boldsymbol{\nabla}\Phi_\alpha. \tag{10.66}$$

If a photon is absorbed rather than emitted in the transition $|\alpha\rangle \to |\beta\rangle$, Eq. (10.64) reads $E_\alpha + E_\gamma = E_\beta$. The first exponential in Eq. (10.65) is then unity, and the second one does not contribute. The transition rate for spontaneous emission is now obtained with the golden rule, Eq. (10.19), which we write as

$$dw_{\beta\alpha} = \frac{2\pi}{\hbar}|\langle\beta|H_{em}|\alpha\rangle|^2\rho(E_\gamma). \tag{10.67}$$

With $p_\gamma = E_\gamma/c$, the density-of-states factor $\rho(E_\gamma)$ is given by Eq. (10.28) as

$$\rho(E_\gamma) = \frac{E_\gamma^2 V d\Omega}{(2\pi\hbar c)^3}. \tag{10.68}$$

Here $dw_{\beta\alpha}$ is the probability per unit time that the photon is emitted with momentum $\boldsymbol{p}_\gamma$ into the solid angle $d\Omega$. With the matrix element Eq. (10.66), the transition rate becomes

$$dw_{\beta\alpha} = \frac{e^2 E_\gamma}{2\pi m^2 c^3}|\hat{\boldsymbol{\epsilon}}\cdot\int d^3x\Phi_\beta^*\boldsymbol{\nabla}\Phi_\alpha|^2\, d\Omega. \tag{10.69}$$

If the wave functions Φ_α and Φ_β are known, the transition rate can be computed. However, the integral containing the wave functions can be changed into a form that expresses the salient facts more clearly. Assume that the Hamiltonian H_0 describing the decaying system, but not the electromagnetic interaction, is

$$H_0 = \frac{p^2}{2m} + V(\boldsymbol{x}),$$

where $V(\boldsymbol{x})$ does not depend on the momentum and hence commutes with $\boldsymbol{x}$. H_0 satisfies the eigenvalue equations

$$H_0\Phi_\alpha = E_\alpha\Phi_\alpha, \qquad H_0\Phi_\beta = E_\beta\Phi_\beta. \tag{10.70}$$

With the commutation relation,

$$xp_x - p_x x = i\hbar, \tag{10.71}$$

and the corresponding relations for the y and z components, the commutator of $\boldsymbol{x}$ and H_0 becomes

$$\boldsymbol{x}H_0 - H_0\boldsymbol{x} = \frac{i\hbar}{m}\boldsymbol{p} = \frac{\hbar^2}{m}\boldsymbol{\nabla}. \tag{10.72}$$

With this expression, the gradient operator in Eq. (10.69) can be replaced, and, with Eq. (10.70), the integral becomes

$$\begin{aligned}\int d^3x\,\Phi_\beta^*\boldsymbol{\nabla}\Phi_\alpha &= \frac{m}{\hbar^2}\int d^3x\,\Phi_\beta^*(\boldsymbol{x}H_0 - H_0\boldsymbol{x})\Phi_\alpha \\ &= \frac{m}{\hbar^2}(E_\alpha - E_\beta)\int d^3x\,\Phi_\beta^*\boldsymbol{x}\Phi_\alpha \\ &= \frac{m}{\hbar^2}E_\gamma\int d^3x\,\Phi_\beta^*\boldsymbol{x}\Phi_\alpha.\end{aligned}$$

The integral is the matrix element of the vector $\boldsymbol{x}$, and it is written as

$$\int d^3x\Phi_\beta^*\boldsymbol{x}\Phi_\alpha \equiv \langle\beta|\boldsymbol{x}|\alpha\rangle. \tag{10.73}$$

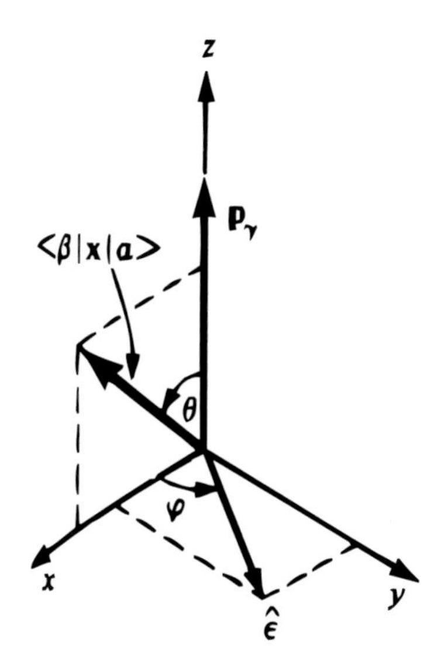

Figure 10.6: The polarization vector $\hat{\boldsymbol{\epsilon}}$ of a photon emitted along the z axis lies in the xy plane. The vector $\langle\beta|\boldsymbol{x}|\alpha\rangle$, describing the decaying system, is taken to lie in the xz plane.

The transition rate into the solid angle $d\Omega$ is thus

$$dw_{\beta\alpha} = \frac{e^2}{2\pi\hbar^4c^3}E_\gamma^3|\hat{\boldsymbol{\epsilon}}\cdot\langle\beta|\boldsymbol{x}|\alpha\rangle|^2\,d\Omega. \tag{10.74}$$

For a moment, we can place e^2 into the matrix element, which then becomes $\langle\beta|e\boldsymbol{x}|\alpha\rangle$. Since $e\boldsymbol{x}$ is the electric dipole moment, the radiation described by Eq. (10.74) is called electric dipole radiation, as mentioned above. The vector $\langle\beta|\boldsymbol{x}|\alpha\rangle$ characterizes the decaying system; the energy E_γ and the polarization vector $\hat{\boldsymbol{\epsilon}}$ describe the emitted photon. For a free photon, the unit vector $\hat{\boldsymbol{\epsilon}}$ is perpendicular to the photon momentum $\boldsymbol{p}_\gamma$ (Section 5.5). The vectors $\langle\beta|\boldsymbol{x}|\alpha\rangle$, $\boldsymbol{p}_\gamma$, and $\hat{\boldsymbol{\epsilon}}$ are shown in Fig. 10.6. Without loss of generality the coordinate system can be so chosen that $\boldsymbol{p}_\gamma$ points into the z direction and $\langle\beta|\boldsymbol{x}|\alpha\rangle$ lies in the xz plane; the polarization vector

$\hat{\boldsymbol{\epsilon}}$ must be in the xy plane. With the angles θ and φ as defined in Fig. 10.6, the components of $\langle\beta|\boldsymbol{x}|\alpha\rangle$ and $\hat{\boldsymbol{\epsilon}}$ are $\langle\beta|\boldsymbol{x}|\alpha\rangle = |\langle\beta|\boldsymbol{x}|\alpha\rangle|(\sin\theta, 0, \cos\theta)$, $\hat{\boldsymbol{\epsilon}} = (\cos\varphi, \sin\varphi, 0)$. Performing the scalar product in Eq. (10.74) then gives

$$dw_{\beta\alpha} = \frac{e^2}{2\pi\hbar^4 c^3} E_\gamma^3 |\langle\beta|\boldsymbol{x}|\alpha\rangle|^2 \sin^2\theta \cos^2\varphi \, d\Omega. \tag{10.75}$$

If the polarization of the emitted photon is not observed, $dw_{\beta\alpha}$ must be integrated over the angle φ and summed over the two polarization states. The sum introduces a factor 2; with

$$d\Omega = \sin\theta \, d\theta \, d\varphi \quad \text{and} \quad \int_0^{2\pi} d\varphi \cos^2\varphi = \pi,$$

the transition rate for an unpolarized photon becomes

$$dw_{\beta\alpha} = \frac{e^2}{\hbar^4 c^3} E_\gamma^3 |\langle\beta|\boldsymbol{x}|\alpha\rangle|^2 \sin^3\theta \, d\theta. \tag{10.76}$$

The total transition rate $w_{\beta\alpha}$ is obtained by integration over $d\theta$,

$$w_{\beta\alpha} = \int_0^\pi dw_{\beta\alpha} = \frac{4}{3} \frac{e^2}{\hbar^4 c^3} E_\gamma^3 |\langle\beta|\boldsymbol{x}|\alpha\rangle|^2. \tag{10.77}$$

The lifetime (mean life) is the reciprocal of $w_{\beta\alpha}$.

The physical content of the expression (10.77) for the total transition rate becomes more transparent if appropriate units are introduced. If the decaying system or particle has a mass m, then the characteristic length associated with it is the Compton wavelength, $\lambda\!\!\!^{-}_c = \hbar/mc$, and $E_0 = mc^2$ is the characteristic energy. The time that it takes light to move the distance $\lambda\!\!\!^{-}_c$ is given by $t_0 = \hbar/mc^2$, and the inverse of this time, $w_0 = 1/t_0 = mc^2/\hbar$, is the characteristic transition rate. With $\lambda\!\!\!^{-}_c$, $E_0 = mc^2$, and w_0, the transition rate is rewritten as

$$\frac{w_{\beta\alpha}}{w_0} = \frac{4}{3}\left(\frac{e^2}{\hbar c}\right)\left(\frac{E_\gamma}{mc^2}\right)^3 \frac{|\langle\beta|\boldsymbol{x}|\alpha\rangle|^2}{\lambda\!\!\!^{-}_c{}^2}. \tag{10.78}$$

The transition rate, expressed in terms of the "natural" rate w_0, becomes a product of three dimensionless factors, each of which has a clear physical interpretation. The last term, $|\langle\beta|\boldsymbol{x}|\alpha\rangle|^2/\lambda\!\!\!^{-}_c{}^2$, contains the information about the structure of the decaying system. If the wave functions Φ_α and Φ_β are known, the electric dipole matrix element $\langle\beta|\boldsymbol{x}|\alpha\rangle$ can be computed. Even without calculation, however, some properties can be deduced. For instance, the states $|\alpha\rangle$ and $|\beta\rangle$ must have opposite parities; otherwise $\langle\beta|\boldsymbol{x}|\alpha\rangle$ vanishes, and no electric dipole radiation can be emitted.

The term $(E_\gamma/mc^2)^3$ gives the dependence of the electric dipole radiation on the energy of the emitted photon. Equation (10.68) shows that two of the three powers of E_γ are contributed by the density-of-states factor: With increasing photon energy, the accessible volume in phase space becomes larger, and the decay consequently becomes faster. The third factor E_γ is introduced by the matrix element $\langle\beta|\boldsymbol{\nabla}|\alpha\rangle$, and it is said to be of dynamical origin.

The factor

$$\frac{e^2}{\hbar c} \equiv \alpha \approx \frac{1}{137} \tag{10.79}$$

characterizes the strength of the interaction between the charged particle and the photon, and it is usually called the fine structure constant. A number of remarks concerning α are in order here. The first one concerns the fact that α, formed from three natural constants, is a dimensionless number. Since α is a pure number, it must have the same value everywhere, even on Trantor or Terminus.[7] Moreover, its value should be calculable in a truly fundamental theory. At the present time, no such theory exists that is generally accepted and understood. The second remark concerns the magnitude of α. Fortunately, α is small compared to 1, and this fact makes the application of perturbation theory successful. The expression (10.78) for the transition rate has been computed with the first-order expression, Eq. (10.1), and the result is proportional to α. The second-order term, Eq. (10.21), involves H_{em} twice, and its contribution will therefore be of order α^2 and considerably smaller than the first-order term. An example of this rapid convergence has already been presented in the discussion of the g factor of the electron, Eq. (6.32). As the third remark we note that the electric charge e plays two different roles. In Section 7.2, the charge appeared as an additive quantum number; in the present section, the strength of the electromagnetic interaction was shown to be proportional to e^2; e is therefore called a *coupling constant.*

10.5 Multipole Radiation

In the previous section, a simple example of the action of the electromagnetic interaction, namely the emission of electric dipole radiation, has been computed in some detail. In the present section, the decay of actual subatomic systems will be discussed, and it will turn out that the previous considerations must be generalized. Two subatomic electromagnetic decays are shown in Fig. 10.7.

[7] I. Asimov, *Foundation*, Avon Books, New York, 1951.

In the nuclear example, the nuclide ^{170}Tm decays with a half-life of 129 d to an excited state of ^{170}Yb, which then decays to its ground state with emission of a gamma ray of 0.084 MeV. The second example is the decay of the neutral sigma; in the transition $\Sigma^0 \xrightarrow{\gamma} \Lambda^0$, a 77-MeV gamma ray is emitted.

The lifetime of the neutral sigma is 7×10^{-20}sec; the half-life of the 84 keV state in ^{170}Yb, on the other hand, has been determined as 1.61 nsec. (It is customary to quote mean lives in particle physics and half-lives in nuclear physics; see Eq. (5.33) for the relationship.) The basic idea underlying the half-life measurement is shown in Fig. 10.8.[8] The radioactive source, in the example ^{170}Tm, is placed between two counters. The beta counter detects the beta ray that populates the 2^+ state in ^{170}Yb. After some delay, the excited state decays with the emission of a 0.084 MeV photon. This photon has a certain probability of being delayed by a time D, and the coincidence rate between the delayed beta pulse and the gamma pulse is detected with an AND circuit (Section 4.9). The coincidence count rate $N(D)$ is recorded on a semilogarithmic plot against D, and the slope of the resulting curve gives the desired half-life. The corresponding ideas have already been discussed in Section 5.7, and the plot shown in Fig. 10.8 is a specific example of an exponential decay as sketched in Fig. 5.15.

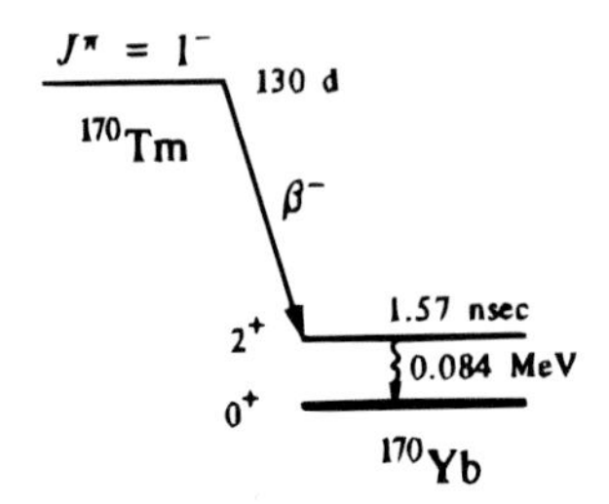

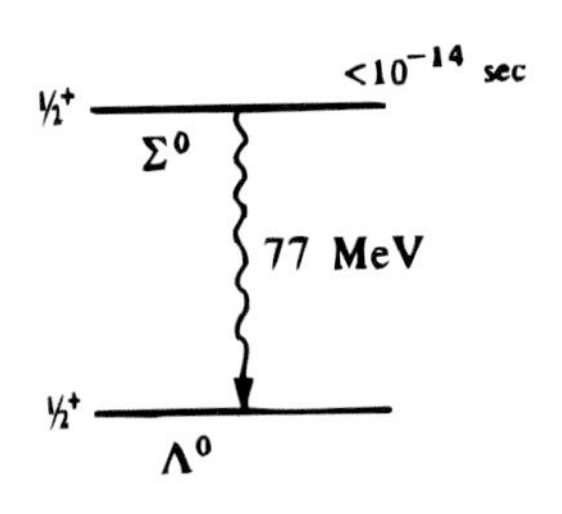

Figure 10.7: Two examples of subatomic gamma decays. Note that the energy scales differ by about a factor 100.

The method shown here, in which the decay curve is measured point by point, is only one possible approach. Many other techniques for investigating decay lifetimes have been evolved[8] and at present the half-lives of more than 1500 states are known.

After this brief excursion into the experimental aspects of electromagnetic transitions of subatomic particles, we return to theory and ask: can the decays shown as examples in Fig. 10.8 be explained by the treatment given in Section 10.4? It can be seen immediately that the transition $\Sigma^0 \to \Lambda^0$ cannot be caused by electric dipole transitions: The matrix element that appears in the electric dipole transition rate, Eq. (10.77), has the form

$$\langle\beta|\boldsymbol{x}|\alpha\rangle \equiv \langle\Lambda^0|\boldsymbol{x}|\Sigma^0\rangle \equiv \int d^3x\psi_\Lambda^*\boldsymbol{x}\psi_\Sigma.$$

[8] The measurement of short mean lives is discussed by R. E. Bell, in *Alpha-, Beta- and Gamma-Ray Spectroscopy*, Vol. 2 (K. Siegbahn, ed.), North-Holland, Amsterdam, 1965; T.K. Alexander and J.S. Foster, *Adv. Nucl. Phys.* **10**, 197 (1979); G. Bellini et al., *Phys. Rept.* **83**, 1 (1982).

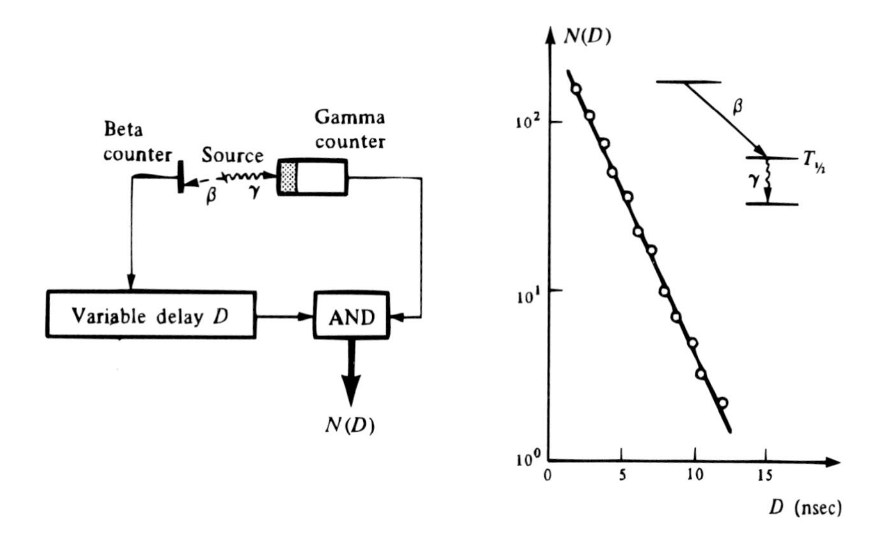

Figure 10.8: Determination of the half-life of a short-lived nuclear state, decaying by gamma emission. The block diagram is shown at the left; a typical curve of coincidence counting rate $N(D)$, taken as a function of the delay time D, is given at the right.

The wave functions ψ_Λ and ψ_Σ have the same parity, and their product is even under the parity operation. The vector $\boldsymbol{x}$, however, is odd under parity, and the integrand is therefore also odd; the integral consequently must vanish. Similarly, it can be shown that dipole radiation cannot explain the $2^+ \rightarrow 0^+$ transition in ^{170}Yb. The treatment given in the previous section must therefore be generalized if it is to explain all electromagnetic radiation emitted by subatomic systems.

The approximation that leads to electric dipole radiation is introduced by keeping only the first term in the expansion (10.61). Removal of this restriction is straightforward but lengthy, and we shall quote only the final result.(9) The emitted radiation can be characterized by its parity, η_P, and by its angular momentum quantum number, j. For any given value of j, the photon can carry away even or odd parity. It is customary to call one of these two an electric and the other a magnetic transition. Parity and angular momentum are related by

$$\begin{aligned} \text{electric radiation: } \eta_P &= (-1)^j \\ \text{magnetic radiation: } \eta_P &= -(-1)^j. \end{aligned} \tag{10.80}$$

[9] Introductions to the theory of multipole radiation can be found in the following references: G. Baym, *Lectures on Quantum Mechanics*, Benjamin, Reading, Mass., 1959, pp. 281, 376; Jackson, Chapter 9; Blatt and Weisskopf, Chapter 12 and Appendix; S. A. Moszkowski, in *Alpha-, Beta- and Gamma-Ray Spectroscopy*, Vol. 2, (K. Siegbahn, ed.), North-Holland, Amsterdam, 1965, Chapter 15; T. W. Donnely and J. D. Walecka, *Ann. Rev. Nucl. Sci.* **25**, 329 (1975).

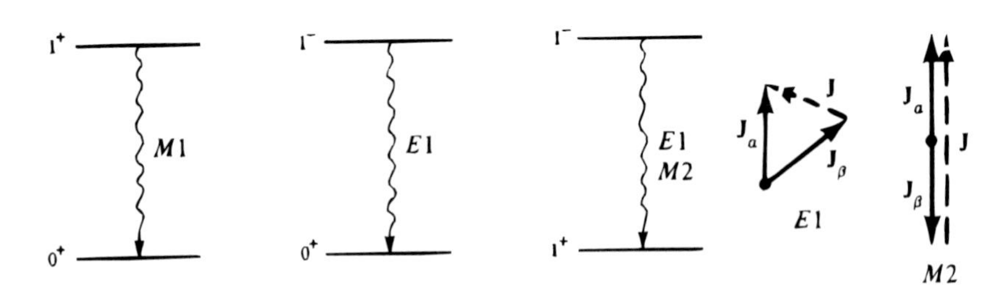

Figure 10.9: A few examples of the possible values of angular momentum and parity emitted in a given transition. The vector diagrams for the transition $1^- \to 1^+$ are shown at the right.

As an example, the electric dipole radiation carries an angular momentum $j = 1$ and, according to Eq. (10.80), a negative parity; it is written as $E1$. More generally, an electric (magnetic) radiation with quantum number j is written as $Ej(Mj)$. [We remind the reader that the quantum number j is defined by Eq. (5.4): If $\boldsymbol{J}$ is the photon angular momentum operator, $j(j+1)\hbar^2$ is the eigenvalue of J^2.]

The values of j and η_P of the photons emitted in a transition $\alpha \to \beta$ are limited by the conservation of angular momentum and parity

$$\boldsymbol{J}_\alpha = \boldsymbol{J}_\beta + \boldsymbol{J}, \qquad \eta_P(\alpha) = \eta_P(\beta)\eta_P. \tag{10.81}$$

A few examples of possible values of j and η_P are given in Fig. 10.9. Note that initial and final spins are vectors. The various values of the angular momentum of the emitted radiation are obtained by vector addition, as also shown in Fig. 10.9.

The selection rules equation (10.81) state which transitions are allowed in a given decay, but they do not give information about the rate with which they occur. To find the rate, dynamical computations must be performed. In the previous section, the transition rate for $E1$ radiation was found, and Eq. (10.77) expresses this rate in terms of the matrix element $\langle\beta|\boldsymbol{x}|\alpha\rangle$. Expressions similar to Eq. (10.77) can be found for all multipole orders Ej and Mj. The real problem then begins: The relevant matrix elements must be evaluated, and this step requires a knowledge of the wave functions ψ_α and ψ_β. Finding the correct wave functions for a particular subatomic system is usually a long and tedious process, and only in a few cases has it come to a satisfactory conclusion. For an estimate of the transition rate, a crude model is therefore a necessity; it will provide at least an approximate value with which observed half-lives can be compared.

For nuclei, the single-particle model is often used to get estimates for the half-lives of the various multipole orders. In the single-particle model it is assumed that the transition of one nucleon gives rise to the radiation. (We shall treat the single-particle model in Chapter 17.) Using a simple form for the single-nucleon wave function, the transition rates can be computed;[10] a result is shown in Fig. 10.10. The curves in Fig. 10.10 are calculated for a single proton in a nucleus with $A = 100$. Under these assumptions it is seen that the lowest multipole allowed by parity and by angular momentum selection rules dominates. Care must be taken in using the single-particle transition rates; in actual nuclei, deviations of one or even more orders of magnitude occur.

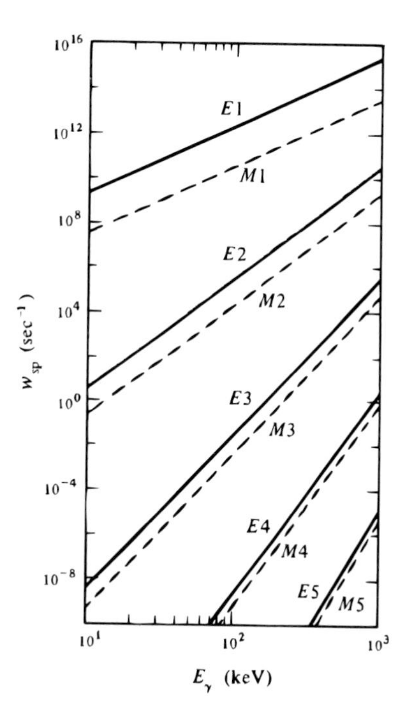

Figure 10.10: Single proton transition rate (in sec^{-1}) as a function of the gamma-ray energy (in keV) for various multipolarities. [After S. A. Moszkowski, in *Alpha-, Beta- and Gamma-Ray Spectroscopy*, Vol. 2 (K. Siegbahn, ed.), North-Holland, Amsterdam, 1965, Chapter 15, p. 882.]

10.6 Electromagnetic Scattering of Leptons

Electromagnetic processes that involve only leptons and photons have been encountered a few times. Photoeffect, Compton scattering, pair production, and bremsstrahlung were mentioned in Sections 3.3 and 3.4. The g factor of the leptons, discussed in Section 6.5, also involves only the electromagnetic interaction of leptons. In the present section we shall outline some of the aspects of the electromagnetic interaction of leptons without performing computations. The process to be discussed is the scattering of electrons. The diagrams for the scattering of electrons by electrons (Møller scattering) or electrons by positrons (Bhabha scattering) are shown in Fig. 10.11. The two electrons in Møller scattering are indistinguishable, and the graphs shown in Fig. 10.11(a) and (b) must both be taken into account.

[10] J. J. Sakurai, *Ann. Phys.* (New York) **11,** 1 (1960); J. J. Sakurai, *Currents and Mesons,* University of Chicago Press, Chicago, 1969.

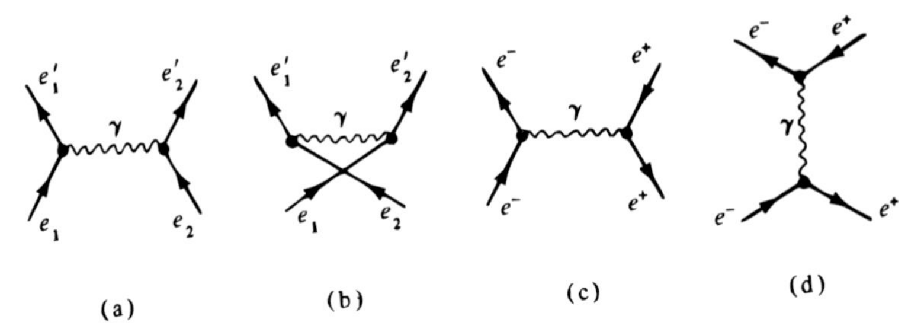

Figure 10.11: Diagrams for the scattering $e^-e^- \to e^-e^-$ and $e^+e^- \to e^+e^-$.

Since it is impossible to tell which process has taken place, the *amplitudes* for the two diagrams in Fig. 10.11(a) and (b) must be added, not the intensities. The particles in Bhabha scattering can be distinguished by their charge. Nevertheless, two graphs appear, and it is impossible to tell through which one scattering has occurred. Again the amplitudes for the two processes must be added. The contribution from Fig. 10.11(c) is called the photon-exchange term, and the one from Fig. 10.11(d) the annihilation term.

The annihilation term, Fig. 10.11(d), deserves closer attention. It appears because the additive quantum numbers of an electron–positron pair are the same as those of the photon, namely $A = q = S = L = L_e = L_\mu = L_\tau = 0$. Once the virtual photon has been "formed" it no longer remembers where it came from, and it can give rise to a number of processes:

$$\begin{aligned} e^+e^- \longrightarrow\ & 2\gamma \\ & e^+e^-,\ \mu^+\mu^-,\ \tau^+\tau^-, \\ & \pi^+\pi^-,\ \pi^+\pi^-\pi^0,\ K^+K^-, \\ & \bar{p}p, \quad \bar{n}n, \\ & \bar{c}c, \quad \bar{b}b, \quad \bar{t}t \\ & \vdots \end{aligned}$$

Only the first four involve the electromagnetic interaction exclusively, and only the second one is shown in Fig. 10.11.

The computation of the cross section for Møller and Bhabha scattering requires knowledge of quantum electrodynamics and Dirac theory. The cross sections depend on the total energy of the two electrons and on the scattering angle θ. If E is the energy of one of the two leptons in the c.m., then the cross section for Møller scattering for large energies ($E \gg m_e c^2$) is of the form

$$\frac{d\sigma}{d\Omega} = \frac{\alpha^2}{E^2}(\hbar c)^2 f(\theta). \tag{10.82}$$

where α is the fine structure constant and $f(\theta)$ is a function of θ that is given explicitly in various texts on quantum electrodynamics. We note that $\alpha = e^2/\hbar c$ occurs *squared* in Eq. (10.82), in agreement with the fact that *two* vertices appear in all graphs in Fig. 10.11. The form of Eq. (10.82) follows unambiguously from dimensional arguments. At very high energies, the electron mass can no longer play a role, and the only quantities that can enter the cross section are the coupling constant, in the dimensionless form α, and the energy, E. From these two quantities and the natural constants $\hbar$ and c, the only combination with the dimension of a cross section (area) is as given in Eq. (10.82). Only the dimensionless function $f(\theta)$ is dependent on the theory.

Experimentally, Møller and Bhabha scattering can be studied in two different ways. The straightforward approach is to employ a beam of electrons or positrons and observe the scattering from the electrons in a metal foil, as indicated in Fig. 10.12. One difficulty of this approach turns up when the cross sections of Møller and Rutherford scattering are compared. For a material with atomic number Z, the ratio of cross sections is approximately $1/Z^2$. For most reasonable target materials, Rutherford scattering will be much more frequent than Møller scattering. How can the two processes be separated? For simplicity we assume the incoming

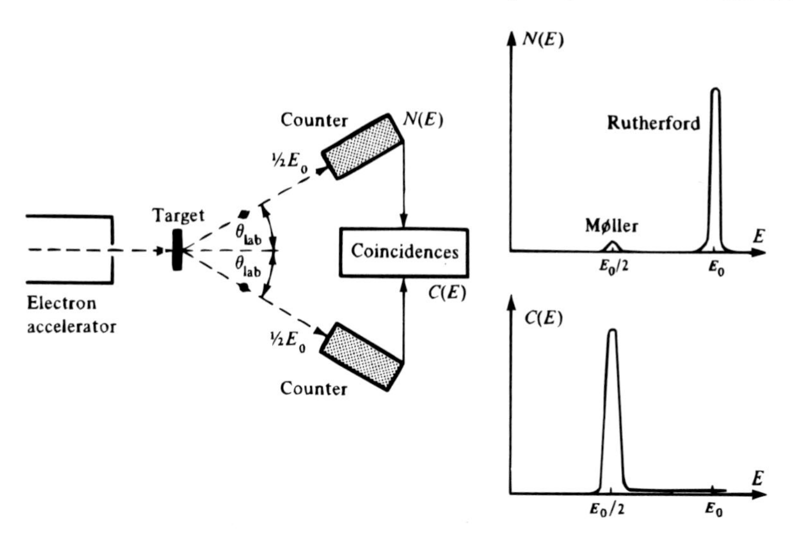

Figure 10.12: Detection of Møller and Bhabha scattering by observing collisions with electrons in matter. $N(E)$ denotes the number of electrons with energy E observed in one counter. $C(E)$ denotes the number of coincidences in which both electrons have the energy E.

energy, E_0, to be much larger than the binding energy of the electrons in the atom. The electrons in the target are thus essentially free. In symmetric scattering, shown in Fig. 10.12, both outgoing electrons make the same angle θ_{lab} with the beam axis, have energies $E_0/2$, and are simultaneous. If two counters are set at the proper angles, accepting only electrons with energies $E_0/2$, and if the signals are required to be simultaneous, Møller and Bhabha scattering can be separated cleanly from Rutherford scattering. A second disadvantage of the approach just outlined is not so easily overcome: The energy available in the c.m. to explore the structure of the electromagnetic interaction is small because of the small electron rest mass. We have studied this problem in Section 2.7; in Eq. (2.32) we found the total energy available in the c.m.,

$$W \approx (2E_0 m_e c^2)^{1/2}. \tag{10.83}$$

With $E_0 = 10$ GeV, the total energy available in the c.m. becomes

$$W \approx 100 \text{ MeV}.$$

Even at 10-GeV incident energy there is not enough c.m. energy to even produce a muon pair. The path around this difficulty has already been shown in Section 2.8; it is the use of colliding beams. As e^+e^- collisions have yielded some of the most beautiful results and promise to continue to do so, we will discuss a few of the experiments and data in the following sections.

One interesting concept occurs in connection with Bhabha scattering. The virtual photons in the photon-exchange and in the annihilation diagram (Fig. 10.11(c) and (d)) have very different properties. Both photons are virtual and do not satisfy the relation $E = pc$. Consider both reactions in the c.m. In the exchange diagram, the incoming and the outgoing electrons have the same energies but opposite momenta. Consequently, energy and momentum of the virtual photon are given by

$$E_\gamma = E_e - E_e' = 0, \quad \boldsymbol{p}_\gamma = \boldsymbol{p}_e - \boldsymbol{p}_e' = +2\boldsymbol{p}_e. \tag{10.84}$$

If we *define* a "mass" for the virtual photon through the relation $E^2 = (pc)^2 + (mc^2)^2$, we find[11]

$$(mc^2)^2 = -(2p_e c)^2 < 0. \tag{10.85}$$

The virtual photon in the exchange diagram carries only momentum—no energy. The square of its mass is negative. Such a photon is called *spacelike*. In the annihilation diagram, the situation is reversed,

$$E_\gamma = E_{e^-} + E_{e^+} = 2E, \quad \boldsymbol{p}_\gamma = \boldsymbol{p}_{e^-} + \boldsymbol{p}_{e^+} = 0. \tag{10.86}$$

The virtual photon carries only energy—no momentum. The square of its mass is given by

$$(mc^2)^2 = (2E)^2 > 0; \tag{10.87}$$

it is positive and the photon is called *timelike*. In electron-positron scattering, both spacelike and timelike photons enter. The agreement of experiment with theory indicates that these concepts are correct, even if they sound strange at first.

[11]The "mass" defined here is related to the four-momentum transfer, q, by $m^2 = (q/c)^2$. It is equal to the actual particle mass only for free particles.

10.7 Vector Mesons as Mediators of the Photon–Hadron Interaction

> The changing of bodies into light, and light into bodies, is very conformable to the course of nature, which seems delighted with transmutations.
>
> Newton, *Opticks*

The previous sections and Section 6.5 have dealt with quantum electrodynamics and the interaction of photons and leptons. Before turning to the electromagnetic interaction involving hadrons, we shall review one of the central assumptions of quantum electrodynamics, namely the form of the interaction Hamiltonian. As pointed out in Section 10.3, the Hamiltonian is obtained from the principle of minimal electromagnetic interaction, Eq. (10.39). The principle introduces only the electric charge as a fundamental constant, and currents are assumed to be due to the motion of charges. Leptons are pictured as point particles and the probability current density of a lepton with velocity $\boldsymbol{v}$ is given by Eq. (10.47).

We already know that the electromagnetic current of *hadrons* is not as simple as the one of leptons. The g factor and the elastic form factor of nucleons, both discussed in Section 6.6, indicate that the interaction of nucleons with the electromagnetic field is not directly given by the minimal electromagnetic interaction. Consequently, we write the total electromagnetic current density of a system as

$$e\boldsymbol{j}_{em} = e\boldsymbol{j}_{em}(\text{leptons}) + e\boldsymbol{j}_{em}(\text{hadrons}) \qquad (10.88)$$

and ask: What experiments will tell us about the hadronic contribution? Since it is assumed that the electromagnetic interaction is mediated by photons, the question can be rephrased: What experiments give information about the interactions of photons with hadrons? How does the photon interact with hadrons? The interaction of the photon with a hadron does not occur through the electric charge alone, as is evidenced by the electromagnetic decay of the *neutral* pion into two photons. One possible way in which a photon can interact with a hadron current is indicated in Fig. 10.13.

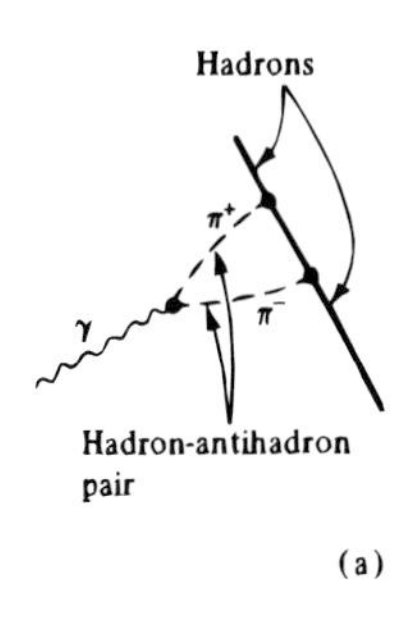

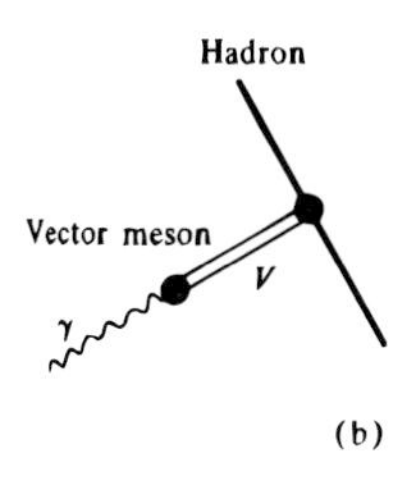

Figure 10.13: Interaction of a photon with a hadron. (a) The photon can produce a hadron–antihadron pair. (b) The photon can produce a vector meson which then interacts with the hadron.

In Fig. 10.13(a) the photon produces a hadron-antihadron pair, and the partners of the pair interact strongly with the hadron current. As early as 1960, Sakurai suggested that the two hadrons of the pair should be strongly coupled and form a vector meson, as shown in Fig. 10.13(b).[10] The photon thus would transform into a vector meson, as already anticipated in Fig. IV.2. Sakurai made his suggestion long before the vector mesons were discovered experimentally. Theoretical suggestions can be useful guides for planning experiments, but only the result of experiments can provide the clues as to the nature of the interaction between the photon and hadrons. Three types of experiments that can provide information about the photon-hadron interaction are illustrated by the Feynman diagrams shown in Fig. 10.14. Two of these involve virtual photons; the third one is performed with real ones. In all three cases the object of interest is the photon–hadron vertex. In the present section, we discuss timelike photons in electron–positron scattering; in Section 10.10, real and spacelike photons will be treated.

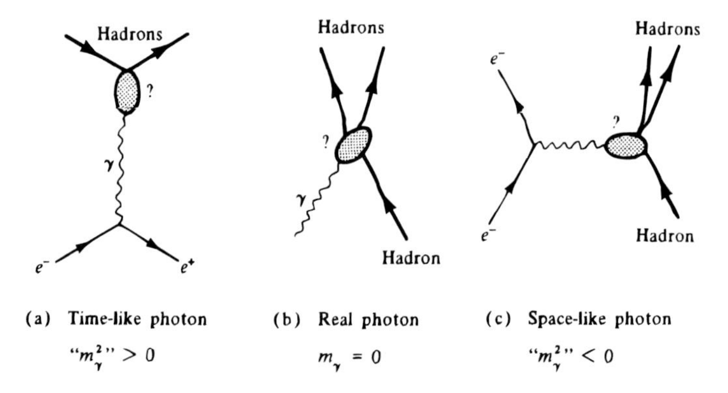

Figure 10.14: Diagrams of three experimental possibilities to study the interaction of photons with hadrons. Details are discussed in the text.

The virtual photon produced in electron–positron collisions is timelike, as follows from Eqs. (10.86) and (10.87); in the $e^- - e^+$ c.m., it has energy but no momentum. The system of hadrons produced by timelike photons must possess quantum numbers that are determined by those of the photon. Since the electromagnetic interaction conserves strangeness, parity, and charge conjugation, only final states with strangeness 0, negative parity, and negative charge parity can be produced. In addition, angular momentum conservation requires the final state to have angular momentum unity. Are there such final states that are produced copiously? The experiments indicate that hadrons satisfying all conditions are indeed produced. Consider first Fig. 10.15. It shows the number of pion pairs observed at a given total energy of the colliding electrons, normalized by division by the number of electrons observed at the same energy. A pronounced peak appears at about 770 MeV, with a width of about 100 MeV.

The reader with a good memory will say "Aha" and will turn back to Fig. 5.12 where a similar peak is shown at the same energy and with the same width. This

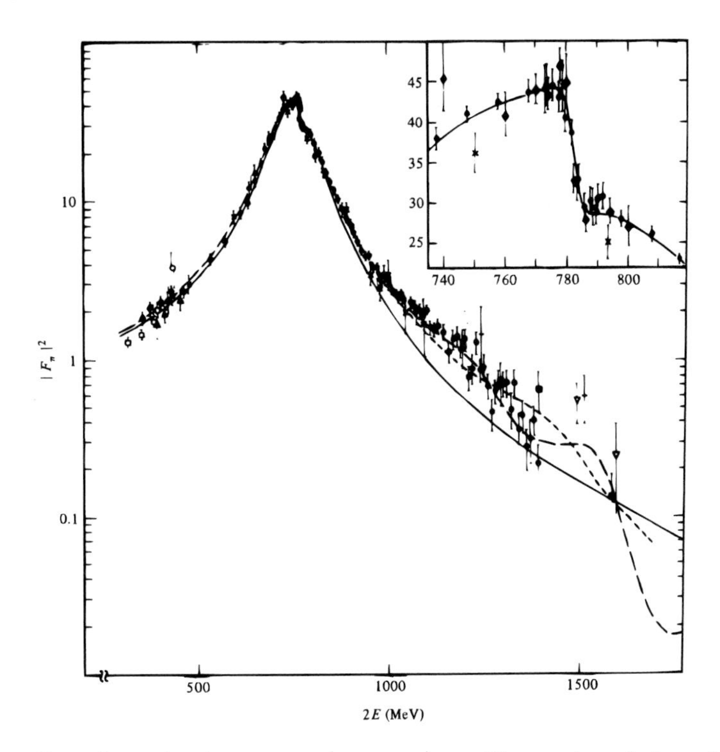

Figure 10.15: Form factor for the process $e^+e^- \to \pi^+\pi^-$. The number of pions observed at a given energy $2E$ is normalized by division through the number of electron and muon pairs observed at the same energy and converted to a squared pion form factor. Unlike form factors in Chapter 6, this form factor is obtained in the time-like region in which the squared momentum transfer $q^2 > 0$. The energy $2E$ is that of the colliding beams. The inset shows the rapid drop at $2E$ close to 780 MeV due to interference of the ω and ρ mesons. The curves are theoretical calculations that include such interference effects. [From L. M. Barkov et al., *Nucl. Phys.* **B256,** 365 (1985).]

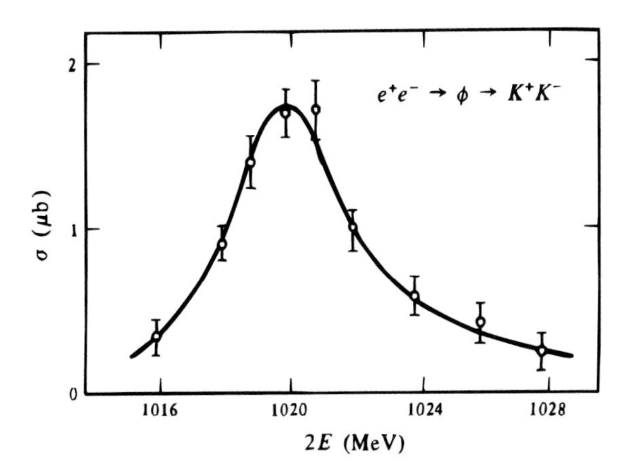

Figure 10.16: Cross section for the process $e^+e^- \to K^+K^-$. [From V. A. Sidorov (NOVOSIBIRSK), *Proceedings of the 4th International Symposium on Electron and Photon Interactions*, (D. W. Braben, ed.), Daresbury Nuclear Phys. Lab, 1969.]

Table 10.1: VECTOR MESONS. η_P is the parity and η_c the charge parity of the vector mesons.

Meson	I	J	η_P	η_c	Y	Rest Energy (MeV)	Width (MeV)	Dominant Decay Mode
ρ^0	1	1	-1	-1	0	770	153	$\pi\pi$
ω^0	0	1	-1	-1	0	783	10	$\pi^+\pi^-\pi^0$
ϕ^0	0	1	-1	-1	0	1020	4	$K\overline{K}$

peak was identified with the rho meson. Why does the rho turn up here? Before answering this question, two more experiments will be discussed to provide additional information. In Fig. 10.16, the cross section for the process $e^+e^- \rightarrow K^+K^-$ is shown as a function of the total energy $2E_0$ at energies near 1 GeV. Again a resonance peak appears but this time with a peak energy of about 1020 MeV and a width of about 4 MeV. The ϕ^0 meson has these two properties. Observation of the reaction $e^+e^- \rightarrow \pi^+\pi^-\pi^0$ yields a peak at about 780 MeV (see inset of Fig. 10.15) with a width of about 10 MeV. These values point to the ω^0. The virtual photon in the reaction $e^+e^- \rightarrow$ hadrons produces resonances at the positions of the ρ^0, the ω^0, and the ϕ^0. To see what these three mesons have in common, we list their properties in Table 10.1.

The three mesons in Table 10.1 satisfy the conditions set out above: They have spin $J = 1$, negative parity, negative charge parity, and strangeness 0. Since a vector has negative parity and the same number of independent components as a spin-1 particle, the mesons are called *vector mesons.* The rho has isospin 1 and is an isovector, whereas the two others are isoscalars.

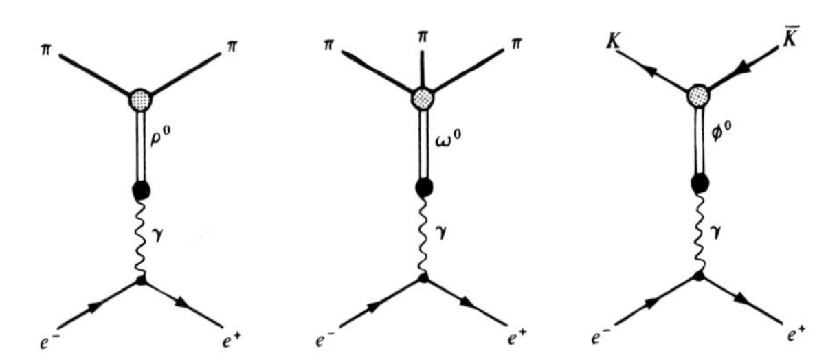

Figure 10.17: The transformation of a virtual photon into a vector meson gives rise to the resonances and their decays observed in colliding beam experiments.

As pointed out in Section 8.6, after Eq. (8.30), the electric charge operator is composed of an isoscalar and the third component of an isovector. The photon, as carrier of the electromagnetic force, should have the same transformation properties, and it matches the vector mesons in their isospin properties. The diagrams for the production of the three vector mesons listed in Table 10.1 are given in Fig. 10.17.

10.8 Colliding Beams

We have already discussed colliding beams in Section 2.8; in Section 7.6 we indicated that e^+e^- experiments were important in the discovery of the new quantum numbers charm and bottom. Actually, the first e^+e^- experiments were done to test QED at high momentum transfers, but emphasis soon changed to studies of hadron production via the photon annihilation channel, Fig. 10.11(d). The virtual photon has spin 1 and negative parity; the hadrons are consequently produced in a unique and well defined state of total angular momentum and parity. Despite this simplicity, electron–positron collisions have been an unexpectedly rich source of new information and surprises. They are ideally suited to search for new leptons and quarks; in addition, they allow tests of the standard model. In the following sections, we describe some of the results.

Two technical achievements are responsible for the outpouring of results from collider experiments: well-designed accelerators and new detectors. We have treated these developments already in Chapters 2 and 4 and add here only some specific information. In Table 10.2, we list some of the existing and planned high-energy colliders. The largest e^+e^- collider built, LEP at CERN in Geneva, is shown in Fig. 2.12. A different arrangement of colliding beams, the Stanford Linear Collider (SLC) is sketched in Fig. 10.18.

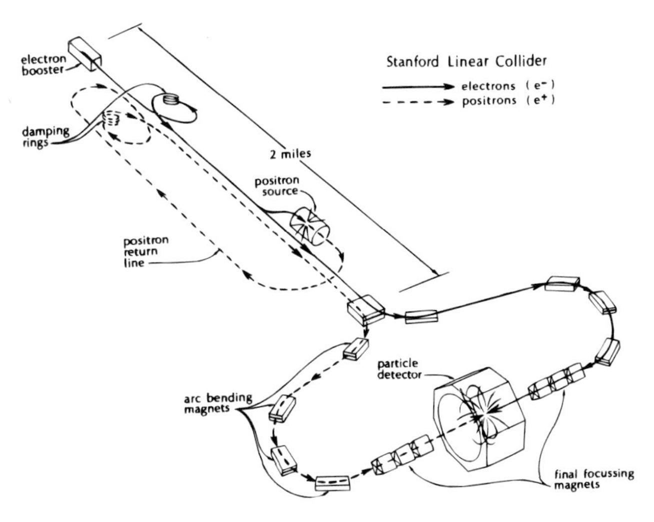

Figure 10.18: Artist's conception of the SLC. Electrons and positrons were accelerated to almost 50 GeV in the linear part, then guided and focused by magnets until they collide head-on. [Figure drawn by Walter Zawojski and reproduced courtesy of SLAC.]

As discussed in Section 2.2, the event rates in a colliding beam experiment are several orders of magnitude smaller than in a typical stationary target experiment.

Table 10.2: SOME EXISTING AND PLANNED COLLIDERS.

Ring	Location	Start of operations	Particles collided	Max Beam Energy (GeV)
CESR	Cornell	1979	e^+e^-	6/6
VEPP-4M	Novosibirsk	1994	e^+e^-	6/6
BEPC	Beijing	1989	e^+e^-	2/2
LEP	CERN	1989	e^+e^-	105/105
DAΦNE	Frascati	1999	e^+e^-	1/1
KEKB	Tokyo	1999	e^+e^-	4/8
PEP-II	Stanford	1999	e^+e^-	3/9
ILC	Undecided		e^+e^-	2000/2000
HERA	Hamburg	1992	ep	30/920
Tevatron	Batavia	1987	$\bar{p}p$	980/980
LHC	CERN	2007	pp	7000/7000

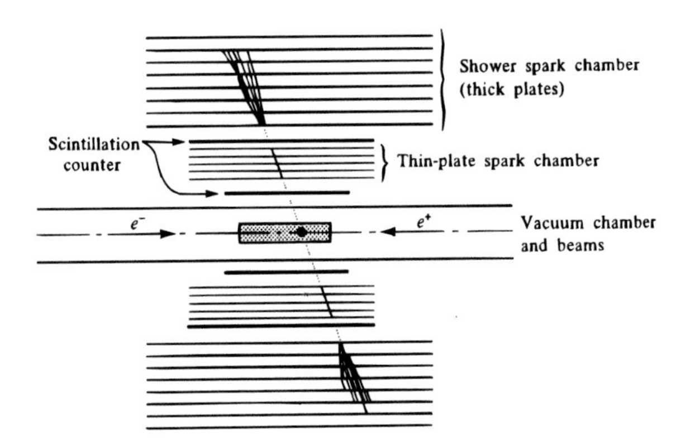

Figure 10.19: Basic arrangement for detectors at colliders.

Consequently, detectors are designed to observe essentially all events. The basic arrangement is given in Fig. 10.19; the detector that was crucial for the discovery of the ψ and thus of charm at SPEAR is shown in Fig. 10.20.[12] Fig. 10.19 illustrates another interesting feature of e^+e^- collisions: the resonances can decay by emitting $q\bar{q}$ pairs that subsequently appear mostly as back-to-back jets.

Table 10.2 shows that the highest energies are achieved in collisions of hadrons. Thus the Tevatron was built and the LHC is being built to search for new physics in $\bar{p}p$ and pp collisions, respectively. However, e^+e^- colliders produce experiments that are much easier to interpret and yield cleaner probes of new physics: the International Linear Collider (ILC) is presently in the planning stage even though its energy will be a fraction of the energy that LHC will reach.

[12]Other detectors are described in Chapter 4.

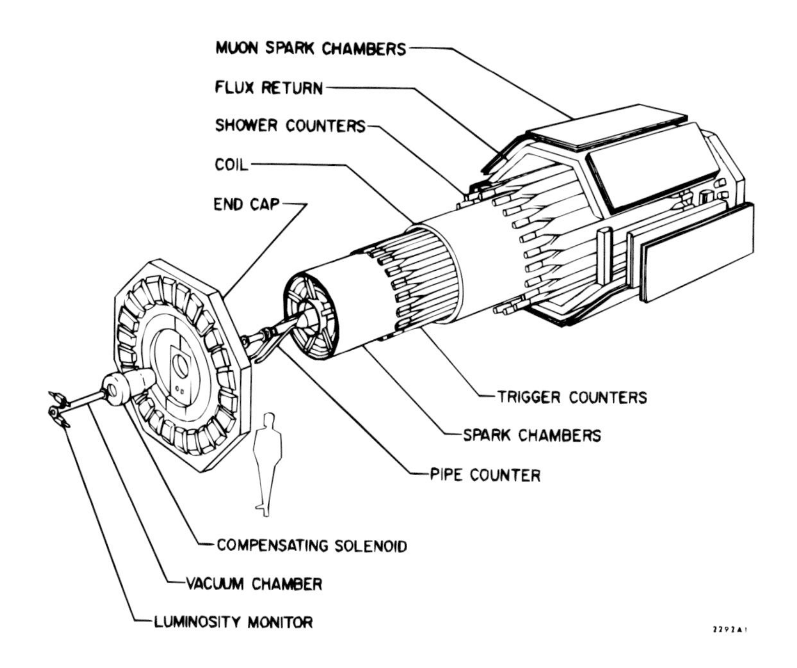

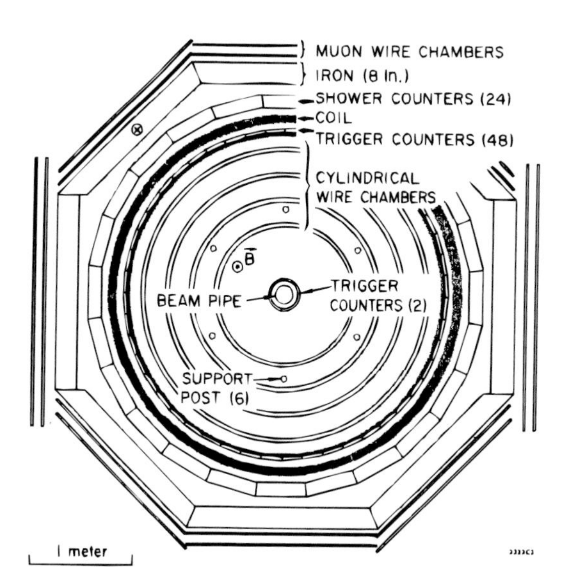

Figure 10.20: Magnetic detector at SPEAR. The detector included spark chambers, scintillation counters, and a solenoid. The solenoid was 3 m in diameter, 3 m long, and produced a 4 kG field parallel to the beam. [Figure courtesy of SLAC.]

10.9 Electron–Positron Collisions and Quarks

In Section 7.6, we mentioned the "1974 November Revolution" in which a new long-lived particle, J/ψ was discovered simultaneously in pp scattering at the Brookhaven National Laboratory and in e^+e^- collisions at SLAC.

In 1977, another long-lived particle, the Υ, was found in p-nucleus collisions at Fermilab. The J/ψ is interpreted as a $c\bar{c}$ state of a charmed quark with its antiquark. Similarly, the upsilon is a $b\bar{b}$ state and there is a $t\bar{(t)}$ state as well. The detailed investigations of these particles and of some closely related states in e^+e^- collisions yield strikingly simple and profound results.

As an example, we show in Fig. 10.21 the total cross section for the production of hadrons in e^+e^- collisions as a function of the total c.m. energy W near 3.1 GeV.[13] Two features stand out, the very large cross section and the narrowness of the resonance peak.

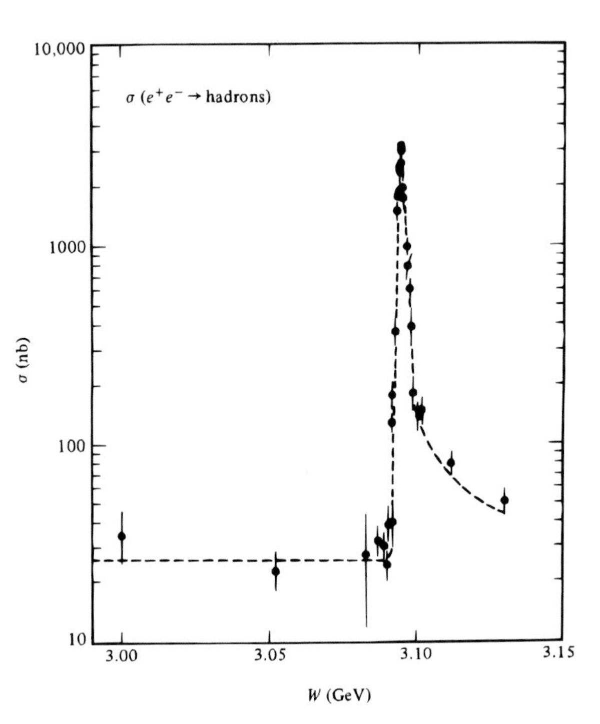

Figure 10.21: Total hadron production cross section in e^+e^- collisions near 3.1 GeV and the J/ψ peak. [From A. M. Boyarski et al., *Phys. Rev. Lett.* **34**, 1357 (1975).]

Muon pair production is described very well by QED. The cross section for $e^+e^- \to \mu^+\mu^-$ (neglecting the muon mass) is given by an equation similar but not identical to Eq. 10.82 because here the particles in the final state are distinguishable from the ones in the initial state:

$$\frac{d\sigma}{d\Omega} = \frac{\alpha^2}{4s}(\hbar c)^2(1 + \cos^2\theta), \tag{10.89}$$

where $s = W^2 = (2E_e)^2$ is the square of the c.m. energy, Eq. (10.83), and θ is the c.m. scattering angle. The total cross section is

$$\sigma = \frac{4\pi\alpha^2}{3s}(\hbar c)^2. \tag{10.90}$$

[13] A. M. Boyarski et al., *Phys. Rev. Lett.* **34,** 1357 (1975); R. F. Schwitters and K. Strauch, *Ann. Rev. Nucl. Sci.* **26**, 89 (1976).

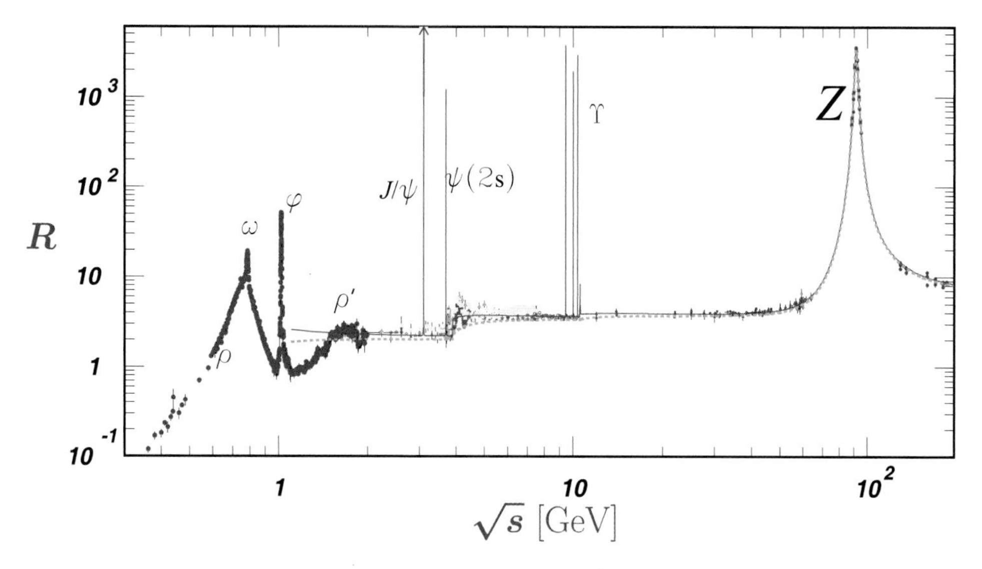

Figure 10.22: The ratio R of the total cross section for e^+e^- annihilation into hadrons to the muon pair production cross section. [From PDG.]

It is therefore convenient to refer all other cross sections to that of muon pair production by introducing the ratio R, defined by

$$R = \frac{\sigma(e^+e^- \longrightarrow \text{hadrons})}{\sigma(e^+e^- \longrightarrow \mu^+\mu^-)}. \tag{10.91}$$

This ratio is shown as a function of W in Fig. 10.22. A number of resonances stand out like beanpoles above a flat landscape. The resonances and the flat background can be described in terms of simple diagrams, as in Fig. 10.23. The resonances (particles) have an energy dependence that is given by a Breit–Wigner shape, Eq. (5.45), and they have large total cross sections. The photon's quantum numbers imply that the resonances have spin and parity $J^\pi = 1^-$. As will be discussed later the resonances are "bound" (confined) quark–antiquark pairs that appear as vector mesons. The flat "background" between resonances is ascribed to nonresonant quark–antiquark pair production. Since quarks are confined, the nonresonant quark–antiquark pair produced by the photon must encounter at least one other quark–antiquark pair and combine with it before emerging as free particles. This process is shown in Fig. 10.23(b).

If quarks are indeed spin-$\frac{1}{2}$ point particles, as postulated in Section 5.11, the cross section for the production of a $q\overline{q}$ pair should also be given by Eq. 10.89 multiplied by the square of the ratio of the quark-to-electron charge. If we denote the electric charge of quark i as a multiple of e by q_i, the assumption of point charges immediately gives for the ratio R,

$$R = \sum_i {q_i}^2, \tag{10.92}$$

because all other factors cancel. The sum in Eq. (10.92) extends over all quark species with mass less than $W/(2c^2)$. For $W < 6$GeV, three quarks can be produced, u, d, and s, with charges 2/3, −1/3, and −1/3, respectively (See Table 5.7), so that Eq. (10.92) gives $R = 2/3$. However, Fig. 10.22 shows that $R_{\text{exp}} = 2$! The discrepancy is explained through the introduction of color, as stated in Section 5.11. If each quark comes in three colors, R is given by $3q_i^2 = 2$ for u, d, and s quarks in agreement with experiment. Above the threshold for J/ψ production there is a fourth quark of charge 2/3, and above the upsilon threshold five known quark flavor u, d, s, c, and b are present; with color, we then expect R to be

$$R = 3\left[2\left(\frac{2}{3}\right)^2 + 3\left(\frac{-1}{3}\right)^2\right] = \frac{11}{3}.$$

Above the threshold for $\bar{t}t$ production, we have to include the charge of this quark. The data in Fig. 10.23 agree approximately with this prediction. The ratio R thus provides strong evidence for two crucial properties of quarks, their point-like nature and their color.

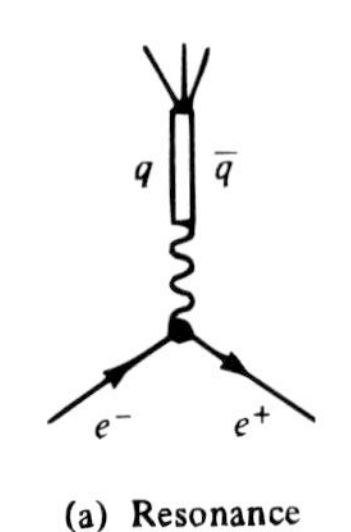

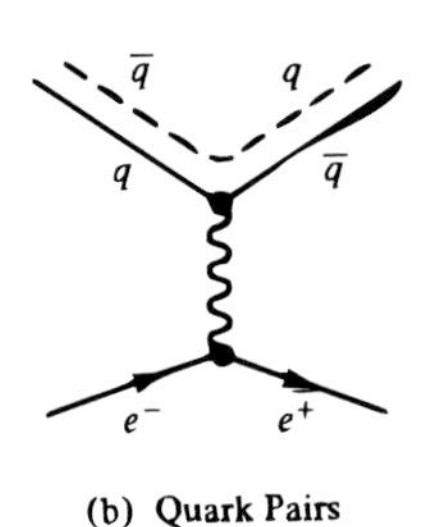

Figure 10.23: Production of (a) resonance and (b) of "individual" quark pair. In (b), the second $q\bar{q}$ pair shown is required to produce observable mesons.

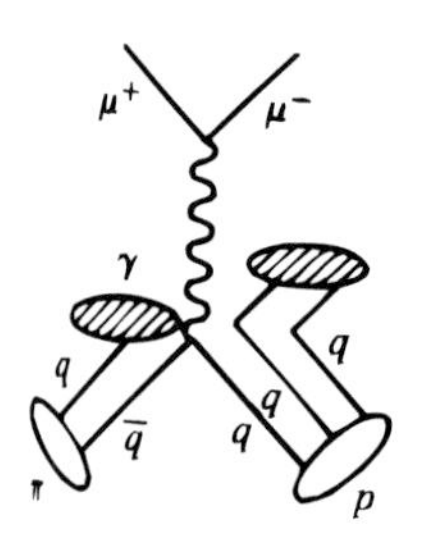

Figure 10.24: The Drell–Yan process for $\mu^+\mu^-$ production in pion–proton scattering.

We have discussed primarily experiments in which lepton pairs annihilate and produce hadrons. The reverse experiment is also feasible, and is called a Drell–Yan reaction.[14] In a typical Drell–Yan process, a high energy pion collides with a proton. The antiquark in the pion annihilates a quark in the proton to produce a virtual photon, which creates a lepton pair, as illustrated in Fig. 10.24. The process has proven useful for studies of QCD.[15]

[14] S.D. Drell and T.M. Yan, *Phys. Rev. Lett.* **25**, 316 (1970); **24**, 181 (1970); *Ann. Phys.* (*New York*) **66**, 578 (1971).

[15] P.N. Harriman, Z. Phys. **C**55, 449 (1992); P.D. Morley, Phys. Rev. C **39**,708 (1989); G.L. Li, J.P. Shen, J.J. Yang, H.Q. Shen, Phys. Rept. **242**, 505 (1994).

10.10 The Photon–Hadron Interaction: Real and Spacelike Photons

> Are there not other original properties of the rays of light, besides those already described?
>
> Newton, *Opticks*

The interaction of *real* photons with hadrons at low and moderate photon energies (say below 20 MeV) has formed a considerable part of nuclear physics for at least 40 years. One example, multipole radiation, was sketched in Section 10.5. Another celebrated case is the photodisintegration of the deuteron,

$$\gamma d \longrightarrow pn,$$

which was discovered in 1934 by Chadwick and Goldhaber[(16)] and used by them for a measurement of the neutron mass. A third example is the exploration of the excited states of nuclei with incident gamma rays. The cross section for gamma-ray absorption shows the existence of individual excited states and the occurrence of the giant dipole resonance.[(17)] The basic features of the resulting cross section have already been given in Fig. 5.34. Such studies produce a great deal of information concerning nuclear structure, but they teach us little new about the nature of the photon–hadron interaction: The photon interacts with the electric charges and currents in the nucleus. The distributions of the charges and currents are determined by the strong force. If they are assumed to be given, then the interaction with the probing photon can be described by the Hamiltonian 10.48. Below, say, 100 MeV incident photon energy, this behavior can be understood: the (reduced) photon wavelength is of the order 2 fm or longer, short enough to probe some details of the nuclear charge and current distributions but not short enough to probe the photon–nucleon interaction.[(18)]

The interaction of high-energy photons ($E \geq$ a few GeV) with hadrons presents a different picture and new aspects emerge: *the photon shows hadron-like properties.*[(19)] The roots of these properties can be understood with concepts that have been introduced earlier. In Section 3.3, the production of real electron–positron pairs by real photons was mentioned. In the previous section, it was found that timelike photons can produce hadrons, as indicated in Figs. 10.21 and 10.22. To describe the high-energy behavior of real photons, we now consider such processes in

[16] J. Chadwick and M. Goldhaber, *Proc. Roy. Soc.* (*London*) **A151**, 479 (1935).

[17] K.A. Snover, *Ann. Rev. Nucl. Part. Sc.* **36**, 545 (1986); J.J. Gaardhoje, *Ann. Rev. Nucl. Part. Sc.* **42**, 483 (1992).

[18] It has been shown by various calculations that the scattering of photons in the limit of zero photon energy is given entirely by the *static* particle properties, mass, charge, and higher moments. The hadron structure dynamics does not enter, and the limit agrees with the classical result. W. Thirring, *Phil. Mag.* **41**, 1193 (1950); F. E. Low, *Phys. Rev.* **96**, 1428 (1954); M. Gell-Mann and M.L. Goldberger, *Phys. Rev.* **96**, 1433 (1954).

[19] L. Stodolsky, *Phys. Rev.* **18**, 135 (1967); S.J. Brodsky and J. Pumplin, *Phys. Rev.* **182**, 1794 (1969); V.N. Gribov, *Sov. Phys. JETP* **30**, 709 (1970); D.R. Yennie, *Rev. Mod. Phys.* **47**, 311 (1975); T.H. Bauer et al., *Rev. Mod. Phys.* **50**, 261 (1978).

more detail. As already stated in Section 3.3 (Problem 3.22), a photon cannot produce a real pair of massive particles in free space. A nucleus must be present to take up momentum in order to satisfy energy and momentum conservation. However, the uncertainty principle permits violation of energy conservation by an amount ΔE during times smaller than $\hbar/\Delta E$. A photon can therefore produce a *virtual* pair or a *virtual* particle with the same quantum numbers as the photon and with total energy ΔE, but such a state can exist only for a time less than $\hbar/\Delta E$. Consider as a simple example the virtual decay of a photon of energy E_γ into a hadron h, with mass m_h. Momentum conservation demands that photon and hadron have the same momentum $p \equiv p_\gamma = E_\gamma/c$.Consequently the energy difference between the photon and the virtual hadron is

$$\Delta E = E_h - E_\gamma = (E_\gamma^2 + m_h^2 c^4)^{1/2} - E_\gamma. \tag{10.93}$$

The time during which the hadron can "virtually exist" is (see Problem 10.31)

$$T = \begin{cases} \dfrac{\hbar}{m_h c^2}, & E_\gamma \ll m_h c^2, \\ \dfrac{2\hbar E_\gamma}{m_h^2 c^4}, & E_\gamma \gg m_h c^2. \end{cases} \tag{10.94}$$

The hadron can travel at most with the velocity of light, and the distance traversed during its virtual existence is limited by

$$L \lesssim \begin{cases} \dfrac{\hbar}{m_h c} & = \lambda\!\!\!^{-}_h, & E_\gamma \ll m_h c^2, \\ \dfrac{2\hbar E_\gamma}{m_h^2 c^3} & = 2\lambda\!\!\!^{-}_h \dfrac{E_\gamma}{m_h c^2}, & E_\gamma \gg m_h c^2, \end{cases} \tag{10.95}$$

where $\lambda\!\!\!^{-}_h$ is the reduced Compton wavelength of the hadron. The quantum numbers of the photon do not allow a decay into one pion; the lowest possible hadron state consists of two pions, and $\lambda\!\!\!^{-}_h$ is consequently limited by

$$\lambda\!\!\!^{-}_h \lesssim \frac{\hbar}{2m_\pi c} \approx 0.7 \text{ fm}. \tag{10.96}$$

The lowest-mass physical particle with $J^{\eta_P} = 1^-$ is the rho meson, for which $\lambda\!\!\!^{-}_h \approx$ 0.3 fm. Equation (10.95, top) then shows that the path length of virtual hadrons associated with low-energy photons is much smaller than nuclear and even smaller than nucleon dimensions. Equation (10.95, bottom) indicates, however, that the path-length can become much larger than nuclear diameters at photon energies exceeding a few GeV.

The argument given so far reveals how far a virtual hadron accompanying the photon can propagate, but it does not predict how often a strong fluctuation arises. To describe the second property, we write for the normalized state function, $|\gamma\rangle$, of the real photon:

$$|\gamma\rangle = c_0|\gamma_0\rangle + c_h|h\rangle. \tag{10.97}$$

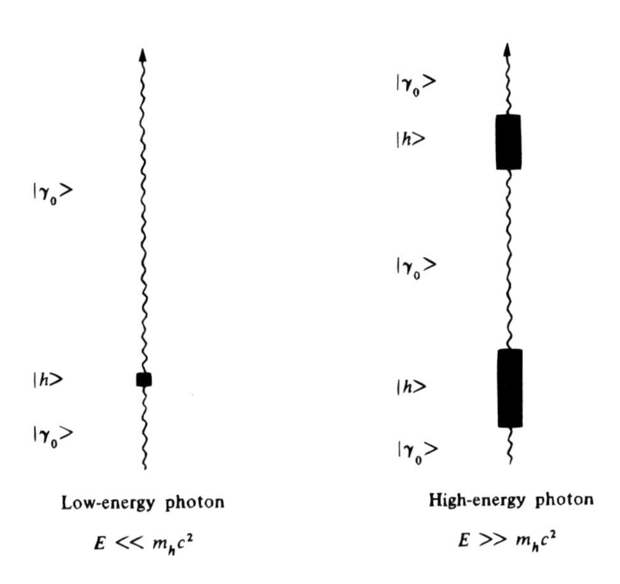

Figure 10.25: Low-energy and high-energy photons. The hadronic contribution for low-energy photons is insignificant. The high-energy photon is accompanied by a hadron cloud that leads to observable effects.

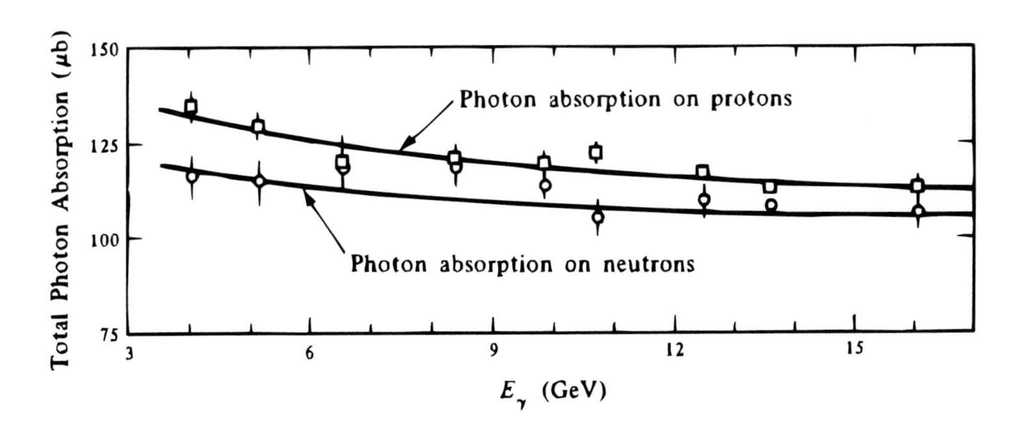

Figure 10.26: Total absorption cross section for photons on nucleons. Very different cross sections are expected if the photon interacts with the electric charge. If the absorption occurs via vector mesons (hadrons), the absorption should be essentially the same for neutron and proton targets. [After D. O. Caldwell et al., *Phys. Rev. Lett.* **25**, 613 (1970).]

Here $c_0|\gamma_0\rangle$ is the purely electromagnetic part of the photon (*bare photon*) and $c_h|h\rangle$ is the hadronic part (*hadron cloud*). The absolute square $c_h^* c_h$ gives the probability of finding the photon in a hadronic state; as we shall see later, it is proportional to α. We shall return to a more detailed discussion of $|h\rangle$ below but note here that we expect, for instance by analogy to the production of real lepton pairs (Fig. 3.7), that the ratio c_h/c_0 increases with increasing energy. Even a small contribution will become experimentally observable because the hadronic force is much stronger than the electromagnetic one. To summarize, we picture low-energy and high-energy photons in Fig. 10.25.

The question as to whether the photon indeed is accompanied by a hadron cloud must be answered by experiments. We shall discuss two examples that demonstrate the existence of a hadronic component. The first one is the scattering of photons from nucleons. The total cross sections for scattering of photons from nucleons have been measured up to center-of-mass energies, $\sqrt{s}$, of 209 GeV for protons and 9 GeV for neutrons. Part of this is shown in Fig. 10.26.[20] As the energy increases above a few GeV, the two cross sections begin to coalesce. If the photons were to interact solely with the electric charge, proton and neutron should have different total cross sections because their electromagnetic properties are different, as indicated by their quark flavor content and the behavior of their form factors G_E and G_M, Eqs. (6.41) and (6.43). The electric form factor of the neutron is small, indicating that the neutron is not only overall neutral but that it contains very little net electric charge at all. The magnetic form factor of the neutron is smaller than that of the proton in the ratio $|\mu_n/\mu_p| \approx 0.7$. If the photon were to interact only with the electric charges and currents, scattering from the neutron would be much smaller than from the proton. The situation is different for the strong component, $c_h|h\rangle$. Proton and neutron form an isospin doublet. According to Eq. (8.15), the strong Hamiltonian commutes with $\vec{I}$ and the hadronic structure is independent of the orientation in isospin space. Proton and neutron consequently have the same hadronic structure. The forces between hadrons are charge-independent and do not depend on the orientation of the nucleon isospin vector. The component $c_h|h\rangle$ therefore should produce equal scattering from protons and neutrons. Indeed, as Fig. 10.26 shows, at energies where $E_\gamma \gg m_h c^2$, the cross sections $\sigma(\gamma, p)$ and $\sigma(\gamma, n)$ approach each other and indicate that the term $c_h|h\rangle$ becomes dominant.

The behavior of the total cross section for photons on nuclei as a function of the scatterer baryon number, A, provides a second striking demonstration of the hadronic traits of high-energy photons. Below an energy of a few GeV, the total cross section is proportional to A,

$$\sigma_{\text{tot}}(\gamma) \propto A, \qquad E < \text{GeV}. \tag{10.98}$$

[20] D. O. Caldwell, et al., *Phys. Rev. Lett.* **25**, 609, 613 (1970); *Phys. Rev.* **D7**, 1362 (1973); Belusov et al., SJNP 21, 289 (1975); ZEUS Collaboration (S. Chekanov et al.) Nucl. Phys. **B627**, 3 (2002).

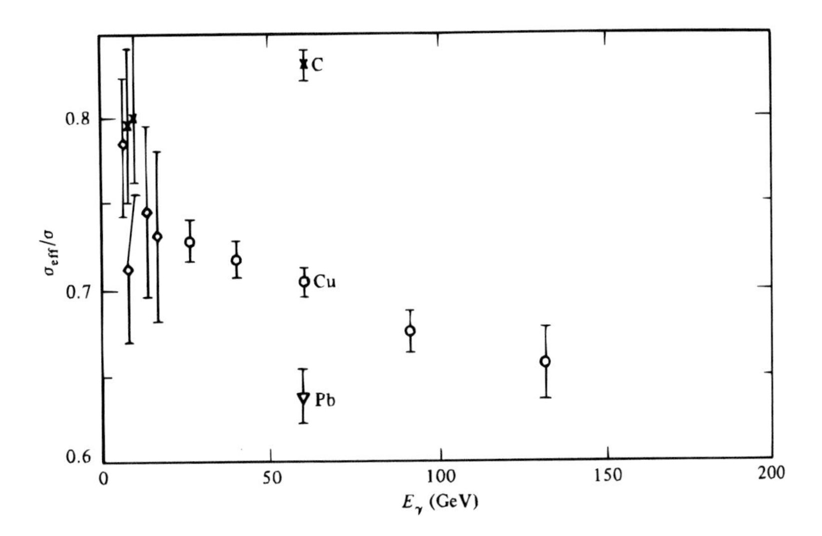

Figure 10.27: Energy dependence of $\sigma_{\text{eff}}/\sigma$ for copper and dependence of $\sigma_{\text{eff}}/\sigma$ on atomic number at 60 GeV photon energy. [After D. O. Caldwell et al., *Phys. Rev. Lett.* **42**, 553 (1979).]

Above a few GeV, however, the total cross section is no longer proportional to A.[(21)] This shadowing effect is displayed in Fig. 10.27 as a plot of $\sigma_{\text{eff}}/\sigma$ against E.[(22)] $\sigma_{\text{eff}}/\sigma$ is the ratio of the photo-production cross section of a nucleus of Z protons and N neutrons to the sum of the individual cross sections of the constituent nucleons. If the high-energy photons were to see all nucleons in a nucleus equally well, $\sigma_{\text{eff}}/\sigma$ would be unity. Fig. 10.27 shows that the ratio $\sigma_{\text{eff}}/\sigma$ at 60 GeV decreases markedly with increasing A; for a fixed A (Cu, $A = 64$), it decreases steadily with increasing energy. To show that this experimental result provides more evidence for the existence of a hadronic contribution to the photon, we shall discuss the behavior of the two components, $|\gamma_0\rangle$ and $|h\rangle$, separately. Consider first the bare photon, $|\gamma_0\rangle$. The mean free path of photons of about 15 GeV energy in nuclear matter (an infinitely large nucleus) is about 600 fm. This number follows with Eq. (2.17) from the values of the photon–nucleon cross section of Fig. 10.26, $\sigma \approx 10^{-2}\text{fm}^2$, and the nuclear density given in Eq. (6.28), $\rho_n \approx 0.17$ nucleon/fm^3. Since the nuclear diameter of even the heaviest nucleus is less than 20 fm, bare photons "illuminate" nuclei uniformly, and the contribution of the term $c_0|\gamma_0\rangle$ to the cross section is proportional to A. The hadronic term, $c_h|h\rangle$, produces two contributions to the total cross section. As will be shown in Chapter 14, the cross section for hadrons is of the order of 3 fm^2, and the mean free path is about 2 fm.

[21] E. M. Henley, *Comm. Nucl. Part. Phys.* **4**, 107 (1970); F. V. Murphy and D. E. Yount, *Sci. Amer.* **224**, 94 (July 1971); D. Schildknecht, hep-ph/0511090, published in *Acta Phys. Polon.* **B37**, 595 (2006).

[22] D. O. Caldwell, V. B. Elings, W. P. Hesse, G. E. Jahn, R. J. Morrison, F. V. Murphy, D. E. Yount, *Phys. Rev. Lett.* **23**, 1256 (1969); D. O. Caldwell et al., *Phys. Rev. Lett.* **42**, 553 (1979); N. Bianchi et al., *Phys. Rev. C* **60**, 064617 (1999).

If the photon transforms to the hadron state *inside* the nucleus, the hadron will interact near the position of production. Since the production can occur anywhere, the contribution to the total cross section is proportional to A, just like that of bare photons. On the other hand, virtual hadrons created *before* striking the nucleus interact with nucleons in the nuclear surface layer because of their short mean free path. The corresponding contribution to the total cross section consequently is proportional to the nuclear area, or to $A^{2/3}$. At a given photon energy, the total cross section is the sum of the three contributions, and it should be of the form

$$\sigma(\gamma A) = aA + bA^{2/3}. \tag{10.99}$$

As stated above, the second term is due to photons that transform into hadrons before striking the nucleus. Such hadrons have a chance to interact if they are produced within a distance L, which at high photon energies is, according to Eq. (10.95, bottom), large compared to nuclear diameters and proportional to E_γ. Other things being equal, the coefficient b should thus be proportional to E_γ, and the surface term should become dominant at energies large compared to $m_h c^2$. The behavior of the cross section as expressed by Eq. (10.99) and Fig. 10.27 therefore can be understood in terms of virtual hadrons.

The expression for the hadron cloud of the photon, $c_h|h\rangle$, can be written in an informative form by using perturbation theory. We assume the states of the various hadrons and of the photon, in the absence of the electromagnetic interaction, to be given by the Schrödinger equations

$$H_h|\gamma_0\rangle = 0, \qquad H_h|n\rangle = E_n|n\rangle. \tag{10.100}$$

H_h is the strong Hamiltonian, $|\gamma_0\rangle$ the state function of the bare photon, and $|n\rangle$ represents a hadronic state. If the electromagnetic interaction is switched on, hadronic states are superimposed onto the bare photon state:

$$|\gamma\rangle = c_0|\gamma_0\rangle + \sum_n c_n|n\rangle, \qquad |c_0|^2 + \sum_n |c_n|^2 = 1. \tag{10.101}$$

Since H_{em} is weaker than H_h, the expansion coefficients c_n are small and $c_0 \approx 1$. The state of the physical photon is a solution of the complete Schrödinger equation,

$$(H_h + H_{em})|\gamma\rangle = E_\gamma|\gamma\rangle. \tag{10.102}$$

Inserting the expansion (10.101) into Eq. (10.102) gives, with Eq. (10.100) and with $\langle n|\gamma_0\rangle = 0, c_n \ll 1$,

$$c_n = \frac{\langle n|H_{em}|\gamma_0\rangle}{E_\gamma - E_n}. \tag{10.103}$$

The energy difference between the photon energy E_γ and the hadron energy E_n is given by Eq. (10.93); for large photon energies, the expansion coefficient becomes, with Eq. (10.94, bottom),

$$c_n = \langle n|H_{em}|\gamma_0\rangle \frac{2E_\gamma}{m_h^2 c^4}. \tag{10.104}$$

The square of the matrix element is of order $\alpha \approx \frac{1}{137}$; if it is constant, the contribution from the hadronic state $|n\rangle$ to the photon state should be proportional to the photon energy. At values of E_γ that are small compared to $m_h c^2$, the photon behaves like an ordinary light quantum.

To compute actual values of c_n and thus find the hadron cloud, the wave functions of the states $|n\rangle$ and H_{em} must be known. At present it is believed that H_{em} is given by the minimal electromagnetic interaction and that all difficulties in computing the matrix elements stem from the absence of a detailed understanding of the structure of the hadron states $|n\rangle$.

Calculations are carried out with simplified models. No one model describes all experiments at the present time, but in the several GeV region the *vector dominance model* (VDM) is reasonably successful in correlating many aspects. This model was introduced by Sakurai,(10) and it is based on the assumption that the only hadronic states of importance in the sum in Eq. (10.101) are the lightest vector mesons ρ, ω, and ϕ. Only three matrix elements, of the form $\langle V|H_{em}|\gamma_0\rangle$, thus appear, and approximate values of these can be obtained from the experiments on vector meson production in colliding beam experiments. (23)

10.11 Magnetic Monopoles

Finally we come to another unsolved aspect of the electromagnetic interaction, the possible existence of *magnetic monopoles*. Classical electrodynamics is based on the observation that electric, but no magnetic, charges exist. The magnetic field is always produced by magnetic dipoles, never by magnetic charges (*monopoles*). This fact is expressed through the Maxwell equation

$$\boldsymbol{\nabla} \cdot \boldsymbol{B} = 0. \tag{10.105}$$

Since this relation states an experimental result, the question as to its validity must be asked. As early as 1931, Dirac proposed a theory with magnetic monopoles.(24) In this theory, Eq. (10.104) is replaced by

$$\boldsymbol{\nabla} \cdot \boldsymbol{B} = 4\pi\rho_m, \tag{10.106}$$

where ρ_m is the magnetic charge density. In an extension of his work, Dirac showed that the rules of quantum mechanics lead to a quantization of the electric charge e and the magnetic charge g(25):

$$eg = \frac{1}{2}n\hbar c, \tag{10.107}$$

[23] For a more detailed description of the VDM see D. Schildknecht, hep-ph/0511090, Proceedings of PHOTON2005 International Conference on the Structure and Interactions of the Photon", Warsaw, 2005.

[24] P. A. M. Dirac, *Proc. Roy. Soc. (London)* **A133**, 60 (1931).

[25] P. A. M. Dirac, *Phys. Rev*, **74**, 817 (1948).

The dimensionless constant describing the interaction between two magnetic monopoles is enormous. Dirac's suggestion led to many unsuccessful searches for magnetic monopoles, both in cosmic rays and at accelerators.[26] The hunt received a new stimulus when it was realized that grand unified theories predict the existence of very heavy monopoles with a mass of about $10^6 \mathrm{GeV}/c^2$,[27] approximately the mass of a bacterium. Obviously such particles cannot be produced with accelerators, but they could have been created in the early universe.[28] No such particles have been seen.

10.12 References

A lucid introduction and history of classical electrodynamics is given in the Feynman Lectures, Vol. II. A more complete and sophisticated treatment can be found in Jackson.

No very easy introduction to quantum electrodynamics exists. There is a nice general introduction to the field in S.S. Schweber, *QED and the Men Who Made It: Dyson, Feynman, Schwinger, and Tomonaga*, Princeton University Press, 1994. However, as already mentioned in footnote 4, the article by Fermi [*Rev. Mod. Phys.* **4**, 87 (1932)] and the book by Feynman (*Quantum Electrodynamics*) can be read by undergraduate students if they do not give up easily. Brief introductions to the quantization of the electromagnetic field can also be found, for instance, in the quantum mechanics texts by Merzbacher, Messiah, or E. G. Harris, *A Pedestrian Approach to Quantum Field Theory*, Wiley-Interscience, New York, 1972.

On a more sophisticated level, a number of excellent books exist: (1) W. Heitler, *The Quantum Theory of Radiation*, Oxford University Press, London, 1954. This book is somewhat old-fashioned, but the physical points are brought out clearly. (2) J. D. Bjorken and S. D. Drell, *Relativistic Quantum Mechanics*, McGraw-Hill, New York, 1964. This book is more modern than Heitler's, and it provides a thorough exposition of the physical ideas and the calculational techniques of relativistic quantum mechanics. (3) J. J. Sakurai, *Advanced Quantum Mechanics*, Addison-Wesley, Reading, Mass., 1967. This book is an excellent companion to Bjorken and Drell. It illuminates many of the same problems from a different point of view. Finally, there are the course lectures by Greiner and Reinhardt *Quantum electrodynamics*, Springer Verlag, New York, 2003. There is also an exposition in I.J.R. Aitchison, and A.J.G. Hey, *Gauge Theories in Particle Physics: A Practical Introduction*, Second Edition; Adam Hilger, 1989.

The classic papers on QED are collected and introduced in J. Schwinger, ed.,

---

[26] See *Phys. Today* **17**, 16 (2006), for searches with accelerators; M. J. Longo, *Phys. Rev.* **D25**, 2399 (1982) for a review on limits; B. Cabrera, *Phys. Rev. Lett.* **48**, 1378 (1982) for an example of searches in cosmic rays.

[27] G. 't Hooft, *Nucl. Phys.* **B79**, 276 (1974); A. Polyakov, *ZhETF Pis. Red.* **20**, 430 (1974) [transl. *JETP Lett.* **20**, 194 (1974)].

[28] J. Preskill, *Phys. Rev. Lett.* **43**, 1365 (1979); P. Langacker, *Phys. Rep.* **72**, 185 (1981).

Quantum Electrodynamics, Dover, New York, 1958. The evidence for the validity of quantum electrodynamics is discussed in S. J. Brodsky and S. D. Drell, *Ann. Rev. Nucl. Sci.* **20**, 147 (1970); R. Gatto, in *High Energy Physics*, Vol. 5 (E. H. S. Burhop, ed.), Academic Press, New York, 1972; P. Dittman and V. Hepp, *Z. Phys* **C10**, 283 (1981); P. Duinker, *Rev. Mod. Phys.* **54**, 325 (1982); *Present Status and Aims of Quantum Electrodynamics*, (G. Gräff, E. Klempt, and G. Werth, eds), Springer Lecture Notes in Physics, Vol. 143, Springer, New York, 1981.

A detailed account of the interaction of photons in the MeV energy range with nuclei is contained in J. M. Eisenberg and W. Greiner, *Nuclear Theory*, Vol. 2: *Excitation Mechanisms of the Nucleus.*, North-Holland, Amsterdam, 3d. revision, 1988. A large number of articles and books review the nuclear and nucleon photoeffect; F. W. K. Firk, *Ann. Rev. Nucl. Sci.* **20**, 39 (1970).

The interaction of high energy photons with hadrons is treated in R. P. Feynman, *Photon–Hadron Interactions*, W. A. Benjamin, Reading, MA, 1972; A. Donnachie and G. Shaw, *Electromagnetic Interactions of Hadrons*, Plenum, New York, 1978, and T. M. Bauer, R. D. Spital, D. R. Yennie, and F. M. Pipkin, *Rev. Mod. Phys.* **50**, 261 (1978). A different point of view is presented in M. Erdmann, *The Partonic Structure of the Photon*, Springer Verlag, New York, 1997.

The photoeffect in the context of nucleon structure is discussed in C.E. Hyde-Wright and K. de Jager, *Ann. Rev. Nucl. Part. Sci.* **54**, 217 (2005).

The subject of e^+e^- collisions is treated in many references, e.g. R. F. Schwitters and K. Strauch, *Ann. Rev. Nucl. Sci.* **26**, 89 (1976); B. Wiik and G. Wolf, *Electron–Positron Interactions*, Springer Tracts in Modern Physics, Springer, New York, Vol. 86, 1979; G. Goldhaber and J. E. Wiss, *Ann. Rev. Nucl. Part. Sci.* **30**, 299, (1980); F. Renard, *Basics of Electron–Positron Collisions*, Editions Frontières, Vol. 32, 1981; *High Energy Electron–Positron Physics*, (A. Ali and P. Söding, eds), World Scientific, Teaneck, NJ, 1988; R. Marshall, *Rep. Prog. Phys.* **52**, 1329 (1989); *Physics opportunities with* $e^+ - e^-$ *colliders* by H. Murayama, M.E. Peskin in *Ann. Rev. Nucl. Part. Sc.* **46**, 533 (1996); *Physics opportunities with a TEV collider* by S. Dawson, M. Oreglia, in *Ann. Rev. Nucl. Part. Sc.* **54**, 269 (2004).

Problems

10.1. Draw the transition probability factor $P_{N\alpha}(T)/4|\langle N|H_{\text{int}}|\alpha\rangle|^2$ of Eq. (10.13) for the following times T:

(a) $T = 10^{-7}\,\text{sec}$

(b) $T = 10^{-22}\,\text{sec}$

10.2. Derive the golden rule No. 1, Eq. (10.21), by developing the approximation involved in Eq. (10.19) to second order.

10.3. Consider the nonrelativistic scattering of a particle with momentum $\boldsymbol{p} = m\boldsymbol{v}$ by a fixed potential $H_{\text{int}} \equiv V(\boldsymbol{x})$ [Fig. 10.1(b)]. Assume that the incident and the scattered particles can be described by plane waves (Born approximation). L^3 is the quantization volume.

(a) Use the golden rule to show that the transition rate into the solid angle $d\Omega$ is given by

$$dw = \frac{v}{L^3}\left|\frac{m}{2\pi\hbar^2}\int d^3x \exp\left[\frac{i(\boldsymbol{p}_\alpha - \boldsymbol{p}_\beta)\cdot\boldsymbol{x}}{\hbar}\right] H_{\text{int}}\right|^2 d\Omega.$$

(b) Show that the connection between cross section $d\sigma$ and transition rate is given by

$$w_{\beta\alpha} = F d\sigma,$$

where F is the incident flux [Eq. (2.11)].

(c) Verify the Born approximation expression, Eq. (6.5), for the scattering amplitude $f(\boldsymbol{q})$.

10.4. Verify Eq. (10.26) by computing the number of states in a three-dimensional box of volume L^3.

10.5. Derive the Lorentz force by starting from the Hamiltonian (10.41).

10.6. Show that the term $q^2\boldsymbol{A}^2/2mc^2$ in Eq. (10.42) can be neglected in realistic situations.

10.7. Verify that $q\rho(\boldsymbol{x})\boldsymbol{v}(\boldsymbol{x})$ in Eq. (10.47) is the charge that traverses unit area per unit time.

10.8. Show that the continuity equation, Eq. (10.51), is a consequence of Maxwell's equations.

10.9. Justify that the total energy in a plane electromagnetic wave in a volume V is given by

$$W = V\frac{|\boldsymbol{E}|^2}{4\pi},$$

where $\boldsymbol{E}$ is the electric field vector.

10.10. Equation (10.69) describes the transition rate for the spontaneous *emission* of dipole radiation in the transition $\alpha \to \beta$.

(a) Compute the corresponding expression for the *absorption* of a photon by dipole radiation inducing the transition $\beta \to \alpha$.

(b) Compare the transition rates for emission and absorption. Compare the ratio with the ratio expected from time-reversal invariance.

10.11. Prove Eqs. (10.71) and (10.72).

10.12. Sketch the radiation pattern predicted by Eqs. (10.75) and (10.76) for dipole radiation, assuming that the vector $\langle \boldsymbol{\beta}|\boldsymbol{x}|\alpha\rangle$ points along the z direction. Compare to the radiation pattern for classical dipole radiation.

10.13. Use Eq. (10.77) to make a crude estimate for the mean life of an electric dipole transition

(a) In an atom, $E_\gamma = 10$ eV.

(b) In a nucleus, $E_\gamma = 1$ MeV.

Find relevant transitions in nuclei and atoms and compare your result with the actual values.

10.14. Discuss an accurate method for determining the fine structure constant.

10.15. Why do nuclei and particles not have permanent electric dipole moments? Why can some molecules have permanent electric dipole moments?

10.16. Why does the transition $\Sigma^0 \to \Lambda^0$ occur through an electromagnetic and not a hadronic decay?

10.17. What kind of multipole transition is involved in the decay $\Sigma^0 \to \Lambda^0$? Use an extrapolation of Fig. 10.10 to estimate the mean life. Compare to the presently known value.

10.18. Discuss time-to-amplitude converters (TACs).

(a) Describe the function of a TAC.

(b) How can a TAC be used to measure lifetimes?

(c) Sketch the block diagram of a TAC.

10.19. Show that a $2^+ \xrightarrow{\gamma} 0^+$ transition, as, for example, shown in Fig. 10.7, cannot occur through dipole radiation.

10.20. Verify that the selection rules of Eq. (10.80) and the conservation laws of Eq. (10.81) together lead to the multipole assignments shown in Fig. 10.9.

10.21. The transition from an excited to a nuclear ground state can usually proceed by two competing processes, photon emission and emission of *conversion electrons*.

(a) Discuss the process of internal conversion.

(b) Assume that a particular decay has a half-life of 1 sec and a conversion coefficient of 10. What is the nuclear half-life for bare nuclei, i.e., nuclei stripped of all their electrons?

(c) The nuclide ^{111}Cd has a first excited state at 247 keV excitation energy. If the electron spectrum of this nuclide is observed, lines appear. Sketch the position of the conversion electron lines produced by the 247 keV transition.

10.22. Consider Møller scattering as shown in Fig. 10.12 (symmetric case).

(a) Assume that the incident electron has a kinetic energy of 1 MeV. Compute the angle θ_{lab}.

(b) Repeat the problem for an incident electron energy of 1 GeV.

(c) Compute the ratio of cross sections for parts (a) and (b) assuming that the angular function $f(\theta)$ in Eq. (10.82) has the same value for both cases.

10.23. Consider Møller scattering. Assume that the electrons in the target foil are completely polarized along the direction of the incident electrons. Use the Pauli principle to get an idea how longitudinally polarized incident electrons will scatter if their spin is (a) parallel and (b) anti-parallel to the target spins. Consider only the symmetric scattering shown in Fig. 10.12.

10.24. To study the high-energy behavior of photons, monoenergetic beams are required. An ingenious way of producing such photons involves an intense laser pulse that collides head-on with a well-focused electron beam. The photons that are scattered by 180° acquire considerable energy. Compute the energy of the photons from a ruby laser that are scattered by 180° from an electron beam of energy

(a) 1 MeV.

(b) 1 GeV.

(c) 100 GeV.

10.25. Estimate the ratio of probabilities for the emission of a rho to that of a gamma ray from a high-energy nucleon that passes close to another one.

10.26. Magnetic monopoles (magnetic charges) would have remarkable properties:

(a) How would a magnetic monopole interact with matter?

(b) How would the track of a monopole look in a bubble chamber?

(c) How could a monopole be detected?

(d) Compute the energy of a monopole accelerated in a field of 20 kG.

10.27. Estimate the mass of a magnetic monopole by using the following, very speculative, approach: The *classical electron radius* r_e is given by

$$r_e = \frac{e^2}{m_e c^2}.$$

Assume that a magnetic monopole has a similar radius, with e replaced by g and m_e by the monopole mass.

10.28. Prove Eq. (10.103).

10.29. Show that a magnetic monopole passing through a superconducting loop of current induces a permanent change of flux, but that a charge or magnetic dipole will not do so.

10.30. Show that the electromagnetic transition from hadronic states of angular momentum and parity 0^+ and 1^- to a state of angular momentum and parity 0^+ are forbidden if both states have isospin zero.

10.31. Consider the virtual decay of a photon of energy E_γ into a hadron with mass m_h. Show that the energy difference between the photon and the virtual hadron is given by Eq. 10.93 and that the uncertainty principle yields Eqs. 10.94 for the time spent as a hadron.

Chapter 11

The Weak Interaction

This chapter explores the weak interaction part of the electro-weak theory. The history of the weak interaction is a series of mystery stories. In each story, a puzzle appears, at first only in a vague form and then more and more clearly. Clues to the solution are present but are overlooked or discarded, usually for the wrong reason. Finally, the hero comes up with the right explanation and everything is clear until the next corpse is unearthed. In the treatment of the electromagnetic interaction, the well-understood classical theory provided an example which, properly translated and reformulated, guided the development of quantum electrodynamics. No such classical analog is present in the weak interaction, and the correct features had to be taken from experiment and from analogies to the electromagnetic interaction. We shall describe some of the puzzles and their solutions. In doing so we are hampered by the self-imposed constraint of not using the Dirac theory. We shall therefore not be able to write the interaction properly but shall use other means to explain the crucial concepts.

At low energies and to lowest order in perturbation theory the weak interaction can be described semi-phenomenologically in a satisfactory way. At high energies, however, problems appear that have no solution if the weak interaction is treated alone. The unification of the weak interaction with the electromagnetic one, however, leads to a deeper understanding and to a solution of these problems. In this chapter, we review some of the experimental knowledge and the basic phenomenology gained from a study of the weak interaction. In the next two chapters we lay the groundwork for, and sketch, the electroweak theory.

11.1 The Continuous Beta Spectrum

> The continuous β-spectrum would then be understandable under the assumption that during β-decay a light neutral particle is emitted with every electron such that the sum of energies of neutrino and electron are constant.
>
> W. Pauli

Radioactivity was discovered in 1896 by Becquerel, and it became clear within a few years that the decaying nuclei emitted three types of radiation, called α, β, and γ rays. The outstanding puzzle was connected with the beta rays. Careful measurements over more than 20 years indicated that the beta particles were electrons and that they were not emitted with discrete energies but as a continuum.

An example of such a beta spectrum is shown in Fig. 11.1. We have discussed nuclear energy levels in chapter 5. The existence of quantized levels was well known in 1920, and the first puzzle posed by the continuous beta spectrum thus was: Why is the spectrum of electrons continuous and not discrete? A second puzzle arose a few years later when it was realized that no electrons are present inside nuclei. Where, then, do the electrons come from?

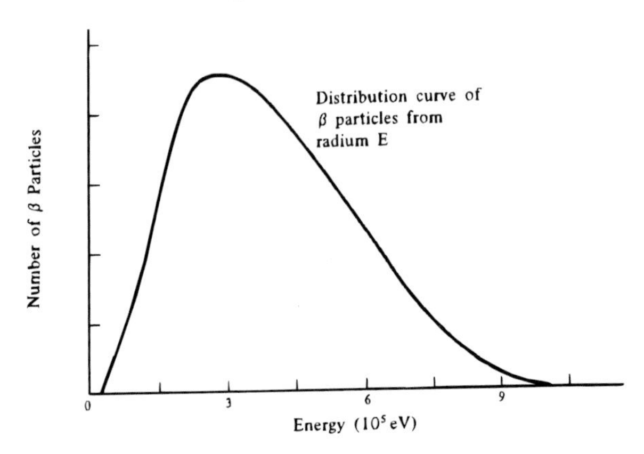

Figure 11.1: Example of a beta spectrum. [This figure is taken from one of the classic papers: C.D. Ellis and W.A. Wooster, *Proc. R. Soc.* (London) **A117**, 109 (1927).] Present experimental techniques yield more accurate energy spectra, but all essential aspects are already contained in the curve reprinted here.

The first puzzle was solved by Pauli, who suggested the existence of a new, very light, uncharged, and penetrating particle, the *neutrino*.[1] Today, with so many particles known, proposing a new particle scarcely raises eyebrows. In 1930, however, it was a revolutionary step. Only two particles were known, the electron and the proton. Destroying the simplicity of the subatomic world by addition of a third citizen was considered to be heresy, and very few people took the idea seriously. One of the ones who did was Fermi; he used Pauli's neutrino hypothesis to solve the second puzzle. Fermi assumed with Pauli that a neutrino is emitted together with the beta particle in every beta decay. Consequently, the simplest nuclear beta decay, the one of the neutron, is written as

$$n \longrightarrow pe^-\bar{\nu}.$$

Since the neutrino is chargeless, it is not observed in a spectrometer. Electron and neutrino share the decay energy, and the observed electrons sometimes have very little of it and sometimes nearly the maximum energy. The spectrum shown in

[1] Pauli first suggested the neutrino in a letter addressed to some of his friends who were attending a physics meeting in Tübingen. He declared that he was unable to be present at the gathering because he wanted to attend the famous annual ball of the Swiss Federal Institute of Technology. The letter is reprinted in R. Kronig and V.F. Weisskopf, eds., *Collected Scientific Papers by Wolfgang Pauli*, Vol. II, Wiley-Interscience, New York 1964, p. 1316. See also L. M. Brown, *Phys. Today* **31**, 23 (September 1978).

Fig. 11.1 is thus qualitatively explained. To avoid the problems posed by electrons inside nuclei, Fermi postulated that the electron and the neutrino were *created* in the decay, just as a photon is created when an atom or a nucleus decays from an excited to the ground state or two photons are created in the decay of the neutral pion.

Fermi did not simply speculate how beta decay could occur; he performed the computations to find the expressions for the electron spectrum and the decay probability. His original treatment[2] is above our level, and it has to be watered down here. In the present section, we shall show that even a crude approach reproduces the shape of the beta spectrum. Since the interaction responsible for beta decay is weak, perturbation theory can be used, and the transition rate is given by the golden rule, Eq. (10.1),

$$dw_{\beta\alpha} = \frac{2\pi}{\hbar}|\langle\beta|H_w|\alpha\rangle|^2\rho(E).$$

Here H_w is the Hamiltonian responsible for beta decay, and we have written $dw_{\beta\alpha}$ rather than $w_{\beta\alpha}$ in order to indicate that we are interested in the transition rate for transitions with electron energies between E_e and $E_e + dE_e$. We first consider the density-of-states factor $\rho(E)$. Three particles are present in the final state, and $\rho(E)$ is given by Eq. (10.34) as

$$\rho(E) = \frac{V^2}{(2\pi\hbar)^6}\frac{d}{dE_{\max}}\int p_e^2\,dp_e\,d\Omega_e p_{\bar{\nu}}^2\,dp_{\bar{\nu}}\,d\Omega_{\bar{\nu}}. \tag{11.1}$$

V is the quantization volume. Since the final results are independent of this volume, it is set equal to 1. The differentiation $d/dE_{\max}$ requires a word of explanation. $E_{\max}$ is constant, and it thus appears at first sight that $d/dE_{\max}$ should vanish. However, it has the meaning of a variation; $(d/dE_{\max})\int\cdots$ indicates how the integral changes under a variation of the maximum energy.

To evaluate $\rho(E)$, we must first decide what we are interested in. Figure 11.1 shows the number of electrons emitted with an energy between E_e and $E_e + dE_e$. To calculate the corresponding transition rate, E_e and consequently also p_e are kept constant. The $d/dE_{\max}$ in Eq. (11.1) then does not affect the terms relating to the electron, and Eq. (11.1) becomes

$$\rho(E) = \frac{d\Omega_e\,d\Omega_{\bar{\nu}}}{(2\pi\hbar)^6}p_e^2\,dp_e p_{\bar{\nu}}^2\frac{dp_{\bar{\nu}}}{dE_{\max}}. \tag{11.2}$$

The next step is simplified by the fact that the nucleon in the final state is much heavier than either lepton and therefore receives very little recoil energy. To a good approximation, electron and neutrino share the total energy:

$$E_e + E_{\bar{\nu}} = E_{\max}. \tag{11.3}$$

[2]E. Fermi, *Z. Physik* **88**, 161 (1934); translated in *The Development of Weak Interaction Theory* (P. K. Kabir, ed.), Gordon and Breach, New York, 1963. See also L. M. Brown and H. Rechenberg, *Am. J. Phys.* **56**, 982 (1988).

For a massless neutrino, $E_{\bar{\nu}} = p_{\bar{\nu}}c$, and for constant E_e,

$$\frac{dp_{\bar{\nu}}}{dE_{\max}} = \frac{1}{c}\frac{dE_{\bar{\nu}}}{dE_{\max}} = \frac{1}{c},$$

so that

$$\rho(E) = \frac{d\Omega_e\, d\Omega_{\bar{\nu}}}{(2\pi\hbar)^6 c}\, p_e^2\, p_{\bar{\nu}}^2\, dp_e. \tag{11.4}$$

As written, $\rho(E)$ is the density-of-states factor for a transition in which the electron has a momentum between p_e and $p_e + dp_e$ and is emitted into the solid angle $d\Omega_e$. With Eq. (11.3), $p_{\bar{\nu}}^2$ is replaced by $(E_{\max} - E_e)^2/c^2$. Moreover, if the matrix element $\langle\beta|H_w|\alpha\rangle$ is averaged over the angle between the electron and the neutrino, $dw_{\beta\alpha}$ can be integrated over $d\Omega_e d\Omega_{\bar{\nu}}$ and with Eq. (11.4) the result is

$$dw_{\beta\alpha} = \frac{1}{2\pi^3 c^3 \hbar^7}\overline{|\langle pe^-\bar{\nu}|H_w|n\rangle|^2}\, p_e^2 (E_{\max} - E_e)^2\, dp_e. \tag{11.5}$$

This expression gives the transition rate for the decay of a neutron into a proton, an electron, and an antineutrino, with the electron having a momentum between p_e and $p_e + dp_e$. Does the expression agree with experiment? Since at this point we know nothing about the matrix element, the simplest approach is to assume that it is independent of the electron momentum and to see how the other factors in Eq. (11.5) fit the observed beta spectra. In principle, then, a function

$$p_e^2 (E_{\max} - E_e)^2 dp_e$$

could be fitted to the experimental data. There exists an easier way: Equation (11.5) is rewritten into the form

$$\left(\frac{dw_{\beta\alpha}}{p_e^2\, dp_e}\right)^{1/2} = \text{const.}\left(\overline{|\langle pe^-\bar{\nu}|H_w|n\rangle|^2}\right)^{1/2} (E_{\max} - E_e). \tag{11.6}$$

If the expression on the left-hand side is determined experimentally and plotted against the electron energy E_e, a straight line results if the matrix element is momentum-independent. Such a plot is called a Fermi or Kurie plot. Figure 11.2 shows the Kurie plot for the neutron decay. It is indeed a straight line over most of the energy range. The deviation at the low-energy end was caused by experimental difficulties in this early experiment: The electron counter had a window 5 mg/cm^2 thick, and it absorbed low-energy electrons. (See Fig. 3.8 and Eq. (3.7).) The number of electrons shown in Fig. 11.2 is not corrected for this loss.

The technique just described can be applied to beta decays other than that of the neutron with a small modification. If a nucleus decays by beta emission, the charged lepton experiences the Coulomb force once it has left the nucleus with charge Ze. This force will decelerate negative and accelerate positive electrons.

The spectrum will be distorted: There will be more positrons of high energy and more electrons of low energy than predicted by Eq. (11.5). The *Coulomb correction* introduces an additional factor in Eq. (11.5), and for a decay $N \to N'e\nu$ it becomes

$$dw_{\beta\alpha} = \frac{1}{2\pi^3 c^3 \hbar^7} \overline{|\langle N'e\nu|H_w|N\rangle|^2} F(\mp, Z, E_e) p_e^2 (E_{\max} - E_e)^2 dp_e. \tag{11.7}$$

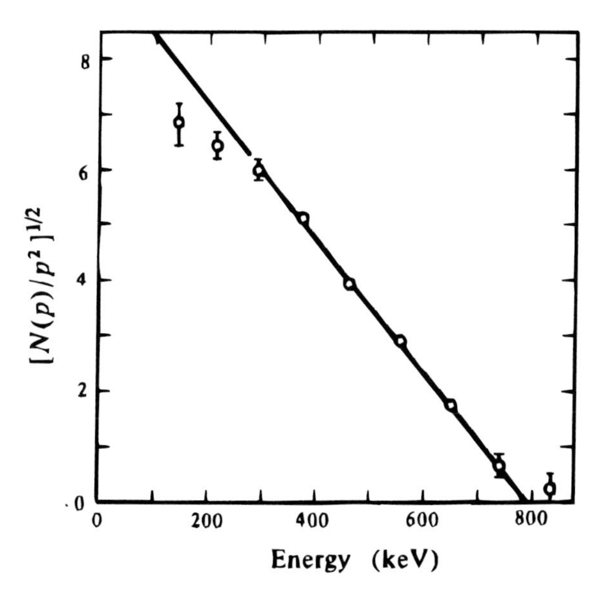

Figure 11.2: Kurie plot for the neutron decay. [From J. M. Robson, *Phys. Rev.* **83**, 349 (1951).] Here $N(p)$ corresponds to dw/dp_e in Eq. (11.6).

$F(\mp, Z, E_e)$ is called the Fermi function;[3] the sign indicates whether it applies to electrons or positrons. The Fermi function also corrects Kurie plots for the Coulomb distortion, and the momentum dependence of the matrix element can be tested in many decays. It turns out that the matrix element is essentially momentum-independent in all cases of interest, for decay energies up to a few MeV. The shape of the electron spectrum in beta decay is dominated by phase–space considerations and not by properties of the matrix element.

However, the high-energy end of the beta spectrum can provide information about the mass of the neutrino. In deriving Eq. (11.7) we have assumed a massless electron neutrino. If the mass is not zero, the Kurie plot will deviate from a straight line at the upper limit; the deviation will be most pronounced for decays with small maximum energy, for instance $^3\text{H} \to e^-\bar{\nu}_e{}^3\text{He}$. Searches with this nucleus indicate that the electron neutrino rest energy is smaller than approximately 3 eV.[4]

11.2 Beta Decay Lifetimes

Information about the magnitude of the matrix element can be obtained from the lifetimes of beta emitters. If the matrix element is momentum-independent, the

[3] H. Behrens and J. Jänecke, *Numerical Tables for Beta Decay and Electron Capture, Landolt-Börnstein*, New Series, Vol. I/4, Springer, Berlin, 1969. H. Behrens and W. Bühring, *Electron Radial Wave Functions and Nuclear Beta-Decay*, Clarendon Press, Oxford, 1982.

[4] J. Bonn et al., *Nucl. Phys. Proc. Suppl.* **110**, 395 (2002); see also http://www-ik.fzk.de/ katrin/index.html for an experiment under preparation.

total transition rate $w_{\beta\alpha}$ and the mean life τ can be obtained from Eq. (11.7) by integration over the momentum:

$$w = \frac{1}{\tau} = \frac{1}{2\pi^3 c^3 \hbar^7} \overline{|\langle N' e\nu | H_w | N \rangle|^2} \int_0^{p_{\rm max}} dp_e F(\mp, Z, E_e) p_e^2 (E_{\rm max} - E_e)^2. \quad (11.8)$$

For very large energies, where $E_{\rm max} \approx cp_{\rm max}$, and for small Z, where $F \approx 1$, the integral becomes

$$\int_0^{p_{\rm max}} dp_e p_e^2 (E_{\rm max} - E_e)^2 \simeq \frac{1}{30c^3} E_{\rm max}^5. \quad (11.9)$$

While this relation is useful for estimates, accurate values of the integral are needed for a meaningful treatment of the data. Fortunately the integral has been tabulated(3) as

$$\int_0^{p_{\rm max}} dp_e F(\mp, Z, E_e) p_e^2 (E_{\rm max} - E_e)^2 = m_e^5 c^7 f(E_{\rm max}). \quad (11.10)$$

The factor $m_e^5 c^7$ has been inserted in order to make f dimensionless. With Eqs. (11.10) and (11.8), the matrix element becomes

$$\overline{|\langle N' e\nu | H_w | N \rangle|^2} = \frac{2\pi^3}{f\tau} \frac{\hbar^7}{m_e^5 c^4}. \quad (11.11)$$

If τ is measured and f computed(3) then the square of the matrix element can be obtained from Eq. (11.11). It is customary to use $ft_{1/2}$ and not $f\tau$ in tabulations. $ft_{1/2}$ is called the *comparative half-life.* The name stems from the fact that all beta-decaying states would have the same value of $ft_{1/2}$ if all matrix elements were equal. Nature provides an enormous range of values of $ft_{1/2}$, from about 10^3 to 10^{23} sec. If such a variation were caused by the fact that the weak interaction, H_w, were not universal but would change from decay to decay, an understanding of the weak processes would be hopeless. It is assumed that H_w is the same for all decays and that the nuclear wave functions that enter the calculation of $\langle N' e\nu | H_w | N \rangle$ are responsible for the variations. The most fundamental decays have the "best" wave functions and give rise to the largest matrix elements. A few cases are listed in Table 11.1.

With $ft_{1/2} = (\ln 2) f\tau$ (Eq. (5.34)) and with the numerical values of the constants, Eq. (11.11) becomes

$$\overline{|\langle N' e\nu | H_w | N \rangle|^2} = \frac{43 \times 10^{-6}\ \text{MeV}^2\ \text{fm}^6\ \text{sec}}{ft_{1/2}(\text{in sec})}. \quad (11.12)$$

Now consider the decay of the neutron. With the value of $ft_{1/2}$ given in Table 11.1, the magnitude of the matrix element of H_w becomes

$$\overline{|\langle pe\bar{\nu} | H_w | n \rangle|} \approx 2 \times 10^{-4}\ \text{MeV fm}^3. \quad (11.13)$$

Table 11.1: Comparative Half-Lives of a Few Beta Decays.

Decay	Spin-parity Sequence	$t_{1/2}$	$E_{\max}$ (MeV)	$ft_{1/2}$ (sec)
$n \to p$	$\frac{1}{2}^+ \to \frac{1}{2}^+$	10.2 min	0.782	1054
$^6\text{He} \to {}^6\text{Li}$	$0^+ \to 1^+$	0.807 sec	3.50	800
$^{14}\text{O} \to {}^{14}\text{N}$	$0^+ \to 0^+$	70.6 sec	1.752	3039

The matrix element (11.13) gives an energy times a volume. The volume of the proton follows from Eq. (6.46) as approximately 2 fm^3. The weak energy, distributed over the volume of the proton, is of the order

$$H_w \approx 10^{-4} \text{ MeV}. \tag{11.14}$$

This number demonstrates the weakness of the weak interaction: Presumably the mass of the proton, about 1 GeV, is given by the hadronic interaction. The weak interaction is consequently about a factor of 10^7 smaller.

11.3 The Current–Current Interaction of the Standard Model

Two facts have become clear in the previous two sections: The dominant feature of the beta spectrum is given by the phase–space factor, and the beta decay interaction is so weak that perturbation theory can be used. However, we have learned very little about the Hamiltonian responsible for beta decay. Is it nevertheless possible to make a stab at the construction of a *weak Hamiltonian*? We have said above that the first successful theory of beta decay was formulated by Fermi(2) and that even less was known about beta decay in 1933 than we have described so far. It is therefore only proper to show how Fermi's genius led to a profound understanding of the weak interaction. We shall follow Fermi's reasoning but use more modern language.

Fermi assumed that electron and neutrino were created during the process of beta decay. This act of creation is similar to the process of photon emission. By 1933 the quantum theory of radiation was well understood, and Fermi patterned his theory after it. The result was incredibly successful and withstood all assaults for nearly 25 years. When parity fell in 1957, Fermi's theory finally required modification. The most successful extension was put forward by Feynman and Gell-Mann and, in somewhat different form, by Marshak and Sudarshan.(5) It can be said that the weak interaction tries as hard as possible to look like its stronger cousin, the electromagnetic one.

[5] R. P. Feynman and M. Gell-Mann, *Phys. Rev.* **109**, 193 (1958); E. C. G. Sudarshan and R. E. Marshak, *Phys. Rev.* **109**, 1860 (1958).

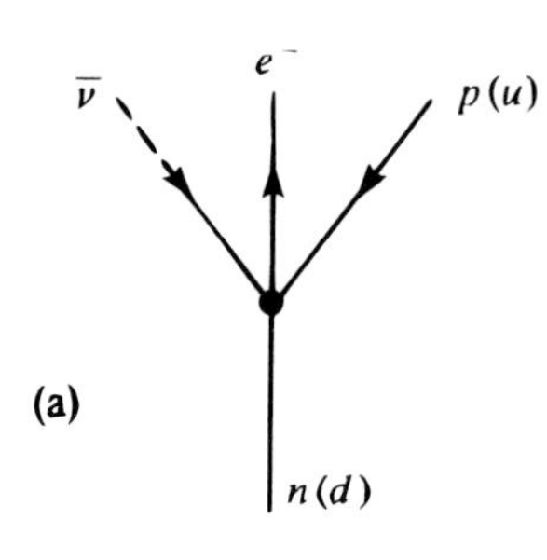

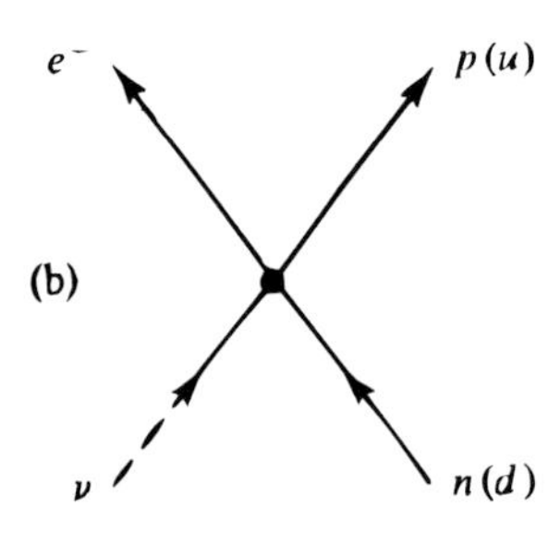

Figure 11.3: Neutron decay and neutrino absorption. It is assumed that the absolute values of the matrix elements for the two processes are the same. The diagrams apply to quarks with u and d substituting for p and n.

Figure 11.3(a) shows a diagram for the decay of the neutron. Such a decay is not the most convenient one for writing down an interaction, because one particle comes in and three particles leave. It is easier to see the analogy to the electromagnetic force in a case where two particles are destroyed and two are created. We learned in Section 5.10 that antiparticles can be looked at as particles going backward in time. One of the outgoing antiparticles, say the antineutrino, is therefore replaced by an incoming particle, in this case a neutrino. The process then appears as in Fig. 11.3(b). It is assumed that the matrix elements for the two processes in Fig. 11.3(a) and (b) have the same magnitude. (The transition rates are different because of unequal phase space factors $\rho(E)$.)

It is helpful, especially for later developments, to build the same figures in terms of quarks. To convert a neutron to a proton requires the change of a down (d) to an up (u) quark. The diagrams in Fig. 11.3 thus apply to quarks through the substitution of d and u for n and p, as shown in parentheses.

In the next step, the electromagnetic and the weak interaction of the standard model are compared (Fig. 11.4). The electromagnetic interaction has the familiar form where the force is transmitted by a virtual photon. The weak interaction has been changed from Fig. 11.3(b), and the *intermediate boson* or W (for *weak*), has been inserted. This force-carrying gauge particle makes the analogy to electromagnetism more obvious.

Consider first the electromagnetic case where two currents, each produced by a particle of charge e, interact via a virtual photon. The interaction energy is given by Eq. (10.50):

$$\begin{aligned} H_{em} &= -\frac{e^2}{c^2}\int d^3x\, d^3x'\, \boldsymbol{j}(\boldsymbol{x})\cdot \boldsymbol{j}'(\boldsymbol{x}')\frac{1}{|\boldsymbol{x}-\boldsymbol{x}'|} \\ &= -\frac{e^2}{c^2}\int d^3x d^3x' \boldsymbol{j}(\boldsymbol{x})\cdot \boldsymbol{j}'(\boldsymbol{x}') f_{\rm em}(r), \end{aligned} \tag{11.15}$$

where $r = |\boldsymbol{x}-\boldsymbol{x}'|$, and $f_{\rm em}$ gives the dependence of $H_{\rm em}$ on the separation of $\boldsymbol{j}(\boldsymbol{x})$ and $\boldsymbol{j}'(\boldsymbol{x}')$. The long range of the interaction, given by $|\boldsymbol{x}-\boldsymbol{x}'|^{-1}$, is caused by the vanishing mass of the photon.

The weak interaction, as shown in the graph in Fig. 11.4, is assumed to arise from a *weak current–current interaction*, and the form of H_w is patterned after $H_{\rm em}$.

Lepton conservation in the weak case corresponds to charge conservation in the electromagnetic interaction, and each weak current retains its lepton number.

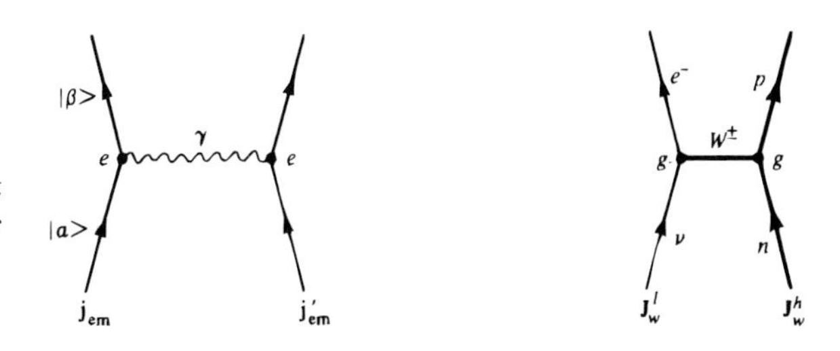

Figure 11.4: Comparison of the electromagnetic and the weak interactions. The superscripts l and h indicate the weak currents of leptons and hadrons, respectively.

Consequently, the lepton number of the W must be zero. (Had we, in going from Fig. 11.3(a) to (b), replaced the outgoing proton by an incoming antiproton, the currents would not satisfy such a conservation law.) The charged weak currents shown in Fig. 11.4 change the value of the electric charge by one unit at the vertex; the neutrino, for instance, changes into an electron. Since the electric charge must be conserved, the W must be charged at each vertex in Fig. 11.4. In analogy to the electromagnetic interaction, Eq. (11.15), the weak Hamiltonian can now be written as

$$H_w = -\frac{g_w^2}{c^2}\int d^3x\, d^3x' \mathbf{J}^l_w(\boldsymbol{x})\cdot \boldsymbol{J}^h_w(\boldsymbol{x}')f(r), \tag{11.16}$$

where g_w is a coupling constant and $f(r)$ gives the dependence of the weak interaction on distance. The range R_W of $f(r)$ must be very short: The mass of the W is about 80 GeV/c^2 (Table 5.9); Eq. (5.52) gives

$$R_W = \frac{\hbar}{m_W c} \approx 2.5 \text{ am}. \tag{11.17}$$

It is customary to describe such short-range forces by a Yukawa shape,

$$f(r) = \frac{\exp(-r/R_W)}{r}. \tag{11.18}$$

We shall return to this form in chapter 14. Here it is sufficient to note that $f(r)$ is a function that is appreciably different from zero only for distances of the order of, or less than, R_W. If we further assume that the weak currents vary very little over distances of the order of R_W, then $\boldsymbol{J}^h_w(\boldsymbol{x}') \approx \boldsymbol{J}^h_w(\boldsymbol{x})$, Eq. (11.18) can be inserted into Eq. (11.16), and the integral over d^3x' can be performed. The result is

$$H_w = -4\pi\frac{g_w^2 R_W^2}{c^2}\int d^3x\, \boldsymbol{J}^l_w(\boldsymbol{x})\cdot\boldsymbol{J}^h_w(\boldsymbol{x}). \tag{11.19}$$

Eq. (11.19) is often rewritten as

$$H_w = -\frac{G_F}{\sqrt{2}\,c^2}\int d^3x\, \boldsymbol{J}^l_w(\boldsymbol{x})\cdot\boldsymbol{J}^h_w(\boldsymbol{x}), \tag{11.20}$$

with

$$G_F = \sqrt{2}\,4\pi g_w^2 R_W^2 = \sqrt{2}\,4\pi \left(\frac{\hbar}{m_W c}\right)^2 g_w^2. \tag{11.21}$$

The factor $1/\sqrt{2}\,c^2$ in Eq. (11.20) is introduced by convention. G_F is a new weak coupling constant called the *Fermi coupling constant* that no longer has the same dimension as the electric charge e.

As Eq. (11.20) stands, it is not yet correct for the following reason: H_w is an operator that must be Hermitian. If the currents $\boldsymbol{J}_w^l$ and $\boldsymbol{J}_w^h$ were Hermitian, H_w would be Hermitian. In the electromagnetic interaction, Hermiticity of $\boldsymbol{j}_{em}$ is guaranteed because the electromagnetic current can be observed; the photon is neutral. No such guarantee exists in the weak interaction, and, in fact, as already indicated, the charged weak current is *not* Hermitian. H_w must therefore be made Hermitian. There are two ways of achieving this goal. One is to add the Hermitian conjugate expression to Eq. (11.20). The second one is again patterned in analogy to the electromagnetic case. In Eq. (10.88) the electromagnetic current was written as the sum of two contributions, one from leptons and the other from hadrons. Similarly, it is assumed that the total weak current is the sum of two contributions, one from leptons and the other from hadrons,

$$\boldsymbol{J}_w = \boldsymbol{J}_w^l + \boldsymbol{J}_w^h. \tag{11.22}$$

The weak Hamiltonian is then Hermitian if Eq. (11.20) is generalized to

$$H_w = -\frac{G_F}{\sqrt{2}\,c^2}\int d^3x \boldsymbol{J}_w(\boldsymbol{x})\cdot\boldsymbol{J}_w^\dagger(\boldsymbol{x}). \tag{11.23}$$

This form is not yet complete. Our starting point, the electromagnetic interaction in the form of Eq. (11.15), describes only the energy due to two currents but leaves out the Coulomb interaction. The Coulomb energy between two charges described by electric charge densities $e\rho(\boldsymbol{x})$ and $e\rho'(\boldsymbol{x}')$ is given by

$$H_c = e^2 \int d^3x d^3x' \frac{\rho(\boldsymbol{x})\rho'(\boldsymbol{x}')}{|\boldsymbol{x}-\boldsymbol{x}'|}.$$

If *weak charges* $g_w\rho_w$ exist, then the arguments leading to Eq. (11.23) can be repeated, and the complete weak Hamiltonian becomes

$$H_w = \frac{G_F}{\sqrt{2}\,c^2}\int d^3x[c^2\rho_w(\boldsymbol{x})\rho_w^\dagger(\boldsymbol{x}) - \boldsymbol{J}_w(\boldsymbol{x})\cdot\boldsymbol{J}_w^\dagger(\boldsymbol{x})]. \tag{11.24}$$

It is possible to treat weak interactions due to charged currents by using H_w in this form. However, relativistic notation makes arguments simpler and more transparent. The probability density and the probability current together form a four-vector, as already indicated in Eq. (1.11):

$$J_w = (c\rho_w, \boldsymbol{J}_w).$$

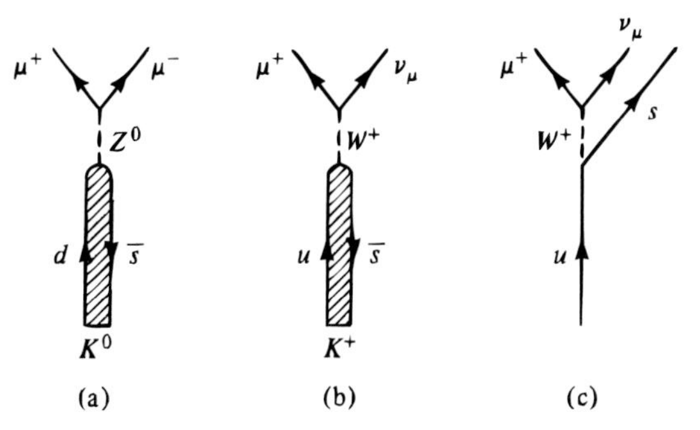

Figure 11.5: Leptonic decays of K^0 and K^+ and quark analogue; the decay shown for the K^0 is forbidden.

For the rest of this chapter we denote four-vectors with ordinary letters. The scalar product of two four-vectors is defined by Eq. (1.10); the product $J_w \cdot J_w^\dagger$ is

$$g_{\mu\,\nu} J_w J_w^\dagger = c^2 \rho_w \rho_w^\dagger - \boldsymbol{J}_w \cdot \boldsymbol{J}_w^\dagger,$$

and the weak Hamiltonian becomes

$$H_w = \frac{G_F}{\sqrt{2}\, c^2} \int d^3x\, J_w(\boldsymbol{x}) \cdot J_w^\dagger(\boldsymbol{x}). \tag{11.25}$$

This equation makes it obvious that H_w is a Lorentz invariant. So far, we have taken the weak current J_w and the intermediate boson W to be charged, as shown in Fig. 11.4. This assumption was generally held to be true until about 1979 and was based on experimental data. It was known, for instance, that the decay $K^0 \to \mu^+\mu^-$, shown in Fig. 11.5a, was absent or greatly suppressed relative to the primary decay mode of the $K^+, K^+ \to \mu^+\nu_\mu$, shown in Fig. 11.5b. Such two-body weak decay modes can be understood more readily in terms of quarks, as illustrated in Fig. 11.5c. The composition of the K^+ is $(u\bar{s})$ and that of the K^0 is $(d\bar{s})$. The analogy to Fig. 11.3 now becomes apparent, even more so if the initial $\bar{s}$ leg is turned into an s in the final state, as shown in Fig. 11.5c. A neutral weak current, mediated by a neutral intermediate vector boson Z^0 would allow processes such as $K^0 \to \mu^+\mu^-$ and the elastic scattering of neutrinos on leptons and protons, $\nu_\mu e \to \nu_\mu e, \nu_\mu p \to \nu_\mu p$, illustrated in Fig. 11.6. Around 1968, Weinberg[6] and Salam[7] independently predicted the existence of weak neutral currents in a theory that unified the weak and electromagnetic interactions. The absence of the decay $K^0 \to \mu^+\mu^-$ was a major hurdle in the acceptance of the Weinberg–Salam theory until 1970 when Glashow, Iliopoulos, and Maiani[8] (GIM) showed that the absence of the missing K^0 decay could be understood by postulating the existence of charmed quarks (Sections 7.6 and 13.2), which permitted a cancelation to occur.

[6] S. Weinberg, *Phys. Rev. Lett.* **19**, 1264 (1967), **27**, 1688 (1977); *Phys. Rev.* **D5**, 1962 (1972).
[7] A. Salam, in *Elementary Particle Theory*, (N. Swartholm ed.) Almqvist and Wiksells, Stockholm, 1969, p. 367.
[8] S. L. Glashow, J. Iliopoulos, and L. Maiani, *Phys. Rev.* **D2**, 1285 (1970).

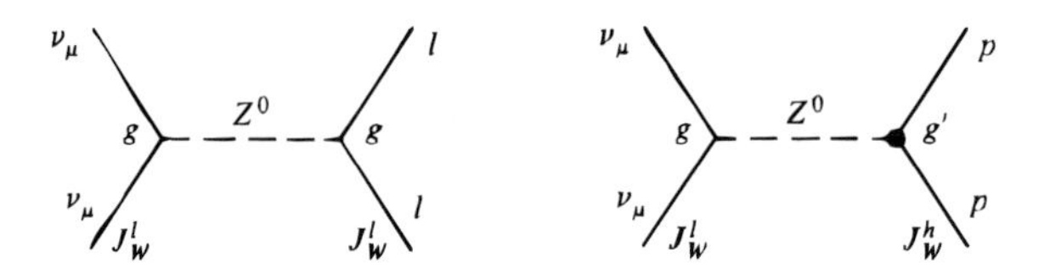

Figure 11.6: Weak neutral currents mediated by Z^0.

Weak neutral currents were discovered at CERN through the observation of the elastic scattering of neutrinos and antineutrinos on electrons, $\nu_\mu e \to \nu_\mu e$ and $\bar{\nu}_\mu e \to \bar{\nu}_\mu e$.[9]

These reactions are forbidden by muon number conservation if only charged weak currents exist. Weak neutral currents now have been verified in many other experiments.[10]

The concepts of a weak current and a weak charge require some reassuring remarks. We are used to electric charges and currents: They can be observed and measured, and they form part of our everyday surroundings. Weak currents and weak charges, on the other hand, have no classical analog. The only way to become familiar with them is to assume their existence and explore the consequences. Since all experiments agree with the predictions of the standard model based on a weak current–current interaction, confidence in the existence of weak charges and currents is justified. In the following sections, we shall inquire into three questions related to H_w: (1) What phenomena are described by H_w? (2) What is the form of the weak current J_w? (3) What is the value of the coupling constant G_F?

11.4 A Variety of Weak Processes

The discussion so far has been restricted to beta decay, the oldest and best known example of a weak interaction. If it were the only manifestation of the weak force, interest would be limited. However, a surprising variety of weak processes is known. Weak reactions have been a rich source of unexpected new phenomena, such as the violation of parity and CP conservation as well as numerous other phenomena associated with the neutral kaons and other systems. Moreover, the unification of the weak and electromagnetic interactions (chapter 13) has had a profound influence on our understanding of fundamental forces. In the present section, we shall categorize the weak processes, list a few examples, and state why they all are called weak.

A *classification* of weak processes can be based on the separation of the weak current into a leptonic and a hadronic part, as in Eq. (11.22). Inserting Eq. (11.22) in the form $J_w = J^l_w + J^h_w$ into the weak Hamiltonian (11.25) produces four scalar products; one involves only leptons and one only hadrons, and two couple lepton and hadron currents. The classification is performed according to these terms:

---

[9]F. J. Hasert et al., *Phys. Lett.* **B46,** 121 (1973); H. Faissner et al., *Phys. Rev. Lett.* **41,** 213 (1978); R.C. Allen, *Phys. Rev. Lett.* **22**,2401 (1985).

[10]V. Nguyen-Khac and A. M. Lutz, eds *Neutral Currents: 20 Years Later*, World Scientific, Singapore, 1993.

$$\begin{aligned} \text{leptonic processes}: &\quad J_w^l \cdot J_w^{l\dagger} \\ \text{semileptonic processes}: &\quad J_w^l \cdot J_w^{h\dagger} + J_w^h \cdot J_w^{l\dagger} \\ \text{hadronic processes}: &\quad J_w^h \cdot J_w^{h\dagger}. \end{aligned} \tag{11.26}$$

Weak processes of each of these three classes are known. In chapter 10, in the treatment of the electromagnetic interaction, we have learned that life is easy as long as only leptons are present. The story repeats itself in the weak interaction: Leptonic processes can be calculated, and theory and experiment agree. Semileptonic processes produce difficulties, and the weak processes involving only hadrons cannot yet be calculated in detail from first principles. We shall now list processes in each of the three classes.

Leptonic Processes The leptonic processes that are easiest to study are the decay of the muon and tau

$$\mu^+ \longrightarrow e^+\bar{\nu}_\mu\nu_e, \quad \tau^+ \longrightarrow l^+\bar{\nu}_\tau\nu_l. \tag{11.27}$$

We will use muons here; muon decay also will be discussed in the following section, where it will be seen that the maximum energy of the emitted electrons is about 53 MeV, the lifetime is 2.2 μsec, and parity is not conserved. Investigations of the decay of the tau are more difficult because the tau is mainly produced through electromagnetic processes such as $e^+e^- \to \tau^+\tau^-$, and not through the decay of a heavier meson as in the case of the muon, where copiously produced pions give rise to the muons.

The scattering of neutrinos with charged leptons also involves only leptons: The processes

$$\nu_e e^- \longrightarrow \nu_e e^-, \quad \nu_\mu e^- \longrightarrow \nu_e \mu^-, \quad \nu_\tau e^- \longrightarrow \nu_e \tau^- \tag{11.28}$$

are without electromagnetic or hadronic complications, and they, and the corresponding ones involving antineutrinos, are ideal for exploring the weak interaction at high energies. Indeed, such reactions have been studied both at accelerators(9,10) and at reactors.(11)

Semileptonic Processes In semileptonic processes, one current is leptonic and the other one hadronic. Three semileptonic decays are listed in Table 11.2. The $\pi^\pm$ decays are similar to that of ^{14}O, Table 11.1, and the $ft_{1/2}$ values are closely related.

Can these decays give sufficient information to study the semi-leptonic weak interaction thoroughly? The maximum energy listed in Table 11.2 is 81 MeV, but the electromagnetic interaction taught us that energies of the order of many GeV

[11] J.M. Conrad, M.H. Shaevitz, and T. Bolton, *Reviews of Modern Physics*, **70**, 1341 (1998).

Table 11.2: DECAY PROPERTIES OF THREE SEMILEPTONIC DECAYS.

Decay	Spin-parity Sequence	$t_{1/2}{}^{\dagger}$ (sec)	$E_{\max}$ (MeV)	$ft_{1/2}$ (sec)
$\pi^{\pm} \to \pi^0 e\nu$	$0^- \to 0^-$	1.76	4.1	3.1×10^3
$n^0 \to pe\bar{\nu}$	$\frac{1}{2}^+ \to \frac{1}{2}^+$	612	0.78	1.1×10^3
$\Sigma^- \to \Lambda^0 e^- \bar{\nu}$	$\frac{1}{2}^+ \to \frac{1}{2}^+$	1.8×10^{-6}	81	6×10^3

†Partial half-life.

are necessary to explore some of the properties. Weak decays with such energies are very difficult to observe because a state with very high excitation generally decays hadronically or electromagnetically, so that the weak interaction cannot compete. An example is the ψ/J and its excited states with energies in excess of 3 GeV. Even though selection rules slow down the decay into hadrons, the contribution from the weak interaction to the decay is so small that it has not yet been observed. At much higher energies, the situation is even more unfavorable.

One of the best ways of studying the high energy behavior of the weak interaction is through semileptonic neutrino-induced reactions such as

$$\begin{aligned} \nu_\mu n &\longrightarrow \mu^- p, \qquad & \nu_\mu p &\longrightarrow \nu_\mu p \\ \bar{\nu}_\mu p &\longrightarrow \mu^+ n, \qquad & \nu_\mu p &\longrightarrow \nu_\mu n \pi^+, \end{aligned} \tag{11.29}$$

and deep inelastic scattering

$$\begin{aligned} \nu_\mu p &\longrightarrow \nu_\mu X, \\ &\longrightarrow \mu^- X, \end{aligned} \tag{11.30}$$

where X is any particle or particles.

The reactions in the first column of Eq. (11.29) involve charged weak currents and the exchange of a $W^{\pm}$, the ones in the right column require neutral weak currents and the exchange of a Z^0. The reactions of the types shown in Eqs. (11.29) and (11.30) have helped to validate the Weinberg–Salam (WS) theory, and have been used to obtain structure functions. They will be discussed in more detail below and in Section 11.14.

In the semileptonic processes listed so far, the weak decays have not involved a change of strangeness. True, the decay $\Sigma^+ \to \Lambda^0 e^+ \nu$ in Table 11.2 involves strange particles, but the hadrons in the initial and final states have the same strangeness. We have, however, mentioned in Section 7.5 that strangeness or hypercharge is not necessarily conserved in the weak interaction. Indeed, strangeness-changing weak decays exist, and three are listed in Table 11.3. They are all mediated by charged currents. No strangeness-changing decays or reactions that occur through neutral weak currents have been observed; for instance, the decay $\Lambda^0 \to ne^+e^-$ is absent.

Table 11.3: STRANGENESS-CHANGING SEMILEPTONIC DECAYS.

Decay	Spin-parity Sequence (of hadron)	$t_{1/2}$† (sec)	$E_{\max}(e)$ (MeV)	$ft_{1/2}$ (sec)
$K^+ \to \pi^0 e^+ \nu_e$	$0^- \to 0^-$	1.8×10^{-7}	358	1×10^6
$\Lambda^0 \to p e^- \bar{\nu}_e$	$\frac{1}{2}^+ \to \frac{1}{2}^+$	2.2×10^{-7}	177	2×10^4
$\Sigma^- \to n e^- \bar{\nu}_e$	$\frac{1}{2}^+ \to \frac{1}{2}^+$	1.0×10^{-7}	257	1×10^5

†Partial half-life.

Hadronic Processes Examples of weak decays in which only hadrons are involved are

$$\begin{aligned} K^+ &\longrightarrow \pi^+\pi^0 \\ &\longrightarrow \pi^+\pi^+\pi^- \\ &\longrightarrow \pi^+\pi^0\pi^0 \end{aligned} \tag{11.31}$$

and

$$\begin{aligned} \Lambda^0 &\longrightarrow p\pi^- \\ &\longrightarrow n\pi^0 \end{aligned} \tag{11.32}$$

Other weak decays involving only hadrons can be found in the tables of PDG. All of these obey the strangeness selection rule

$$|\Delta S| = 1.$$

The absence of observed $\Delta S = 0$ transitions is easily explained: transitions without change of strangeness can proceed by hadronic or electromagnetic decays, and the weak branch is hidden.

Why are all the processes listed in the present section called weak, regardless of whether they involve leptons, hadrons, or both? The justification comes from the fact that the strength of the interaction responsible for the various processes appears to be the same. Additional support comes from considerations of selection rules and from the observation that all processes that are weak according to the strength classification also show violations of parity and charge conjugation invariance.

The *strength* of the interaction responsible for a decay expresses itself in the lifetime, other things being equal. The decays in Table 11.2 are of the type $A \to Be\nu$. While the decay energies vary by about a factor of 100 and the density-of-states factors by a factor of 10^{10}, the ft values are approximately the same. It is therefore likely that the three very different decays in Table 11.2 are caused by the same force. A discrepancy appears when the ft values in Table 11.2 and 11.3 are compared. While the decays appear to be similar, the ft values for hypercharge-changing decays are between one and three orders of magnitude larger than the corresponding ones for hypercharge-conserving decays. We shall return to this discrepancy in Section 11.9 and show that it has an explanation within the framework of the weak current–current interaction.

Parity violation has already been treated in Section 9.3; the electromagnetic and the hadronic force conserve parity, but a violation appears in the weak one. The example discussed in Section 9.3 was a semileptonic decay. The original evidence for parity nonconservation came from the decay of the charged kaons into two and three pions; these weak decays involve hadrons. In the next section we shall show that the purely leptonic decay of the muon also does not conserve parity. These examples indicate that the various processes all violate parity conservation. This fact alone would not justify classing them all into one category. However, it indicates a similarity in the *form* of the interaction that causes these decays, and it supports the conclusion already reached from a consideration of the lifetimes.

Conservation of strangeness or hypercharge in the hadronic and the electromagnetic interaction was postulated in Eq. (7.42). The examples of weak decays discussed in Section 7.5 and in the present section indicate that many cases are known where the strangeness changes by one unit; no case has been found where a change of two units occurs. The *selection rule* for strangeness,

$$\begin{aligned}&\Delta S = 0 \qquad \text{in hadronic and electromagnetic interaction}\\&\Delta S = 0, \pm 1 \quad \text{in the weak interaction,}\end{aligned} \tag{11.33}$$

thus establishes another characteristic feature of the weak interaction.

11.5 The Muon Decay

In the previous section we have surveyed weak processes, and we have partially answered the first question posed at the end of Section 11.3, namely what phenomena are described by H_w. The form of the weak current and the value of the weak coupling constant remain to be studied. We can expect that the fundamental features of the weak interaction will be easiest to explore in purely leptonic processes because no serious interference from the hadronic force is present there. In this section, the salient features of the much studied muon decay will be described. The decay of the tau is very similar, but it can decay into either muons or electrons.

Muons and taus do not interact strongly, and it is consequently not possible to produce them directly and copiously through a reaction. However, the decay of charged pions is a convenient source of muons. Assume, for instance, that positive pions are produced at an accelerator. The pions are selected in a pion channel and slowed down in an absorber (Fig. 11.7). If their energy is not too high they usually come to rest before decaying through the mode

$$\pi^+ \longrightarrow \mu^+ \nu_\mu. \tag{11.34}$$

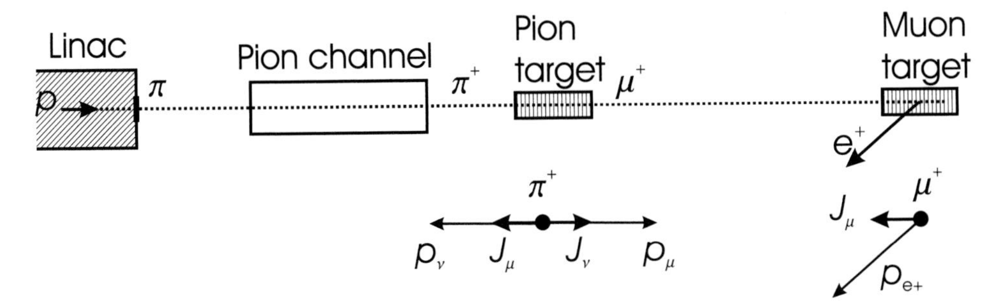

Figure 11.7: A positive pion is selected in the pion channel and comes to rest in the pion target. The pion decay results in a fully polarized muon. The muon escapes from the pion target and comes to rest in the muon target. Its spin points in the direction from which it came. The decay electron is then observed.

Conservation laws determine much of what happens: Conservation of the lepton and muon numbers requires the neutral particle to be a muon neutrino. Momentum conservation demands that the muon and the muon neutrino have equal and opposite momenta in the c.m. of the decaying pion. The muon neutrino has its spin opposite to its momentum, as shown in Fig. 7.2. Since the pion has spin 0, angular momentum conservation insists that the positive muon must be fully polarized, with its spin pointing opposite to its momentum. The muons escape from the pion target; some are stopped in the muon target, and their decay positron can be detected. With proper choice of the muon target, the decaying muon is still polarized, and its spin $\boldsymbol{J}$ points into the direction from which it came.

The processes just described and shown in Fig. 11.7 permit a number of measurements that all give information concerning the weak interaction. We shall discuss three aspects here, parity nonconservation, the lifetime of the muon, and the spectrum of the decay electrons.

Parity Nonconservation As Fig. 11.7 is drawn, it shows the breakdown of parity in two different places. The muon is expected to be polarized because the neutrino emitted together with it is polarized. A longitudinally polarized muon violates parity conservation, as was explained in Section 9.3. A measurement of the polarization of the muon thus demonstrates that parity is not conserved in the weak decay of the pion. Such a polarization has been detected.(12) The second place where parity nonconservation shows up is in the decay of the muon. As sketched in Fig. 11.7, the muon spin points into a well-defined direction, and the probability of positron emission can now be determined with respect to this direction. This experiment is analogous to the one discussed in Section 9.3 and shown in Fig. 9.6. Indeed, as in the Wu–Ambler experiment, it was found that the positron is prefer-

[12] G. Backenstoss, B. D. Hyams, G. Knop, P. C. Marin, and U. Stierlin, *Phys. Rev. Lett.* **6**, 415 (1961); M. Bardon, P. Franzini, and J. Lee, *Phys. Rev. Lett.* **7**, 23 (1961); TWIST collaboration, *Phys. Rev. D* **71**, 071101 (2005).

entially emitted parallel to the spin of the incoming muon, indicating that parity is also violated in the muon decay.[13]

Muon Lifetime The experimental arrangement for determining the muon lifetime has already been described in chapter 4. In Fig. 4.22, the logic elements are shown, and it is easy to see how they fit into the setup of Fig. 11.7. Observation of the number of electrons detected in counter D as a function of the delay time between counters B and D gives a curve of the form shown in Fig. 5.15, and the slope of the curve determines the muon lifetime. For most estimates it is sufficient to remember that the muon mean life is 2.2 μsec.

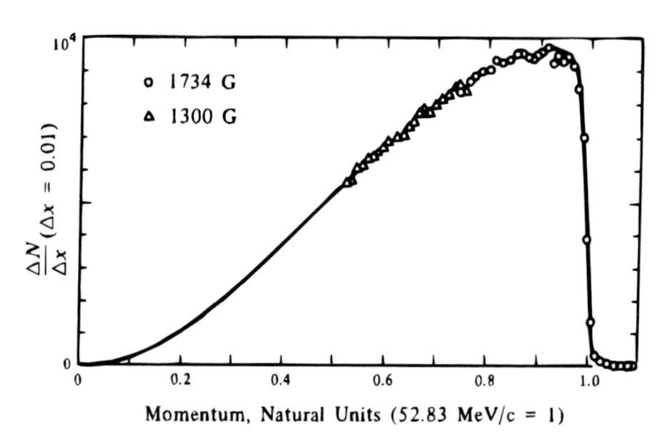

Figure 11.8: Electron spectrum from unpolarized muons. [B. A. Sherwood, *Phys. Rev.* **156**, 1475 (1967).] The momentum is measured in units of the maximum electron momentum.

Electron Spectrum To investigate the electron spectrum, the number of electrons is measured as a function of momentum. To determine the momentum, the electron path in a magnetic field is observed. One possibility to detect the electrons is to use wire spark chambers the result of which is shown in Fig. 11.8. Another detection scheme using drift chambers has recently led to spectacular precision and was shown in chapter 4 (see Fig. 4.16).[14] Some similarity to the electron spectrum in nuclear beta decay, Fig. 11.1, exists but the drop-off at high electron momenta is much steeper. The electron spectrum is no longer determined by the phase-space factor alone and comparison with theory provides information on the form of the weak Hamiltonian.

11.6 The Weak Current of Leptons

In the previous section, some of the salient features of the muon decay have been discussed. The τ decay is similar, but there are many more open channels. These data and some additional information will now be used to construct the weak Hamiltonian, Eq. (11.25), in more detail. In particular, we shall have to find the form of the weak current, J^l_w, as far as we can with our limited tools. The first fact to be used is the uncanny similarity between electron and muon, a fact often stated by

[13] R. L. Garwin, L. M. Lederman, and M. Weinrich, *Phys. Rev.* **105**, 1415 (1957); J. L. Friedman and V. L. Telegdi, *Phys. Rev.* **105**, 1681 (1957).

[14] TWIST Collaboration, *Phys. Rev. Lett.* **94**, 101805 (2005).

the words *muon-electron universality.*[15] This universality is expressed by writing the total weak current of leptons as the sum of an electron and a muon current,

$$J_w^l = J_w^e + J_w^\mu \tag{11.35}$$

and assuming that both behave alike. The leptonic part of the weak Hamiltonian H_w is found by inserting Eq. (11.35) into Eq. (11.25):

$$H_w = \frac{G_F}{\sqrt{2}\,c^2} \int d^3x (J_w^e \cdot J_w^{e\dagger} + J_w^e \cdot J_w^{\mu\dagger} + J_w^\mu \cdot J_w^{e\dagger} + J_w^\mu \cdot J_w^{\mu\dagger}). \tag{11.36}$$

For the explicit construction of the weak current J_w^e, we use the analogy to electromagnetism. In chapter 10, we systematically went from the classical Hamiltonian, Eq. (10.48),

$$H_{em} = \frac{e}{c} \int d^3x \boldsymbol{j}_{em} \cdot \boldsymbol{A}$$

to the matrix element, Eq. (10.60),

$$\langle\beta|H_{em}|\alpha\rangle = -i\frac{e\hbar}{mc} \int d^3x\, \psi_\beta^* \boldsymbol{\nabla} \psi_\alpha \cdot \boldsymbol{A}.$$

Comparison of these two expressions shows that the substitution

$$\boldsymbol{j}_{em} = -i\frac{\hbar}{m}\psi_\beta \boldsymbol{\nabla}\psi_\alpha = \psi_\beta^* \left(\frac{\boldsymbol{p}_{op}}{m}\right)\psi_\alpha = \psi_\beta^* \boldsymbol{v}_{op} \psi_\alpha \tag{11.37}$$

provides the transition from the classical Hamiltonian to the quantum mechanical matrix element. The analogous substitution for the probability density is

$$\rho_{em} = \psi_\beta^* \psi_\alpha. \tag{11.38}$$

Equations (11.37) and (11.38) are valid for nonrelativistic electrons. To allow for generalizations, we introduce two operators, V_0 and $\boldsymbol{V}$, and write

$$\rho_{em} = \psi_\beta^* V_0 \psi_\alpha, \quad \boldsymbol{j}_{em} = c\psi_\beta^* \boldsymbol{V} \psi_\alpha.$$

The velocity of light, c, has been inserted in order to make $\boldsymbol{V}$ dimensionless. Charge density and current density combine to form a four-vector,

$$j_{em} = (c\rho, \boldsymbol{j}),$$

or, with the operators V_0 and $\boldsymbol{V}$,

$$j_{em} = c\psi_\beta^* V \psi_\alpha. \tag{11.39}$$

[15] Since three charged leptons are known, electron–muon universality should be replaced by electron–muon–tau universality and Eq. (11.35) should be generalized to read $J_w^l = J_w^e + J_w^\mu + J_w^\tau$. In order to keep the equations manageable, we retain the form Eq. (11.35); the generalization to take the τ into account is straightforward.

The notation $\mathcal{V} \equiv (V_0, \boldsymbol{V})$ is a reminder that the "sandwich" $\psi^* \mathcal{V} \psi$ transforms like a four-vector. With Eqs. (11.37) and (11.38), the explicit form of $\mathcal{V}$ for a nonrelativistic electron is

$$\mathcal{V} \equiv (V_0, \boldsymbol{V}), \quad V_0 = 1, \quad \boldsymbol{V} = \frac{\boldsymbol{p}}{mc}. \tag{11.40}$$

There are a number of differences between the electromagnetic and weak currents. Whereas the electromagnetic current is always a neutral one that conserves charge, the weak current has a charge-changing part, $J_w^{(-)}$, in addition to the neutral one, $J_w^{(0)}$. For electrons, the corresponding weak current densities are written in analogy to the electromagnetic ones as

$$\begin{aligned} J_w^{e(-)} &= c\psi_e^* \mathcal{V} \psi_{\nu_e}, \\ J_w^{e(0)} &= c\psi_e^* \mathcal{V} \psi_e, \quad J_w^{\nu(0)} = c\psi_{\nu_e}^* \mathcal{V} \psi_{\nu_e}. \end{aligned} \tag{11.41}$$

The weak current is more complicated than the electromagnetic one in other ways. We have seen in chapter 9 and earlier in this chapter that the weak interaction does not respect parity. The operator $\mathcal{V} = (V_0, \boldsymbol{V})$ behaves under the parity operation as

$$V_0 \xrightarrow{P} V_0 \quad \boldsymbol{V} \xrightarrow{P} -\boldsymbol{V}. \tag{11.42}$$

The fact that the vector part changes sign follows from Eq. (9.1). V_0, on the other hand, is a probability density, and it remains unchanged under the parity operation. According to the golden rule, the transition rate for a reaction from a polarized or unpolarized source is proportional to the square of a matrix element, or

$$w_\mu \propto \left| \int d^3x \psi_e^* \mathcal{V} \psi_{\nu_e} \cdot \psi_{\nu_\mu}^* \mathcal{V} \psi_\mu \right|^2 .$$

The vector product $\mathcal{V} \cdot \mathcal{V} = V_0 V_0 - \boldsymbol{V} \cdot \boldsymbol{V}$ remains unchanged under P; if w_μ^P denotes the transition rate after the parity operation, it is equal to w_μ:

$$w_\mu^P = w_\mu .$$

This result disagrees with the electron asymmetry observed in beta and muon decays. How can the expression for the weak current be generalized in such a way that the analogy to the electromagnetic current is not completely destroyed but that parity nonconservation is included? A hint to the answer comes from comparing linear and angular momentum. Under ordinary rotations, both behave in the same way. We have not demonstrated this fact explicitly, but the proof is straightforward if the arguments given in Section 8.2 are used. Under the parity operation, the polar vector $\boldsymbol{p}$ and the axial vector $\boldsymbol{J}$ reveal their difference: $\boldsymbol{p}$ changes sign, whereas $\boldsymbol{J}$ does not. These properties remain true for general operators $\mathcal{V}$ and $\mathcal{A}$: $\mathcal{V}$ and $\mathcal{A}$

behave identically under ordinary rotations but differently under space inversion. The properties of a general axial four-vector $\mathcal{A}$ under P are given by

$$\mathcal{A}_0 \xrightarrow{P} -\mathcal{A}_0 \quad \boldsymbol{\mathcal{A}} \xrightarrow{P} \boldsymbol{\mathcal{A}}. \tag{11.43}$$

The behavior of the *axial probability density* cannot be visualized as easily as the one for the ordinary probability density: The electric charge provides an example for the properties of V_0, but no classical example for an axial charge exists.[(16)] The suggested generalization of the weak current, Eq. (11.41), is for instance

$$J_w^{e(-)} = c\psi_e^*(\mathcal{V} - \mathcal{A})\psi_{\nu_e}. \tag{11.44}$$

Let us next use the simplicity of Eq. (11.41) to learn more about the physics that is hidden in it. To do so, we consider the Hermitian conjugate of the current J_w^e. The operators $\mathcal{V}$ and $\mathcal{A}$ are Hermitian; noting that for a one-component wave function $\psi^\dagger = \psi^*$ then gives

$$\begin{aligned} J_w^{e(-)\dagger} &= c[\psi_e^*(\mathcal{V} - \mathcal{A})\psi_{\nu_e}]^\dagger \\ &= c\psi_{\nu_e}^*(\mathcal{V} - \mathcal{A})\psi_e = J_w^{e(+)} \\ J_w^{e(0)\dagger} &= J_w^{e(0)}, \quad J_w^{\nu(0)\dagger} = J_w^{\nu(0)}. \end{aligned} \tag{11.45}$$

Comparison with J_w^e and with Fig. 11.4 shows that $J_w^{e(-)\dagger} = J_w^{e(+)}$ describes the destruction of an electron and the creation of an electron neutrino. The four vector product $J_w^e \cdot J_w^{e\dagger}$ in H_w^e is thus responsible in part for the scattering of electron neutrinos by electrons, $\nu_e e^- \rightarrow \nu_e e^-$, a process that has already been listed in Eq. (11.28). Weak neutral currents, however, also contribute to this scattering through the products $J_w^{e(0)} \cdot J_w^{\nu(0)\dagger}$ and $J_w^{\nu(0)\dagger} \cdot J_w^{e(0)}$. The various currents and scattering processes are displayed in Fig. 11.9. The operator $J_w^{e(-)} \cdot J_w^{e(-)\dagger}$ can also induce antineutrino scattering on electrons or positrons, e.g.,

$$e^+\bar{\nu}_e \longrightarrow e^+\bar{\nu}_e. \tag{11.46}$$

The other terms in the Hamiltonian (11.36) similarly give rise to weak processes involving only leptons. One term that is responsible for muon decay is easily seen to be

$$J_w^e \cdot J_w^{\mu\dagger} = c^2\psi_e^*(\mathcal{V} - \mathcal{A})\psi_{\nu_e} \cdot \psi_{\nu_\mu}^*(\mathcal{V} - \mathcal{A})\psi_\mu. \tag{11.47}$$

In the previous section, the muon decay has been discussed and predictions based on the scalar product, Eq. (11.47), must now be compared to the experimental facts. With Eqs. (10.1) and (11.25) the transition rate for the muon decay then becomes

[16] If magnetic monopoles exist, they provide an example for an axial charge. The magnetic charge density ρ_m, introduced in Eq. (10.106), changes sign under the parity operation. This fact can be proved by considering the energy of a magnetic monopole in a magnetic field and assuming invariance of the corresponding Hamiltonian under P.

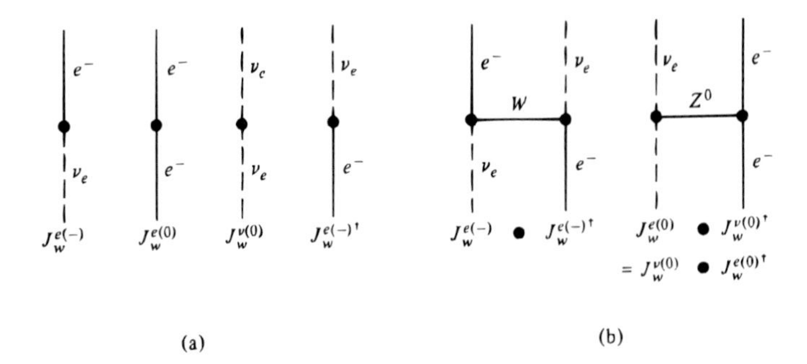

Figure 11.9: (a) Interpretation of some leptonic currents, and (b) their products.

$$w_\mu = \frac{\pi G_F^2}{\hbar} \left| \int d^3x \psi_e^* (\mathcal{V} - \mathcal{A}) \psi_{\nu_e} \cdot \psi_{\nu_\mu}^* (\mathcal{V} - \mathcal{A}) \psi_\mu \right|^2 \rho(E),$$

or

$$w_\mu = \frac{\pi G_F^2}{\hbar} |M_{\text{even}} - M_{\text{odd}}|^2 \rho(E), \tag{11.48}$$

with

$$M_{\text{even}} = \int d^3x (\psi_e^* \mathcal{V} \psi_{\nu_e} \cdot \psi_{\nu_\mu}^* \mathcal{V} \psi_\mu + \psi_e^* \mathcal{A} \psi_{\nu_e} \cdot \psi_{\nu_\mu}^* \mathcal{A} \psi_\mu)$$

$$M_{\text{odd}} = - \int d^3x (\psi_e^* \mathcal{V} \psi_{\nu_e} \cdot \psi_{\nu_\mu}^* \mathcal{A} \psi_\mu + \psi_e^* \mathcal{A} \psi_{\nu_e} \cdot \psi_{\nu_\mu}^* \mathcal{V} \psi_\mu).$$

Under the parity operation, M_{even} remains unchanged, M_{odd} changes sign, and the transition rate becomes

$$w_\mu^P = \frac{\pi G_F^2}{\hbar} |M_{\text{even}} + M_{\text{odd}}|^2 \rho(E). \tag{11.49}$$

Comparison of Eqs. (11.48) and (11.49) shows that

$$w_\mu^P \neq w_\mu.$$

The presence of both a vector *and* an axial vector operator in the weak current permits the description of the observed violation of parity invariance. The violation becomes maximal if $\mathcal{V}$ and $\mathcal{A}$ have equal magnitudes.

The detailed computation of a transition rate or cross section can be performed only if the explicit form of the operators $\mathcal{V}$ and $\mathcal{A}$ is known. This form depends on the type of particles that carry the weak current. For *nonrelativistic electrons*, the operators V_0 and $\boldsymbol{V}$ are given in Eq. (11.40). The axial vector current is usually not treated in introductory quantum mechanics. We establish its form by using invariance arguments. An electron is described by its energy, its momentum $\boldsymbol{p}$, and its spin $\boldsymbol{J}$. For spin 1/2, it is customary to use instead of the spin $\boldsymbol{J}$ the dimensionless Pauli spin operator $\boldsymbol{\sigma}$; it is connected to $\boldsymbol{J}$ by

$$\boldsymbol{\sigma} = \frac{2\boldsymbol{J}}{\hbar}. \tag{11.50}$$

The only axial vector available is $\boldsymbol{J}$, or $\boldsymbol{\sigma}$. The operator $\mathcal{A}$ must therefore be proportional to $\boldsymbol{\sigma}$. The axial charge operator, $\mathcal{A}_0$, changes sign under the parity operation as indicated by Eq. (11.46); since $\boldsymbol{\sigma}\cdot\boldsymbol{p}$ has this property, we set

$$\mathcal{A} = (\mathcal{A}_0, \boldsymbol{\mathcal{A}}), \quad \mathcal{A}_0 = \frac{\boldsymbol{\sigma}\cdot\boldsymbol{p}}{mc}, \quad \boldsymbol{\mathcal{A}} = \boldsymbol{\sigma}. \tag{11.51}$$

The factor $1/mc$ in $\mathcal{A}_0$ is chosen to make the operator dimensionless.

The nonrelativistic operators, as given in Eqs. (11.40) and (11.51), cannot be used for the evaluation of the muon and tau[on] decays because there all particles in the final state must be treated relativistically. The generalization of the operators $\mathcal{V}$ and $\mathcal{A}$ to relativistic leptons is well known.[17] Calculations with the relativistic operators are, however, beyond our means here, and we therefore give the transition rate for the muon decay without proof. The rate $dw_\mu(E_e)$ for the emission of an electron with energy between E_e and $E_e + dE_e$ becomes, for $E_e \gg m_e c^2$,

$$dw_\mu(E_e) = G_F^2 \frac{m_\mu^2}{4\pi^3\hbar^7 c^2} E_e^2 \left(1 - \frac{4}{3}\frac{E_e}{m_\mu c^2}\right) dE_e. \tag{11.52}$$

This expression, after replacing the electron energy by the electron momentum, agrees very well with the spectrum shown in Fig. 11.7.

11.7 Chirality versus Helicity

• In Eq. (9.33) we gave a definition for the *helicity* of particles. For massive particles this quantity is not frame independent as can be seen from the fact that the dot product involves only the space-like components of the momentum so a Lorentz-transformation can clearly change it. In other words, an observer moving faster than the particle would see the opposite helicity as one moving slower than the particle.

In the relativistic treatment of quantum mechanics another observable emerges which is called *chirality*. It plays a central role in the proper definition of the currents. We don't have the room to properly define it here, but we can make a connection to limiting cases. In particular, for highly relativistic particles ($p \gg mc$) it can be shown that:[17]

$$\text{chirality} \rightarrow \text{helicity}. \tag{11.53}$$

[17] See Halzen and Martin, *Quarks and Leptons*, John Wiley & Sons (1984); Chapter 5.

So when the term *chirality* arises, one can try to visualize its implications by assuming particles are massless and use *helicity* as a synonym of *chirality*.

For leptons, the fact that the charged current of the weak interaction is purely $\mathcal{V}-\mathcal{A}$ is usually expressed by stating that only the left-handed chirality states participate in the interaction. As examples, neglecting the mass of neutrinos and taking into account Eq. (11.53) implies that neutrinos should all have negative helicity. Massless right-handed neutrinos, if they existed, would not interact via the weak interaction. On the other hand, for low-energy electrons from beta decay, where the mass cannot be neglected, the wave function with well-defined chirality will be a combination of eigen-states of helicity.

Another important result that we can't prove here is that for anti-particles the chirality is expected to be opposite to the one for particles and thus anti-neutrinos, for example, are expected to be right handed. •

11.8 The Weak Coupling Constant G_F

The electromagnetic coupling constant e can be determined by observing the force on a charged particle in a known field, by measuring the Rutherford or Mott cross section (Eqs. (6.9) or (6.11)) from a point scatterer, or by determining the lifetime of a decay with well-known matrix element $\langle f|\boldsymbol{x}|i\rangle$ [Eq. (10.77)]. What is the best way of determining the weak coupling constant G_F? Again there are a number of possibilities, but the total lifetime of the muon is a good choice. The reason is twofold: the muon decay involves no hadrons so that complications due to the hadronic interaction do not have to be considered, and the muon lifetime has been measured very accurately.

The total transition rate for the muon decay is obtained by integrating Eq. (11.52), with $E_{\text{max}} \approx m_\mu c^2/2$,

$$\begin{aligned} w_\mu &= \int_0^{E_{\text{max}}} dw_\mu(E_e) \\ &= G_F^2 \frac{m_\mu^2}{4\pi^3\hbar^7 c^2} \int_0^{E_{\text{max}}} dE_e E_e^2 \left(1 - \frac{4}{3}\frac{E_e}{m_\mu c^2}\right) = \frac{G_F^2 m_\mu^5 c^4}{192\pi^3\hbar^7}. \end{aligned} \tag{11.54}$$

With the muon lifetime, $\tau = 1/w_\mu$, the (Fermi) coupling constant becomes[18]

$$\begin{aligned} G_F &= (1.16637 \pm 0.00001) \times 10^{-5}\ \text{GeV}^{-2}(\hbar c)^3 \\ &= 0.896 \times 10^{-4}\ \text{MeV-fm}^3 \\ &= 1.435 \times 10^{-49}\ \text{erg-cm}^3. \end{aligned} \tag{11.55}$$

[18]PDG.

In the electromagnetic case we have expressed the strength of the interaction by making e^2 dimensionless as in Eq. (10.79):

$$\alpha = \frac{e^2}{\hbar c} \approx \frac{1}{137}.$$

Comparison of Eqs. (11.15) and (11.16) makes it clear that the weak analog to the electric charge is g_w, not G_F. Like e^2, g_w^2 is made dimensionless by division by $\hbar c$. The connection with G_F, as given in Eq. (11.21), then permits us to write $g_w^2/\hbar c$ in terms of G_F and the mass m,

$$\frac{g_w^2}{\hbar c} = \frac{1}{\sqrt{2}\, 4\pi} \frac{1}{\hbar c} \left(\frac{m_W c}{\hbar}\right)^2 G_F.$$

With $m_W \approx 80$ GeV$/c^2$, we find

$$\frac{g_w^2}{\hbar c} \approx \frac{1}{240}. \tag{11.56}$$

The coupling constants g_w and e are of the same order of magnitude, suggesting that the weak and electromagnetic interactions are related. The observed weakness of the weak interactions at low energies is not caused by a small coupling constant, but rather by a short range [Eq. (11.17)]. Actually, when these arguments were first made, the mass of the W was not known and the formulation of a unified electroweak theory led to the prediction of the correct mass of the W.

11.9 Weak Decays of Quarks and the CKM Matrix

In chapter 5 we introduced quarks which can be arranged in pairs:

$$\begin{pmatrix} u \\ d \end{pmatrix} \begin{pmatrix} c \\ s \end{pmatrix} \begin{pmatrix} t \\ b \end{pmatrix}. \tag{11.57}$$

Here the top row contains the $q = 2e/3$ members and the bottom row contains the $q = -e/3$ ones, grouped by family in order of increasing mass from left to right. It turns out that all charged weak decays of quarks can be explained by assuming that all transitions that change row are allowed. Thus, a down quark can change into an up quark and emit a W^- which may then decay into an electron and an anti-neutrino. This is what happens in neutron beta decay. A different example is a decay like: $K^+ \to \pi^+ e \nu_e$, where a strange quark decays to an up quark. The situation can be better summarized by listing allowed parents and daughters together:

$$\begin{pmatrix} u \\ d' \end{pmatrix} \begin{pmatrix} c \\ s' \end{pmatrix} \begin{pmatrix} t \\ b' \end{pmatrix}, \tag{11.58}$$

where now decays can take place only within a column and the primed states are linear combinations of the mass eigenstates:

$$\begin{pmatrix} d' \\ s' \\ b' \end{pmatrix} = \begin{pmatrix} V_{ud} & V_{us} & V_{ub} \\ V_{cd} & V_{cs} & V_{cb} \\ V_{td} & V_{ts} & V_{tb} \end{pmatrix} \begin{pmatrix} d \\ s \\ b \end{pmatrix}. \tag{11.59}$$

Here we have used matrix multiplication to simplify our notation. The matrix is called the Cabibbo-Kobayashi-Maskawa (CKM) mixing matrix.[19] A consequence of this scheme is that nuclear weak decays have an *effective coupling constant* $V_{ud}G_F$ and decays involving a $u \leftrightarrow s$ quark transitions have a different *effective coupling constant*, $V_{us}G_F$. Experimental determinations yield[18]:

$$V_{ud} \approx 0.97, \qquad V_{us} \approx 0.22 \tag{11.60}$$

so the effective coupling constant for nuclear beta decays is much larger than that for decays involving $u \leftrightarrow s$ transitions. The CKM matrix as introduced above should only produce a rotation from the mass eigenstates to the weak eigenstates and consequently should be unitary. This implies, for example, $|V_{ud}|^2 + |V_{us}|^2 + |V_{ub}|^2 = 1$. Finding that the sum of the squares of a row or a column don't add up to unity could thus be an indication of new physics. For this reason much effort has been dedicated to checking the unitarity of the matrix.[20]

11.10 Weak Currents in Nuclear Physics

In this section we will discuss a particular example, the decay of ^{14}O,

$$^{14}\text{O} \xrightarrow{\beta^+} {}^{14}\text{N}$$

as a means of getting a better understanding on how the ideas we have discussed so far work when applied to real cases.

Figure 8.5 displays the $A = 14$ isobars ^{14}C,^{14}N, and ^{14}O. The ground states of ^{14}C and ^{14}O and the first excited state of ^{14}N form an isospin triplet. The positron decay of interest leads from the ground state of ^{14}O to the first excited state of ^{14}N. The maximum positron energy is 1.81 MeV, the half-life of ^{14}O is 71 sec, and the ft value is 3072 sec (Table 11.1). There are two reasons why this decay is useful: (1) The transition occurs between members of an isospin multiplet. Apart from electromagnetic corrections, the wave functions of the initial and final states of the decay consequently describe the same hadronic state and thus are identical in their spin and space properties. Matrix elements involving them can be computed accurately. Such transitions are called *superallowed*. (2) Initial and final states have

[19] N. Cabibbo, *Phys. Rev. Lett.* **10**, 531 (1963); M. Kobayashi and T. Maskawa, *Prog. Theor. Phys.* **49**, 652 (1973). We will give more details on Eqs. (11.58) and (11.59) and explain the historical facts that established this logic in Chapter 13.

[20] J.C. Hardy and I.S. Towner, Phys. Rev. Lett. **94**, 092502 (2005).

spin-parity $J^\pi = 0^+$. Parity and angular momentum selection rules then severely restrict the matrix elements.

Using Eq. (11.59) for writing down the weak current of hadrons, and taking into consideration that in nuclear beta decay there is is not enough energy available for transformations involving quarks other than u and d:

$$J_w^h(\text{nuclear physics}) = V_{ud} J_w. \tag{11.61}$$

Denoting the wave functions of the initial and final nuclear states by $\psi_{0+\alpha}$ and $\psi_{0+\beta}$ and writing the weak current J_w in the same form as J_w^e, Eq. (11.44), $\boldsymbol{J}_w^h$ becomes

$$J_w^h(0^+ \longrightarrow 0^+) = cV_{ud}\psi_{0+\beta}^*(\mathcal{V} - \mathcal{A})\psi_{0+\alpha}.$$

With Eqs. (11.25) and (11.44), the matrix element of H_w then becomes

$$\langle\beta|H_w|\alpha\rangle = \frac{1}{\sqrt{2}} G_F V_{ud} \int d^3x \psi_{e+}^*(\mathcal{V} - \mathcal{A})\psi_{\bar{\nu}_e} \cdot \psi_{0+\beta}^*(\mathcal{V} - \mathcal{A})\psi_{0+\alpha}.$$

The positron and the neutrino are leptons, and they do not interact hadronically with the nucleus. After emission, they can therefore be described by plane waves, like free particles:

$$\psi_{e+} = u_e \exp\left(\frac{i\boldsymbol{p}_e \cdot \boldsymbol{x}}{\hbar}\right), \qquad \psi_{\bar{\nu}} = u_{\bar{\nu}} \exp\left(\frac{i\boldsymbol{p}_\nu \cdot \boldsymbol{x}}{\hbar}\right). \tag{11.62}$$

Here the spin wave functions u_e and $u_{\bar{\nu}}$ are no longer functions of $\boldsymbol{x}$. (The plane wave for the electron is slightly distorted by the Coulomb field of the nucleus. This distortion results in a small correction that has been discussed in Section 11.2 and is given by the function F introduced there.) The energies of the leptons are less than a few MeV, the reduced wavelengths $\lambda\!\!\!^- = \hbar/p$ are long compared to the nuclear radius, and the lepton wave functions can be replaced by their values at the origin, u_e and $u_{\bar{\nu}}$. The matrix element then becomes

$$\langle\beta|H_w|\alpha\rangle = \frac{1}{\sqrt{2}} G_F V_{ud} u_e^*(\mathcal{V} - \mathcal{A})u_{\bar{\nu}} \cdot \int d^3x \psi_{0+\beta}^*(\mathcal{V} - \mathcal{A})\psi_{0+\alpha}. \tag{11.63}$$

Parity and angular momentum conservation simplify this expression. Consider parity first.[21] Under P, the nuclear wave functions $\psi_{0+\alpha}$ and $\psi_{0+\beta}$ remain unchanged. According to Eqs. (11.42) and (11.43), $\boldsymbol{V}$ and $\mathcal{A}_0$ change sign. Consequently, the corresponding integrands are odd under P and the integrals vanish. The term involving $\boldsymbol{A}$ also vanishes because the wave functions are scalars under rotation, whereas $\boldsymbol{A}$ behaves like a vector. The average of a vector over a spherical surface

[21] At first sight, the parity argument seems inappropriate, because the weak interaction does not conserve parity. However, the parity of the initial and the final nuclear states is given by the hadronic interaction, which, due to the non-relativistic nature of the motion of the hadrons, does conserve parity. The argument is therefore correct.

vanishes: Scalars transform like Y_0, vectors like Y_1, and the integral $\int d^3x Y_0^* Y_1 Y_0$ vanishes. The only term left under the integral is V_0, and the matrix element takes on the form

$$\langle\beta|H_w|\alpha\rangle = \frac{1}{\sqrt{2}} G_F V_{ud} u_e^* (V_0 - A_0) u_{\bar{\nu}} \langle 1\rangle, \tag{11.64}$$

where $\langle 1\rangle$ is the symbol used in nuclear physics for the integral

$$\langle 1\rangle = \int d^3x \psi_{0^+\beta}^* V_0 \psi_{0^+\alpha}. \tag{11.65}$$

The recoil energy imparted to the decaying nucleus is very small so that the nuclear matrix element $\langle 1\rangle$ can be computed nonrelativistically; the result is

$$\langle 1\rangle = \sqrt{2}, \tag{11.66}$$

if the states β and α have the same isospin and are part of the same multiplet.

• To verify Eq. (11.66), we use the nonrelativistic operator $V_0 = 1$ from Eq. (11.40) so that

$$\langle 1\rangle = \int d^3x \psi_{0^+\beta}^* \psi_{0^+\alpha}.$$

A new problem arises here: the wave functions ψ_β and ψ_α belong to different isobars and hence are orthogonal. As written, the integral vanishes. The solution to the problem is simple if the isospin formalism is introduced. The states in ^{14}O and ^{14}N belong to the same $I = 1$ isospin multiplet, with I_3 values of 1 and 0, respectively. They have the same spatial wave function so that the total wave functions can be written

$$^{14}\text{O}: \qquad \psi_\alpha = \psi_0(\boldsymbol{x})\boldsymbol{\Phi}_{1,1}$$

$$^{14}\text{N}: \qquad \psi_\beta = \psi_0(\boldsymbol{x})\boldsymbol{\Phi}_{1,0}.$$

where, $\boldsymbol{\Phi}_{1,1}$ and $\boldsymbol{\Phi}_{1,0}$ denote the normalized isospin functions. The weak current changes ^{14}O into ^{14}N; it lowers the I_3 value by one unit. This lowering is expressed by the operator I_-, given in Eq. (8.26). In the isospin formalism the complete matrix element $\langle 1\rangle$ thus becomes

$$\langle 1\rangle = \int d^3x \psi_0^*(x)\psi_0(x)\boldsymbol{\Phi}_{1,0}^* I_- \boldsymbol{\Phi}_{1,1}.$$

The isospin part is evaluated with Eq. (8.27):

$$\boldsymbol{\Phi}_{1,0}^* I_- \boldsymbol{\Phi}_{1,1} = \sqrt{2}\boldsymbol{\Phi}_{1,0}^* \boldsymbol{\Phi}_{1,0} = \sqrt{2}.$$

The spatial wave function is normalized to 1 so that the final result, $\langle 1\rangle = \sqrt{2}$, verifies Eq. (11.66). •

With Eq. (11.66), the square of the matrix element of H_w becomes

$$|\langle\beta|H_w|\alpha\rangle|^2 = G_F^2 V_{ud}^2 |u_e^*(V_0 - A_0)u_{\bar{\nu}}|^2.$$

The magnitude of the lepton matrix element can be obtained by assuming spinless nonrelativistic electrons and by first considering only the vector term, proportional to V_0. Equation (11.40) then gives

$$u_e^* V_0 u_{\bar{\nu}} = u_e^* u_{\bar{\nu}} \qquad \text{and} \qquad |u_e^* V_0 u_{\bar{\nu}}|^2 = u_e^* u_e u_{\bar{\nu}}^* u_{\bar{\nu}}.$$

If the leptons are normalized to one particle per unit volume, Eq. (11.62) gives $u_e^* u_e = u_{\bar{\nu}}^* u_{\bar{\nu}} = 1$. The matrix element of A_0 vanishes nonrelativistically, as is evident from Eq. (11.51) with $p/m \to 0$. For highly relativistic electrons, $p/mc \to pc/E \to 1$, and the matrix element of A_0 approaches that of V_0. There is no interference between A_0 and V_0 in this case so that the square of the lepton matrix element becomes

$$|u_e^*(V_0 - A_0)u_{\bar{\nu}}|^2 = |u_e^* V_0 u_{\bar{\nu}}|^2 = |u_e^* A_0 u_{\bar{\nu}}|^2 = 2. \tag{11.67}$$

The square of the matrix element for a weak $0^+ \to 0^+$ transition thus is

$$|\langle\beta|H_w|\alpha\rangle|^2 = 2G_F^2 V_{ud}^2. \tag{11.68}$$

With Eq. (11.11) and $ft_{1/2} = f\tau \ln 2$, the final result becomes

$$G_F^2 V_{ud}^2 = \pi^3 \ln 2 \frac{\hbar^7}{m_e^5 c^4} \frac{1}{ft_{1/2}}. \tag{11.69}$$

The ft value of ^{14}O is given in Table 11.1. A number of other $0^+ \to 0^+$ superallowed transitions have been investigated carefully. Taking into account some small corrections, the value of $G_F V_{ud}$ becomes[22]

$$G_F^V V_{ud} = (1.400 \pm 0.002) \times 10^{-49} \text{ erg cm}^3. \tag{11.70}$$

The superscript V on G_F indicates that the constant has been determined from decays involving only the vector interaction in the hadronic matrix element. Investigations of decays to which the axial vector interaction contributes, for instance that of the neutron, yield a value for the corresponding coupling constant G_F^A. The ratio $|G_F^A/G_F^V|$ is found to have the value[22]

$$\left|\frac{G_F^A}{G_F^V}\right| = 1.267 \pm 0.003. \tag{11.71}$$

In many mystery stories, the essential clues are hidden in aspects that appear, at first sight, completely normal, and the obviously guilty party often turns out to

[22]See PDG for the latest value.

be innocent. We now have $G_F, V_{ud}, G_F^V V_{ud}$, and $|G_F^A/G_F^V|$, given in Eqs. (11.55), (11.60), (11.70), and (11.71). Within the given limits of error, the following relations hold:

$$G_F^V = G_F, \qquad G_F^A \neq G_F. \tag{11.72}$$

What do these relations tell us about the weak interaction? At first sight it appears that the equal coupling constants for the vector current (G_F^V) and for the purely leptonic current (G_F) simply express the *universality* of the weak interaction and that $G_F^A \neq G_F$ requires an explanation. However, the situation is not so straightforward. A proton, for instance, is not just a simple point particle. At small distance it is made up of three quarks confined by gluons, and at distances $\gtrsim$1 fm it is aptly described as clothed by a meson cloud (Fig. 6.8). Why should the physical proton have the same vector current as a point lepton? There is no a priori reason why G_F^V and G_F should be identical. The result $G_F^A \neq G_F$ appears to be more in agreement with intuitive arguments, and the primary puzzle is the explanation of $G_F^V = G_F$. The solution to the puzzle is the *conserved vector current hypothesis* (CVC). It was first proposed in a tentative way by Gershtein and Zeldovich[(23)] and put into a powerful form by Feynman and Gell-Mann.[(5)] To explain CVC, consider first the electromagnetic case. In Section 7.2, it was pointed out that the electromagnetic charge is conserved. The positron and proton have the same electric charge despite the structure of the proton. In other words, the coupling constant e, which characterizes the interaction with the electromagnetic field, is the same for particles of the same charge regardless of their structural properties. The hadronic force responsible for the confinement of the quarks does not change the coupling constant e. The classical expression for this fact is current conservation, Eq. (10.51). The CVC hypothesis postulates that the *weak vector* current is also conserved:

$$\frac{1}{c}\frac{\partial V_0}{\partial t} + \boldsymbol{\nabla} \cdot \boldsymbol{V} = 0. \tag{11.73}$$

The equality of the coupling constants G_F^V and G_F then follows: whenever a hadron virtually decomposes into another set of hadrons (for instance, a proton into a neutron and a negative pion), the weak vector current is conserved. The equality of G_F^V and G_F is not the only evidence for CVC; many additional experiments support Eq. (11.73).[(24,20)]

An example is the comparison of the beta decay rates for ^{14}O and π^+. The systems are quite different; however, they have some common features. Both are decays from and to states of spin zero and isospin 1. Since the final and initial hadronic states are within an isospin multiplet, the decays are superallowed with matrix elements given by Eq. (11.66). The $ft_{1/2}$ for both ^{14}O and π^+ should thus be identical. Tables 11.1 and 11.2 show that they are almost identical; indeed, they

[23] S. S. Gershtein and Y. B. Zeldovich *ZhETF* **29**, 698(1955) [Transl. *Sov. Phys. JETP* **2**, 576 (1957)].
[24] L. Grenacs, *Annu. Rev. Nucl. Part. Sci.* **35**, 455 (1985) and references therein.

are equal to each other within experimental errors, after radiative corrections have been made.[(20,25)]

The hypothesis of the conservation of the *vector* current is based on the analogy to the electromagnetic current, which is also a vector current. No electromagnetic *axial* vector current exists, and it is thus not possible to refer to a well-known theory for guidance. Indeed, $G_F^A \neq G_F$ shows that the axial vector current is not conserved. The fact, however, that G_F^A does not differ from G_F by more than about 25% shows that the axial current is almost conserved. The detailed description of this fact is called the PCAC hypothesis or the *partially conserved axial vector current hypothesis.*[(24)]

11.11 Inverse Beta Decay: Reines and Cowan's Detection of Neutrinos

We now turn to neutrinos: they had been hypothesized by Pauli to save the law of conservation of energy in 1931, but Pauli thought that they were so weakly interacting that they would never be detected, so he considered his hypothesis somewhat *sinful.* In order to understand how neutrinos were detected, we consider the "elastic" scattering of neutrinos or antineutrinos due to the charged weak currents, e.g.,

$$\bar{\nu} p \longrightarrow l^+ n, \tag{11.74}$$

where l^+ is a positive lepton. The transition rate for this semileptonic process is given by the golden rule,

$$dw = \frac{2\pi}{\hbar} |\langle n l^+ | H_w | p \bar{\nu} \rangle|^2 \rho(E).$$

The transition rate gives the number of particles scattered per unit time by one scattering center. Equation (2.14) then shows that cross section and transition rates are connected by

$$d\sigma = \frac{dw}{F}. \tag{11.75}$$

Antineutrinos move close to the velocity of light; with the normalization of one particle per unit volume, the flux F is equal to the velocity, $F = c$. Consequently, the cross section becomes

$$d\sigma = \frac{2\pi}{\hbar c} |\langle n l^+ | H_w | p \bar{\nu} \rangle|^2 \rho(E). \tag{11.76}$$

The density-of-states factor for two particles in the final state, in their c.m., is given by Eq. (10.31). With $V = 1$, $\rho(E)$ is given by

$$\rho(E) = \frac{E_n E_l p_l}{(2\pi\hbar)^3 c^2 (E_n + E_l)} \, d\Omega_l,$$

[25] P. DePommier et al., *Nucl. Phys.* **B4**, 189 (1968); D. Počanić et al., Phys. Rev. Lett. **93**, 181803 (2004).

where $d\Omega_l$ is the solid-angle element into which the lepton is scattered. The differential cross section for antineutrino capture in the c.m. becomes

$$d\sigma_{\rm cm}(\bar{\nu}p \longrightarrow ln) = \frac{1}{4\pi^2\hbar^4c^3}\frac{E_nE_lp_l}{E_n+E_l}|\langle nl|H_w|p\bar{\nu}\rangle|^2 d\Omega_l. \tag{11.77}$$

Considering *low-energy electron anti-neutrinos* we can relate it to our earlier development. There we pointed out that the magnitude of the matrix element $\langle ne^+|H_w|p\bar{\nu}\rangle$ is the same as that for the neutron decay, $\langle pe^-\bar{\nu}|H_w|n\rangle$. The neutron decay matrix element is connected to the neutron $f\tau$ value by Eq. (11.11). Integrating Eq. (11.77) over $d\Omega_l$, inserting Eq. (11.11) into Eq. (11.77), and noting that for low electron energies $E_n \approx m_nc^2, E_e \ll m_nc^2$, we find

$$\sigma(\bar{\nu}_e p \longrightarrow e^+n) = \frac{2\pi^2\hbar^3}{m_e^5c^7}\frac{p_eE_e}{(f\tau)_{\rm neutron}}. \tag{11.78}$$

With the numerical values of the constants and the observed $f\tau$ (Table 11.1) and with convenient energy and momentum units, the cross section is

$$\sigma(\text{cm}^2) = 2.3\times10^{-44}\frac{p_e}{m_ec}\frac{E_e}{m_ec^2}.$$

At the antineutrino energies occurring at a reactor, the recoil energy of the neutron in the reaction $\bar{\nu}p \to e^+n$ can be neglected, and the total energy of the positron is connected to the antineutrino energy by $E_{e^+} = E_{\bar{\nu}}+(m_p-m_n)c^2 = E_{\bar{\nu}}-1.293$ MeV. For an antineutrino energy of 2.5 MeV, the cross section becomes 12×10^{-44} cm^2.

Antineutrino capture was first observed by Reines, Cowan, and co-workers at Los Alamos in 1956.[26] They set up a large and well-shielded liquid scintillation counter near a reactor. A reactor emits an intense stream of antineutrinos, in the Los Alamos experiment about $10^{13}\bar{\nu}/\text{cm}^2$ sec. A few of these are captured in the liquid and give rise to a neutron and a positron. These produce a characteristic signal, and the Los Alamos group was able to determine the cross section as

$$\sigma_{\rm exp} = (11\pm4)\times10^{-44}\ \text{cm}^2.$$

To compare this number to the one expected from Eq. (11.74), the antineutrino spectrum must be known. It can be deduced from the beta spectrum of the fission fragments of ^{238}U,[27] and a cross section of about 10×10^{-44} cm^2 is computed, in good agreement with the actually observed value. The agreement is reassuring; it indicates that the low-energy features of the weak interaction theory are capable of describing neutrino reactions.

[26] F. Reines and C. L. Cowan, *Science* **124**, 103 (1956); *Phys. Rev.* **113**, 273 (1959); F. Reines, C.L. Cowan, F.B. Harrison, A.D. McGuire, and H.W. Kruse, *Phys. Rev.* **117**, 159 (1960).
[27] R. E. Carter, F. Reines, J. J. Wagner, and M.E. Wyman, *Phys. Rev.* **113**, 280 (1959).

11.12 Massive Neutrinos

In our treatment so far we have assumed massless neutrinos. In Section 11.2, we pointed out that the high energy end of the beta spectrum can be used to search for a finite electron neutrino mass and yields a limit of $m(\nu_e)c^2 < 2.2$ eV. Similarly, searches for masses of ν_μ and ν_τ give $m(\nu_\mu)c^2 \leq 0.19$ MeV and $m(\nu_\tau)c^2 \leq 18.2$ MeV. Because the masses of neutrinos are so much smaller than the masses of the other particles, they were assumed to carry no mass in the Standard Model (Chapter 13.)

Here the mystery story starts with two experiments that were not motivated by measuring neutrino masses: one was the IMB-collaboration detector that was mounted to search for proton decay and the other was the Homestake-mine Cl detector set up to detect neutrinos from the Sun and confirm the mechanisms for production of solar energy (the intensity of light produced by the Sun is directly related to the intensity of neutrinos, see Problem 11.46.) The IMB-collaboration detector did not find any evidence for proton decay but proved able to detect and identify the flavor of neutrinos that are produced in the upper atmosphere (called *atmospheric neutrinos*). Both detectors found something unexpected: the IMB detector[28] determined that the ratio of muon neutrinos to electron neutrinos from the upper atmosphere was approximately a factor of 2 too small compared to expectations and the Homestake-mine Cl detector found only $\approx 1/3$ of the electron neutrinos expected from the Sun.[29] For many years it was thought that the solutions to these problems were unrelated.[30] Many scientists thought, for example, that the solar neutrino problem was due to lack of proper understanding of the solar physics. Other detectors were built to confirm the findings and better understand them and eventually it became clear that neutrinos do have mass and undergo flavor oscillations. A detector built in Japan, Super-Kamiokande, showed clear evidence for atmospheric neutrino oscillations[31] and a Canadian-American collaboration, the SNO detector, showed clear evidence for solar neutrino oscillations.[32]

Assuming that there are 3 kinds of neutrinos, ν_1, ν_2, and ν_3 with corresponding masses m_1, m_2 and m_3, and that weak decays produce neutrinos not in a pure mass eigenstate, but in a general linear combination of all possible states, we have:

$$\begin{pmatrix} \nu_e \\ \nu_\mu \\ \nu_\tau \end{pmatrix} = \begin{pmatrix} V_{e1} & V_{e2} & V_{e3} \\ V_{\mu 1} & V_{\mu 2} & V_{\mu 3} \\ V_{\tau 1} & V_{\tau 2} & V_{\tau 3} \end{pmatrix} \begin{pmatrix} \nu_1 \\ \nu_2 \\ \nu_3 \end{pmatrix}. \tag{11.79}$$

This should be reminiscent of Eq. 11.59 but here the matrix is called the Pontecorvo-Maki-Nakagawa-Sakata matrix.[33] To simplify our equations we assume only two

[28] D. Casper et al., *Phys. Rev. Lett.* **66**, 2561 (1991).
[29] J.N. Bahcall and R. Davis, jr. *Science* **191**, 264 (1976).
[30] For a very nice description of the history see J.N. Bahcall, posted at the Nobel prize web site: http://nobelprize.org/physics/articles/bahcall/index.html .
[31] Y. Fukuda et al., *Phys. Rev. Lett.* **81**, 1562 (1998).
[32] Q.R. Ahmad et al. *Phys. Rev. Lett.* **89**, 011301 (2002).
[33] V.N. Gribov and B.M. Pontecorvo, *Phys. Lett.* **B28**, 493 (1969); Z. Maki, M. Nakagawa, and

neutrino flavors (this turns out to be a very good approximation in most of the experimental situations); then the problem is a two-state one, similar to that of the neutral kaon system. In this case

$$\begin{aligned} \nu_e &= \cos\theta_{12}\nu_1 + \sin\theta_{12}\nu_2 \\ \nu_\mu &= -\sin\theta_{12}\nu_1 + \cos\theta_{12}\nu_2 \end{aligned} \tag{11.80}$$

and the time evolution of an electron neutrino born in the Sun will be:

$$|\nu_e(t)\rangle \quad = e^{-iE_1t/\hbar}\cos\theta_{12}\,|\nu_1\rangle + e^{-iE_2t/\hbar}\sin\theta_{12}\,|\nu_2\rangle\,. \tag{11.81}$$

It is ν_1 and ν_2 that have well-defined time dependences. So, the probability amplitude of finding a muon neutrino at time t can be obtained by using Eqs. (11.80) and (11.81) (compare also to Eq. (9.73))

$$\mathcal{P}_{\nu_\mu}(t) = |\langle\nu_\mu|\nu_e(t)\rangle|^2 = \sin^2 2\theta_{12}\sin^2\left[\frac{1}{2}\frac{(E_1-E_2)t}{\hbar}\right]. \tag{11.82}$$

Since the masses of the neutrinos are very small, we have $m_\nu \ll p/c$, where p is the neutrino momentum, and

$$E_i \approx pc + \frac{m_i^2c^3}{2p} \quad \text{and } t \sim L/c$$

where L is the propagation length.[34] With this approximation, Eq. (11.82) can be written as

$$\mathcal{P}_{\nu_\mu}(t) = \sin^2 2\theta_{12}\sin^2\left(\Delta m^2\frac{c^3L}{4pc\hbar}\right). \tag{11.83}$$

In summary, neutrinos change flavors as they move away from the point of production. This has been shown to be the explanation of both the atmospheric neutrino and solar neutrino puzzles.[35] For solar neutrinos, the SNO collaboration was able to show that, while the number of electron neutrinos from the Sun was significantly reduced from the number expected (in agreement with the Homestake-mine experiment) the missing neutrinos had changed flavors by the time they arrived on Earth and had consequently been missed by the Homestake-mine experiment (which could only see electron neutrinos.) A confirmation of the oscillations using neutrinos from reactors was performed by the Kamland collaboration.[36] The Kamland collaboration observed a similar phenomenon as had been observed with the solar neutrinos, but using a shorter distance and looking at lower-energy neutrinos.

S. Sakata, *Prog. Theor. Phys.* **28**, 870 (1962); note that these papers were written before the one by Kobayashi and Maskawa.

[34] For an understandable but more sophisticated treatment, see H.J. Lipkin, *Phys. Lett. B* **642**, 366 (2006).

[35] W.C. Haxton and B.R. Holstein, *Am. Jour. Phys.* **72**, 18 (2004).

[36] K. Eguchi et al., *Phys. Rev. Lett.* **90**, 021802 (2003).

• Because the Sun has a large number of electrons that interact with ν_e's but not with ν_μ's or ν_τ's, there is an additional phase shift between the wave function of the electron neutrinos with respect to the other neutrinos. The consequence is that oscillations can get enhaced under the proper conditions. These are called matter-enhanced or MSW oscillations[37] as opposed to the vacuum oscillations described by Eqs. (11.80) and (11.81). •

11.13 Majorana versus Dirac Neutrinos

A question that remains un-answered is why the neutrino masses are so small compared to those of the other particles. Theorists have come up with a mechanism called "see-saw" that involves an extremely heavy neutrino with mass M_{RH} postulated in a grand unified theory. In this theory, the neutrinos of the electron, muon and tau turn out to have masses of the order of $m_\nu \sim m_L^2/M_{RH}$, where m_L is the mass of a lepton or that of the $W^\pm$. The small mass could then be a signal of grand unification. However, in order for this mechanism to exist neutrinos should be identical to their anti-particles. In this case they would be called Majorana particles.[38] Fermions that are distinguishable from their anti-particles are called Dirac particles. Clearly all charged fermions are Dirac particles, but neutral particles can in principle have either identity. The π^0 is indistinguishable from its anti-partner, but the K_0 is distinct from the $\overline{K_0}$. Fig. 11.10 shows a scheme of the helicity components of fermions under the two scenarios.[39]

How can we determine whether neutrinos are Majorana or Dirac particles? Because neutrinos are massive, their helicity depends on the frame of reference. If neutrinos were Majorana particles one could consider electron anti-neutrinos from, say, nuclear beta decays then take a frame of reference that moves faster than the anti-neutrinos and see if they behave just like neutrinos. However, performing such an experiment is impractical. There is one practical way of experimentally finding out whether neutrinos are Majorana particles: It is the observation of *zero-neutrino double-beta decay*. Some nuclei do not have enough energy for an ordinary beta decay, but the energy difference of nuclei with Z and $Z+2$ protons may be sufficient to allow a decay with the emission of two electrons and two anti-neutrinos:

$$(Z, N) \longrightarrow (Z+2, N-2) + 2e^- + 2\overline{\nu_e}. \tag{11.84}$$

In this decay two neutrons from the original nucleus simultaneously undergo beta decay. This decay is very slow, but it has been observed for many cases.[40] If neutrinos are Majorana particles then the double-beta decay becomes possible without

[37] L. Wolfenstein, *Phys. Rev. D* **17**, 2369 (1978); S.P. Mikheyev and A.Y. Smirnov, *Soviet Journal Nuclear Physics* **42**, 913 (1985). The phenomenon is nicely described in H.A. Bethe, *Phys. Rev. Lett.* **56**, 1305 (1986).

[38] E. Majorana, *Nuovo Cimento* **14**, 171 (1937).

[39] B. Kayser et al., World Sci. Lecture Notes in Physics, Vol. 25, (1989).

[40] The first direct observation of 2ν decay came from ^{82}Se with a mean lifetime of about 10^{20} years, S.R. Elliott, A.A. Hahn, and M.K. Moe, *Phys. Rev. Lett.* **59**, 2020 (1987).

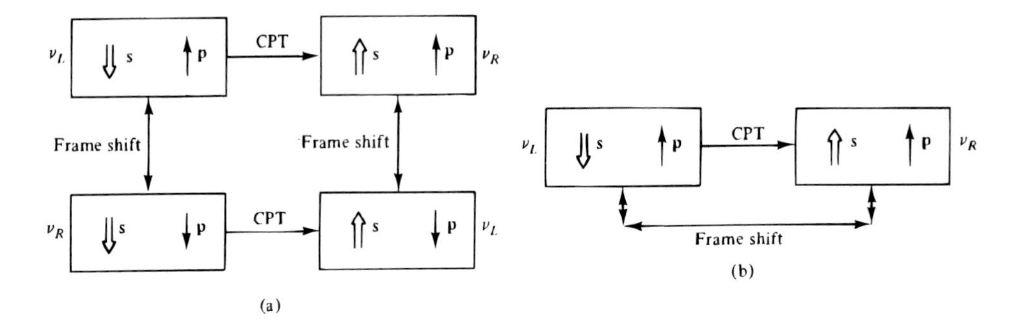

Figure 11.10: (a) Four states of a Dirac neutrino; (b) two states of a Majorana neutrino.

the emission of neutrinos:

$$(Z, N) \longrightarrow (Z+2, N-2) + 2e^-. \tag{11.85}$$

In simple terms, virtual neutrinos from the decays of the two neutrons would anihilate each other in this case. This can only happen if neutrinos are Majorana particles. However, one more condition has to be met: the helicities of the neutrinos have to be opposite each other. This can happen because neutrinos are massive and consequently not in an eigenstate of helicity (see Section 11.7) and the decay amplitude ends up being proportional to the neutrino masses. Because presently there is only an upper limit on the masses of neutrinos (see Table 5.7), it is not possible to use an upper limit on the rate found in a particular experiment to exclude the possibility of Majorana neutrinos. Rather, the upper limits for the decaying rates can be used to put upper limits on the neutrino masses.

Nevertheless, because the two electrons take up almost all the energy, the experimental signature is very clear: one should observe a spike at the endpoint in the electron energy spectrum. The observation of such decays would be a clear signal of Majorana neutrinos and, of course, would require lepton number violation.[41]

11.14 The Weak Current of Hadrons at High Energies

High energies are important for the exploration of two aspects of the weak interaction: (1) Nucleons and nuclei have weak charges and weak currents as well as electromagnetic ones. To investigate their distributions (weak form factors), weakly interacting probes with wavelengths smaller than the dimensions of interest are required. The problem is similar to the study of the electromagnetic structure of subatomic particles discussed in Chapter 6. If the weak form factors have about

[41] There is presently a controversial claim of observation of such neutrino-less double beta decay: H.V. Klapdor-Kleingrothaus et al., *Mod. Phys. Lett.* **A16**, 2409 (2001); C.E. Aalseth et al., *ibid.* **A17**, 1475 (2002). Detectors that should significantly improve the sensitivity are under consideration so there is hope that this issue may be cleared up in the next few years.

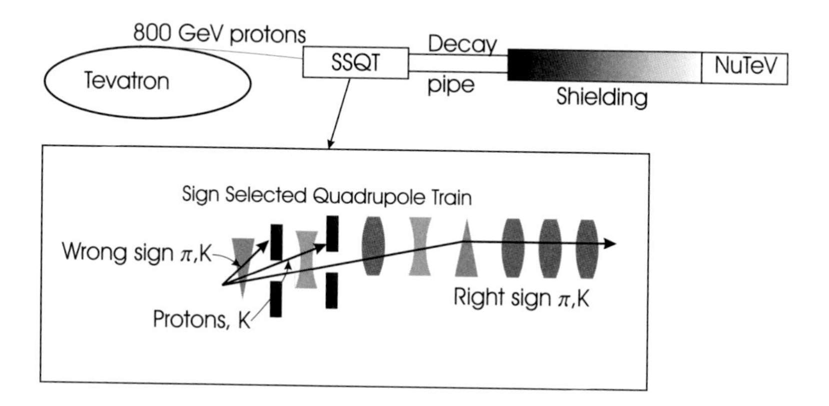

Figure 11.11: Sketch of the NuTev experiment at Fermilab. The device labelled SSQT allowed selection of mesons with the approriate sign. The 'shielding' acted as a filter that enhanced the ratio of neutrinos to charged particles.

the same behavior as the electromagnetic ones, then the discussion in Chapter 6 shows that weak probes with energies upward of a few GeV are needed. (2) The range of the weak interaction is given by Eq. (11.17) as about 2.5 am. To study characteristics of the weak interaction, energies that approach or exceed $m_W c^2$ are required.

In the electromagnetic case, structure investigations use charged leptons (electrons or muons) and photons. In the weak case, both neutrinos and charged leptons provide information. Since neutrinos interact only through the weak interaction they are an obvious choice for structure studies. Even though interaction cross sections are small, existing and planned accelerators provide large neutrino fluxes from the decays of pions and kaons; huge detectors are required for meaningful studies. In the following, we shall discuss some theoretical and experimental aspects of neutrino scattering. The charged leptons also interact via the weak interaction which can be separated from the much stronger electromagnetic one because the former violates parity and charge conjugation invariance. The interference of the electromagnetic and weak parity- or charge conjugation-violating amplitudes in the cross section can be observed, as has been pointed out in Chapter 9 and will be seen again in Chapter 13.

The feasibility of experiments to detect *neutrino reactions at high energies* was pointed out by Pontecorvo and by Schwartz.[42] The theoretical possibilities were first explored by Lee and Yang.[43] As so often in physics, the basic idea is simple, and it is sketched in Fig. 11.11: protons from a high-energy accelerator strike a target and produce high-energy pions and kaons. Mesons of one charge and one

[42] B. Pontecorvo, *Sov. Phys. JETP* **37**, 1751 (1959); M. Schwartz, *Phys. Rev. Lett.* **4**, 306 (1960). For a fascinating personal account, see B. Maglich, ed., *Adventures in Experimental Physics*, Vol. α, World Science Communications, Princeton, N.J., 1972, p. 82.

[43] T. D. Lee and C. N. Yang, *Phys. Rev. Lett.* **4**, 307 (1960); *Phys. Rev.* **126**, 2239 (1962).

variety, for instance, π^+, are selected and focused into the desired direction. If no material is placed in their path, they decay in flight and create positive muons and muon neutrinos. In the c.m. of the pion, muon and neutrino are emitted with opposite momenta. Because of the large momentum of the decaying pion, in the laboratory, most of the decay products move forward in a small cone. Further focusing is possible, or a "narrow-band" of neutrinos may be selected by restricting the momenta of the pions selected, as shown in Fig. 11.11. A typical flux of "wide-band" neutrinos is shown in Fig. 11.12. The detector is placed at a fairly large distance (e.g., >300 m) from the target and is so well shielded that mostly neutrinos can reach it. The small cross section requires a large detector so as to have a reasonable rate of events. The detector should be able to distinguish neutral and charged current events as well as measure the kinematic variables of produced particles. A typical detector is shown in Fig. 11.13.

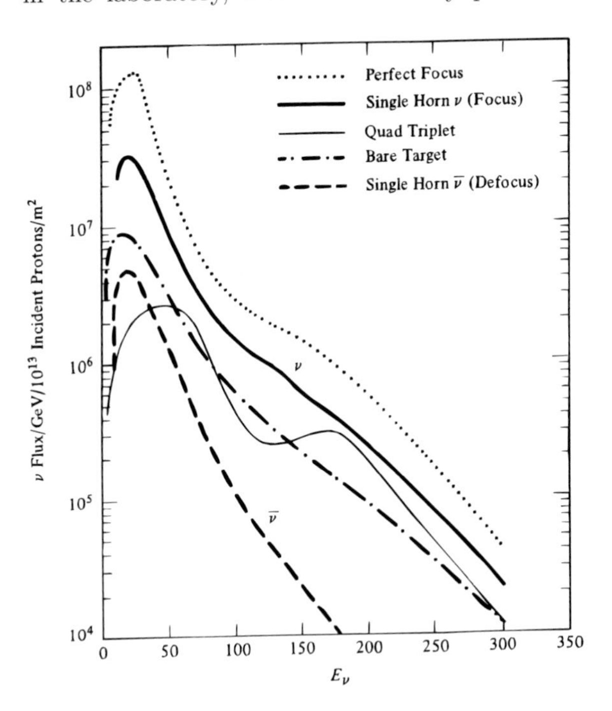

Figure 11.12: The calculated flux of neutrinos from various broad-band focusing devices used at Fermilab compared to that which would result from a perfect focusing device. The proton beam energy has been taken to be 400 GeV. [Courtesy H. E. Fisk and F. Sciulli.]

At first sight, neutrino experiments at high-energy accelerators appear to be hopeless because the neutrino flux is much smaller than at reactors. Fortunately, the cross section increases rapidly with energy: For energies such that $m_pc^2 \ll E_{cm} \ll m_wc^2$, where E_{cm} is the center-of-mass energy, there is no other dimension than the energy $E_{cm} = \sqrt{s}$, to set the scale. Thus, we can use dimensional arguments, as for Eq. (10.82), to obtain the energy dependence of the cross section.

In the present case the coupling constant G_F has the dimension energy-volume so that the cross section is given by

$$\sigma = CG_F^2 s/(\hbar c)^4 = 2CG_F^2 m_p E_{\rm lab}/(\hbar c)^4, \tag{11.86}$$

where C is a dimensionless constant, and $E_{\rm lab}$ is the laboratory energy of the neutrinos. The linear dependence of the total neutrino and antineutrino cross sections

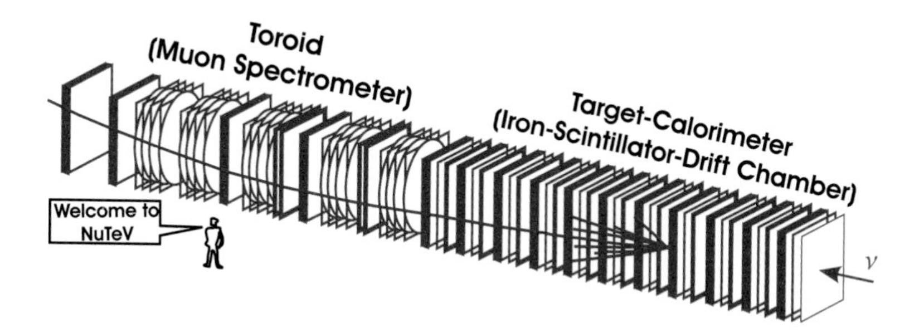

Figure 11.13: The NuTev detector in Fermilab. The steel target region is instrumented with counters and spark chambers to detect the interaction point and to track muons downstream. The toroids permit measurement of final state muon momenta. The large size is apparent by comparison with the individual shown.

on the laboratory energy is shown in Fig. 11.14.

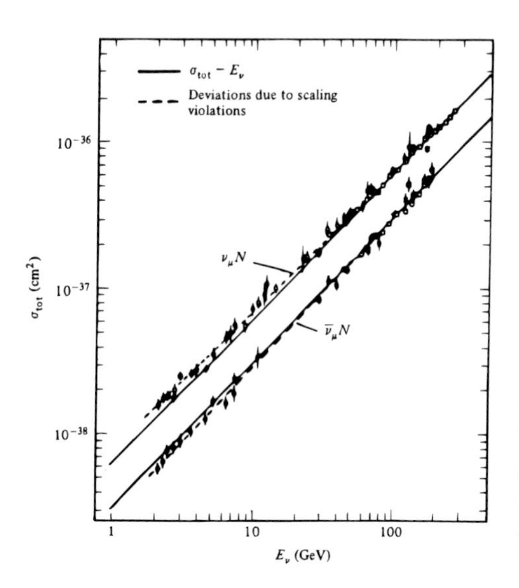

Figure 11.14: Total charged current neutrino and antineutrino cross sections plotted against energy. [From F. Eisele, *Rep. Prog. Phys.* **49**, 233 (1986).]

The factor of 3 difference between σ_ν and $\sigma_{\bar{\nu}}$ in Fig. 11.14 can be understood from angular momentum conservation. Neutrinos are purely left-handed, antineutrinos right-handed. For massless quarks and leptons, only the left-handed components of these particles participate in charged current weak interactions, as we will detail in Chapter 13. Then, as shown in Fig. 11.15, angular momentum can be conserved for backward scattering of neutrinos, but not of antineutrinos. The consequence is that the angular distribution of neutrinos is isotropic, but that of antineutrinos is $[(1+\cos\theta)/2]^2$. The resulting decrease in the integral of the differential cross section accounts for the smaller antineutrino total cross section.

For elastic scattering, form factors are important and the effective size of the target particle provides a scale. Consequently, the cross section as a function of the laboratory energy flattens out after an initial rise. Lee and Yang used the conserved vector current hypothesis of Gell-Mann and Feynman to compute the expected cross sections; their result is shown in Fig. 11.16.

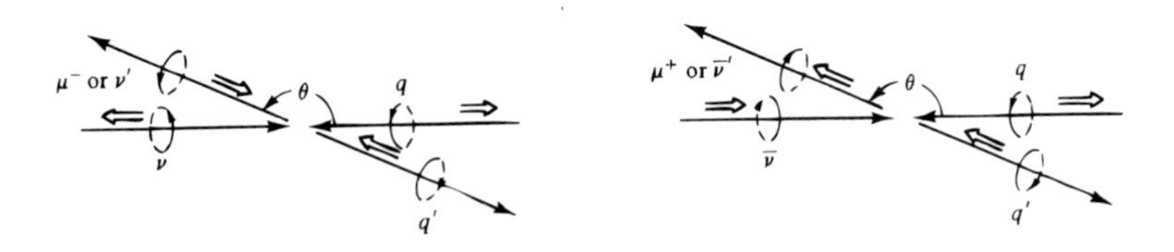

Figure 11.15: Illustration of angular momentum conservation for backward (180°) neutrino and antineutrino scattering on quarks. For the sake of clarity, θ is shown as close to, but not equal to 180°.

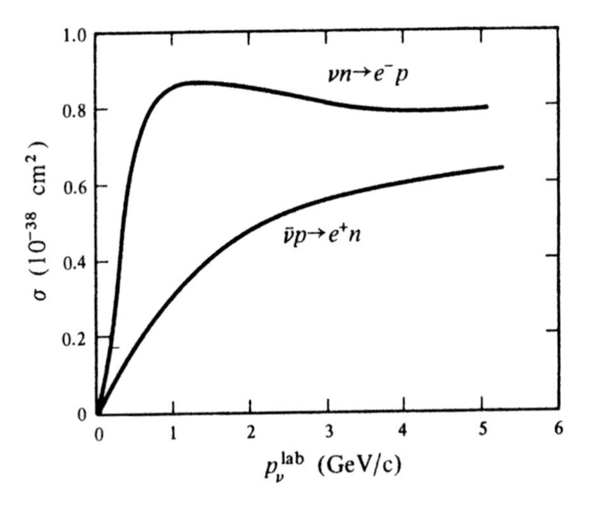

Figure 11.16: Cross section for the reaction $\nu n \to l^- p$, as predicted by T. D. Lee and C. N. Yang, *Phys. Rev. Lett*, **4**, 307 (1960).

The cross section increases very steeply up to laboratory neutrino energies of about 1 GeV and then levels off. The maximum cross section is of the order of 10^{-38} cm^2, about five orders of magnitude larger than the one observed in the Los Alamos neutrino experiment. The larger cross section made it possible for the Columbia group to perform the memorable experiment that revealed the existence of two kinds of neutrinos (Section 7.4).

We now turn our attention to the behavior of the matrix element of H_w at high energies. We shall first evaluate the cross section for the reaction $\nu_\mu \mathrm{N} \to \mu^- N'$, where N and N' are spinless hypothetical nucleons.

We shall discuss the modifications required to describe real nucleons later. The cross section for this reaction is given by Eq. (11.77) with small changes in notation. At high energies, the lepton mass can be neglected and E_μ can be replaced by $p_\mu c$. Equation (11.77) then reads

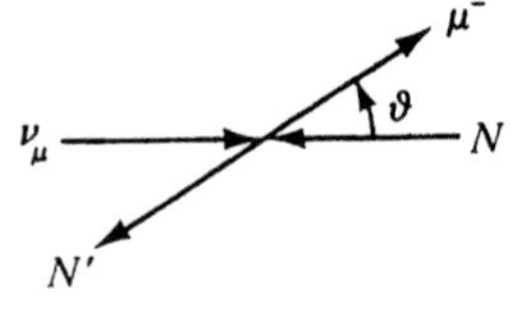

Figure 11.17: "Elastic" reaction $\nu_\mu N \to \mu^- N'$ in the c.m.

$$d\sigma_{\text{c.m.}}(\nu_\mu N \longrightarrow \mu^- N') = \frac{1}{4\pi^2\hbar^4c^2}\frac{E}{W}p_\mu^2|\langle\mu^- N'|H_w|\nu N\rangle|^2 d\Omega,$$

where E is the energy of N' and W the total energy in the c.m. The reaction $\nu_\mu N \to \mu^- N'$ is shown in Fig. 11.17.

In the c.m. all momenta have the same magnitude so that the square of the momentum transfer becomes

$$-\boldsymbol{q}^2 = (\boldsymbol{p}_\nu - \boldsymbol{p}_\mu)^2 = 2p_\nu^2(1-\cos\vartheta), \qquad (11.87)$$

where ϑ is the c.m. scattering angle. With Eq. (11.87), the solid-angle element $d\Omega = 2\pi \sin\vartheta d\vartheta$ can be written as

$$d\Omega = -\frac{\pi}{p_\nu^2} dq^2,$$

so that

$$d\sigma = \frac{-1}{4\pi\hbar^4 c^2} \frac{E}{W} |\langle \mu^- N' | H_w | \nu N \rangle|^2 dq^2. \qquad (11.88)$$

The central problem is now the matrix element. At low energies, where the structure of the particles can be neglected, we have already considered weak $0^+ \to 0^+$ transitions caused by charged weak currents.

The matrix element is given by Eq. (11.68), and the differential cross section in this case is

$$d\sigma = -\frac{G_F^2 V_{ud}^2}{2\pi\hbar^4 c^2} \frac{E}{W} dq^2. \qquad (11.89)$$

The total cross section is obtained by integrating over dq^2. The minimum squared momentum transfer is $-4p_\nu^2$, the maximum as given by Eq. (11.87) is 0, and the integration from 0 to $-4p_\nu^2$ yields

$$\sigma_{\text{tot}} = \frac{2G_F^2 V_{ud}^2}{\pi\hbar^4 c^2} \frac{E}{W} p_\nu^2 \qquad (11.90)$$

For the case of spin zero considered here, the cross section is modified by a weak form factor, F_w, and Eq. (11.89) becomes

$$d\sigma = -\frac{G_F^2 V_{ud}^2}{2\pi\hbar^4 c^2} \frac{E}{W} |F_w(q^2)|^2 dq^2. \qquad (11.91)$$

The weak form factor F_w is predicted by the CVC hypothesis. Feynman and Gell-Mann postulated that the vector form factors appearing in the electromagnetic and in the weak currents must have the same form. For our simplified example CVC states that for the vector interaction

$$F_w(q^2) = F_{em}(q^2). \qquad (11.92)$$

No spinless nucleons exist, and the form factor F_{em} for our specific example cannot be determined. However, we can assume that F_{em} has the same form as the form factors that appear in the nucleon structure. In particular we can identify F_{em} with G_D as given in Eq. (6.42): The weak cross section then becomes, with Eq. (11.92),

$$d\sigma = -\frac{G_F^2 V_{ud}^2}{2\pi\hbar^4 c^2} \frac{E}{W} \frac{dq^2}{(1+|q^2|/q_0^2)^4}. \qquad (11.93)$$

The total cross section is obtained by integration from 0 to $-4p_\nu^2$,

$$\sigma = \frac{G_F^2 V_{ud}^2 E q_0^2}{6\pi\hbar^4 c^2 W}\left(1 - \frac{1}{(1+4p_\nu^2/q_0^2)^3}\right). \tag{11.94}$$

This expression displays the essential features of the theoretical cross sections shown in Fig. 11.16: At low energies, the term in the large parentheses can be expanded; the result is identical to Eq. (11.90), and the cross section increases as p_ν^2. At higher energies, the term in the large parentheses becomes unity, and the cross section is a constant.

The cross section, Eq. (11.94) has been derived for a superallowed $0^+ \to 0^+$ transition, for which only a single vector form factor enters. Nucleons have spin 1/2, and at least three form factors are required to describe the cross section. Two of these form factors are predicted from the CVC hypothesis to be identical to those for the electromagnetic scattering of electrons, G_E and G_M introduced in Eq. (6.38). The weak current, however, also contains an axial part, $\mathcal{A}$, and a single form factor is sufficient to describe it. It is assumed that it has the same *form* as G_D, Eq. (6.42). Thus only one free parameter is left, $q_0^2 \equiv M_A^2 c^2$. Figure 11.18 presents data for the elastic scattering $\nu_\mu n \to \mu^- p$ and neutral current elastic scatterings on protons. The theoretical curves are cross sections computed with three form factors, G_E, G_M, and G_F^A. G_E and G_M are given in Eq. (6.43) and G_F^A by Eq. (6.42), with $q_0^2 \equiv M_A^2 c^2$ and M_A as indicated in Fig. 11.18. The data show that the experimental results are compatible with these form factors and with an *axial mass* $M_A = 1.06$ GeV/c^2, somewhat larger than the *vector mass* $M_V \equiv q_0/c = \sqrt{0.71}$ GeV/c^2. This result is expected because axial vector mesons have higher masses than their vector counterparts; the lowest axial vector meson is the h_1 with a mass of 1190 MeV/c^2.

So far the discussion has been restricted to the elastic scattering due to charged currents. The cross section for the true elastic scattering due to neutral currents

$$\nu_\mu p \longrightarrow \nu_\mu p$$

is more difficult to measure, but has been studied[44] to test the standard model [Weinberg–Salam theory] (Chapter 13).

Both charged and neutral current weak interactions of neutrinos induce many other reactions such as

$$\nu_\mu p \longrightarrow \mu^- \pi^+ p.$$

Of particular interest are the *inclusive reactions*

$$\begin{aligned} \nu_\mu p &\longrightarrow \mu^- X, & \bar{\nu}_\mu p &\longrightarrow \mu^+ X, \\ \nu_\mu p &\longrightarrow \nu_\mu X, & \bar{\nu}_\mu p &\longrightarrow \bar{\nu}_\mu X, \end{aligned}$$

[44] G. P. Zeller et al. *Phys. Rev. Lett.* **88**, 091802 (2002).

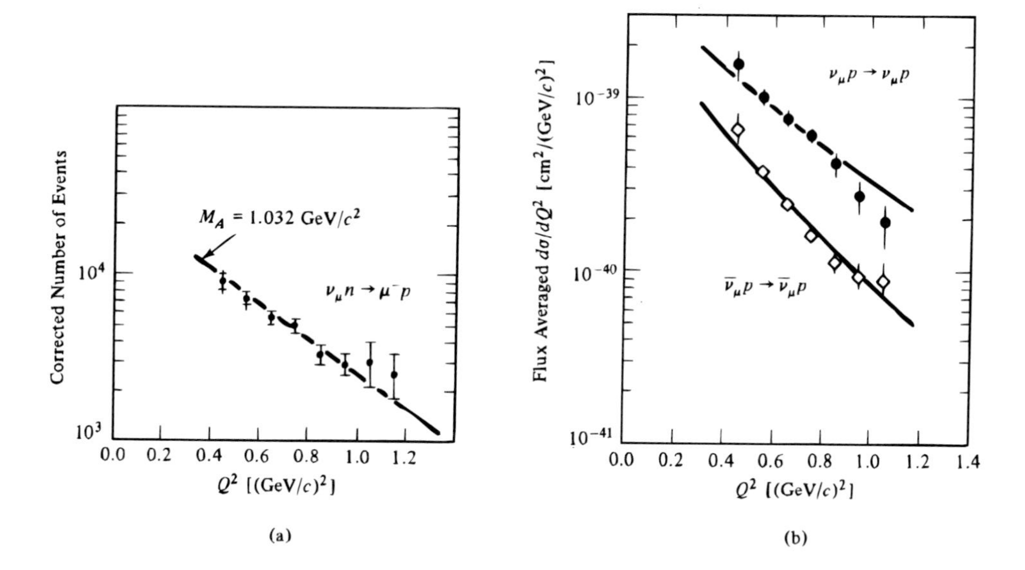

Figure 11.18: (a) The flux-averaged differential cross section for quasielastic events obtained from two-prong events. The smooth curve is for $M_A = 1.032$ GeV/c^2. (b) The flux-averaged differential cross sections for neutrino and antineutrino scattering on protons. The solid curves correspond to $M_A = 1.06$ GeV/c^2. [From L. A. Ahrens et al., *Phys. Rev.* **D35**, 785 (1987).]

where X stands for any number of particles. As for inclusive electron scattering, discussed in Sections 6.9 and 6.10, these reactions have been employed to explore the quark–parton model and obtain quark distribution functions. We have already shown the total charged current cross section for neutrinos and antineutrinos as a function of laboratory energy. The linear dependence of the cross section provides evidence for the point parton substructure of the proton. As shown by Eq. (11.90) for scattering from point particles, the cross section is proportional to the square of the c.m. momentum or energy; this squared energy, in turn, is proportional to the laboratory energy (see Eq. [11.86]).

The deep inelastic scattering of neutrinos or antineutrinos complements that of electrons. The charged current inclusive reactions are easier to study, since a charged lepton is detected in the final state rather than a neutrino. The development is similar to that of Sections 6.9 and 6.10. For instance, the scattering of the neutrinos from the quark–partons is elastic and incoherent, as for electrons, and scaling occurs.

There are also differences between inclusive deep inelastic electron and neutrino scattering. For ν scattering, the interaction is of very short range rather than $1/r$. This difference requires that the electromagnetic α/q^2 be replaced by $G_F^2/8\pi$ or by $(g_w^2/\hbar c)/(q^2 + m_{W,Z}^2 c^2)$ at higher energies in Eq. (6.67).

Also, the charged current weak interaction with up and down quarks is not proportional to their electric charges, and Eq. (6.65) must be modified to

$$\mathcal{P}(x) = \sum_i \mathcal{P}(x_i) \qquad (11.95)$$

where i is a sum over quarks. Furthermore, the weak interaction may also occur with gluons (neglected below) and does not conserve parity. The axial current gives rise to a third, parity-violating, structure function, F_3, from the interference of the vector and axial vector current matrix elements in the cross section.

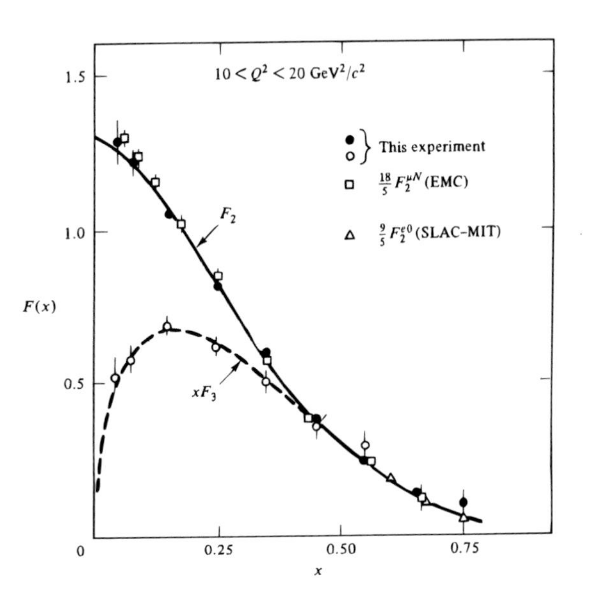

Figure 11.19: The structure functions $F_2(x)$ and $xF_3(x)$ for a fixed value of the squared momentum transfer, Q^2, for neutrino scattering. Also shown is the renormalized $F_2(x)$ obtained from muon and electron scattering. [After F. Eisele, *Rep. Prog. Phys.* **49**, 233 (1986).]

This term changes sign for neutrino and antineutrino scattering from nucleons. Thus, the structure function F_3 can be determined from the difference of the charged current (cc) (or neutral current) neutrino and antineutrino inclusive cross sections,[45]

$$F_3 \propto \sigma_{\rm cc}(\nu) - \sigma_{\rm cc}(\bar{\nu}). \qquad (11.96)$$

F_3 is shown for protons in Fig. 11.19. The other two structure functions, F_1 and F_2 are identical to the electromagnetic ones to within a constant because of the CVC relation. In Fig. 11.19, F_2 from ν and $\bar{\nu}$ scattering is compared to F_2 from charged lepton scattering. From Eqs. (6.65) and (11.95), and the relation $F_2(x) = x\mathcal{P}(x)$, Eq. (6.68), we obtain for an isoscalar target with equal numbers of protons and neutrons and therefore up and down quarks

$$\frac{F_2(e)}{F_2(\nu)} = \frac{5}{18}. \qquad (11.97)$$

Since F_3 measures the probability of finding a quark (if the presence of antiquarks is neglected), it follows that $\int F_3 dx = 3$.

[45] J. V. Allaby et al., *Phys. Lett.* **213B**, 554 (1988).

11.15 References

General theory and experimental facts on beta decay are describe in: H. J. Lipkin, *Beta Decay for Pedestrians*, North-Holland, Amsterdam, 1962; C. S. Wu and S. A. Moszkowski in *Beta Decay*, Wiley-Interscience, New York, 1966; and E.J. Konopinski, *The Theory of Beta Radioactivity*, International Series of Monographs on Physics, Oxford, 1966, cover nuclear beta decay lucidly and in depth up to ~1965. Enphasis on modern issues can be found in: B. Holstein, *Weak Interactions in Nuclei*, Princeton University Press, Princeton, NJ, 1989; K. Grotz and H.V. Klapdor, *The weak interaction in nuclear, particle, and astrophysics*, Bristol; New York: Hilger, 1990; W.C. Haxton and E.M. Henley, eds, *Symmetries and Fundamental Interactions in Nuclei*, World Sci., Singapore, 1995; L. Wolfenstein, *The strength of the weak interactions*, in *Annual Review of Nuclear and Particle Science* **54**, 1 (2004).

The weak interaction is discussed in more general terms in L.B. Okun, *Leptons and Quarks*, North-Holland, Amsterdam, 1982; E.D. Commins and P.H. Bucksbaum, *Weak Interactions of Leptons and Quarks*, Press Syndicate University of Cambridge, New York, 1983; R.N. Cahn and G. Goldhaber, *The Experimental Foundations of Particle Physics*, Cambridge University Press, 1988. The latter includes copies of the original papers of many of the milestone discoveries.

A summary of the development of the weak interactions over a thirty year stretch can be found in *Thirty Years Since Parity Nonconservation*, A Symposium for T.D. Lee, (R. Novick, ed.) Birkhäuser, Boston, 1988. For a *History of Weak Interactions*, see T.D. Lee in *Elementary Processes at High Energy*, (A. Zichichi, ed.) Academic Press, New York, 1971, p. 828. Many of the influential papers on weak interactions, including a translation of Fermi's classic work, are reprinted in P.K. Kabir, *The Development of Weak Interaction Theory*, Gordon and Breach, New York, 1963. Original papers, with introductory discussions, are also collected in C. Strachan, *The Theory of Beta Decay*, Pergamon, Elmsford, N. Y., 1969.

There have been numerous reviews on various aspects of the weak interactions. For instance, muon decay properties are discussed in detail in *Muon Physics*, (V. Hughes and C.S. Wu, eds) Academic Press, New York, 1975; and R. Engfer and H.K. Walter, *Annu. Rev. Nucl. Part. Sci.* **36**, 327 (1986).

Neutrinos and Double-Beta Decay: F. Boehm, P. Vogel, *Physics of massive neutrinos*, Cambridge University Press, 1992; E. Kearns, T. Kajita, and Y. Totsuka, *Detecting Massive Neutrinos*, *Sci. Amer.* **281**, 48 (February 1999); P. Fisher, B. Kayser, and K. McFarland, *Annu. Rev. Nucl. Part. Sci.* **49**, 481 (1999); *Neutrino Physics*, K. Winter, ed., 2nd ed., Cambridge Univ. Press, Cambridge, 2000; A. Franklin, *Are there Really Neutrinos?*, Perseus, Cambridge, 2001; *Current Aspects of Neutrino Physics*, D.O. Caldwell, ed., Springer, Heidelberg 2001; M. Koshiba, *Birth of Neutrino Astrophysics* in *Rev. Mod. Phys.* **75**, 1011 (2003); M. C. Gonzales-Garcia and Y. Nix, *Rev. Mod. Phys.* **75**, 345 (2003). W.C. Haxton and B.R. Holstein, *Neutrino Physics: An update, Am. J. Phys* **72**, 18 (2004); R.N. Mohapatra and P.B. Pal, *Massive Neutrinos in Physics and Astrophysics*, 3rd ed., World Sci., Singapore, 2004. Detailed reviews include W.C. Haxton and G.J. Stephenson, Jr., *Prog. Part. Nucl. Phys.* (D. Wilkinson, ed.) **12**, 409 (1984); M. Doi, T. Kotani and E. Takasugi, *Prog. Theor. Phys.* Suppl. 83, 1 (1985); F. T. Avignone III and R. L. Brodzinski *Prog. Part. Nucl. Phys.*, (A. Faessler, ed.) **21**, 99 (1988); F. Reines (Nobel Prize address), *Rev. Mod. Phys.* **68**, 317 (1996); J.N. Conrad, M.H. Shaevitz, and T. Bolton, *Rev. Mod. Phys.* **70**, 1341 (1998); K. Hagiwara, *Annu. Rev. Nucl. Part. Sci.* **48**, 463 (1998); S.R. Elliott and P. Vogel, *Annu. Rev. Nucl. Part. Sci.* **52**, 65 (2002); J. Suhonen and O. Civitarese, *Phys. Rep.* **300**, 123 (1998); Yu. Zdesenko, *Rev. Mod. Phys.* **74**, 663 (2002).

Reviews of neutrinos and their high energy interactions, including nucleon structure information, are: D. Cline and W. F. Fry, *Annu. Rev. Nucl. Sci.* **27**, 209 (1977); H.E. Fisk and F. Sciulli, *Annu. Rev. Nucl. Part. Sci.* **32**, 499 (1982); P. Eisele, *Rep. Prog. Phys.* **49**, 233 (1986); M. Diemoz, F. Ferroni, and E. Longo *Phys. Rep.* **130**, 293 (1986).

Problems

11.1. Verify that the proton recoil energy can be neglected in the discussion of neutron beta decay.

11.2. Plot the phase–space distribution, Eq. (11.4), and check that a typical beta spectrum is well represented by it.

11.3. (a) Discuss how the upper end of the beta spectrum and the Kurie plot are distorted if the neutrino has a finite rest mass.

(b) Show the deviation of the Kurie plot for the beta decay of ^{3}H if the electron neutrino has a mass of 50 eV/c^2. What are some of the background problems that can plague a measurement of this deviation?

11.4. Discuss the beta decay of the neutron:

(a) Sketch the measurement of the mean life.

(b) Discuss the measurement of the spectrum.

(c) Use Eqs. (11.9) and (11.10) to compute the value of f for the neutron decay. Assume that $F(-, 1, E) = 1$. Compare the resulting value of ft with the one given in Table 11.1.

(d) In what observables does parity-nonconservation show up in neutron decay? How can it be observed experimentally? Discuss the results of such measurements.

11.5. Beta spectra can be measured in a variety of instruments. Two that are often used are magnetic beta spectrometers and solid-state detectors.

(a) Discuss both methods. Compare momentum resolution and counting statistics for a given source strength.

(b) What are the advantages and disadvantages of either method?

11.6. Assume that the mass difference between the charged and the neutral pions is caused by the electromagnetic interaction. Compare the corresponding energy to the weak energy given in Eq. (11.14).

11.7. Verify the integration leading to Eq. (11.19).

11.8. List three nuclear beta decays, one with a very small, one with an average, and one with a very large ft value. Consider the spin and parities involved and discuss why the variation in ft is not an argument against the universal Fermi interaction.

11.9. Compute the ratio of lifetimes for the decays

$$\Sigma^+ \to \Lambda^0 e^+ \nu \quad \text{and} \quad \Sigma^- \to \Lambda^0 e^- \bar{\nu}.$$

Compare your value to the experimental ratio.

11.10. Verify the f values in Table 11.3.

11.11. Consider the branching (intensity) ratio

$$\frac{\pi \to e\nu}{\pi \to \mu\nu}.$$

(a) How were the two decay modes observed?

(b) Compute the branching ratio expected if the matrix elements for both decays are assumed to be equal. Compare the result with the experimental ratio.

(c) Discuss the helicities of the charged leptons emitted in the pion decay; assume that neutrinos and antineutrinos are fully polarized, as shown in Fig. 7.2. Sketch the helicities of the e^+ and e^-.

(d) Experiment indicates that the helicity of negative leptons emitted in beta decay is given by $-v/c$, where v is the lepton velocity. Use this fact, together with the result of part (c), to explain the low branching ratio that is found experimentally.

11.12. Why do positive muons in matter usually come to rest before they decay? Describe the processes involved and give approximate values for the characteristic times that enter the considerations. Why do negative muons behave differently?

11.13. * Discuss the experimental determination of the polarization of the muon emitted in the decay of the pion.

11.14. Discuss the experimental determination of the electron spectrum in muon decay:

(a) Sketch a typical arrangement.

(b) How thin should the target be (in g/cm^2) in order not to affect the spectrum appreciably?

(c) How does one guarantee that the spectrum observed is that of an unpolarized muon source?

(d) How can the spectrum at low electron momenta be found?

11.15. Use the spectrum of Fig. 11.6 to construct an approximate Kurie plot for the muon decay. Show that a simple phase-space spectrum does not fit the observed data.

11.16. List reactions and decays that are described by the leptonic Hamiltonian, Eq. (11.36).

11.17. Show that the linear momentum and the angular momentum have the same transformation properties under ordinary rotations.

11.18. Show that neutral currents cannot contribute to beta decay in lowest order of G_F.

11.19. Show that the electron spectrum in Fig. 11.8 can be fitted with Eq. (11.52), after proper change of the variable.

11.20. (a) Determine the value of E_{max} in Eq. (11.54). Assume that $m_e = 0$.

(b) Verify the result of the integration in Eq. (11.54).

(c) Use the value of the muon mean life listed in PDG to verify the value of G_F given in Eq. (11.55).

11.21. Verify Eq. (11.56).

11.22. (a) What are the properties of W as predicted by the arguments in Sections 11.3 and 11.6?

(b) * Discuss experiments that could give information about W.

11.23. Find some examples other than the ones given in Table 11.3 to demonstrate that strangeness-changing weak decays are systematically slower than the corresponding strangeness-conserving ones. Use your examples to find a value for the sine of the Cabibbo angle, V_{us}.

11.24. Verify that the wave functions of the neutrino and the electrons, given in Eq. (11.62), are essentially constant over the nuclear volume.

11.25. Prove in detail that the integral containing $\mathcal{A}$ in Eq. (11.63) vanishes.

11.26. The computation of the lepton matrix element in Eq. (11.67) gives

$$|u_e^*(\mathcal{V}_0 + \mathcal{A}_0)u_{\bar{\nu}}|^2 = 2\left(1 + \frac{v}{c}\cos\theta_{e\nu}\right),$$

where v is the positron velocity and $\theta_{e\nu}$ the angle between positron and neutrino momenta.

(a) How can the positron–neutrino correlation be measured? Discuss the principle of the method and a typical experiment.

(b) Show that the observed positron–neutrino (and electron–antineutrino) correlations are in agreement with a $\mathcal{V} - \mathcal{A}$ interaction.

11.27. * List some superallowed $0^+ \to 0^+$ transitions and show that their ft values are all closely identical.

11.28. High-energy neutrinos have been observed in bubble chambers (propane and hydrogen) and in spark chambers.

(a) Compare typical count rates.

(b) What are the advantages and the disadvantages of the various detectors?

11.29. Plot a few numerical values of the cross section equation (11.94) as a function of the neutrino momentum

(a) In the c.m.

(b) In the laboratory.

Compare your curves with the ones shown in Figs. 11.16 and 11.18.

11.30. * Consider the strangeness-changing weak current of hadrons, for instance, in the case $\Lambda^0 \rightarrow p$ in semi-leptonic processes. Such a current satisfies the selection rule

$$\Delta S = \Delta Q,$$

where ΔS is the change of strangeness and ΔQ the change in charge.

(a) Give a few additional currents that have been observed and that satisfy this selection rule.

(b) Have currents with $\Delta S = -\Delta Q$ been observed? (The quantum numbers S and Q always refer to the hadrons.)

11.31. Discuss the isospin selection rules that are satisfied by the weak interaction

(a) In nonstrange decays, and

(b) In decays involving a change of strangeness.

(c) What experiments can be used to test these selection rules?

11.32. Discuss the evidence for and against the existence of neutral currents.

11.33. Show that the maximum cross section for a point interaction is given by the so-called unitary limit

$$\sigma_{\text{max}} = 4\pi\hbar^2/p^2,$$

where p is the c.m. momentum.

11.34. What experiments can be carried out to test the absence of $\Delta S \geq 2$ weak currents?

11.35. Show that the reaction $\nu_\mu e \rightarrow \nu_\mu e$ is forbidden if only charged currents exist.

11.36. Determine and briefly discuss one or more tests of the conserved vector current hypothesis.

11.37. Use the lifetime for the beta decay of ^{14}O and Eq. (11.9) to determine the beta decay lifetime of the positive pion (see Tables 11.1 and 11.2). Compare with experiment.

11.38. How can the decay $\Lambda^0 \rightarrow n\pi^0$ occur despite the absence of strangeness-changing neutral currents?

11.39. Use Eq. 11.80 to show that $m_2, m_1 \neq m_e, m_\mu$.

11.40. Find the probability $\mathcal{P}_{\nu_e}$ that an electron neutrino is still an electron neutrino after time t, rather than having turned into a muon neutrino, as in Eq. (11.82).

11.41. For nucleons composed of point quarks,

(a) show that the total cross section for high-energy neutrino scattering varies linearly with the laboratory energy E.

(b) How is the total cross section modified as the center-of-mass energy approaches $m_W c^2$?

(c) What is the laboratory energy of neutrinos for which the center-of-mass energy is equal to $m_W c^2$?

11.42. Verify Eq. (11.97).

11.43. Electron neutrinos can interact with electrons via charged and neutral weal currents.

(a) Draw Feynman diagrams showing these possibilities.

(b) Explain why muon neutrinos can interact with electrons only via neutral currents and not charged currents.

11.44. Electron neutrinos produced in beta decays are actually a mixture of two mass eigenstates, ν_1 and ν_2, $\mid \nu_e\rangle = cos\theta \mid \nu_1\rangle + sin\theta \mid \nu_2\rangle$.

(a) Deduce the equation that gives the probability of observing the neutrino as a muon neutrino for vacuum oscillations.

(b) Consider electron neutrinos from the decays of ^{8}B in the sun. Assume $\theta = 45\,\text{deg}$ and $m_2^2 - m_1^2 = 5 \times 10^{-5}$ eV^2. Plot the electron neutrino energy spectrum on earth assuming vacuum oscillations.

11.45. (a) Explain why the total cross section for high energy muon neutrinos to scatter off hadronic targets is three times larger than for antineutrinos.

(b) Would you expect electron neutrinos to have approximately the same, larger, or smaller cross sections than those for muon neutrinos?

(c) Repeat for electron antineutrinos.

11.46. Assume that the Sun obtains its energy from the transformation of 4 protons into a doubly ionized He atom, liberating ≈ 26 MeV: $4p \to {}^4\text{He}^{++} + 2e^+ + 2\nu_e$ and use the solar luminosity on Earth, 1.4 kWatt/m^2, to derive the expected intensity of neutrinos (number per unit time and area) on Earth.

11.47. Estimate V_{ud} using the ft values for the decays $\Lambda_0 \to pe^-\nu_e$ and $n \to pe^-\nu_e$ from Tables 11.2 and 11.3.

Chapter 12

Introduction to Gauge Theories

12.1 Introduction

In chapter 7 we introduced both global and local gauge transformations. In this chapter we continue the discussion of gauge invariance and its applications. This invariance has emerged as the primary underpinning of all fundamental subatomic interactions. It is now believed that all forces are described by *gauge theories*, theories for which local gauge invariance holds. The importance of gauge theories became obvious with the development of the unified electroweak theory; the Standard Model is based on gauge theories for the strong, electromagnetic and weak interactions. In the present chapter we discuss the ideas underlying modern gauge theories. The material is somewhat more difficult than what we have treated so far, but is necessary for understanding the Standard Model.

In chapter 7 we saw that additive conservation laws, including charge conservation, follow from a global gauge transformation, Eq. (7.21). We also showed that a local gauge transformation, Eq. (7.27), allows us to identify the charge as the electric one. The development in chapter 7 was for a static charge. However, the Schrödinger equation (7.1)

$$i\hbar\frac{\partial\psi}{\partial t} = H\psi \tag{7.1}$$

with the Hamiltonian of Section 10.3, for a particle of charge q under the influence of an electromagnetic field,

$$H = \frac{1}{2m}\left(\boldsymbol{p} - \frac{q}{c}\boldsymbol{A}\right)^2 + qA_0, \tag{12.1}$$

is also invariant under the combined local gauge transformation,

$$\psi_q' = e^{iQ\epsilon(\boldsymbol{x},t)}\psi_q \equiv U_Q(\epsilon)\psi_q, \tag{12.2}$$

where Q is the charge operator and

$$A_0' = A_0 - \hbar \frac{\partial \epsilon(\boldsymbol{x}, t)}{\partial t}$$
$$\boldsymbol{A}' = \boldsymbol{A} + \hbar c \boldsymbol{\nabla} \epsilon(\boldsymbol{x}, t), \tag{12.3}$$

or in four-vector notation

$$A_\mu' = A_\mu - \hbar c \nabla_\mu \epsilon(\boldsymbol{x}, t). \tag{12.4}$$

The local gauge invariance of Maxwell's equations for classical electricity and magnetism has been known for many decades. In classical electromagnetism, only the electromagnetic fields $\boldsymbol{E}$ and $\boldsymbol{B}$ have physical meaning, and gauge invariance is associated with the partial freedom of choice of the electromagnetic potentials A_0 and $\boldsymbol{A}$ of Eqs. (10.37) and (10.38). As we shall show in Section 12.2 the same is not true in quantum mechanics. With the advent of general relativity, which employs local gauge invariance, Weyl in 1919 tried to generalize the electromagnetic local gauge invariance as a geometrical means to unify electromagnetism and gravity.[(1)] His attempts were unsuccessful and the development lay dormant for over 30 years. However, in the past several decades, local gauge invariance was successfully extended and applied to the unification of electromagnetic and weak interactions. The invariance also underlies the basic theory of all interactions of the Standard Model, grand unified theories, as well as supersymmetric theories that include gravitation. Indeed, all modern descriptions of basic forces are gauge theories.

Gauge invariance is a powerful tool. We shall show that it dictates the form of the interaction, and requires massless vector fields, as for instance the electromagnetic field with its massless photon; Table 5.9 shows that the quanta of all subatomic forces have spin 1, and thus correspond to vector fields.

In chapter 7, we demonstrated that the form of Eqs. (12.2) and (12.4) leads to invariance under a local gauge transformation. Here we reverse the argument. If the Schrödinger equation (7.1) with H that for a free particle, $H = \boldsymbol{p}^2/2m$, is to remain invariant under the local gauge transformation (12.2), then a compensating *four-vector* field, with time and space components which can be called A_0 and $\boldsymbol{A}$, abbreviated as $(A_0, \boldsymbol{A})$ or simply A_μ *must* be introduced. Its concommitant transformations must be given by Eq. (12.4). In the following development we sometimes shall use the shorthand notation for four-vectors. More elaborate manipulations will be shown in brackets and bullets.

The requirement for a compensating vector field to maintain invariance of the Schrödinger equation under a local gauge transformation can be seen most easily by introducing the *covariant gauge* (sometimes simply called *covariant derivatives*) $D_\mu = (D_0, \boldsymbol{D})$

[1] H. Weyl, *Ann. Physik* **59**, 101 (1919).

$$D_0 \equiv \frac{1}{c}\frac{\partial}{\partial t} + \frac{iqA_0}{\hbar c},$$
$$\boldsymbol{D} \equiv \boldsymbol{\nabla} - \frac{iq\boldsymbol{A}}{\hbar c}. \tag{12.5}$$

If these derivatives replace the normal ones, $(1/c)\partial/\partial t$ and $\boldsymbol{\nabla}$, it follows with Eqs. (7.1) and (12.2) that

$$D'_0\psi'_q = D'_0 U_Q \psi_q = U_Q D_0 \psi_q,$$
$$\boldsymbol{D}'\psi'_q = \boldsymbol{D}' U_Q \psi_q = U_Q \boldsymbol{D} \psi_q, \tag{12.6}$$

where D'_0 and $\boldsymbol{D}'$ have A'_0 and $\boldsymbol{A}'$ as dependent variables. It is important to note that if U_Q stands to the left of D_0 and $\boldsymbol{D}$, it is a simple phase factor, since the derivatives only act on quantities to their right. With the introduction of the gauge covariant derivatives, D_0 and $\boldsymbol{D}$ transform under local gauge transformations just like $\frac{1}{c}\frac{\partial}{\partial t}$ and $\boldsymbol{\nabla}$ do under a global gauge transformation (ϵ = constant). The vector nature of the compensating field which appears in the covariant derivative is determined by the vector property of the momentum $\boldsymbol{p}$ for the free Hamiltonian and the time dependence of Eq. (7.1). When the covariant derivative is introduced in the Schrödinger equation, including the compensating field, the resulting particle Hamiltonian has the form Eq. (12.1). Thus, the requirement of local gauge invariance *generates* the qA_0 and $\boldsymbol{j} \cdot \boldsymbol{A}$ interaction of a charged particle with the electromagnetic field. We note, in addition, that space and time transformations are tied together.

So far, we have neglected the equation of motion for the vector field $(A_0, \boldsymbol{A})$. In the case of the electromagnetic field, it is given by Maxwell's equations ($i = x, y, z$)

$$\frac{1}{c^2}\frac{\partial^2 A_0}{\partial t^2} - \boldsymbol{\nabla}^2 A_0 = \rho = \psi^* q \psi,$$
$$\frac{1}{c^2}\frac{\partial^2 A_i}{\partial t^2} - \boldsymbol{\nabla}^2 A_i = \frac{j_i}{c} = \psi^* \frac{q\boldsymbol{v}_i}{c}\psi, \tag{12.7}$$

if we use the Lorentz condition

$$\frac{1}{c}\frac{\partial A_0}{\partial t} + \boldsymbol{\nabla} \cdot \boldsymbol{A} = 0. \tag{12.8}$$

The equations (12.7) are invariant under the gauge transformations, Eq. (12.4), if we impose the condition

$$\frac{1}{c^2}\frac{\partial^2 \epsilon(\boldsymbol{x}, t)}{\partial t^2} - \boldsymbol{\nabla}^2 \epsilon(\boldsymbol{x}, t) = 0. \tag{12.9}$$

- In four-vector notation Eq. (12.7) becomes

$$\left[\Box A_\mu \equiv g_{\alpha\,\nu} \nabla_\alpha \nabla_\nu A_\mu = \frac{j_\mu}{c}.\right]$$

If we do not impose the Lorentz condition, Maxwell's equations are

$$\begin{aligned}\frac{1}{c}\frac{\partial}{\partial t}\boldsymbol{\nabla}\cdot\boldsymbol{A}+\boldsymbol{\nabla}^2 A_0&=-\rho\\ \frac{1}{c}\frac{\partial^2\boldsymbol{A}}{\partial t^2}+\boldsymbol{\nabla}\frac{\partial A_0}{\partial t}&=\boldsymbol{j}\end{aligned}\tag{12.7a}$$

or

$$\Box A_\mu - g_{\alpha\,\nu}\nabla_\mu\nabla_\alpha A_\nu = g_{\alpha\,\nu}\left(\nabla_\nu\nabla_\alpha A_\mu - \nabla_\mu\nabla_\alpha A_\nu\right)=\frac{j_\mu}{c}$$

where we follow the convention that the $\sum$ sign is omited when indices are repeated. These equations are invariant under the gauge transformations, Eq. (12.4) for an arbitrary function $\epsilon(\boldsymbol{x},t)$. •

If the electromagnetic field quantum had a mass m_γ, Eq. (12.7) would be changed to

$$\begin{aligned}\frac{1}{c^2}\frac{\partial^2 A_0}{\partial t^2}-\nabla^2 A_0+\frac{m_\gamma^2c^2A_0}{\hbar^2}&=\rho,\\ \frac{1}{c^2}\frac{\partial^2\boldsymbol{A}}{\partial t^2}-\nabla^2\boldsymbol{A}+\frac{m_\gamma^2c^2\boldsymbol{A}}{\hbar^2}&=\frac{\boldsymbol{j}}{c},\\ \left[\text{or}\qquad \Box A_\mu+\frac{m_\gamma^2c^2}{\hbar^2}A_\mu\right.&=\left.\frac{j_\mu}{c},\right]\end{aligned}\tag{12.10}$$

and this additional mass term spoils the invariance under the gauge transformation. Thus, gauge invariance of the full theory, including the electromagnetic gauge field, only holds for massless photons or gauge particles.

Alternatively, we can write Maxwell's equations in terms of the electric and magnetic field strengths $\boldsymbol{E}$ and $\boldsymbol{B}$, defined in Eqs. (10.37) and (10.38). These field strengths are invariant under the gauge transformation (12.4), as is well known from classical electricity and magnetism.

12.2 Potentials in Quantum Mechanics—The Aharonov–Bohm Effect

The local gauge transformations clearly contain global ones as a special case. For the latter, we can say that the phase of a wavefunction is arbitrary and can be changed at will; however, the phase must be identical at all points in space and time. That this restriction is not essential was not fully appreciated for many years. For a local gauge invariance, the phase becomes a degree of freedom that varies with space and time, but its dependence is connected to the (vector) potentials A_0 and $\boldsymbol{A}$. The potentials thus acquire a physical meaning that they did not have in classical electricity and magnetism and that was not realised till several decades ago.[2] Their effect can be determined experimentally, as will now be shown.

[2] Y. Aharonov and D. Bohm, *Phys. Rev.* **115**, 485 (1959).

In the absence of the electromagnetic field, the stationary nonrelativistic wave equation for a free electron is

$$-\frac{\hbar^2}{2m}\nabla^2\psi_0 = E\psi_0. \tag{12.11}$$

The solution is a plane wave with a phase given by $\boldsymbol{p}\cdot\boldsymbol{x}/\hbar$,

$$\psi_0 = \exp\left(\frac{i\boldsymbol{p}\cdot\boldsymbol{x}}{\hbar}\right).$$

In the presence of a static electromagnetic vector potential $\boldsymbol{A}$, the stationary Schrödinger equation becomes, with Eq. (12.5)

$$\begin{aligned}-\frac{\hbar^2}{2m}\boldsymbol{D}^2\psi &= -\frac{\hbar^2}{2m}\left(\boldsymbol{\nabla}+\frac{ie\boldsymbol{A}(\boldsymbol{x})}{\hbar c}\right)^2\psi \\ &= E\psi.\end{aligned} \tag{12.12}$$

If the field $\boldsymbol{B}=0$, i.e. $\boldsymbol{\nabla}\times\boldsymbol{A}=0$ in the region where ψ is to be obtained, the solution to this equation can be written as

$$\psi = \psi_0 e^{i\varphi} \tag{12.13}$$

with the change of phase φ equal to

$$\varphi = \frac{e}{\hbar c}\int_{\text{path}}\boldsymbol{A}\cdot d\boldsymbol{x}. \tag{12.14}$$

Consider then the experimental arrangement shown in Fig. 12.1, where an electron beam from a source S is diffracted by two slits behind which there is a solenoid of sufficient length that we can neglect external fringing magnetic fields in the region where the electrons will be found.

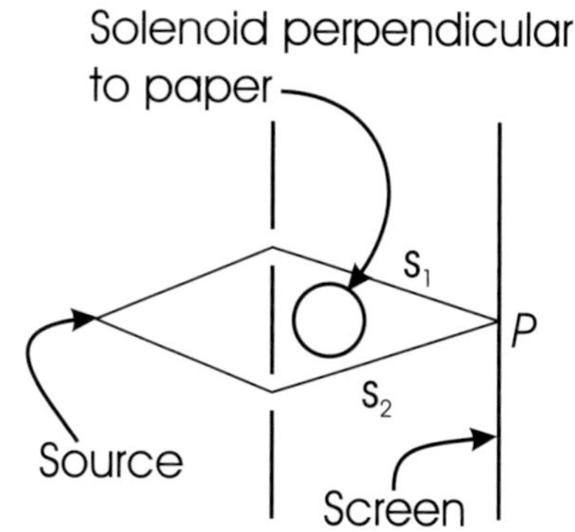

Figure 12.1: Two slit arrangement for observing the Aharonov–Bohm effect.

Thus, for the experimental arrangement shown, the wavefunction at P is expected to be ψ_0', representing the superposition of two free spherical waves emanating from slits 1 and 2 with phases shifted by $\boldsymbol{p}\cdot\boldsymbol{s}_1/\hbar$ for the wave from slit 1 and by $\boldsymbol{p}\cdot\boldsymbol{s}_2/\hbar$ for that from slit 2. However, even though the magnetic field $\boldsymbol{B}$ is confined to the solenoid, the vector potential $\boldsymbol{A}$ cannot be zero everywhere outside the solenoid, since the flux through any loop surrounding the solenoid is given by

$$\Phi = \int\boldsymbol{B}\cdot d\boldsymbol{S} = \oint_{\text{path}}\boldsymbol{A}\cdot d\boldsymbol{x}, \tag{12.15}$$

where $d\boldsymbol{S}$ is an element of area of the loop. Thus, there are additional phase shifts given by Eq. (12.14) for the two different paths,

$$\varphi_1 = \frac{e}{\hbar c}\int_{s_1} \boldsymbol{A}\cdot d\boldsymbol{x}, \qquad \varphi_2 = \frac{e}{\hbar c}\int_{s_2} \boldsymbol{A}\cdot d\boldsymbol{x}. \tag{12.16}$$

The interference pattern observed on the screen is determined by the phase difference of the two waves. If $|s_1| = |s_2|$, so that P is located at equal distances from the two slits, the phase difference $\delta\varphi$ is

$$\begin{aligned}\delta\varphi = \varphi_1 - \varphi_2 &= \frac{e}{\hbar c}\left(\int_{s_1} \boldsymbol{A}\cdot d\boldsymbol{x} - \int_{s_2} \boldsymbol{A}\cdot d\boldsymbol{x}\right)\\ &= \frac{e}{\hbar c}\oint \boldsymbol{A}\cdot d\boldsymbol{x} = \frac{e}{\hbar c}\,\Phi.\end{aligned} \tag{12.17}$$

Thus, even though there is no magnetic field along the paths of the electrons, they show interference effects that depend on, and vary with the vector potential $\boldsymbol{A}$, which therefore acquires a physical reality that was absent in classical mechanics. The effect occurs because the local phase at two space–time points is connected by the potential. The importance of potentials in quantum theory was stressed by Aharonov and Bohm(2) and the phase difference dependence on the vector potential $\boldsymbol{A}$ has been observed.(3)

It was shown by Berry(4) that the Aharonov–Bohm effect is a special case of a geometric phase present for any system transported adiabatically (slowly) around a closed circuit. The phase can be made visible by beating the system that is made to go around the circuit with the same system made to go straight to the detector; another way is to examine the superposition of stationary spin states of a system of particles, such as neutrons, before and after they have completed a closed path, as in a helical magnetic field. Berry makes a classical analogy to a body moving around a closed path on a curved surface. Thus, if a matchstick is taken around a closed path on a plane, without rotating it, it points in the same direction at the end as at the start. If, however, it is taken around a path on a sphere, such as from the North pole of the Earth to the equator, then taken to a different longitude and returned to the North pole, it ends up pointing along a different longitude at the end than at the start. Like the quantum mechanical effect, the change in direction only depends on geometrical factors.

12.3 Gauge Invariance for Non-Abelian Fields

The electromagnetic field is a simple example of a gauge field. If we are to include the weak interactions, then there are two problems that need to be solved. The first one is that both neutral and charged vector bosons are required. The second one is that the weak bosons W^+ and Z^0 are massive, whereas we showed that gauge invariance requires massless fields. We tackle the first problem in this section.

[3] R. G. Chambers, *Phys. Rev. Lett.* **5**, 3 (1960).
[4] V. M. Berry, *Proc. R. Soc. London* **A392**, 45 (1984); *Sci. Amer.* **259**, 46 (December 1988).

How can we generalize the gauge invariance of the single vector field (*Abelian* case) to theories of several non-commuting (*non-Abelian*) massless vector fields? An example would be a vector field with internal degrees of freedom, such as charge; suppose the photon had isospin unity and came in three charge states. In chapter 8 we saw that this generalization is possible for a global gauge transformation with the introduction of isospin. However, there we used a constant phase rotation, $U = \exp(-i\omega\hat{\boldsymbol{\alpha}}\cdot\boldsymbol{I})$, Eq. (8.20). The extension to a space–time dependent phase was formulated by Yang and Mills.[5] Their result lay dormant for many years because the strong interactions were described by the exchange of massive bosons (e.g., π, ρ) only some of which are vector particles; the weak interaction also requires very massive bosons, but no theory with such bosons was available.

Consider a vector field V, with three internal (not space) components, $V^{(a)}$ (e.g., isospin $= 1$, with $a = 1\ldots3$). In analogy to Eq. (8.20), we generalize Eq. (12.2) by introducing a different function $\xi^{(a)}$ for each internal (isospin) component of the vector field, $V^{(a)}$

$$\psi' \equiv U\psi = \exp[igI^{(a)}\xi^{(a)}(\boldsymbol{x},t)]\psi = \exp[ig\vec{I}\cdot\vec{\xi}(\boldsymbol{x},t)]\psi, \tag{12.18}$$

where a sum over repeated indices is assumed and the quantities $\vec{I}$ and $\vec{\xi}$ are vectors in the internal space. Thus there are now three separate space-and time-dependent phase angles $\xi^{(a)}$ and three non-commuting isospin vectors $I^{(a)}$. It is this non-commuting property that makes the theory non-Abelian. The difference between the local gauge invariance and the global one, described in chapter 8, can be stated in terms of a neutron and proton. These particles represent two states of different I_3. The choice of phase, $I_3 = +1/2$ for the protons, is a matter of convention, but it is the same everywhere in space. Since we deal with local actions and forces carried by fields, rather than actions at a distance, Yang and Mills[5] questioned whether two nucleons separated by a large distance can communicate their phase instantaneously. Stated another way, could the proton have $I_3 = 1/2$ in one place and $I_3 = -1/2$ in another one? They went on to investigate the consequences of a local invariance as we are doing here.

In analogy to the electromagnetic case, where the interaction can be obtained from the replacement of

$$\nabla_\mu = \left(\frac{1}{c}\frac{\partial}{\partial t}, -\boldsymbol{\nabla}\right) \quad \text{by} \quad D_\mu = (D_0, \boldsymbol{D}),$$

we define a generalized operator D_μ by

$$D_\mu = \nabla_\mu + ig\vec{I}\cdot\vec{V}_\mu \tag{12.19}$$

with $V_\mu = (V_0, \boldsymbol{V})$. Arrows are used for internal vectors. This equation is similar to Eq. (12.5) with different dimensions and with q replaced by g, and is part of the required generalization for a non-Abelian theory.

[5]C.N. Yang and R.L. Mills, *Phys. Rev.* **96**, 191 (1954).

We still need to generalize the gauge transformation, Eq. (12.4), to the fields V_μ. The isospin components of the vector fields, $V^{(a)}$, do not commute with each other, thus we rewrite Eq. (8.22) as

$$[I^{(a)}, I^{(b)}] = i\epsilon_{abc} I^{(c)}. \tag{12.20}$$

The symbol ϵ_{abc} is +1 if abc are normal-ordered or a cyclic variation thereof, and −1 otherwise.

To derive the appropriate gauge transformation, we ask that $D'_\mu \psi' = U D_\mu \psi$, since this condition assures invariance of the equations of motion under the gauge transformation, as we discussed earlier. With Eq. (12.19) we have

$$\begin{aligned}
D'_\mu \psi' &= (\nabla_\mu + ig\vec{I}\cdot\vec{V}'_\mu)\psi' \\
\psi' &= \exp(ig\vec{I}\cdot\vec{\xi}\,)\psi. \\
D'_\mu \psi' &= \psi[\nabla_\mu \exp(ig\vec{I}\cdot\vec{\xi}\,)] + \exp(ig\vec{I}\cdot\vec{\xi}\,)\nabla_\mu \psi \\
&\quad + ig\vec{I}\cdot\vec{V}'_\mu \exp(ig\vec{I}\cdot\vec{\xi}\,)\psi \\
&= \exp(ig\vec{I}\cdot\vec{\xi}\,)\{ig\vec{I}\cdot(\nabla'_\mu\vec{\xi}\,) + \nabla_\mu + ig\vec{I}\cdot\vec{V}'_\mu \\
&\quad + [ig\vec{I}\cdot\vec{V}'_\mu, \exp(ig\vec{I}\cdot\vec{\xi}\,)]\}\psi.
\end{aligned} \tag{12.21}$$

In order to make the evaluation of the commutator simpler in Eq. (12.21), we assume $\xi^{(a)}(\boldsymbol{x}, t)$ to be an infinitesimal and keep only linear terms in ξ. The commutator in Eq. (12.21) is then

$$\begin{aligned}
[ig\vec{I}\cdot\vec{V}'_\mu, 1 + ig\vec{I}\cdot\vec{\xi}\,] &= -ig^2 V'^{(a)}_\mu \epsilon_{abc}\xi^{(b)} I^{(c)} \\
&= -ig^2 \vec{V}'_\mu \cdot \vec{\xi} \times \vec{I} \approx -ig^2 \vec{V}_\mu \times \vec{\xi}\cdot\vec{I}.
\end{aligned} \tag{12.22}$$

Since we only keep linear terms in ξ and Eq. (12.22) already is linear, we have set $V'_\mu = V_\mu$ in the last equality of this equation. The equality $D'_\mu \psi' = U D_\mu \psi$ then leads to[6]

$$\vec{I}\cdot\vec{V}_\mu = \vec{I}\cdot\vec{V}'_\mu + \vec{I}\cdot\vec{\nabla}_\mu \xi - g\vec{V}_\mu \times \vec{\xi}\cdot\vec{I},$$

or

$$\vec{V}'_\mu = \vec{V}_\mu - \nabla_\mu \vec{\xi} + g\vec{V}_\mu \times \vec{\xi}\,. \tag{12.23}$$

This is the desired generalization of Eqs. (12.4); note the appearance of the coupling constant g in the additional term of Eq. (12.23).

[6]It is relatively straightforward to generalize this expression to massless particles with higher degrees of freedom. As shown earlier, a mass term would break gauge invariance. It is therefore important that the gauge field quanta represented by $V^{(a)}$ retain zero mass.

Once again, we see that the requirement of gauge invariance specifies the "minimal" interaction. In the quantum equation of motion we replace ∇_μ by D_μ. For instance, for the nonrelativistic Schrödinger equation, we have

$$\left(i\hbar\frac{\partial}{\partial t} - g\hbar c\vec{I}\cdot\vec{V}_0\right)\psi = \frac{1}{2m}(-i\hbar\boldsymbol{\nabla} - g\hbar\vec{I}\cdot\vec{V})^2\psi. \tag{12.24}$$

An important difference from charge couplings, which can vary for different members of an isomultiplet, is that the coupling strength g is the *same* for all isospin components of the vector field $\vec{V}_\mu$. Local gauge invariance imposes global gauge invariance and, in our example, requires isospin invariance of the theory with a *single* strength of coupling, g.

It is also of interest to examine the free field equations for the massless vector fields V_μ. We have already examined briefly the case of the electromagnetic field. There, the electric and magnetic fields $\boldsymbol{E}$ and $\boldsymbol{B}$ are invariant under the gauge transformations (12.3), so that the equations for the free fields also do not depend on the choice of gauge. Here, in contrast, the electric and magnetic fields of the non-Abelian theory cannot be gauge invariant. If we define them as in Eqs. (10.37) and (10.38), then under the gauge transformation of the vector field, $V^{(a)}$, we find with Eq. (12.23)

$$\begin{aligned}
\boldsymbol{E}'^{(a)} &= -\frac{1}{c}\frac{\partial \boldsymbol{V}'^{(a)}}{\partial t} - \boldsymbol{\nabla}V_0'^{(a)} \\
&= -\frac{1}{c}\frac{\partial \boldsymbol{V}^{(a)}}{\partial t} - \boldsymbol{\nabla}V_0^{(a)} - g\epsilon_{abc}\left[\frac{1}{c}\frac{\partial}{\partial t}(\boldsymbol{V}^{(b)}\xi^{(c)}) + \boldsymbol{\nabla}V_0^{(b)}\xi^{(c)}\right], \\
\boldsymbol{B}'^{(a)} &= \boldsymbol{\nabla}\times\boldsymbol{V}'^{(a)} = \boldsymbol{B}^{(a)} + g\epsilon_{abc}\boldsymbol{\nabla}\times(\boldsymbol{V}^{(b)}\xi^{(c)}).
\end{aligned} \tag{12.25}$$

Thus, the definitions of the non-Abelian fields $\boldsymbol{E}^{(a)}$ and $\boldsymbol{B}^{(a)}$ must be modified. The new definitions will involve the coupling constant g, and we will show that the appropriate expressions are

$$\begin{aligned}
\boldsymbol{E}^{(a)} &= -\frac{1}{c}\frac{\partial \boldsymbol{V}^{(a)}}{\partial t} - \boldsymbol{\nabla}V_0^{(a)} - g\epsilon_{abc}\boldsymbol{V}^{(b)}V_0^{(c)} \\
\boldsymbol{B}^{(a)} &= \boldsymbol{\nabla}\times\boldsymbol{V}^{(a)} - g\epsilon_{abc}\boldsymbol{V}^{(b)}\times V^{(c)}.
\end{aligned} \tag{12.26}$$

With these definitions, the relationship between the original and the gauge transformed fields are given by

$$\begin{aligned}
\boldsymbol{E}'^{(a)} &= \boldsymbol{E}^{(a)} + g\epsilon_{abc}\boldsymbol{E}^{(b)}\xi^{(c)} \\
\boldsymbol{B}'^{(a)} &= \boldsymbol{B}^{(a)} + g\epsilon_{abc}\boldsymbol{B}^{(b)}\xi^{(c)},
\end{aligned} \tag{12.27}$$

with an extra term similar to that in Eq. (12.23). $\boldsymbol{E}$ and $\boldsymbol{B}$ are not only vectors in space; they are also vectors in the internal (isospin) space. We denote this fact by superscripts or arrows, $\vec{\boldsymbol{E}}$ and $\vec{\boldsymbol{B}}$. The proof of Eq. (12.27) is straightforward, even if somewhat tedious. We shall show it for the electric field $\boldsymbol{E}^{(a)}$ and leave it as an exercise for the reader to prove it for the magnetic field $\boldsymbol{B}^{(a)}$. We use the isospin indices (a) here in order not to confuse isospin and normal vectors. With the definition (12.26), and Eq. (12.23), we find

$$\begin{aligned} \boldsymbol{E}'^{(a)} = \boldsymbol{E}^{(a)} &- \frac{g}{c}\frac{\partial}{\partial t}(\epsilon_{abc}\boldsymbol{V}^{(b)}\xi^{(c)}) - g\boldsymbol{\nabla}(\epsilon_{abc}V_0^{(b)}\xi^{(c)}) \\ &- g\boldsymbol{V}^{(b)}\epsilon_{abc}V_0^{(c)} + g\left[(\boldsymbol{V}^{(b)} + \boldsymbol{\nabla}\xi^{(b)} + \epsilon_{bde}g\boldsymbol{V}^{(d)}\xi^{(e)}\right. \\ &\left.\times\epsilon_{abc}\left((V_0^{(c)} - \frac{1}{c}\frac{\partial}{\partial t}\xi^{(c)} + \epsilon_{cfg}V_0^{(f)}\xi^{(g)}\right)\right]. \end{aligned} \tag{12.28}$$

The last term in square brackets can be simplified, especially because we are only keeping terms of first order in ξ, in accord with the derivation of Eq. (12.23). Together with the second-to-last term, this term becomes (we use vector signs for isospin)

$$\begin{aligned} &-\frac{g}{c}\boldsymbol{V}\times\frac{\partial\vec{\xi}}{\partial t} + g^2\boldsymbol{V}\times(\vec{V}_0\times\vec{\xi}) + g\boldsymbol{\nabla}\vec{\xi}\times\vec{V}_0 + g^2(\vec{\boldsymbol{V}}\times\vec{\xi}\,)\times\vec{V}_0 \\ &\quad = -g\left(\boldsymbol{V}\times\frac{1}{c}\frac{\partial\vec{\xi}}{\partial t} - \boldsymbol{\nabla}\vec{\xi}\times\vec{V}_0\right) + g^2(\vec{\boldsymbol{V}}\times\vec{V}_0)\times\vec{\xi}\,. \end{aligned} \tag{12.29}$$

Combining Eqs. (12.28) and (12.29), we find that the unwanted terms in Eq. (12.25) cancel and the relationship between $\vec{\boldsymbol{E}}'$ and $\vec{\boldsymbol{E}}$ becomes

$$\boldsymbol{E}'^{(a)} = \boldsymbol{E}^{(a)} + g\epsilon_{abc}\boldsymbol{E}^{(b)}\xi^{(c)}$$

or

$$\vec{\boldsymbol{E}}' = \vec{\boldsymbol{E}} + g\vec{\boldsymbol{E}}\times\vec{\xi}\,. \tag{12.30}$$

Thus, Eq. (12.27) is obtained for the electric field $\boldsymbol{E}^{(a)}$. The last term, proportional to $\vec{\boldsymbol{E}}\times\vec{\xi}$ is required by the non-Abelian nature of the theory, and comes about because of the non-commuting nature of the various isospin components. The similarity of the last term of Eq. (12.27) with that of Eq. (12.23) for the transformation of the vector field $V^{(a)}$ thus is not surprising, and is required. Moreover, the last term in Eq. (12.26), which is required to cancel the unwanted ones in Eq. (12.25) is quadratic in the vector field V_μ. The theory thus becomes nonlinear and this additional term has drastic consequences, which we now shall examine.

The coupling constant g in Eqs. (12.23) and (12.25) is similar to the charge e in quantum electrodynamics and is thus sometimes referred to as the "charge". The non-Abelian theory thus describes a "charged" field, in contrast to the "uncharged" or neutral electromagnetic field A_μ. To examine the consequences of this "charge", which is related to isospin, we look at the energy of the field, corresponding to the Hamiltonian. The energy density u is given by

$$u = \frac{1}{2}(\vec{\boldsymbol{E}}^2 + \vec{\boldsymbol{B}}^2) \qquad (12.31)$$

If we substitute Eqs. (12.26) into (12.31), we observe that the extra term in Eq. (12.26) leads to cubic and quartic self-interactions of the non-Abelian "free" field; examples are

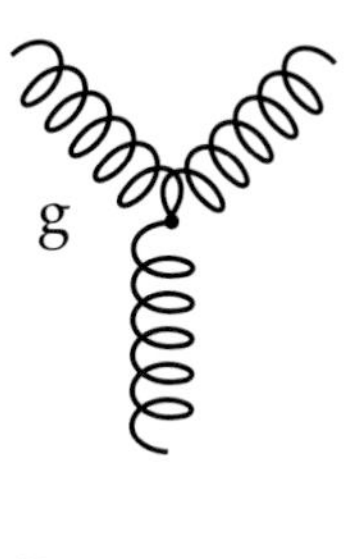

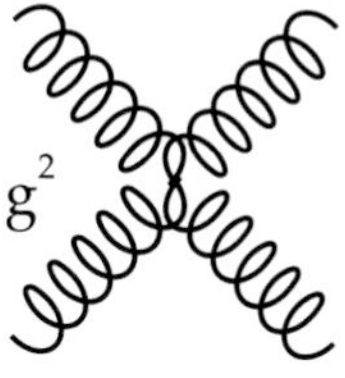

Figure 12.2: Feynman diagrams for self-interactions of a "charged" field.

$$\begin{aligned} &\text{cubic terms} \propto g\left(\frac{1}{c}\frac{\partial \vec{\boldsymbol{V}}}{\partial t} + \boldsymbol{\nabla}\vec{V}_0\right)\cdot(\vec{V}_0 \times \vec{\boldsymbol{V}}) \\ &\text{quartic terms} \propto g^2(\vec{V}_0 \times \vec{V})^2. \end{aligned} \qquad (12.32)$$

There is no free field! The gauge field $V^{(a)}$ and its quanta are "charged", thus the quanta interact directly with each other, unlike photons. The self-interactions are the cubic terms, proportional to the "charge" g, and the quartic ones proportional to g^2 in Eq. (12.32). Feynman diagrams for these interactions are shown in Fig. 12.2. The strengths of these interactions are given in terms of the unique coupling g. If g is the "charge" of the matter field, as q was in the electromagnetic case, then the gauge vector fields are seen to carry this "charge"; they are not "neutral."

Quantum chromodynamics (QCD) is a theory quite analogous to, but somewhat more general than what we have developed in this section. In QCD, the charge is called "color charge" and the massless vector gauge bosons are the colored gluons. The gluons, however, come in eight colors, not just three charges. The self-interactions are present and there is no free gluon field. Since the gluons are color-charged, they always interact with each other. Our model can be generalized to this situation.

12.4 The Higgs Mechanism; Spontaneous Symmetry Breaking

We saw in Section 12.1 that gauge theories require massless vector bosons. Any connection to a theory of weak interactions, where the vector bosons are very massive, therefore appears to be lost. A non gauge-invariant theory, however, leads to a multitude of problems, including infinities for physical quantities in second order of perturbation theory. The solution to this dilemma lies in "spontaneously" broken symmetries.

There are two kinds of symmetry-breaking. We have discussed the first one, namely an approximate symmetry, at some length. In this case, a small part of the Hamiltonian spoils the exact symmetry. An example is the breaking of exact isospin invariance by the electromagnetic (and weak) interaction(s), as discussed in Section 8.5. The second kind of symmetry breaking, often called "spontaneous", was not studied seriously until the 1960s.[(7)] Here the Hamiltonian that describes the dynamics of the system retains the full symmetry, but the ground state breaks it. This phenomenon can occur if the ground state of the Hamiltonian is degenerate; the choice of a particular state among the degenerate ones then breaks the symmetry. A well-known example is a ferromagnet. Although the Hamiltonian which describes the ferromagnet is rotationally invariant, a gnome walking along the domains of a given ferromagnet, with its spins aligned in a given direction, would certainly not realize it. For this reason the symmetry also is sometimes referred to as a "hidden symmetry." It is only when the gnome realizes that the spins of a ferromagnet could point in any direction of space that the rotational symmetry becomes apparent. For a given ferromagnet, the rotational symmetry is broken.

We have not yet discussed how a hidden symmetry can explain massive gauge bosons, and it seems at first sight that such an explanation is not possible. Goldstone[(8)] pointed out that a hidden symmetry will always have associated with it a massless field because no energy is required to shift from the chosen ground state to another degenerate one. In a ferromagnet these zero-mass excitations are (long wavelength) spin waves.

The appearance of zero-mass "Goldstone bosons" in a theory with spontaneous symmetry breaking might suggest that such theories have no connection to the weak interactions. However, through the efforts of Higgs,[(9)] Kibble[(10)], Weinberg, Salam, and others who persisted in their belief that hidden symmetries could be used, we now have a viable theory of electroweak interactions within the Standard Model. Before describing this theory in the following chapter, we explain how hidden symmetries can generate masses.

It is helpful to consider a specific example. To this end we introduce globally gauge invariant complex scalar (Higgs) fields ϕ and ϕ^*, which might represent scalar mesons H^+ and H^-. These fields can be considered to be combinations of two real fields, ϕ_1 and ϕ_2.

$$\phi = \frac{1}{\sqrt{2}}(\phi_1 + i\phi_2), \quad \phi^* = \frac{1}{\sqrt{2}}(\phi_1 - i\phi_2). \tag{12.33}$$

[7]M. Baker and S. L. Glashow, *Phys. Rev.* **128**, 2462 (1962).

[8]J. Goldstone, *Nuovo Cim.* **19**, 154 (1961); J. Goldstone, A. Salam, and S. Weinberg, *Phys. Rev.* **127**, 965 (1962).

[9]P.W. Higgs, *Phys. Lett.* **12**, 132 (1964), *Phys. Rev.* **145**, 1156 (1966).

[10]T.W.B. Kibble, *Phys. Rev.* **155**, 1554 (1967).

These scalar fields obey the Klein–Gordon equation, the relativistic generalization of the Schrödinger equation. For a free particle of mass m, this equation is the quantum mechanical translation of

$$E^2 = (pc)^2 + (mc^2)^2, \tag{12.34}$$

with

$$E \longrightarrow i\hbar\frac{\partial}{\partial t} \quad \boldsymbol{p} \longrightarrow -i\hbar\boldsymbol{\nabla}. \tag{12.35}$$

The Klein–Gordon equation for ϕ thus becomes

$$\left(\frac{1}{c^2}\frac{\partial^2}{\partial t^2} - \nabla^2 + \frac{m^2c^2}{\hbar^2}\right)\phi(\boldsymbol{x}, t) = 0. \tag{12.36}$$

The same equation holds for ϕ^*, so that there is an obvious symmetry here between the fields and quanta represented by ϕ and ϕ^* or ϕ_1 and ϕ_2. The solutions of Eq. (12.36) are plane waves

$$\phi \propto \exp\left(\frac{i\boldsymbol{p}\cdot x - Et}{\hbar}\right) \tag{12.37}$$

with $E = \pm\sqrt{(p^2c^2 + m^2c^4)}$.

The Hamiltonian from which the free Klein-Gordon equation can be obtained is

$$H_0 = \frac{1}{2}\int(\hbar^2\frac{\partial\phi^*}{\partial t}\frac{\partial\phi}{\partial t} + \hbar^2c^2\boldsymbol{\nabla}\phi^*\cdot\boldsymbol{\nabla}\phi + m^2c^4\phi^*\phi)d^3x, \tag{12.38}$$

The state of lowest absolute energy, which we will call the ground state, has $\boldsymbol{p} = 0$ and $E = mc^2$. If $m = 0$, then this state is a constant in both space and time with zero momentum and energy. In the presence of a (scalar) potential V the Klein–Gordon Hamiltonian and equation for ϕ can be written as

$$H = H_0 + \int V\phi d^3x, \tag{12.39a}$$

and

$$\left(\frac{1}{c^2}\frac{\partial^2}{\partial t^2} - \nabla^2 + \frac{V'}{(\hbar c)^2} + \frac{m^2c^2}{\hbar^2}\right)\phi(\boldsymbol{x}, t) = 0, \tag{12.39b}$$

where V' stands for differentiation of V with respect to ϕ, if V depends on ϕ. We can consider the mass as a constant potential and write

$$\left(\frac{1}{c^2}\frac{\partial^2}{\partial t^2} - \nabla^2 + U\right)\phi = 0, \tag{12.39c}$$

where U has the dimension of (length)$^{-2}$. The state of lowest energy occurs for $\phi = \text{constant} = 0$ if U is not zero.

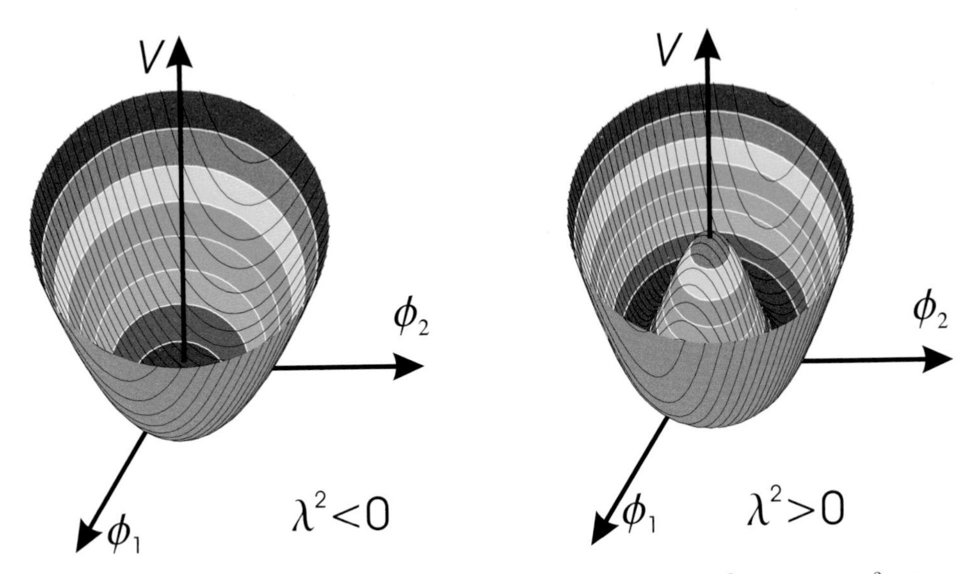

Figure 12.3: Potential V as a function of the complex field ϕ. Left: $\lambda^2 < 0$; Right: $\lambda^2 > 0$

We now assume that the masses of the quanta of the fields ϕ and ϕ^* are zero, but that the particles move in a potential which depends on the fields themselves. As a specific example, consider the Hamiltonian with a potential $V = -\lambda^2\phi^*\phi + \eta^2(\phi^*\phi)^2$, in which case the Klein–Gordon equation for ϕ is

$$\left(\frac{1}{c^2}\frac{\partial^2}{\partial t^2} - \nabla^2 - \lambda^2 + 2\eta^2(\phi^*\phi)\right)\phi = 0. \tag{12.40}$$

We show the potential V as a function of ϕ for both $\lambda^2 < 0$ and $\lambda^2 > 0$ in Fig. 12.3. In the latter case, the potential is often referred to as a Mexican hat.

If λ were imaginary or $\lambda^2 = -u^2 < 0$, then the state of lowest energy would occur when $\phi = \phi^* = 0$, as above; it is unique. For small deviations from this minimum we could expand ϕ and ϕ^* about $\phi = \phi^* = 0$; if we kept only linear terms, the term proportional to η^2 would not contribute and the Klein–Gordon equation for ϕ becomes

$$\left(\frac{1}{c^2}\frac{\partial^2}{\partial t^2} - \nabla^2 + u^2\right)\phi = 0. \tag{12.41}$$

This is just Eq. (12.35) for a free particle of mass $\hbar u/c$.

Since the quantities λ^2 and η^2 in Eq. (12.40) are positive definite, λ cannot be interpreted as being proportional to a mass. For this case, the minimum kinetic energy still occurs when the magnitude of ϕ, i.e. $|\phi| = constant$,

$$|\phi_{min}| \equiv \sqrt{\frac{\lambda^2}{2\eta^2}} = \frac{v}{\sqrt{2}}, \tag{12.42}$$

where $v = \lambda/\eta$. The minimum in momentum and energy is now degenerate. It lies anywhere on the circle of radius $v/\sqrt{2}$, or $\phi_{min} = \frac{v}{\sqrt{2}}e^{i\alpha}$, where α is an arbitrary phase. Since ground states are expected to be unique, we assume that nature picks out a particular one of these solutions; this choice "spontaneously" breaks the symmetry, that is, it hides the symmetry inherent in the equation of motion (12.40) and its counterpart for ϕ^*. This symmetry-breaking is similar to the ferromagnet, where the choice of lining up the magnet in a particular direction in space hides the symmetry. A particularly simple choice for the ground state is $\phi_1 = v/\sqrt{2}$ and $\phi_2 = 0$, or

$$\phi = \phi^* = \frac{v}{\sqrt{2}}. \tag{12.43}$$

We have taken a simple ground state; other choices can be made, but once made, the symmetry is lost. For small excitations in the continuum, we assume that ϕ and ϕ^* can be expanded about the "ground state" solutions,

$$\begin{aligned} \phi &= \frac{1}{\sqrt{2}}(v+R)e^{i\theta/v} \approx \frac{1}{\sqrt{2}}(v+R+i\theta) \\ \phi^* &= \frac{1}{\sqrt{2}}(v+R)e^{-i\theta/v} \approx \frac{1}{\sqrt{2}}(v+R-i\theta). \end{aligned} \tag{12.44}$$

The new fields are called R and θ. With the expansion about the asymmetric solution v, we have lost the symmetry between ϕ_1 and ϕ_2. The reason for choosing the exponential form for one of the fields will become clear shortly. If we substitute Eq. (12.44) into the Klein–Gordon Eq. (12.40), we obtain to first order in R and θ,

$$\begin{aligned} \left(\frac{1}{c^2}\frac{\partial^2}{\partial t^2} - \nabla^2 + 2\lambda^2\right) R(\boldsymbol{x},t) &= 0, \\ \left(\frac{1}{c^2}\frac{\partial^2}{\partial t^2} - \nabla^2\right) \theta(\boldsymbol{x},t) &= 0. \end{aligned} \tag{12.45}$$

Comparison with Eq. (12.36) shows that the particle corresponding to the field R has acquired a mass m,

$$m = \frac{\sqrt{2}\lambda\hbar}{c}, \tag{12.46}$$

whereas the field θ remains massless. The massless quantum of this field is called a Goldstone boson[8]. Such a (zero spin) boson always occurs when a global symmetry is broken spontaneously, as is done here by the choice of a specific ground state. On the other hand, the particle corresponding to the field R has acquired a mass. The mass is associated with the minimum energy required to reach an excited state for a radial oscillation in Fig. 12.4. The simple model used here began with two massless bosons described by the fields ϕ_1 and ϕ_2, or ϕ and ϕ^*, but the spontaneous symmetry-breaking led to a new field with a nonvanishing mass and a field that remains massless. We now have the background to incorporate local gauge invariance by coupling the massless spin-zero boson fields, ϕ and ϕ^*, to the electromagnetic field.

This coupling is completely specified by the requirement of gauge invariance, as discussed in Section 12.1. We will see that the Goldstone theorem is evaded in the example we are considering and that the Goldstone boson assists in giving a mass to the photon. First, we show that a charge and current can be associated with the Klein–Gordon equation (12.36). Indeed, we can define ρ and $\boldsymbol{j}$ by

$$\begin{aligned} \rho &= \frac{i}{c}q\left(\phi^*\frac{\partial\phi}{\partial t} - \phi\frac{\partial\phi^*}{\partial t}\right), \\ \boldsymbol{j} &= -iqc(\phi^*\boldsymbol{\nabla}\phi - \phi\boldsymbol{\nabla}\phi^*), \end{aligned} \tag{12.47}$$

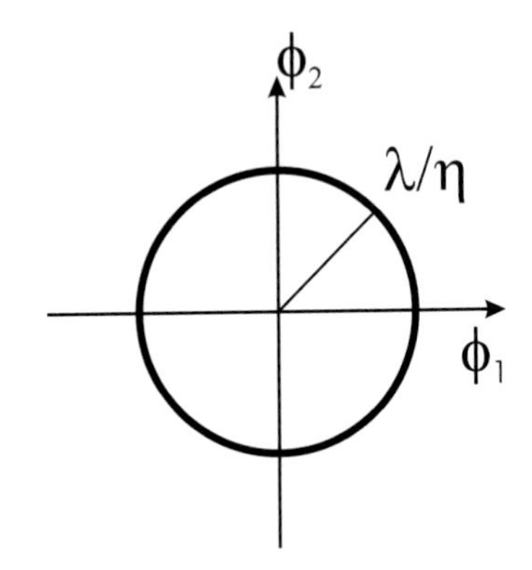

Figure 12.4: The minimum energy condition, Eq. (12.43).

such that the continuity equation, Eq. (10.51), holds

$$\frac{\partial\rho}{\partial t} + \boldsymbol{\nabla}\cdot\boldsymbol{j} = 0. \tag{10.51}$$

From Eq. (12.47) it follows that ϕ has the dimension of $(\text{length})^{-1}$. We assume that the charge associated with the field ϕ is the electrical charge q. The equation of motion is then given by Eq. (12.40) with D_0 and $\boldsymbol{D}$ substituted for $c^{-1}\partial/\partial t$ and $\boldsymbol{\nabla}$,

$$[D_0^2 - \boldsymbol{D}^2 - \lambda^2 + 2\eta^2(\phi^*\phi)]\phi = 0. \tag{12.48}$$

This equation and that for ϕ^* are invariant under the local gauge transformations

$$\begin{aligned} \phi &\longrightarrow \phi' = \exp[iQ\epsilon(\boldsymbol{x},t)]\phi \\ \phi^* &\longrightarrow \phi^{*\prime} = \exp[-iQ\epsilon(\boldsymbol{x},t)]\phi^* \\ A_\mu &\longrightarrow A'_\mu = A_\mu - \hbar c\nabla_\mu\epsilon(\boldsymbol{x},t). \end{aligned} \tag{12.49}$$

Again, if λ^2 were negative, the solution with zero momentum would occur for $\phi = 0$, and $-\lambda^2$ would be proportional to the mass of the spin-zero boson. However, with $\lambda^2 > 0$, such an interpretation is not possible. The spin-zero particles have no mass, and the lowest absolute value of energy is shifted as shown in Fig. 12.4. By picking one of the degenerate "ground" states as that of choice, we break the symmetry of the equations of motion and give masses to both the gauge field (i.e., to the photon) and one of the spin-zero field particles.

The ground state is shifted from that of the globally gauge invariant example. The lowest energy now occurs when

$$\phi^*\phi \equiv \frac{v'^2}{2} = \frac{\lambda^2}{2\eta^2} + \frac{q^2}{\hbar^2 c^2}\frac{1}{2\eta^2}(A_0^2 - \mathbf{A}^2), \tag{12.50}$$

and the expansion (12.44) can be used. Thus, for low excitations, where we keep only linear terms in R and θ, the equation of motion for ϕ becomes

$$\begin{aligned} \left(D_0^2 - \boldsymbol{D}^2 + 2\lambda^2 + \frac{3q^2}{\hbar^2c^2}(A_0^2 - \boldsymbol{A}^2)\right) R = 0, \\ \left(D_0^2 - \boldsymbol{D}^2 + \frac{q^2}{\hbar^2c^2}(A_0^2 - \boldsymbol{A}^2)\right)\theta + \frac{q}{\hbar c}\left(\frac{1}{c}\frac{\partial A_0}{\partial t} + \boldsymbol{\nabla}\cdot\boldsymbol{A}\right) = 0. \end{aligned} \tag{12.51}$$

The equation for R is somewhat more complicated than Eq. (12.45), but otherwise there are no surprises. As anticipated, the R field has acquired a mass of $\sqrt{2}\lambda\hbar/c$ and the θ field remains massless. Although there is an additional term that involves the electromagnetic field, it can be eliminated by invoking the Lorentz condition, Eq. (12.8). However, there is another change that has occurred, namely the electromagnetic field quantum has acquired a mass. To see this fact explicitly, we return to the charge and current, given by Eq. (12.47). With the substitution of $\partial/\partial t$ by cD_0 and $\boldsymbol{\nabla}$ by $\boldsymbol{D}$, we find, for instance

$$\rho = iq[\phi^* D_0\phi - \phi(D_0\phi)^*]. \tag{12.52}$$

If we substitute Eq. (12.44) and keep only first-order terms in R and θ, we obtain

$$\begin{aligned} \rho &= \frac{iq}{c}\left(\phi^*\frac{\partial\phi}{\partial t} - \phi\frac{\partial\phi^*}{\partial t} + \frac{2iq}{\hbar}A_0\phi^*\phi\right) \\ &\approx -v'\frac{q}{c}\frac{\partial\theta}{\partial t} - \frac{2q^2}{\hbar c}A_0 v'^2 - \frac{4q^2}{\hbar c}A_0 v' R. \end{aligned} \tag{12.53}$$

When this charge density is used in the equation of motion of A_0, Eq. (12.7), we find

$$\begin{aligned} \frac{1}{c^2}\frac{\partial^2 A_0}{\partial t^2} - \nabla^2 A_0 + \frac{2q^2}{\hbar c}v'^2 A_0 &= \frac{iq}{c}\left(\phi^*\frac{\partial\phi}{\partial t} - \phi\frac{\partial\phi^*}{\partial t}\right) \\ &\approx \frac{v'q}{c}\frac{\partial\theta}{\partial t} - \frac{4q^2}{\hbar c}A_0 v' R. \end{aligned} \tag{12.54}$$

By comparison with Eq. (12.10), we see that the new term $(2q^2/\hbar c)v'^2 A_0$ corresponds to a mass for the gauge "photon" of the electromagnetic field. The mass is

$$m = \sqrt{2\frac{\hbar}{c}\frac{q}{c}}v'.$$

This mass does not spoil gauge invariance because the original equation (12.48) is gauge invariant.

If we do not make the linear approximation, we can simplify the equation of motion for the scalar fields R and θ by taking advantage of the local gauge invariance, namely by choosing the transformation (local phase rotation)

$$\begin{aligned} \phi \longrightarrow \phi' &= \exp\left(\frac{-i\theta(\boldsymbol{x},t)}{v}\right)\phi \approx \frac{1}{\sqrt{2}}(v+R) \\ \phi^* \longrightarrow \phi^{*\prime} &= \exp\left(\frac{i\theta(\boldsymbol{x},t)}{v}\right)\phi^* \approx \frac{1}{\sqrt{2}}(v+R) \\ A_\mu \longrightarrow A'_\mu &= A_\mu + \frac{\hbar c}{qv}\nabla_\mu\theta. \end{aligned} \tag{12.55}$$

This choice fixes the gauge, called the unitary or U-gauge; in this gauge, the θ field has been eliminated and ϕ and ϕ^* are equal. The equation of motion (12.48) is gauge invariant. Hence, in the U-gauge in terms of the new field ϕ', Eq. (12.48) holds for ϕ' if we simultaneously change A_μ to A'_μ,

$$[{D_0'}^2 - \boldsymbol{D}'^2 - \lambda^2 + \eta^2(v'+R')^2](v'+R') = 0. \tag{12.56}$$

The θ field no longer appears in the equation of motion through our choice of gauge. Where has it gone? To answer this question, we examine the degrees of freedom in our problem. To begin with there were two internal degrees of freedom for the spin-zero boson, namely ϕ and ϕ^* with charges $\pm e$ and two directions of polarization (helicity) for the photon. At the end, we have one R field and a massive photon of spin one which has three degrees of polarization. Thus, there are again four degrees of freedom. When the photon acquired a mass it also gained a longitudinal degree of polarization that was not present originally. The zero mass Goldstone boson, θ, has been used up to provide this extra degree of freedom, and it is said that the gauge field "ate up" the Goldstone boson to become massive and gain its longitudinal polarization. Before eliminating θ by the choice of gauge, we had an extra, spurious degree of freedom.

This type of model, generalized to a non-Abelian vector gauge field, is successful in describing the weak interaction in conjunction with the electromagnetic one. We will consider this case in the next chapter.

In summary, we have shown that the imposition of gauge invariance determines the form of the interaction. Although gauge invariance requires massless fields or gauge quanta, spontaneous symmetry-breaking permits the introduction of masses in the theory, but at the cost of introducing extra (Higgs) fields, ϕ_1 and ϕ_2 or ϕ and ϕ^*. This method of generating masses is often called a "Higgs mechanism."[11] In our simple Abelian "toy model" the photon acquired a mass. Since the real photon is massless, this example is not realized in nature.

[11] M. J. Veltman, *Sci. Amer.* **255**, 76 (November 1986).

12.5 General References

A simple introduction and a good set of references can be found in a Resource letter by T.P. Cheng and L.-F. Li, *Am. J. Phys.* **56**, 586 (1988). Good general introductions to gauge theories can be found in Merzbacher, Ch. 17; R. Mills, *Am. J. Phys.* **57**, 493 (1989); and J.D. Jackson and L. B. Okun, *Rev. Mod. Phys* **73**, 663 (2001). There are also a number of books on the subject, of varying degrees of difficulty. In approximate order of increasing difficulty, they are: C. Quigg, *Gauge Theories of the Strong, Weak, and Electromagnetic Interactions*, 2nd ed., Westview Press, Boulder, Co, 1997; I. J. R. Aitchison and A. J. G. Hey, *Gauge Theories in Particle Physics, A Practical Introduction*, Institute of Physics Pub., 2003; M.W. Guidry, *Gauge Field Theories: An Introduction with Applications*, Wiley, (1991). U. Mosel, *Fields, Symmetries, and Quarks*, 2nd. ed, Springer Verlag, New York, 1999; E. Leader and E. Predazzi, *An Introduction to Gauge Theories and Modern Particle Physics*, Vol. I, Cambridge University Press, Cambridge, UK 1996. A complete listing of references up to 1988 can be found R. H. Stuewer, "Resource Letter GI-1, Gauge Invariance," *Am. J. Phys.* **56**, 586 (1988). A more recent introduction is R. Barlow, *Eur. J. Phys.* **11**, 45 (1990). A recent review and historical notes can be found in L. O/Raifeartaigh, N. Straumann, *Rev. Mod. Phys.* **72**, 1 (2000).

The Aharanov–Bohm effect is discussed in Y. Imry and R.A. Webb, *Sci. Amer.* **260**, 56 (April 1989). A more general treatment of geometric phases can be found in *Geometric Phases in Physics*, (A. Shapere and F. Wilczek, eds) World Sci., Teaneck, NJ, 1989; M. Peshkin and R. Tonomura, *The Aharonov-Bohm Effect*, Springer Verlag, New York, 1989.

Problems

12.1. Show that the Schrödinger equation with the Hamiltonian, Eq. (12.1) is invariant under the local gauge transformations, Eqs. (12.2) and (12.4).

12.2. Show that

$$
\begin{aligned}
D_0'\psi_q' &= U_Q D_0 \psi_q, \\
\boldsymbol{D}'\psi_{\mathrm{q}}' &= U_Q \boldsymbol{D}\psi_q.
\end{aligned}
$$

12.3. Show that Eqs. (12.10) are not invariant under local gauge transformations due to the mass term, but that Eq. (12.7) does satisfy this invariance.

12.4. Show that the definitions of $\boldsymbol{E}^{(a)}$ and $\boldsymbol{B}^{(a)}$, given by Eq. (12.26) lead to the gauge transformations given by Eq. (12.27).

12.5. Repeat the theoretical derivation of Section 12.3 for a non-Abelian vector field of isospin 1/2 rather than 1, as given in the text. Use Pauli matrices $\vec{\tau}$ instead of $\vec{I}$.

12.6. Derive the relationship between $\boldsymbol{B}'^{(a)}$ and $\boldsymbol{B}^{(a)}$, Eq. (12.27).

12.7. Choose a different ground state than that given by Eq. (12.43), and show that the physical consequences are the same.

12.8. Consider the example of a match carried without rotation in a closed loop along a line of fixed longitude on the surface of the Earth from the North pole to the equator, to a different longitude, and back to the North pole. Determine the geometrical factors which determine the change of orientation of the match from the beginning to the end of the loop.

12.9. * Outline an experiment to determine the Berry phase.

12.10. Show the correctness of Eq. (12.14) when $\boldsymbol{B} = 0$.

12.11. Show that the local gauge invariance of Eq. (12.7) requires Eq. (12.9).

12.12. Show that the Hamiltonian of Eqs. 12.40 and 12.41 is gauge invariant.

Chapter 13

The Electroweak Theory of the Standard Model

13.1 Introduction

In this chapter, we provide an introduction to the "standard model of the electroweak interactions." The subject is complex; for details, we refer the reader to texts and reviews listed at the end of the chapter.

The phenomenological current–current interaction as described in chapter 11 gives excellent agreement with low energy experiments. It is not, however, a well-defined theory. All calculations are performed to lowest order in the effective coupling constant, G_F, i.e., to lowest (first) order in perturbation theory. Computations of higher order, or of radiative corrections lead to physically meaningless infinities which we do not know how to remove. On the other hand, it is experimentally known that the higher order weak processes are extremely small. For instance, the mass difference between K_L and K_S is of second order in G_F and is tiny (Section 9.7). Consequently, the "theory" in the form given in chapter 11 is unsatisfactory. No adequate theory of the weak interaction, *alone*, has been discovered. This shortcoming was a challenge to solve a wider problem, and produced a more fundamental theory that describes the weak interactions unified with the electromagnetic one.

The electroweak theory is a major triumph. In 1879 James Clerk Maxwell formulated a unified theory of electricity and magnetism; and exactly one hundred years later Sheldon Glashow, Abdus Salam, and Steven Weinberg received the Nobel prize for a comparable achievement, the unification of the weak and electromagnetic forces.[(1)] As we saw in chapter 11, the two interactions are of the current–current form and require vector (also axial vector for the weak interactions) currents. Both the weak vector and electromagnetic currents are conserved. Despite these and other similarities, the two forces appear at first sight to have little in common. The electromagnetic force has an infinite range and is carried by massless photons, whereas the weak force has a very short range and is mediated by very heavy vector bosons.

[1] S. Weinberg, *Rev. Mod. Phys.* **52**, 515 (1980); A. Salam, *Rev. Mod. Phys.* **52**, 525 (1980); S. L. Glashow, *Rev. Mod. Phys.* **52**, 539 (1980).

Furthermore, the finiteness of the results of higher order electromagnetic processes requires gauge invariance, which in turn requires zero mass particles. How, then, can the massive bosons be incorporated? Glashow discussed this problem in 1961,[(2)] realized that it was a "principal stumbling block," and suggested neutral currents. Both Salam and Weinberg believed that the spontaneous symmetry breaking introduced in chapter 12 could provide masses to the "intermediate" bosons in a gauge theory which begins with massless particles, and that the resultant theory would be finite.[(3)] The mathematical proof of the fact that a finite theory, to all orders in the appropriate coupling constant, could be constructed in this manner did not come until later.[(4)] The proof makes use of the important point that the symmetry breaking does not spoil the gauge invariance of the theory. The electroweak theory was formulated and predicted the masses of the W^+ and Z^0 *before these particles were found experimentally.*

13.2 The Gauge Bosons and Weak Isospin

If a theory is to combine electromagnetism and the weak interactions, it must include the photon as well as the massive intermediate bosons. A gauge theory, as described in Section 12.4, requires that the charged bosons be supplemented by a neutral one in order to make an isospin multiplet and to have current conservation. The massive gauge bosons do not have strong interactions, thus the relationship between them is called "weak isospin". Since there are three charge states, corresponding to charged and neutral currents, the intermediate boson must have weak isospin 1. These particles are not necessarily those observed in nature. Nevertheless, we call the three gauge bosons W^+, W^-, and W^0; they have zero mass to begin with, as required by a gauge theory. In addition, there is a neutral "electromagnetic" field with a weak isospin singlet particle we shall call the B^0. Then, in the theory of Weinberg and Salam, the neutral particles associated with the weak and electromagnetic fields, and observed in nature, the Z^0 and the photon, are mixtures of the B^0 and the W^0,

$$\gamma = \cos\theta_W B^0 - \sin\theta_W W^0, \quad Z^0 = \cos\theta_W W^0 + \sin\theta_W B^0. \tag{13.1}$$

The mixing angle θ_W is called the Weinberg angle and can be determined from experiment, as we shall see. The photon and Z^0 are mixtures of a weak isospin singlet, B^0, and a component, W^0, of the isospin triplet W bosons. They are not simple particles, even though the photon has zero mass. We have seen in chapter 10 that the photon is a mixture of isospin zero and one for strong isospin; now we see that the photon is also a mixture of weak isospin zero and one. The Higgs mechanism is responsible for giving the W and Z bosons their masses. The masses of the Z^0

[2]S. L. Glashow, *Nucl. Phys.* **22**, 579 (1961).

[3]S. Weinberg, *Phys. Rev. Lett.* **19**, 1264 (1967); A. Salam, *Nobel Symposium*, No. 8 (N. Svartholm, ed.), Almqvist and Wiksell, Stockholm, 1968, p. 367.

[4]G. 't Hooft, *Nucl. Phys.* **B33**, 173 (1971), *B35*, 167 (1971); G. 't Hooft and M. Veltman, *Nucl. Phys.* **B44**, 189 (1972), **B50**, 318 (1972).

and the W^+ (or W^-) need no longer be the same because of the mixing of neutral particles.

In chapter 7 we saw that each type of lepton (electron, muon, tau) and its associated neutrino are separately conserved to a very good approximation. It thus makes sense to discuss separately each lepton pair, that is each "family" consisting of a charged lepton and its associated neutrino. The basic theory contains massless leptons, so that the charged lepton and its neutrino have the same mass. Weinberg introduced[5] weak isospin doublets to characterize each family of leptons. For instance, we can write for the electron and its neutrino

$$|I, I_3\rangle = |1/2, 1/2\rangle = \nu_{e_L}, \quad |I, I_3\rangle = |1/2, -1/2\rangle = e_L, \tag{13.2}$$

where the subscript L will be explained shortly. This formalism is analogous to that used for the neutron and proton, Eq. (8.13). We use the same notation, because we do not believe that it will create any confusion. As for the nucleon, the particle with the largest charge (zero here) is taken to be that of $I_3 = 1/2$. Furthermore, as in Eq. (8.14) or (8.30), we can write

$$q = I_{3l} + \frac{Y_l}{2}, \tag{13.3}$$

where Y_l is the "weak hypercharge" of the lepton; it follows that $Y_l = -1$. Actually, Weinberg introduced the "doublet" above for the combination of vector and axial vector currents that occurs in beta decay and the other weak interactions studied in chapter 11. This combination is $\mathcal{V} - \mathcal{A}$ for leptons and $\mathcal{V} - g_A\mathcal{A}$ for hadrons; it is usually called "left-handed", and this handedness is maintained by the weak interaction. According to the present evidence (chapter 11) the neutrino is almost purely left-handed, because the mass of the neutrino is very small. The right handed component plays no role in a first order interactions, since the neutrino only has weak interactions. Thus, it has its spin aligned antiparallel to its momentum. Similarly, only the left-handed component of the electron enters in first order weak interactions. Earlier, in Section 9.3, Eq. (9.34), we saw that the electron emitted in beta decay and the negative muon emitted in π^- decay (Problem 9.16) are polarized in a direction opposite to their momenta, like the neutrino in Fig. 7.2. These particles are said to be left-handed because an angular momentum opposite in direction to the velocity is like a left-handed screw-type motion. Both the electron and the neutrino have non-zero mass; thus they cannot be purely left-handed and cannot be fully polarized; their polarization for a velocity v is $-v/c$, as given by Eq. (9.34). For the neutrino, this is just about unity, since the mass is (probably) less than $1eV$. The right-handed (R) partner of the electron does not couple to the weak current, but only to the electromagnetic current. The electromagnetic current conserves parity and therefore couples to both left-handed and right-handed electrons with equal strength.

[5] S. Weinberg, *Phys. Rev. Lett.* **19**, 1264 (1967), *Phys. Rev.* **D11**, 3583 (1975).

For the neutrino, we have to distinguish between Dirac and Majorana types. For a Dirac neutrino, the right-handed partner takes no part in the weak interaction. On the other hand, for a Majorana neutrino, the right-handed partner is, in many ways, like a Dirac antineutrino and thus does participate, just like an antineutrino.. For the weak interactions, the appropriate weak isospin doublet is E_L with

$$E_L = \begin{pmatrix} \nu_e \\ e \end{pmatrix}_L \tag{13.4}$$

and with the isospin components for ν_{e_L} and e_L given by Eq. (13.2). Similar pairings are made for the muon and its Dirac neutrino and for the tau and the tau Dirac neutrino. For isospin invariance to hold, the masses of the charged leptons must be equal to those of the associated neutrinos. Moreover, the electromagnetic interaction does not favor the left over the right, so that there must also be right-handed (spin parallel to momentum) components for the charged leptons, e_R, μ_R, τ_R. The masses of the right- and left-handed components must be identical since we can turn a massive left-handed electron or neutrino into a right-handed one by a frame transformation (Fig. 11.10). On the other hand, since $I_{3R} = 0$, for the Gell–Mann–Nishijima charge relation, Eq. (13.3), to hold we need $Y_{lR} = -2$, whereas $Y_{lL} = -1$.

For the quarks, a weak isospin can also be introduced. Again, we have three families, and in a parallel fashion to the leptons, we write

$$f_L = \begin{pmatrix} u \\ d' \end{pmatrix}_L, \quad m_L = \begin{pmatrix} c \\ s' \end{pmatrix}_L, \quad h_L = \begin{pmatrix} t \\ b' \end{pmatrix}_L, \tag{13.5}$$

with weak third components of isospin $= \frac{1}{2}$ and $-\frac{1}{2}$ for upper and lower components, respectively. In Eq. (13.5) we have used the notation f, m, h for feather-, medium-, and heavy-weight quarks. Since all quarks have finite masses, there are also right-handed isospin singlets,

$$u_R, d_R, c_R, s_R, t_R, b_R. \tag{13.6}$$

In order to have the usual charge relation, Eq. (13.3), we require

$$\begin{aligned} Y_{fL} &= Y_{mL} = Y_{hL} = \tfrac{1}{3}, \\ Y_{uR} &= Y_{cR} = Y_{tR} = \tfrac{4}{3}, \\ Y_{dR} &= Y_{sR} = Y_{bR} = -\tfrac{2}{3}. \end{aligned} \tag{13.7}$$

A bit of history: Initially only the u, d, and s quarks were hypothesized to explain the known mesons and baryons. Cabibbo noted[6] that the weak currents of hadrons could be interpreted in terms of an isospin doublet:

$$\vec{J}_{\text{quark}} \sim g(\overline{u}, \overline{d'})\vec{I}\begin{pmatrix} u \\ d' \end{pmatrix} \tag{13.8}$$

[6] N. Cabibbo, *Phys. Rev. Lett.* **10**, 531 (1963).

where $\vec{I}$ represents the three-component isospin matrix (see Sect. 8.5),

$$d' = d\cos\theta_C + s\sin\theta_C, \tag{13.9}$$

and θ_C is called the Cabibbo angle.

Cabibbo showed that assuming $\theta_C \approx 0.22$ allowed for a natural description of the relative strengths of weak decays involving $u \leftrightarrow d$ versus $u \leftrightarrow s$ quark transitions (see Eq. 11.60.) In the notation of Eq 11.59:

$$V_{ud}^2 = \cos^2\theta_C. \tag{13.10}$$

As example, take the beta decay of the neutron. The weak current J_w^h of Eq. (11.61) that changes a neutron into a proton or a down quark into an up one conserves strangeness, $\Delta S = 0$. In terms of quarks, we can identify this current as that operating between the two members of the featherweight isospin $\frac{1}{2}$ doublet. This identity is analogous to that for leptons, where the weak current also stays within a single family. On the other hand, the decay

$$K^-(\overline{u}s) \longrightarrow \pi^0(\overline{u}u)e^-\nu_e,$$

which connects an s to a u quark, is also within the f family.

A main issue was left unexplained by the scheme proposed by Cabibbo: The decay

$$K^0(\overline{d}s) \longrightarrow \mu^+\mu^-$$

appears much more supressed than other weak decays. This supression is common to all flavor changing decays driven by the neutral part of the weak current. Glashow, Iliopoulos, and Maiani (GIM)[7] introduced the charmed quark as the left-handed partner of the strange quark in order to eliminate the strangeness-changing weak neutral currents. With the presence of charm, the s and c quarks belong to a weak isospin $\frac{1}{2}$ doublet and the medium-weight family must be orthogonal to the light-weight one, so that s' is

$$s' = s\cos\theta_C - d\sin\theta_C. \tag{13.11}$$

The cancellation of strangeness-changing neutral currents now can be shown as follows: The neutral current occurs for the left-handed doublet and therefore for d' and s' and not for d and s. In any reaction, it is the *sum* of the matrix elements

[7]S. L. Glashow, J. Iliopoulos, and L. Maiani, *Phys. Rev* **D2**, 1285 (1970).

for the two families that contributes. For the neutral current, J^{nc}, this sum is

$$\begin{aligned}
&\langle d\cos\theta_C + s\sin\theta_C|J^{nc}|d\cos\theta_C + s\sin\theta_C\rangle \\
&\qquad + \langle s\cos\theta_C - d\sin\theta_C|J^{nc}|s\cos\theta_C - d\sin\theta_C\rangle \\
&\quad = \langle d|J^{nc}|d\rangle(\cos^2\theta_C + \sin^2\theta_C) \\
&\qquad + \langle s|J^{nc}|s\rangle(\cos^2\theta_C + \sin^2\theta_C) \\
&\qquad + \langle s|J^{nc}|d\rangle(\sin\theta_C\cos\theta_C - \sin\theta_C\cos\theta_C) \\
&\qquad + \langle d|J^{nc}|s\rangle(\cos\theta_C\sin\theta_C - \sin\theta_C\cos\theta_C) \\
&\quad = \langle d|J^{nc}|d\rangle + \langle s|J^{nc}|s\rangle.
\end{aligned} \tag{13.12}$$

There is thus no contribution for any process of a neutral current connecting the strange and down quarks, or a neutral strangeness-changing current. The contribution of this current is canceled by the symmetry between the second and first families introduced by the charmed quark.

The existence of the c, t, and b quarks was later confirmed and their properties are now fairly well known. Moreover, presently all direct evidence for CP violation can be explained by a phase in the CKM matrix.[8]

13.3 The Electroweak Interaction

In this section, we concentrate on the interaction terms of the electroweak theory to demonstrate the unity of the weak and electromagnetic interactions. We first need to write the currents in a transparent manner. In the notation of chapter 11 [see e.g., Eq. (11.37)], we can write,

$$\begin{aligned}
j_{\mu,\text{em}}(e) &= \psi_e^* \mathcal{V}_{\mu,\text{em}} \psi_e \\
&= \psi_{e_L}^* \mathcal{V}_{\mu,\text{em}} \psi_{e_L} + \psi_{e_R}^* \mathcal{V}_{\mu,\text{em}} \psi_{e_R},
\end{aligned} \tag{13.13}$$

where we have generalized the operators 1 and $\boldsymbol{p}/m$ or $\boldsymbol{v}_{\text{op}}$ by the relativistic $\mathcal{V}_{\mu,\text{em}}$ with $\mu = 0\ldots3$;

$$V_{0,\text{em}} = 1 \quad \text{and} \quad V_{i,\text{em}} = v_{i,\text{op}}$$

for $i = 1, 2, 3$ or x, y, z in the nonrelativistic limit. We have also replaced the wavefunctions ψ_α and ψ_β of Eq. (11.37) by ψ_e, ψ_{e_L} or ψ_{e_R}. The break-up into left-handed (L) and right-handed (R) currents in Eq. (13.13) is just a formal change without importance for the electromagnetic interaction. For the weak interaction, however, the break-up becomes useful. As discussed earlier, the weak current contains both a vector and an axial vector operator in the combination $\mathcal{V}_\mu - \mathcal{A}_\mu$, with $A_0 = \boldsymbol{\sigma}\cdot\boldsymbol{p}/m$ and $\mathcal{A}_i = \sigma_i$ in the nonrelativistic approximation. Instead of using the two operators between the complete wavefunction ψ_e, we can sandwich the operator

[8] T.E. Browder, R. Faccini, *Annu. Rev. Nucl. Part. Sci.* **53**, 353 (2003).

$\mathcal{V}_\mu$ alone between the left-handed and the right-handed wavefunctions:

$$\psi^*_{e_L}\mathcal{V}_\mu\psi_{e_L} = \tfrac{1}{2}\psi^*_e(\mathcal{V}_\mu - \mathcal{A}_\mu)\psi_e, \tag{13.14}$$
$$\psi^*_{e_R}\mathcal{V}_\mu\psi_{e_R} = \tfrac{1}{2}\psi^*_e(\mathcal{V}_\mu + \mathcal{A}_\mu)\psi_e. \tag{13.15}$$

Both sides of Eqs. (13.14) and (13.15) are equivalent, but the left-hand side provides a convenient description of the weak currents of leptons. The weak charged currents of the leptons are purely left-handed. Equation (13.14) consequently permits us to write this current of the electron and its neutrino as

$$j^{ch}_{\mu,\mathrm{wk}} = \psi^*_{e_L}\mathcal{V}_\mu\psi_{v_L} + \psi^*_{v_L}\mathcal{V}_\mu\psi_{e_L}, \tag{13.16}$$

These currents can be expressed more succinctly in terms of the weak isospin notation, which also brings out their symmetry under this operation; we use matrices for this description.

In terms of the isospin and matrix notation, the currents for the doublet E_L, Eq. (13.16), can be written as

$$\begin{aligned}
j_{\mu,\mathrm{em}} &= \psi^*_{E_L}\mathcal{V}_\mu\left(I_3 + \frac{\mathcal{Y}}{2}\right)\psi_{E_L} + \psi^*_{E_R}\mathcal{V}_\mu\frac{\mathcal{Y}}{2}\psi_{E_R}\\
&= \psi^*_{E_L}\mathcal{V}_\mu I_3\psi_{E_L} + \psi^*_E\mathcal{V}_\mu\frac{\mathcal{Y}}{2}\psi_E\\
j^{ch}_{\mu,\mathrm{wk}} &= \psi^*_{E_L}\mathcal{V}_\mu 2I_-\psi_{E_L} + \psi^*_{E_L}\mathcal{V}_\mu 2I_+\psi_{E_L},\\
j^{nc}_{\mu,\mathrm{wk}} &= \psi^*_{E_L}2I_3\psi_{E_L}
\end{aligned} \tag{13.17}$$

with $E_R = e_R$ and $E = E_L + E_R$. In this equation $\mathcal{Y}$ is a weak hypercharge operator with eigenvalues given by Y_l, i.e. $\mathcal{Y}|l\rangle = Y_l|l\rangle$, where $|l\rangle$ is a lepton. We have also introduced the isospin raising and lowering operators and matrices I_+ and I_-, and I_3

$$\begin{aligned}
I_\pm &= \frac{1}{2}(I_1 \pm iI_2),\\
I_+ &= \frac{1}{2}\begin{pmatrix}0 & 1\\ 0 & 0\end{pmatrix}, \quad I_- = \frac{1}{2}\begin{pmatrix}0 & 0\\ 1 & 0\end{pmatrix},\\
I_3 &= \frac{1}{2}\begin{pmatrix}1 & 0\\ 0 & -1\end{pmatrix}.
\end{aligned} \tag{13.18}$$

The properties of these raising and lowering operators are

$$\begin{aligned}
I_+|\nu_L\rangle = 0, &\quad 2I_+|e_L\rangle = |\nu_L\rangle,\\
I_-|e_L\rangle = 0, &\quad 2I_-|\nu_L\rangle = |e_L\rangle.
\end{aligned} \tag{13.19}$$

The coefficients in front of the isospin and the hypercharge operators in Eq. (13.17) are fixed by Eq. (13.3) and the properties of the electron (charge $= -e$) and neutrino (charge $=$ zero).

These currents can now be introduced in the equation of motion. As we learned in chapter 12, the form of this equation is dictated by gauge invariance and the interaction is generated thereby. The leptons e and ν are light, so that a non-relativistic equation of motion cannot be used except at very low energies for the electron. The Schrödinger equation must be modified, since it has first-order time derivatives but second-order space derivatives; it is, consequently, not relativistically invariant. The problem was solved by Dirac, who invented an equation that is first order in both space and time. We will not introduce the Dirac equation, but use the fact that the electroweak theory is most important at high energies where the lepton masses can be neglected. The generalization of the Schrödinger equation for a particle of zero mass was found by Weyl, and it can be written down easily for a massless electron by noting that the only observables are spin and momentum. The simplest equation consequently is

$$\frac{i\hbar\partial\psi}{\partial t} = \boldsymbol{\sigma} \cdot \boldsymbol{p}\psi = -i\hbar\boldsymbol{\sigma} \cdot \boldsymbol{\nabla}\psi.$$

This equation has the right form but is not general enough for the electroweak theory. We generalize it by introducing the vector $\mathcal{V}_\mu = (V_0, \boldsymbol{V})$ as in Eq. (13.13) and writing[9]

$$i\hbar V_0 \frac{\partial\psi}{\partial t} = \boldsymbol{V} \cdot \boldsymbol{p}c\psi = -i\hbar c\boldsymbol{V} \cdot \boldsymbol{\nabla}\psi. \tag{13.20}$$

The vector $\mathcal{V}_\mu$ is related to the spin of the fermion. In the presence of the electromagnetic field, gauge invariance dictates that the derivatives $\partial/\partial t$ and $\boldsymbol{\nabla}$ be replaced by D_0 and $\boldsymbol{D}$, so that Eq. (13.20) becomes

$$\begin{aligned} i\hbar c V_0 D_0 \psi &= i\hbar V_0 \left(\frac{\partial}{\partial t} + \frac{ieA_0}{\hbar}\right)\psi \\ &= -i\hbar c\boldsymbol{V} \cdot \boldsymbol{D}\psi = -i\hbar c\boldsymbol{V} \cdot \left(\boldsymbol{\nabla} - \frac{ie\boldsymbol{A}}{\hbar c}\right)\psi. \end{aligned} \tag{13.21}$$

This equation applies to particles of charge e. In the electroweak theory, we need gauge invariance with respect to both the isoscalar B field and the isovector W fields. The latter are non-commuting and thus non-Abelian. We also have both the neutrino and electron to consider. The left-handed components of the electron and the neutrino couple to both the isovector weak, W, and isoscalar, B, fields, whereas the right-handed component of the electron couples only to the isosinglet field B, since it does not participate in the weak interaction. Consider the equation of motion for e_R first. For a free electron, Eq. (13.20) can be used. In the presence of the $\boldsymbol{B}$ field, we have from Eq. (13.21)

$$i\hbar V_0 \left(\frac{\partial}{\partial t} - i\frac{g'}{\hbar}B_0\frac{\mathcal{Y}}{2}\right)\psi_{e_R} = -i\hbar c\boldsymbol{V} \cdot \left(\boldsymbol{\nabla} + i\frac{g'}{\hbar c}\boldsymbol{B}\frac{\mathcal{Y}}{2}\right)\psi_{e_R}, \tag{13.22}$$

[9]Merzbacher, Ch. 24.

with a coupling strength g' to the B field. If we multiply this equation on the left by ψ^*_{eR}, we can write the expectation value of the interaction terms as

$$\begin{aligned} H_{\text{int}}(e_R) &= \frac{-g'}{2}\psi^*_{eR}(V_0 B_0 - \boldsymbol{V}\cdot\boldsymbol{B})\mathcal{Y}\psi_{eR} \\ &= -\frac{g'}{2}\psi^*_{eR}g_{\mu\nu}\mathcal{V}_\mu B_\nu \mathcal{Y}\psi_{eR}, \end{aligned} \tag{13.23}$$

where a sum is implied over repeated indices. For the left-handed component of the electron and neutrino we use the isospin doublet ψ_{E_L}; it gets coupled to both the $\boldsymbol{W}$ fields with strength g and the $\boldsymbol{B}$ field with strength g', so that equation (13.21) becomes

$$\begin{aligned} i\hbar V_0 &\left(\frac{\partial}{\partial t} - i\frac{g'}{\hbar}B_0\frac{\mathcal{Y}}{2} - i\frac{g}{\hbar}\vec{I}\cdot\vec{W}_0\right)\psi_{E_L} \\ &= -i\hbar c\boldsymbol{V}\cdot\left(\boldsymbol{\nabla} - i\frac{g'}{\hbar c}\boldsymbol{B}\frac{\mathcal{Y}}{2} - i\frac{g}{\hbar c}\vec{I}\cdot\vec{\boldsymbol{W}}\right)\psi_{E_L}. \end{aligned} \tag{13.24}$$

Again, we multiply on the left by $\psi^*_{E_L}$ and isolate the interaction terms, which, in the shorthand notation of Eq. (13.23), are

$$H_{\text{int}}(E_L) = -\psi^*_{E_L}g_{\mu\nu}\mathcal{V}_\mu\left(\frac{g'}{2}B_\nu\mathcal{Y} + g\vec{I}\cdot\vec{W}_\nu\right)\psi_{E_L}. \tag{13.25}$$

Table 13.1: Eigenvalues of the Weak Hypercharge. The eigenvalues can be translated to more massive families.

Particle or Multiplet	E_L	e_R	f_L	u_R	d_R
Y	-1	-2	$1/3$	$4/3$	$-2/3$

The hypercharge operator $\mathcal{Y}$ commutes with the isospin operator I, and has eigenvalues Y given by Table 13.1. At this stage it appears that we have introduced two new coupling constants, g and g'.

However, because we know the strength of the coupling of the electron to the electromagnetic field, only one of them is unknown. To see the relationship between the coupling constants g, g' and e, we write out the two interaction terms Eqs. (13.23) and (13.25) in terms of the physical fields $W^\pm$, Z^0 and A. The charged current interaction part is

$$H_{\text{int}}(\text{charged currents}) = \frac{-g}{\sqrt{2}}(\psi^*_{\nu L}g_{\mu\nu}\mathcal{V}_\mu W_\nu{}^{(+)}\psi_{eL} + \psi^*_{eL}g_{\mu\nu}\mathcal{V}_\mu W_\nu{}^{(-)}\psi_{\nu L}), \tag{13.26}$$

with $W^{(\mp)} = \frac{1}{\sqrt{2}}(W_1 \pm iW_2)$. The neutral current interaction is

$$\begin{aligned}H_{\text{int}}(\text{neutral currents}) = \frac{1}{2} g_{\mu\nu} \mathcal{V}_\mu [&\psi^\dagger_{E_L} (g \sin\theta_W 2I_3 - g' \cos\theta_W \mathcal{Y}) A_\nu \psi_{E_L} \\ &- \psi^*_{eR} g' \cos\theta_W A_\nu \mathcal{Y} \psi_{eR} \\ &- \psi^\dagger_{E_L} (g \cos\theta_W 2I_3 + g' \sin\theta_W \mathcal{Y}) Z_\nu \psi_{E_L} \\ &- \psi^*_{eR} g' \sin\theta_W Z_\nu \mathcal{Y} \psi_{eR}].\end{aligned} \tag{13.27}$$

The terms multiplied by A_ν represent the electromagnetic interaction. Since we know that the electromagnetic coupling constant is e, it follows that

$$g \sin\theta_W = -g' \cos\theta_W = e. \tag{13.28}$$

Therefore the electroweak neutral current interaction is

$$\begin{aligned}H_{\text{int}}(\text{neutral currents}) = -g_{\mu\nu} \Big[&e\psi^*_e \mathcal{V}_\mu A_\nu \psi_e + \frac{g}{2\cos\theta_W} (\psi^*_{\nu_L} Z_\nu \mathcal{V}_\mu \psi_{\nu_L} \\ &- \psi^*_{e_L} \mathcal{V}_\mu Z_\nu \psi_{e_L} + 2\psi^*_e \mathcal{V}_\mu Z_\nu \sin^2\theta_W \psi_e) \Big].\end{aligned} \tag{13.29}$$

It is totally determined by the requirement of gauge invariance. Eq. (13.28) connects g, g', θ_W, and e; since e is known, there remains but a single unknown parameter, which we take to be $\sin^2\theta_W$. Not only is a neutral current required by the global isospin symmetry, but the strengths of both the neutral and charged current interactions are given in terms of only two constants, e, and θ_W. Indeed, the weak interaction coupling is proportional to e, as advertised.

So far, the leptons and gauge bosons are massless, but we know from chapter 12 how to remedy this situation. There are four field quanta, the W^+, W^-, Z^0, and γ, three of which must acquire masses. Thus we introduce a doublet of scalar fields, of isospin 1/2, with charges +1 and 0

$$\mathbf{\Phi} \equiv \begin{pmatrix} \phi^{(+)} \\ \phi^{(0)} \end{pmatrix}. \tag{13.30}$$

Both $\phi^{(0)}$ and $\phi^{(+)}$ are complex fields,

$$\phi^{(0)} = \frac{1}{\sqrt{2}}(\phi_1 + i\phi_2) \quad \text{and} \quad \phi^{(+)} = \frac{1}{\sqrt{2}}(\phi_3 + i\phi_4),$$

so that there are four real fields. It follows from Eq. (13.3) that the hypercharge of these fields must be $Y_\phi = 1$. The equation of motion for these fields is Eq. (12.40),

$$\left(\frac{1}{c^2}\frac{\partial^2}{\partial t^2} - \nabla^2 - \lambda^2 + 2\eta^2 (\mathbf{\Phi}^\dagger \mathbf{\Phi}) \right) \mathbf{\Phi} = 0. \tag{13.31}$$

As in Section 12.5, we expand about the minimum, $\mathbf{\Phi}_0$,

$$\mathbf{\Phi}_0 = \begin{pmatrix} 0 \\ v/\sqrt{2} \end{pmatrix} \tag{13.32}$$

$$\mathbf{\Phi} \approx \begin{pmatrix} 0 \\ \frac{1}{\sqrt{2}}(v+H) \end{pmatrix} \exp(i\vec{\theta} \cdot \vec{I}/v). \tag{13.33}$$

There are no linear terms in the expansion of $\phi^{(+)}$. The reason for choosing $\phi^{(0)}$ to have a non-vanishing vacuum expectation value is connected to having the photon remain massless. The choice of ground state, Eq. (13.32), breaks both isospin and Y-symmetry, since

$$e^{i\epsilon\mathcal{Y}}\mathbf{\Phi}_0 \approx (1 + i\epsilon\mathcal{Y} + \cdots)\mathbf{\Phi}_0 \neq \mathbf{\Phi}_0 \tag{13.34}$$

(i.e., $Y_v \neq 0$). However, the ground state maintains charge conservation and is invariant under the combined isospin and hypercharge transformation for which the generator is the charge operator, that is

$$\begin{aligned} e^{i\epsilon Q}\mathbf{\Phi}_0 &= (1 + i\epsilon Q + \cdots)\mathbf{\Phi}_0 \\ &= [1 + i(I_3 + \mathcal{Y}/2) + \cdots]\mathbf{\Phi}_0 = \mathbf{\Phi}_0, \end{aligned} \tag{13.35}$$

since the charge of $\mathbf{\Phi}^{(0)}, q_v = 0$. It follows that the photon will remain massless, whereas the three degrees of freedom associated with the other three ($\vec{\theta}$) fields will be "eaten up" to supply the extra longitudinal polarizations of the three gauge fields, $W^{\pm}, W^0$, which acquire masses. This procedure is similar to the Abelian case.

As in Section 12.4, we transform to the U-gauge

$$\begin{aligned} \mathbf{\Phi} &\longrightarrow \mathbf{\Phi}' = \exp\left(\frac{-i\vec{\theta}\cdot\vec{I}}{v}\right)\mathbf{\Phi} = \begin{pmatrix} 0 \\ \frac{1}{\sqrt{2}}(v+H) \end{pmatrix}, \\ B_\mu &\longrightarrow B'_\mu = B_\mu, \\ W_\mu &\longrightarrow W'_\mu = W_\mu, \\ \psi_{E_L} &\longrightarrow \psi'_{E_L} = \exp\left(\frac{-i\vec{\theta}\cdot\vec{I}}{v}\right)\psi_{E_L}, \\ \psi_{eR} &\longrightarrow \psi'_{eR} = \psi_{eR}. \end{aligned} \tag{13.36}$$

As a result, we find that the (Higgs) field, H, and the W fields have acquired masses; that of the Higgs boson, the quantum of the H field, is

$$m_H = \frac{\sqrt{2}\lambda\hbar}{c}. \tag{13.37}$$

The Higgs boson has yet to be discovered. The charged W fields have a mass

$$m_{W^\pm} = \frac{gv}{2\hbar c}. \tag{13.38}$$

Finally, from the combinations given by Eq. (13.1), we obtain for the neutral Z^0 meson a mass

$$m_{Z^0} = m_{W^\pm}\sqrt{1+\frac{g'^2}{g^2}} = \frac{m_{W^\pm}}{\cos\theta_W}. \tag{13.39}$$

We also find from Eq. (11.20)

$$\frac{G_F}{\sqrt{2}} = \frac{e^2\hbar^2}{8m_W^2\sin^2\theta_W c^2} \tag{13.40}$$

or

$$m_W c^2 = \frac{37.3\text{GeV}}{\sin\theta_W}. \tag{13.41}$$

What we have done here can be repeated for the muon and tau lepton families, for which the couplings will be identical.

There are other aspects of the electroweak theory we shall not discuss here. For instance, the leptons must still acquire masses. These masses can be provided by Yukawa-type couplings to the Higgs boson. We shall also not treat the hadron sector. The development in terms of quarks parallels that of the leptons, except that all quarks have right-handed components since they all have masses. The interaction for hadrons such as protons can then be obtained by adding the contributions from the quarks that make up their structure.

13.4 Tests of the Standard Model

The most crucial and direct test of the Weinberg-Salam (WS) theory was the discovery of the $W^+(W^-)$ and Z^0. As pointed out in the last section, the standard model predicts not only the existence, but also relationships between the couplings and the masses of these gauge bosons.(10) They can be produced in reactions such as $p+\overline{p}\to W^+ + \cdots$ or $\to Z^0 + \cdots$, and detected through their decays, such as

$$\begin{aligned} W^+ &\longrightarrow e^+\nu_e & \qquad Z^0 &\longrightarrow e^+e^- \\ &\longrightarrow \mu^+\nu_\mu & &\longrightarrow \mu^+\mu^-. \end{aligned}$$

The $p\overline{p}$ experiment was carried out in the 1980's and, although the production rate was very small, the detection signal was clean. In the case of the W^+, a single charged lepton was detected at a large transverse momentum ($\lesssim m_W c/2$) relative to the production axis and with a large energy, $m_W c^2/2$, and no other high energy particle was observed.

For the Z^0 two charged particles were detected at an energy $m_Z c^2/2$. A particularly clean event is shown in Fig. 13.1. This figure shows the energy deposited

[10] See a resource letter by J.L. Rosner, *Am. Jour. Phys.* **71**, 302 (2003).

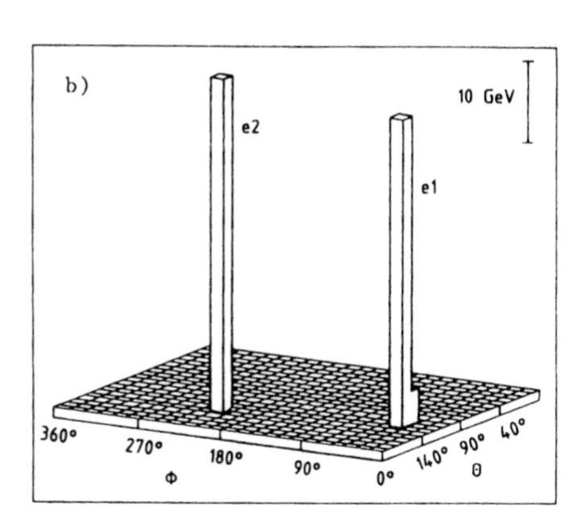

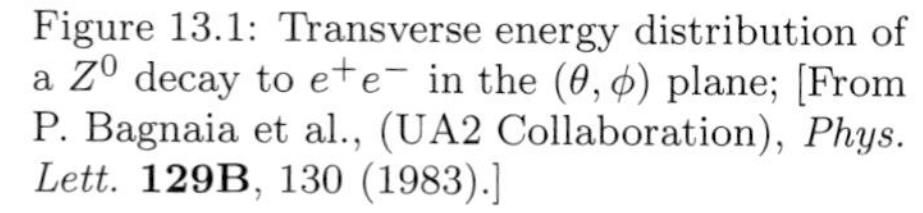
Figure 13.1: Transverse energy distribution of a Z^0 decay to e^+e^- in the (θ, ϕ) plane; [From P. Bagnaia et al., (UA2 Collaboration), *Phys. Lett.* **129B**, 130 (1983).]

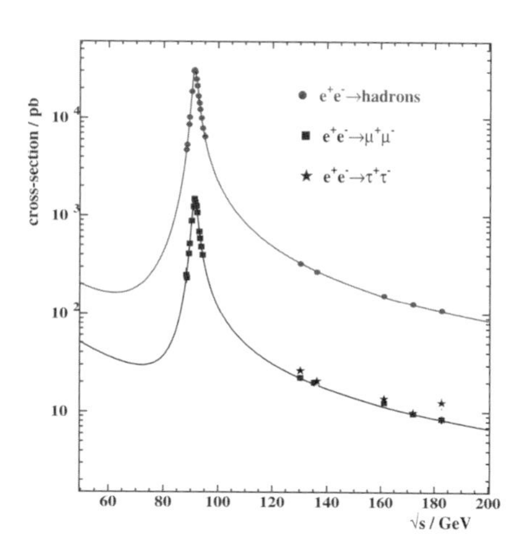

Figure 13.2: Cross section versus invariant mass for e^+e^- to hadrons and lepton pairs. [From *Rev. Mod. Phys.* **71**, s96 (1999).]

in a calorimeter as a function of polar and azimuthal angles relative to the proton axis. Both W^+ and Z^0 were discovered at the predicted masses. In the 1990's with data from e^+e^- colliders from SLAC and particularly LEP at CERN the masses and the widths of the W^+ and Z^0 were determined with high accuracy. The best measurements at present give $m_W c^2 = 80.425 \pm 0.038$ GeV, and $m_z c^2 = 91.1876 \pm 0.0021$ GeV.[11] Figure 13.2 shows measurements of the cross sections for e^+e^- to hadrons and lepton pairs produced by the experiments at the LEP collider. The Z resonance is clearly visible.

In chapter 11, we focused on charged current weak interaction processes; these are incorporated into the electroweak theory. In the last section we stressed the neutral current parts of the electroweak theory because it is here that we find important tests of the theory. Unlike their charged current counterparts, the neutral current weak interactions are not simply left-handed or of the form $\mathcal{V} - \mathcal{A}$, but, because of the mixing of weak and electromagnetic interactions (B^0 and W^0), involve a mixture of $\mathcal{V} - \mathcal{A}$ and $\mathcal{V} + \mathcal{A}$. The mixture is determined by θ_W. For instance, for the electron and its neutrino, the effective weak neutral current J^n_μ times its coupling to leptons can be written as (see Eq. (13.29))

$$g_{\text{eff}} J^n_\mu = \frac{g}{4\cos\theta_W}[\psi^*_{\nu_e}(\mathcal{V}_\mu - \mathcal{A}_\mu)\psi_{\nu_e} - \psi^*_e \mathcal{V}_\mu (1 - 4\sin^2\theta_W)\psi_e + \psi_{\text{e}}{}^* \text{A}_\mu \psi_{\text{e}}]. \quad (13.42)$$

The interference between the right- and left-handed components can be observed by measuring, for example, the forward-backward asymmetry in $e^+e^- \to \mu^+\mu^-$. As shown in Fig. 13.3, the agreement with the prediction of the Standard Model is very good.

[11] PDG.

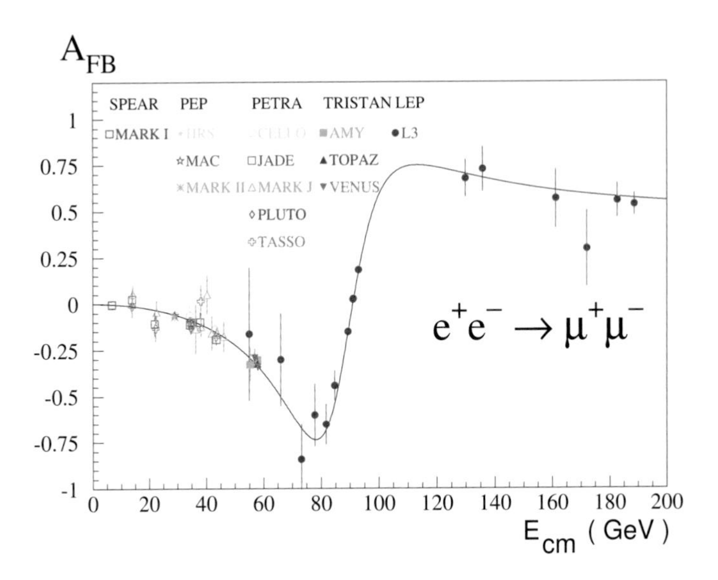

Figure 13.3: Forward-backward asymmetry in $e^+e^- \to \mu^+\mu^-$ and $e^+e^- \to \tau^+\tau^-$ as a function of energy from different experiments. The interference of γ and Z contributions gives the asymmetry variation with energy, as indicated by the standard model curve. [Courtesy of P. Grannis; see *Rev. Mod. Phys.* **71**, s96 (1999).]

As another example, consider the purely leptonic process, depicted in Fig. 13.4, $\nu_\mu e \to \nu_\mu e$, for energies $m_e c^2 \ll E \ll m_W c^2$. The cross section calculated in the Born approximation is

$$\sigma = \frac{G_F^2 m_e E_{\nu,\mathrm{Lab}}}{2\pi\hbar^4 c^2}\left[(1-2\sin^2\theta_W)^2 + \frac{4}{3}\sin^4\theta_W\right]. \tag{13.43}$$

By contrast, that for $\overline{\nu}_\mu e \to \overline{\nu}_\mu e$, is

$$\sigma = \frac{G_F^2 m_e E_{\nu,\mathrm{Lab}}}{2\pi\hbar^4 c^2}\left[4\sin^4\theta_W + \frac{1}{3}(1-2\sin^2\theta_W)^2\right]. \tag{13.44}$$

Clearly, these cross sections can be used to determine the Weinberg angle θ_W and test the theory.

For electron neutrino scattering from electrons, there occurs an interference between the neutral and charged weak current interactions, since both diagrams of Fig. 13.5 can contribute. The calculated cross sections are

$$\begin{aligned}\sigma(\nu_e e \to \nu_e e) &= \frac{G_F^2 m_e E_{\nu,\mathrm{Lab}}}{2\pi\hbar^4 c^2}\left[(1+2\sin^2\theta_W)^2 + \frac{4}{3}\sin^4\theta_W\right],\\ \sigma(\overline{\nu}_e e \to \overline{\nu}_e e) &= \frac{G_F^2 m_e E_{\nu,\mathrm{Lab}}}{2\pi\hbar^4 c^2}\left[4\sin^4\theta_W + \frac{1}{3}(1+2\sin^2\theta_W)^2\right].\end{aligned} \tag{13.45}$$

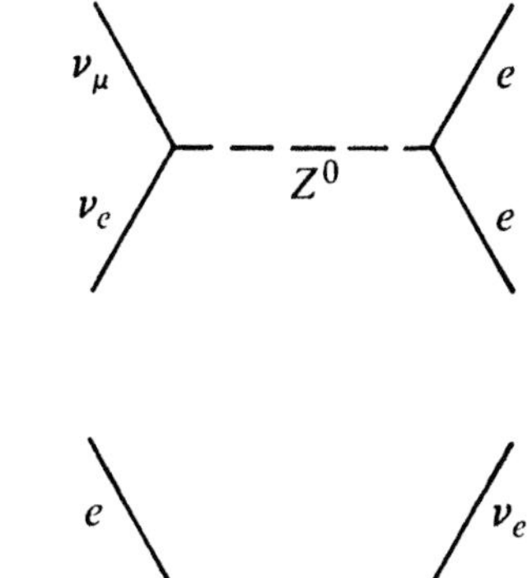

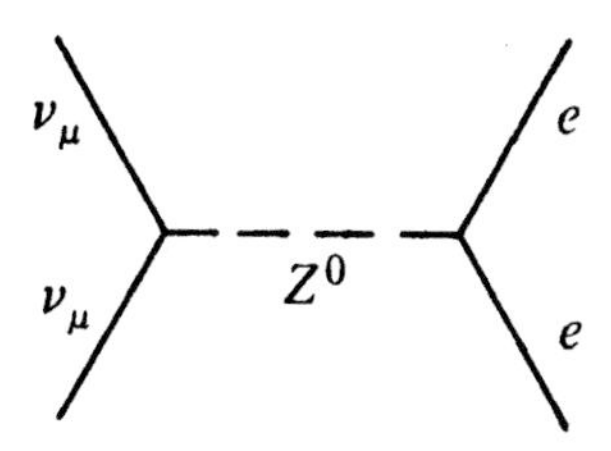

Figure 13.4: Feynman diagram for the elastic scattering of muon neutrinos on electrons.

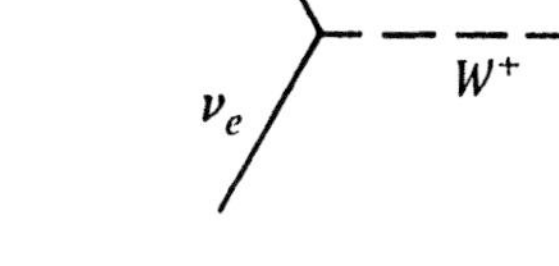

Figure 13.5: Feynman diagrams for elastic scattering of electron neutrinos from electrons.

Similarly, we can compare elastic neutrino and antineutrino scattering on the proton. At zero momentum transfer (forward scattering) there is no interference between the vector and axial vector currents and the form factors are unity, so that the ratio of antineutrino or neutrino scattering cross section due to neutral currents to that for charged currents becomes

$$\frac{\sigma(\overline{\nu}_\mu p \to \overline{\nu}_\mu p)}{\sigma(\overline{\nu}_\mu p \to \mu^+ n)} = \frac{\sigma(\nu_\mu p \to \nu_\mu p)}{\sigma(\nu_\mu n \to \mu^- p)} = \frac{1}{4\cos^2\theta_C}\left(\frac{(1-4\sin^2\theta_W)^2 + g_A^2}{1+g_A^2}\right), \tag{13.46}$$

where $g_A \approx 1.27$.

The Weinberg angles obtained from the different processes described above have been extracted and found to agree with the expectations from the standard model.[(12)]

The Higgs structure of the theory can be tested by measuring the parameter ρ_0,

$$\rho_0 = M_W^2/(M_Z^2 \cos^2\theta_W \hat{\rho}), \tag{13.47}$$

which is unity in the WS theory, but could be different if the Higgs bosons were not in a doublet. The factor $\hat{\rho}$ corresponds to calculated radiative corrections and differs from unity by about one percent. Experimentally, the parameter ρ_0 is found to be 1 to within less than one tenth of one percent.[11] A major confirmation of

[12]Measurements have become precise enough that radiative corrections need to be taken into consideration. See SLAC E158 collaboration, *Phys. Rev. Lett.* **95**, 081601 (2005). However, measurements from scattering of ν and $\overline{\nu}$ off nuclei disagree by more than 2σ with the expectations; see NuTev collaboration, *Phys. Rev. Lett.* **88**, 091802 (2002).

the scheme presented in this chapter would be the observation of the Higgs particle itself. This is expected to happen at the LHC experiments.

We have described experiments performed at the highest energies available in accelerators, but many tests of the standard model have been performed at lower energies as well.[13] Many other tests of the electroweak sector of the standard model, that we don't have room to present here, have been performed, and, in brief, excellent agreement has been found over a broad range of phenomena and energies. This is a remarkable achievment. However, after enjoying this triumph for a short while, physicists are now again searching for discrepancies with the model. This is based on a conviction that there must be a theory deeper than the standard model that could explain some of its apparently arbitrary parameters (like the values of the masses of the particles and the Weinberg angle[14]) and give a deeper insight on how nature works.

13.5 References

A somewhat technical, but complete book devoted to the weak interactions and the electroweak theory is E. D. Commins and P. H. Bucksbaum, *Weak Interactions of Leptons and Quarks*, Cambridge University Press, New York, 1983; for up-to-date information, see PDG. A recent resource letter by J.L. Rosner, *Am. Jour. Phys.* **71**, 302 (2003) provides a good list of references for different subtopics.

There are also a number of historical reviews. The Nobel lectures given by S. L. Glashow, *Rev. Mod. Phys.* **52**, 539 (1980), A. Salam, *Rev. Mod. Phys.* **52**, 525 (1980), and S. Weinberg, *Rev. Mod. Phys.* **52**, 515 (1980) give insight into the development of the theory. Others are S. Weinberg, *Sci. Amer.* **231**, 50 (July 1974); P.Q. Hung and C. Quigg, Science, **210**, 1205, (1980); M.K. Gaillard, *Comm. Nucl. Part. Phys.* **9**, 39 (1980); G. 't Hooft, *Sci. Amer.* **242**, 104 (June 1980); H. Georgi, *Sci. Amer.* **244**, 40 (April 1981), M.A.B. Beg and A. Sirlin, *Phys. Rep.* **88**, 1 (1982); S. Weinberg, *Phys. Today*, **39**, 35 (August 1986); P. Langacker and A.K. Mann, *Phys. Today*, **42**, 22 (Dec. 1989). An historical perspective on the discovery of neutral currents is given by P. Galison, *Rev. Mod. Phys.* **55**, 477 (1983) and by F. Sciulli, *Prog. Part. Nucl. Phys.* (D. H. Wilkinson, ed.) **2**, 41 (1979). Other reviews of neutral currents include T. W. Donnely and R.D. Peccei, *Phys. Rep.* **50**, 1 (1979); C. Baltay, *Comm. Nucl. Part. Phys.* **8**, 157 (1979); J.E. Kim et al., *Rev. Mod. Phys.* **53**, 211 (1981), P. Q. Hung and J. J. Sakurai, *Annu. Rev. Nucl. Part. Sci.* **31**, 375 (1981); L.M. Sehgal, *Prog. Part. Nucl. Phys.*, (A. Faessler, ed.) **14**, 1 (1985); D.B. Cline, ed. *Weak Neutral Current; the Discovery of the Electro-Weak Force.*Addison-Wesley, Reading, MA .

Tests of the electroweak theory are described in C. Kiesling, *Recent Experimental Tests of the Standard Theory of Electroweak Interactions*, Springer Tracts in Modern

[13]See for example, N. Severijns and M. Beck, *Rev. Mod. Phys.* **77**, (2006).
[14]R. N. Cahn, *Rev. Mod. Phys.* **68**, 951 (1996).

Physics. **112**, Springer, New York, 1988, in P. Langacker, *Comm. Nucl. Part. Phys.* **19**, 1 (1989); M. Martinez, R. Miquel, L. Rolandi, and R. Tenchini, *Rev. Mod. Phys.* **71**, 575 (1999).

The search for the W and Z gauge bosons is reviewed in the Nobel lectures of C. Rubbia and S. van der Meer, *Rev. Mod. Phys.* **57**, 689 and 699 (1985); D. B. Cline, C. Rubbia, and S. van der Meer, *Sci. Amer.* **246**, 48 (March 1982); J. Ellis et al., *Annu. Rev. Nucl. Part. Sci.* **32**, 443 (1982); and E. Radermacher, *Prog. Part. Nucl. Phys.*, (A. Faessler, ed.) **14**, 23 (1985); a book on the subject is P. Watkins, *Story of the W and Z*, Cambridge University Press, cambridge, 1986;

Measurements of $\sin^2\theta_W$ are discussed in PDG. Recent measurements are reported in the references quoted under footnote 12.

Measurements of ρ and other quantities related to the Higgs sector are discussed in PDG. See also P.Q. Hung and J.J. Sakurai, *Annu. Rev. Nucl. Part. Sci.* **31**, 375 (1981); M.K. Gaillard, *Comm. Nucl. Part. Phys.* **9**, 39 (1980); R. N. Cahn, *Rep. Prog. Phys.* **52**, 389 (1989).

Problems

13.1. Verify the assignments in Eq. (13.7).

13.2. Generalize the Cabibbo mixing formalism to three families and show that no neutral "bottomness"-changing current occurs.

13.3. Verify Eq. (13.17).

13.4. Verify Eq. (13.19).

13.5. (a) Determine the matrices I_1 and I_2. (See Eq. (13.18.))

(b) Show that $[I_i, I_j] = i\epsilon_{ijk}I_k$, where $\epsilon_{ijk} = +1$ for 1,2,3 or a cyclic permutation thereof, but -1 for an aniticyclic permutation.

(c) Find $I_1|E_L\rangle$.

13.6. Verify Eq. (13.27).

13.7. Show that $m_H = \sqrt{2}\hbar\lambda/c$, Eq. (13.37).

13.8. Show that $m_W = gv/2\hbar c$, Eq. (13.38).

13.9. Determine the charged and neutral weak current interactions of the featherweight quark family $\begin{pmatrix} u \\ d_W \end{pmatrix}$, u_R, and d_R by analogy to the development for the lepton family E_L and e_R. Make use of Table 13.1. That is, find the analogue of Eqs. (13.26) and (13.27) or (13.29) for the quark sector.

13.10. Make use of the quark structure of the proton and neutron and the solution to problem 13.9 to determine

(a) The weak coupling of the Z^0 to the proton and neutron at zero momentum transfer. Assume that the vector current is conserved, but that the axial current is "renormalized" by the multiplicative factor g_A of chapter 11.

(b) Express your answer to part (a) in terms of isospin operators or matrices for the isospin doublet $\begin{pmatrix} p \\ n \end{pmatrix}$. Hint: The interaction can be written in terms of an isoscalar and the third component, I_3, of an isovector.

13.11. Determine the ratios (for $\sin^2\theta_W = 0.225$)

(a) $\dfrac{\sigma(\nu_\mu e \to \nu_\mu e)}{\sigma(\overline{\nu_\mu} e \to \overline{\nu_\mu} e)}$,

(b) $\dfrac{\sigma(\nu_e e \to \nu_e e)}{\sigma(\overline{\nu_\mu} e \to \overline{\nu_\mu} e)}$, and compare to experiments.

13.12. Verify Eq. (13.46).

13.13. Estimate the order of magnitude of the parity-admixture in the $1S$ atomic hydrogen wavefunction due to the neutral weak current interaction between the electron and proton.

Chapter 14

Strong Interactions

All good things must come to an end. In chapters 10 and 11 we have seen that the electromagnetic and the weak interactions of leptons at low energies were characterized each by a single coupling constant. Furthermore, nature uses only one type of current for the electromagnetic interaction, a vector; and two for the weak interaction, a vector and an axial vector. The situation with the strong interaction at low energies is much more complicated. In the nucleon-nucleon interaction, for instance, almost every term allowed by general symmetry principles appears to be required to fit the experimental data. In addition, at energies $\lesssim$1 GeV the phenomenological strong interactions do not provide any evidence that they are goverened by one universal coupling constant. Consider, for instance, Figs. IV.1 and IV.3. The strength of the interaction of the pion with the baryon is described by the constant $f_{\pi NN^*}$ in the first case, and by $f_{\pi NN}$ in the second one. The two constants are not identical. The interaction of the pion with pions is characterized by yet another constant. Since many hadrons exist, a large number of coupling constants occur. The corresponding interactions are all called strong because they all are about two orders of magnitude stronger than the electromagnetic one. However, they are not exactly alike. While some connections among the coupling constants can be derived by using symmetry arguments, these relations are only approximate, and many constants appear at present to be unrelated. The situation resembles a jigsaw puzzle in which it is not known if all pieces are present and in which the shape of some pieces cannot be seen clearly.

We found in chapter 13 how a clever idea led to a simplification and a unification of the weak and electromagnetic interactions into a single interaction with only one coupling constant. Is it possible that the strong interaction at these lower energies masks simplicity that sets in at higher energies? In the past several decades a theory has been developed in which the strong interactions at sufficiently high energies (more precisely, high momentum transfers or short distances) are just as simple as the electroweak theory and are described by a single coupling constant. This theory is called quantum chromodynamics[1] (QCD) and has received over-

[1] I.R. Aitchison, *An Informal Introduction to Gauge Field Theories*, Cambridge University Press, Cambridge, 1982; I.R. Aitchison and A.J.G. Hey, *Gauge theories in particle physics:*

Table 14.1: ANALOGIES OF QCD AND QED.

	QED	QCD
Fundamental particles	Charged leptons	Quarks
Gauge quanta	Photon	Gluons
Source of interaction	Charge	Color charge
Basic strength	$\alpha = e^2/\hbar c$	α_s

whelming support from experiments carried out at the highest energies available. QCD has features that are analogous to those of the theory for the electron and the electromagnetic field (quantum electrodynamics), but has also important differences therefrom. The analogy is shown in Table 14.1. The fundamental particles of QED are the leptons, those of QCD the quarks. The gauge quanta of the electromagnetic field is the photon and that of the QCD field the gluon. The strength of the electromagnetic field is determined by the electric charge, that of QCD by the color charge. QCD, like QED, is described by a single coupling constant the square of which, α_S, is the analogue of the fine structure constant $\alpha = e^2/\hbar c$. There are also differences between QED and QCD. Whereas the photon is electrically neutral and therefore transfers no charge, the gluon carries color. This difference is crucial. The strong squared coupling constant, α_S, depends on momentum transfer or distance probed, and becomes progressively weaker as the former is increased and the latter becomes smaller. By contrast, the squared coupling constant of QED, α, has only a weak dependence on momentum transfer, and increases as that momentum grows. For QCD at very high momentum transfers perturbation theory becomes practical so that the theory can be easily tested in this realm. The theory is said to be "asymptotically free" in that the coupling constant is predicted to vanish as the distance probed shrinks to zero. On the other hand, the squared coupling constant α_S becomes very large for large distances, which leads to quark confinement or what is often called "infrared slavery." The word infrared connotes large wavelengths or distances. This feature of QCD implies that neither single quarks nor gluons can be observed as free particles. The theory is thus highly nonlinear, and it is the large-distance behavior that is probed at low energies where it is depicted more effectively by meson exchanges and their couplings to baryons. In this limit the theory is solved numerically by 'lattice' calculations (section 14.9.) We will describe some features of the low-energy theory in Section 14.1.

14.1 Range and Strength of the Low-Energy Strong Interactions

Some features are common to all low-energy strong interactions, and in this section we shall describe two of the most important ones, range and strength. The range is the distance over which the force is effective. Historically, much of the information

A practical introduction Philadelphia, Institute of Physics Pub., 2003; C. Quigg, *Gauge Theories of the Strong, Weak and Electromagnetic Interactions*, Benjamin, Reading, Ma 1983; K. Gottfried and V.F. Weisskopf, *Concepts of Particle Physics*, Oxford University Press, New York, 1984.

on strong forces was gleaned from studying nuclei, and the force between nucleons therefore enters heavily into the discussion here. It also serves as an introduction of how dynamical information can be deduced from experiments.

Range The early alpha-particle scattering experiments by Rutherford indicated that the nuclear force must have a range of at most a few fm. In 1933, Wigner pointed out that a comparison of the binding energies of the deuteron, the triton, and the alpha particle leads to the conclusion that nuclear forces must have a range of about 1 fm and be very strong.[2] The argument goes as follows. The binding energies of the three nuclides are given in Table 14.2. Also listed are the binding energies per particle and per "bond." The increase in binding energy cannot be due only to the increased number of bonds. However, if the force has a very short range, the increase can be explained: The larger number of bonds pulls the nucleons together, and they experience a deeper potential; the binding energies per particle and per bond increase correspondingly.

Strength The strength of a strong force is best described by a coupling constant. However, to extract a coupling constant from experimental data, a definite form of the strong Hamiltonian must be assumed. We shall do this in later sections.

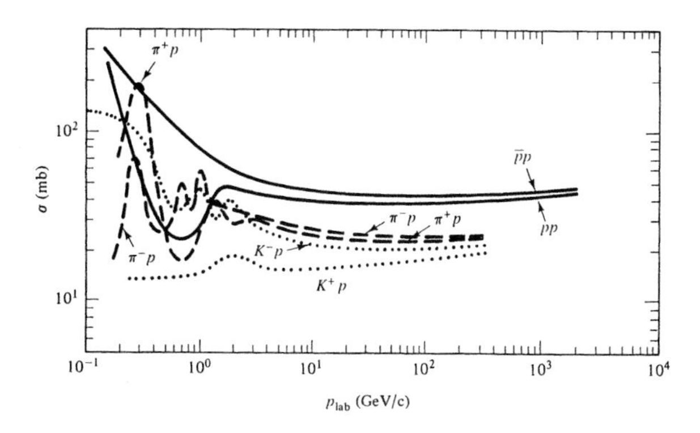

Figure 14.1: Total cross sections for various strong collisions.

Here we compare the strength of the strong forces to that of the electromagnetic and the weak ones from scattering total cross sections. This comparison is somewhat arbitrary because the energy dependence of the cross sections are different. The total cross section for the scattering of neutrinos from nucleons at high energies increases linearly with laboratory energy as shown in Fig. 11.14; it is of the order of $5 \times 10^{-39} E_{\text{Lab}}(\text{GeV})\text{cm}^2$. The cross section for electron scattering from protons is of the order of magnitude of the Mott cross section, Eq. (6.11), at high energies, as discussed in Section 6.8. We take the total cross section to be approximately $90\mu\text{b}/(E_{\text{cm}} \text{in GeV})^2$. In Fig. 14.1, we compare various strong cross sections as a function of laboratory momentum; in all cases, the cross section is of the order of several times 10^{-26} cm^2, or approximately geometric. In Fig. 14.2 we compare the total cross sections for the strong, electromagnetic, and weak processes.

[2]E. P. Wigner, *Phys. Rev.* **43**, 252 (1933).

Table 14.2: Binding Energies of ^{2}H, ^{3}H, and ^{4}He.

Nuclide	Number of Bonds	Binding Energy (MeV)		
		Total	Per Particle	Per Bond
^{2}H	1	2.2	1.1	2.2
^{3}H	3	8.5	2.8	2.8
^{4}He	6	28	7	4.7

To obtain the relative strengths of the three interactions, we compare cross sections somewhat arbitrarily at the approximate border between low and high energies, namely 1 GeV kinetic energy in the laboratory. For the order of magnitude of the relative strengths we take the ratios of the square root of the cross sections, since the strengths appear in the scattering amplitudes; from Fig. 14.2 it then follows that

$$\text{strong / electromagnetic / weak} \approx 1/10^{-3}/10^{-6}. \tag{14.1}$$

The electromagnetic strength is somewhat small because of the comparison energy; a more widely accepted value would be closer to 10^{-2}, or $e^2/\hbar c = 1/137$. Since the coupling constant of the electromagnetic interaction in dimensionless units is of the order of 10^{-2}, as indicated in Eq. (10.79), the corresponding coupling constant for the strong force is of the order of unity. Consequently, the perturbation approach is at best of limited use in the theory of strong interactions at these energies.

The fact that the absolute strength of the strong interactions is characterized by a coupling constant of the order of 1 can be seen in a different manner in Fig. 14.2. At the energies where the comparison of the coupling strengths was performed, namely at a GeV, the strong cross section is of the order of the geometrical cross section of the proton, which is about 3 fm^2. If the proton were *transparent* to the incident hadrons, we would expect the cross section to be much smaller than the geometrical cross section. However, the size of the total cross section, of the order of a few fm^2, indicates that nearly every incident hadron that comes within "reach" of a scattering center suffers an interaction. In this sense, the strong interaction is indeed strong. Even if it were much stronger, it could not scatter appreciably more. On the other hand, it appears that at sufficiently high energy and momentum transfer, the strong interaction becomes weaker and may be accessible to a perturbative treatment. This observation follows from QCD, where the coupling constant decreases at short distances.

A close examination of Fig. 14.2 shows that the weak, electromagnetic, and (perhaps) strong cross sections may be approaching each other as the energy increases. We will point out in Section 14.8 that recent "grand unified" theories of these three interactions predict that measurements at presently available laboratory energies are a low-energy manifestation of a single force; the scale at which the three subatomic forces become equal is predicted to be of the order of 10^{16} GeV, very much higher than any energy available today.

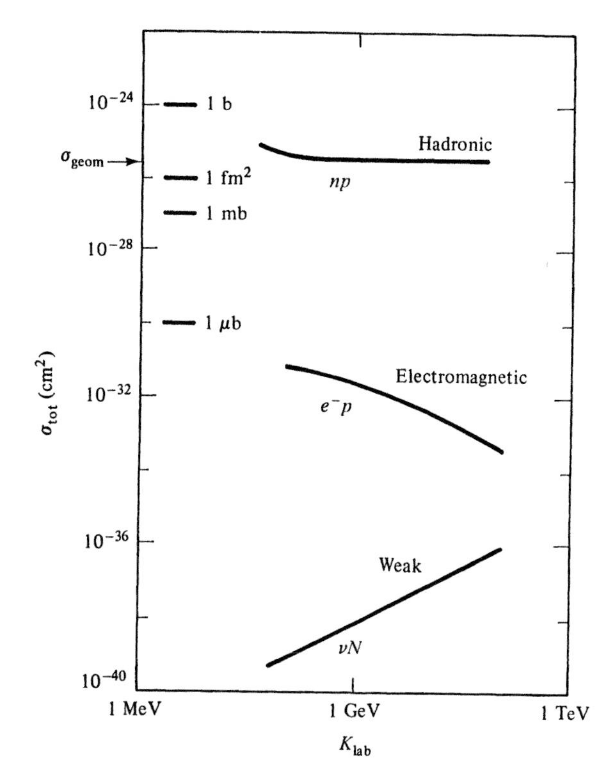

Figure 14.2: Comparison of the total cross sections for strong, electromagnetic, and weak processes on nucleons. σ_{geom} indicates the geometrical cross section of a nucleon, and K is the kinetic energy.

14.2 The Pion–Nucleon Interaction—Survey

Explaining the nuclear forces was one of the main goals of subatomic physicists during most of the last century. We have already pointed out in Section 5.8 that there was almost complete ignorance as to the nature of the nuclear force before Yukawa postulated the existence of a heavy boson in 1934.[3] Yukawa's revolutionary step did not solve the nuclear force problem completely because no calculation reproduced the experimental data well and because it was not even clear what properties the proposed quantum should have.[4] When the pion was discovered, identified with the Yukawa quantum, and found to be a pseudoscalar isovector particle, some of the uncertainties were removed, but it was still not possible to describe the nuclear force satisfactorily. Today we know that, in terms of a meson basis for describing nuclear forces, many more carriers exist and must be taken into account.

[3]H. Yukawa, *Proc. Phys. Math. Soc. Japan* **17**, 48 (1935).
[4]W. Pauli, *Meson Theory of Nuclear Forces*, Wiley-Interscience, New York, 1946.

Nevertheless, the pion and its interaction with nucleons play a special role. First, the pion lives long enough so that intense pion beams can be prepared and the interaction of pions with nucleons can be studied in detail. Second, the pion is the lightest meson; it is more than three times lighter than the next heavier one. In the energy range up to 500 MeV the pion–nucleon interaction can be studied without interference from other mesons. Moreover, because the range of a force, $R = \hbar/mc$, is inversely proportional to the mass of the quantum, the pion alone is responsible for the long-range part of the nuclear force. In principle, the properties of the nuclear force beyond a distance of about 1.5 fm can be compared with the theoretical predictions without severe complications from other mesons.

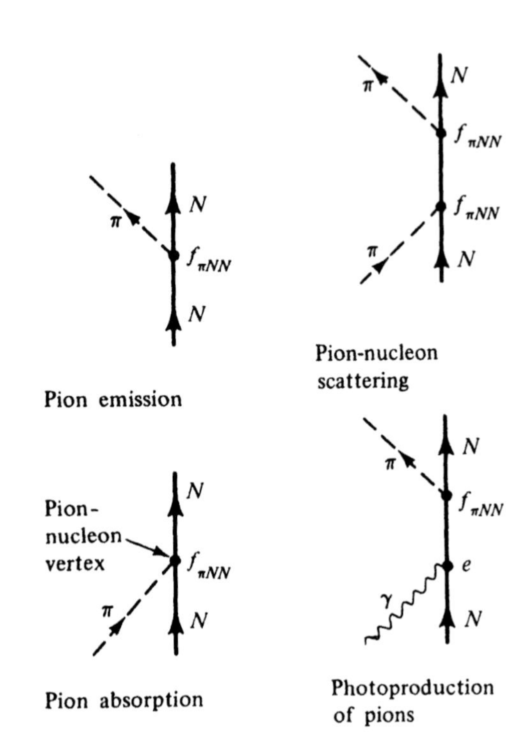

Figure 14.3: Pions can be emitted and absorbed singly. The strength of the pion–nucleon interaction is characterized by the coupling constant $f_{\pi NN}$.

Figure 14.4: Typical diagrams for pion–nucleon scattering and for pion photoproduction.

Experimentally and theoretically, then, the pion–nucleon force plays the role of a test case, and we shall therefore discuss some of the important aspects here. Pions, being bosons, can be emitted and absorbed singly, as shown in Fig. 14.3.

The actual experimental exploration of the pion–nucleon force is performed, for instance, through studies of pion–nucleon scattering and of the photoproduction of pions. Two typical diagrams are shown in Fig. 14.4. In principle, many different pion–nucleon scattering processes can be observed, but only the following three can be readily investigated at low energies:

$$\pi^+ p \longrightarrow \pi^+ p \tag{14.2}$$

$$\pi^- p \longrightarrow \pi^- p \tag{14.3}$$

$$\pi^- p \longrightarrow \pi^0 n. \tag{14.4}$$

[5] V. Flaminio et al., *Compilation of Cross Sections I: π^+ and π^- Induced Reactions*, CERN/HERA Report 83-01, 1983. See also G. Höhler, *Pion–Nucleon Scattering*, (H. Schopper, ed.) Landoldt–Bernstein New Series *I/9 b1* (1982) and *I/9 b2* (1983).

The total cross sections for the scattering of positive and negative pions have been displayed in Fig. 5.34. The cross sections for the elastic processes, Eqs. (14.2) and (14.3), and for charge exchange, Eq. (14.4), are sketched in Fig. 14.5 up to a pion kinetic energy of about 500 MeV.[(5)]
The best-known photoproduction processes are

$$\gamma p \longrightarrow \pi^0 p \qquad (14.5)$$

$$\gamma p \longrightarrow \pi^+ n. \qquad (14.6)$$

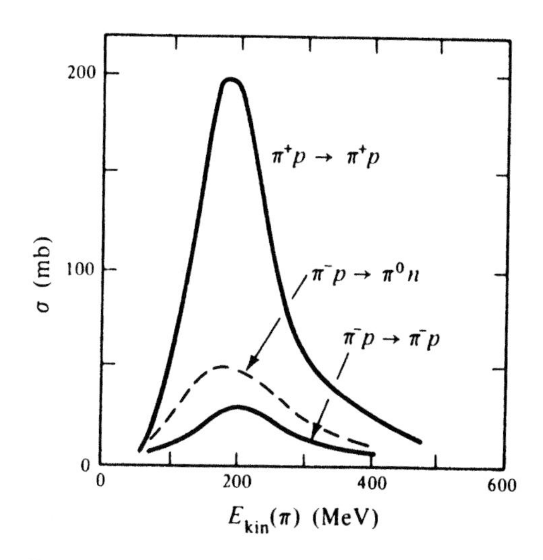

Figure 14.5: Cross sections for the low-energy elastic and charge-exchange pion–proton reactions.

The reaction γn can be studied by using deuterium targets and subtracting the contribution of the proton. The cross sections for the processes in Eqs. (14.5) and (14.6) are shown in Fig. 14.6. The feature that dominates Eqs. (14.2)–(14.6) is the appearance of a resonance. In pion scattering, it occurs at a pion kinetic energy of about 170 MeV; in photoproduction, the photon energy at the peak is about 300 MeV. Despite this difference in kinetic energies, the peaks in pion scattering and pion photoproduction can be interpreted by one phenomenon, the formation of an excited nucleon state, Δ, as indicated in Fig. 14.7.

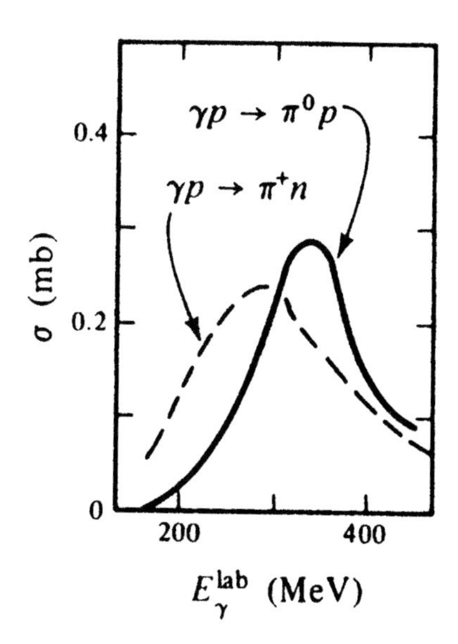

Figure 14.6: Total cross sections for the photoproduction of neutral and charged pions from hydrogen, as a function of the incident photon energy.

The mass of this resonance particle is approximately given by $m_{N^*} \approx m_N + m_\pi + E_{\rm kin}/c^2 = 1260\ {\rm MeV}/c^2$ in pion scattering, and by $m_{N^*} \approx m_N + E_\gamma = 1240\ {\rm MeV}/c^2$ in photoproduction. Proper computation, taking into account the recoil of the N^*, gives a mass of 1232 MeV$/c^2$ for both processes, and it is appealing to assume that they represent the same resonance. The discovery of this resonance, called $\Delta(1232)$, was already discussed in Section 5.12. The cross sections in Figs. 14.5 and 14.6 show that the interaction of pions with nucleons at energies below about 500 MeV is dominated by this resonance.

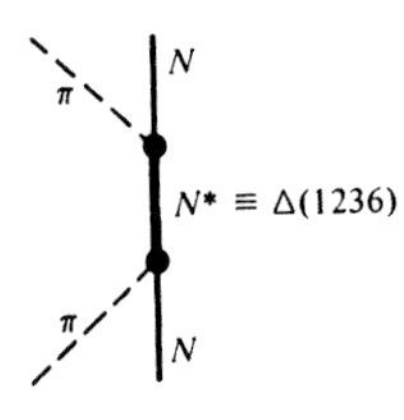

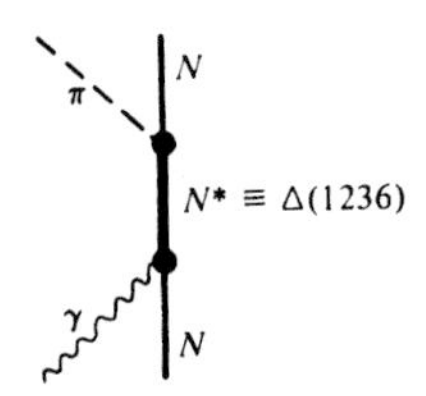

Figure 14.7: Pion scattering and pion photoproduction at low energies are dominated by the formation of an excited nucleon, N^*, usually called $\Delta(1232)$.

Isospin and spin of $\Delta(1232)$ can be established by simple arguments. Pion ($I=1$) and nucleon ($I=\frac{1}{2}$) can form states with $I=\frac{1}{2}$ and $I=\frac{3}{2}$. If $\Delta(1232)$ had $I=\frac{1}{2}$, only two charge states of the resonance would occur. According to the Gell-Mann–Nishijima relation, Eq. (8.30), they would have the same electric charges as the nucleons, namely 0 and 1. These two resonances, $\Delta^0(1232)$ and $\Delta^+(1232)$, are indeed observed. In addition, however, the $\Delta^{++}(1232)$ appears in the process $\pi^+p \to \pi^+p$, and Δ consequently must have $I=\frac{3}{2}$. The fourth member of the isospin multiplet, $\Delta^-(1232)$, cannot be observed with proton targets; deuteron targets permit the investigation of the reaction $\pi^-n \to \pi^-n$, where Δ^- shows up. To establish the spin of $\Delta(1232)$, we note that the maximum cross section for the scattering of unpolarized particles is given by[6]

$$\sigma_{\max} = 4\pi\lambda\!\!\!^{-2}\frac{2J+1}{(2J_\pi+1)(2J_N+1)} = 4\pi\lambda\!\!\!^{-2}\left(J+\frac{1}{2}\right). \tag{14.7}$$

J, J_π, and J_N are the spins of the resonance and of the colliding particles, and $\lambda\!\!\!^{-}$ is the c.m. reduced pion wavelength at resonance. $4\pi\lambda\!\!\!^{-2}$ at 155 MeV is almost 100 mb, and $\sigma_{\max}$ is about 200 mb, so that $J+\frac{1}{2}\approx 2$ or $J=\frac{3}{2}$. To form a state with spin $\frac{3}{2}$ in pion–nucleon scattering, the incoming pions must carry one unit of orbital angular momentum. Pion–nucleon scattering at low energies occurs predominantly in p waves.

• The fact that pion–nucleon scattering at low energies occurs predominantly in the state $J=\frac{3}{2}, I=\frac{3}{2}$ (the so-called 3–3 resonance) can be verified by a spin–isospin phase-shift analysis. We shall not present the complete analysis here, but we shall outline its isospin part because it provides an example for the use of isospin invariance. We first note that experimental states are prepared with well-defined charges. Theoretically, however, it is more appropriate to use well-defined values of the total isospin. It is therefore necessary to express the experimentally prepared

[6]The maximum cross section for the scattering of spinless particles with zero orbital angular momentum is given by $4\pi\lambda\!\!\!^{-2}$. A particle with spin J is $(2J+1)$-fold-degenerate. By assuming that the above cross section holds for each substate, Eq. (14.7) follows.

states in terms of eigenstates of I and I_3, denoted by $|I, I_3\rangle$. Starting with the $|\pi^+p\rangle = \left|\frac{3}{2}, \frac{3}{2}\right\rangle$ state and applying twice the isospin-lowering operator:[7]

$$
\begin{aligned}
|\pi^+p\rangle &= \left|\frac{3}{2}, \frac{3}{2}\right\rangle \\
|\pi^-p\rangle &= \sqrt{\frac{1}{3}}\left|\frac{3}{2}, -\frac{1}{2}\right\rangle - \sqrt{\frac{2}{3}}\left|\frac{1}{2}, -\frac{1}{2}\right\rangle \\
|\pi^0n\rangle &= \sqrt{\frac{2}{3}}\left|\frac{3}{2}, -\frac{1}{2}\right\rangle + \sqrt{\frac{1}{3}}\left|\frac{1}{2}, -\frac{1}{2}\right\rangle.
\end{aligned}
\tag{14.8}
$$

To describe pion–nucleon scattering, a scattering operator S is introduced. The operator S is not as frightening as it usually appears to the beginner, and all we have to know about it are two properties. (1) The scattering amplitude f for a collision $ab \to cd$ is proportional to the matrix element of S,

$$f \propto \langle cd|S|ab\rangle.$$

The cross section is related to f by Eq. (6.2), or $d\sigma/d\Omega = |f|^2$. (2) The pion–nucleon force is strong and assumed to be charge-independent. Thus the Hamiltonian $H_{\pi N}$ must commute with the isospin operator,

$$[H_{\pi N}, \vec{I}] = 0.$$

Since pion–nucleon scattering occurs through the pion–nucleon force as shown in Fig. 14.4, the scattering operator can be constructed from $H_{\pi N}$. It therefore must also commute with $\vec{I}$,

$$[S, \vec{I}\,] = 0, \tag{14.9}$$

and with I^2,

$$[S, I^2] = 0. \tag{14.10}$$

Thus, if $|I, I_3\rangle$ is an eigenstate of I^2 with eigenvalues $I(I+1)$, so is $S|I, I_3\rangle$. Consequently, the state $S|I, I_3\rangle$ is orthogonal to the state $|I', I_3'\rangle$, and the matrix element $\langle I', I_3'|S|I, I_3\rangle$ vanishes unless $I' = I, I_3' = I_3$. Moreover, S does not depend on I_3, as is indicated by Eq. (14.9); the matrix element is independent of I_3 and can simply be written as $\langle I|S|I\rangle$. With the abbreviations

$$f_{1/2} = \left\langle \frac{1}{2}\,|S|\,\frac{1}{2}\right\rangle, \quad f_{3/2} = \left\langle \frac{3}{2}\,|S|\,\frac{3}{2}\right\rangle$$

[7] Merzbacher, Section 17.6; see also problem 15.7.

and with Eqs. (14.8), the matrix elements for the elastic and the charge exchange processes become

$$\begin{aligned}\langle\pi^+p|S|\pi^+p\rangle &= f_{3/2}\\ \langle\pi^-p|S|\pi^-p\rangle &= \frac{1}{3}f_{3/2}+\frac{2}{3}f_{1/2}\\ \langle\pi^0n|S|\pi^-p\rangle &= \frac{\sqrt{2}}{3}f_{3/2}-\frac{\sqrt{2}}{3}f_{1/2}.\end{aligned} \tag{14.11}$$

The matrix elements are complex numbers, and three reactions are not sufficient to determine $f_{1/2}$ and $f_{3/2}$. However, if the resonances shown in Fig. 14.5 occur in the $I=\frac{3}{2}$ state, then $f_{3/2}$ should dominate at the resonance energy. With $|f_{3/2}|\gg|f_{1/2}|$ and with $\sigma\propto|f|^2$, Eq. (14.11) predicts for the ratios of cross sections at resonance

$$\begin{aligned}&\sigma(\pi^+p\to\pi^+p)/\sigma(\pi^-p\to\pi^-p)/\\ &\quad\sigma(\pi^-p\to\pi^0n)=9/1/2.\end{aligned} \tag{14.12}$$

The agreement of this prediction with experiment provides additional support for the hypothesis of charge independence of the pion–nucleon force. •

14.3 The Form of the Pion–Nucleon Interaction

In this section, we shall construct a possible form for the Hamiltonian $H_{\pi N}$ at low pion energies by using invariance arguments and the properties of pions and nucleons. The pion is a pseudoscalar boson with isospin 1; consequently, the wave function $\vec{\Phi}$ of the pion is a pseudoscalar in ordinary space but a vector in isospace. The nucleon is a spinor in ordinary space and in isospace. The Hamiltonian $H_{\pi N}$ must be a scalar in ordinary and in isospace. In the nonrelativistic case (static limit), the nucleon recoil is neglected, and the building blocks available for the construction of $H_{\pi N}$ are

$$\vec{\Phi}, \qquad \vec{\boldsymbol{\tau}}, \qquad \boldsymbol{\sigma}. \tag{14.13}$$

Here, $\vec{\Phi}$ is the pion wave function, $\vec{\boldsymbol{\tau}}=2\vec{I}$ is related to the nucleon isospin operator, and $\boldsymbol{\sigma}=2\boldsymbol{J}/\hbar$ is related to the nucleon spin operator. The Hamiltonian is a scalar in isospace if it is proportional to the scalar product of the two isovectors listed in (12.13),

$$H_{\pi N}\propto\vec{\tau}\cdot\vec{\Phi}.$$

$H_{\pi N}$ is a scalar in ordinary space if it is proportional to the scalar product of two vectors or two axial vectors. The list (14.13) contains only one axial vector, $\boldsymbol{\sigma}$, and a pseudoscalar, Φ. The easiest way to create a second axial vector is to form the gradient of $\vec{\Phi}$ so that

$$H_{\pi N}\propto\boldsymbol{\sigma}\cdot\boldsymbol{\nabla}\vec{\Phi}.$$

Combining the ordinary and isoscalars gives

$$H_{\pi N} = F_{\pi N}\boldsymbol{\sigma}\cdot(\vec{\tau}\cdot\boldsymbol{\nabla}\vec{\Phi}(\boldsymbol{x})), \tag{14.14}$$

where $F_{\pi N}$ is a coupling constant. This Hamiltonian describes a point interaction: Pion and nucleon interact only if they are at the same point. However, the interaction is known to occur over an extended region. To smear out the interaction, a weighting (source) function $\rho(\boldsymbol{x})$ is introduced; $\rho(\boldsymbol{x})$ can, for instance, be taken to represent the nucleon probability density, $\rho = \psi^*\psi$. The function $\rho(\boldsymbol{x})$ falls rapidly to zero beyond about 1 fm and is normalized so that

$$\int d^3x\rho(\boldsymbol{x}) = 1. \tag{14.15}$$

The Hamiltonian between a pion and an extended nucleon fixed at the origin of the coordinate system becomes

$$H_{\pi N} = F_{\pi N}\int d^3x\rho(\boldsymbol{x})\boldsymbol{\sigma}\cdot(\vec{\tau}\cdot\boldsymbol{\nabla}\vec{\Phi}(\boldsymbol{x})). \tag{14.16}$$

This interaction is the simplest one that leads to single emission and absorption of pions. It is not unique; additional terms such as $F'\vec{\Phi}^2$ may be present. Moreover, it is nonrelativistic and therefore limited in its range of validity. However, at higher energies, where Eq. (14.16) is no longer valid, other particles and processes complicate the situation so that consideration of the pion–nucleon force alone becomes meaningless anyway.

The integral in Eq. (14.16) vanishes for a spherical source function $\rho(r)$ unless the pion wave function describes a p wave ($l = 1$). This prediction is in agreement with the experimental data described in the previous section.

The first successful description of pion–nucleon scattering and pion photoproduction was due to Chew and Low,[8] who used the Hamiltonian (14.16). Because of the angular momentum barrier present in the $l = 1$ state, the low-energy pion–nucleon scattering cross section (below about 50 MeV) can be computed in perturbation theory. At higher energies, the approach is more sophisticated, but it can be shown that the Hamiltonian (14.16) leads to an attractive force in the state $I = \frac{3}{2}, J = \frac{3}{2}$ and can explain the observed resonance.[9] At still higher energies, the nonrelativistic approach is no longer adequate.

The numerical value of the pion–nucleon coupling constant $F_{\pi N}$ is determined by comparing the measured and computed values for the pion–nucleon scattering cross section. It is customary not to quote $F_{\pi N}$ but rather the corresponding dimensionless and rationalized coupling constant, $f_{\pi NN}$. The dimension of $F_{\pi N}$ in

[8] G. F. Chew, *Phys. Rev.* **95**, 1669 (1954); G. F. Chew and F. E. Low, *Phys. Rev.* **101**, 1570 (1956); G. C. Wick, *Rev. Mod. Phys.* **27**, 339 (1955).

[9] Detailed descriptions of the Chew-Low approach can be found in G. Källen, *Elementary Particle Physics*, Addison-Wesley, Reading, Mass., 1964; E. M. Henley and W. Thirring, *Elementary Quantum Field Theory*, McGraw-Hill, New York, 1962; and J. D. Bjorken and S. D. Drell, *Relativistic Quantum Mechanics*, McGraw-Hill, New York, 1964. While these accounts are not elementary, they contain more details than the original papers.

Eq. (14.16) depends on the normalization of the pion wave function $\vec{\Phi}$. Since the pion should be treated relativistically, the probability density is normalized not to unity, but to E^{-1}, where E is the energy of the state. This normalization gives the probability density the correct Lorentz transformation properties; the probability density is not a relativistic scalar, but transforms like the zeroth component of a four-vector. With this normalization, $\vec{\Phi}$ has the dimension of $E^{-1/2}L^{-3/2}$ and the dimensionless rationalized coupling constant has the value[10]

$$f_{\pi NN}^2 = \frac{m_\pi^2}{4\pi\hbar^5 c} F_{\pi N}^2 \approx 0.08. \tag{14.17}$$

When the pion was the only known meson, the subject of the pion–nucleon interaction played a dominant role in theoretical and experimental investigations. It was felt that a complete knowledge of this interaction would be the clue to a complete understanding of strong physics. However, attempts to explain, for instance, the nucleon–nucleon force and the nucleon structure in terms of the pion alone were never successful. Other mesons were postulated, and these and some unexpected ones were found. It became clear that the pion–nucleon interaction is not the only problem of interest and that an interaction-by-interaction approach would not necessarily solve the entire problem. At present, in this energy domain the field is very complicated and far beyond a brief and low-brow description. Our discussion here is therefore limited; we shall not treat other interactions but shall turn to the nucleon–nucleon force because it plays an important role in nuclear *and* particle physics.

14.4 The Yukawa Theory of Nuclear Forces

We have stated at the beginning of Section 14.2 that Yukawa introduced a heavy boson for the explanation of nuclear forces in 1934. The fundamental idea thus antedates the discovery of the pion by years. The role of mesons in nuclear physics was not discovered experimentally; it was predicted through a brilliant theoretical speculation. For this reason we shall first sketch the basic idea of Yukawa's theory before expounding the experimental facts. We shall introduce the Yukawa potential in its simplest form by analogy with the electromagnetic interaction.

The interaction of a charged particle with a Coulomb potential has been discussed in chapter 10. The scalar potential A_0 produced by a charge distribution $q\rho(\boldsymbol{x}')$ satisfies the wave equation[11]

$$\boldsymbol{\nabla}^2 A_0 - \frac{1}{c^2}\frac{\partial^2 A_0}{\partial t^2} = -4\pi q\rho. \tag{14.18}$$

[10] O. Dumbrajs et al., *Nucl. Phys.* **B216**, 277 (1983).

[11] The inhomogeneous wave equation can be found in most texts on electrodynamics, for instance, in Jackson, Eq. (6.73). As in chapter 10, our notation differs slightly from Jackson; here ρ is not a charge but a probability distribution.

If the charge distribution is time-independent, the wave equation reduces to the Poisson equation,

$$\nabla^2 A_0 = -4\pi q\rho. \tag{14.19}$$

It is straightforward to see that the potential (10.45),

$$A_0(\boldsymbol{x}) = \int d^3x' \frac{q\rho(\boldsymbol{x}')}{|\boldsymbol{x}-\boldsymbol{x}'|}, \tag{14.20}$$

solves the Poisson equation.[12] For a point charge q located at the origin, A_0 reduces to the Coulomb potential,

$$A_0(r) = \frac{q}{r}. \tag{14.21}$$

When Yukawa considered the interaction between nucleons in 1934, he noticed that the electromagnetic interaction could provide a model but that it did not fall off rapidly enough with distance. To force a more rapid decrease, he added a term $k^2\Phi$ to Eq. (14.19):

$$(\nabla^2 - k^2)\Phi(\boldsymbol{x}) = 4\pi\frac{g}{(\hbar c)^{1/2}}\rho(\boldsymbol{x}). \tag{14.22}$$

Equation (14.22) is the Klein–Gordon equation introduced in Eq. (12.35). The electromagnetic potential A_0 has been replaced by the field $\Phi(\boldsymbol{x})$, and the strength of the field is determined by the strong source $g\rho(\boldsymbol{x})$, where g determines the dimensionless strength, and ρ is a probability density. The sign of the source term has been chosen opposite to the electromagnetic case.[3] The solution of Eq. (14.22) that vanishes at infinity is

$$\Phi(\boldsymbol{x}) = \frac{-g}{(\hbar c)^{1/2}}\int \frac{\exp(-k|\boldsymbol{x}-\boldsymbol{x}'|)}{|\boldsymbol{x}-\boldsymbol{x}'|}\rho(\boldsymbol{x}')d^3x'. \tag{14.23}$$

For a strong point source, placed at position $\boldsymbol{x}' = 0$, this solution becomes the *Yukawa potential*,

$$\Phi(r) = -\frac{g}{(\hbar c)^{1/2}}\frac{\exp(-kr)}{r}. \tag{14.24}$$

The constant k can be determined by considering Eq. (14.22) for the free case ($\rho(\boldsymbol{x}) = 0$) and comparing it to the corresponding quantized equation. The substitution

$$E \longrightarrow i\hbar\frac{\partial}{\partial t} \quad \boldsymbol{p} \longrightarrow -i\hbar\nabla, \tag{14.25}$$

changes the energy–momentum relation,

$$E^2 = (pc)^2 + (mc^2)^2,$$

[12] See, for instance, Jackson, Section 1.7. The important step can be summarized in the relation $\nabla^2(1/r) = -4\pi\delta(\boldsymbol{x})$, where δ is the Dirac delta function.

into the Klein–Gordon equation,

$$\left[\frac{1}{c^2}\frac{\partial^2}{\partial t^2} - \nabla^2 + \left(\frac{mc}{\hbar}\right)^2\right]\Phi(\boldsymbol{x}) = 0. \qquad (14.26)$$

For a time-independent field and for $\rho(\boldsymbol{x}) = 0$, comparison of Eqs. (14.26) and (14.22) yields

$$k = \frac{mc}{\hbar}. \qquad (14.27)$$

The constant k in the Yukawa potential is just the inverse of the Compton wavelength of the field quantum. The mass of the field quantum determines the range of the potential. We have thus regained the result already expressed in Section 5.8. In addition, we have found the radial dependence of the potential for the case of a point source. The simple form of the Yukawa theory thus provides a description of the strong potential produced by a point nucleon in terms of the mass of the field quantum. It "explains" the short range of the strong forces. Before delving deeper into meson theory, we shall describe in more detail what is known about the forces between nucleons.

14.5 Low-Energy Nucleon–Nucleon Force

The properties of the forces between nucleons at energies where its constituent substructure can be neglected, has been studied *directly* in collision experiments or *indirectly* by extracting them from the properties of bound systems, namely the nuclei. In the present section, we shall first discuss the properties of the nuclear force as deduced from nuclear characteristics and then sketch some of the results obtained in scattering experiments below a few hundred MeV.

From the observed characteristics of nuclei, a number of conclusions about the nuclear force, that is, the strong force between nucleons, can be drawn. The most important ones will be summarized here.

Attraction The force is predominantly attractive; otherwise stable nuclei could not exist.

Range and Strength As explained in Section 14.1, comparison of the binding energies of ^{2}H,^{3}H, and ^{4}He indicates that the range of the nuclear force is of the order of 1 fm. If the force is represented by a potential with such a width, a depth of about 50 MeV is found (Section 16.2).

Charge Independence As discussed in chapter 8, the strong force is charge-independent. After correction for the "electromagnetic interaction,"[(13)] the pp, nn,

[13] We have placed "electromagnetic interaction" in quotes because there is an additional effect of the same order, which is not electromagnetic in origin: the masses of the up and down quarks are not identical. This mass difference, which is not believed to be primarily electromagnetic in origin, affects charge independence.

and np forces between nucleons in the same states are identical.

Saturation If every nucleon interacted attractively with every other one, there would be $A(A-1)/2$ distinct interacting pairs. The binding energy would be expected to be proportional to $A(A-1) \approx A^2$, and all nuclei would have a diameter equal to the range of the nuclear force. Both predictions, binding energy proportional to A^2 and constant nuclear volume, disagree violently with experiment for $A > 4$. For most nuclei, the volume and the binding energy are proportional to the mass number A. The first fact is expressed in Eq. (6.26); the second one will be discussed in Section 16.1. Consequently, the nuclear force exhibits saturation: One particle attracts only a limited number of others; additional nucleons are either not influenced or are repelled. A similar behavior occurs in chemical bonding and with van der Waals' forces. Saturation can be explained in two ways; through exchange forces[14] or through strongly repulsive forces at short distances (hard core).[15] Exchange forces lead to saturation in chemical binding, and hard cores account for it in classical liquids. In the strong case, the decision between the two cannot be made by considering nuclear properties, but scattering experiments indicate that both contribute. We shall return to both phenomena later.

The next two properties require a somewhat longer discussion; after stating the properties, they will be treated together.

Spin Dependence The force between two nucleons depends on the orientation of the nucleon spins.

Noncentral Forces Nuclear forces contain a noncentral component.

The two properties follow from the quantum numbers of the deuteron and from the fact that it has only one bound state. The deuteron consists of a proton and a neutron. The spin, parity, and magnetic moment are found to be

$$J^\pi = 1^+, \qquad \mu_d = 0.85742\mu_N. \tag{14.28}$$

The total spin of the deuteron is the vector sum of the spins of the two nucleons and of their relative orbital angular momentum,

$$\boldsymbol{J} = \boldsymbol{S}_p + \boldsymbol{S}_n + \boldsymbol{L}.$$

[14] W. Heisenberg, *Z. Physik* **77**, 1 (1932).
[15] R. Jastrow, *Phys. Rev.* **81**, 165 (1951).

The even parity of the deuteron implies that L must be even. There are then only two possibilities for forming total angular momentum 1, namely $L = 0$ and $L = 2$. In the first case, shown in Fig. 14.8(a), the two nucleon spins add up to the deuteron spin; in the second, shown in Fig. 14.8(b), orbital and spin contributions are antiparallel. In the s state, where $L = 0$, the expected magnetic moment is the sum of the moments of the proton and the neutron, or

$$\mu(s \text{ state}) = 0.879\,634\mu_N.$$

The actual deuteron moment deviates from this value by a few percent,

$$\frac{\mu_d - \mu(s)}{\mu_d} = -0.026. \tag{14.29}$$

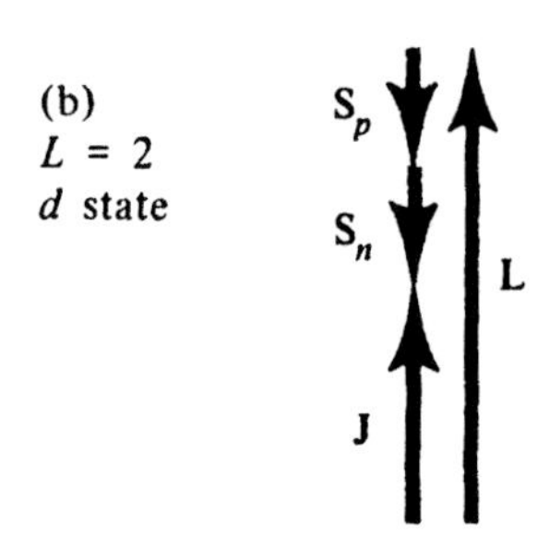

Figure 14.8: The two possible ways in which spin and orbital contribution can form a deuteron of spin 1.

The approximate agreement between μ_d and $\mu(s)$ implies that the deuteron is predominantly in an s state, with the two nucleon spins adding up to the deuteron spin. If the nuclear force were spin-independent, proton and neutron could also form a bound state with spin 0. The absence of such a bound state is evidence for the *spin dependence* of the nucleon–nucleon force. The deviation of the actual deuteron moment from the s-state moment can be explained if it is assumed that the deuteron ground state is a superposition of s and d states. Part of the time, the deuteron has orbital angular momentum $L = 2$. Independent evidence for this fact comes from the observation that the deuteron has a small, but finite, *quadrupole* moment. The electric quadrupole moment measures the deviation of a charge distribution from sphericity. Consider a nucleus with charge Ze to have its spin $\boldsymbol{J}$ point along the z direction, as shown in Fig. 14.9. The charge density at point $\boldsymbol{r} = (x, y, z)$ is given by $Ze\rho(\boldsymbol{r})$. The classical quadrupole moment is *defined* by

$$\mathcal{Q} = Z\int d^3r(3z^2 - r^2)\rho(\boldsymbol{r}) = Z\int d^3r r^2(3\cos^2\theta - 1)\rho(\boldsymbol{r}). \tag{14.30}$$

For a spherically symmetric $\rho(\boldsymbol{r})$, the quadrupole moment vanishes. For a cigar-shaped (prolate) nucleus, the charge is concentrated along z, and $\mathcal{Q}$ is positive. The quadrupole moment of a disk-shaped (oblate) nucleus is negative. As defined here, $\mathcal{Q}$ has the dimension of an area and is given in cm^2, or barns (10^{-24} cm^2), or fm^2.

In an external inhomogeneous electric field, a nucleus with quadrupole moment acquires an energy that depends on the orientation of the nucleus with respect to the field gradient.(16) This interaction permits the determination of $\mathcal{Q}$; for the deuteron, a nonvanishing value was found.(17) The present value is

$$\mathcal{Q}_d = 0.282 \text{ fm}^2. \tag{14.31}$$

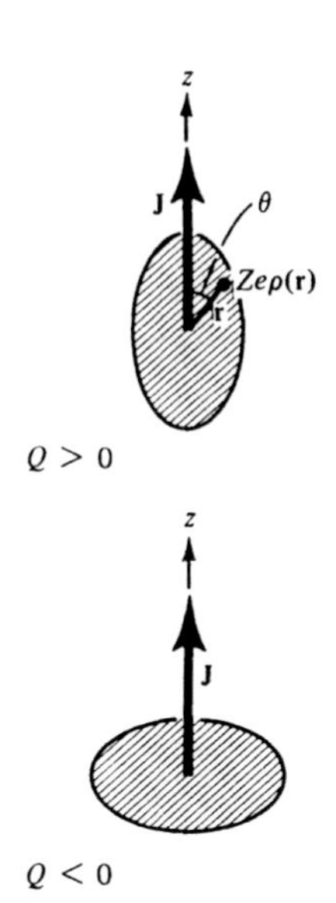

Figure 14.9: Prolate and oblate nuclei, with spins pointing in the z direction. The nuclei are assumed to be axially symmetric; z is the symmetry axis.

s states are spherically symmetric and have $\mathcal{Q} = 0$. The nonvanishing value of $\mathcal{Q}_d$ thus verifies the conclusion drawn from the nonadditivity of the magnetic moments: The deuteron ground state must possess a d-state admixture. (See also Section 6.8, in particular Fig. 6.35.) The presence of a d-state component implies that the nuclear force cannot be purely central, because the ground state in a central potential is always an s state; the energies of states with $L \neq 0$ are pushed higher by the centrifugal potential. The *noncentral force* giving rise to the deuteron quadrupole moment is called the *tensor force*. Such a force depends on the angle between the vector joining the two nucleons and the deuteron spin.

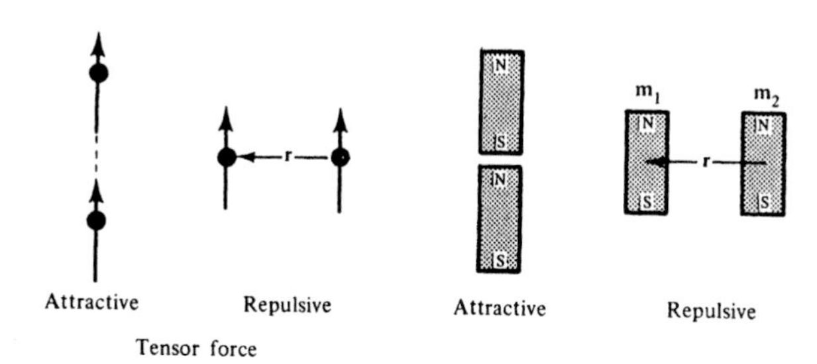

Figure 14.10: The tensor force in the deuteron is attractive in the cigar-shaped configuration and repulsive in the disk-shaped one. Two bar magnets provide a classical example of a tensor force.

Figure 14.10 shows two extreme positions. Since the deuteron quadrupole moment is positive, comparison of Figs. 14.9 and 14.10 indicates that the tensor force must be attractive in the prolate and repulsive in the oblate configuration.

A simple and well-known example of a classical tensor force is also shown in Fig. 14.10. Two bar magnets, with dipole moments m_1 and m_2, attract each other in the cigar-shaped arrangement but repel each other in the disk-shaped one.

[16] Careful discussions of the quadrupole moment are given in E. Segrè, *Nuclei and Particles*, Benjamin, Reading, Mass., Section 6.8; and Jackson, Section 4.2.

[17] J. M. B. Kellog, I. I. Rabi, and J. R. Zacharias, *Phys. Rev.* **55**, 318 (1939).

The interaction energy between two dipoles is well known[18]; it is

$$E_{12} = \frac{1}{r^3}(\boldsymbol{m}_1 \cdot \boldsymbol{m}_2 - 3(\boldsymbol{m}_1 \cdot \hat{\boldsymbol{r}})(\boldsymbol{m}_2 \cdot \hat{\boldsymbol{r}})). \tag{14.32}$$

The vector $\boldsymbol{r}$ connects the two dipoles; $\hat{\boldsymbol{r}}$ is a unit vector along $\boldsymbol{r}$. In analogy to this expression, a tensor operator is introduced to describe the noncentral part of the force between two nucleons.[19] This operator is defined by

$$S_{12} = 3(\boldsymbol{\sigma}_1 \cdot \hat{\boldsymbol{r}})(\boldsymbol{\sigma}_2 \cdot \hat{\boldsymbol{r}}) - \boldsymbol{\sigma}_1 \cdot \boldsymbol{\sigma}_2, \tag{14.33}$$

where $\boldsymbol{\sigma}_1$ and $\boldsymbol{\sigma}_2$ are the spin operators for the two nucleons [Eq. (11.50)]. E_{12} and S_{12} have the same dependence on the orientation of the two components. S_{12} is dimensionless; the term $\boldsymbol{\sigma}_1 \cdot \boldsymbol{\sigma}_2$ makes the value of S_{12} averaged over all angles equal to zero and thus eliminates components of the central force from S_{12}. The exchange of a pion between two nucleons gives rise to just such a tensor force as we shall show in the next Section, and this interaction is the longest range part of the nucleon–nucleon force.

The arguments given so far show that the properties of nuclei allow many conclusions concerning the nucleon–nucleon interaction. However, it is hopeless to extract the strength and the radial dependence of the various components of the nuclear force from nuclear information. *Collision experiments* with nucleons are required for a more complete elucidation of the nucleon–nucleon interaction. Here we shall show that collision experiments provide evidence for *exchange* and *spin-orbit* forces.

Exchange Forces The existence of exchange forces is readily apparent in the angular distribution (differential cross section as a function of the scattering angle) of *np* scattering at energies of a few hundred MeV. The expected angular distribution can be obtained with the help of the Born approximation. This approximation is reasonable here because the kinetic energy of the incident nucleon is much larger than the depth of the potential. The particle therefore crosses the potential region rapidly and barely feels the interaction. The differential cross section for a scattering process is given by Eqs. (6.2) and (6.5) as

$$\frac{d\sigma}{d\Omega} = |f(\boldsymbol{q})|^2,$$

where

$$f(\boldsymbol{q}) = -\frac{m}{2\pi\hbar^2}\int V(\boldsymbol{x}) \exp\left(\frac{i\boldsymbol{q}\cdot\boldsymbol{x}}{\hbar}\right) d^3x. \tag{14.34}$$

---

[18] Jackson, Section 4.2.

[19] A good description of the tensor force and its effects is given in J. M. Blatt and V. F. Weisskopf, *Theoretical Nuclear Physics*, John Wiley, New York, 1952, ch. 2. The *d*-state admixture and tensor force in the deuteron are reviewed in T. E. O. Ericson and M. Rosa–Clot, *Annu. Rev. Nucl. Part. Sci.* **35**, 271 (1985).

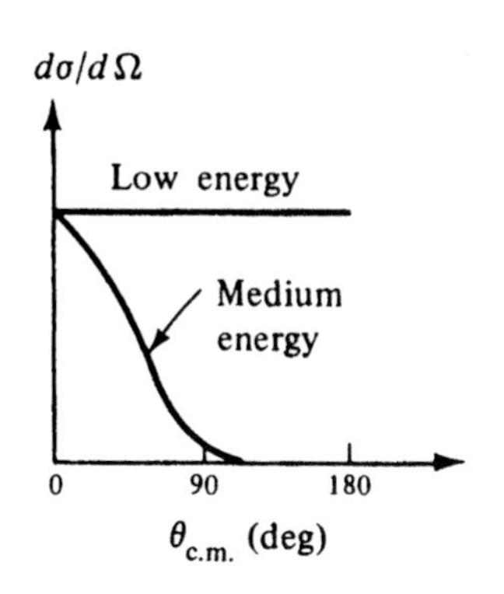

Figure 14.11: Predicted shape of the differential cross section for np scattering at low and medium energies. The curves follow from the first Born approximation using an ordinary potential.

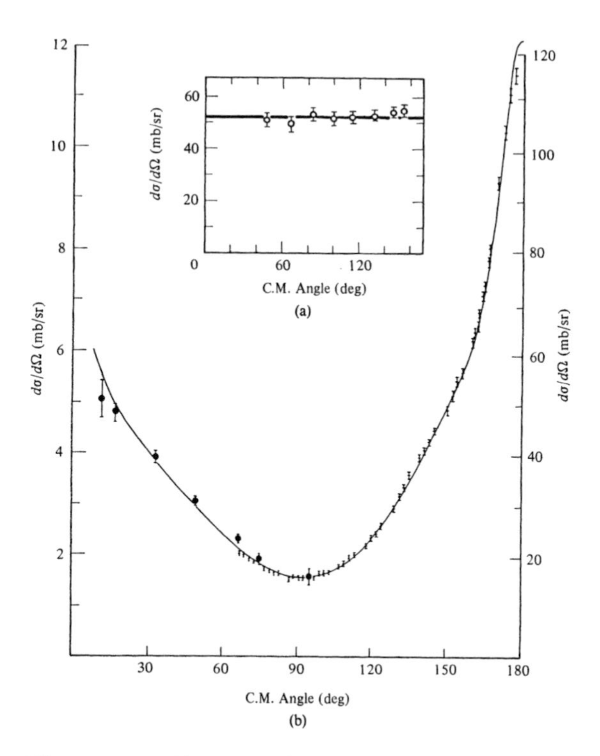

Figure 14.12: Observed differential cross sections for np scattering. (a) The angular distribution at 14 MeV neutron energy is isotropic. [J. C. Alred et al., *Phys. Rev.* **91**, 90 (1953).] (b) At a neutron energy of 425 MeV, a pronounced backward peak is present. [Courtesy D. V. Bugg; see also, D. V. Bugg, *Prog. Part. Nucl. Phys.* (D. H. Wilkinson, ed.) **7**, 47 (1981)].

Here $V(\boldsymbol{x})$ is the interaction potential and $\boldsymbol{q} = \boldsymbol{p}_i - \boldsymbol{p}_f$ is the momentum transfer. For elastic scattering in the c.m., $p_i = p_f = p$, and the magnitude of the momentum transfer becomes

$$q = 2p \sin \tfrac{1}{2}\theta.$$

The maximum momentum transfer is given by $q_{\text{max}} = 2p$. At low energies, $2pR/\hbar \ll 1$, where R is the nuclear force range.

Equation (14.34) then predicts isotropic scattering. At higher energies, where $2pR/\hbar \gg 1$, the situation is different. For forward scattering, at a sufficiently small scattering angle θ, q is small, and the cross section will remain large. For backward scattering, $q \approx q_{\text{max}} = 2p$, the exponent in Eq. (14.34) oscillates rapidly, and the integral becomes small. The predicted behavior, isotropy at low energies and forward scattering at higher energies, is shown in Fig. 14.11. The two features do not depend on the Born approximation; they are more general. Low-energy scattering in a short-range potential is always isotropic, and the high-energy scattering usually acquires a diffractionlike character where small angles (low momentum transfers) are preferred. Experiments at low energies indeed give an isotropic c.m. differential cross section. Even at a neutron energy of 14 MeV, the angular distribution is isotropic, as displayed in Fig. 14.12(a).[20] At higher energies, however, the behavior is very different from the one sketched in Fig. 14.11.

A measurement at a neutron energy of 418 MeV is reproduced in Fig. 14.12(b).[21] The differential cross section displays a pronounced peak in the backward direction. Such a behavior cannot be understood with an ordinary potential that leaves the neutron a neutron and the proton a proton.

It is evidence for an *exchange force* that changes the incoming neutron into a proton through the exchange of a charged meson with the target proton. The forward-moving neutron now has become a proton, and the recoiling target proton a neutron. In effect, then, the neutron is observed in the backward direction after scattering. The exchange nature of the nucleon–nucleon force can also be understood simply from the Yukawa meson exchange theory. As shown in Fig. 14.13, the exchange of a charged meson transfers the charge from the proton to the neutron and *vice versa*, so that an *exchange force* results.

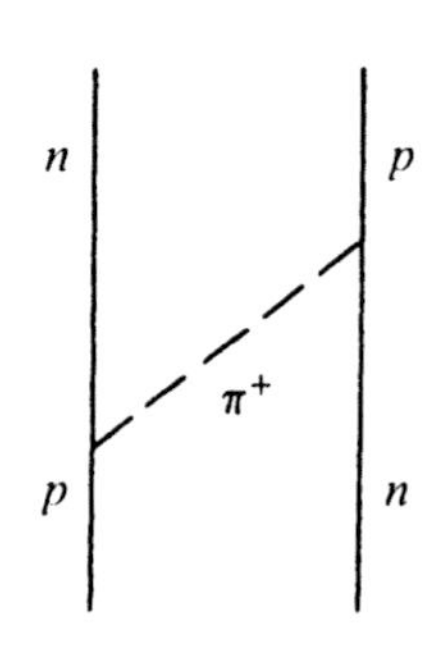

Figure 14.13: Charged-pion-exchange force between a neutron and proton.

Spin–Orbit Force The existence of a spin–orbit interaction can be seen in scattering experiments involving either polarized particles or polarized targets.[22] The idea underlying such experiments can be explained with a simple example, the scattering of polarized nucleons from a spinless target nucleus, for instance, ^{4}He or ^{12}C. Assume that the nucleon–nucleus force is attractive; it then gives rise to trajectories as shown in Fig. 14.14(a). Assume further that the two incoming protons are fully polarized, with spins pointing "up," perpendicular to the scattering plane. Proton 1, scattered to the right, has an orbital angular momentum $\boldsymbol{L}_1$ with respect to the nucleus that is pointing "down." Proton 2, scattered to the left, has its orbital angular momentum $\boldsymbol{L}_2$ "up." Assume that the nuclear force consists of two terms, a central potential, V_c, and a *spin–orbit potential* of the form $V_{LS}\boldsymbol{L}\cdot\boldsymbol{\sigma}$,

$$V = V_c + V_{LS}\boldsymbol{L}\cdot\boldsymbol{\sigma}. \tag{14.35}$$

Figure 14.14(b) implies that the scalar product $\boldsymbol{L}\cdot\boldsymbol{\sigma}$ has opposite signs for nucleons 1 and 2. Consequently, the total potential V is larger for one nucleon than for the other, and more polarized nucleons will be scattered to one side than to the other.

[20] J. C. Alred, A. H. Armstrong, and L. Rosen, *Phys. Rev.* **91**, 90 (1953).

[21] D. V. Bugg, *Prog. Part. Nucl. Phys.*, (D. H. Wilkinson, ed.) **7**, 47 (1981).

[22] For a nice introduction see H.H. Barshall, *Am. Jour. Phys.* **35**, 119 (1967); for current issues see Proceedings of the 16th International Spin Physics Symposium and Workshop on Polarized Electron Sources and Polarimeters, SPIN2004, F. Bradamante, A. Bressan, A. Martin, K. Aulenbacher eds., World Sci. (2005).

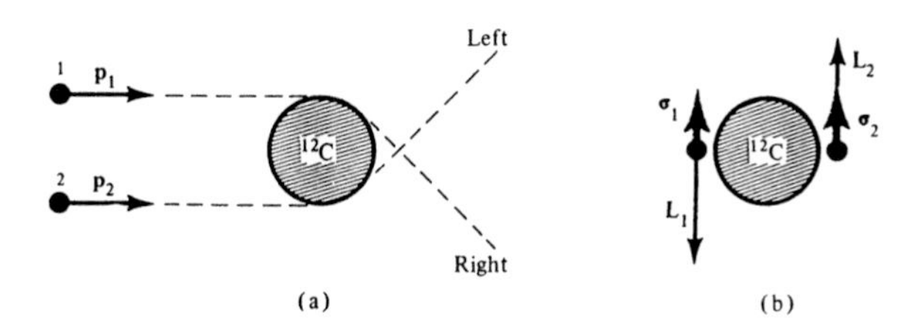

Figure 14.14: Scattering of polarized protons from a spinless nucleus. (a) The trajectories in the scattering plane. (b) The spins and the orbital angular momenta of nucleons 1 and 2.

Experimentally, such left–right asymmetries are observed[(22)] and provide evidence for the existence of a spin–orbit force.

The information obtained in the present section can be summarized by writing the potential energy between two nucleons 1 and 2 as

$$V_{NN} = V_c + V_{sc}\boldsymbol{\sigma_1} \cdot \boldsymbol{\sigma}_2. + V_T S_{12} + V_{LS}\boldsymbol{L} \cdot \tfrac{1}{2}(\boldsymbol{\sigma}_1 + \boldsymbol{\sigma}_2), \tag{14.36}$$

where $\boldsymbol{\sigma}_1$ and $\boldsymbol{\sigma}_2$ are the spin operators of the two nucleons and $\boldsymbol{L}$ is their relative orbital angular momentum,

$$\boldsymbol{L} = \tfrac{1}{2}(\boldsymbol{r}_1 - \boldsymbol{r}_2) \times (\boldsymbol{p}_1 - \boldsymbol{p}_2). \tag{14.37}$$

V_c in Eq. (14.36) describes the ordinary central potential energy, V_{sc} is the spin-dependent central term discussed above. V_T gives the tensor force; the tensor operator S_{12} is defined in Eq. (14.33). V_{LS} characterizes the spin–orbit force introduced in Eq. (14.35). V_{NN} in Eq. (14.36) is nearly the most general form allowed by invariance laws.[(23)]

Charge independence of the strong force implies invariance under rotation in isospin space. The two isospin operators $\vec{I}_1$ and $\vec{I}_2$ of the two nucleons can only occur in the combinations

$$1 \quad \text{and} \quad \vec{I}_1 \cdot \vec{I}_2.$$

Thus each coefficient V_i in V_{NN} can still be of the form

$$V_i = V_i' + V_i'' \vec{I}_1 \cdot \vec{I}_2, \tag{14.38}$$

where V' and V'' can be functions of $r \equiv |\boldsymbol{r}_1 - \boldsymbol{r}_2|, p = \frac{1}{2}|\boldsymbol{p}_1 - \boldsymbol{p}_2|$, and $|\boldsymbol{L}|$.

The coefficients V_i are determined by a mixture of theory and phenomenology. The features that are reasonably well understood are incorporated in the potential to begin with. An example is the one-pion exchange potential. Other features are added to reach agreement with experiment.[(24)] A large[(25)] number

[23] S. Okubo and R. E. Marshak, *Ann. Physik* **4**, 166 (1958). Actually one term allowed by invariance arguments, the quadratic spin-orbit term, is missing in Eq. (12.37).

[24] K. Holinde, *Phys. Rep.* **68**, 121 (1981); S.-O. Backman, G. E. Brown, and J. A. Niskanen, *Phys. Rep.* **124**, 1 (1985).

of pp and np collision experiments have been performed.[26] In addition to total cross sections and angular distributions, collisions with polarized projectiles and polarized targets have been studied. The potential has the general appearance shown in Fig. 14.15. The essential features of V_{NN} are common to the various fits. In particular, the presence of all the terms listed in Eq. (14.36) is required. The coefficients depend on the total spin and isospin of the pair. At large radii ($r \gtrsim 2$ fm), V_{NN} falls off as predicted by the Yukawa potential, Eq. (14.24), with a range $\hbar/m_\pi c$ equal to the Compton wavelength of the pion. The potential is attractive at medium distances, and a common feature is a strong repulsion for distances shorter than about 0.5 fm in all states.

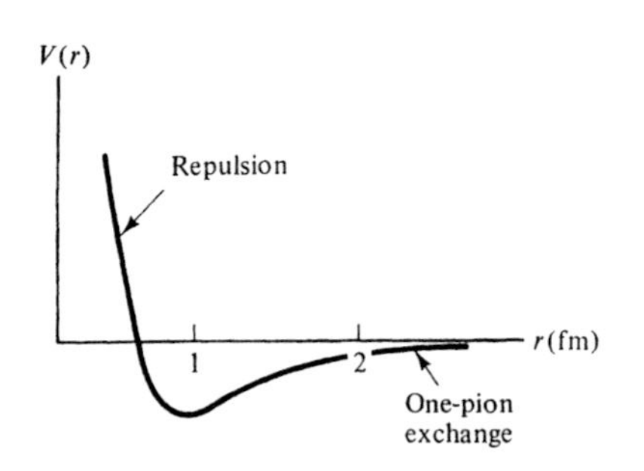

Figure 14.15: Sketch of the nucleon-nucleon potential.

At short distances the potential is believed to arise primarily from the quark structure of the nucleon and the effects of the quark–quark forces. The short distance repulsion between nucleons can be explained in this manner, and reasonable fits to the scattering data are obtained with quark–quark interactions at short distances and single meson exchanges at larger ones.[27]

14.6 Meson Theory of the Nucleon–Nucleon Force

Potentials that use one and two (and even more) pion exchanges, as shown in Fig. 14.16, are used[28] to describe the nucleon-nucleon force with nucleons or Δ's in intermediate states of the Feynman diagrams. The potentials also incorporate more massive meson exchanges, up to masses of the order of 1 GeV/c^2. In these models, the exchange of the ω vector meson is responsible for a large part of the short-range repulsion. As stated earlier, the longest range part of the interaction between two nucleons is due to one pion exchange.

• In Section 12.4, the Yukawa potential was introduced in analogy to electromagnetism by finding the solution to a Poisson equation with a mass term. In the

[25] G. E. Brown and A. D. Jackson, *The Nucleon–Nucleon Interaction*, North-Holland, Amsterdam, 1976.

[26] G. J. M. Austen, T. A. Rijken, and P. A. Verhoeven, in *Few Body Systems and Nuclear Forces*, (J. Ehlers et al., eds.) Vols. 82 and 87, Springer Verlag, New York, 1987; D. V. Bugg, *Comm. Nucl. Part. Phys.* **12**, 287 (1984); D. V. Bugg, *Annu. Rev. Nucl. Part. Sci.* **35**, 295 (1985); C. R. Newsom et al., *Phys. Rev.* **C39**, 965 (1989).

[27] K. Maltman and N. Isgur, *Phys. Rev. D* **29**, 952 (1984); A. Faessler, *Prog. Part. Nucl. Phys.*, (A. Faessler, ed.) **13**, 253 (1985).

[28] R. Vinh Mau, *Nucl. Phys.* **A328**, 381 (1979); in *Mesons in Nuclei*, (M. Rho and D. H. Wilkinson, ed.) Vol. 1, Ch. 12, North-Holland, Amsterdam, 1979; M. Lacombe et al., *Phys. Rev. C* **21**, 861 (1980); S.-O. Bäckman, G. E. Brown, and J. A. Niskanen, *Phys. Rep.* **124**, 1 (1985); R.B. Wiringa et al., *Phys. Rev. C* **51**, 38 (1995); R. Machleidt, *Phys. Rev. C* **63**, 024001 (2001).

present section we shall establish the expression for the interaction energy between two nucleons.

We begin with the simplest case, where the interaction is mediated by the exchange of a *neutral scalar* meson. The emission and absorption of such a meson is described by an interaction Hamiltonian. For the pseudoscalar case, the corresponding Hamiltonian $H_{\pi N}$ has been discussed in Section 12.3.

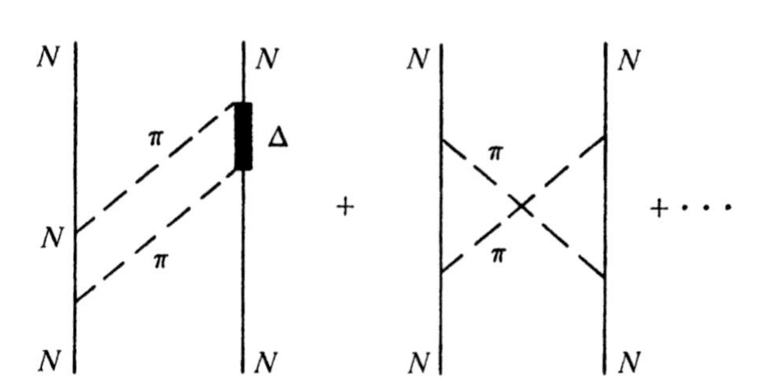

Figure 14.16: Typical two pion-exchange potential diagrams.

The Hamiltonian, H_s, for the scalar interaction can be obtained by similar invariance arguments: Φ is now a scalar in ordinary and in isospin space, and the simplest expression for the energy of interaction between a scalar meson and a fixed nucleon characterized by a source function $\rho(\boldsymbol{x})$ is

$$H_s = g(\hbar c)^{3/2} \int d^3x \Phi(\boldsymbol{x}) \rho(\boldsymbol{x}). \tag{14.39}$$

Between emission and absorption, the meson is free. The wave function of a free spinless meson satisfies the Klein–Gordon equation, Eq. (14.26). In the time-independent case, it reads

$$\left[\boldsymbol{\nabla}^2 - \left(\frac{mc}{\hbar}\right)^2\right] \Phi(\boldsymbol{x}) = 0. \tag{14.40}$$

Together with Hamilton's equations of motion,(25,29) Eqs. (14.39) and (14.40) lead to

$$\left[\boldsymbol{\nabla}^2 - \left(\frac{mc}{\hbar}\right)^2\right] \Phi(\boldsymbol{x}) = \frac{4\pi g \rho(\boldsymbol{x})}{(\hbar c)^{1/2}}. \tag{14.41}$$

This expression is identical to Eq. (14.22). In Section 14.4, we constructed it by starting from the corresponding one in electromagnetism and adding a mass term. Here it follows logically from the wave equation for the scalar meson together with the simplest form for the interaction Hamiltonian. The solution to Eq. (14.42) has

[29] A brief derivation is given in W. Pauli, *Meson Theory of Nuclear Forces*, Wiley-Interscience, New York, 1946. The elements of Lagrange and Hamiltonian mechanics can be found in most texts on mechanics. The application to wave functions (fields) is described in E. M. Henley and W. Thirring, *Elementary Quantum Field Theory*, McGraw-Hill, New York, 1962, p. 29, or F. Mandl, *Introduction to Quantum Field Theory*, Wiley-Interscience, New York, 1959, chapter 2.

already been given in Section 14.4. In particular, for a point nucleon at position $\boldsymbol{x} = 0$, it is the Yukawa potential, Eq. (14.24). The nucleon acts as a source of the meson field and

$$\Phi(\boldsymbol{x}) = -\frac{g}{(\hbar c)^{1/2} \mathrm{r}} \exp(-kr), \quad r = |\boldsymbol{x}|, \quad k = \frac{mc}{\hbar}, \tag{14.42}$$

is the field produced at $\boldsymbol{x}$ by a point nucleon sitting at the origin. The interaction energy between this and a second point nucleon at position $\boldsymbol{x}$ is found by inserting Eq. (14.42) into Eq. (14.39) and by using the fact that $\rho(\boldsymbol{x})$ now also describes a point nucleon. The interaction energy then becomes

$$V_s = -g^2 \hbar c \frac{\exp(-kr)}{r}. \tag{14.43}$$

The negative sign means attraction and two nucleons consequently attract each other if the force is produced by a neutral scalar meson.

Pions are pseudoscalar and not scalar particles, although the latter appear in the meson exchange potentials used to fit data.[25] As the next step, we therefore consider the contribution from a *neutral pseudoscalar meson*. Looking through the list at PDG indicates that η, with a mass of 549 MeV/c^2, is such a particle. The interaction Hamiltonian is very similar to the one given in Eq. (14.16); for an isoscalar particle, this relation simplifies to

$$H_p = F \int d^3x \rho(\boldsymbol{x}) \boldsymbol{\sigma} \cdot \boldsymbol{\nabla} \Phi. \tag{14.44}$$

The free pseudoscalar meson is also described by the Klein–Gordon equation, Eq. (14.44), because it is not possible to distinguish between free scalar and pseudoscalar particles. For the meson field in the presence of a nucleon, Eqs. (14.44) and (14.40) together yield

$$\left[\boldsymbol{\nabla}^2 - \left(\frac{mc}{\hbar} \right)^2 \right] \Phi = -\frac{4\pi}{\hbar^2 c^2} F \boldsymbol{\sigma} \cdot \boldsymbol{\nabla} \rho(\boldsymbol{x}). \tag{14.45}$$

This equation is solved as in Section 14.4. Inserting the solution into Eq. (14.44) then gives, for the potential energy due to the exchange of the neutral pseudoscalar meson between point nucleons A and B,

$$V_p = \frac{F^2}{\hbar^2 c^2} (\boldsymbol{\sigma}_A \cdot \boldsymbol{\nabla})(\boldsymbol{\sigma}_B \cdot \boldsymbol{\nabla}) \frac{\exp(-kr)}{r}. \tag{14.46}$$

The differentiations can be performed, and the final result is[25,30]

$$V_p = \frac{F^2}{\hbar^2 c^2} \left[\frac{1}{3} \boldsymbol{\sigma}_A \cdot \boldsymbol{\sigma}_B + S_{AB} \left(\frac{1}{3} + \frac{1}{kr} + \frac{1}{(kr)^2} \right) \right] \times k^2 \frac{\exp(-kr)}{r}, \tag{14.47}$$

[30] Details can be found in L. R. B. Elton, *Introductory Nuclear Theory*, 2nd ed, Saunders, Philadelphia, 1966, Section 10.3. V_p, as given in Eq. (14.47), is not complete; a term proportional to $\delta(\boldsymbol{r})$ is missing. The omission is unimportant because the short-range repulsion between nucleons makes the term ineffective.

where k is given in Eq. (14.42) and S_{AB} is the tensor operator defined in Eq. (14.33). V_p can be generalized immediately to the pion: The only modification is a factor $\vec{\tau}_A \cdot \vec{\tau}_B$ multiplying (14.47)

$$V_\pi = 4\pi f_{\pi NN}^2 \hbar c \vec{\tau}_A \cdot \vec{\tau}_B \left[\frac{1}{3} \boldsymbol{\sigma}_A \cdot \boldsymbol{\sigma}_B + S_{AB} \left(\frac{1}{3} + \frac{1}{kr} + \frac{1}{(kr)^2} \right) \right] \frac{\exp(-kr)}{r}, \quad (14.48)$$

where use has been made of Eq. (14.17).

It is remarkable that the exchange of a pseudoscalar meson leads to the experimentally observed tensor force. Even before the pion was discovered and its pseudoscalar nature established, Eq. (14.47) was known and was taken as a hint as to the properties of the Yukawa quantum.(3) However, it turned out to be impossible to explain all features of the nucleon–nucleon force in terms of the exchange of pions only. Today we know the reason for the failure: the pion is only one of many mesons; it leads to the longest-range part of the nucleon–nucleon force. •

Evidence for the longest-range role of the pion exchange interaction can be found, for instance, the d/s ratio of the deuteron. This ratio can be measured accurately in the asymptotic region of the wave function, and is a good test of the existence and correctness of the description of the long range nucleon–nucleon force in terms of the pion exchange theory.(31)

The use of meson theory to calculate the nucleon-nucleon force began over 50 years ago. Its major problem is that it is semi-phenomenological and that it is difficult to estimate errors. We will come back to discuss the nucleon–nucleon force in connection with QCD in Section 14.9.

14.7 Strong Processes at High Energies

Early explorers of the Earth faced an uncertain fate. They did not know if they would fall off into the unknown when they reached the end of the disk-shaped world. Bounds were placed on the possible disasters when it was realized that the earth was approximately a sphere. Further exploration led to more bounds, and the presently existing topographic maps leave little room for major surprises. A few decades ago the situation in high-energy physics resembled that of the early explorers. At present, far more is known. The immediate neighborhood, the strong interaction at energies below, say, 1 TeV, is reasonably well explored experimentally. Much remains to be explained, but it is possible that no major new feature will emerge in this energy region in future experiments. At higher energies, however, a new world may be waiting for us. Experiments, with the scarce cosmic rays and at DESY in Hambourg and the Tevatron at FNAL provide some glimpses into the ultrahigh-energy region, but it is very likely that much more will be learned when

[31] T. E. O. Ericson, *Comm. Nucl. Part. Phys.* **13**, 157, (1984).

the LHC at CERN will be completed. In this section, we shall sketch three aspects of ultrahigh-energy collisions.

Inelastic Collisions[32] Most of the discussions so far have been restricted to elastic collisions. These are dominant at low energies. As the energy increases, more and more particles can be created.

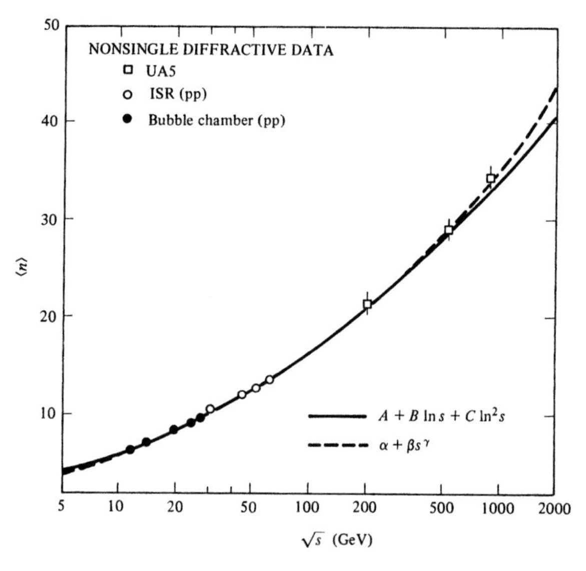

Figure 14.17: Multiplicity, $\langle n_{\rm Ch} \rangle$ of charged secondaries in pp and $\overline{p}p$ collisions as a function of c.m. energy. [From C. Geich-Gimbel, *Int. J. Mod. Phys.* **A4**, 1527 (1989).]

At ultrahigh energies, the interaction of two nucleons can indeed be a spectacular event. The experimental data, obtained at various high-energy accelerators and in cosmic ray studies, display the following prominent features: (i) Small transverse momenta. The pp elastic differential cross sections reproduced in Fig. 6.29 decrease exponentially with t ($|t| = |q|^2$): collision events with large perpendicular momentum transfer are rare. The reluctance of particles to transfer momentum perpendicular to its motion persists in inelastic events.

A different way of stating this finding, which relates it to our earlier discussion, is that the interaction at high momentum transfers or small distances becomes weak; perturbation theory is therefore applicable in this region. The number of particles produced falls off very rapidly as a function of p_T, the momentum transverse to the incident beam. The average value of p_T is of the order of 0.3 GeV/c and nearly independent of the incoming energy. (ii) Low multiplicity. The multiplicity, the number n of secondary particles, can be compared with the maximum allowed by energy conservation. By this criterion, n increases only slowly with energy. The average multiplicity of charged secondaries, $\langle n_{ch} \rangle$, is shown in Fig. 14.17 for both pp and $\overline{p}p$ collisions as a function of the c.m. energy $W = \sqrt{s}$. The curves represent two possible fits, one a logarithmic increase favored by QCD, and one a power law fit favored by statistical, thermodynamic, or hydrodynamic models.[34] However, these last models predict a power law proportional to s^{γ} with $\gamma = \frac{1}{4}$, whereas

[32] D. Green, *High Pt Physics at Hadron Colliders* (Cambridge Monographs on Particle Physics, Nuclear Physics and Cosmology)(2005).

experimentally γ is considerably smaller, $\gamma = 0.127 \pm 0.009$.(33)

The slower logarithmic increase predicted by QCD indicates that not all the energy is distributed statistically, but that a disproportionate amount goes to a few "leading" particles.(35)

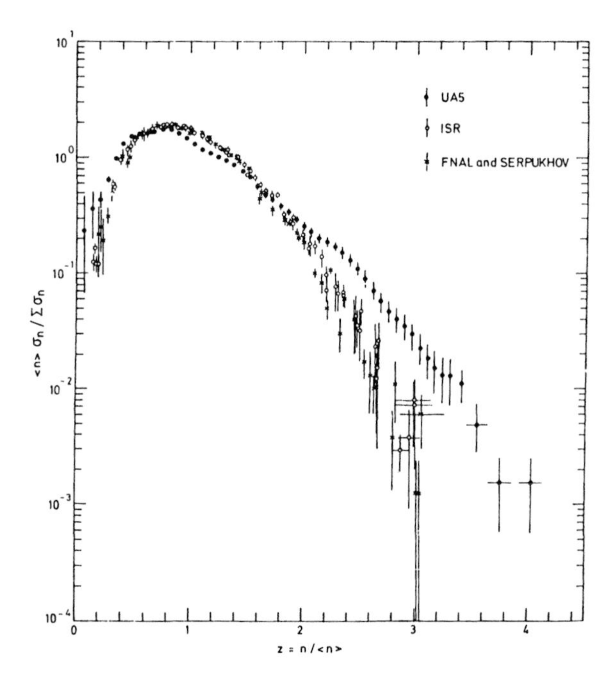

Figure 14.18: Normalized distributions in charged multiplicity in the range of c.m. energies of 11.3–62.2 GeV (ISR, FNAL and Serpukhov) and at 546 GeV (UA5). [From J. G. Alner et al., (UA5 Collaboration), *Phys. Lett.* **138B**, 304 (1984).]

(iii) Poisson-like distributions. The cross sections for the production of events with n prongs are shown for two energies in Fig. 14.18.(33,36) The distributions are plotted as a function of $z = n/\langle n \rangle$. Fig. 14.18 shows that the normalized distributions resemble a Poisson distribution, Eq. (4.3), but are somewhat broader. On the basis of scaling, it was predicted by Koba, Nielsen, and Olesen(37) that the normalized charged particle multiplicity should become independent of energy for asymptotically large energies; this is often referred to as KNO scaling. This scaling behavior appears to hold over a region of c.m. energies of about 10–70 GeV.

However, at higher energies we observe in Fig. 14.18 that the tail of the distribution function broadens, so that "asymptotia" has not yet been reached.

High-Energy Theorems (Asymptotia) Processes at ultrahigh energies can be extremely complex. It is nevertheless possible to extrapolate lower-energy data to predict features of cross sections that should emerge as the total energy in the c.m.,

[33] C. Geich–Gimbel, *Int. J. Mod. Phys.* **A4**, 1527 (1989).

[34] E. Fermi, *Phys. Rev.* **81**, 683 (1951); L. D. Landau, *Izv. Akad. Nauk SSSR* **17**, 51 (1953) [transl. *Collected Papers of L. D. Landau*, (D. ter Haar, ed.)] Pergamon Press and Gordon and Breach, New York, 1965; M. Kretzschmar, *Annu. Rev. Nucl. Sci.* **10**, 765 (1958); D. Kharzeev, E. Levin and M. Nardi, *Nucl. Phys.* **A 747**, 609(2005); .

[35] E. M. Friedlander and R. M. Weiner, *Phys. Rev.* **D28**, 2903 (1983).

[36] G. J. Alner et al., (UA5 Collaboration), *Phys. Lett.* **B138**, 304 (1984).

[37] Z. Koba, H. B. Nielsen, and P. Olesen, *Nucl. Phys.* **B40**, 317 (1972); T. Renk, S.A. Bass and D.K. Srivastava, *Phys. Lett.* **B 632**, 632 (2006).

W, tends toward infinity. This energy region is usually called *asymptotia*, and it is not yet clear if and where this strange land begins.

In Section 9.8, we stated that the TCP theorem can be proved with very general arguments. These are based on axiomatic quantum field theory, which is an extension of quantum mechanics into the relativistic region. This theory can also be used to derive theorems on high-energy collisions.(38) Quantum field theory lies far outside the scope of this book, but we shall state two theorems because they are typical of the results that can be expected from this approach. The first theorem follows rigorously from quantum field theory,(38) and it gives an upper bound on the total cross section as $s = W^2$ tends to infinity:

$$\sigma_{\text{tot}} < \text{ const.}(\log s)^2. \tag{14.49}$$

This bound was discovered by Froissart(39) and it limits the rise of the total cross section with increasing energy regardless of the type of interaction involved. An example of cross sections that increase with increasing energy is shown in Fig. 6.30, namely the pp and $\bar{p}p$ total cross sections at values of s greater than about 1000 GeV^2. This increase follows the maximum rate allowed by the Froissart bound, Eq. (14.49). The growth of the cross section with energy is thought to be due to two reasons, an increase in the effective interaction radius, R, of the two nucleons or nucleon–antinucleon and a decrease in the transparency or increase in the *opacity*, or blackness.(33) For a black target, every wave that passes through it would be absorbed and we would have $\sigma_{el} = \sigma_{abs} = \pi R^2$, so that $\sigma_{\text{tot}} = 2\pi R^2$. Will the increase in cross section continue indefinitely or will the cross section flatten out again? The observed increases indicate that, even at the highest energies available so far, the asymptotic region has not yet been reached. The second theorem follows from quantum field theory if it is additionally assumed that the total cross sections at asymptotic energies become constants. The Pomeranchuk theorem(40) then predicts that the total cross sections for the particle–target and antiparticle–target collisions approach the same value as the energy tends toward infinity:

$$\frac{\sigma_{\text{tot}}(\overline{A}+B)}{\sigma_{\text{tot}}(A+B)} \longrightarrow 1 \text{ in asymptotia.} \tag{14.50}$$

In a simplified geometrical interpretation, the Pomeranchuk theorem can be understood as follows: As the energy approaches infinity, so many reactions are possible that the collision can almost be thought of as one between two totally absorbing black disks. The cross section is thus essentially geometric (the radii of the two objects are not well defined, but we are only providing a qualitative argument). Since the geometrical structures of the positive and negative pions are identical (the charge is certainly not important), the cross sections for π^+p and π^-p would

[38] A. Martin, *Nuovo Cim.* **42**, 930 (1966); R.J. Eden, *Rev. Mod. Phys.* **43**, 15 (1971).
[39] M. Froissart, *Phys. Rev.* **123**, 1053 (1961).
[40] I.Ya. Pomeranchuk, *Sov. Phys. JETP* **7**, 499 (1958).

be expected to be identical. The fact that π^+p can only be in an isospin state $I=\frac{3}{2}$ whereas π^-p can scatter in both $I=\frac{3}{2}$ and $\frac{1}{2}$ is of no importance because there is a huge (infinite) number of possible final states in both cases.

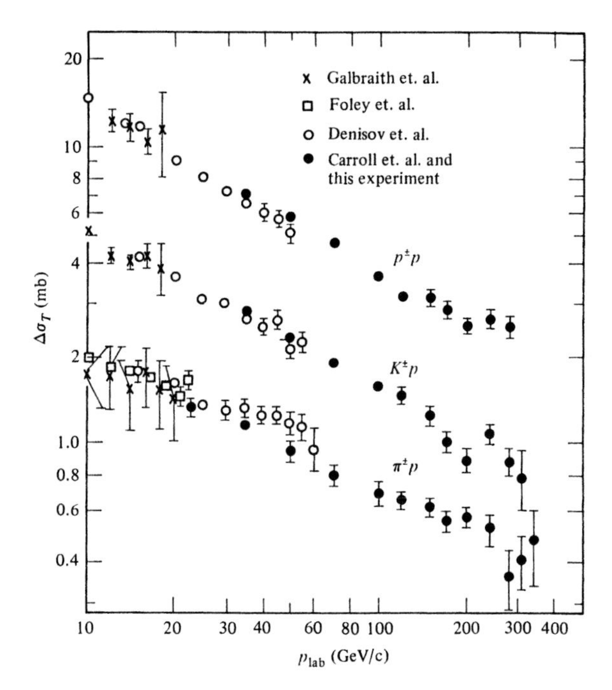

Figure 14.19: The differences between particle–target and antiparticle–target cross section. [After A.S. Carroll et al., *Phys. Lett.* **80B**, 423 (1979). See PDG for more recent measurements.]

The same argument can be made, for instance, for $\bar{p}p$ and pp scatterings, where the additional annihilation for $\bar{p}p$ scattering is a very small fraction of the total cross section. Experimentally, it is found that

$$\sigma^+ - \sigma^- \approx \text{const } p_{\text{lab}}^{-1/2}. \tag{14.51}$$

The experimental data appear to bear out the Pomeranchuk theorem. Figure 14.19 shows some results.[41] The relevant cross sections indeed tend toward a common constant value, and the differences $\Delta\sigma$ tend towards zero.

Scale Invariance[42] Where is asymptotia? At the present time, this question is not settled, but some insight can be obtained with simple arguments. Consider first a world in which only the electron and positron exist. The bound system in such a world is positronium, an "atom" in which an electron and a positron revolve around the common c.m. The energy levels of positronium are given by the Bohr formula,

$$E_n = -\alpha^2 m_e c^2 \frac{1}{(2n)^2}, \quad n = 1, 2, \ldots, \tag{14.52}$$

[41] A. S. Carroll et al., *Phys. Lett.* **80B**, 423 (1979); see also R. E. Breedon et al., UA6 Collaboration, *Phys. Lett.* **216B**, 459 (1989) and PDG.

[42] T. D. Lee, *Phys. Today* **25**, 23 (April 1972); R. Jackiw, *Phys. Today* **25**, 23 (January 1972); J. D. Bjorken, *Phys. Rev.* **179**, 1547 (1969); I. Mishustin, J. Bondorf, M. Rho, *Nucl. Phys.* **A555**, 215 (1993).

where $\alpha = e^2/\hbar c$ is the fine structure constant. Apart from the factor $(2n)^{-2}$, the energy levels are determined by two factors, α^2 and $m_e c^2$. The first describes the strength of interaction, and the second sets the *scale*. At energies of the order of, or smaller than, the scale energy $m_e c^2$, the physical phenomena are dominated by the existence of discrete energy levels. At energies large compared to $m_e c^2$, asymptotia has been reached in the positronium world, and physical phenomena satisfy simple laws in which m_e does not appear. Consider Bhabha scattering,

$$e^+e^- \longrightarrow e^+e^-. \tag{14.53}$$

The total cross section for electron–positron scattering in asymptotia can depend only on W, the total c.m. energy, and on the strength factor α^2 but not on m_e. The cross section has the dimension of an area, and the only possible form *not* containing m_e is

$$\sigma = \text{const.}\frac{\alpha^2}{W^2}, \quad \text{in asymptotia.} \tag{14.54}$$

This form expresses *scale invariance.* It is not possible from the measured cross section to determine the mass of the colliding particles.

Now consider e^+e^- scattering in the real world. Equation (14.54) is valid at c.m. energies greater than a few MeV. At energies of a few hundred MeV, deviations begin to occur, and a peak appears at $W = 760$ MeV, as indicated in Fig. 10.15. The deviation and the observed resonance reveal that m_e is not the only mass that sets a scale but that higher-mass particles exist, in this case the pions and their resonances. In addition to Bhabha scattering, processes such as

$$e^+e^- \longrightarrow \text{hadrons} \tag{14.55}$$

become possible, and σ depends on the masses of the various hadrons. The departure of the total cross section from the form of Eq. (14.54) indicates that a new basic energy scale has appeared. The energy scale is now given by

$$E_h = m_h c^2, \tag{14.56}$$

where m_h is the mass of a suitably chosen quark or hadron. Usually, m_h is taken to be the nucleon mass, $m_h = m_N$. What could have been considered asymptotia for Bhabha scattering has turned out to be nothing but a transition region. However, the game can now be replayed. At energies large compared to the new scale energy, E_h, we again expect independence of the total cross section on the hadron masses as discussed in Section 10.9. Dimensional arguments then show that σ_{tot} must again be of the form of Eq. (14.54); see also Eq. (10.90),

$$\sigma_{\text{tot}} = \text{const.}\frac{\alpha^2}{W^2}, \quad \text{for } W \gg m_h c^2. \tag{14.57}$$

The constant can be different from the one given in Eq. (14.54), but the energy dependence is the same.

It is important to note that the high energy results presented earlier dealt with total cross sections. For hadrons, these appear to be dominated by large distance phenomena. It is not clear that this feature will persist at ever higher energies. Moreover, high momentum transfer, or short distance, collisions are different, and serve as tests of the underlying theory, quantum chromodynamics or QCD, since the interaction is predicted to become ever weaker.

14.8 The Standard Model, Quantum Chromodynamics

There is now good evidence that the theory of the strong forces is quantum chromodynamics (QCD), so named because of its analogy to quantum electrodynamics, the quantum theory of electricity and magnetism. The term "chromodynamics" refers to the key ingredient of color in the theory. In Section 10.9 we saw that the experimental production of hadrons in e^+e^- collisions provides evidence that quarks must come in three colors. This additional degree of freedom is responsible for the forces between quarks.

Table 14.1 presents the analogy between QCD and QED; the gauge field quantum, the gluon, like its counterpart, the photon, is massless and has a spin of $1\hbar$; thus there are color electric and color magnetic forces.

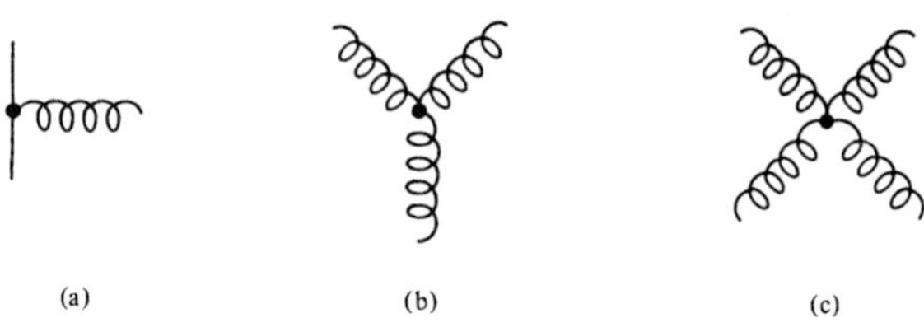

Figure 14.20: (a) Gluon coupling to quarks and (b), (c) gluon self-couplings.

However, there are also crucial differences between QCD and QED. The gluons themselves are "color charged" and not neutral as is the photon. Indeed, the gluons can be considered to be bicolored, that is, to be made up of a color and an anticolor. The gluon color leads to a non-Abelian (noncommuting) theory. There are eight colored gluons. Out of three colors and their anticolors, we can make up nine possible combinations; one of these, $r\overline{r}+g\overline{g}+b\overline{b}$ is colorless and the remaining eight correspond to the gluons.

Because the gluons themselves are color-charged, they can interact with each other and there are not only quark–gluon couplings as shown in Fig. 14.20(a), but also gluon–gluon couplings as shown in Figs. 14.20(b) and 14.20(c). The source of the gluon fields need not be quarks, but can be other gluons! This self-coupling gives rise to a highly nonlinear theory with no "free" gluon field. There also arises the possible existence of mesons made up of gluons only. Such objects are called glueballs; they have been sought but have not yet been found, and may not exist in pure form. The color combinations carried by the gluons can be described in terms of the three colors of the quarks. In Fig. 14.21 we show two ways of drawing the exchange of a gluon between a quark and an antiquark. The exchange leads from a red–antired to a blue–antiblue combination. The red quark is changed to a blue quark because the gluon carries away the color $r\bar{b}$; similarly $\bar{r}$ is changed to $\bar{b}$. The colors of the gluons cause the theory to be non-commuting: if a red quark emits a $r\bar{b}$ gluon it becomes blue as shown in Fig. 14.22; the subsequent emission of a $b\bar{g}$ gluon leads to a green quark; however, the reverse order of emission of the gluons cannot occur; a red quark cannot emit a $b\bar{g}$ gluon.

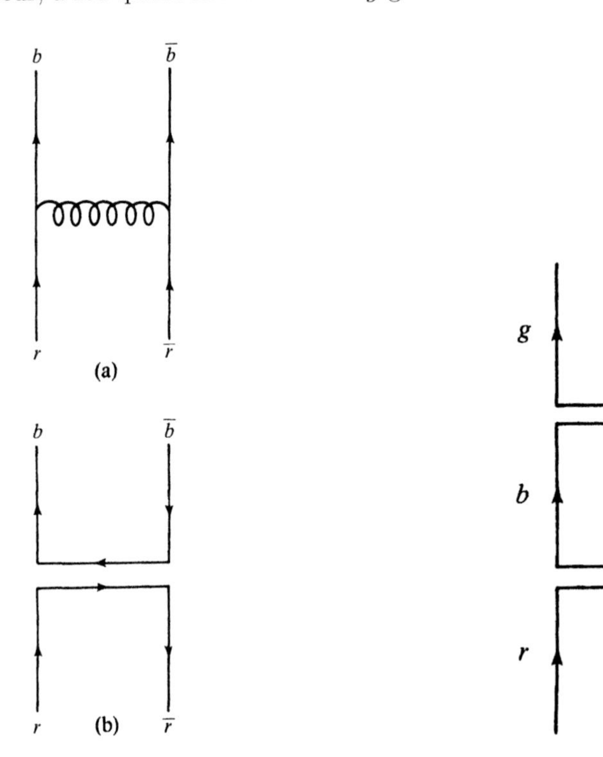

Figure 14.21: Two ways of depicting the exchange of a gluon between a quark and antiquark: (a) Standard way, and (b) bicolored way.

Figure 14.22: The emission of two gluons by a red quark.

What are the features of the QCD force that we expect and/or require? The theory should conserve charge, strangeness, charm and other flavor quantum numbers as well as the other additive and multiplicative quantum numbers discussed in chapters 7 to 9. We expect the theory to lead to the confinement of color: No colored objects, made up of gluons or quarks can exist freely, but must be combined and confined into colorless (white) hadrons. Evidence for confinement comes from the fact that only particles corresponding to white (colorless) quark combinations, such as $q\bar{q}$ or qqq are observed. Colored combinations such as qq or $qq\bar{q}$ have never been seen. The forces should thus be strongly attractive for colorless states and repulsive for other ones—indeed infinitely repulsive for colored objects since they do not appear in nature. We expect the long-range confining force to be universal and thus flavor-independent (independent of the quark type). This feature gives rise to isospin conservation, for instance, even though the theory says nothing about the equality of the masses of the up and down quarks! It also means that we can relate the energy levels of the bound $u\bar{u}, d\bar{d}, s\bar{s}, c\bar{c}, b\bar{b}$, and $t\bar{t}$ systems. Another property that is expected for the light quarks is chiral symmetry and will be discussed in Sec. 14.9.

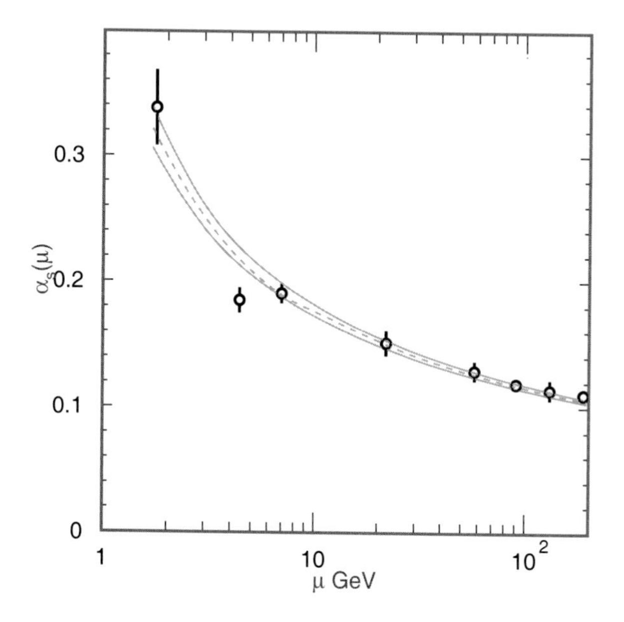

Figure 14.23: Running of α_s with the mass scale parameter μ. [From PDG.]

The theory also has the property that the force becomes weak at short distances. This "asymptotic freedom" of the theory has been tested at high energies and momentum transfers. Thus, QCD predicts that the analogue of the fine structure constant, proportional to the square of the strong coupling constant, $\alpha_s = g_s^2/\hbar c$, varies with momentum transfer,[43]

$$\alpha_s(q^2) = \frac{\alpha_s(\mu^2 c^2)}{1 + \frac{\alpha_s(\mu^2 c^2)}{12\pi}(33 - 2n_f)\ln\left(\frac{q^2}{\mu^2 c^2}\right)}, \tag{14.58}$$

where μ is a mass that sets the scale (renormalization mass), q is the four-momentum transfer with $q^2 = q_0^2/c^2 - \boldsymbol{q}^2, q_0$ is the energy transfer, and n_f is the number of flavors (six). Figure 14.23 shows a comparison between measurements of α_s and Eq. 14.58 versus the theoretical prediction.

The fine structure constant of electrodynamics also changes with momentum transfer, but much more slowly and in the opposite direction, it becomes slightly larger at high momentum transfers. This distinction between QCD and QED rests on the self-interaction of the gluons due to their color charge. We can illustrate the difference with the help of a thought-experiment. In Fig. 14.24(a) an external electron is shown; although it cannot create real electron–positron pairs, it can do so virtually as long as the pair lives for a time less that about $\hbar/mc^2$, where m

[43] See PDG.

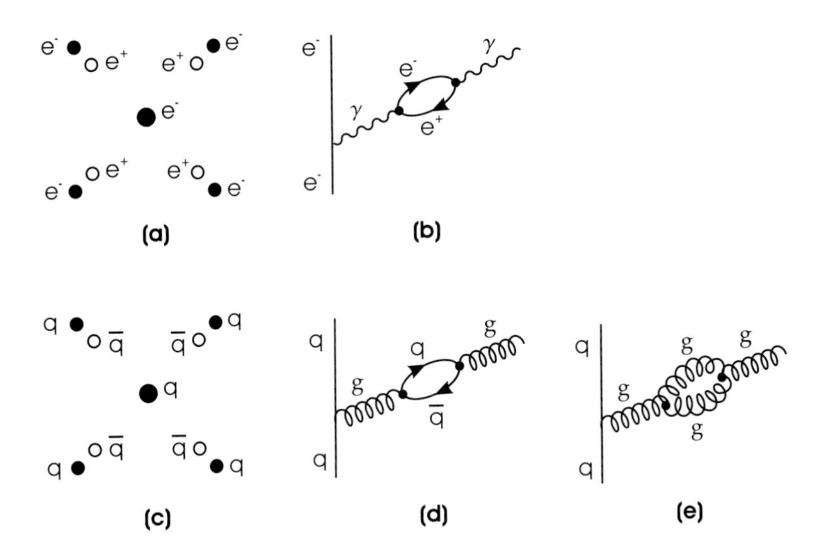

Figure 14.24: Shielding and antishielding. (a) An external charge $-e$ is shown surrounded by electron–positron pairs; (b) Feynman diagram corresponding to Fig. (a); (c) and (d) + (e) are diagrams similar to (a) and (b), respectively, but for QCD and a quark q.

is the mass of the electron. A Feynman diagram corresponding to Fig. 14.24(a) is shown in Fig. 14.24(b). Since the external electron attracts positive charges, the positron of the virtual pair will be closer to the test electron than the electron of the pair. Consequently the effective charge (strength of interaction) of the real electron seen by a very small test charge is reduced when the test charge is some distance away. But this effective charge increases in magnitude as the test charge approaches the electron, since its screening by the positron of the e^-e^+ pair is reduced. The effective interaction strength, $\alpha = e^2/\hbar c$, increases slightly at small distances or high momentum transfers. The situation is different in QCD, because in addition to the effect of screening from quark–antiquark pairs (Fig. 14.24(d)), the gluons can interact with themselves (Fig. 14.24(e)). These gluons carry away color so that, if not too many types of quark–antiquarks pairs can be created, there is antishielding and the color charge decreases as we approach the colored quark, as shown in Fig. 14.24(e). Eq. (14.58) implies that this decrease holds for $n_f < 33/2$. Thus α_s, which measures the strength of the interaction, is reduced at short distances or large momentum transfers, quite the opposite from QED. At these momentum transfers, which require very high energies, QCD can be and has been tested. If the effective strength, as measured by α_s, is sufficiently weak, then perturbation theory can be used.

As an example, consider the production of quark pairs in e^+e^- collisions at very high energies, as shown in Fig. 10.23(b). By analogy to Eq. 10.89, $e^+e^- \rightarrow \mu^+\mu^-$, the production of $q\overline{q}$, Fig. 10.23(b), should show the same angular distribution, namely $(1 + \cos^2\theta)$. In the colliding frame, which is the c.m. frame, the $\mu^+\mu^-$ or

$q\overline{q}$ must emerge back-to-back. Single quarks, however, cannot appear; the quarks create further $q\overline{q}$ pairs. This process goes on until insufficient energy is left for further $q\overline{q}$ production. The gluon thus creates "jets" of back-to-back mesons.[44] At energies above 10–20 GeV such two-jet events predominate and the hadrons have an angular distribution proportional to $(1 + \cos^2\theta)$. This angular distribution also shows that the quarks have spin 1/2, just as the muon.

As a result of confinement the lines of force between a quark and antiquark are different than those between a positive and negative charge (Fig. 14.25). In the case of QCD, the lines of force are compressed into a cylindrical bundle because for a linear confining potential the force is constant.

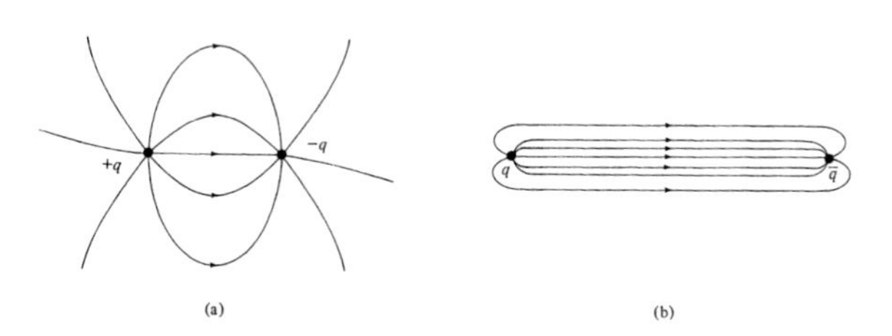

Figure 14.25: (a) Lines of force for charges $\pm q$; (b) lines of force for quarks q and $\overline{q}$.

Thus, as the quark and antiquark are separated, the energy required to do so increases linearly with the separation, and it takes an infinite energy to "liberate" the particles. Therefore, they are confined.

The theoretical study of confinement is difficult because QCD is highly nonlinear. It has been examined for a discrete space–time, namely on a lattice, by means of numerical (Monte Carlo type) techniques pioneered for this type of problem by Wilson.[45] Thanks to improved and faster computers, continuous progress has occurred. There has been continual progress and refinements in lattice QCD computations (see Section 14.9), so that agreement with experiment can be obtained for more and more strong interaction phenomena.

These numerical approaches hint, but do not prove that confinement will result from the theory. Often, especially for heavy quarks, confinement is modeled by a linearly rising or similar potential.

Although the detailed nature of the "long-range" confinement force is not known, it is expected to be like a string or spring, i.e., a restitutative force that is independent of spin, color, and flavor. One way to examine this force theoretically and experimentally is in a heavy quark–antiquark system, where the quarks can be considered as nonrelativistic.

---

[44] G. Kramer, *Theory of Jets in Electron–Positron Annihilation*, Springer Tracts in Modern Physics No. 102, (G. Höhler, ed.) Springer, New York, 1984.

[45] K.G. Wilson, *Phys. Rev.* **D10**, 2455 (1974) and in *New Phenomena in Subnuclear Physics*, (A. Zichichi, ed.) Plenum, New York, 1977.

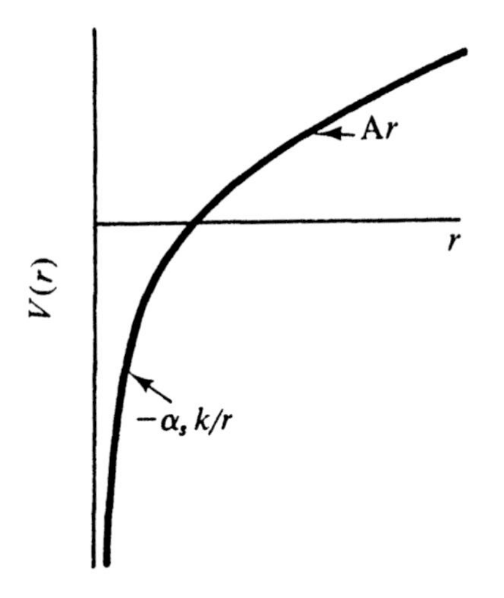

Figure 14.26: The potential V of Eq. (14.59).

We expect the short-distance one-gluon exchange force between the heavy quarks to be primarily like a Coulomb force. The distance dependence then is r^{-2}, just as between two fixed (heavy) electrical charges. At large distances the confining force should predominate. Good results in fitting the spectrum of $c\bar{c}(J/\psi)$ and $b\bar{b}(\Upsilon)$ systems have been obtained with a potential of the form

$$V = -\frac{\alpha_s k}{r} + Ar \tag{14.59}$$

where k and A are constant coefficients. This potential is illustrated in Fig. 14.26.

14.9 QCD at Low Energies

In Section 14.6 we discussed phenomenological approaches to extracting the nucleon-nucleon force. Ideally one would deduce this force from QCD, but, as we have already mentioned, this is a complicated problem that has not been solved.

Chiral Perturbation QCD-inspired systematic methods have been introduced at low energies. Use is made of the symmetries of QCD, particularly chirality (see Section 11.7.) The left and right handed light quarks (up, down, and strange) are decoupled from each other in the QCD Hamiltonian if their masses can be neglected; it is the mass term which connects them. Since the masses are small, but not zero, this symmetry is only approximate. At low energies a systematic expansion can be carried out by constructing the most general Hamiltonian which incorporates all terms with the symmetries of QCD, primarily chirality. For this reason, the method is called *chiral perturbation theory.*(46) In addition, it is possible to carry out an expansion in powers of p^2/χ^2, where p is the relative momentum of the nucleons and χ is the chiral perturbation theory limit, of order ~ 1 GeV, where the strong fine structure constant become of order unity.(47)

[46] S. Scherer, Introduction to Chiral Perturbation in *Advances in Nuclear Physics* **27**, 277 (2003).

[47] Nucleon-nucleon chiral potentials were developed in C. Ordoñez and U. VanKolck, *Phys. Lett.* **B291**, 459 (1992); and brought to a fine point in D.R. Entem and R. Machleidt, *Phys. Lett.* **B524**, 93 (2002).

An expansion in terms of the light quark masses (alternatively the pion mass) can also be made. Infinities which may occur are absorbed order by order in unknown constants. These are fixed from experiments or calculated in QCD-inspired models.

In so-called effective field theories, heavy mesons (the ρ, ω, etc...) in the low-energy potential are replaced by a zero-range (δ) function. At higher energies ($E \sim m_\pi c^2$) pions should be incorporated. A systematic expansion in powers of p^2/χ^2 and m_π^2/χ^2 can then be carried out. Thus, effective field theories lead to a systematic treatment, where it is known what the next order correction will be, so that errors can be estimated.(48) The expansion can also be carried out directly for the scattering amplitude (see Chapter 6.) In that case , it corresponds to the effective range expansion:(49)

$$\sigma = \frac{4\pi}{k^2} \frac{1}{1 + \cot^2 \delta_0} \tag{14.60}$$

where $\hbar k$ is the relative momentum of the nucleons and δ_0 is the s-wave phase shift, given by

$$\cot \delta_0 = -\frac{1}{ka} + \frac{1}{2} k r_0 + ... \tag{14.61}$$

Here a is called the *scattering length* and r_0 the *effective range.* For the nucleon-nucleon problem, the expansion needs to be applied separately to the singlet and triplet states.

Lattice QCD Although no analytic solutions of QCD have been found, the improvement of computers have permitted numerical solutions. R. Wilson developed a way of numerically solving the evolution of a state in the presence of strong interactions without the need of perturbative approximations and preserving the gauge invariance of the theory.(50) Calculations are carried out in a finite *lattice* representing space and time. Generally, studies are carried out changing the total size and the number of points of the lattice to observe stability. Considerable success has been achieved, especially over the past few years. Earlier calculations used a "quenched" approximation in which fermion loops, or vacuum polarization effects were omitted. These approximations are much less costly, but it is impossible to estimate the errors made, and they are no longer the norm. Now complete QCD calculations are carried out and accuracies of a few % have been achieved, e.g., in fitting heavy quark masses and decay constants. A comparison is shown in Fig. 14.27,(51) where both quenched and full calculations are compared. It is now

---

[48] P.F. Bedaque and U. Van Klock, *Annu. Rev. Nucl. Part. Sci* **52**, 339 (2002).
[49] *Nuclear Physics with Effective Field Theory* ed. R. Seki, U. Van Kolck, M.J. Savage, World Sci., Singapore, 1998.
[50] R. Wilson, *Phys. Rev. D* **10**, 2445 (1974).
[51] C.T.H. Davies et al, *Phys. Rev. Lett* **92**, 022001 (2004).

possible to make predictions as well as fit measured quantities. Nucleon form factors, quark distributions in the nucleon, contributions to the nucleon spin, the axial vector weak coupling constant, g_A, and other physical quantities can be computed.

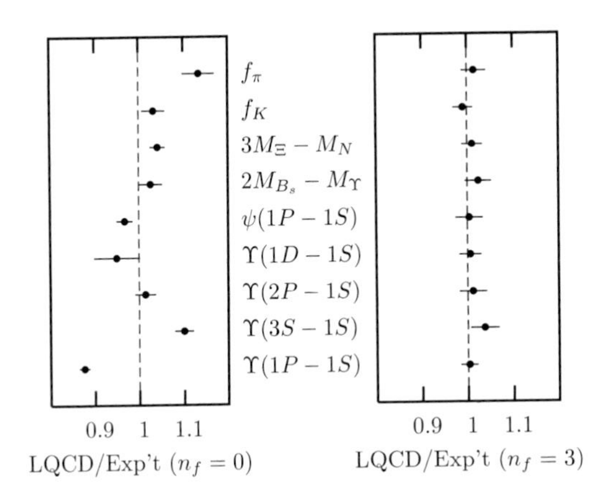

Figure 14.27: Ratio of lattice calculations to experimental values for the decay constants f_π and f_K and for mass splittings. Left: 'quenched'; right: 'full'. n_f indicates the number of light-quark flavors included in the calculations. [From ref.(51).]

However, the pion (and light quark) masses used are still large compared to experimental values, e.g., 350 MeV/c^2 for the pion mass. Costs escalate rapidly as m_π decreases (proportional to m_π^{-9}). It has been shown how to incorporate chiral symmetry in lattice calculations[52] and it is sometimes possible to compare results with chiral perturbation theory ($m_\pi = 0$), in order to extrapolate lattice results to more realistic pion masses.

14.10 Grand Unified Theories, Supersymmetry, String Theories

The success of the unification of the weak and electromagnetic interactions has led to attempts to include the strong forces, and even gravity. The first type of theories were "grand unified theories" or GUTs. The theories predict that the strengths of the three interactions only differ at "low" energies, but approach each other at energies of the order of 10^{15} to 10^{17} GeV. These energies are not far removed from the Planck mass,

$$\sqrt{\frac{\hbar c}{G}} = 1.22 \times 10^{19}\ \mathrm{GeV}/c^2,$$

where G is the gravitational constant.

In GUTs, quarks and leptons occur symmetrically in a single multiplet, thus "explaining" why there are as many lepton as quark families (i.e., three), and also predicting that quarks and leptons can be interchanged. Thus, there is no longer a reason why the proton should be stable and the theories predict its decay lifetime to be of the order of 10^{31} to 10^{33} years. The long lifetime of the proton stems from the

[52] D.B. Kaplan, *Phys. Lett.* **B288**, 342 (1992).

high unification energy. Experimentally, the lifetime of the proton to the expected dominant mode $p \to e^+\pi^0$ is $\geq 1.6 \times 10^{33}$ y[(43)] rules out the simplest GUTs.[(53)] In GUT models, not only is baryon number not conserved, but lepton number and muon number as well, and thus $\mu^\pm \to e^\pm\gamma, \mu^\pm \to e^\pm e^+ e^-$ are allowed, although at a very low rate. These decays have been sought but, as described in Section 7.4, have not yet been found.

GUTs also predict that the various neutrinos are massive and convert to other flavors, as has been found experimentally. In addition, the theories predict the existence of massive monopoles, which so far have not been found.

On the other side of the coin, GUTs have had a number of successes. Many of them yield predictions of $\sin^2\theta_W$ and masses of the heavy quarks that are close to the experimental values.[(54)] GUTs also make interesting connections to cosmology. They are a possible explanation for the "missing mass", the reasons for the low ratio of baryon to photon density ($\sim 10^{-9}$) in the universe, and the cause of the baryon asymmetry, i.e., why we have many many more baryons than antibaryons.[(55)] This requires the three Sakharov conditions[(56)]: CP/T violation, baryon/lepton nonconservation, and non-equilibrium conditions.(See Chapter 19.)

An important component in modern theories is supersymmetry (SUSY) It was originally introduced by Wess and Zumino[(57)] to develop a quantum field theory of gravity and remove infinities. According to it, every particle in nature has a 'superpartner' of the opposite statistics, spin $0 \leftrightarrow$ spin $1/2$; for instance, 'squarks' of spin 0 and 'photinos' of spin 1/2 are the superpartners of the quarks and photon, respectively. If the symmetry were exact, the superparticles would have the same mass as the ordinary ones. This is clearly not the case so this symmetry is broken. In Grand Unified Theories the 's-particles' are expected to have mases $\gtrsim 100$ GeV/c^2 and could be observed when the LHC starts running. Supersymmetry could explain peculiarities that otherwise seem capricious.[(58)]

The latest addition to the theories of nature are so-called superstring theories of particles.[(59)] The most natural of these theories are based on a universe which is more than four dimensional (three space and one time dimension), generally ten-dimensional (nine space and one time dimension); six of these dimensions are then collapsed. Such theories have a number of appealing features. They include

---

[53] D.V. Nanopoulos, *Comm. Nucl. Part. Phys.* **15**, 161 (1985); H. Georgi, *Sci. Amer.* **244**, 48 (April 1981); M. Goldhaber and W.J. Marciano, *Comm. Nucl. Part. Phys.* **16**, 23 (1986).

[54] A.J. Buras et al., *Nucl. Phys.* **B135**, 66 (1978); a nice review is presented in J. Ellis and M. Jacob, *Phys. Rep.* **403**, 445 (2004).

[55] E.W. Kolb and M.S. Turner, *Annu. Rev. Nucl. Part. Sci.* **33**, 645 (1983).

[56] A.D. Sakharov, *Pis'ma Z. Eksp. Teor. Fiz.* **5**, 32 (1967); English Translation: *JETP Lett.* **5**, 24 (1967); L.B. Okun, Ya.B. Zeldovich, *Comments Nucl. Part. Phys.* **6**, 69 (1976).

[57] J. Wess and B. Zumino, *Nucl. Phys.* **B70**, 31 (1974); see also H. E. Haber and G. L. Kane, *Sci. Amer.* **255,** 52 (June 1986).

[58] S. Dimopoulos, S. Raby, F. Wilczek, *Phys. Tod.*, pg. 25 (October 1991).

[59] M.B. Green, *Sci. Amer.* **255,** 48 (September 1986); B. Zwiebach, *A First Course in String Theory*, Cambridge University Press, Cambridge, (2004); B.R. Greene, *The Elegant Universe*, Norton, WW & Co. (1999).

quantum gravity, produce gauge theories with spin-1 gauge quanta, spin-2 massless particles (gravitons), and eliminate many of the infinities which plague the quantum theory of gravity. The theories have as their basis that fundamental particles, e.g., quarks and leptons, are strings and not points, but the string dimensions are very much smaller than we can measure at present, of the order of the Planck length, $\sqrt{G\hbar/c^3}$, about 10^{-33} cm. Although there is no experimental support for these theories, they are being pursued avidly. One reason is that they can predict why gravity is so much weaker than the other forces: It is spread over more dimensions than the other forces so that the part in our 4-dimensional universe is weaker.

14.11 References

The literature covering the field of strong interactions is immense. Most texts and reviews, and particularly the original theoretical papers, are rather sophisticated. In the following we list some reviews and books which are either simpler, or from which, with some effort, information can be extracted even at the level assumed here.

The book *Pions and Nuclei* by T. Ericson and W. Weise, Clarendon Press, Oxford, (1988) presents the essential experimental data and provides the necessary theoretical background for a discussion of the data.

Both the meson theory and a general description of nuclear forces can be found in K. S. Krane, *Introductory Nuclear Physics*, John Wiley, New York, 1987. A historical paper on Yukawa's discovery is L.M. Brown, *Phys. Today* **39**, 55 (December 1986). An interesting introduction to nuclear forces and a collection of some of the pioneering papers can be found in D.M. Brink, *Nuclear Forces*, Pergamon, Elmsford, N.Y., 1965. The nucleon–nucleon interaction is also discussed in: *The NN Interaction and Many Body Problems*, (S. S. Wu et al., eds.), World Sci., Teaneck, N.J., 1984; A good introduction to Chiral theories is given by H. Leutwyler in*Chiral Dynamics, Theory and Experiment III*, ed. A.M. Bernstein, J.L. Goity, and U-G Meissner, World Sci. Singapore, 2001. Effective field theory is introduced in *Nuclear Physics with Effective Field Theory*, ed. R. Seki, U. van Kolck, and M.J. Savage, World Sci. Singapore, 1998; S. Scherer *Introduction to Chiral Pertrubation Theory*, in *Advances in Nuclear Physics* **27**, 277 (2003); B. Holstein, *Introduction to Chiral Perturbation Theory* in *Hadronic Structure*, ed. J. Goity, World Sci., Singapore, 2001, p. 204; hep-ph/0210398; P.F. Bedaque and U. van Kolck, *Annu. Rev. Nucl. Part. Sci.* **52**, 339 (2002).

High energy nucleon–nucleon and nucleon–antinucleon experiments and theory are reviewed in M. Block and R. N. Cahn, *Rev. Mod. Phys.* **57**, 563 (1985); *The Nucleon–Nucleon and Nucleon–Antinucleon Interactions*, (H. Mitter and W. Plessas, eds.) Springer, New York, 1985; H.G. Dosch, P. Gauron, and B. Nicolescu, LSANL arch., hep-ph 0206214. Plots can be found in PDG.

Introductions to QCD can be found in F. E. Close, *An Introduction to Quarks and Partons*, Academic Press, New York, 1979; Y. Nambu, *Quarks*, World Sci., Singapore, 1981; C. Quigg, *Gauge Theories of the Strong, Weak, and Electromagnetic Interactions*, Benjamin-Cummings, Reading, MA., 1983; K. Gottfried and V. F. Weisskopf, *Concepts of Particle Physics*, Oxford University, New York, Vol. I, 1984, Vol. II, 1986; I. S. Hughes, *Elementary Particles*, 2nd. ed., Cambridge University Press, Cambridge, 1985; D. H. Perkins, *Introduction to High Energy Physics*, 4th. ed., Cambridge University Press, Cambridge, 2000. F. Wilczek has written two articles that are suited for the level of this book and beautifully written: *Rev. Mod. Phys.* **77**, 857 (2005); and *QCD Made Simple, Phys. Today* **53**, 22 (2000); more advanced treatises can be found in F. Wilczek, *Annu. Rev. Nucl. Part. Sci.* **32**, 177 (1982); G. Altarelli, *A QCD Primer*, LANL arch. hep-ph/0204179. Experimental tests of QCD are reviewed in S. Bethke and J.E. Pilcher *Annu. Rev. Nucl. Part. Sci.* **42**, 251 (1992) and J.E. Huth and M.L. Mangano, *Annu. Rev. Nucl. Part. Sci* **43**, 585 (1993).

Glueballs are reviewed in P. M. Fishbane and S. Meshkov, *Comm. Nucl. Part. Phys.* **13,** 325 (1984); J. Ishikawa, *Sci. Amer.* **247,** 142 (November 1984); J. F. Donoghue, *Comm. Nucl. Part. Phys.*, **16,** 277 (1986); F. E. Close, *Rep. Prog. Phys.* **51,** 833 (1988); F.E. Close, Sci. Amer. **279**, 80 (1998).

QCD studies on a lattice are discussed in "Lattices for Laymen" by D. J. E. Callaway in *Contemp. Phys.* **26,** 23, (1985); C. Rebbi, *Sci. Amer.* **248,** 54 (February 1983); A. Hasenfratz and P. Hasenfratz, *Annu. Rev. Nucl. Part. Sci.* **35,** 559 (1985); A.S. Kronfeld and S.P.B. MacKenzie, *Annu. Rev. Nucl. Part. Sci.* **43**, 793 (1993); A.M. Green, ed., *Hadronic Physics from Lattice QCD*, World Sci., Singapore (2004); H. Neuberger, , *Annu. Rev. Nucl. Part. Sci*, **53**, 23 (2001).

There are numerous review articles and books on GUTs. Reasonably accessible ones are H. Georgi and S.L. Glashow, *Phys. Today* **33**, 30 (September 1980); H. Georgi, *Sci. Amer.* **244,** 48 (April 1981); L. B. Okun, *Leptons and Quarks*, North-Holland, Amsterdam, 1982; L.B. Okun, *Particle Physics The Quest for the Substance of Substance*, Harwood Academic, New York, 1985; M. Jacob and P. V. Landshoff, *Rep. Prog. Phys.* **50,** 1387 (1987).

Superstrings and Supersymmetry are really beyond the level of this text; however, we list some books and reviews for the interested reader: B. Zwiebach, *A First Course in String Theory*, Cambridge University Press, Cambridge, (2004); M.B. Green, *Sci. Amer.* **255,** 48 (Sept. 1986); H.E. Haber and G. L. Kane, *Sci. Amer.* **255,** 52 (June 1986); P.G.O. Freund, *An Introduction to Supersymmetry*, Cambridge University Press, New York, 1986; *Supersymmetry and Supergravity*, A Reprint Volume from Phys. Rep., (M. Jacob, ed.) World Sci., Teaneck, N.J., 1985; *Supersymmetry, A Decade of Development*, (P. C. West, ed.) Adam Hilger, Boston, 1986; *String Theory is Testable, Phys. Today* **50**, 40 (February 1999); J. Hewett and M. Spiropulu, *Annu. Rev. Nucl. Part. Sci.*, **52**397 (2002); M.J. Duff and P.R. Page, *Sci. Amer*, **278**, 64 (Feb. 1998); J. Jolie, *Sci. Amer.* **297**, 70 (July 2002); G. Kane *Supersymmetry*, Perseus Publ., Cambridge, MA, 2000. Popularized texts include J. Gribbin, *The Search for Superstrings, Symnmetry, and the Theory of Everything*, Little, Brown and Co., New York, 1998; B. Greene, *The Fabric of the Cosmos*, A.A. Knopf, New York, 2004; D. Falk, *Universe on a T-Shirt: The Quest for the Theory of Everything*, Arcade Publ. Co., New York, 2004.

Problems

14.1. (a) List 10 possible pion–nucleon scattering processes, with, at most, one pion and one nucleon.

(b) Which of these processes are related by time-reversal invariance?

(c) Express all cross sections in terms of $M_{3/2}$ and $M_{1/2}$.

14.2. * Sketch an experimental arrangement used to study pion–nucleon scattering.

(a) How is the total cross section observed?

(b) How is the charge-exchange reaction cross section determined?

14.3. Use the observed cross sections to show that the peaks of the first resonance in pion–nucleon and in photonucleon reactions occur at the same mass of the Δ. Take recoil into account.

14.4. Treat the pion–nucleon scattering at the first resonance classically: Compute the classical distance from the center of the nucleon at which a pion with angular momentum $l = 0, 1, 2, 3$ (in units of $\hbar$) will strike. Which partial waves will contribute significantly according to this argument? Use a parity argument to rule out the values $l = 0$ and $l = 2$.

14.5. Justify Eq. (14.7) by a crude (nonrigorous) argument.

14.6. Verify the expansions (14.8).

14.7. Consider $H_{\pi N}$, Eq. (14.16). Assume a spherical source function $\rho(r)$. Assume the pion wave function to be a plane wave. Show that only the p-wave part of this plane wave leads to a nonvanishing integral.

14.8. Consider Fig. 5.35. The second and third resonances in the $\pi^- p$ system have no counterpart in the $\pi^+ p$ system. What is the isospin of these resonances?

14.9. (a) Do conservation laws permit terms in the pion–nucleon interaction that are quadratic in the pion wave function $\vec{\Phi}$? If so, give an example.

(b) Repeat part (a) for terms cubic in $\vec{\Phi}$. If your answer is yes, give an example.

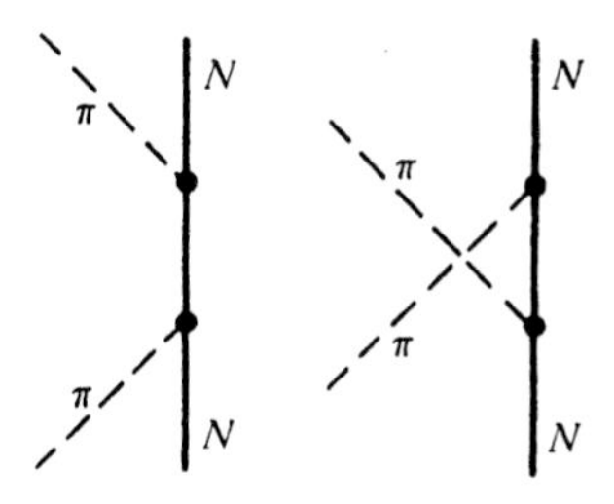

Fig. 14.28

14.10. Use second-order nonrelativistic perturbation theory and the two diagrams in Fig. 14.28 to compute the low-energy pion–nucleon scattering cross section. * Compare with the experimental data.

14.11. Use the decay $\Delta \to \pi N$ to compute a crude value for the coupling constant $f_{\pi N \Delta}$. Compare to $f_{\pi NN}$.

14.12. Assume that particles of 1 GeV kinetic energy are produced at the center of a lead nucleus. Estimate the fraction of particles that escape from the nucleus without interaction if the particles interact

(a) Strongly.

(b) Electromagnetically.

(c) Weakly.

14.13. Show that the Coulomb potential, Eq. (14.21), solves the Poisson equation, Eq. (14.19).

14.14. Show that the Yukawa potential, Eq. (14.24), is a solution of Eq. (14.22).

14.15. Assume attractive spherically symmetric nuclear forces with a range R and point nucleons. Show that the most stable nucleus has a diameter about equal to the force range R. (*Hint:* Consider the total binding energy, the sum of the kinetic and the potential energy, as a function of the nuclear diameter. The nucleus is in its ground state; the nucleons obey Fermi statistics. The arguments in chapter 16 may be helpful.)

14.16. * *Deuteron—Experimental.* Describe how the following deuteron characteristics have been determined:

(a) The binding energy.

(b) The spin.

(c) The isospin.

(d) The magnetic moment.

(e) The quadrupole moment.

14.17. Show that the ground state of a two-body system with central force must be an s state, that is, have orbital angular momentum zero.

14.18. *Deuteron—Theory.* Treat the deuteron as a three-dimensional square well, with depth $-V_0$ and range R.

(a) Write the Schrödinger equation. Justify the value of the mass used in the Schrödinger equation.

(b) Assume the ground state to be spherically symmetric. Find the ground-state wave function inside and outside the well. Determine the binding energy in terms of V_0 and R. Show that B fixes only the product V_0R^2.

(c) Sketch the ground-state wave function. Estimate the fraction of time that the neutron and proton spend outside each other's force range. Why does the deuteron not disintegrate when the nucleons are outside the force range?

14.19. Dineutrons and diprotons, that is, bound states consisting of two neutrons or two protons, are not stable. Explain why not in terms of what is known about the deuteron.

14.20. At one time evidence for a bound state consisting of an antiproton and a neutron had been found, and the binding energy of this $\bar{p}n$ system was noted to be 83 MeV. [L. Gray, P. Hagerty, and T. Kalogeropoulos, *Phys. Rev. Lett.* **26,** 1491 (1971).] Describe this system by a square well with radius $b = 1.4$ fm and depth V_0. Compute V_0 and compare the numerical value with that of the deuteron.

14.21. * Antideuterons have been observed. How were they identified? [D. E. Dorfan et al., *Phys. Rev. Lett.* **14,** 1003 (1965); T. Massam et al., *Nuovo Cim.* **39,** 10 (1965).]

14.22. Verify that a cigar-shaped nucleus, with the nuclear symmetry axis parallel to the z axis, has a positive quadrupole moment.

14.23. Show that the quadrupole moment of a nucleus with spin $\frac{1}{2}$ is zero.

14.24. Show that the quadrupole moment of the deuteron is "small," i.e., that it corresponds to a small deformation.

14.25. The lowest-lying singlet state of the neutron–proton system, with quantum numbers $J = 0, L = 0$, is sometimes called the *singlet deuteron.* It is not bound, and scattering experiments indicate that it occurs just a few keV above *zero* energy; it is just slightly *unbound.* Assume that the singlet state occurs at zero energy, and find the relation between well depth and well radius for a square well. Assume equal singlet and triplet well radii, and show that the singlet well depth is smaller than the triplet one.

14.26. Show that the tensor operator, Eq. (14.33), vanishes if it is averaged over all directions $\hat{\boldsymbol{r}}$.

14.27. Prove that the operator $L = \frac{1}{2}(\boldsymbol{r}_1 - \boldsymbol{r}_2) \times (\boldsymbol{p}_1 - \boldsymbol{p}_2)$ [Eq. (14.37)] is the orbital angular momentum of the two colliding nucleons in their c.m.

14.28. Show that hermiticity of V_{NN}, Eq. (14.36), demands that the coefficients V_i be real.

14.29. Show that translational invariance implies that the coefficients V_i in Eq. (14.36) can depend only on the relative coordinate $\boldsymbol{r} = \boldsymbol{r}_1 - \boldsymbol{r}_2$ of the two colliding nucleons and not on $\boldsymbol{r}_1$ or $\boldsymbol{r}_2$ separately.

14.30. Galilean invariance demands that the transformation

$$\boldsymbol{p}_i' = \boldsymbol{p}_i + m\boldsymbol{v}$$

leaves the V_i in Eq. (14.36) unchanged. Show that this condition implies that V_i can depend only on the relative momentum $\boldsymbol{p} = \frac{1}{2}(\boldsymbol{p}_1 - \boldsymbol{p}_2)$.

14.31. Show that the spin operators $\boldsymbol{\sigma}_1$ and $\boldsymbol{\sigma}_2$ satisfy the relations

$$\begin{aligned}
&\boldsymbol{\sigma}_x^2 = \boldsymbol{\sigma}_y^2 = \boldsymbol{\sigma}_z^2 = 1 \\
&\boldsymbol{\sigma}_x\boldsymbol{\sigma}_y + \boldsymbol{\sigma}_y\boldsymbol{\sigma}_x = 0 \\
&\boldsymbol{\sigma}^2 = 3 \\
&(\boldsymbol{a} \cdot \boldsymbol{\sigma})^2 = a^2 \\
&(\boldsymbol{\sigma}_1 \cdot \boldsymbol{\sigma}_2)^2 = 3 - 2\boldsymbol{\sigma}_1 \cdot \boldsymbol{\sigma}_2.
\end{aligned}$$

14.32. Show that the following eigenvalue equations hold:

$$\begin{aligned}
\boldsymbol{\sigma}_1 \cdot \boldsymbol{\sigma}_2|t\rangle &= 1|t\rangle \\
\boldsymbol{\sigma}_1 \cdot \boldsymbol{\sigma}_2|s\rangle &= -3|s\rangle.
\end{aligned}$$

Here $|s\rangle$ and $|t\rangle$ are the spin eigenstates of the two-nucleon system: $|s\rangle$ is the singlet and $|t\rangle$ is the triplet state.

14.33. Show that the operator

$$P_{12} = \frac{1}{2}(1 + \boldsymbol{\sigma}_1 \cdot \boldsymbol{\sigma}_2)$$

exchanges the spin coordinates of the two nucleons in the two-nucleon system.

14.34. At which energy in the laboratory system does pp scattering become inelastic, i.e., can pions be produced?

14.35. Show that Hamilton's equations of motion, together with Eqs. (14.39) and (14.40), lead to Eq. (14.41).

14.36. Verify Eq. (14.46).

14.37. Show that Eq. (14.47) follows from Eq. (14.46).

14.38. (a) Compute the expectation value of the single-pion exchange potential energy in the s states of two nucleons.

(b) Compute the effective force in any even angular momentum state with spin-1 and with spin-0.

14.39. Explain why, at low energies, the $\overline{p}p$ and the $\overline{p}n$ cross sections are much larger than the pp and the pn ones. [J. S. Ball and G. F. Chew, *Phys. Rev.* **109,** 1385 (1958).]

14.40. Verify Eq. (14.52).

14.41. Show that dimensional analysis leads to Eq. (14.54). Determine the dimension of the constant.

14.42. Show that the total cross section for the scattering of neutrinos and nucleons in asymptotia is given by

$$\sigma_{\text{tot}} = \text{const. } G^2 W^2,$$

where G is the weak coupling constant and W the total energy in the c.m. Compare this result with experiment.

14.43. (a) Can the total photon absorption cross section of Fig. 10.26 be used to obtain the relative strength of the electromagnetic interaction, as outlined in Section 14.1?

(b) What is the appropriate method for making the comparison in this case? Use it to determine the ratio of the electromagnetic and strong strengths.

14.44. What are the spins and parities of the four lowest energy states of glueballs?

14.45. (a) What are the possible eight bi-colored combinations orthogonal to $r\overline{r} + g\overline{g} + b\overline{b}$?

(b) What combinations can be emitted by a red quark?

(c) What combinations can be emitted by a single gluon joining to two other gluons as shown in Fig. 14.20b?

14.46. (a) In attempts at unification of the subatomic forces with gravity, a maximum mass scale and a minimum length sometimes appear; they are the Planck mass and Planck length. Use dimensional arguments to obtain these two measures in terms of $\hbar$, c, and the gravitational constant G.

(b) Evaluate the values of the Planck mass in GeV/c^2 and the Planck length in cm.

14.47. What is the range of the one-gluon exchange force?

14.48. In chapter 10 we demonstrated that color was important in understanding the cross sections of hadrons produced in high-energy electron–positron collisions. What other instances can you think of where experimental evidence for three colors can be obtained?

14.49. If the couplings of the ρ and ω mesons to a nucleon are similar to that of a photon, determine the nucleon–nucleon potentials due to the exchange of these particles.

Part V

Models

> "A model is like an Austrian timetable. Austrian trains are always late. A Prussian visitor asks the Austrian conductor why they bother to print timetables. The conductor replies: If we didn't, how would we know how late the trains are?"
>
> V. F. Weisskopf

Atomic physics is very well understood. A simple model, the Rutherford model, describes the essential structure: A heavy nucleus gives rise to a central field, and the electrons move primarily in this central field. The force is well known. The equation describing the dynamics is the Schrödinger equation or, if relativity is taken into account, the Dirac equation. Historically, this satisfactory picture is not the end result of one single line of research, but it is the confluence of many different streams of discoveries, streams that at one time appeared to have nothing in common. The Mendeleev table of elements, the Balmer series, the Coulomb law, electrolysis, black-body radiation, cathode rays, the scattering of alpha particles, and Bohr's model all were essential steps and milestones. What is the situation with regard to particles and nuclei? We have described the elementary particle zoo and the nature of the forces. Are the known facts sufficient to build a coherent picture of the subatomic world? The theoretical description of *nuclei* is in good shape: There exist successful models, and most aspects of the structure and the interaction of nucleons and nuclei can be described reasonably well. Although many nuclear properties can be obtained from first principles (e.g., through a time-dependent Hartree–Fock treatment), the complexity of the many-body problem usually leads to the replacement of such a description by specific models. They involve the known properties of the nuclear forces but focus on simple modes of motion. Much remains to be done until nuclear theory is as complete and as free from assumptions as atomic physics. The *particle* situation is in about the same shape. Many properties of the particle zoo can be explained rather well in terms of quarks and gluons. The so-called *standard model*, which includes QCD for the strong interactions and the electroweak theory of Chapter 13, can be used to fit much data.

In the following chapters, we shall briefly outline the quark model of particles and some of the most successful nuclear models. The discussion in these chapters is restricted to *hadrons*. Only brief reference will be made to leptons; in particular, the symmetry between leptons and quarks will be pointed out and described.

Photo 6: Sky and Water I, (1938). From *The Graphic Work of M. C. Escher*, Hawthorn Books, New York. [Courtesy of M. C. Escher Foundation, Gemeente Museum, The Hague.] Compare this illustration to Figure 15.3.

Chapter 15

Quark Models of Mesons and Baryons

> Consider all substances; can you find among them any enduring "self"? Are they not all aggregates that sooner or later will break apart and be scattered?
>
> The Teaching of Buddha

15.1 Introduction

The number of subatomic particles is at least as large as the number of elements. To find out how progress in understanding the particle zoo could occur, it is a good idea to take a look at the history of chemistry and atomic physics. The discovery of the periodic table of elements was an essential cornerstone for the development of a systematic chemistry. Rutherford's model of the atom brought a first understanding of the atomic structure, and it formed the basis on which the periodic system of elements could be explained. Quantum mechanics then provided a deeper understanding of Bohr's atom and of the periodic system. Progress in atomic theory thus started from the empirical observation of regularities, proceeded via a model, and it came to a conclusion with the discovery of the dynamical equations.

The time delay between recognizing regularities and explaining them fully was long. The Balmer formula was proposed in 1885; the Schrödinger equation made its appearance 40 years later. The periodic table of elements was discovered in 1869; its explanation in terms of the exclusion principle came 55 years later. Where do we stand in particle physics? The recent developments parallel those just described, but at a much faster rate. Impressive progress has been made, regularities have been found and explained, and QCD is providing a theoretical under-pinning and deeper understanding.

15.2 Quarks as Building Blocks of Hadrons

In 1964, Gell–Mann, and independently Zweig suggested a triplet of hypothetical particles with remarkable properties.[1] Gell–Mann called his particles quarks after *Finnegan's Wake*,[2] whereas Zweig called his particles aces. The name quark has stuck. It is now generally accepted that hadrons are made up of quarks, the properties of which were discussed in Section 5.11. Here, we review some basic properties quarks must have if hadrons are to be made up from them. Quarks must be fermions; it is only with fermion building blocks that both fermions and bosons can be constructed. Quarks have spin 1/2, positive parity, and come in three colors. Mesons are composed primarily of a quark–antiquark pair, and a baryon of three quarks. It is not ruled out that additional mesons and baryons containing one or more additional $q\bar{q}$ pairs and gluons exist.[3]

First, we discuss the structure and relationships of hadrons below a mass of the order of 1 GeV/c^2 and include all quarks below this mass, namely the up, down, and strange quarks. Can symmetry considerations guide us in developing relationships among the low mass hadrons?

Isospin, an internal rotational symmetry, is known to be helpful, as pointed out in Chapter 8. In terms of quarks, this symmetry neglects the mass difference between the up and down quarks and treats them as two species with the same hadronic properties and differing only in charge.

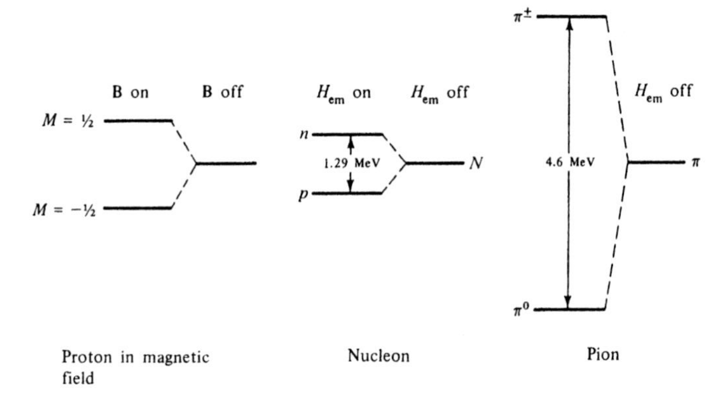

Figure 15.1: Mass (energy) splitting produced by a field. The magnetic field can be switched off; the two magnetic sublevels of the proton then become degenerate. The electromagnetic interaction, however, can be switched off only in a gedanken-experiment.

The strange quark, however, is an isosinglet as far as the strong interactions are concerned, even though it is a member of a weak interaction iso-doublet. Can the strong isospin symmetry be enlarged? Would additional simplicity result if certain parts of the strong interaction were switched off, say the mass difference between the u or d and s quarks? Is this neglect of the order of 150 MeV/c^2 reasonable? To answer these questions we look at particles with the same spin and parity within

[1] M. Gell–Mann, *Phys. Lett.* **8**, 214 (1964); G. Zweig, CERN Report 8182/Th401 (1964).

[2] James Joyce, *Finnegan's Wake*, Viking, New York, 1939, p. 38.

[3] Existing evidence for a pentaquark is very controversial. See CLAS collaboration, *Phys. Rev. Lett.* **97**, 032001 (2006).

a reasonable mass range. To estimate the *reasonable mass range*, we note that the mass splitting due to the electromagnetic interaction is of the order of a few MeV, as indicated in Fig. 15.1. Since the hadronic interaction is about 100 times stronger than the electromagnetic one, a mass splitting of the order of a few hundred MeV can be expected. Since the pion is the lightest hadron, it is tempting to look first at the low-lying 0^- bosons. There are nine such particles below 1 GeV: three pions, two kaons, two antikaons, the eta, and the eta-prime.

These particles are shown to the left in Fig. 15.2. In nature, only the positive and negative members of the same isomultiplet are degenerate, and all other particles possess different masses. If the weak interaction is switched off, the very small splitting between K^0 and $\overline{K^0}$ disappears. If in addition H_{em} is turned off, the neutral and the charged members of the same isospin multiplet become degenerate.

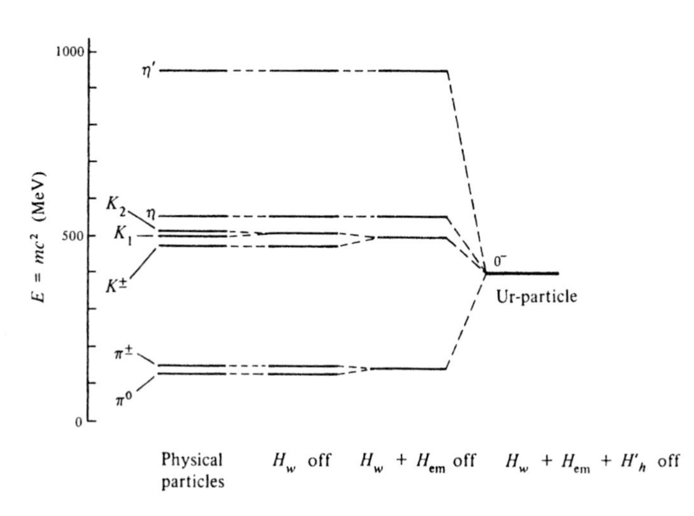

Figure 15.2: The nine pseudoscalar mesons with mass below 1 GeV. At the left the masses are given as they occur in nature. Going to the right, first the weak interaction is switched off, then the electromagnetic interaction, and finally part of the hadronic interaction. The mass splittings caused by H_w and H_{em} are exaggerated. The position of the 0^- particle is unknown.

Finally, it is assumed that all nine pseudoscalar mesons become degenerate if part of the hadronic interaction is turned off. We call the resulting nine-fold degenerate pseudoscalar state the 0^- particle. The mass of the 0^- particle is determined by the part of the hadronic interaction that has not been switched off. According to Fig. 15.2, the 0^- particle gives rise to a native family of nine different particles.

Closer inspection shows that three other particle multiplets can be discerned in the region below a few GeV. The characteristics of the four multiplets are summarized in Table 15.1.

The crucial question is now: Is this scheme useful and can it be brought into more precise form? Does it then yield new predictions? To make the classification more quantitative, we discuss it in terms of quarks.

Table 15.1: HADRONS. The four lowest-lying multiplets of hadrons are listed. They give rise to a total of 36 particles. The rest energy is arbitrarily taken as the central energy of the multiplet.

Spin-parity, J^π	Rest Energy (GeV)	Type	Members of the Multiplet	Number of Members
0^-	0.5	Boson	$\pi K\overline{K}\eta\eta'$	9
1^-	0.8	Boson	$\rho K^*\overline{K}^*\omega\phi$	9
$\frac{1}{2}^+$	1.1	Fermion	$N\Lambda\Sigma\Xi$	8
$\frac{3}{2}^+$	1.4	Fermion	$\Delta\Sigma^*\Xi^*\Omega$	10

15.3 Hunting the Quark

Do quarks exist? Considerable effort has been spent by many experimental groups since 1964 to find quarks in nature, but no conclusive positive evidence has yet been uncovered. Fortunately the fractional electric charges would make quark signatures in careful experiments unambiguous.

In principle, quarks can be produced by high-energy protons through reactions of the type

$$pN \longrightarrow NNq\overline{q} + \text{bosons}, \quad pN \longrightarrow Nqqq + \text{bosons}. \tag{15.1}$$

The thresholds of these reactions depend on the mass m_q of the quarks; the magnitudes of the cross sections are determined by the forces between the hadrons and the quarks. (Since neither the forces nor the quark masses are known, the search is an uncertain affair. If quarks are not found, one never knows if it is because they do not exist, because their mass is too high, or because the production cross section is too low.)

The high energies required to produce massive particles are available in the biggest accelerators, in high-energy colliding beams, and in cosmic rays. Moreover, if free quarks exist, and if the world was created in a "big bang," it is likely that quarks were produced during a very early stage when the temperature was exceedingly high. Some of these original quarks could still be around; searches in sedimentary rocks have not found any.

Quarks can be hunted at accelerators[4] and in cosmic rays. Moreover, since at least one quark must be stable, they should have accumulated in the earth's crust, in meteorites, or in moon rocks. Quarks can be distinguished from other particles either by their fractional charge or by their mass. If the mass is studied, stability is taken as an additional criterion. If the charge is used as signature, the idea is simple. Equation (3.2) shows that the energy loss of a particle in matter is proportional to the square of its charge. A quark of charge $e/3$ would produce one ninth of the ionization of a singly charged particle of the same velocity. If the particle is relativistic,

[4]M. Banner et al., (UA2 Collaboration) *Phys. Lett.* **121B**, 187 (1983).

it produces approximately minimum ionization (see Fig. 3.5). A relativistic quark of charge $e/3$ would therefore show only one ninth of the minimum ionization, and it should have a very different appearance than an ordinary charged particle. A quark with charge $2e/3$ would yield four ninths of the standard ionization.

In reality, experiments are more complicated because it is difficult to find faint tracks. We shall not discuss any of the various experiments here because all the reliable ones have produced negative results.(5) If quarks are found, the excitement will be so enormous that the relevant experiment will be well advertised.

15.4 Mesons as Bound Quark States

According to Eq. (5.66), mesons are bound quark–antiquark pairs. Since the long-range confining part of the QCD force is primarily central, the lowest meson state has zero relative orbital angular momentum, l, between the pair. The intrinsic parity of a fermion–antifermion pair is negative and the two spin-1/2 quarks can form two states(6) with $l = 0$:

$$\begin{array}{lll} {}^1S_0 & J^\pi = 0^- & \text{pseudoscalar mesons,} \\ {}^3S_1 & J^\pi = 1^- & \text{vector mesons.} \end{array}$$

To see how the observed mesons can be understood with these assignments, we consider the quark properties listed in Table 15.2. In "low energy" models of hadrons, the quarks are taken to be dressed by their interactions with gluons. These dressed quarks, particularly useful in hadronic structure calculations, are called constituent quarks; the additional inertia of the virtual gluons makes the light (u, d, and s) quarks in Table 15.2 considerably more massive than their undressed counterparts, the "current" quarks in Table 5.7. The name of these quarks stems from their role in quark currents, as in the electroweak theory.

The mass values in Table 15.2 imply that only the three light quarks u, d, and s and their antiparticles need be considered for the mesons with masses less than 1GeV/c^2, listed in Table 15.1. The relevant flavor combinations are

$$\begin{array}{ccc} u\bar{u} & d\bar{u} & s\bar{u} \\ u\bar{d} & d\bar{d} & s\bar{d} \\ u\bar{s} & d\bar{s} & s\bar{s} \end{array} \tag{15.2}$$

In writing these combinations it must be remembered that each quark comes in three colors (red, green, blue) and that the observed particles are color neutral

[5] L. W. Jones, *Rev. Mod. Phys.* **49**, 717 (1977); L. Lyons, *Phys. Rep.* **129**, 226 (1985).

[6] The ordinary spectroscopic notation is used where the capital letter gives the orbital angular momentum, the subscript indicates the value of the total angular momentum, and the left superscript is equal to $2S + 1$, where S is the spin. ${}^3S_1$ thus denotes a state with $l = 0, J = 1, S = 1, 2S + 1 = 3$.

Table 15.2: SOME PROPERTIES OF CONSTITUENT QUARKS.

Quark	Charge (e)	Spin	I	I_3	S	C	B	T	Mass (MeV/c^2)†
u	2/3	1/2	1/2	1/2	0	0	0	0	330
d	−1/3	1/2	1/2	−1/2	0	0	0	0	336
s	−1/3	1/2	0	0	−1	0	0	0	540
c	2/3	1/2	0	0	0	1	0	0	1,550
b	−1/3	1/2	0	0	0	0	−1	0	4,800
t	2/3	1/2	0	0	0	0	0	1	178,000

†The masses cannot be measured directly; they are model-dependent, and hence approximate.

Table 15.3: REORDERING THE $q\bar{q}$ STATES ACCORDING TO STRANGENESS AND ISOSPIN COMPONENT I_3.

	$I_3 = -1$	$-1/2$	0	$1/2$	1
S = 1		$d\bar{s}$		$u\bar{s}$	
S = 0	$d\bar{u}$		$u\bar{u}, d\bar{d}, s\bar{s}$		$u\bar{d}$
S = −1		$s\bar{u}$		$s\bar{d}$	

(color singlets). Quarks of all three colors and three anticolors must appear with equal probability so that, for instance, the product $u\bar{u}$ should really be written as

$$\frac{u_r\bar{u}_r + u_g\bar{u}_g + u_b\bar{u}_b}{\sqrt{3}}, \tag{15.3}$$

where the subscripts denote the colors.

The matrix, (15.2), implies the existence of nine different mesons, in agreement with the numbers listed in Table 15.1. However, the arrangement in Eq. (15.2) is not made according to quantum numbers, and comparison with the experimentally observed mesons is thus not obvious. In Table 15.3, the nine combinations are reordered according to the values of the strangeness S and the isospin component I_3. Table 15.2 is helpful in such rearrangements. The states in Table 15.3 can now be compared with the nine pseudoscalar and the nine vector mesons. For the *pseudoscalar mesons*, arranging these in the same scheme gives

$$\begin{array}{ccccc} & K^0 & & K^+ & \\ \pi^- & & \pi^0\eta^0\eta' & & \pi^+ \\ & K^- & & \overline{K^0} & \end{array} \tag{15.4}$$

and for the *vector mesons*

$$\begin{array}{ccccc} & K^{*0} & & K^{*+} & \\ \rho^- & & \rho^0\omega^0\phi^0 & & \rho^+ \\ & K^{*-} & & \overline{K^{*0}} & \end{array} . \tag{15.5}$$

Table 15.4: NONSTRANGE NEUTRAL MESONS.

Meson	$I(J^\pi)$	Rest Energy (MeV)	Meson	$I(J^\pi)$	Rest Energy (MeV)
π^0	$1(0^-)$	135	ρ^0	$1(1^-)$	770
η^0	$0(0^-)$	549	ω^0	$0(1^-)$	782
η'	$0(0^-)$	958	ϕ^0	$0(1^-)$	1019

In both cases, the assignments for the six states in the *outer ring* are unambiguous. The three states in the center, however, have the same quantum numbers S and I_3. How are the states $u\overline{u}, d\overline{d}$ and $s\overline{s}$ related to the corresponding mesons with $S = I_3 = 0$? Since any linear combination of states $u\overline{u}, d\overline{d}$ and $s\overline{s}$ has the same quantum numbers, it is not possible to identify one quark combination with one meson. To get more information, we summarize the properties of the nonstrange neutral mesons in Table 15.4, which shows that it will be straightforward to find the quark content of the neutral pion and the neutral rho: these two particles are members of isospin triplets. Knowing the quark assignment of the other members of the isotriplet should help. Consider, for instance, the three rho mesons:

$$\rho^+ = u\overline{d} \quad \rho^0 = ?, \quad \rho^- = d\overline{u}. \tag{15.6}$$

The charged members of the rho do not contain a contribution from the strange quark in their wave function. The neutral rho forms an isospin triplet with its two charged relatives and thus should also not contain a strange component. Of the three products listed in the $I_3 = 0, S = 0$ entry in Table 15.3, only the first two can appear, and the wave function must have the form

$$\rho^0 = \alpha u\overline{u} + \beta d\overline{d}.$$

Normalization and symmetry give

$$|\alpha|^2 + |\beta|^2 = 1, \quad |\alpha| = |\beta|, \quad \text{or} \quad \alpha = \pm\beta = \frac{1}{\sqrt{2}}.$$

If we were to add up two ordinary spin-1/2 particles to get a spin-1 system, it would be easy to select the correct sign: The linear combination must be an eigenfunction of J^2, with eigenvalue $j(j+1)\hbar^2 = 2\hbar^2$. This condition determines that the sign is positive.(7) The situation here is different, because we are dealing with particle-antiparticle pairs and the antiparticle introduces a minus sign. We shall not justify the appearance of this minus sign because it will not occur in any measurable quantity in our discussion. The wave functions of the three rho mesons in terms of

[7] Park, Eq. (6.35); Merzbacher, Eq. (16.85); G. Baym, *Lectures in Quantum Mechanics*, Benjamin, Reading, Mass., 1969, Chapter 15.

their quark constituents are

$$\begin{aligned}\rho^+ &= u\bar{d} \\ \rho^0 &= \frac{d\bar{d} - u\bar{u}}{\sqrt{2}} \\ \rho^- &= d\bar{u}.\end{aligned} \tag{15.7}$$

These quark combinations also apply to the pions; the difference between the rho and the pion lies in the ordinary spin. The rho is a vector meson ($J^\pi = 1^-$), while the pion is a pseudoscalar meson ($J^\pi = 0^-$). The other neutral mesons will be discussed in Section 15.6.

If masses beyond 1 GeV/c^2 are considered, mesons of orbital angular momentum $l = 1\hbar$ begin to appear, with $J^\pi = 0^+, 1^+$, and 2^+; they correspond to $q\bar{q}$ states $^1P_1, ^3P_0, ^3P_1$, and $^3P_2$; the isospin can be zero or one, as for the lower mass mesons.

15.5 Baryons as Bound Quark States

Three quarks form a baryon. Since quarks are fermions, the overall wave function of the three quarks must be antisymmetric; the wave function must change sign under any interchange of two quarks:

$$|q_1q_2q_3\rangle = -|q_2q_1q_3\rangle. \tag{15.8}$$

To explain why the wave function of the three quarks must be antisymmetric, the ideas expounded in Chapter 8 are generalized. There, with the introduction of isospin, proton and neutron were considered to be two states of the same particle. The total wave function, including isospin, of a two-nucleon system then must be antisymmetric under exchange of the two nucleons. Here it is assumed that the three quarks are three states of the same particle, and Eq. (15.8) is then the expression of the Pauli principle. The simplest situation arises when the three quarks have no orbital angular momentum between any pairs and have their spins parallel. The resultant baryon then has spin 3/2 and positive parity. As in the case of the mesons, it is straightforward to find the quantum numbers of the various quark combinations. Consider, for instance, the combination uuu.

$$uuu: \quad A = 1, S = 0, I_3 = \frac{3}{2}, q = 2e, J = \left(\frac{3}{2}\right),$$

where S is the strangeness. These are just the quantum numbers of the Δ^{++}, the doubly charged member of the $\Delta(1232)$. For a Δ^{++}, however, with all spin components parallel ($J = 3/2, J_z = 3/2$) and all quarks in S^+ states, or no orbital angular momentum, the wavefunction is symmetric under interchange of any pair of quarks.

The lack of antisymmetry of the wavefunction was a large impediment for the development of the quark model until the idea of an extra degree of freedom appeared. This new degree, color, was introduced initially to solve the antisymmetry puzzle.[8] Its effect on the meson wavefunction is given in Eq. (15.3). With three colors, an antisymmetric colorless (white) wavefunction can be formed. If the three colors were three unit vectors along the x, y and z axes in color space, the colorless (scalar) combination would be $\hat{x} \cdot \hat{y} \times \hat{z}$. If we denote the three colors by a, b, c, the unnormalized color singlet combination of quarks can be written as

$$\sum_{a,b,c} \epsilon_{abc} q_a q_b q_c, \tag{15.9}$$

with a, b, c running over the three colors red, green, and blue; ϵ_{abc} is the antisymmetric tensor which is $+1$ for even permutations of a, b, c (r, g, b) and -1 for odd ones. Three colors are the minimum required to form an antisymmetric state of three quarks. Although color was introduced in an *ad hoc* fashion, it has become all important in our understanding of the strong interactions through QCD, as discussed in Section 14.8. The evidence for color includes saturation of the lowest mass baryons by three quarks and mesons by $q\bar{q}$, an explanation of the decay width of the π^0 to two photons, and the magnitude of the cross section for reactions such as $e^+e^- \rightarrow$ hadrons, as discussed in Section 10.9.

The three quarks u, d and s can be combined to form 10 combinations, and particles exist for all 10. The quark combinations and the corresponding baryons are shown in Fig. 15.3.

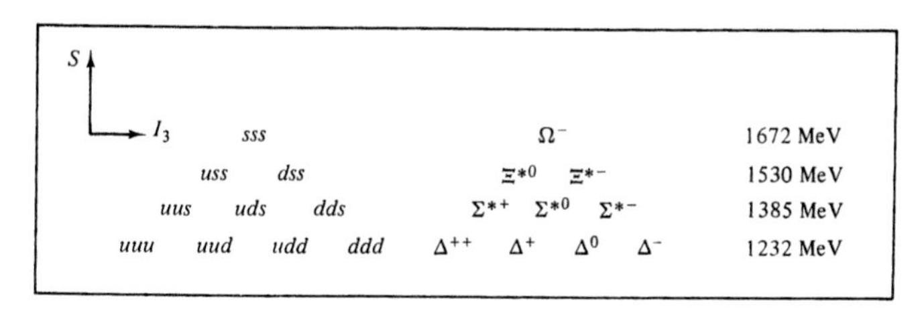

Figure 15.3: Quarks and the $(3/2)^+$ decimet. The states and the particles are arranged so that the x axis gives I_3, and the y axis S. The rest energies are given at the right.

Also indicated are the rest energies of the isomultiplets. Since there are 10 particles, the array is called the $(3/2)^+$ *decimet* (or decuplet). The similarity to Escher's "Sky and Water I" on p. 470 is impressive, particularly if it is noted that the decimet of the antiparticles also exists.

Three spin-1/2 fermions in an S-state can also be coupled to form a state with spin 1/2 and positive parity. Examples in nuclear physics are ^{3}H and ^{3}He. Table 15.1 indicates that only eight members of the $(1/2)^+$ family are known. The eight particles and the corresponding quark combinations are shown in Fig. 15.4.

[8] O. W. Greenberg, *Phys. Rev. Lett.* **13,** 598 (1964); M. Y. Han and Y. Nambu, *Phys. Rev.* **139,** B1006 (1965).

	uss		*dss*		Ξ^0		Ξ^-	1318 MeV
uus		*uds*		*dds*	Σ^+	Σ^0	Σ^-	1192 MeV
		uds				Λ^0		1116 MeV
	uud		*udd*		p		n	939 MeV

Figure 15.4: The $(1/2)^+$ baryon octet and the corresponding quark combinations. The rest energies of the isomultiplets are given at the right. All states are antisymmetric in color.

Two questions are raised by the comparison of existing particles and quark combinations in Fig. 15.4: (1) Why are the corner particles uuu, ddd, and sss present in the $(3/2)^+$ decimet but absent in the $(1/2)^+$ octet? (2) Why does the combination uds appear twice in the octet but only once in the decimet? Both questions have a straightforward answer:

1. No symmetric (or antisymmetric) state with spin 1/2 and zero angular momentum can be formed from three identical fermions. (Try!) The "corner particles" in the $(1/2)^+$ octet are therefore forbidden by the Pauli principle, Eq. (15.8), and indeed are not found in nature.

2. If the z component of each quark spin is denoted with an arrow, a state with $L = 0$ and $J_z = +1/2$ can be formed in three different ways:

$$u\uparrow d\uparrow s\downarrow, \quad u\uparrow d\downarrow s\uparrow, \quad u\downarrow d\uparrow s\uparrow. \tag{15.10}$$

   From these three states, three different linear combinations can be formed that are orthogonal to each other and have a total spin J. Two of these combinations have spin $J = 1/2$ and one has spin $J = 3/2$. The one combination with $J = 3/2$ turns up in the decimet; the two others are members of the octet.

15.6 The Hadron Masses

A remarkable regularity appears if the masses of the particles are plotted against their *quark content*. In the last two sections we have found definite assignments of quark combinations to all of the hadrons that comprise the set of the four multiplets listed in Table 15.1. A careful look at the mass values of the various states shows that the mass depends strongly on the number of strange quarks. In Fig. 15.5, the rest energies of most of the particles are plotted, and the number of strange quarks is indicated for each level. The masses of the various states can be understood if it is assumed that the nonstrange constituent quarks are approximately equally heavy but that the strange quark is heavier by an amount Δ (see Table 15.1)

$$m(u) = m(d), \quad m(s) = m(u) + \Delta. \tag{15.11}$$

Figure 15.5 implies that the value of Δ is of the order of two hundred MeV$/c^2$, in agreement with Table 15.2. The fact that the observed levels are not all equally

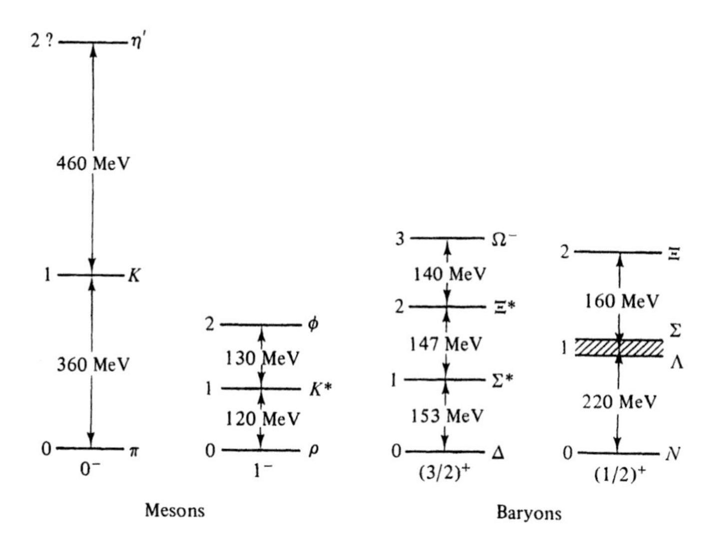

Figure 15.5: Particle rest energies. Each level is labeled with the number of s quarks that it contains.

spaced is not surprising. The mass of a meson built from quarks q_1 and $\overline{q}_2$ is given by

$$m = m(q_1) + m(\overline{q}_2) - \frac{B}{c^2}.$$

It is too much to hope that the binding energy B is exactly the same for all mesons and baryons. B depends on the nature of the forces between quarks and on the state of the quarks. Figure 15.5 therefore provides only a crude value for the mass difference Δ.

A few observations follow directly from the simple arguments made so far. The first one concerns the Ω^-. When Gell–Mann introduced strangeness he conjectured that a particle with strangeness -3 should exist and called it Ω^-. In terms of the quark structure of hadrons and Fig. 15.3, the conjecture is easy to understand. With u, d, and s quarks as units, a baryon composed of three quarks and mesons made of quark–antiquark pairs, a baryon can have any strangeness between 0 and -3 and a meson strangeness 0, and ± 1. The possible isospins and charge characteristics also follow easily from this picture. Gell–Mann used group symmetry arguments to predict the mass of the Ω^-,[9] but the prediction can be understood by looking at Fig. 15.5. Once Gell–Mann had written down all particles except Ω^-, the top of the pyramid followed logically. Fig. 15.5 shows that the energy differences between the three lower layers of the pyramid are 153 and 147 MeV, respectively. Consequently, the top of the pyramid should be about 140 MeV above the rest energy of the Ξ^* and that was where the Ω^- was found.[10]

[9] See M. Gell–Mann and Y. Ne'eman, *The Eightfold Way*, Benjamin, Reading, MA, 1964.
[10] The first Ω^- was probably seen in a cosmic-ray experiment in 1954 [Y. Eisenberg, *Phys. Rev.*

The masses of the mesons are somewhat more difficult to obtain. There is no pure $s\bar{s}$ state for the pseudoscalar mesons (although the ϕ is almost so); also the pion has an abnormally low mass because it may not be a pure $q\bar{q}$ state; it may be a partial Goldstone boson.[11] In Fig. 15.5 we have arbitrarily taken the mass of the pseudoscalar $s\bar{s}$ state to lie half-way between the masses of the η and η' mesons. The pseudoscalar nonet masses cannot be obtained in this approximate manner.

The second observation concerns the mass splitting within a multiplet of a given spin and parity. This splitting could be caused by the fact that the force between a strange and nonstrange quark differs from that between two strange or two nonstrange ones, but it is much simpler to interpret it as due to the mass difference between the constituent strange and nonstrange quarks. Indeed, a study of $c\bar{c}$ and $b\bar{b}$ systems shows that, for a given spin and orbital angular momentum, the dominant longrange QCD confining force is independent of flavor (quark type).[12] Thus, we can say that if we neglect the mass splitting between the s, and u or d quarks, we would have degenerate multiplets of 0^- and 1^- mesons and also $1/2^+$ and $3/2^+$ baryons, as shown in Table 15.1. The u, d, and s quarks thus also form a multiplet that is a generalization of an isospin multiplet.

The third observation is that the QCD force depends on spin. The mass splitting, primarily due to a spin–spin force, Eq. (15.16), is of the order of 300 MeV/c^2 between both the 0^- and 1^- and between the $1/2^+$ and $3/2^+$ multiplets, as shown in Table 15.1.

The last observation leads us back to the problem of the neutral mesons. This problem was only partially solved in Section 15.4. In Eq. (15.7) the quark composition of the ρ^0 was given, but ω^0 and ϕ^0 were left without assignment. Figure 15.5 implies that ϕ^0, which is about 130 MeV above K^*, is almost solely composed of two strange quarks:

$$\phi^0 = s\bar{s}. \tag{15.12}$$

The state function of ω^0 can now be found by setting

$$\omega^0 = c_1 u\bar{u} + c_2 d\bar{d} + c_3 s\bar{s}. \tag{15.13}$$

The state representing ω^0 should be orthogonal to the states representing ρ^0 and ϕ^0. With Eqs. (15.12) and (15.7), the state of ω^0 then becomes

$$\omega^0 = \frac{1}{\sqrt{2}}(u\bar{u} + d\bar{d}) \tag{15.14}$$

and the mass of ω^0 should satisfy

$$m_{\omega^0} \approx m_{\rho^0}. \tag{15.15}$$

This prediction is in approximate agreement with reality.

96, 541 (1954)]. The unambiguous discovery, however, occurred in 1964 [Barnes et al., *Phys. Rev. Lett.* **12,** 204 (1964)]. See also W. P. Fowler and N. P. Samios, *Sci. Amer.* **211,** 36 (April 1964).

[11] W. Weise, *Nucl. Phys.* **A434,** 685 (1985) and *Prog. Part. Nucl. Phys.*, (A. Faessler, ed) **20,** 113 (1988); C.P. Burgess, *Phys. Rep.* **330**, 193 (2000).

[12] N. Isgur and G. Karl, *Phys. Today* **36**, 36 (November 1983).

15.7 QCD and Quark Models of the Hadrons

The basic theory of hadronic forces is now known to be quantum chromodynamics or QCD. Some of the features of QCD were developed in Section 14.8. The potential between massive quarks was given by Eq. 14.59 and shown in Fig. 14.26. It is approximately proportional to $1/r$ at short distances and becomes linear at large separations. The short distance behavior suggests that we examine the spectrum of positronium, an atom of e^+ and e^- bound by Coulomb and magnetic forces. It is a relativistic problem, but we can be guided by a nonrelativistic approximation. In that case, the problem reduces to that of a hydrogen atom with an effective reduced mass of the electron $= m_e/2$, but with states of opposite parities to those of hydrogen because the parity of the particle–antiparticle system is negative.

The lowest energy levels of the spectrum are shown in Fig. 15.6. The ground state is the usual $1S^-$ state. Because the problem is relativistic, we cannot neglect the magnetic force between the electron and positron. This force is spin-dependent, of the form

$$V = \kappa \boldsymbol{\sigma_1} \cdot \boldsymbol{\sigma_2} \delta^3(\mathbf{r}), \tag{15.16}$$

where κ is a constant, $\delta^3(\mathbf{r})$ is a Dirac-delta function which vanishes everywhere except at $\mathbf{r} = 0$ and has the property that $\int \delta^3(\mathbf{r}) d^3r = 1$.

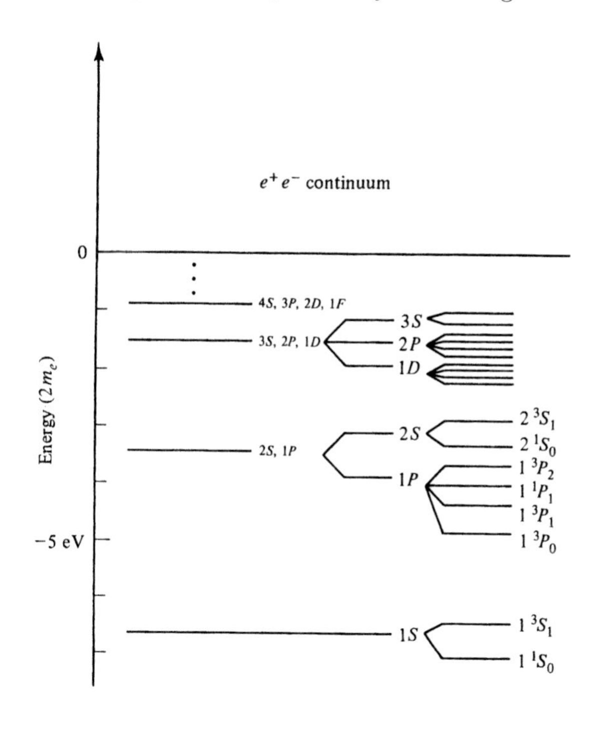

Figure 15.6: Energy levels of the positronium atom. The splitting of the Coulomb energy levels, shown at the left, is schematic and magnified. The splitting is due to spin–orbit and spin–spin forces. [From N. Isgur and G. Karl, *Phys. Today* **36**, 36 (November 1983).]

The spin dependence causes a splitting between the $^3S_1$ and $^1S_0$ states, the latter one lying lower in energy. A similar splitting is observed in atomic systems.

Features similar to the e^+e^- spectrum can be seen in the spectrum of the light mesons.[12] The lowest state, comprising the mesons, corresponds to the $^1S_0$ ground state; the spin-dependent splitting gives rise to the $^3S_1$ vector mesons. The $1P$

mesons correspond to the $1P$ states of e^+e^-. Calculations based on a potential like that of Eq. (14.59) give reasonable agreement with the experimental masses of both mesons and baryons.[13] We will see these features in more detail in Section 15.8.

The QCD long-range confining force is approximately constant in space, corresponding to a linear potential. A well-known confining potential in physics is that of a spring or harmonic oscillator. Although this potential differs from a linear one, it may provide guidance in establishing energy levels or masses and other properties of hadrons. Indeed, one of the qualitatively successful quark models for describing hadrons and their interactions uses a harmonic potential at large distances and one-gluon exchange to describe the short-distance force between quarks.[14] We sketch some of the relevant ideas, and we start by discussing the energy levels of the three-dimensional harmonic oscillator. Since these energy levels will reappear in the nuclear shell model in Chapter 17, the harmonic oscillator is treated here in more detail than would otherwise be necessary.[15] The physical facts are simple, but the complete mathematics is somewhat involved; only the parts needed here and in Chapter 17 are given.

A particle attracted toward a fixed point by a force proportional to the distance r' from the point has a potential energy

$$V(r') = \frac{1}{2}\kappa r'^2. \tag{15.17}$$

The Schrödinger equation for such a three-dimensional harmonic oscillator is

$$\boldsymbol{\nabla}^2\psi + \frac{2m}{\hbar^2}\left(E - \frac{1}{2}\kappa r'^2\right)\psi = 0. \tag{15.18}$$

With the substitutions

$$\kappa = m\omega^2, \quad r' = \left(\frac{\hbar}{m\omega}\right)^{1/2} r, \quad E = \frac{1}{2}\hbar\omega\lambda, \tag{15.19}$$

the Schrödinger equation reads

$$\boldsymbol{\nabla}^2\psi + (\lambda - r^2)\psi = 0. \tag{15.20}$$

[13] A. de Rújula, H. Georgi, and S. L. Glashow, *Phys. Rev.* **D12**, 147 (1975); N. Isgur and G. Karl, *Phys. Rev.* **D18**, 4187 (1978), **D19**, 2653 (1979) **D20**, 1191 (1979); N. Isgur in *Particles and Fields–1981: Testing the Standard Model*, (C. A. Heusch and W. T. Kirk, eds) *AIP Conf. Proc. 81, Amer. Inst. Phys.*, New York, 1982, p. 7; S. Godfrey and N. Isgur, *Phys. Rev.* **D32**, 189 (1985).

[14] M. G. Huber and B. C. Metsch, *Prog. Part. Nucl. Phys.*, (A. Faessler, ed.) **20**, 187 (1988) M. Oka and K. Yazaki, *Prog. Theor. Phys.* **66**, 556, 572, (1981); A. Faessler et al., *Nucl. Phys.* **A402**, 555 (1983).

[15] The one-dimensional harmonic oscillator is treated, for instance, in Merzbacher. The three-dimensional oscillator can be found in Messiah, Section 12.15, or in detail in J. L. Powell and B. Crasemann, *Quantum Mechanics*, Addison-Wesley, Reading, Mass., 1961, Section 7.4.

Since the harmonic oscillator is spherically symmetric, it is advantageous to write the Schrödinger equation in spherical polar coordinates, r, θ, and φ. In these coordinates, the operator $\boldsymbol{\nabla}^2$ becomes

$$\boldsymbol{\nabla}^2 = \frac{1}{r^2}\frac{\partial}{\partial r}\left(r^2\frac{\partial}{\partial r}\right) - \frac{1}{r^2\hbar^2}\boldsymbol{L}^2, \tag{15.21}$$

where $\boldsymbol{L}^2$ is the operator of the square of the total angular momentum,

$$\boldsymbol{L}^2 = -\hbar^2\left[\frac{1}{\sin\theta}\frac{\partial}{\partial\theta}\left(\sin\theta\frac{\partial}{\partial\theta}\right) + \frac{1}{\sin^2\theta}\frac{\partial^2}{\partial\varphi^2}\right]. \tag{15.22}$$

An ansatz of the form

$$\psi = R(r)Y_l^m(\theta,\varphi) \tag{15.23}$$

solves Eq. (15.20), where the Y_l^m are spherical harmonics. Y_l^m is an eigenfunction of L^2 and L_z (compare Eq. (5.7)),

$$\boldsymbol{L}^2 Y_l^m = l(l+1)\hbar^2 Y_l^m, \quad L_z Y_l^m = m\hbar Y_l^m. \tag{15.24}$$

The radial wave function $R(r)$ satisfies

$$\frac{1}{r^2}\frac{d}{dr}\left(r^2\frac{dR}{dr}\right) + \left(\lambda - r^2 - \frac{l(l+1)}{r^2}\right)R = 0. \tag{15.25}$$

This equation can be solved in a straightforward way(15) and the results of interest here can be summarized as follows.[16] Equation (15.25) has acceptable solutions only if

$$E_N = \left(N + \frac{3}{2}\right)\hbar\omega, \tag{15.26}$$

where N is an integer, $N = 0, 1, 2, \ldots$. The potential and the energy levels are shown in Fig. 15.7.

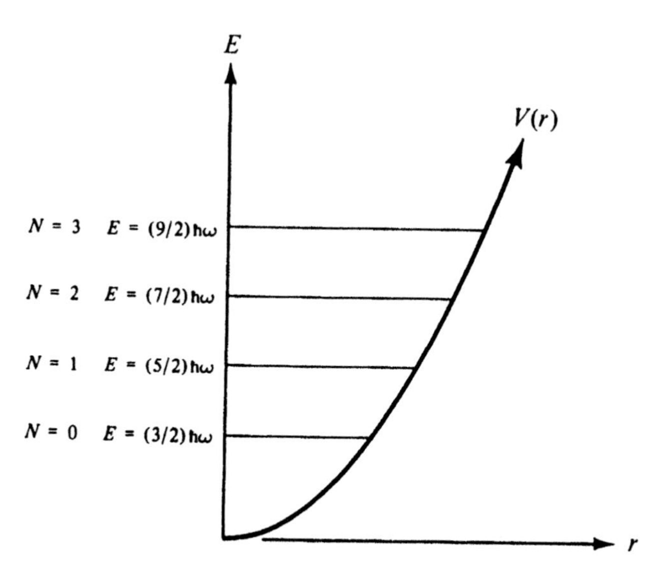

Figure 15.7: Three-dimensional harmonic oscillator and its energy levels.

[16] Various definitions of the quantum numbers are in use. Our notation agrees with A. Bohr and B. R. Mottelson, *Nuclear Structure*, Vol. I. Benjamin, Reading, Mass. 1969, p. 220.

The complete wave function is given by

$$\psi_{Nlm} = \left(\frac{2}{r}\right)^{1/2} \Lambda_k^{l+1/2}(r^2) Y_l^m(\theta, \varphi), \quad k = \frac{1}{2}(N - l), \tag{15.27}$$

where $\Lambda(r^2)$ is a Laguerre function. It is related to the more familiar Laguerre polynomials $L_k^\alpha(r)$ by

$$\Lambda_k^\alpha(r^2) = \left[\Gamma(\alpha + 1)\left(k + \alpha k\right)\right]^{-1/2} \exp\left(-\frac{r^2}{2}\right) r^\alpha L_k^\alpha(r^2). \tag{15.28}$$

At first, these functions appear terrifying. However, they become docile if one simply looks up their properties and behavior in one of the many books on mathematical physics.[17] The radial wave functions of the first three levels ($N = 0, 1, 2$) are shown in Fig. 15.8. What is the physical meaning of the indices N, l, and m? N has already been defined in Eq. (15.26); it labels the energy levels. Equation (15.24) shows that l is the orbital angular momentum quantum number; it is restricted to values $l \le N$.

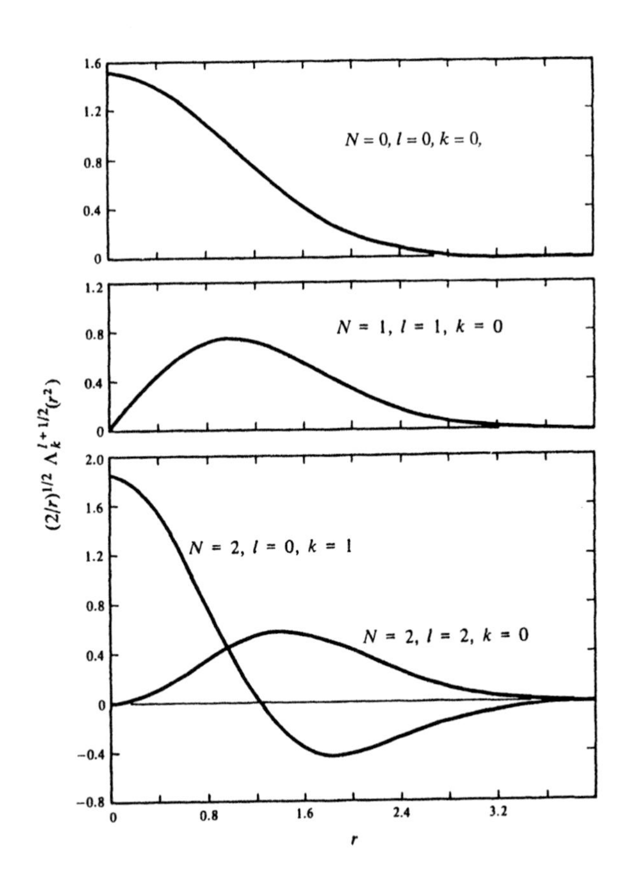

Figure 15.8: Normalized radial wave functions $(2/r)^{1/2}\Lambda$ for the three-dimensional harmonic oscillator. The distance r is measured in units of $(\hbar/m\omega)^{1/2}$.

[17] For example, P. M. Morse and H. Feshbach, *Methods of Theoretical Physics*, McGraw-Hill, New York, 1953, Section 12.3, Eq. (12.3.37).

For each value of l, the magnetic quantum number m can assume the $2l+1$ values from $-l$ to l. The parity of each state is given by Eq. (9.10) as

$$\pi = (-1)^l.$$

States of even and odd parity exist, and consequently the possible orbital angular momenta for a state with quantum number N are given by

$$\begin{array}{lll} N\text{even} & \pi\text{even} & l = 0, 2, \ldots, N \\ N\text{odd} & \pi\text{odd} & l = 1, 3, \ldots, N. \end{array} \tag{15.29}$$

The degeneracy of each level N can now be obtained by counting: The possible angular momenta are determined by Eq. (15.29); each angular momentum contributes $2l+1$ substates, and the total degeneracy becomes

$$\frac{1}{2}(N+1)(N+2). \tag{15.30}$$

The radial wave function $R(r) = (2/r)^{1/2}\Lambda$ is characterized by the number, n_r, of its nodes. It is customary to exclude nodes at $r = 0$ and include nodes at $r = \infty$ in counting. The examples in Fig. 15.8 then show that

$$n_r = 1 + k = 1 + \frac{1}{2}(N - l). \tag{15.31}$$

This relation is valid for all radial wave functions $R(r)$.

After this long preparation we return to our goal, connecting the properties of the harmonic oscillator to particle models. A state of a particle can be characterized by its mass (energy) and its angular momentum. In Fig. 15.9, we show the lowest few levels of the harmonic oscillator, labeled by the quantum numbers N, the radial quantum numbers n_r, and the angular momenta in units of $\hbar$, and the corresponding levels of the e^+e^- system without the magnetic force effects.

We expect the $q\overline{q}$ bound states to lie somewhere between these two extremes, as shown in the figure; we have also included the effect of the short-range spin–spin force, Eq. (15.16), for the lowest several states. The first two states correspond to the 0^- and 1^- multiplets, the next two to ^3P and $^1P_1$ states. The ^3P state is split by spin–orbit forces into $0^+, 1^+$, and 2^+ meson multiplets, most of the members of which have masses above 1 GeV/c^2. The center of the $^1P_1$ state is at about 1240 MeV/c^2, as is that of the $^3P_0$ state. The $^3P_1$ state lies at about 1350 MeV/c^2 and the $^3P_2$ state at 1400 MeV/c^2.

The harmonic oscillator shows another feature of the particle spectrum, namely, some general relationships between particles of different spins but the same parity. In some cases these particles appear to be rotationally excited states of the particle of lowest mass; we then expect to find a multiplet with the same number of components as in the lowest mass state.

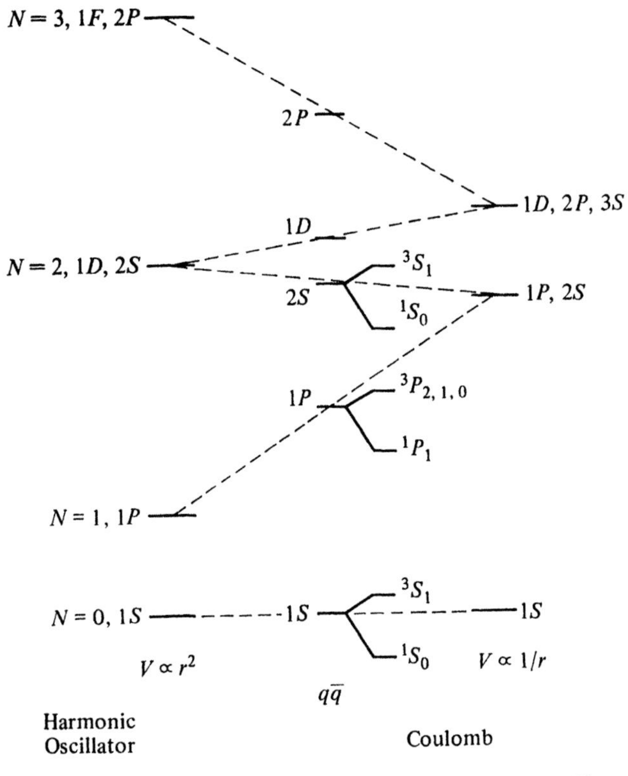

Figure 15.9: Lowest few energy levels for a harmonic oscillator potential, Coulomb potential, and $q\bar{q}$. The latter are taken to lie about halfway between the levels of the other two potentials and the spin–spin splitting is shown. The labels, in addition to N, are the radial quantum numbers and the orbital angular momenta.

Here we sketch some ideas relevant to these higher mass states and to the long-range confining force of quarks, taken to be a harmonic oscillator potential. A plot of the angular momentum, l, against energy is shown in Fig. 15.10. The states in Fig. 15.10 can be ordered into families in a variety of ways: states with equal values of N, of l, or of n_r can be connected. In Fig. 15.10, the last possibility has been chosen, and the result is a series of straight-line trajectories that rise with increasing energy. The straightness is a property of the harmonic oscillator; if a different potential shape is chosen, the trajectories will in general no longer be straight, but the general appearance remains. Why have levels with equal n_r rather than equal l been connected? The quantum numbers l and n_r have a different physical origin. We can, in principle, take a quantum mechanical system and spin it with various values of its angular momentum without changing its internal structure. The quantum number l describes the behavior of the system under rotations in space, and it can be called an *external quantum number*. The number of radial nodes, however, is a property of the structure of the state, and n_r (like intrinsic parity) can be called an *internal quantum number*. In this sense the states on one trajectory have a similar structure. Actually, the particles lying on a given trajectory can be further subdivided: States with the same parity recur at intervals $\Delta l = 2$. State B in Fig. 15.9 can be considered to be state A recurring with a higher angular momentum.

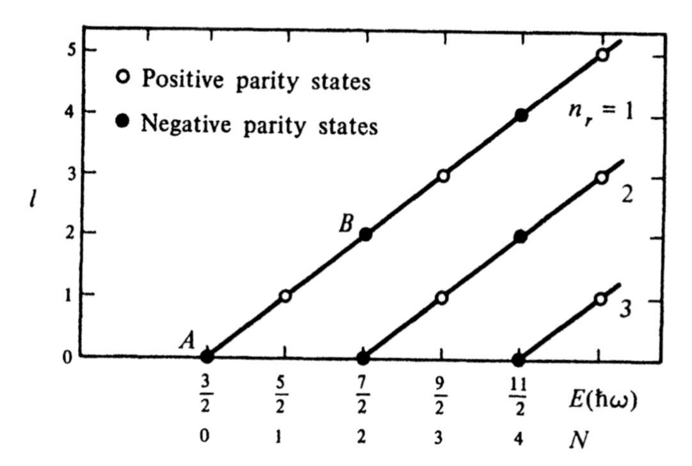

Figure 15.10: Plot of the angular momentum against energy for the states of the three-dimensional harmonic oscillator.

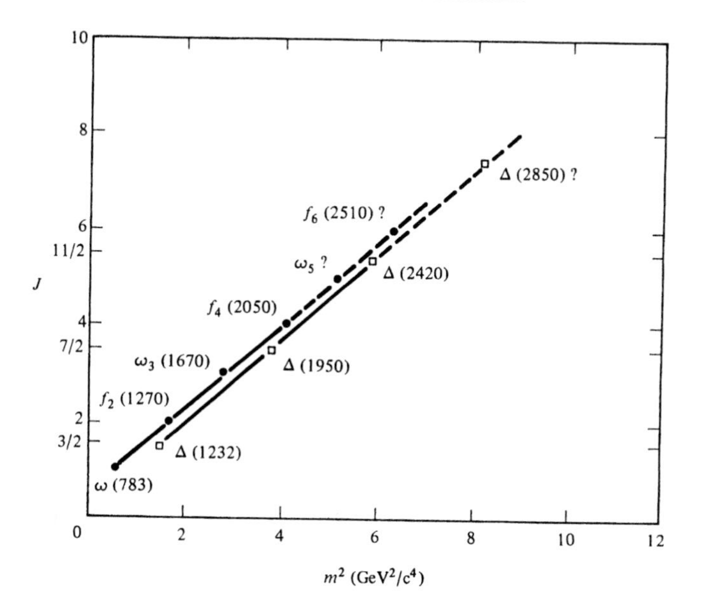

Figure 15.11: Plot of the spin against squared mass for isospin zero, negative parity mesons and for isospin 3/2, positive parity baryons.

The question now arises: Do particles show a similar behavior if the particle mass is plotted against particle spin for particles that have the same internal quantum numbers? Indeed, pronounced regularities appear, and we present an example in Fig. 15.11. Here we show the spins of the negative parity isospin 0 mesons and of the positive parity isospin 3/2 baryons as a function of the square of the particle masses. The appearance of a family for each case is clear, and the similarity to Fig. 15.10 is evident. The graph is called a Chew–Frautschi plot and also Regge trajectory; the higher mass particles are called Regge recurrences of the lowest mass state.(9)

[18] V.D. Barger and D.B. Cline, *Phenomenological Theories of High Energy Scattering, An Experimental Evaluation*, Benjamin, Reading, Mass., 1969.

It is remarkable that the slopes of the two Regge trajectories are quite similar, about 0.9 GeV$^2/c^4$. This slope is related to the spring constant in the case of the harmonic oscillator.

The QCD confining forces for quarks are strong, highly nonlinear and difficult to deal with. So far, it has only been done succssfully on a lattice. It is thus not surprising that a number of other models have been constructed which allow one to obtain quark wavefunctions and static and spectroscopic properties of hadrons, particularly baryons.

One of the earliest successful models was the MIT bag model.(20) The baryon is visualized as a bag or bubble of radius R which confines the quarks. If constituent quarks are used, then the momenta of the quarks are of order $\hbar/R$, or of the same order of magnitude as the mass of the quarks multiplied by c, the speed of light. A nonrelativistic treatment then causes concern, and the MIT group decided to take current quarks, approximated by massless ones. In its simplest form, the three massless quarks move freely inside the bag. The boundary condition at the bag surface is chosen to prevent color flux from leaving the confining region. A constant pressure B exerted radially inward on the bag counter-acts the kinetic energy of the quarks inside the bag. The model is remarkably successful in obtaining some properties of nucleons such as magnetic moments, and radii. However, the bag radius has to be of order 1–1.2 fm to fit the data, so that there is little room for pions and other mesons. This shortcoming was removed in the "little" or chiral bag model, in which pions are coupled to the surface of an MIT-like bag.(21) The coupling to pions allows the bag to have a smaller radius, about 0.5 fm. This model was further improved by treating the dynamics of the pions and quarks together in a "cloudy" bag, where the pions are allowed to penetrate into the bag.(22) The pions can be coupled to the surface only, or can have an equivalent coupling throughout the volume of the bag.(23) In all of these models the quarks are taken to be the current quarks with a small mass of the order of 4–10 MeV, in contrast to the potential models which use constituent quarks. There are still other bag models, including so-called "solitons".(24) It remains to be seen which of these many models, if any, remains successful in the long run as more data becomes available.

---

[19] Regge trajectories are based on much more general grounds (analytic properties) than the derivation given here. See, e.g., T. Regge, *Nuovo Cim.* **14**, 951 (1959); **18**, 947 (1960).

[20] A. Chodos et al., *Phys. Rev.* **D9**, 3471 (1974); **D10**, 2599 (1974); T. De Grand et al., *Phys. Rev.* **D12**, 2060 (1975); K. Johnson, *Acta Phys. Polon.*, **B6**, 865 (1975).

[21] G. E. Brown and M. Rho, *Phys. Lett.* **82B** 177 (1979); G. E. Brown, M. Rho, and V. Vento, *Phys. Lett.* **84B**, 383 (1979).

[22] S. Theberge, A. Thomas, and G. A. Miller, *Phys. Rev.* **D22**, 2838 (1980); **D23**, 2106(E) (1981).

[23] A. W. Thomas, *Adv. Nucl. Phys.* **13**, 1 (1983); G. A. Miller in *Quarks and Nuclei*, (W. Weise, ed.) Ch. 3, World Scientific, Singapore, 1984.

[24] L. Wilets, *Nontopological Solitons*, World Sci., Teaneck, NJ, 1989; I. Zahed and G.E. Brown, *Phys. Rep.* **129**, 226 (1986); N.S. Manton and P. Sutcliffe, *Topological Solitons*, Cambridge, 2004; R. Alkofer, H. Reinhardt and H. Weisel, *Phys. Rept.* **265**, 139 (1996).

15.8 Heavy Mesons: Charmonium, Upsilon, ...

The existence of quarks heavier than the strange one was predicted on the basis of the electroweak theory introduced in Chapters 11–13. The absence of weak strangeness-changing neutral currents required a new quark, charm. A whole new era of physics was ushered in when Richter at SLAC (Stanford) and Ting at Brookhaven and their collaborators almost simultaneously discovered the J/ψ.[25] This meson, composed of $c\bar{c}$, was found at SLAC in e^+e^- collisions as described in Section 10.9, and at Brookhaven in the study of hadronically produced e^+e^-. There could be little doubt that a new chapter of physics had been opened since

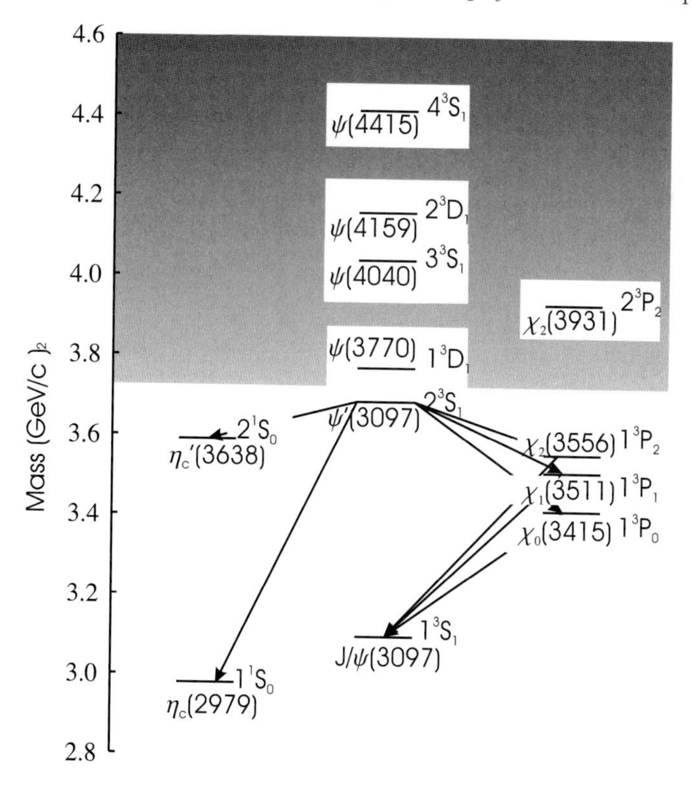

Figure 15.12: Spectrum of charmonium. The lines connecting energy levels represent photon transitions. The shaded region is the continuum for decays into $D\overline{D}$ mesons. [From PDG.]

the decay width of the J/ψ is only about 70 keV and not of the order of a hundred MeV. The decay width to the specific channel of e^+e^- is only of the order of 5 keV, as expected for a vector meson. It is now known that the J/ψ is a $^3S_1(1^-)$ state of $c\bar{c}$. The excitement of the physics community over the new state of matter was heightened further by the discovery of excited states of $c\bar{c}$.

[25] J. J. Aubert et al., *Phys. Rev. Lett.* **33**, 1404 (1974), J. E. Augustin et al., *Phys. Rev. Lett.* **33**, 1406 (1974).

The level structure of charmonium, the bound $c\bar{c}$ system, is shown in Fig. 15.12. It is similar to that of positronium; the deviations from that spectrum can be understood by using in the Schrödinger equation the masses $m(c) = m(\bar{c}) = 1550\ \text{MeV}/c^2$ and a central potential of the form[26]

$$V = -\frac{\alpha_s k}{r} + Ar, \tag{14.59}$$

shown in Fig. 14.26. The energies for the potential of Eq. (14.59) are intermediate between those in a Coulomb potential and a harmonic oscillator potential, as shown in Fig. 15.9. With the addition of spin–orbit and spin–spin potentials, Eqs. (14.35) and (15.16), as given by one-gluon exchange,[13] the spectrum and the gamma-ray decay rates can be reproduced. The widths of the $c\bar{c}$ states are small.

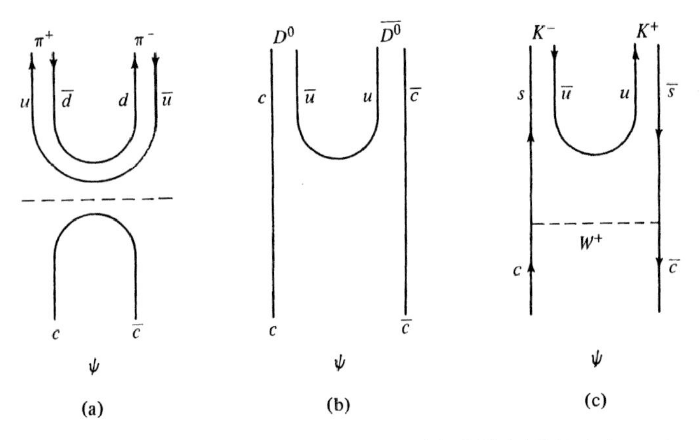

Figure 15.13: J/ψ decays into hadrons: (a) OZI inhibited decay to nonstrange mesons; (b) a preferred decay above the $D\overline{D}$ threshold; (c) a preferred weak decay.

Below approximately 3.7 GeV/c^2 the widths of the states broaden with increasing energy from about 10 keV to a few MeV, but above this energy the widths increase to several tens of MeV; the $c\bar{c}$ system has moved from bound to continuum states.

The continuum feature can be understood if we postulate that above 3.7 GeV, the $c\bar{c}$ system can decay into two charmed mesons, D and $\overline{D}$, e.g. $c\bar{u}$ and $\bar{c}u$, but that below this threshold these channels are closed and the system is quasi-stable or bound. The small widths of the bound $c\bar{c}$ states suggest a selection rule. Such a rule had already been postulated by Okubo, Zweig, and Iizuka.[27] The OZI rule states that transitions described by diagrams with quark lines that are disconnected, i.e., diagrams which can be cut by a line that does not intersect any quark lines, are severely suppressed. An example of such an OZI-inhibited decay and of the cutting line (dashed) for the decay of the $c\bar{c}$ system into two pions is shown in Fig. 15.13(a),

[26] See, T. Appelquist, R. M. Barnett, and K. Lane, *Annu. Rev. Nucl. Part. Sci.* **28**, 387 (1978); E.S. Swanson, *Phys. Rep.* **429**, 243 (2006).

[27] S. Okubo, *Phys. Lett.* **5**, 163 (1963); G. Zweig, CERN report No. 8419/Th 412 (unpubl.); J. Iizuka, *Prog. Theor. Phys. Suppl.* No. 37–38, 21 (1966).

whereas the allowed decay into charmed mesons, $D\overline{D}$ above the 3.7 GeV/c threshold is shown in Fig. 15.13(b). Below this threshold, the decay into hadrons is primarily electromagnetic, or via three gluons, in order to be colorless. For charmed mesons with $c = 1$, the preferred virtual decay is weak, $c \rightarrow sW^+$, shown in Fig. 15.13(c).

The charmed quark was postulated on theoretical grounds, and the discovery of the $c\bar{c}$ meson, the J/ψ, was a theoretical and experimental triumph. In contrast, the discovery of the bottom quark was unexpected. Preliminary evidence for a fifth quark had been obtained as early as 1968, but only the observation of a narrow dimuon resonance in the scattering of 400 GeV protons from nuclei at Fermilab in 1977 provided the conclusive evidence for a new particle, the upsilon.[28] The upsilon is the bound state of a bottom (or beauty) quark with its antiquark, $b\bar{b}$. Similar to charmonium, "bottonium" possesses a spectrum of excited positronium-like resonances. The spectrum of $b\bar{b}$ resonances can be understood on the basis of Eq. (14.59) with the same constant A and with k as predicted by the one-gluon exchange potential. Thus the long range confining force is flavor-independent. As seen in Table 15.2 the partner of the bottom quark, called top, has a very large mass of ~ 179 GeV/c^2 and toponium can be studied as well.

15.9 Outlook and Problems

We have only scratched the surface of particle models. The detailed discussion goes far deeper and involves more than the composition of hadrons in terms of quarks and gluons. It includes particle properties such as static moments, decays, form factors, and couplings of mesons to baryons.

The description of hadrons in terms of quarks is very successful. The success leads to a number of questions; a few of these are listed here:

1. The indirect evidence for quarks is overwhelming, but quark confinement is not yet understood fully. It is expected to follow from QCD, and numerical calculations (on a lattice rather than for a continuum space) suggest that it does occur.[29]

2. Are the quarks themselves structureless particles? Since there are at least 18 quarks, composed of six flavors and three colors, we must wonder whether the quarks are really the fundamental constituents of hadrons.

3. What is the relationship of the strong multiplets and the electroweak families? Is there a relationship, as suggested by grand unified theories (GUTs)?)

4. Color is the important attribute of quarks for the strong interactions. Flavor is more important for the electroweak interaction. Why is this so? What is

[28] S. W. Herb et al., *Phys. Rev. Lett.* **39**, 252 (1977); W. R. Innes et al., *Phys. Rev. Lett.* **39**, 1240 (1977); L. Lederman, *Sci. Amer.* **239**, 72, (October, 1978).

[29] N.A. Campbell, L.A. Huntley, and C. Michael, *Nucl. Phys.* **B306,** 51 (1978).

the connection between flavor and color, between weak and strong isospin? What is the relationship between the strong and electroweak interactions?

5. What is the cause of isospin symmetry? The masses of the up and down current quarks are quite different, as shown in Table 5.7. On the other hand, the constituent up and down quark masses are almost the same. What is the relationship between the current and constituent quarks?

6. The mesons are composed primarily of $q\overline{q}$ and the baryons of three (valence) quarks. However, there is evidence of a background "sea" of quark–antiquark pairs. How important are these sea quarks and what is their role? Are there other constituents of the known hadrons? What is the role of the gluons? The pion does not easily fit into the mass scheme of the mesons. Why is its mass so small? Is the pion partially a Goldstone boson, as described in Section 12.5, rather than a $q\overline{q}$ meson?[(11)]

7. What is the connection between the quark–gluon and baryon–meson degrees of freedom? What is the role of the pion and other mesons in the structure of the baryons?

8. What is the relationship between hadrons and leptons? Whereas an earlier problem was the *raison d'être* of the muon, at present the question is the relationship of quarks and leptons. Finite results for the electroweak theory require an equal number of leptons and quark flavors so that the sums of the charges over all leptons and quarks is zero. Is this equality related to the reason that quark charges are fractional multiples of e? Or is the fractional electric charge related to the role of color?

9. Although we have only touched briefly on the subject of Regge poles, there exist a number of problems here as well. For instance, do Regge poles really describe all particles?

The questions we have posed here are but a small sample of those that attract the attention of particle theorists. The success of the Weinberg–Salam theory and the apparent success of QCD have led to speculative theories that attempt to combine these forces, as we described in Chapter 14. Will these grand unified theories and their successors be able to answer the above questions?

15.10 References

The following two books provide very readable introductions to the role of quarks in the structure of hadrons:

Y. Nambu, *Quarks*, World Scientific, Singapore, 1985; L. B. Okun, *Particle Physics, The Quest for the Substance of Substructure*, Harwood Academic, New York, 1985, especially Chapter 3.

On a more advanced level, quark models are discussed in a number of books. Particularly informative are:

F. E. Close, *An Introduction to Quarks and Partons*, Academic Press, New York, 1979; F. Halzen and A. D. Martin, *Quarks and Leptons*, John Wiley, New York, 1984; K. Gottfried and V. F. Weisskopf, *Concepts of Particle Physics*, Oxford University Press, New York, Vol. I, 1984, Vol. II, 1986; I. S. Hughes, *Elementary Particles*, 2nd Edition, Cambridge Univ. Press, Cambridge, 1985; M. Jacob, *The Quark Structure of Matter*, World Scientific, Singapore, 1992; A. Hosaka and H. Toki, *Quarks, Baryons, and Chiral Symmetry*, World Scientific, Singapore, 2000.

There are also numerous reviews on various aspects of quark models. Among them are: A. W. Hendry and D. B. Lichtenberg, *Properties of Hadrons in the Quark Model, Fortschr. Phys.* **33,** 139 (1985); F. Wilczek, *QCD Made Simple, Phys. Today*, **53**, 22, Aug. 2000.

Reviews of light hadron spectroscopy can be found in A. J. G. Hey and R. L. Kelly, *Phys. Rep.* **96,** 72 (1983); B. Diekmann, *Phys. Rep.* **159,** 100 (1988); *Hadron Spectroscopy 1985*, (S. Oneda, ed.) *Amer. Inst. Phys. Confer. Proc. 132*, AIP, New York, 1985;

Bag models of hadrons, with emphasis on the MIT model, are reviewed in G. E. De Tar and J. F. Donoghue, *Annu. Rev. Nucl. Part. Sci.* **33,** 235 (1983). The little bag is described in G. E. Brown and M. Rho, *Phys. Today* **36**, 24 (February 1983). The cloudy bag and other models are reviewed in A. Thomas, *Adv. Nucl. Phys.* **13,** 1 (1983) and G. A. Miller in *Quarks and Nuclei*, (W. Weise, ed.) World Scientific, Singapore, 1984, Ch. 3. Soliton bag models are described in L. Wilets, *Nontopological Solitons*, World Scientific, Teaneck, NJ, 1989, and I. Zahed and G. E. Brown, *Phys. Rep.* **129,** 226 (1986); N.S. Manton and P. Sutcliffe, *Topological Solitons*, Cambridge 2004; R. Alkofer, H. Reinhardt and H. Weisel, *Phys. Rept.* **265**, 139 (1996).

Searches for free quarks are reviewed by L. Lyons, *Phys. Rep.* **129,** 226 (1985); P. F. Smith, *Annu. Rev. Nucl. Part. Sci.* **39,** 73 (1989).

A discussion of $c\bar{c}$ can be found in E.D. Bloom and G.J. Feldman, *Sci. Amer.* **246,** 66 (May 1982). Reviews of charmonium are in K. Königsmann, *Phys. Rep.* **139,** 244 (1986); D.G. Hitlin and W.H. Toki, *Annu. Rev. Nucl. Part. Sci.* **38**, 497 (1988); R.M. Barnett, H. Muehry, and H.R. Quin, *The Charm of Strange Quarks*, Springer Verlag, New York, 2000; J.A. Appel, *Annu. Rev. Nucl. Part. Sci.*, **42**, 367 (1992)

Both $c\bar{c}$ and $b\bar{b}$ mesons are reviewed in W. Kwong and J.L. Rosner, *Annu. Rev. Nucl. Part. Sci.* **37,** 325 (1987). The mesons are reviewed in P. Franzini and J. Lee-Franzini, *Annu. Rev. Nucl. Part. Sci.* **33,** 1 (1983); K. Berkelman, *Phys. Rep.* **98,** 146 (1983). A more popular article is N.B. Mistry, R.A. Poling, and E. M. Thorndyke, *Sci. Amer.* **249,** 106 (July 1983); A. Manohar and M. Wise, *Heavy Quark Physics*, Cambridge Univ. Press, New York, 2000; D. Berson and T. Sharmicki, *Annu. Rev. Nucl. Part. Sci.* **43**, 333 (1993).

The discovery of the top quark and the top quark are discussed by P.L. Tipton, *Sci. Amer.* **277**, 54 (Sept. 1997); S. Willenbrock, *Rev. Mod. Phys.* **72**, 1141 (2000); K. Tollefson and E.W. Varnes, *Annu. Rev. Nucl. Part. Sci.* **49**, 435 (1999). Toponium is discussed by J. H. Kuhn and P. M. Zerwas, *Phys. Rep.* **167,** 321 (1988); C. Quigg *Phys. Today* **50**, 19 (May 1997).

Asymptotic freedom and the confinement of quarks is described by F. Wilczek, *Rev. Mod. Phys.* **77**, 857 (2005), and C. Rebbi, *Sci. Amer.* **248,** 54 (February 1983). Lattice work on this aspect of QCD is reviewed in a recent very readable article by S.R. Sharpe in *Glueballs, Hybrids, and Exotic Hadrons*, (S-W. Chung, ed.) *Amer. Inst. Phys. Confer. Proc.* **185**, AIP, New York, 55 (1989).

Regge phenomenology is reviewed by A. C. Irving and R. P. Worden, *Phys. Rep.* **34,** 144 (1977); L. Caneshi, ed., *Regge Theory and Low* p_T *Hadronic Interactions*, North Holland, Amesterdam, 1989. Descriptions can also be found in the texts listed at the beginning of this Section.

Problems

15.1. Assume that the nonstrange quarks u and d are stable as free particles. Describe their fate upon entering a solid. What will the ultimate fate of either one be, and where do you expect them to come to rest?

15.2. Describe possible ways to search for quarks at accelerators. How are quarks distinguished from other particles? What limits the mass of the quark that can be found?

15.3. Could quarks be seen in a Millikan-type (oil drop) experiment? Estimate the lower limit of the concentration that can be observed in an ordinary oil droplet experiment. How can the approach be improved?

15.4. Use the quark model to compute the ratio of the magnetic moment of the proton to that of the neutron.

15.5. Use a simple potential well with range given by the proton radius to discuss the validity of the nonrelativistic treatment of the constituent quarks in the simple quark model.

15.6. Justify that only one baryon state can be formed from three identical quarks with $L = 0$; verify that this state corresponds to a particle with spin 3/2.

15.7. (a) Prove that the square of the sum of two angular momentum operators, $\mathbf{J}$ and $\mathbf{J}'$, can be written as

$$\begin{aligned}(\mathbf{J}+\mathbf{J}')^2 &= J^2 + J'^2 + 2\mathbf{J}\cdot\mathbf{J}' \\ &= J^2 + J'^2 + 2J_zJ'_z + J_+J'_- + J_-J'_+,\end{aligned}$$

where

$$J_\pm = J_x \pm iJ_y, \quad J'_\pm = J'_x \pm iJ'_y$$

are raising and lowering operators with properties as given in Eq. (8.27).

(b) Consider the two quark states

$$|\alpha\rangle = |u\uparrow\rangle|d\downarrow\rangle$$
$$|\beta\rangle = |u\downarrow\rangle|d\uparrow\rangle$$

where, for instance, $|u\uparrow\rangle$ denotes an up quark with spin up ($J_z = 1/2$), and $|d\downarrow\rangle$ denotes a down-quark with spin down ($J'_z = -1/2$). Use the result of part (a) to find the linear combinations of the states $|\alpha\rangle$ and $|\beta\rangle$ that correspond to values $J_{\text{tot}} = 1$ and $J_{\text{tot}} = 0$ of the total angular momentum quantum number of the two quarks.

15.8. Assume the u and d quarks to be massless and to belong to an isomultiplet with total isospin of 1/2.

(a) Show that the spin and isospin assignments of the proton, neutron and delta are those that can be reached with 3 quarks.

(b) Show that the lowest mass mesons that can be made with u and d quarks (and their antiquarks) have isospin 0 and 1.

15.9. Verify Eqs. (15.14) and (15.15). Why should the various particle states be orthogonal to each other?

15.10. Apply the argument that leads to Eq. (15.15) to the neutral pseudoscalar mesons. Try to find possible explanations why the agreement with experiment is much less satisfactory than for the vector mesons.

15.11. Instead of the assignments made in Table 15.2, one could choose

	J	A	S	I	I_3
u	1/2	1	0	1/2	1/2
d	1/2	1	0	1/2	−1/2
s	1/2	1	−1	0	0

(a) What is q/e for each quark in this case?

(b) Mesons would be constructed from $q'\overline{q}'$, where $\overline{q}'$ is an antiquark. Would this assignment work? Explain any difficulties that are encountered.

(c) Why is this model not used? [S. Sakata, *Prog. Theor. Phys.* **16,** 686 (1956)].

15.12. If a real quark is ever seen, how could it be caught and kept in captivity? To what uses could it be put?

15.13. (a) Show that "normal" quark configurations for bosons, $q\bar{q}$, must satisfy the conditions

$$|S| \leq 1, \quad |I| \leq 1, \quad \left|\frac{q}{e}\right| \leq 1.$$

(b) Have "exotic" mesons, i.e., mesons that do not satisfy these conditions, been found?

15.14. Verify Eq. (15.20).

15.15. Show that L^2, Eq. (15.22), is indeed the operator of the square of the orbital angular momentum.

15.16. Show that $R(r)$ satisfies Eq. (15.25).

15.17. Prove Eq. (15.30).

15.18. Prepare a plot similar to Fig. 15.10 for the energy levels of the hydrogen atom.

15.19. (a) Find the Regge trajectory for baryons of isospin 1/2 and positive parity; repeat for those of isospin 1/2 and negative parity. Determine the slopes and compare to those of Fig. 15.11.

(b) Discuss the occurrence of parity doubling in the baryon spectrum in light of the above and other data. [See F. Iachello, *Phys. Rev. Lett.* **62,** 2440 (1989)].

15.20. What is the evidence, if any, for the presence of gluons and sea quarks inside hadrons?

15.21. (a) Compare the approximate predicted and experimental spectra of the $b\bar{b}$ and $c\bar{c}$ systems if the primary potential is given by Eq. (14.59) with the *same* constants in both cases. Take the average energy for the states split by spin–orbit and spin–spin forces.

(b) Predict the spectrum of $t\bar{t}$ for a top quark mass of $170\text{GeV}/c^2$ on the basis of the results of (a).

15.22. (a) Assume that the spin-spin potential. Eq. (15.16), is proportional to the inverse square of the quark masses. Write the potential in terms of a dimensionless constant of proportionality, k. Determine $k|\psi(0)|^2$ from the splitting of the $^3S_1$ and $^1S_0$ states of the $c\bar{c}$ system, and predict the splitting for the $b\bar{b}$ system. Assume that $|\psi(0)|^2$ is independent of the quark mass. Compare with experimental data, if feasible.

(b) Apply the results of part (a) to the splitting of the light pseudoscalar and vector mesons of Table 15.1. How well does the application work here?

(c) If the spin–orbit potential is proportional to the inverse square of the quark masses, repeat (a) for the splitting of the ${}^3P_2, {}^3P_1$, and ${}^3P_0$ states. Can you explain the discrepancy with experiment?

15.23. (a) For a harmonic oscillator how does the energy splitting between S-states depend on the mass of the bound particle?

(b) Repeat (a) for a Coulomb ($1/r$) potential.

(c) Compare these energy spacings to those for charmonium and bottomium.

15.24. Table 15.2 indicates that the masses of the up and down quarks differ by about 6MeV$/c^2$. Show that, to lowest order in $m_u - m_d$, this mass difference contributes to the mass difference of the neutron and proton, but not to that of the π^+ and π^0.

15.25. (a) Show that there is no symmetric total spin-1/2 wavefunction in spin space for three quarks of spin 1/2.

(b) Repeat part (a) for an antisymmetric wavefunction.

15.26. (a) If the mass differences between the light pseudoscalar and vector mesons is due to a difference in the forces between nonstrange quarks, a strange and a nonstrange quark, and between strange quarks, determine the nature of the difference.

(b) Apply (a) to the low lying $1/2^+$ and $3/2^+$ baryons.

15.27. Determine the boundary conditions in the MIT bag model if color is to be confined inside a bag of radius R, and if the quarks are free to move inside the bag.

15.28. Explain how saturation of the lowest mass baryon states by three quarks and mesons by $q\overline{q}$ is evidence for color.

15.29. Show that the mean square radii of the K^0 and $\overline{K^0}$ with a simple nonrelativistic central quark–antiquark potential are such that $\langle r(K^0)^2\rangle$ is negative and $\langle r(\overline{K}^0)^2\rangle$ is positive.

Chapter 16

Liquid Drop Model, Fermi Gas Model, Heavy Ions

Computations of nuclear properties *ab initio* are very difficult and have only been carried out for light nuclei. The force is very complicated, and nuclei are many-body problems. It is therefore necessary with most nuclear problems to simplify the approach and use specific nuclear models combined with simplified nuclear forces.

Nuclear models generally can be divided into independent particle models (IPM) in which the nucleons are assumed, in lowest order, to move nearly independently in a common nuclear potential, and strong interaction (collective) models (SIM) in which the nucleons are strongly coupled to each other. The simplest SIM is the liquid drop model; the simplest IPM is the Fermi gas model. Both of these will be treated in this chapter. In the following two chapters we shall discuss the shell model (IPM) in which nucleons move nearly independently in a static spherical potential determined by the nuclear density distribution, and the collective model (SIM) in which collective motions of the nucleus are considered. The unified model combines features of the shell and of the collective model: The nucleons are assumed to move nearly independently in a common, slowly changing, nonspherical potential, and excitations of the individual nucleons and of the entire nucleus are considered.

16.1 The Liquid Drop Model

One of the most striking facts about nuclei is the approximate constancy of nuclear density: the volume of a nucleus is proportional to the number A of constituents. The same fact holds for liquids, and one of the early nuclear models, introduced by Bohr(1) and von Weizsäcker,(2) was patterned after liquid drops; nuclei are considered to be nearly incompressible liquid droplets of extremely high density. The model leads to an understanding of the trend of binding energies with atomic number, and it also gives a physical picture of the fission process. We shall sketch the simplest aspects of the liquid drop model in the present section.

[1] N. Bohr, *Nature* **137**, 344 (1936).
[2] C.F. von Weizsäcker, *Z. Physik* **96**, 431 (1935).

In Section 5.3, nuclear mass measurements were introduced, and in Section 5.4 some basic features of nuclear ground states were mentioned. Fig. 5.20 represents a plot of the stable nuclei in a NZ plane. We return to the nuclear masses here, and we shall describe their behavior in more detail than in Chapter 5. Consider a nucleus consisting of A nucleons, Z protons, and N neutrons. The total mass of such a nucleus is somewhat smaller than the sum of the masses of its constituents because of the binding energy B which holds the nucleons together. For bound states, B is positive and represents the energy that is required to disintegrate the nucleus into its constituent neutrons and protons. B is given by

$$\frac{B}{c^2} = Zm_p + Nm_n - m_{nuclear}(Z, N). \tag{16.1}$$

Here, $m_{\text{nuclear}}(Z, N)$ is the mass of a nucleus with Z protons and N neutrons. It is customary to quote *atomic* and not *nuclear* masses and use atomic mass units (see Eq. (5.23)). In terms of the atomic mass $m(Z, N)$, the binding energy can be written as

$$\frac{B}{c^2} = Zm_H + Nm_n - m(Z, N). \tag{16.2}$$

A small term due to atomic binding effects is neglected in Eq. (16.2); m_H is the mass of the hydrogen atom. The difference between the atomic rest energy $m(Z, N)c^2$ and the nucleon or mass number times uc^2 is called the mass excess (or mass defect),

$$\Delta = m(Z, N)c^2 - A\, uc^2. \tag{16.3}$$

Comparison between Eqs. (16.2) and (16.3) shows that $-\Delta$ and B measure essentially the same quantity but differ by a small energy. Tables usually list Δ because it is the quantity that follows from mass-spectroscopic measurements. The average binding energy per nucleon, B/A, is plotted in Fig. 16.1. The binding energy curve exhibits a number of interesting features:

1. Over most of the range of stable nuclei, B/A is approximately constant and of the order of 8–9 MeV. This constancy results from the saturation of nuclear forces discussed in Section 14.5. If all nucleons inside a nucleus were within each other's force range, the total binding energy would be expected to increase proportionally to the number of bonds or approximately proportionally to A^2. B/A would then be proportional to A.

2. B/A reaches its maximum in the region of iron ($A \approx 60$). It drops off slowly toward large A and more steeply toward small A. This behavior is crucial for the synthesis of the elements and for nuclear power production. One consequence is that the abundance of elements around iron is especially large. Also if a nucleus of say, $A = 240$, is split into two roughly equal parts, the binding of the two parts is stronger than that of the original nuclide, and energy is liberated.

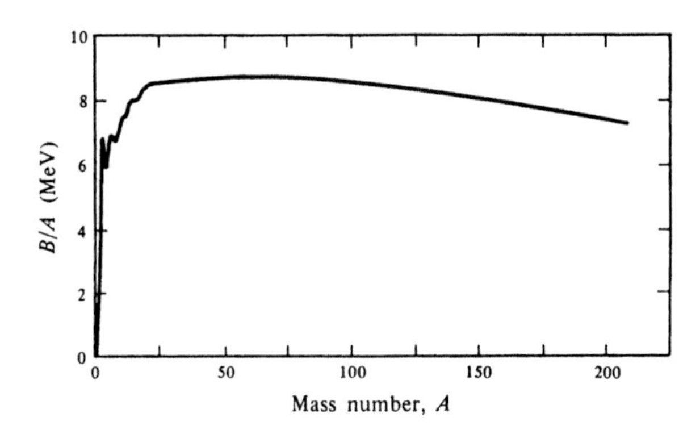

Figure 16.1: Binding energy per particle for nuclei.

This process is responsible for energy production in fission. At the other end, if two light nuclides are fused, the binding of the fused system will be stronger, and energy will again be liberated. This energy release is the base for energy production in fusion.

The smooth variation of the binding energy B/A as a function of mass number A suggests that it should be possible to express the nuclear masses by a simple formula. The first semiempirical mass formula was obtained by von Weizsäcker, who noted that the constant average binding energy per particle and the constant nuclear density suggested a liquid drop model.(2) The primary fact needed to arrive at a mass formula is the tendency of B/A *to be approximately constant for* $A \gtrsim 50$. The binding energy per particle for an infinite nucleus without surface thus should have a constant value, a_v, the binding energy of nuclear matter. Since there are A particles in the nucleus, the *volume contribution* E_v to the binding energy is

$$E_v = +a_v A. \tag{16.4}$$

Nucleons at the *surface* have fewer bonds and the finite size of a real nucleus leads to a contribution E_s to the energy that is proportional to the surface area and decreases the binding energy,

$$E_s = -a_s A^{2/3}. \tag{16.5}$$

Volume and surface terms correspond to a liquid drop model. If only these two terms were present, isobars would be stable regardless of the value of N and Z. Figure 5.20, however, demonstrates that only nuclides in a narrow band are stable. For lighter nuclides, the self-conjugate isobars ($N = Z$ or $A = 2Z$) are the most stable ones, whereas heavier stable isobars have $A > 2Z$. These features are explained by two additional terms, a symmetry term and the Coulomb energy.

The *Coulomb energy* is caused by the repulsive electrical force acting between any two protons; this energy favors isobars with a neutron excess. For simplicity we assume that the protons are uniformly distributed throughout a spherical nucleus of radius $R = R_0 A^{1/3}$; with Eq. (8.37), the Coulomb energy becomes

$$E_c = -a_c Z^2 A^{-1/3}. \tag{16.6}$$

The fact that only nuclides in a small band are stable is explained by another term, the symmetry energy. The effect of the *symmetry energy* is best seen if the mass excess Δ is plotted against Z for all isobars characterized by a given value of A. As an example, such a plot is shown in Fig. 16.2 for $A = 127$.

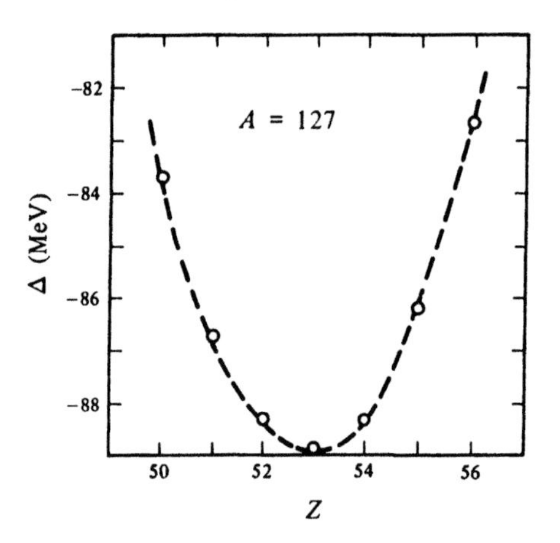

Figure 16.2: Mass excess Δ as a function of Z for $A = 127$.

The figure appears like a cross section through a deep valley; the isobar at the bottom is the only stable one, and the ones clinging to the steep sides tumble down toward the bottom of the valley, usually by emission of electrons or positrons. The isobars with $A = 127$ are not an isolated case; the mass excesses for all other isobars also are shaped like cross sections through a valley. Figure 5.20 can therefore be brought into a more informative form by adding a third dimension to the plot: the binding energy or the mass excess.

Such a plot is analogous to a topographic map, and Fig. 16.3 presents the contour map of the binding energy in an $N - Z$ plane. Figure. 16.2 is the cross section through the valley at the position indicated in Fig. 16.3. The sides of the valley are steep, and it is consequently difficult experimentally to explore the valley to the "top" because the nuclei are shortlived. Some of these nuclei can be produced in accelerators, separated and re-accelerated in special facilities for short lived ions.[3] The dashed countour lines in Fig. 16.3 indicate the shorter lifetimes. The limits of the region of stability are called "neutron and proton *drip lines*". Beyond these the nuclei decay by the strong interactions and the lifetimes are shorter than $\sim 10^{-18}$ sec.

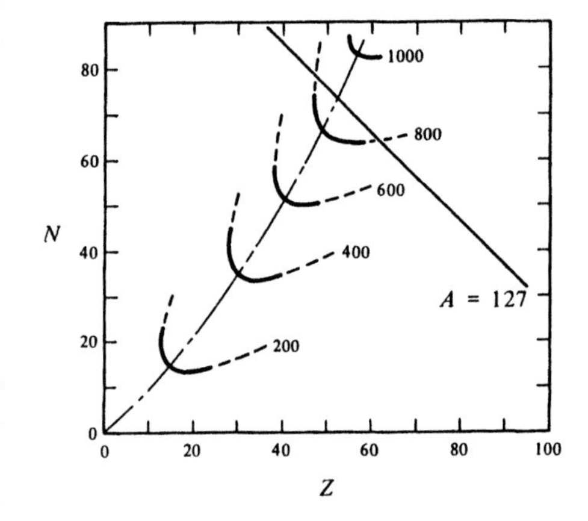

Figure 16.3: Sketch of binding energy B plotted in form of a contour map in an $N - Z$ plane. The *energy valley* appears clearly; it forms a canyon in the $N - Z$ plane. The numbers on the contour lines give the total binding energy in MeV.

[3]W. Henning, *Nucl. Phys.* **A746**, 3c (2004); J.A. Nolen, *Nucl. Phys.* **A746**, 9c (2004).

Beyond the drip lines nuclei are called "unstable" because they can decay by emitting particles via the strong interaction and their lifetimes are extremely short.

The symmetry energy arises because the exclusion principle makes it more expensive in energy for a nucleus to have more of one type of nucleon than the other. In the following section we shall derive an approximate expression for the symmetry energy; it is of the form

$$E_{\text{sym}} = -a_{\text{sym}} \frac{(Z-N)^2}{A}. \tag{16.7}$$

Collecting the terms gives the Bethe–Weizsäcker relation for the binding energy of a nucleus (A, N),

$$B = a_v A - a_s A^{2/3} - a_{\text{sym}}(Z-N)^2 A^{-1} - a_c Z^2 A^{-1/3}. \tag{16.8}$$

The binding energy per particle becomes

$$\frac{B}{A} = a_v - a_s A^{-1/3} - a_{\text{sym}} \frac{(Z-N)^2}{A^2} - a_c Z^2 A^{-4/3}. \tag{16.9}$$

The constants in these relations are determined by fitting the experimentally observed binding energies; a typical set is

$$\begin{aligned} a_v &= 15.6 \text{ MeV}, \quad a_{\text{sym}} = 23.3 \text{ MeV}, \\ a_s &= 16.8 \text{ MeV}, \quad a_c = 0.72 \text{ MeV}. \end{aligned} \tag{16.10}$$

With these values, the general trend of the curves shown in Figs. 16.1 and 16.2 is reproduced well. Of course, finer details are not given, and relations with many more terms are employed when small deviations from the smooth behavior are studied.[4] Two remarks concerning the binding energy relation are in order. (1) Here, we have assumed that the coefficients in Eq. (16.8) are adjustable parameters to be determined by experiment. In a more thorough treatment of nuclear physics, the coefficients are derived from the characteristics of nuclear forces. In particular the calculation of the most important coefficient, a_ν, has occupied theoretical physicists for a long time because it is intimately related to the properties of *nuclear matter.* Nuclear matter is the state of matter that would exist in an infinitely large nucleus. The closest approximation to nuclear matter presumably exists in neutron stars (Chapter 19.) (2) Sophisticated versions of the Bethe–Weizsäcker relation, or some of its updated forms, can be used to explore the stability properties of matter by extrapolating to regions that are not well known. Such studies are important, for instance, in the investigation of very heavy artificial elements (see Section 16.3), in the treatment of nuclear explosions, and in astrophysics.

---

[4]See, for instance, D.N. Basu, nucl-th/0309045 and R.C. Nayak and L. Satpathy, *At. Data and Nucl. Data Tables* **73**, 213 (1999).

16.2 The Fermi Gas Model

The semiempirical binding-energy relation obtained in the previous section is based on treating the nucleus like a liquid drop. Such an analogy is an oversimplification, and the nucleus has many properties that can be explained more simply in terms of independent-particle behavior rather than in terms of the strong-interaction picture implied by the liquid drop model. The most primitive independent-particle model is obtained if the nucleus is treated as a degenerate Fermi gas of nucleons. The nucleons are assumed to move freely, except for effects of the exclusion principle, throughout a sphere of radius $R = R_0 A^{1/3}, R_0 \approx 1.2$ fm. The situation is represented in Fig. 16.4 by two wells, one for neutrons and one for protons.

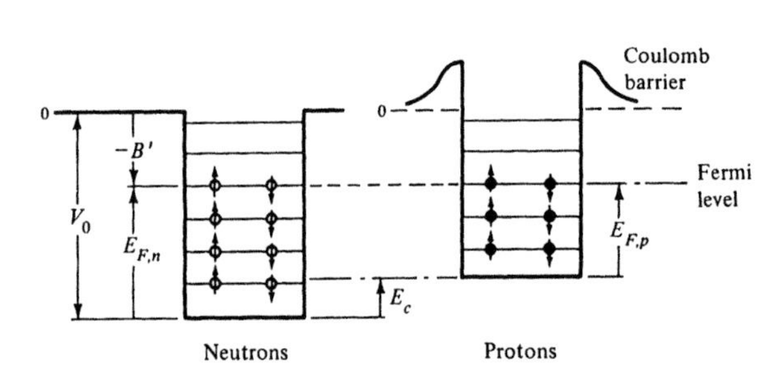

Figure 16.4: Nuclear square wells for neutrons and protons. The well parameters are adjusted to give the observed binding energy B'.

Free neutrons and free protons, far away from the wells, have the same energy, and the zero levels for the two wells are the same. The two wells however, have different shapes and different depths because of the Coulomb energy, Eq. (8.37): The bottom of the proton well is higher than the bottom of the neutron well by an amount E_c, and the proton potential has a *Coulomb!barrier*.

Protons that try to enter the nucleus from the outside are repelled by the positive charge of the nucleus; they must either "tunnel" through the barrier or have enough energy to pass over it.

The wells contain a finite number of levels. Each level can be occupied by two nucleons, one with spin up and one with spin down. It is assumed that, under normal conditions, the *nuclear temperature* is so low that the nucleons occupy the lowest states available to them. Such a situation is described by the term *degenerate Fermi gas*. The nucleons populate all states up to a maximum kinetic energy equal to the Fermi energy E_F. The total number, n, of states with momenta up to p_{max} follows from Eq. (10.25), after integration over d^3p, as

$$n = \frac{V p_{\text{max}}^3}{6\pi^2 \hbar^3}. \tag{16.11}$$

Each momentum state can accept *two* nucleons so that the total number of one species of nucleons with momenta up to p_{max} is $2n$. If neutrons are considered, then $2n = N$, the number of neutrons, and N is given by

$$N = \frac{Vp_N^3}{3\pi^2\hbar^3}, \tag{16.12}$$

where p_N is the maximum neutron momentum, and V is the nuclear volume. With $V = 4\pi R^3/3 = 4\pi R_0^3 A/3$, the maximum neutron momentum follows from Eq. (16.12) as

$$p_N = \frac{\hbar}{R_0}\left(\frac{9\pi N}{4A}\right)^{1/3}. \tag{16.13}$$

Similarly, the maximum proton momentum is obtained as

$$p_Z = \frac{\hbar}{R_0}\left(\frac{9\pi Z}{4A}\right)^{1/3}. \tag{16.14}$$

The appropriate value of the Fermi energy can be found by considering self-conjugate nuclei for which $N = Z$. Equation (16.13), after inserting the numerical values and using the nonrelativistic relation between energy and momentum, then yields

$$E_F = \frac{p_F^2}{2m} \approx 40 \text{ MeV}. \tag{16.15}$$

The average kinetic energy per nucleon can also be calculated, and it is given by

$$\langle E \rangle = \frac{\displaystyle\int_0^{p_f} E d^3p}{\displaystyle\int_0^{p_F} d^3p} = \frac{3}{5}\left(\frac{p_F^2}{2m}\right) \approx 24 \text{ MeV}. \tag{16.16}$$

This result justifies the nonrelativistic approximation for nuclei. With Eqs. (16.13) and (16.14) the total average kinetic energy becomes

$$\langle E(Z,N) \rangle = N\langle E_N \rangle + Z\langle E_Z \rangle = \frac{3}{10m}(Np_N^2 + Zp_Z^2)$$

or

$$\langle E(Z,N) \rangle = \frac{3}{10m}\frac{\hbar^2}{R_0^2}\left(\frac{9\pi}{4}\right)^{2/3}\frac{(N^{5/3} + Z^{5/3})}{A^{2/3}}. \tag{16.17}$$

Equal masses for proton and neutron and equal radii for the proton and neutron wells have been assumed. Moreover, neutrons and protons move independently of each other. The interaction between the various particles has been replaced by the boundary of the nucleus, represented by the potential well.

For a given value of A, $\langle E(Z,N) \rangle$ has a minimum for equal numbers of protons and neutrons, or $N = Z = A/2$. To study the behavior of $\langle E(Z,N) \rangle$ around this minimum we set

$$Z - N = \epsilon, \qquad Z + N = A \qquad \text{(fixed)}$$

or $Z = \frac{1}{2}A(1 + \epsilon/A), N = \frac{1}{2}A(1 - \epsilon/A)$, and assume that $(\epsilon/A) \ll 1$. With

$$(1 + x)^n = 1 + nx + \frac{n(n-1)}{2}x^2 + \cdots ,$$

and after reinserting $Z - N$ for ϵ, Eq. (16.17) becomes, near $N = Z$,

$$\langle E(Z, N)\rangle = \frac{3}{10m}\frac{\hbar^2}{R_0^2}\left(\frac{9\pi}{8}\right)^{2/3}\left(A + \frac{5}{9}\frac{(Z-N)^2}{A} + \cdots\right). \tag{16.18}$$

The first term is proportional to A, and it contributes to the volume energy. The leading-order deviation is of the form assumed in Eq. (16.7) for the symmetry energy, and the coefficient of $(Z - N)^2/A$ can be evaluated numerically:

$$\frac{1}{6}\left(\frac{9\pi}{8}\right)^{2/3}\frac{\hbar^2}{mR_0^2}\frac{(Z-N)^2}{A} \approx 11\ \text{MeV}\frac{(Z-N)^2}{A}. \tag{16.19}$$

The evaluation has produced the expected form for the symmetry energy, but the coefficient is only about half as big as a_{sym} in Eq. (16.10). We shall now briefly describe where the missing contribution to the symmetry energy comes from.[5]

In the discussion leading to Eq. (16.18) it was tacitly assumed that the potential depth V_0 (Fig 16.4) does not depend on the neutron excess $(Z-N)$. This assumption is not very good because the average interaction between like nucleons is less than it is between neutron and proton, mainly because of the exclusion principle. The Pauli principle weakens the interaction between like particles by forbidding some of the two body states, while the interaction between neutron and proton is allowed in all states. The change in the potential well depth has been determined, and it is of the order

$$\Delta V_0(\text{in MeV}) \approx 30\frac{(Z-N)}{A}. \tag{16.20}$$

This decrease in depth of the potential well accounts for the missing contribution to the symmetry energy.[6]

16.3 Heavy Ion Reactions

In the last few decades, heavy ion reactions have become a significant tool for investigating nuclei under extreme conditions. Heavy ion reactions permit one to create new species of nuclei away from the stable valley, Fig. 16.3, and also to study nuclei under conditions of higher than normal densities and excitations. It is thus possible to explore the *nuclear equation of state*, related to the dependence of the energy on density and temperature.[7] This equation is essential for an understanding of star collapse and what remains afterwards (Chapter 19).

[5] K.A. Brueckner, *Phys. Rev.* **97**, 1353 (1955).
[6] B.L. Cohen, *Am. J. Phys.* **38**, 766 (1970).
[7] A. Akmal, V.R. Pandharipande, and D.G. Ravenhall, *Phys. Rev. C* **58**, 1804 (1998); H.A. Bethe, *Ann. Rev. Nucl. Part. Sci.* **38**, 1 (1988).

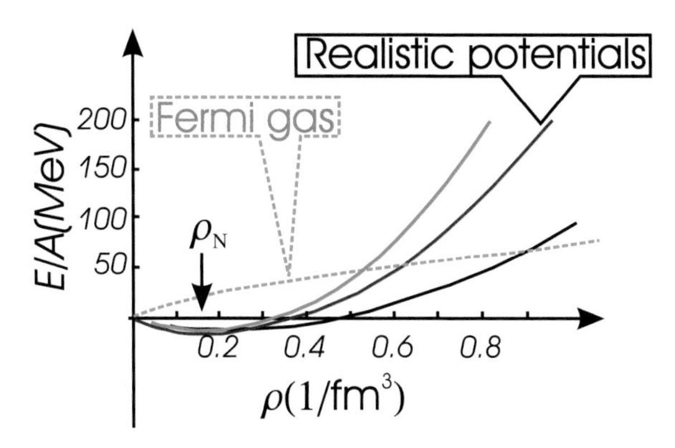

Figure 16.5: Sketch of calculations of energy per nucleon as a function of density at a temperature of 0 K. The 'Fermi gas' curve assumes no interaction appart from Pauli blocking. The other curves that show minima around the observed nuclear density come from realistic potentials with different ingredients. [After A. Akmal, V.R. Pandharipande, and D.G. Ravenhall, *Phys. Rev. C* **58**, 1804 (1998).

Some typical plots of the energy dependence on density are shown in Fig. 16.5 for nuclear matter. Nuclear matter consists of an equal and infinite number of neutrons and protons distributed uniformly throughout space, but with the neglect of Coulomb forces. At low densities, nuclear matter is unbound because nuclear forces are only felt at short range. A minimum energy is reached at normal nuclear matter density, $\rho_n \approx 0.17$ nucleons/fm^3, the central density of finite nuclei.

The minimum energy corresponds to the volume energy of Eq. (16.4), about -15.6 MeV per nucleon. The curvature at the minimum, $\delta^2 E/\delta\rho^2$, is related to the incompressibility of nuclear matter,

$$K = 9\left(\rho^2 \frac{\delta^2 E/A}{\delta\rho^2}\right)_{\min} . \tag{16.21}$$

The value of K is ~ 210 MeV and can be obtained from the excitation energy of the collective 0^+ "breathing" mode[(8)] (Section 18.6) and from kaon production in heavy ion collisions.[(9)]

The series of drawings in Fig. 16.6 show typical events in heavy ion collisions as the energy is increased. The dynamics are determined by the competition between the Coulomb force, the centrifugal barrier, and the nuclear force. Owing to these forces, the shapes of the nuclei change as they approach each other and surface modes of motion are excited (see Chapter 18). For energies below the Coulomb barrier, Coulomb excitation dominates the interaction. Above the Coulomb barrier, many nuclear processes occur. Examples are particle transfers, fusion reactions, and nuclear excitations, often with large angular momenta, particularly for grazing collisions. To see that very high angular momenta can be reached and to study the collisions in more detail, we note that semi-classical approaches can be used because $pr_c/\hbar \gg 1$, where p is the relative momentum of the two colliding ions and r_c is the approximate distance of closest approach. For energies close to the Coulomb barrier, this distance can be found by assuming that the nuclei remain undistorted.

[8] Youngblood et al., *Phys. Rev. Lett.* **82**, 691 (1999).
[9] Hartnack et al, *Phys. Rev. Lett.* **96**, 012302 (2006).

In that case, the classical distance of closest approach is given approximately by

$$r_c \approx \frac{Z_1 Z_2 e^2}{E_K}, \tag{16.22}$$

where Z_i is the atomic number of ion i and E_K is the relative (c.m.) kinetic energy. As an example, consider the reaction $^{36}\mathrm{S}(^{122}\mathrm{Sn}, 4n)^{154}\mathrm{Dy}$ at incident energies ~ 170 MeV.[10] The c.m. energy is ~ 130 MeV and the distance of closest approach ~ 9 fm.

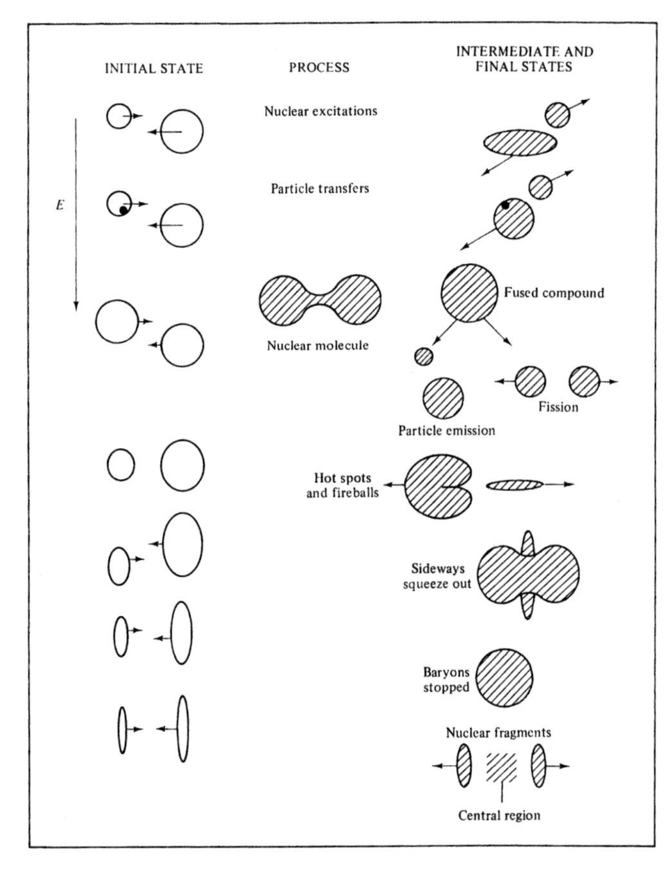

Figure 16.6: Sketches of some heavy ion reactions as a function of increasing energy.

The parameter $pr_c/\hbar > 10^2$ is much larger than 1, as required by the semi-classical approximation. Angular momenta up to this magnitude are hard to excite because the nuclei having peripheral collisions don't fuse, but there is clear experimental evidence that angular momenta up to $\sim 50 - 60\hbar$ were reached. Thus, heavy ion reactions allow one to study nuclei at very high rotations and orbital angular momenta. Fusion reactions are also likely in central collisions. The Coulomb barrier slows down the nuclei so that fusion is more probable.

If the nuclei fuse, then a "compound nucleus" is formed that can be heavier than known stable nuclei. After the evaporation of a few neutrons, the ground state of a heavier system emerges which may be characterized by its decay products. The latter can include fragmentation into several pieces or fission into two fragments. Transfermium elements of $Z = 105$ to 118, for example, were found in this

[10] W.C. Ma et al., *Phys. Rev. C* **65**, 034312 (2002).

manner.[11] Superheavy nuclei of special stability ($Z = 114 - 120$ $N = 172 - 184$; see Chapter 17) have been sought, but have not been found. In some cases nuclear quasi-molecules may be formed; they are like a dumb-bell with two centers separated by a short neck and bound by valence nucleons. Resonances seen, e.g., in $^{12}C + ^{12}C$ collisions have been ascribed to such molecules;[12] they have rotational and vibrational states like ordinary molecules and may act like "doorway" states to complete fusion.

As the collision energy increases, the number of possible reaction products grows rapidly and the reaction becomes increasingly complex; the production of pions and other mesons increases in importance. The nucleus can be treated as a quantum fluid.[13] When the velocity of the projectile is much larger than the average speed of the nucleons in the nucleus ($\sim c/4$), then these nucleons cannot move aside to accommodate the projectile; this leads to a high density and the disorganization leads to heating. Head-on collisions may produce shock waves if the mean-free path of the nucleons is much smaller than the nuclear radii; but no direct evidence exists. Particularly in off-center collisions, nuclear material may be squeezed out sideways as shown in Fig. 16.6.

16.4 Relativistic Heavy Ion Collisions

Highly relativistic collisions ($E_K \gtrsim 10$ GeV/nucleon in the c.m.) between heavy ions are of great interest, partly because of their connection to processes that occurred in the early universe. In Chapter 14 we described how quarks are confined inside nucleons and how QCD implies that, at high momentum transfers, there should be asymptotic freedom. It is believed that the early universe (a few microseconds after the Big Bang) consisted of a partially equilibrated system composed of a large number of quarks with densities and temperatures high enough that quarks were deconfined. As the universe expanded it cooled down and the baryons were "frozen out" (formed). Can something similar to the deconfined system be created in the laboratory? Over the last 30 years physicists have searched for it using relativistic heavy ion collisions. Expectations based on QCD predicted that at energy densities of ~ 1 GeV/fm^3 (corresponding to temperatures of $kT \sim 170$ MeV; see Problem 16.27) a state of weakly interacting hadronic matter, called the quark–gluon plasma (QGP), could be formed. The quarks from a particular nucleon would be shielded from each other by gluons and quarks from other nucleons so that the color binding to a particular nucleon would be dissolved and the quarks would move throughout the whole system. In order to detect the phase transition

[11] Yu.Ts. Oganessian; *Phys. Scr.* **T125,** 57 (2006); S. Hofman and G. Münzenberg, *Rev. Mod. Phys.* **72**, 733 (2000); Yu.Ts. Oganessian et al., *Phys. Rev. C* **74**, 044602 (2006).

[12] D.A. Bromley, J.A. Kuehner, and A. Almqvist, *Phys. Rev. Lett.* **4**, 365 and 515 (1960); E.R. Cosman et al., *Phys. Rev. Lett.* **35**, 265 (1975); T. A. Cormier et al., *Phys. Rev. Lett.* **38**, 940 (1977).

[13] J.M. Eisenberg and W. Greiner, *Nuclear Theory*, Vol. 1, *Nuclear Models*, 3d. Ed., Ch. 17, North Holland, Amsterdam, 1987.

from the regular hadronic matter to the QGP one could consider following similar steps as one would with, say, water. Thus one may observe how a sample changes

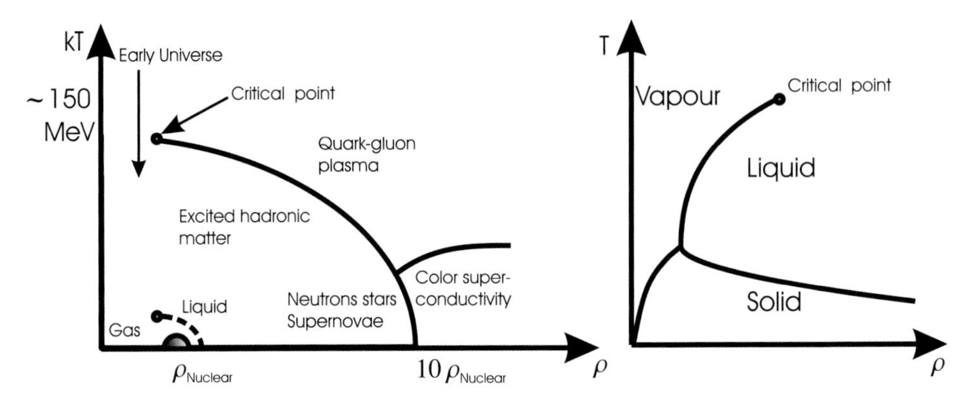

Figure 16.7: Sketches of phase diagrams. Left: expected for hadronic matter. Right: measured for water.

its temperature as one compresses it from one phase to the other. Fig. 16.7 presents the theoretical phase diagram for water and for hadronic matter up to the extreme conditions we are discussing. The normal nucleus occupies a tiny region of this phase diagram. However, determining the properties of the QGP is difficult primarily because its life time is only $\sim 10^{-23}$ seconds and all that is observed are the particles emitted from it. In order to use data to test theoretical expectations, calculations which take into account the following steps are needed: 1) the initial collision between the incoming nuclei; 2) the formation of the QGP; 3) the final emission of particles and their interactions with the QGP.

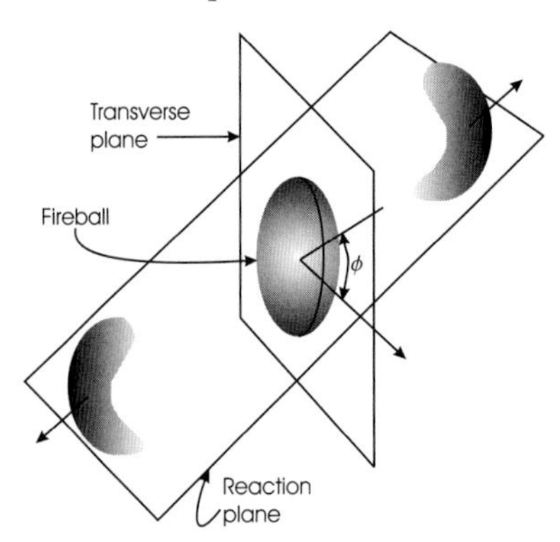

Figure 16.8: Sketch of a Au-Au collision showing the formation of the hot asymmetric *fireball*.

The Relativistic Heavy Ion Collider (RHIC) accelerator at Brookhaven (see Chapter 2) was built to study these collisions. Beams of ions ranging from protons to gold can be accelerated to energies up to 100 GeV/nucleon ($\gamma \sim 100$). In what follows we will describe what has been learned from the RHIC collisions. To fix ideas we refer to Fig. 16.8. After the collision, the colliding nuclei leave behind a non-spherical *fireball*. An important issue is to what extent this *fireball* is in thermodynamic equilibrium. After a typical RHIC collision thousands of hadrons, mainly mesons, are observed.

In a hot equilibrated system many different particles get produced in pairs, as long as there is enough energy to produce them. Since the different particles have different masses the relative particle abundances reflect the energy distribution in the original system and can probe whether particles are emitted from a hot equilibrated source or from individual collisions. Fig. 16.9 shows that the data is in good agreement with the hypothesis of emission from a thermally equilibrated source. However,

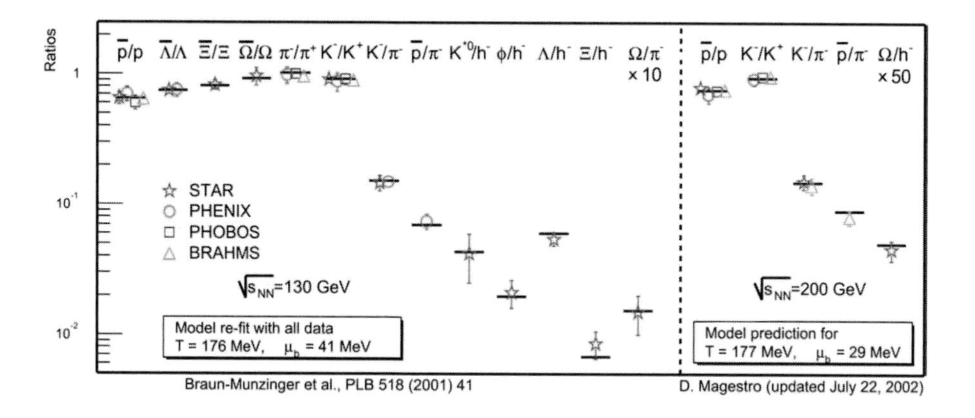

Figure 16.9: Measured particle abundance ratios at RHIC show good agreement with a statistical thermal model. [From T. Ludlam, *Nucl. Phys.* **A750**, 9 (2005).]

presently all evidence indicates that the *fireball* contains a medium that is *strongly* coupled, rather than the expected weakly interacting medium. Part of the evidence for the fact that the *fireball* is not weakly coupled comes from observation of two-particle distributions. Pairs of particles that are emitted from the *fireball* in the manner sketched in Fig. 16.10 can be observed with tracking detectors (e.g. the STAR time-projection chamber) and Fig. 16.11 presents the measured distribution of events versus the angle between the pair of particles.

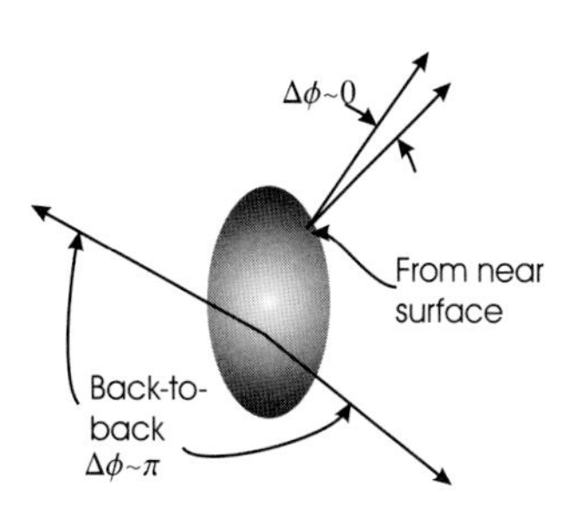

Figure 16.10: Sketch of two-particle emission from fireball.

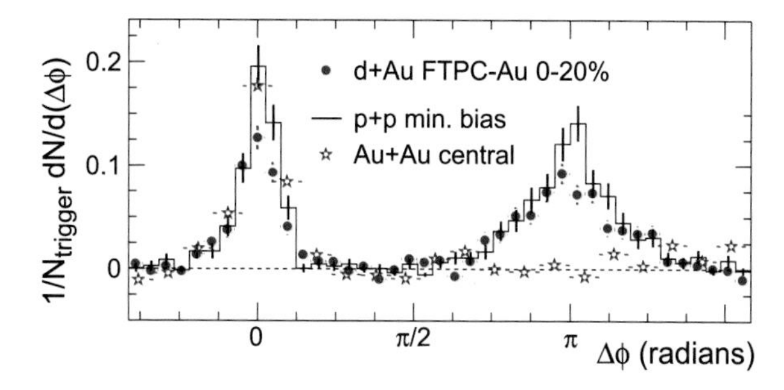

Figure 16.11: Two-particle azimuthal distributions. [From STAR collaboration, *Phys. Rev. Lett.* **91**, 072304 (2003).]

Particles originating from single jets come at angles $\Delta\phi$ close to zero while di-jet events yield particles coming at $\Delta\phi \sim \pi$. The significant suppression (called *jet quenching*) observed at $\Delta\phi \sim \pi$ in the Au+Au data can be understood assuming a strongly interacting medium that absorbs particles that move through it. The $\Delta\phi \sim 0$ events are dominated by particles emitted from the surface, but in the $\Delta\phi \sim \pi$ events one of the particles is obliged to go through the *fireball*. Fig. 16.11 shows a comparison between Au-Au and d-Au data where it is clear that the jet-quenching effect occurs only in heavy-ion collisions, presumably because here this equilbrated phase of matter is present, but not in $d + \mathrm{Au}$ collisions or in $p + p$ collisions, where it is not expected. The amount of absorption observed implies that this phase of matter has not been observed before.

Additional evidence for the strong coupling in the *fireball* comes from observations of the distribution of particles as a function of the azimuthal angle, ϕ, measured with respect to the reaction plane shown in Fig. 16.8. For convenience the distributions are described by a function of the form:

$$\frac{dN}{d\phi} = A(1 + 2v_1 \cos\phi + 2v_2 \cos 2\phi + ..) , \tag{16.23}$$

where v_1 is called *the direct flow* and v_2 *the elliptic anisotropy* or *elliptic flow*.

Fig. 16.12 shows a plot of measurements of elliptic flow from recent experiments at RHIC versus the observed particle multiplicity (n_{ch} in the figure.) The larger multiplicities correspond to more *central* collisions, because the two incoming nuclei have larger overlap and consequently more nucleons participate in the collision. As the centrality increases the initial *fireball* looks more circular in the transverse plane and the values of v_2 should get smaller, as observed.

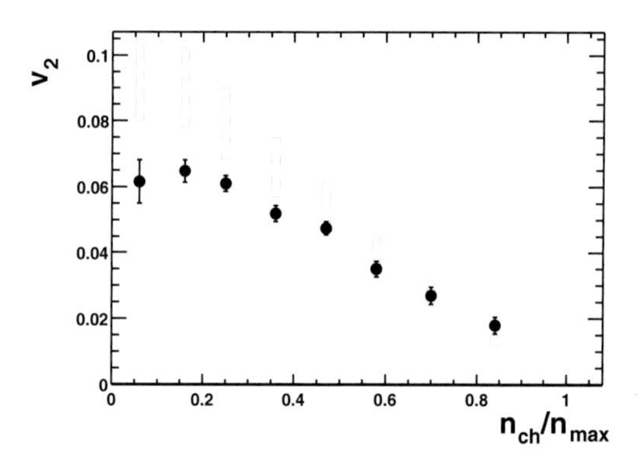

Figure 16.12: Elliptic flow measured at RHIC versus particle multiplicity. The bars show calculations assuming a zero-viscosity fluid. [From T. Ludlam, *Nucl. Phys.* **A750**, 9 (2005).]

The measurements are sensitive to the viscosity in the *fireball*: in a medium with high viscosity the pressure gradients from the initial space anisotropy don't translate efficiently into flow anisotropies in the final state, yielding low values of v_2. For a weakly interacting system, as the QGP was expected to be, the viscosity should be large because it is proportional to the mean-free path (see Problem 16.28.) The large values of v_2 observed can be explained only by assuming a strongly coupled medium (i.e. short mean-free path, low viscosity.) The bars in Fig. 16.12 show calculations that assume a *fireball* with zero viscosity. Although the agreement is far from perfect, the fact that the medium is strongly coupled is well established.

Much remains to be understood about the properties of the matter that has been observed at RHIC and its relationship to the early universe.

16.5 References

A derivation of the semiempirical mass formula based on a nucleon–nucleon interaction is given in J. P. Wesley and A. E. S. Green, *Am. J. Phys*, **36**, 1093 (1968).

An authoritative series of seven volumes on heavy ion physics is *Treatise on Heavy Ion Science*, (D. A. Bromley, ed.) Plenum, New York, 1984.

Heavy Ion Reactions are reviewed in *Nuclear Structure and Heavy Ion Dynamics*, Int. School E. Fermi, 1982, Varenna, (L. Moretto and R. A. Ricci, eds.) North-Holland, New York, 1983. A conference on *Heavy Ion Collisions*, (Cargèse 1984) (P. Bonche et al. eds.), Plenum Press, Elmsford, NY, 1984 reviews the subject up to that time. Resonances in heavy ion reactions are found in T. M. Cormier, *Ann. Rev. Nucl. Part. Sci.* **33**, 271 (1983). Heavy ion molecular phenomena are described in *Heavy Ion Collisions, Nuclear Molecular Phenomena*, (N. Cindro, eds.) North-Holland, New York, 1978. Fusion at and below the Coulomb barrier is described by S.G. Steadman and M.J. Rhoades–Brown, *Ann. Rev. Nucl. Part, Sci.* **36**, 649 (1986), by P. Frobrich, *Phys. Rep.* **116**, 338 (1984), and by M. Beckerman, *Rep. Prog. Phys.* **51**, 1047 (1988). Higher energy heavy ion reactions are reviewed by S. Nagamiya, J. Randrup, and T. J. M. Symons in *Ann. Rev. Nucl. Part. Sci.* **34**, 155 (1984). Shock compression is discussed by K–H. Kampart, *J. Phys.* **G15**, 691 (1989). Reaction mechanisms are reviewed in *Heavy Ion Reaction Mechanisms*, (M. Martinet, C. Ngô, and F. LePage, eds.) *Nucl. Phys.* **428A**, (1984). A useful reference is the International School of High Energy Physic, 3d. Course, *Probing the Nuclear Paradigm with Heavy Ion Reactions*, ed. R.A. Broglia, P. Kienle, and P.F. Bortignon, World Scientific, Singapopre, 1994.

The equation of state is discussed by S.H. Kahana, *Ann. Rev. Nucl. Part. Sci.* **39**, 231 (1989).

A nice introduction to RHIC and the connection with the early universe is given in M. Riordan and W.A. Zajc, *The First Few Microseconds*, *Sci. Amer.* **294**, 34 (2006); more details can be found in T. Ludlam, *Nucl. Phys.* **A750**, 9 (2005); B. Müller, J.L. Nagle, *Ann. Rev. Nucl. and Part. Sci.* **56**, 93 (2006). The four collaborations from RHIC have recently presented summaries on their experiments in *Nucl. Phys.* **A757**, 1-283(2005).

Problems

16.1. Estimate the magnitude of the correction that must be applied to Eq. (16.2) in order to take into account atomic binding effects.

16.2. Find the relation between the binding energy B and the mass excess Δ. Can either quantity be used if, for instance, the stability of isobars is studied?

16.3. Discuss the decays of the nuclides shown in Fig. 16.2.

16.4. Use the Bethe–Weizsäcker relation to estimate the position in Fig. 16.2 of the $A = 127$ isobars with $Z = 48$, 49, 67, and 58. How would these isobars decay? With what decay energies? Estimate very crudely the lifetimes that you would expect.

16.5. Prepare a plot similar to Fig. 16.2 for the $A = 90$ isobars. Show that *two* parabolic curves appear. Explain why. How could the appearance of two such curves be introduced into the binding energy relation?

16.6. Consider possible decays $(A, Z) \to (A', Z')$. Write down criteria that involve the corresponding atomic masses $m(A,\ Z)$ and that indicate when a nucleus $(A,\ Z)$ is stable against

(a) Alpha decay.

(b) Electron decay.

(c) Positron decay.

(d) Electron capture.

16.7. Derive Eq. (16.6) and find an expression for the coefficient a_c in terms of R_0. Compute a_c and compare with the empirical value quoted in Eq. (16.10).

16.8. Use Figs. 16.1–fig16.3 to find approximate values for the coefficients in the Bethe–Weitzsäcker relation. Compare with the values in Eq. (16.10).

16.9. Verify that nucleons in the ground state of a nucleus indeed form a degenerate Fermi gas, i.e., occupy the lowest available levels, at all temperatures obtainable in the laboratory. At what temperature (approximately) would a fair fraction of nucleons be excited?

16.10. What would the ratio Z/A be for a nucleus if the exclusion principle were inoperative?

16.11. Consider a nucleus with $A = 237$. Use the semiempirical mass formula to:

(a) Find Z for the most stable isobar.

(b) Discuss the stability of this nuclide for various likely decay modes.

16.12. Symmetric fission is the splitting of a nucleus (A, Z) into two equal fragments $(A/2, Z/2)$. Use the Bethe–Weizsäcker relation to derive a condition for fission instability:

(a) Find the dependence on Z and A.

(b) For what values of A is fission possible for nuclides lying along the line of stability (Fig. 5.20)?

(c) Compare the result obtained in part (b) with reality.

(d) Compute the energy released in the fission of ^{238}U and compare with the actual value.

16.13. (a) Consider isobars with A odd. How many stable isobars would you expect for a given value of A? Why?

(b) Consider isobars with N and Z even. Explain why more than one even stable isobar can occur. Discuss an actual example.

16.14. Verify Eq. (16.18).

16.15. B/A gives the *average* binding energy of a nucleon in a nucleus. The separation energy is the energy required to remove the nucleon that is easiest to remove from a nucleus.

(a) Give an expression for the separation energy in terms of binding energies.

(b) * Use a table of mass excesses, (http://www.nndc.bnl.gov/masses/), to find the neutron separation energies for ^{113}Cd and ^{114}Cd.

16.16. Compare the ratio of the binding energy to the mass of the system for atoms, nuclei, and elementary particles. (Assume that elementary particles are built from constituent quarks.)

16.17. Use the dependence of the potential depth V_0 on $N - Z$, as expressed by Eq. (16.20), to compute the corresponding contribution to the symmetry energy.

16.18. Discuss the symmetry energy due to

(a) An ordinary central force.

(b) A space and spin exchange potential (Heisenberg force). If s_1 and s_2 are the spins of particles 1 and 2, this potential is given by

$$V\psi(\mathbf{r}_1, s_1; \mathbf{r}_2, s_2) = f(r)\psi(\mathbf{r}_2, s_2; \mathbf{r}_1, s_1),$$

where $\mathbf{r} = \mathbf{r}_1 - \mathbf{r}_2$.

16.19. (a) How can the incompressibility of nuclei be measured?

(b) Should there be an excited nuclear state or resonance corresponding to a spherical compression and decompression (breathing)? If so, relate its excitation energy to the incompressibility.

16.20. Consider a collision of ^{32}S on ^{208}Pb at an energy corresponding to the height of the Coulomb barrier. Find this energy in the laboratory system.

16.21. For a collision of ^{64}Cu on ^{208}Pb at a laboratory energy of twice the Coulomb barrier:

(a) What is the c.m. energy?

(b) What is the approximate maximum angular momentum state that can be excited?

16.22. Estimate the (lowest) vibrational and rotational excitation energies for a nuclear molecule formed in ^{12}C collisions on ^{12}C. Compare to experiments.

16.23. In a relativistic heavy ion collision of 250 GeV/nucleon ^{32}S ions on ^{208}Pb:

(a) Find the c.m. energy available.

(b) Estimate the density of nuclear matter formed if the nuclei fuse into a compound of radius equal to that of the Pb nucleus. Neglect relativistic contractions.

(c) What would be the energy density under the conditions of part (b) if all the energy is available?

16.24. CERN has built a lead source for heavy ion collisions. Consider the collision of a 2 TeV/nucleon ^{208}Pb beam colliding with a stationary ^{238}U target.

(a) Find the c.m. energy and the relativistic $\gamma = (1 - v^2/c^2)^{-1/2}$ factor for the collision, where v is the c.m. velocity.

(b) Determine the approximate volume, energy density, and particle density of both the beam and target if the relativistic contraction is taken into account.

(c) If the central collision contains 100 particles from both the beam and target and fills a region corresponding to the contracted volume of 100 particles in (b), find the nucleon and energy densities in this fused compound.

16.25. At RHIC Au + Au collisions occur with a total energy of 200 GeV/A.

(a) Find the relativistic γ and v/c of the ions.

(b) If the radius of a gold nucleus is $R \sim 8$ fm, what is the longitudinal dimension of the colliding beams?

(c) What is the total energy available?

(d) What laboratory energy would be required for the same ceneter-of-mass energy for a Au beam on a fixed Au target?

16.26. Consider the energy density of a fermion and a boson relativistic gases and show that the number of degrees of freedom expected for a quark-gluon plasma with N_q number of quarks is $N = 2(N_c^2 - 1) + 2\ N_c N_q 7/4$, where N_c is the number of colors.

16.27. (a) Find an expression that relates the energy density to the temperature of black-body *photon* radiation. Hint: use Stefan-Boltzman's law.

(b) Now assume the quark-gluon plasma can be treated as a gas of bosons plus fermions with $N_f \sim 3$ (the three lightest quarks) and calculate the temperature corresponding to an average energy density of $\sim 1\text{GeV/fm}^3$. You may use the result from problem 16.26.

16.28. At the end of Sec. 16.4 it is claimed that shear viscosity increases with the mean-free path, and consequently that low viscosity is an indication of a strongly-coupled medium. Use the classical kinetic theory to explain this behavior. [Hint: in the latter context the force of shear viscosity is proportional to the average velocity *difference* between layers separated by the mean-free path.]

Chapter 17

The Shell Model

The liquid drop and the Fermi gas models represent the nucleus in very crude terms. While they account for gross nuclear features, they cannot explain specific properties of excited nuclear states. In Section 5.11 we have given some aspects of the nuclear energy spectrum, and we have also pointed out that progress in atomic physics was tied to an unraveling of the atomic spectra. In atomic physics, solid-state physics, and quantum electrodynamics, unraveling began with the independent-particle model. It is therefore not surprising that this approach was tried early in nuclear theory also. Bartlett, and also Elsasser,(1) pointed out that nuclei display particularly stable configurations if Z or N (or both) is one of the *magic numbers*

$$2,\ 8,\ 20,\ 28,\ 50,\ 82,\ 126. \tag{17.1}$$

The main evidence at that time consisted of the number of isotopes, alpha-particle emission energies, and elemental abundance. Elsasser tried to understand this stability in terms of the neutrons and protons moving independently in a single particle potential well, but he was unable to account for the stability of N or Z at 50 and 82 and N at 126. Scant attention was paid to his work for two reasons. One was that the model had no apparent theoretical basis. Unlike atoms, nuclei have no fixed center, and the short range of nuclear forces seems to imply that one cannot use a smooth average potential to represent the actual potential felt by a nucleon. The second reason was the meager experimental evidence available.

However, the evidence for the existence of magic numbers continued to increase. As in the case of atoms, such magic numbers try to tell us that some kind of *shells* exist in nuclei. Finally, in 1949, the magic numbers were explained in terms of single-particle orbits by Maria Goeppert Mayer(2) and by J. H. D. Jensen.(3) The crucial element was the realization that spin–orbit forces are essential for an understanding of the closed shells at 50, 82, and 126. Moreover, the suggestion was

[1] J. H. Bartlett, *Phys. Rev.* **41,** 370 (1932); W. M. Elsasser, *J. Phys. Radium* **4,** 549 (1933); **5,** 625 (1934).

[2] M. G. Mayer, *Phys. Rev.* **74**, 235 (1948); **75**, 1969 (1949); **78**, 16 (1950).

[3] O. Haxel, J. H. D. Jensen, and H. Suess, *Phys. Rev.* **75**, 1766 (1949); *Z. Physik* **128**, 295 (1950).

made that the Pauli principle strongly suppresses collisions between nucleons and thereby provides for nearly undisturbed orbits for the nucleons in nuclear matter.[4]

The naive shell model assumes that nucleons move independently in a spherical potential. The assumptions of independence and sphericity are oversimplifications. Interactions between nucleons are present that cannot be described by an average central potential, and the nuclear shape is known to not always be spherical. The shell model can be refined by taking some of the *residual interactions* into account and by studying orbits in a deformed well.

In the following sections we shall first exhibit some of the experimental evidence for the existence of magic numbers. We shall then discuss shell closures and the single-particle shell model and finally sketch some refinements.

17.1 The Magic Numbers

In this section, we shall review some experimental evidence for the fact that nuclides with either Z or N equal to one of the magic numbers 2, 8, 20, 28, 50, 82, or 126 are particularly stable. Of course, these numbers are now well explained by the shell model but the adjective *magic* is so descriptive that it is retained.

Clear evidence for the magic numbers comes from the *separation energies of the last nucleons.* To explain the concept, consider atoms. The separation energy or ionization potential is the energy needed to remove the least tightly bound (the *last*) electron from a neutral atom. The separation energies of the elements are shown in Fig. 17.1. The atomic shells are responsible for the pronounced peaks: if the last electron fills a major shell, it is particularly tightly bound, and the separation energy reaches a peak. The next electron finds itself outside a closed shell, has very little to hold onto, and can be removed easily. The nuclear quantity that is analogous to the ionization potential is the separation energy of the last nucleon. If, for instance, a neutron is removed from a nuclide (Z, N), a nuclide $(Z, N-1)$ results. The energy needed for removal is the difference in binding energies between these two nuclides,

$$S_n(Z,N) = B(Z,N) - B(Z,N-1). \tag{17.2}$$

An analogous expression holds for the proton separation energy. With Eqs. (16.2) and (16.3), the separation energy can be written in terms of the mass excesses,

$$S_n(Z,N) = m_n c^2 - u + \Delta(Z,N-1) - \Delta(Z,N) \tag{17.3}$$

or with the numerical values of the neutron mass and the atomic mass unit

$$S_n(Z,N) = 8.07\text{ MeV} + \Delta(Z,N-1) - \Delta(Z,N).$$

[4] E. Fermi, *Nuclear Physics*, University of Chicago Press, Chicago, 1950; V. F. Weisskopf, *Helv. Phys. Acta* **23**, 187 (1950); *Science* **113**, 101 (1951).

The result can be presented in two different ways: Either Z can be kept fixed, or the neutron excess $N-Z$ can be kept constant. The first situation is easier to visualize: We start with a certain nuclide, continue adding neutrons, and record the energy with which each one is bound. Such a plot is shown in Fig. 17.2 for the isotopes

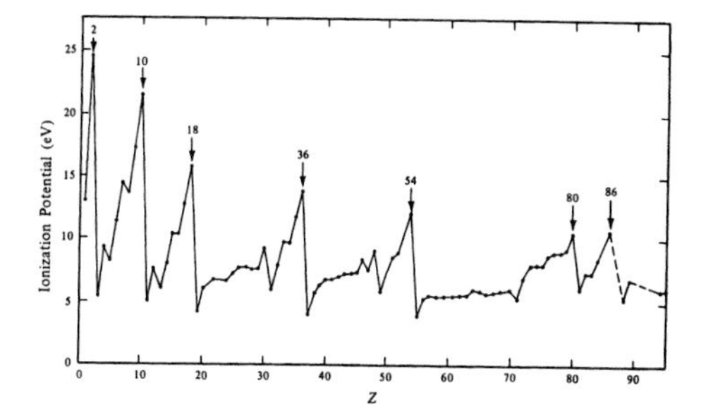

Figure 17.1: Separation energies of the neutral atoms (ionization potentials). [Based on data from C. E. Moore, "Ionization Potentials and Ionization Limits Derived from the Analyses of Optical Spectra," *NSRDS—NBS 34*, 1970.]

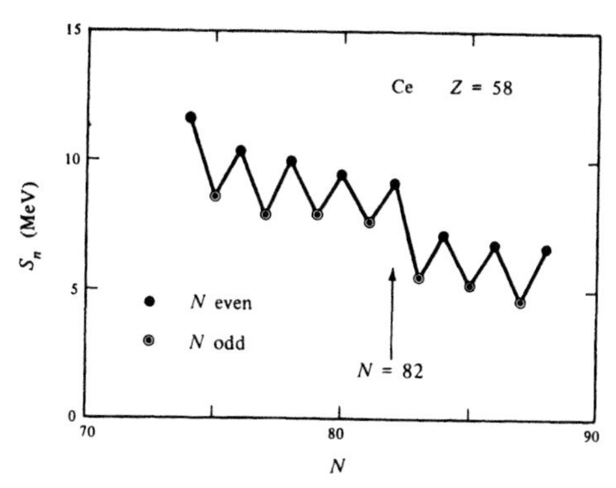

Figure 17.2: Separation energy of the last neutron for the isotopes of cerium.

of cerium, $Z = 58$. Two effects are apparent, an even–odd difference and a closed-shell discontinuity. The even–odd behavior indicates that neutrons are more tightly bound when N is even than when N is odd. The same holds for protons. This fact, together with the empirical observation that all even–even nuclei have spin zero in their ground states, shows that an extra attractive interaction occurs when two like particles pair off to zero angular momentum. This *pairing interaction* is important for understanding nuclear structure in terms of the shell model, and we shall explain it later. Here we note that a similar effect occurs in superconductors where two electrons of opposite momenta and spins form a Cooper pair.[(5)] In nuclei, the interacting boson model replaces the Cooper pairs; this model will be discussed in Chapter 18. From Fig. 17.2 follows that the pairing energy is of the order of 2 MeV in cerium. Once this pairing is corrected for, for instance by only considering isotopes with even N, the second effect, namely the influence of the closed shell at $N = 82$, stands out. Neutrons after a closed shell are less tightly bound by about 2 MeV than just before the closed shell. Figures similar to Fig. 17.2 can be prepared for other regions, and shell closure at all magic numbers can be observed.

Closed shells should be spherically symmetric, have a total angular momentum of zero, and be especially stable. The stability of closed shells can be seen from the energies of the first excited states; a pronounced stability means that it will be hard to excite a closed shell, and consequently the first excited state should lie especially high. An example of this behavior is given in Fig. 17.3 where the ground states and first excited states of the Pb isotopes with even A are shown. ^{208}Pb, with $N = 126$,

[5]See, for instance, G. Baym, *Lectures on Quantum Mechanics*, Benjamin, Reading, Mass., 1969, Chapter 8; E. Moya de Guerra in J.M. Arias and M. Lozano, eds.,*An Advanced Course in Modern Nuclear Physics*, Springer, New York, 2001, p. 255.

has an excitation energy that is nearly 2 MeV larger than that of the other isotopes.

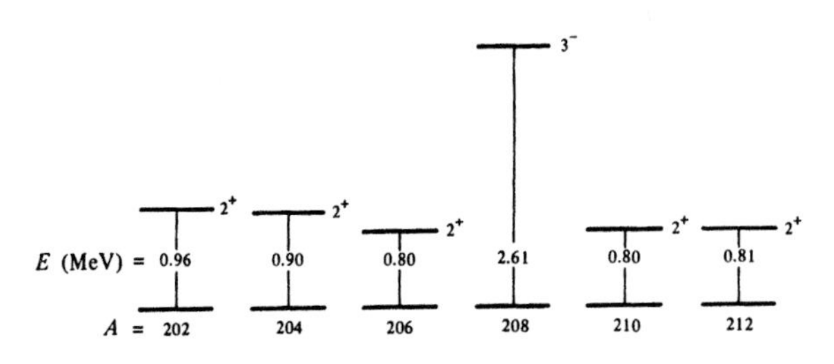

Figure 17.3: Ground and first-excited states of the even-A isotopes of Pb.

Furthermore, unlike all the other isotopes for which the spins and parities of the first excited states are 2^+, that of ^{208}Pb is 3^-. The closed shell affects not only the energy of the first excited state but also its spin and parity.

17.2 The Closed Shells

The first task in the construction of the shell model is the explanation of the magic numbers. In the independent-particle model it is assumed that the nucleons move independently in the nuclear potential. Because of the short range of the nuclear forces, this potential resembles the nuclear density distribution. To see the resemblance explicitly, we consider a two-body force of the type

$$V_{12} = V_0 f(\boldsymbol{x}_1 - \boldsymbol{x}_2), \tag{17.4}$$

where V_0 is the central depth of the potential and f describes its shape. The function f is assumed to be smooth and of very short range. A crude estimate of the strength of the central potential acting on nucleon 1 in the nucleus can be obtained by averaging over nucleon 2. Such an averaging represents the action of all nucleons (except 1) on 1. Averaging is performed by multiplying V_{12} by the density distribution of nucleon 2 in the nucleus, $\rho(\boldsymbol{x}_2)$,

$$V(1) = V_0 \int d^3\boldsymbol{x}_2 f(\boldsymbol{x}_1 - \boldsymbol{x}_2)\rho(\boldsymbol{x}_2).$$

If f is of sufficiently short range, $\rho(\boldsymbol{x}_2)$ can be approximated by $\rho(\boldsymbol{x}_1)$, and $V(1)$ becomes

$$V(1) = CV_0\rho(\boldsymbol{x}_1), \quad C = \int d^3\boldsymbol{x} f(\boldsymbol{x}). \tag{17.5}$$

The potential seen by a particle is indeed proportional to the nuclear density distribution. The density distribution, in turn, is approximately the same as the charge distribution. The charge distribution of spherical nuclei was studied in Section 6.4, and it was found that it can be represented in a first approximation by the Fermi distribution, Fig. 6.4. It would therefore be appropriate to start the investigation of the single-particle levels by using a potential that has the form of a Fermi distribution but is attractive. The Schrödinger equation for such a potential cannot be solved in closed form. For many discussions the realistic potential is consequently replaced by one that can be treated easily, either a square well or a harmonic oscillator potential. We have encountered the latter in Section 15.7, and we can now use the relevant information with very minor changes. The nuclear potential and its approximation by the harmonic oscillator are shown in Fig. 17.4.

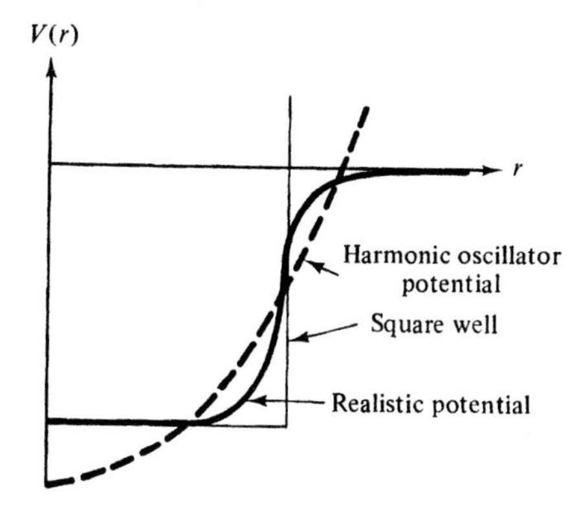

Figure 17.4: The more realistic potential resembling the actual nuclear density distribution is replaced by a harmonic oscillator potential or a square well.

Consider first the harmonic oscillator whose energy levels are shown in Fig. 15.7. The group of degenerate levels corresponding to one particular value of N is called an oscillator shell. The degeneracy of each shell is given by Eq. (15.30). In the application to nuclei, each level can be occupied by two nucleons, and consequently the degeneracy is given by $(N+1)(N+2)$. In Table 17.1 the oscillator shells, their properties, and the total number of levels up to the shell N are listed. The orbitals are denoted by a number and a letter; $2s$, for instance, means the second level with an orbital angular momentum of zero.

Table 17.1: Oscillator Shells for the Three-Dimensional Harmonic Oscillator.

N	Orbitals	Parity	Degeneracy	Total Number of Levels
0	$1s$	+	2	2
1	$1p$	−	6	8
2	$2s$, $1d$	+	12	20
3	$2p$, $1f$	−	20	40
4	$3s$, $2d$, $1g$	+	30	70
5	$3p$, $2f$, $1h$	−	42	112
6	$4s$, $3d$, $2g$, $1i$	+	56	168

Table 17.1 shows that the harmonic oscillator predicts shell closures at nucleons numbers 2, 8, 20, 40, 70, 112, and 168. The first three agree with the magic numbers, but after $N = 2$, the real shell closures differ from the predicted ones. One of two conclusions is forced on us: either the agreement of the first three numbers is fortuitous or an important feature is still missing. Of course by now it is well understood that the second conclusion is correct. To introduce the missing feature, we turn again to the level diagram.

The energy levels of the harmonic oscillator are degenerate for two different reasons. Consider, for example, the level with $N = 2$, which contains the orbitals

$2s$ and $1d$. The $2s$ state has $l = 0$, and it can accept two particles because of the two possible spin states. Rotational symmetry gives the d state ($l = 2$) a $(2l+1)$-fold degeneracy, and, considering the two spin states, this degeneracy leads to $2(4+1) = 10$ states. The fact that the $2s$ and the $1d$ state have the same energy is a feature peculiar to the harmonic oscillator. It is somewhat unfortunate that the harmonic oscillator, which is otherwise so straightforward to understand, possesses this dynamical degeneracy. What happens to the degeneracy in a more realistic potential, such as the one shown in Fig. 17.4? The wave functions of the harmonic oscillator in Fig. 15.8 indicate that particles in states with higher angular momenta are more likely to be found at larger radii than particles in states with small or zero orbital angular momenta.

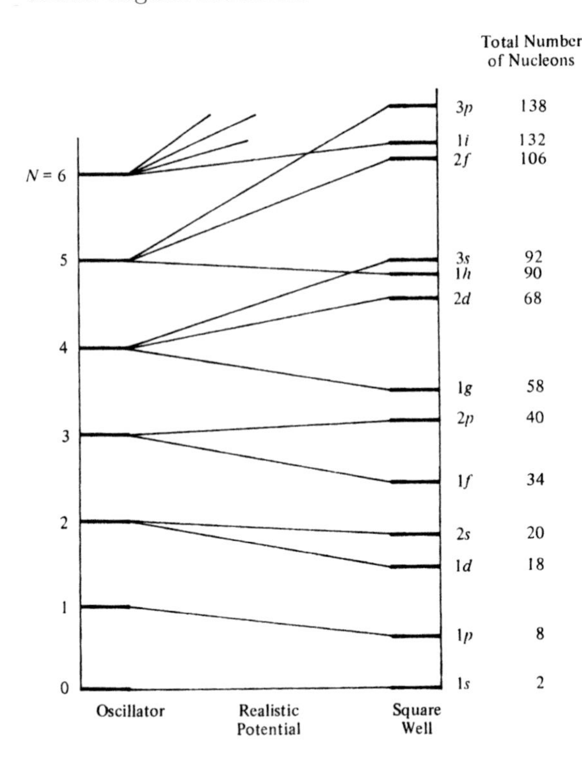

Figure 17.5: Single particle shells. At the left are the harmonic oscillator levels. If the accidental degeneracy in each oscillator shell is lifted by a change in potential shape to a square well, a level diagram as given at the right appears. The total number of nucleons that can be placed into the well up to, and including, the particular shell are also given.

Figure 17.4 shows that the Fermi potential has a flat bottom and for identical central depth is thus deeper at large radii than the oscillator potential. Consequently, the states with higher angular momenta see a deeper potential in the realistic case, the degeneracy will be lifted, and the high-l states will move to lower energies. The lifting of the degeneracy by this feature can be shown explicitly for the square well; the levels for a square well with infinitely high walls are shown on the right side of Fig. 17.5. The realistic case lies somewhere between the square well and the harmonic oscillator, shown on the left side of Fig. 17.5. The magic numbers 50, 82, and 126 are still not explained.

So far, the energy levels have been labeled only with n and l, but the nucleon spin has been neglected. A nucleon with orbital angular momentum l can be in two states, with total angular momenta $l \pm \frac{1}{2}$. As an example, consider the oscillator shell $N = 1$. A nucleon in state $1p$ can have total angular momentum $\frac{1}{2}$ or $\frac{3}{2}$, and

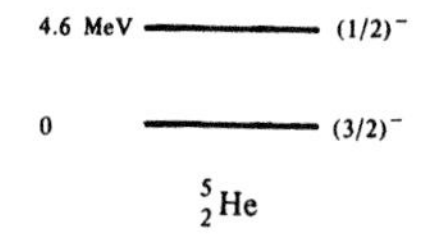

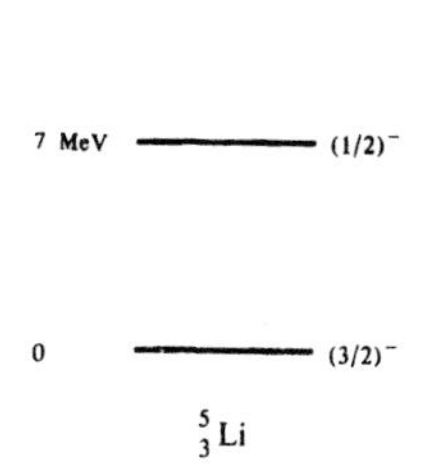

Figure 17.6: Lowest energy levels of ^{5}He and ^{5}Li. Actually, the states have very short half-lives and consequently very large widths. Since the widths are not germane to our arguments, they are not shown.

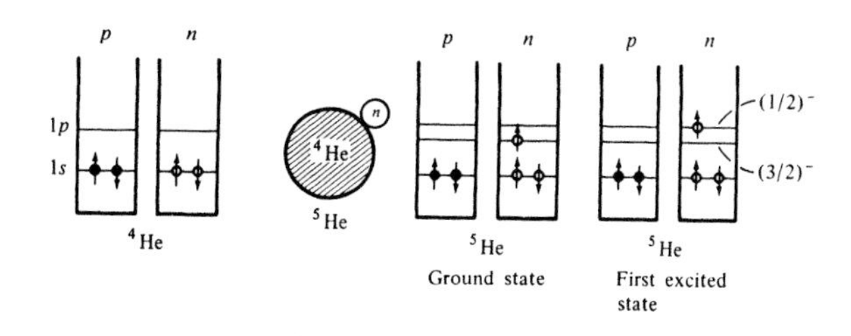

Figure 17.7: Occupation of the nucleon energy levels in ^{4}He, ^{5}He, and ^{5}He*. For simplicity the Coulomb interaction has been neglected, and the neutron and proton wells have been drawn identically. Moreover, only the two lowest energy levels are shown.

the corresponding states are denoted with $1p_{1/2}$ and $1p_{3/2}$. In the central harmonic oscillator potential and in the square well potential, these states are degenerate. The situation is altered by *spin-dependent* forces. Consider, for instance, the lowest energy levels of ^{5}He and ^{5}Li, as given in Fig. 17.6. The ground states of these nuclides have spin $\frac{3}{2}$ and negative parity, and the first excited states spin $\frac{1}{2}$ and negative parity. These quantum numbers are explained by considering ^{5}He(^{5}Li) as a closed shell core of ^{4}He plus one neutron (proton). In ^{4}He, the $1s$ levels for neutrons and protons are filled, and it is the first doubly magic nucleus. The next nucleon, neutron or proton, must go into one of the $1p$ levels, either $1p_{1/2}$ or $1p_{3/2}$. The spins of the observed levels (Fig. 17.7) tell us that the $1p_{3/2}$ level has the lower energy. If the nucleon outside the closed shell, the so-called valence nucleon, is lifted to the next higher level, the first excited state of ^{5}He results.

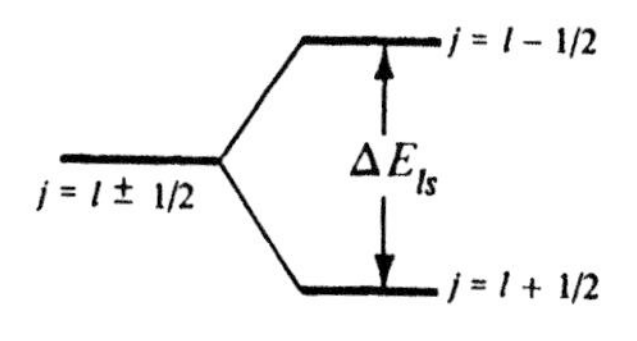

Figure 17.8: Splitting of the states with a given value of l into two states. The spin–orbit interaction depresses the state with total angular momentum $j = l + \frac{1}{2}$ and raises the one with $j = l - \frac{1}{2}$.

The spin and parity values, $(\frac{1}{2})^-$, of this state indicate that it is a $1p_{1/2}$ single-particle level. The degeneracy of the $1p_{1/2}$ and $1p_{3/2}$ levels is lifted in actual nuclei, and the energy splitting is of the order of a few MeV in the light nuclei. This conclusion can be tentatively generalized by assuming that the degeneracy between the levels $l + \frac{1}{2}$ and $l - \frac{1}{2}$ is always lifted in real nuclei, as shown in Fig. 17.8.

The splitting between states $l+\frac{1}{2}$ and $l-\frac{1}{2}$ is now known to be caused primarily by the interaction between the nucleon spin and its orbital angular momentum. Such a spin–orbit force is well known in atomic physics,[(6)] but it was not expected that it would be so strong in nuclei. Since the orbital angular momentum increases with A, so does the importance of the spin–orbit force. We return to the spin–orbit force in the next section but show here that the magic numbers can be explained if its effects are taken into account. A nucleon, moving in the central potential of the nucleus with orbital angular momentum $\boldsymbol{l}$, spin $\boldsymbol{s}$, and total angular momentum $\boldsymbol{j}$,

$$\boldsymbol{j} = \boldsymbol{l} + \boldsymbol{s}, \tag{17.6}$$

acquires an additional energy

$$V_{ls} = C_{ls}\boldsymbol{l}\cdot\boldsymbol{s}. \tag{17.7}$$

We must find the effect of this potential-energy operator on a state $|\alpha; j, l, s\rangle$. Here α denotes all quantum numbers other than j, l, and s. (The reason that j, l, and s can be specified simultaneously is that states of $l = j \pm \frac{1}{2}$ have opposite parities, and parity is conserved in the hadronic force.) With the square of Eq. (17.6), the operator $\boldsymbol{l}\cdot\boldsymbol{s}$ is written as

$$\boldsymbol{l}\cdot\boldsymbol{s} = \tfrac{1}{2}(j^2 - l^2 - s^2). \tag{17.8}$$

The actions of the operators j^2, l^2, and s^2 on $|\alpha; j, l, s\rangle$ are given by Eq. (5.7) so that

$$\boldsymbol{l}\cdot\boldsymbol{s}|\alpha; j, l, s\rangle = \tfrac{1}{2}\hbar^2\{j(j+1) - l(l+1) - s(s+1)\}|\alpha : j, l, s\rangle. \tag{17.9}$$

For a nucleon, with spin $s = \frac{1}{2}$, only two possibilities exist, namely $j = l + \frac{1}{2}$ and $j = l - \frac{1}{2}$, and for these Eq. (17.9) yields

$$\boldsymbol{l}\cdot\boldsymbol{s}|\alpha; j, l, \tfrac{1}{2}\rangle = \begin{cases} \tfrac{1}{2}\hbar^2 l|\alpha; j, l, \tfrac{1}{2}\rangle & \text{for } j = l + \tfrac{1}{2} \\ -\tfrac{1}{2}\hbar^2(l+1)|\alpha; j, l, \tfrac{1}{2}\rangle & \text{for } j = l - \tfrac{1}{2}. \end{cases} \tag{17.10}$$

The energy splitting ΔE_{ls}, shown in Fig. 17.8, is proportional to $l + \frac{1}{2}$:

$$\Delta E_{ls} = (l + \tfrac{1}{2})\hbar^2 C_{ls}. \tag{17.11}$$

The spin–orbit splitting increases with increasing orbital angular momentum l. It consequently becomes more important for heavier nuclei, where higher l values appear. For a given value of l, the level with higher total angular momentum, $j = l + \frac{1}{2}$, lies lower, and it has a degeneracy of $2j + 1 = 2l + 2$. The upper level, with $j = l - \frac{1}{2}$, is $2l$-fold degenerate.

[6] Tipler-Llewellyn, Chapter 7; H. A. Bethe and R. Jackiw, *Intermediate Quantum Mechanics*, 2nd ed. Benjamin, Reading, Mass., 1968, Chapter 8; Park, Chapter 14; G. P. Fisher, *Am. J. Phys.* **39,** 1528 (1971).

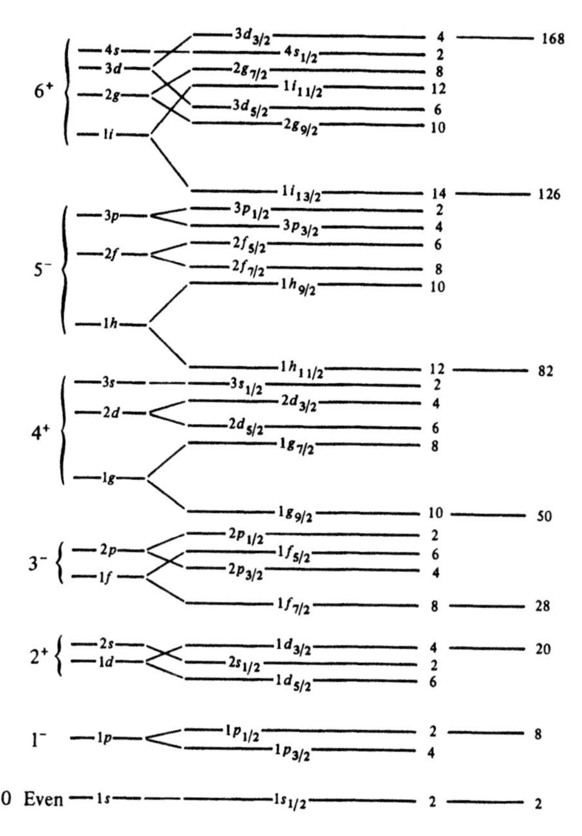

Figure 17.9: Approximate level pattern for nucleons. The number of nucleons in each level and the cumulative totals are shown. The oscillator grouping is shown at the left. Neutrons and protons have essentially the same level pattern up to 50. From then on, some deviations occur. Low neutron angular momenta are more favored than low proton angular momenta.

With these remarks, shell closure at the magic numbers can be understood. Consider Fig. 17.5. The total number of nucleons up to the oscillator shell $N = 3$ is 40; the correct magic number is 50. The $1g_{9/2}$ state has a degeneracy of 10, as shown in Fig. 17.9. This level is depressed by the spin–orbit interaction so that it intrudes into the $N = 3$ oscillator shell, and the total number of nucleons adds up to 50, the correct magic shell closure. Similarly, the $1h_{11/2}$ state has a degeneracy of 12; depressed and added to the $N = 4$ oscillator shell, it produces the number 82. The $1i_{13/2}$, depressed into the $N = 5$ shell, adds 14 nucleons and produces the magic number 126. The situation is summarized in Fig. 17.9, where the level pattern is shown. The details differ slightly for protons and neutrons.

The situation can be summarized by saying that a sufficiently strong spin–orbit interaction which is attractive in the states $j = l + \frac{1}{2}$ can account for the experimentally observed shell closures.

17.3 The Spin–Orbit Interaction

In the previous section it was shown that a spin–orbit interaction, of the form of Eq. (17.7), can produce the experimentally observed shell closures, provided the constant C_{ls} is sufficiently large. Is the evidence from nuclear properties in agreement with what is known about the nucleon–nucleon potential? In Section 14.5 it was shown that the nucleon–nucleon potential energy represented in Eq. (14.36) contains a spin–orbit term,

$$V_{LS}\boldsymbol{L}\cdot\boldsymbol{S}. \tag{17.12}$$

Here $\boldsymbol{L} = \frac{1}{2}(\boldsymbol{x}_1 - \boldsymbol{x}_2) \times (\boldsymbol{p}_1 - \boldsymbol{p}_2)$ is the relative orbital angular momentum of the two nucleons and $\boldsymbol{S} = \boldsymbol{s}_1 + \boldsymbol{s}_2 = \frac{1}{2}(\sigma_1 + \sigma_2)$ is the sum of the spins.

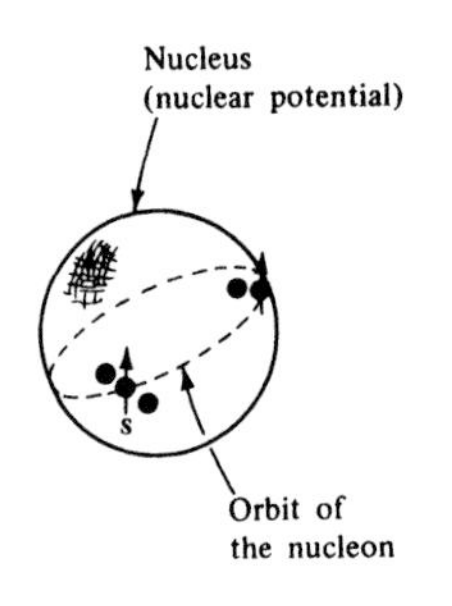

Figure 17.10: Nucleon with orbital angular momentum $\boldsymbol{l}$ and spin $\boldsymbol{s}$ moving in the nuclear potential.

Such a term in the nucleon–nucleon force will produce a term

$$V_{ls} = C_{ls}\boldsymbol{l} \cdot \boldsymbol{s}$$

in the *nuclear* potential, where $\boldsymbol{l}$ is the orbital angular momentum of the nucleon that moves in the nuclear potential and $\boldsymbol{s}$ is its spin. To see the connection, we consider an orbit as shown in Fig. 17.10. In the interior of the nucleus, where the nuclear density is constant, there are an equal number of nucleons on either side of the orbit within reach of the nuclear force. The spin–orbit interaction consequently averages out.

Near the surface, however, nucleons are only on the interior side of the orbit, the relative orbital angular momentum $\boldsymbol{L}$ in Eq. (17.12) always points in the same direction, and the two-body spin–orbit interaction gives rise to a term of the form of Eq. (17.7). To make this argument more precise, the spin–orbit interaction energy [Eq. (17.12)] between two nucleons, 1 and 2, is written as

$$V(1,2) = \tfrac{1}{2}V_{LS}(r_{12})(\boldsymbol{x}_1 - \boldsymbol{x}_2) \times (\boldsymbol{p}_1 - \boldsymbol{p}_2) \cdot (\boldsymbol{s}_1 + \boldsymbol{s}_2). \qquad (17.13)$$

If particle 1 is the nucleon under consideration, an estimate of the nuclear spin–orbit potential can be obtained by averaging $V(1, 2)$ over nucleon 2,

$$V_{ls}(1) = \mathrm{Av} \int d^3\boldsymbol{x}_2 \rho(\boldsymbol{x}_2) V(1,2), \qquad (17.14)$$

where Av indicates that we must average over the spin and the momentum of nucleon 2, and where $\rho(\boldsymbol{x}_2)$ is the probability density of nucleon 2. After inserting $V(1, 2)$ from Eq. (17.13), $V_{ls}(1)$ becomes

$$V_{ls}(1) = \tfrac{1}{2} \int d^3\boldsymbol{x}_2 \rho(\boldsymbol{x}_2) V_{LS}(r_{12})(\boldsymbol{x}_1 - \boldsymbol{x}_2) \times \boldsymbol{p}_1 \cdot \boldsymbol{s}_1; \qquad (17.15)$$

the average of all other terms is zero. The nuclear density at position $\boldsymbol{x}_2$ can be expanded in a Taylor series about $\boldsymbol{x}_1$ because of the short range of the spin–orbit force:

$$\rho(\boldsymbol{x}_2) = \rho(\boldsymbol{x}_1) + (\boldsymbol{x}_2 - \boldsymbol{x}_1) \cdot \boldsymbol{\nabla}\rho(\boldsymbol{x}_1) + \cdots. \qquad (17.16)$$

After inserting the expansion into $V_{ls}(1)$, the integral containing the factor $\rho(\boldsymbol{x}_1)$ vanishes. The remaining integral can be computed; under the assumption that the range of the nucleon spin–orbit interaction is small compared to the nuclear surface thickness, which is the only region wherein $\boldsymbol{\nabla}\rho$ is appreciable, it gives

$$V_{ls}(1) = C \frac{1}{r_1} \frac{\partial \rho(r_1)}{\partial r_1} \boldsymbol{l}_1 \cdot \boldsymbol{s}_1, \qquad (17.17)$$

where $r \equiv |\boldsymbol{x}|$ and

$$C = -\frac{1}{6}\int V_{LS}(r)r^2 d^3r. \tag{17.18}$$

The nucleon–nucleon spin–orbit interaction leads to a spin–orbit interaction for a nucleon moving in the average nuclear potential. As Eq. (17.17) shows, the interaction vanishes where the density is constant, and it is strongest at the nuclear surface. Numerical estimates with Eqs. (17.17) and (17.18) give the correct order of magnitude of V_{ls}.

17.4 The Single-Particle Shell Model

The simplest atomic system is hydrogen because it consists of only one electron moving in the field of a heavy nucleus. Next in simplicity are the alkali atoms which consist of a closed atomic shell plus one electron. In a first approximation they are treated by assuming that the one valence electron moves in the field of the nucleus shielded by the closed shells of electrons which form a spherically symmetric system with zero angular momentum. The entire angular momentum of the atom is provided by the valence electron (and the nucleus). In nuclear physics, the two-body system (deuteron) has only one bound state and does not provide much insight. In analogy to the atomic case, the next simplest cases then are nuclei with closed shells plus one valence nucleon (or nuclides with closed shells minus one nucleon). To discuss such nuclides we first return to closed shells.

What are the quantum numbers of nuclides with closed shells? In the shell model, protons and neutrons are treated independently. Consider first a subshell with a given value of the total angular momentum j, for instance, the proton subshell $1p_{1/2}$ (Fig. 17.9). There are $2j + 1 = 2$ protons in this subshell. Since protons are fermions, the total wave function must be antisymmetric. The spatial wave function of two protons in the same shell is symmetric, and consequently the spin function must be antisymmetric. Only one totally antisymmetric state can be formed from two protons, but a state described by *one* wave function only must have spin $J = 0$. The same argument holds for any closed subshell or shell of protons or neutrons: closed shells always have a total angular momentum of zero. The parity of a closed shell is even because there are an even number of nucleons filling it.

Ground-state spin and parity of nuclides with closed shells plus or minus a single particle are now straightforward to predict. Consider first a single proton outside a closed shell. Because the closed shell has zero angular momentum and even parity, angular momentum and parity of the nucleus are carried by the valence proton. Angular momentum and parity of the proton can be read off from Fig. 17.9. The corresponding level diagram for neutrons is very similar. A first example was already given in Fig. 17.7 from which we deduced that the ground state assignment of ^{5}He should be $p_{3/2}$, or spin $\frac{3}{2}$ and negative parity. A few additional examples are shown in Table 17.2. The agreement between predicted and observed values of spins and

Table 17.2: Ground-State Spins and Parities as Predicted by the Single-Particle Shell Model and as Observed.

Nuclide	Z	N	Shell-Model Assignment	Observed Spin and Parity
^{17}O	8	9	$d_{5/2}$	$\frac{5}{2}^+$
^{17}F	9	8	$d_{5/2}$	$\frac{5}{2}^+$
^{41}Sc	21	20	$f_{7/2}$	$\frac{7}{2}^-$
^{209}Pb	82	127	$g_{9/2}$	$\frac{9}{2}^+$
^{209}Bi	83	126	$h_{9/2}$	$\frac{9}{2}^-$

parities is complete. The quantum numbers of nuclei with a complete shell minus a single particle can also be obtained from Fig. 17.9. Such a *single-hole state* can be described in the language used in Section 5.10 for antiparticles; the hole appears as an antiparticle, and Eq. (5.63) tells us that the angular momentum of the state must be the same as that of the missing nucleon. Similarly, the parity of the hole state must be the same as that of the missing nucleon state.(7) These properties of holes also follow from the remark that a hole, together with the particle that can fill it, couple together to give $J = 0^+$ for the closed shell. As a simple example, consider ^{4}He, shown in Fig. 17.7. Removing one neutron from ^{4}He gives ^{3}He. The removed neutron was in an $s_{1/2}$ state; the absence is indicated by the symbol $(s_{1/2})^{-1}$. The corresponding spin-parity assignment of ^{3}He is $(\frac{1}{2})^+$, in agreement with experiment. Assignments for other single-hole nuclides can easily be given, and they also agree with the experimental values.

Next we turn to *excited states*. In the spirit of the extreme single-particle model, they are described as excitations of the valence nucleon alone; it moves into a higher orbit. The core (closed shell) is assumed to remain undisturbed. Up to what energies can such a picture be expected to hold? Figures 17.2 and 17.3 indicate that the pairing energy is of the order of about 2 MeV. At an excitation energy of a few MeV it is therefore possible that the valence nucleon remains in its ground state but that a pair from the core is broken up and that one of the nucleons of the pair is promoted to the next higher shell. It is also possible that a pair is excited to the next higher shell. In either case, the resulting energy level is no longer describable by the single-particle approach. It is consequently not surprising to find "foreign" levels at a few MeV. Two examples are shown in Fig. 17.11, both doubly magic nuclei plus one valence nucleon. In the case of ^{57}Ni, the single-particle shell-model assignments hold up to about 1 MeV, but above 2.5 MeV, foreign states appear.

[7] A detailed discussion of hole states and of the particle–hole conjugation is given in A. Bohr and B. R. Mottelson, *Nuclear Structure*, Benjamin, Reading, Mass., 1969. See Vol. I, p. 312 and Appendix 3B.

The foreign states are not really foreign. While they cannot be described in terms of the extreme single-particle shell model, they can be understood in terms of the general shell model, through excitations from the core. In the case of ^{209}Pb, the first such state appears at 2.15 MeV. The estimate based on Figs. 17.2 and 17.3 that core excitation will play a role at about 2 MeV is verified.

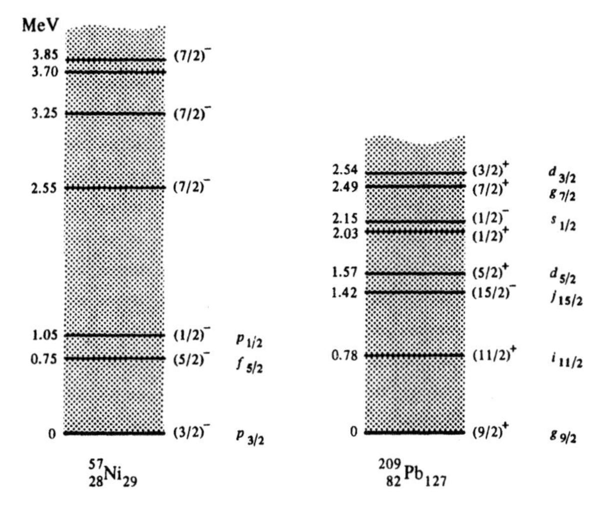

Figure 17.11: Excited states in ^{57}Ni and ^{209}Pb. The states that allow an unambiguous shell-model assignment are labeled with the corresponding quantum numbers.

We have discussed only two properties of nuclei that are well described by the single-particle model, spin and parity of ground states and the level sequence and quantum numbers of the lowest excited states. There are other features that are explained by the extreme single-particle model, for instance the existence of very-long-lived first excited states in certain regions of N and Z, the so-called islands of isomerism. However, the model applies only to a restricted class of nuclei—namely those with only one nucleon outside a closed shell—and an extension to more general conditions is necessary.

17.5 Generalization of the Single-Particle Model

The extreme single-particle shell model, discussed in the previous section, is based on a number of rather unrealistic assumptions: The nucleons move in a spherical fixed potential, no interactions among the particles are taken into account, and only the last odd particle contributes to the level properties. These restrictions are removed in various steps and to various degrees of sophistication; we briefly outline some of the extensions.

1. All particles outside the closed major shells are considered. The angular momenta of these particles can be combined in various ways to get the resulting angular momentum. The two main schemes are the Russell–Saunders, or LS, coupling, and the jj coupling. In the first, the orbital angular momenta are assumed to be weakly coupled to the spins; spin and orbital angular momenta of all nucleons in a shell are added separately to get the resulting $\boldsymbol{L}$ and

$\boldsymbol{S}: \Sigma_i \boldsymbol{l}_i = \boldsymbol{L}, \Sigma_i \boldsymbol{s}_i = \boldsymbol{S}$. The total orbital angular momentum $\boldsymbol{L}$ and the total spin $\boldsymbol{S}$ of all nucleons in a shell are then added to form a given $\boldsymbol{J}$. In the jj coupling scheme, the spin–orbit force is assumed to be stronger than the residual force between individual nucleons so that the spin and the angular momentum of each nucleon are added first to give a total angular momentum $\boldsymbol{j}$; these $\boldsymbol{j}$'s are then combined to the total $\boldsymbol{J}$. In most nuclei, the empirical evidence indicates that the jj coupling is closer to the truth; in the lightest nuclei ($A \lesssim 16$), the coupling scheme appears to be intermediate between the LS and the jj coupling.

2. Residual forces between the particles outside the closed shells are introduced. That such residual interactions are needed can be seen in many ways. Consider, for instance, ^{69}Ga. It has three protons in state $2p_{3/2}$ outside the closed proton shell. These three protons can add their spins to get values of $J = \frac{1}{2}, \frac{3}{2}, \frac{5}{2}, \frac{7}{2}$. (The state $J = \frac{9}{2}$ is forbidden by the Pauli exclusion principle.) In the absence of a residual interaction, these states are degenerate. Experimentally, one state is observed to be lowest—quite often the state $J = j (= \frac{3}{2}$in this case). There must be an interaction that splits these degenerate states. In principle, one should derive the residual interaction as what remains after the nucleon–nucleon interaction is replaced by an average single-nucleon potential. In practice, such a program is too difficult, and the residual interaction is determined empirically. However, many of the features of the residual interaction can be understood on theoretical ground. Consider as an example the *pairing force* described in Section 17.1. We have pointed out there that two like nucleons prefer to be in an antisymmetric spin state, with spins opposed and with a relative orbital angular momentum of zero ($^1S_0$). If the residual force has a very short range and is attractive, this behavior can be understood immediately. Consider for simplicity a zero-range force. The two nucleons can take advantage of such a residual attraction only when they are in a relative s state; the exclusion principle then forces their spins to be opposed, as is observed in reality. Although the true nuclear forces are not of such short range (indeed there is a repulsion at about 0.5 fm), the net effect is unchanged. The energy gained by the action of the *pairing force* is called *pairing energy*, and it is found empirically to be of the order of $12A^{-1/2}$ MeV. The pairing energy leads to an understanding of the energies of the first excited states of even–even nuclei: A pair must be broken, and the corresponding first excited state lies roughly 1–2 MeV above the ground state.

3. Descriptions of nuclei with the inclusion of a dynamic treatment of closed shells have become possible thanks to advanced computers. Such "extended" shell model calculations allow the excitation of closed shell nucleons into open, vacant ones, leaving behind holes. These extended shell models have been successful, for instance in understanding level structure, electromagnetic transi-

tion rates, and weak interaction asymmetry calculations.[8]

4. It is known that many nuclides are permanently deformed and hence cannot be described properly by a spherical potential. For such nuclei, the potential in which the single particles move is assumed to be nonspherical.[9] This deformed-shell or Nilsson model will be described in Section 18.4.

5. The residual interaction between nucleons can also be used to generate new dynamical, collective variables. The importance of such variables in understanding spectra was first noted by Arima and Iachello,[10] and the model they developed has become very useful over the past decades. We will describe the approach in the next chapter.

When the restrictions discussed here are removed, the shell model describes many states very well. However, there remain systematic deviations from properties predicted by the shell model. The two most pronounced ones are quadrupole moments that are much larger than expected and electric quadrupole transitions that are much faster than calculated. These features are most pronounced far away from closed shells, and they point to the existence of collective degrees of freedom that we have not yet considered. We shall turn to the collective model in the following chapter.

17.6 Isobaric Analog Resonances

So far we have discussed states in a given nuclide, without considering neighboring isobars. In Section 8.7 we proved that the charge independence of nuclear forces leads to the assignment of an isospin I to a nuclear state; as long as the Coulomb interaction can be neglected, such a state will show up in $2I + 1$ isobars. Such isobaric analog states have even been found in medium and heavy nuclei[11], [12] and have received attention because of their value for nuclear structure studies.[13]

To describe analog states, we consider the isobars $(Z,\ N)$ and $(Z+1, N-1)$. The energy levels in the absence of the Coulomb interaction are shown in Fig. 17.12. The difference in energy between the two ground states can be computed from the

---

[8] J. B. McGrory and B. H. Wildenthal, *Annu. Rev. Nucl. Part. Sci.* **30**, 383 (1980); B. A. Brown and B. H. Wildenthal, *Annu. Rev. Nucl. Part. Sci.* **38**, 29 (1988); E. Caurier, G. Martnez-Pinedo, F. Nowacki, A. Poves, and A.P. Zuker, *Rev. Mod. Phys.* **77**, 427 (2005).

[9] S. G. Nilsson, *Kgl. Danske Videnskab. Selskab, Mat.-fys. Medd.* **29**, No. 16 (1955).

[10] A. Arima and F. Iachello, *Phys. Rev. Lett.* **35**, 1069 (1975); F. Iachello and A. Arima, *Phys. Lett.* **53B**, 309 (1974); F. Iachello, *Comm. Nucl. Part. Phys.* **8**, 59 (1978).

[11] J. D. Anderson and C. Wong, *Phys. Rev. Lett.* **7**, 250 (1961); J. D. Anderson, C. Wong, and T. W. McClure, *Phys. Rev.* **126**, 2170 (1962).

[12] J. D. Fox, C. F. Moore, and D. Robson, *Phys. Rev. Lett.* **12**, 198 (1964).

[13] H. Feshbach and A. Kerman, *Comm. Nucl. Part. Phys.* **1**, 69 (1967); M. H. Macfarlane and J. P. Schiffer, *Comm. Nucl. Part. Phys.* **3**, 107 (1969). D. Robson, *Science* **179**, 133 (1973).

symmetry term in the semiempirical mass formula. Equation (16.7) gives

$$\Delta_{\text{sym}} = E_{\text{sym}}(Z+1, N-1) - E_{\text{sym}}(Z, N) = -4a_{\text{sym}}\frac{N-Z-1}{A},$$

or

$$\Delta_{\text{sym}}(\text{in MeV}) = -90\frac{N-Z-1}{A}. \tag{17.19}$$

The volume and surface terms are equal for the isobars, and thus the ground state of the isobar with higher Z lies lower by the amount Δ_{sym}. For the pair ^{209}Pb and ^{209}Bi, for instance, $\Delta_{\text{sym}} \sim -19$ MeV.

In the absence of the Coulomb interaction, isospin is a good quantum number. As stated in Section 8.7, the isospin of a nuclear ground state assumes the smallest allowed value. The isospin of the ground state of the isobar (Z, N) is thus given by Eq. (8.34) as

$$I_{>} = \frac{N-Z}{2}, \tag{17.20}$$

whereas for the isobar $(Z+1, N-1)$, the assignment is

$$I_{<} = \frac{N-Z}{2} - 1 = I_{>} - 1. \tag{17.21}$$

Because of charge independence, the levels of the *parent nucleus* (Z, N) also appear with the same energy in the isobar $(Z+1, N-1)$. These analog states are shown in Fig. 17.12. At this point, a crucial difference between light and heavy nuclei appears. To appreciate it we return to Fig. 5.34, Table 5.11, and Eq. (17.3) and note that nuclei have discrete levels (bound states) up to excitation energies of about 8 MeV. Above about 8 MeV, emission of nucleons becomes possible, and the spectrum is continuous.

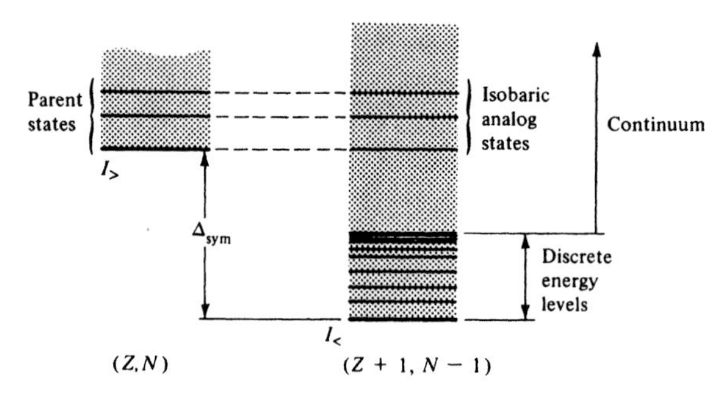

Figure 17.12: Energy level diagram for the isobars (Z, N) and $(Z+1, N-1)$ in the absence of the Coulomb interaction.

In *light nuclei*, where the symmetry energy is small, the isobaric analogs of the ground state and of low-lying excited states of the parent nucleus lie in the discrete part of the spectrum and consequently are *bound states*. An example is shown in Fig. 8.5 where the 0^+ state in ^{14}C is the parent state, and the first excited state in ^{14}N is the isobaric analog state.

In *heavy nuclei*, the situation is as shown in Fig. 17.12: The symmetry energy is larger than the energy at which the continuum begins, and the analog states lie in the continuum. Nevertheless, in the absence of the Coulomb interaction, the analog states would remain bound, as can be seen in the following. Decay by neutron emission will lead to a neutron and a nucleus $(Z+1, N-2)$. The isospin of the ground state and the low-lying excited states of the nuclide $(Z+1, N-2)$ is given by $I = \frac{1}{2}(N-Z-3) = I_> - \frac{3}{2}$. Isospin conservation forbids the decay of the analog state with $I = I_>$ into a state with $I_> - \frac{3}{2}$ and a neutron. In the absence of the Coulomb interaction, the threshold for proton emission is so high that a decay of the analog state by proton emission is not possible.

17.7 Nuclei Far From the Valley of Stability

As we explained in chapter 14 the nucleon-nucleon interaction has not been calculated in an exact form and instead approximate approaches are used. Most of these rely ultimately on comparisons to data for their validation. The data can not simply be that coming from single nucleon-nucleon scattering experiments because the correlations that take place in larger nuclei involve other degrees of freedom (like the three-body force.) As a consequence the models that have been developped have good predictive power close to the line of stability, where there is a wealth of data available, but not for nuclei far from stability, where data is poor. Moreover, it is believed that imperfections of the models close to stability could be resolved from studies of nuclei far from stability, where some features would be exaggerated. For that reason, in recent years interest has focused on studies of exotic nuclei, particularly those far from the valley of stability. The nuclei with large or small Z/N have binding energies close to zero, and often decay by weak interactions with lifetimes larger than 1 ms; they play a role in nucleosynthesis (see Chapter 19.)

Nuclei far from stability can be produced by using short-lived beams. These radioactive beams can be produced by fusion-evaporation reactions (primarily light nuclei), by fission-fragmentation, or by spallation reactions on a primary target. They can then be reaccelerated in a time short compared to their lifetimes in *rare isotope* or *radioactive isotope* accelerators to study reactions (or their by-products) on secondary targets.

The shell model we have discussed earlier applies to nuclei in or near the valley of stability, but far from it, normal shell closures tend to disappear (e.g., $N = 20$ near the neutron drip line) and energy levels become more uniformly spaced. Fig. 17.13 shows how the position of the shells changes from nuclei near stability to those far from it.(14) Doubly magic nuclei will be different than those close to the valley of stability.(15)

---

14 D.F. Geesaman, C.K. Gelbke, R.V.F. Janssens, B.M. Sherrill, *Annu. Rev. Nucl. Part. Sci.* **56**, 53 (2006).

15 R. Schneider et al., *Z. Phys.* **A348**, 241 (1994) and **A352**, 351 (1995).

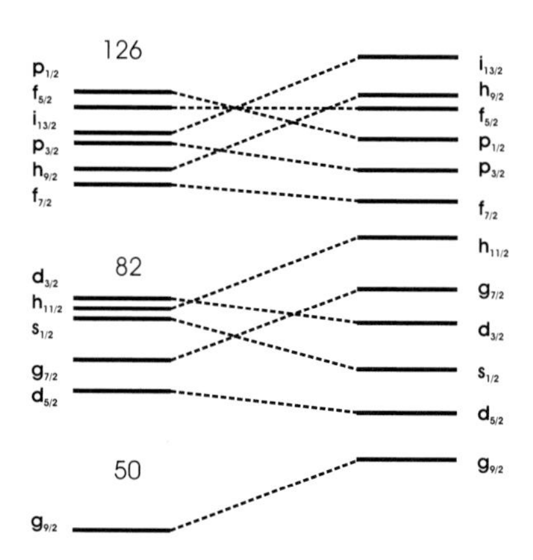

Figure 17.13: Sketch of energies of the shells for nuclei near stability (left) and neutron-rich nuclei (right). [After (14).]

Many-body systems are complicated so it is hard to predict what new phenomena one might observe with nuclei far from stability. An example of an issue that had not been anticipated but was discovered investigating nuclei away from stability are *neutron-halo* nuclei.$^{(16,17)}$ In these nuclei, the last neutron(s) have a reduced wavelength, $\lambda\!\!\!^{-}$, much larger than the average separation of nucleons ($\lambda\!\!\!^{-} = \hbar/\sqrt{2M_nB}$, where B is the binding energy and M_n is the mass of the neutron.)

Because the average shell model potential is weak at these distances, the pairing energy becomes more important so that the addition of two neutrons may be stable, even if the addition of only one neutron is not. In the case of ^{11}Li, for example, one can consider a core of ^{9}Li plus two neutrons: the systems $^9\text{Li} + n$ or $n + n$ are unstable but the combination $^9\text{Li} + n + n$ is stable (with respect to the strong interaction.) The size of the neutron halo in ^{11}Li is about equal to that of the ^{208}Pb nucleus and it has other interesting properties.$^{(18)}$

Are there other surprises in store? Of the ~ 7000 nuclei that are predicted to be stable with respect to decays via the strong interaction we have experimental knowledge of less than approximately half. With the development of efficient techniques to get more intense radioactive beams it is expected that a significant improvement in our knowledge will be achieved over the next decade.

17.8 References

A very careful and easily readable introduction to the shell model, including a thorough discussion of the experimental material, is given by the two founders of the model: M. Goeppert Mayer and J. H. D. Jensen, *Elementary Theory of Nuclear Shell Structure*, Wiley, New York, 1955.

More modern aspects of the shell model and a critical review of many experimental aspects is given in Chapter 3 of A. Bohr and B. R. Mottelson, *Nuclear Structure* Vol. 1. W. A. Benjamin, Reading, Mass. 1969; see also K.L.G. Heyde, *The Nuclear Shell Model*, 2nd edition, Springer, NY, 1994.

[16] I. Tanihata et al., *Phys. Lett.* **B160**, 380 (1985); *Phys. Rev. Lett* **55**, 2676 (1985).
[17] S.M. Austin, G.F. Bertsch, *Sci. Am.* **272**, 90 (1995); A.S. Jensen, K. Riisager, D.V. Fedorov, and E. Garrido, *Rev. Mod. Phys.* **76**, 215 (2004).
[18] T. Nakamura et al., *Phys. Rev. Lett.* **96**, 252502 (2006).

An elementary approach that differs considerably from the one given in the present chapter is given in Chapter 4 of B. L. Cohen, *Concepts of Nuclear Physics*, McGraw-Hill, New York, 1971.

The mathematical problems that appear in the shell model are expounded in detail in A. de-Shalit and I. Talmi, *Nuclear Shell Theory*, Academic Press, New York, 1963.

Descriptions of the nuclear shell model can also be found in G.A. Jones, *The Properties of Nuclei*, 2nd ed., Clarendon Press, Oxford, 1987, Ch. 3, and in more detail in R.D. Lawson, *Theory of the Nuclear Shell Model*, Clarendon Press, Oxford, 1980; K. Heyde, *The Nuclear Shell Model*, Springer, New York, 1994. Reviews can be found in E. Caurier, G. Martnez-Pinedo, F. Nowacki, A. Poves, and A.P. Zuker, *Rev. Mod. Phys.* **77**, 427 (2005); B.A. Brown and B.H. Wildenthal, *Annu. Rev. Nucl. Part. Sci.* **38**, 29 (1988); *Shell Model and Nuclear Structure: Where Do We Stand?*, (A. Covello, ed.) World Sci., Teaneck, NJ, 1989.

Analogue resonances are treated in reviews to be found in *Isospin in Nuclear Physics*, (D. H. Wilkinson, ed.) North-Holland, Amsterdam, 1969; E. G. Bilpuch et al., *Phys. Rep.* **28**, 146 (1976); C. Gaarde, *Annu. Rev. Nuc. Part. Sci.* **41**, 187 (1991); F. Osterfeld, *Rev. Mod. Phys.* **64**, 491 (1992).

Nuclei near the drip lines are described in D.F. Geesaman, C.K. Gelbke, R.V.F. Janssens, B.M. Sherrill, *Annu. Rev. Nucl. Part. Sci.* **56**, 53 (2006); B.A. Brown, *Prog. Part. Nucl. Phys.* **47**, 517 (2001); C. A. Bertulani, L.F. Canto, and M.S. Hussein, eds. *Physics of Unstable Nuclear Beams*, World Sci., Singapore, 1997; J.M. Arias and M. Lozano, eds. *An Advanced Course in Modern Nuclear Physics*, Springer, New York 2001; P. Chomaz, ed. *Comptes Rendus Physique*, **4**, Nos. 4 and 5, 2003.

Problems

17.1. * Use a table of mass excesses given in, e.g., http://www.nndc.bnl.gov/masses/ to discuss the evidence for shell closure as obtained from proton separation energies:

(a) Plot the proton separation energies for some nuclides across the magic numbers while keeping N constant.

(b) Repeat part (a) but keep $N - Z$ constant.

17.2. Discuss additional evidence for the existence of magic numbers by considering the following properties:

(a) The number of stable isotopes and isotones.

(b) Neutron absorption cross sections.

(c) The excitation energies of the first excited states of even–even nuclides.

(d) Beta decay energies.

17.3. Add the following spin–spin term to the two-body force, Eq. (17.4):

$$\boldsymbol{\sigma}_1 \cdot \boldsymbol{\sigma}_2 V_0' g(\boldsymbol{x}_1 - \boldsymbol{x}_2).$$

Assume that g is smooth and very short range. Show that this term gives no contribution to $V(1)$, Eq. (17.5), for closed shell nuclei. Show that the term can be neglected for a nucleus with one particle outside a closed shell.

17.4. Study the level sequence in the infinite three-dimensional square well. Compare the sequence with that obtained from the harmonic oscillator and given in Fig. 17.5.

17.5. Discuss additional evidence for the existence of a strong spin-orbit term in the nucleon–nucleus interaction by considering the scattering of protons from ^{4}He.

17.6. Verify Eq. (17.10).

17.7. Verify Eqs. (17.17) and (17.18).

17.8. (a) Estimate the A dependence of the spin–orbit force.

(b) What is the strength of the two-body spin–orbit force needed to obtain the empirical nuclear spin–orbit splitting? Compare to ^{5}He, ^{5}Li splitting.

(c) What is the sign of the two-body spin–orbit force that gives the correct nuclear spin–orbit term?

17.9. Verify the step from Eq. (17.14) to (17.15). Prove that the terms that are not shown in Eq. (17.15) average to zero.

17.10. Find the spin and parity assignment for the following single-hole nuclear ground states: ^{15}O, ^{15}N, ^{41}K, ^{115}In, ^{207}Pb. Compare your predictions with the measured data.

17.11. Compare the first few excited states of the nuclides ^{15}N, ^{17}O, and ^{39}K with the prediction of the single-particle shell model. Discuss spin, parity, and level ordering.

17.12. Use the single-particle model to calculate for odd-mass nuclei the magnetic moments as a function of spin for:

(a) Z odd, and

(b) N odd.

(c) Compare the result with experimental values.

17.13. What isospin value would you expect for the ground state of an odd-mass nuclide (Z, N) in the single-particle shell model?

17.14. Use the single-particle shell model to explain why the *islands of isomerism* exist. (Traditionally, a long-lived excited nuclear state is called an isomer.) In particular, explain why the nuclide ^{85}Sr has an excited state, at 0.225 MeV, with a half-life of about 70 min.

17.15. Discuss direct nuclear reactions, for instance, $(p, 2p)$, in the shell model and show in the case of a particular example (for instance $p^{16}\text{O} \rightarrow 2p^{15}\text{N}$) that the shell structure is readily apparent in the differential cross section. [For example, see Th. A. Maris, P. Hillman, and H. Tyrèn, *Nucl. Phys.* **7**, 1 (1958).]

17.16. Explain the reaction mechanism for exciting analog states in (d, n) reactions. Find an example in the literature.

17.17. The force acting on a nucleon incident on a nucleus can be represented by a single-particle *optical potential.* Such a potential can contain a term $C\vec{I} \cdot \vec{I}' f(\boldsymbol{r})$, where $\vec{I}$ is the isospin of the incident nucleon and $\vec{I}'$ that of the target nucleus.

(a) Show that such a term is allowed.

(b) Explain how such a term permits excitation of isobaric analog resonances in (p, n) and (n, p) reactions, among others. Are these reactions (either or both) still "allowed" if the electromagnetic interaction is switched off?

(c) Estimate the magnitude and the mass number dependence of the constant C.

17.18. Consider the state of a proton, with small excitation energy, in a heavy nucleus. Explain why the application of the charge-lowering operator, I_-, to such a state gives zero.

17.19. (a) Determine the next shell closures beyond those of $Z = 82$ and $N = 128$. What would be the atomic number and mass number of the next doubly magic nucleus?

(b) Would you expect this nucleus to be stable? Give reasons, or explain.

(c) How would you search for this doubly magic nucleus? Has it been sought and what have been the results of the searches?

17.20. (a) Why do proton drip line nuclei tend to be easier to reach experimentally than neutron ones?

(b) What are some reasons why the usual shell model does not apply for nuclei near the drip lines?

Chapter 18

Collective Model

Although the shell model describes the magic numbers and the properties of many levels very well, it has a number of failures. The most outstanding one is the fact that many quadrupole moments are much larger than those predicted by the shell model.[1] It was shown by Rainwater that such large quadrupole moments can be explained within the concept of a shell model if the closed-shell core is assumed to be deformed.[2] Indeed, if the core is ellipsoidal it acquires a quadrupole moment proportional to the deformation. A deformation of the core is evidence for many-body effects, and collective modes of excitation are possible. The appearance of such modes is not surprising. Lord Rayleigh investigated the stability and oscillations of electrically charged liquid drops in 1877,[3] and Niels Bohr and F. Kalckar showed in 1936 that a system of particles held together by their mutual attraction can perform collective oscillations.[4] A classical example of such collective effects is provided by plasma oscillations.[5] The existence of large nuclear quadrupole moments provides evidence for the possibility of collective effects in nuclei. From about 1950, Aage Bohr and Ben Mottelson started a systematic study of collective motions in nuclei;[6] over the years, they and their collaborators have improved the treatment so that today the model combines the desirable features of shell and collective models and is called the *unified nuclear model.*

The salient facts can be discussed most easily by describing two extreme situations. *Closed shell* nuclei are spherically symmetric and not deformed. The primary collective motions of such nuclei are surface oscillations, like the surface waves on a liquid drop. For small oscillations, harmonic restoring forces are assumed, and equally spaced vibrational levels result. *Far away* from closed shells, the nucleons outside the core polarize the core, and the nucleus can acquire a *permanent*

---

[1] C.H. Townes, H.M. Foley, and W. Low, *Phys. Rev.* **76,** 1415 (1949).

[2] J. Rainwater, *Phys. Rev.* **79**, 432 (1950).

[3] J.W.S. Rayleigh, *The Theory of Sound*, Vol. II, Macmillan, New York, 1877, §364.

[4] N. Bohr, *Nature* **137**, 344 (1936); N. Bohr and F. Kalckar, *Kgl. Danske Videnskab. Selskab. Mat.-fys. Medd.* **14**, No. 10 (1937).

[5] *Feynman Lectures*, II-7-5ff; Jackson, Chapter 7.

[6] A. Bohr, *Phys. Rev.* **81,** 134 (1951); A. Bohr and B.R. Mottelson, *Kgl. Danske Videnskab. Selskab. Mat-fys. Medd.* **27,** No. 16 (1953).

deformation. The entire deformed nucleus can rotate, and this type of collective excitation leads to the appearance of rotational bands. The deformed nucleus acts as a nonspherical potential for the much more rapid single-particle motion; the energy levels of a single particle in such a potential can be investigated, and the result is the *Nilsson model*,[7] already mentioned at the end of the previous chapter.

We shall begin the discussion in the present chapter with deformations and rotational excitations because these two features are easiest to understand and give the most spectacular effects.

18.1 Nuclear Deformations

As early as 1935 optical spectra revealed the existence of nuclear quadrupole moments.[8] We have encountered the quadrupole moment in Section 14.5, and we have seen there that it measures the deviation of the shape of the nuclear charge distribution from a sphere. The existence of a quadrupole moment hence implies nonspherical (deformed) nuclei. For the discussion of nuclear models, the sign and magnitude of the deformation are important. As we shall see below, the quadrupole moments far away from closed shells are so large that they cannot be due to a single particle and thus cannot be explained by the naive shell model. The discrepancy is particularly clear around $A \approx 25$ (Al, Mg), $150 < A < 190$ (lanthanides), and $A > 220$ (actinides).

The classical definition of the quadrupole moment has already been given in Eq. (14.30) as

$$\mathcal{Q} = Z \int d^3r (3z^2 - r^2)\rho(\mathbf{r}). \tag{18.1}$$

Note that the quadrupole moment as defined here has the dimension of an area. In some publications, an additional factor e is introduced in the definition of $\mathcal{Q}$. For estimates, $\mathcal{Q}$ is computed for a homogeneously charged ellipsoid with charge Ze and semiaxes a and b. With b pointing along the z axis, $\mathcal{Q}$ becomes

$$\mathcal{Q} = \frac{2}{5}\, Z(b^2 - a^2). \tag{18.2}$$

If the deviation from sphericity is not too large, the average radius $\overline{R} = \frac{1}{2}(a+b)$ and $\Delta R = b - a$ can be introduced. With $\delta = \Delta R/\overline{R}$, the quadrupole moment becomes

$$\mathcal{Q} = \frac{4}{5}\, Z R^2 \delta. \tag{18.3}$$

Quantum mechanically, the probability density $\rho(r)$ is replaced by $\psi^*_{m=j}\psi_{m=j}$. Here j is the spin quantum number of the nucleus and $m = j$ indicates that the nuclear

[7] S.G. Nilsson, *Kgl. Danske Videnskab, Selskab. Mat.-fys. Medd.* **29,** No. 16 (1955); S.G. Nilsson and I. Ragnarsson, *Shapes and Shells in Nuclear Structure*, Cambridge University Press, New York, 1995.

[8] H. Schüler and T. Schmidt, *Z. Physik* **94,** 457 (1935).

spin is taken to point along the z direction. Thus

$$\mathcal{Q} = Z \int d^3 r \psi^*_{m=j} (3z^2 - r^2) \psi_{m=j}. \tag{18.4}$$

It is customary to introduce a reduced quadrupole moment,

$$\mathcal{Q}_{\text{red}} = \frac{\mathcal{Q}}{ZR^2}. \tag{18.5}$$

For a uniformly charged ellipsoid, Eq. (18.3) shows that the reduced quadrupole moment is approximately equal to the deformation parameter δ:

$$\mathcal{Q}_{\text{red}}(\text{ellipsoid}) = \frac{4}{5}\,\delta. \tag{18.6}$$

After these preliminary remarks we turn to some experimental evidence. Figure 18.1 displays the reduced quadrupole moments as a function of the number of odd nucleons (Z or N); it shows that the nuclear deformation is very small near the magic numbers but assumes values as large as 0.4 between shell closures. The large deformations are all positive. Equation (18.1) then indicates that these nuclei are elongated along their symmetry axes; they are cigar-shaped (prolate).

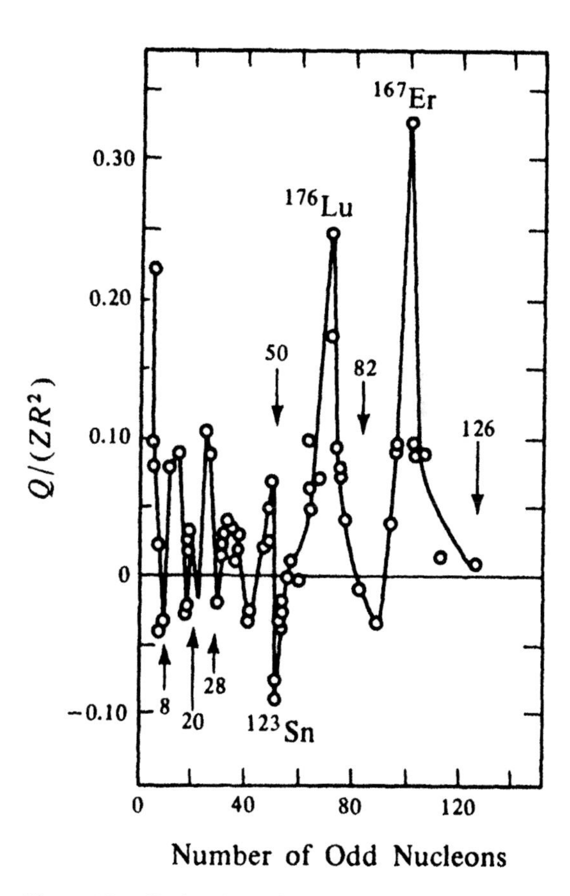

Figure 18.1: Reduced quadrupole moment plotted versus the number of odd nucleons (A or N). Arrows indicate the positions of closed shells, where $\mathcal{Q} = 0$.

The first question is now: Can the observed deformations be explained by the shell model? In the single-particle shell model, the electromagnetic moments are due to the last nucleon; the core is spherically symmetric and does not contribute to the quadrupole moment. The situation for a single proton and a single proton hole are sketched in Fig. 18.2. To compute the quadrupole moment arising from the single particle, a single-particle wave function, for instance as in Eq. (15.27), is inserted

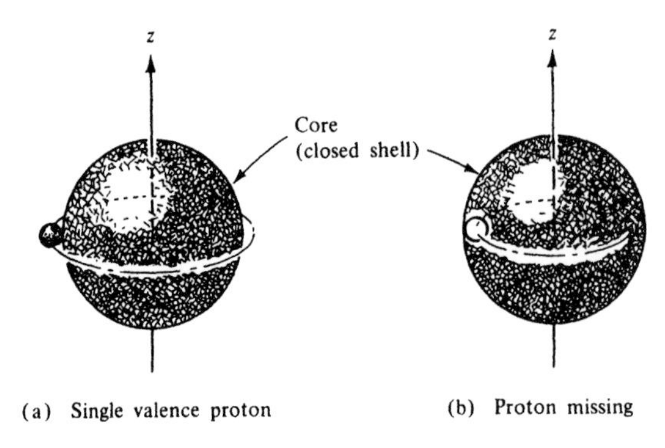

Figure 18.2: Quadrupole moment produced by a closed shell core plus (a) a single proton and (b) a single-proton hole.

into Eq. (18.4); the result is

$$\mathcal{Q}_{sp} = -\langle r^2 \rangle \frac{2j-1}{2(j+1)}. \quad (18.7)$$

Here, j is the angular momentum quantum number of the single particle and $\langle r^2 \rangle$ is the mean-square radius of the single-nucleon orbit. With $\langle r^2 \rangle \approx R^2$, the reduced quadrupole moment for a single proton becomes approximately

$$\mathcal{Q}^p_{\text{red.sp}} \approx -\frac{1}{Z}. \quad (18.8)$$

A single neutron, in first order, produces no quadrupole moment. However, its motion affects the proton distribution by shifting the c.m., and the corresponding value is

$$\mathcal{Q}^n_{\text{sp}} \approx \frac{Z}{A^2} \mathcal{Q}^p_{\text{sp}}. \quad (18.9)$$

For single-hole states, relations similar to Eqs. (18.7) and (18.9) hold, but the sign is positive.

Even a quick glance at Fig. 18.1 shows that many of the observed quadrupole moments are far larger than the estimates given in Eqs. (18.8) and (18.9). A more detailed comparison for four specific cases is given in Table 18.1. For the estimates of the predicted single-particle quadrupole moments, $\langle r^2 \rangle$ has been taken equal to the square of the half-density radius c, given in Eq. (6.27). The values in the table show that in the case of a doubly magic nucleus plus a proton, the single-particle estimate agrees reasonably well with the actual quadrupole moment. In the other cases, the observed values are very much larger than the predicted ones. In the case of ^{175}Lu even the sign is wrong. The features shown in Table 18.1 for a few typical cases hold true when more nuclides are considered. The naive single-particle shell model cannot explain the observed large quadrupole moments.

How can the large quadrupole moments be explained? As stated earlier, the crucial step to a solution of the puzzle was taken by Rainwater. In the naive shell model it is assumed that the closed shells do not contribute to the nuclear moments: the core is assumed to be spherical. Rainwater suggested that the core of nuclides with large quadrupole moments is not spherical but permanently *deformed* by the valence nucleons. Since the core contains most of the nucleons and hence also most of the electric charge, even a small deformation produces a sizable quadrupole moment. An estimate of the deformation necessary to produce a certain reduced quadrupole moment can be obtained from Eq. (18.6). In the case of ^{17}O, for instance, a deformation of only $\delta = 0.07$ is needed to obtain the observed value.

Table 18.1: COMPARISON OF OBSERVED AND PREDICTED SINGLE-PARTICLE QUADRUPOLE MOMENTS.

Nuclide	Z	N	Character	j	Q_{obs} (fm^2)	Q_{sp} (fm^2)	Q_{obs}/Q_{sp}
^{17}O	8	9	Doubly magic + 1 neutron	5/2	−2.6	−0.1	20
^{39}K	19	20	Doubly magic + proton hole	3/2	+5.5	+5	1
^{175}Lu	71	104	Between shells	7/2	+560	−25	−20
^{209}Bi	83	126	Doubly magic + proton hole	9/2	−35	−30	1

The nuclear deformation can be understood by starting from a closed shell nuclide. As discussed in Chapter 17, the short-range pairing force makes such a nucleus spherical, with zero angular momentum. The addition of nucleons outside the closed shell tends to polarize the core through the long-range attractive part of the nuclear force. If only one nucleon is outside the core, the distortion is of the order of $1/A$. Since there are about Z electric charges in the core, such a distortion leads to an induced quadrupole moment of the order of $(Z/A)\,Q_{sp}$. The distortion is about the same for neutrons as for protons, and nuclei with one neutron outside a closed shell thus should have a quadrupole moment of the same sign and about the same magnitude as odd-proton nuclides. The quadrupole moment of ^{17}O, listed in Table 18.1, can consequently be understood in a crude way. When more nucleons are added outside the closed shell, the polarization effect is enhanced, and the observed quadrupole moments can be explained.

The existence of a nuclear deformation makes itself felt not only in the static quadrupole moments but also in a number of other properties. We shall discuss two in the following sections: the appearance of a rotational spectrum and the behavior of shell-model states in a deformed potential.

18.2 Rotational Spectra of Spinless Nuclei

In the previous section we have shown that considerable evidence for the existence of permanently deformed nuclei exists. A nuclear deformation implies that the orientation of such a nucleus in space can be determined and can be described by a set of angles. This possibility leads to a prediction.[9] There exists an uncertainty relation between an angle, φ, and the corresponding orbital angular momentum operator, $L_\varphi = -i\hbar(\partial/\partial\varphi)$,

$$\Delta\varphi\Delta L_\varphi \gtrsim \hbar. \tag{18.10}$$

[9] A. K. Kerman, "Nuclear Rotational Motion," in *Nuclear Reactions*, Vol. I, (P. M. Endt and M. Demeur, eds.), North-Holland, Amsterdam, 1959. The uncertainty relation Eq. (18.10), which underlies the discussion here, gives rise to interesting problems and arguments. If such arguments surface, read M. M. Nieto, *Phys. Rev. Lett.* **18**, 182 (1967), and P. Carruthers and M. M. Nieto, *Rev. Mod. Phys.* **40**, 411 (1968).

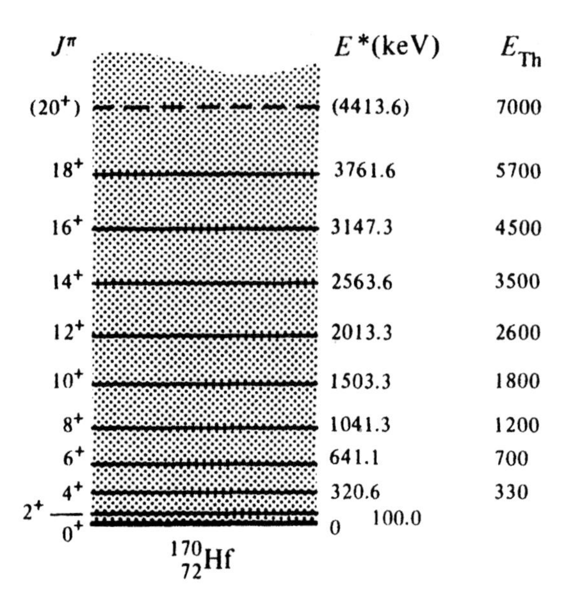

Figure 18.3: Rotational spectrum of the strongly deformed nuclide ^{170}Hf. [After F.S. Stephens, N.L. Lark and R.M. Diamond, *Nucl. Phys.* **63**, 82 (1965).] The levels were observed in the reaction $^{165}\text{Ho}(^{11}\text{B}, 6n)^{170}\text{Hf}$. The values E_{Th} are taken from Eq. (18.14), assuming that $E_2 = 100$ keV.

The levels of ^{170}Hf in Fig. 18.3 show remarkable regularities: All levels have the same parity, the spin increases in units of 2, and the spacing between adjacent levels increases with increasing spin. These properties are very different from those of shell-model states, discussed in Chapter 17. Moreover, ^{170}Hf is an even–even nucleus. We expect that in its ground state all nucleons have their spins paired. The energy needed to break a pair is of the order of 2 MeV (Fig. 17.2), much larger than the energy of the first excited state of ^{170}Hf. The levels therefore do not involve the breaking of a pair. This is a feature common to most of lowest 2^+ states of even-even nuclides.

The angle can be determined to a certain extent, thus the corresponding angular momentum cannot be restricted to one sharp value, but a number of angular momentum states must exist. Such angular momentum states have been observed in many nuclides. They are called *rotational states*, and their physical characteristics will be discussed in more detail below. A particularly beautiful example of a rotational spectrum is shown in Fig. 18.3. A large number of similar spectra have been found in other nuclides.

We shall now show that levels of the type shown in Fig. 18.3 can be explained by collective rotations of deformed nuclei. For simplicity, we assume the deformed nucleus to be axially symmetric (spheroidal), as shown in Fig. 18.4. A Cartesian system of axes, 1, 2, and 3, is fixed in the nucleus, with 3 being chosen as the nuclear symmetry axis. Axes 1 and 2 are equivalent. Naively it could be expected that such a nucleus could rotate about its symmetry axis as well as about any axis perpendicular to it. However, rotation about the symmetry axis is quantum mechanically not a meaningful concept. This fact can be seen as follows: Denote the angle about the symmetry axis 3 by ϕ. Axial symmetry implies that the wave function, ψ, is independent of ϕ,

$$\frac{\partial \psi}{\partial \phi} = 0.$$

R_3, the operator of the component of the orbital angular momentum along the 3 axis, is given by $R_3 = -i\hbar(\partial/\partial\phi)$. Axial symmetry thus implies that the component of the orbital angular momentum along the symmetry axis is zero: No collective rotation about the symmetry axis can occur. Rotation about any axis perpendicular to the symmetry axis, however, can lead to observable results.

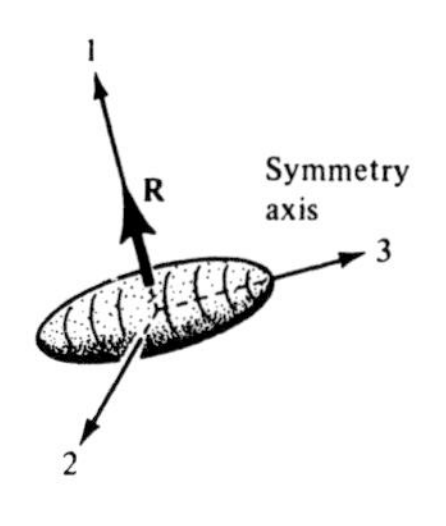

Figure 18.4: Permanently deformed axially symmetric nucleus. **R** is the rotational angular momentum discussed in the text.

For simplicity we first assume a deformed nucleus with zero intrinsic angular momentum and consider rotations about axis 1 (Fig. 18.4). If the nucleus possesses a rotational angular momentum $\boldsymbol{R}$, the energy of rotation is given by

$$H_{\text{rot}} = \frac{\boldsymbol{R}^2}{2\mathcal{I}}, \tag{18.11}$$

where $\mathcal{I}$ is the moment of inertia about axis 1. Translation into quantum mechanics yields the Schrödinger equation

$$\frac{\boldsymbol{R}_{\text{op}}^2}{2\mathcal{I}}\psi = E\psi. \tag{18.12}$$

We have already encountered the operator $\boldsymbol{R}_{\text{op}}^2$ in Chapter 15; we called it $\boldsymbol{L}^2$ there, and it is given by Eq. (15.22). According to Eq. (15.24), the eigenvalues and eigenfunctions of R_{op}^2 are given by

$$\boldsymbol{R}_{\text{op}}^2 Y_J^M = J(J+1)\hbar^2 Y_J^M, \quad J = 0, 1, 2, \dots, \tag{18.13}$$

where Y_J^M is a spherical harmonic. The parity of Y_J^M is given by Eq. (9.10) as $(-1)^J$. The spinless nucleus assumed here is invariant against reflection in the 1–2 plane. Since the spherical harmonics of odd J have odd parity, they change sign under such a reflection and are not admissible eigenfunctions. Only even values of J are allowed; with Eq. (18.12), the rotational energy eigenvalues of the nucleus become

$$E_J = \frac{\hbar^2}{2\mathcal{I}} J(J+1), \quad J = 0, 2, 4, \dots. \tag{18.14}$$

The spin assignments of the levels in Fig. 18.3 agree with these values. If the energy of the first excited state is taken as given, the energies of the higher levels follow from Eq. (18.14) as

$$E_J = \tfrac{1}{6} J(J+1) E_2. \tag{18.15}$$

The values of E_J for ^{170}Hf predicted by this relation are given in Fig. 18.3. The general trend of the experimental spectrum is reproduced, but the computed values are all higher than the observed ones. The deviation can be explained by a centrifugal stretching of the nucleus. Taking stretching into account, the ratios observed for ^{170}Hf can be explained.[10]

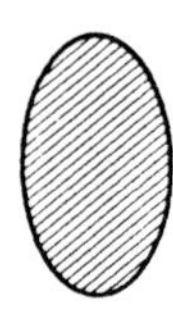

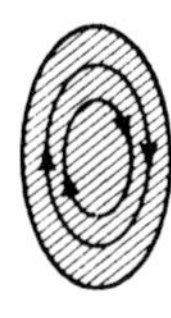

Figure 18.5: Rigid and wavelike (irrotational) rotations. The two rotations are seen from a coordinate system that rotates with the nucleus. For the rigid rotation, the velocities vanish. For the irrotational motion, the streamlines form closed loops. The particles circulate opposite to the rotation of the entire nucleus.

Through Eq. (18.14), the energies of the rotational levels are described in terms of a moment of inertia, $\mathcal{I}$. The experimental value of this parameter for a particular nucleus can be obtained from the observed excitation energies, and this value can then be compared with that computed for a model. Two extreme models suggest themselves, rigid and irrotational motions.

For a uniform rigid spherical body, of radius R_0 and mass Am, the moment of inertia is given by

$$\mathcal{I}_{\text{rigid}} = \frac{2}{5} Am R_0^2. \tag{18.16}$$

In the other extreme, the nuclear rotation is considered as a wave traveling around the nuclear surface; the nuclear shape rotates and the nucleons oscillate. The moment of inertia is given by

$$\mathcal{I}_{\text{irrot}} = \frac{2}{5} Am(\Delta R)^2, \tag{18.17}$$

or

$$\mathcal{I}_{\text{irrot}} = \mathcal{I}_{\text{rigid}} \delta^2. \tag{18.18}$$

Here $\delta = \Delta R/R_0$ is the deformation parameter already encountered in Eq. (18.3). The streamline picture for the two types of rotation, seen from a rotating coordinate system, are given in Fig. 18.5.[11] The empirical values of the moment of inertia lie between the two extremes. The nucleus is certainly not a rigid rotator, but the flow is also not completely irrotational.

Finally, we come to a conceptual problem: A favorite examination question in quantum mechanics is to ask for a proof that a particle with spin J less than 1 cannot have an observable quadrupole moment. Yet we have assumed that a spinless nucleus, as in Fig. 18.4, possesses a permanent deformation. How does this assumption agree with the theorem just mentioned? The solution to the problem lies in a distinction between the *intrinsic quadrupole moment* and the *observed quadrupole moment.*[12]

[10] A. S. Davydov and A. A. Chaban, *Nucl. Phys.* **20,** 499 (1960); R. M. Diamond, F. S. Stephens, and W. J. Swiatecki, *Phys. Rev. Lett.* **11,** 315 (1964).

[11] The two models can be appreciated by playing with a hard-boiled and a raw egg.

[12] K. Kumar, *Phys. Rev. Lett.* **28**, 249 (1972).

A spinless nucleus can have a permanent deformation (intrinsic quadrupole moment), and its effect can be seen in the existence of rotational levels and also in the rates of transitions leading to and from the $J = 0$ level. However, the quadrupole moment cannot be observed directly because the absence of a finite spin does not permit singling out a particular axis. In any measurement, an average over all directions is involved, and the permanent deformation appears only as a particularly large skin thickness.

18.3 Rotational Families

Deformed nuclei with spin zero in their ground state give rise to a rotational band, with spin-parity assignments $0^+, 2^+, \ldots$ Since many deformed nuclei with spins different from zero exist, the treatment of rotations must be extended to this more general case. The situation then becomes considerably more complicated, and we shall only treat the simplest case, namely a nucleus consisting of a deformed, axially symmetric, spinless core and one valence nucleon, and we shall neglect the interaction between the intrinsic and the collective (rotational) motion.

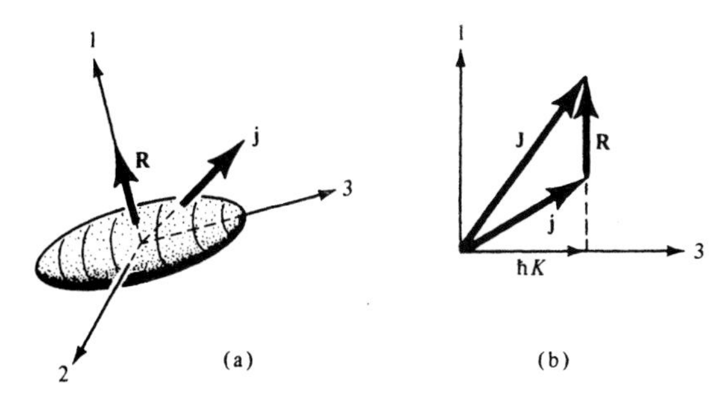

Figure 18.6: (a) The deformed core gives rise to a collective angular momentum **R**; the valence nucleon produces an angular momentum **j**. (b) $\boldsymbol{R}$ and $\boldsymbol{j}$ add up to the total nuclear angular momentum $\boldsymbol{J}$. The eigenvalue of the component of $\boldsymbol{J}$ along the symmetry axis 3 is denoted by $\hbar K$.

We assume that the valence nucleon does not affect the core so that it behaves like the deformed spinless nucleus treated in the previous section. The core then gives rise to a rotational angular momentum **R** perpendicular to the symmetry axis, 3, so that $R_3 = 0$. The valence nucleon produces an angular momentum **j**; **R** and **j** are shown in Fig. 18.6(a); they add up to the total nuclear angular momentum **J**:

$$\boldsymbol{J} = \boldsymbol{R} + \boldsymbol{j}. \tag{18.19}$$

The total angular momentum **J** and its component, J_3, along the nuclear symmetry axis are conserved, and they satisfy the eigenvalue equations

$$\boldsymbol{J}^2_{\text{op}}\psi = J(J+1)\hbar^2\psi, \quad J_{3,\text{op}}\psi = K\hbar\psi. \tag{18.20}$$

Because $R_3 = 0$, the eigenvalue of $j_{3,\text{op}}$ is also given by $\hbar K$.

If, as assumed, the state of the valence nucleon is not affected by the collective rotation, then it is to be expected that each state of the valence nucleon can form the base (head) of a separate rotational band. In the following we shall compute the energy levels of these bands. The Hamiltonian is the sum of the rotational energy and the energy of the valence nucleon,

$$H = H_{\text{rot}} + H_{\text{nuc}},$$

or, with Eqs. (18.11) and (18.19),

$$H = \frac{\boldsymbol{R}_{\text{op}}^2}{2\mathcal{I}} + H_{\text{nuc}} = \frac{1}{2\mathcal{I}}(\boldsymbol{J}_{\text{op}} - \boldsymbol{j}_{\text{op}})^2 + H_{\text{nuc}}.$$

The physical meaning becomes clearer if the Hamiltonian is written as the sum of three terms,

$$\begin{aligned} H &= H_R + H_p + H_c, \quad H_R = \frac{1}{2\mathcal{I}}(\boldsymbol{J}_{\text{op}}^2 - 2J_{3,\text{op}}j_{3,\text{op}}), \\ H_p &= H_{\text{nuc}} + \frac{1}{2\mathcal{I}}\boldsymbol{j}_{\text{op}}^2, \quad H_c = -\frac{1}{\mathcal{I}}(J_{1,\text{op}}j_{1,\text{op}} + J_{2,\text{op}}j_{2,\text{op}}). \end{aligned} \tag{18.21}$$

The third term, H_c, resembles the classical Coriolis force, and it is called the Coriolis, or rotation–particle coupling, term. It can be neglected except for the special case $K = \frac{1}{2}$.[13] The second term, H_p, is independent of the rotational state of the nucleus, and its contribution to the energy can be found by solving

$$H_p\psi = E_p\psi.$$

The first term describes the energy of the rotational motion. With Eq. (18.20), the energy eigenvalues of this term are given by

$$E_R = \frac{\hbar^2}{2\mathcal{I}}[J(J+1) - 2K^2], \quad J \geq K. \tag{18.22}$$

The total energy is then[13]

$$E_{J,K} = \frac{\hbar^2}{2\mathcal{I}}[J(J+1) - 2K^2] + E_p. \tag{18.23}$$

This relation describes a sequence of levels, similar to the one given in Eq. (18.14) for spinless nuclei. Following the terminology in molecular physics, the sequence belonging to a particular value of K is called a *rotational band*, and the state with lowest spin is called the *band head*. Characteristic differences exist between the case $K = 0$ and $K \neq 0$:

1. The spins for the case $K = 0$ are the even integers, while the spins for $K \neq 0$ are given by

$$J = K, K+1, K+2, \ldots, \quad K \neq 0. \tag{18.24}$$

[13] For the treatment of the case $K = \frac{1}{2}$, see ref. (6) or (9).

2. The ratios of excitation energies above the band head are not given by Eq. (18.15). For instance, the ratio of excitation energies of the second to the first excited state is not $\frac{10}{3}$, but

$$\frac{E_{K+2,K} - E_{K,K}}{E_{K+1,K} - E_{K,K}} = 2 + \frac{1}{K+1}. \tag{18.25}$$

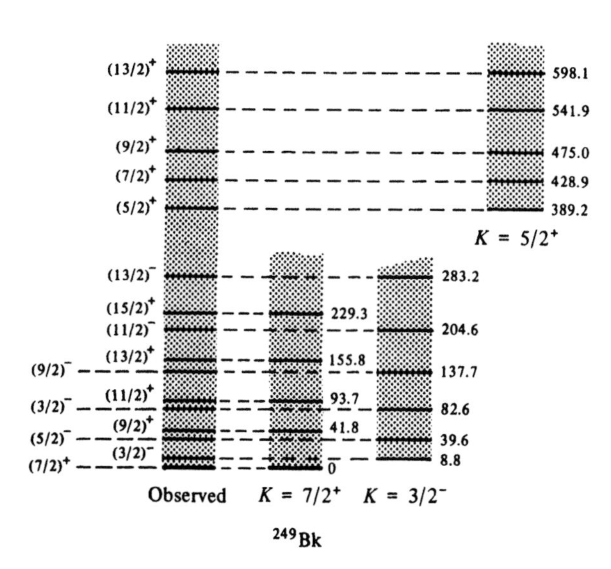

Figure 18.7: Energy levels of ^{249}Bk. All observed energy levels up to an excitation energy of about 600 keV are given at the left. The levels fall into three rotational bands; these are shown at the right. All energies are in keV.

The value of the component K can be determined from this ratio. As an example of the appearance of rotational bands in an odd-A nucleus, the level diagram of ^{249}Bk is shown in Fig. 18.7. The energy levels are drawn at the left, with spins and parities. Three bands can be distinguished; their band heads have assignments $K = (\frac{7}{2})^+, (\frac{3}{2})^-$, and $(\frac{5}{2})^+$. The level sequences satisfy Eq. (18.24), and the energies are reasonably well described by Eq. (18.23). The values of K follow unambiguously from Eq. (18.25).

The rotational families can be represented as trajectories in an angular momentum plot, just as was done for the harmonic oscillator levels in Fig. 15.10 and for some hyperons in Fig. 15.11. Such a plot is shown in Fig. 18.8 for the three families that have emerged from the ^{249}Bk decay scheme of Fig. 18.7. The states on one trajectory have the same internal structure and are distinct only in their collective rotational motion.

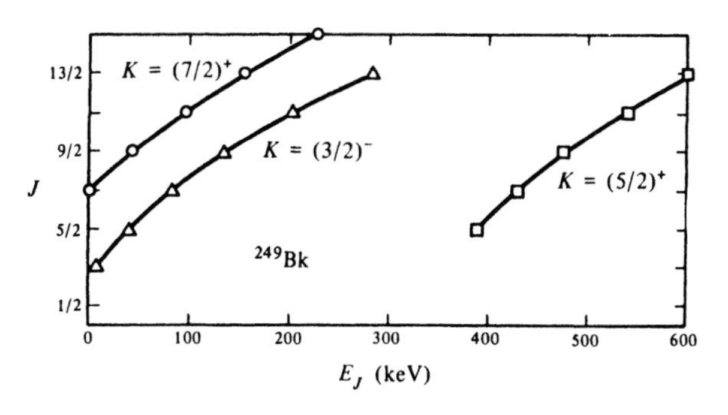

Figure 18.8: Angular momentum plot for the three rotational families of ^{249}Bk displayed as energy levels in Fig. 18.7.

So far we have discussed nuclear deformations and the resulting rotational structure of energy levels. While the treatment has been superficial and many complications and justifications have been omitted, the most important physical ideas have emerged. In the following sections, two more aspects of collective motions must be taken up—the influence of the nuclear deformation on shell-model states (the Nilsson model) and collective vibrations.

18.4 One-Particle Motion in Deformed Nuclei (Nilsson Model)

In Chapter 17, the nuclear shell model is treated; in the previous section, nuclei are considered as collective systems that can rotate. These two models are prototypes of two extreme and opposite points of view. Is there a way to weld the two models into one? In the present section, we shall describe the first step to a unified picture, namely the Nilsson model.[7] This model considers a deformed nucleus as consisting of independent particles moving in a *deformed well.* In Chapter 17, shell-model states in a spherical well were treated. As justified in Section 17.2, the average potential seen by nucleons resembles the nuclear density distribution. With Eqs. (15.12), (15.14), and (17.17), the spherical shell model potential can be written as

$$V(r) = \frac{1}{2}m\omega^2 r^2 - C\mathbf{l}\cdot\mathbf{s}. \tag{18.26}$$

The first term is the central potential, and the second the spin-orbit potential. The factor ω is related to the energy of an oscillator level (Fig. 15.7) through Eq. (15.26), $E = (N + \frac{3}{2})\hbar\omega$. The levels in the potential (18.26) are given, for instance, in Fig. 17.9; they are labeled by the quantum numbers N, l, and j. Rotational and parity invariance mean that the total angular momentum, j, and the orbital angular momentum, l (or the parity), of the nucleon are good quantum numbers, and N, l, and j are used to label the levels.

Since many nuclei possess large permanent deformations, as described in Section 18.1, nucleons do not always move in a spherical potential, and Eq. (18.26) must be generalized. A well-known generalization is due to Nilsson, who wrote instead of Eq. (18.26),

$$V_{\text{def}} = \frac{1}{2}\, m[\omega_\perp^2(x_1^2 + x_2^2) + \omega_3^2 x_3^2] + C\mathbf{l}\cdot\mathbf{s} + Dl^2. \tag{18.27}$$

This potential describes an axially symmetric situation—the one that applies to most deformed nuclei. The coordinates x_1, x_2, and x_3 are fixed in the nucleus: x_3 lies along the symmetry axis, 3 (Fig. 18.4). C determines the strength of the spin–orbit interaction. The term Dl^2 corrects the radial dependence of the potential: the oscillator potential differs markedly from the realistic potential near the nuclear surface, as shown in Fig. 17.4. States with large orbital angular momentum are most sensitive to this difference, and the term Dl^2, with $D < 0$, lowers the energy of these states. Nuclear matter is nearly incompressible: For a given form of the deformation, the coefficients $\omega_\perp$ and ω_3 are thus related. For a pure quadrupole deformation, discussed in the following section, the relation between the coefficients $\omega_\perp$ and ω_3 is expressed in terms of a deformation parameter ϵ:

$$\omega_3 = \omega_0 \left(1 - \frac{2}{3}\,\epsilon\right), \qquad \omega_\perp = \omega_0 \left(1 + \frac{1}{3}\epsilon\right). \tag{18.28}$$

For $\epsilon^2 \ll 1, \omega_\perp^2$ and ω_3 satisfy

$$\omega_\perp^2 \omega_3 = \omega_0^3, \tag{18.29}$$

and this relation expresses the constancy of the nuclear volume on deformation. The parameter ϵ is connected to the deformation parameter δ introduced in Section 18.1 by

$$\delta = \epsilon\ (1 + \tfrac{1}{2}\epsilon). \tag{18.30}$$

With Eqs. (18.3), (18.30), and (6.73), the intrinsic quadrupole moment can be written as

$$\mathcal{Q} = \frac{4}{3}\ Z\langle r^2\rangle\epsilon(1 + \tfrac{1}{2}\epsilon). \tag{18.31}$$

Equations (18.27) and (18.28) show that $V_{\rm def}$ is determined by four parameters, ω_0, C, D, and ϵ. Only ϵ depends strongly on the nuclear shape. For a given nuclide, ϵ is found by measuring $\mathcal{Q}$ and $\langle r^2\rangle$. The first three parameters, ω_0, C, and D, are independent of the nuclear shape for $\epsilon^2 \ll 1$, and they are determined from the spectra and radii of spherical nuclei, where $\epsilon = 0$. Approximate values of these parameters are

$$\hbar\omega_0 \approx 41 A^{-1/3}\ \text{MeV} \tag{18.32}$$

and

$$C \approx -0.1\hbar\omega_0, \qquad D \approx -0.02\hbar\omega_0. \tag{18.33}$$

The choice (18.27) of the potential $V_{\rm def}$ is not unique, and forms other than the one introduced by Nilsson have been studied extensively.[14] Since the salient features of the resulting spectra are unchanged, we restrict the discussion to the Nilsson model.

[14] A detailed investigation of the single-particle levels of nonspherical nuclei in the region $150 < A < 190$ is given by W. Ogle, S. Wahlborn, R. Piepenbring, and S. Fredriksson, *Rev. Mod. Phys.* **43**, 424 (1971).

In the Nilsson model, as in the spherical single-particle model treated in Chapter 17, it is assumed that all nucleons except the last odd one are paired and do not contribute to the nuclear moments. To find the wave function and the energy of the last nucleon, the Schrödinger equation with the potential $V_{\rm def}$ is solved numerically with the help of a computer. A typical result for small A is shown in Fig. 18.9.

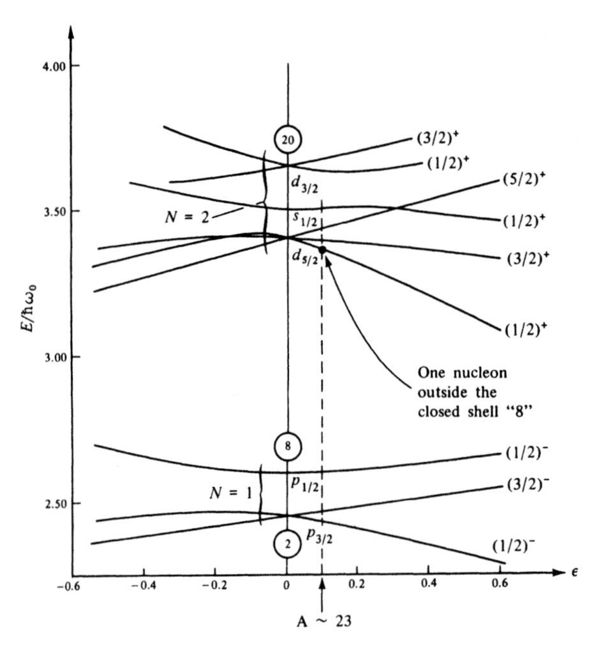

Figure 18.9: Level diagram in the Nilsson model. The notation is explained in the text. Each state can accept two nucleons.

For zero deformation the levels agree with the ones shown in Fig. 17.9, and they can be labeled with the quantum numbers N, j, and l. (N characterizes the oscillator shell and is given in Table 17.1.) In this limit ($\epsilon = 0$), the states are $(2j+1)$-fold degenerate. The deformation lifts the degeneracy, as can be seen from Fig. 18.9. State $p_{3/2}$ splits into two and state $d_{5/2}$ into three levels. A nucleon with total angular momentum j in the spherical case gives rise to $\frac{1}{2}(2j+1)$ different energy levels, with K values $j, j-1, j-2, \ldots, \frac{1}{2}$.

The factor $\frac{1}{2}$ describes a remaining twofold degeneracy which is caused by the symmetry of the nucleus about the 1–2 plane: The states K and $-K$ have the same energy (Fig. 18.10). A state with a given value of K can accommodate two nucleons of a given kind.

Which *quantum numbers* describe the levels in a deformed potential? Rotational symmetry, except about the symmetry axis, is destroyed, and the angular momenta $\mathbf{j}$ and $\mathbf{l}$ are no longer conserved. Only two quantum numbers remain exact in the Nilsson model, the parity, $\pi = (-1)^N$, and the component K.[15]

[15] The fact that a nucleon with total angular momentum $\boldsymbol{j}$ can give rise to the various states K can be understood in the vector model: The angular momentum $\boldsymbol{j}$ precesses rapidly around the symmetry axis 3. Any component perpendicular to 3 is averaged to zero and has no effect.

A state is consequently denoted by K^{π}. Actually, three partially conserved quantum numbers are used to describe a given level further. We shall not need these *asymptotic* quantum numbers here.

As an application of the Nilsson model, we consider the ground states of some nuclides with a neutron or proton number around 11. Figure 18.1 shows that these nuclides are expected to have a deformation of the order of 0.1, and consequently the Nilsson model should be applicable.

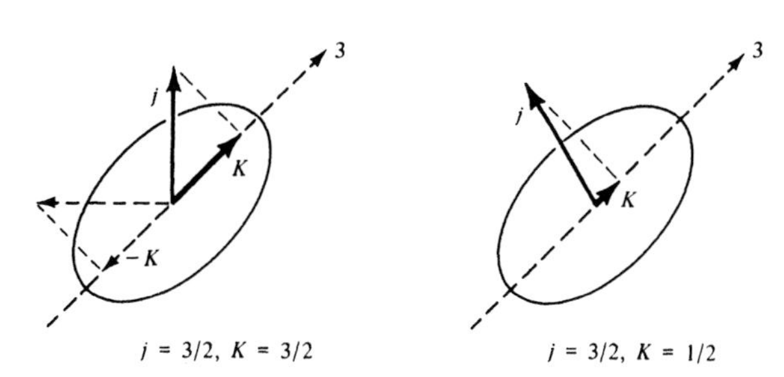

Figure 18.10: In a nonspherical nucleus, the total angular momentum, j, of a nucleon is no longer a conserved quantity. Only its component, K, along the nuclear symmetry axis is conserved. A nucleon with spin j (in the spherical case) gives rise to K values $j, j-1, \ldots, \frac{1}{2}$. States K and $-K$ have the same energy.

The relevant properties of a number of nuclides are summarized in Table 18.2. If it is assumed that the nuclides are described by the single-particle spherical shell model, their ground-state spin-parity assignment can be read from Fig. 17.9: only the last odd nucleon is assumed to determine the moments. The listed nuclides have one or three nucleons outside the closed shell 8: According to Fig. 17.9, they should all have an assignment $(\frac{5}{2})^+$. In reality, the spins are different, even for ^{19}F, which has only one proton outside the magic number 8. The quadrupole moment has been measured for two of the listed nuclides, and $\langle r^2 \rangle$ can be taken from Eq. (6.26); Eq. (18.31) then provides the value of the deformation parameter $\delta(\approx \epsilon)$. In agreement with the estimate from Fig. 18.1, δ is of the order of 0.1. The value $\delta = 0.1$ is indicated in Fig. 18.9. Following this line the predicted assignments can be read: for one nucleon outside the closed shell 8, $(\frac{1}{2})^+$ is predicted. Three nucleons outside the shell lead to an assignment $(\frac{3}{2})^+$. As Table 18.2 shows, these values agree with experiment and demonstrate that the Nilsson model can explain at least some of the properties of deformed nuclei. (In all these assignments it is assumed that the even number of nucleons, for instance, the 10 neutrons in ^{19}F, remain coupled to zero.)

The prediction of ground-state moments is only one of the successes of the Nilsson model. It is also able to correlate a great many other observed properties of deformed nuclei.[16,17]

[16] B.R. Mottelson and S.G. Nilsson, *Kgl. Danske Videnskab. Selskab, Mat-fys. Medd.* **1**, No. 8 (1959).

[17] M.E. Bunker and C.W. Reich, *Rev. Mod. Phys.* **43**, 348 (1971).

Table 18.2: DEFORMED NUCLEI AROUND $A \approx 23$.

					Ground-State Assignment		
						Shell	Nilsson
Nuclide	Z	N	Q	$\delta \approx \epsilon$	Exp.	Model	Model
^{19}F	9	10			$(1/2)^+$	$(5/2)^+$	$(1/2)^+$
^{21}Ne	10	11	9 fm^2	0.09	$(3/2)^+$	$(5/2)^+$	$(3/2)^+$
^{21}Na	11	10			$(3/2)^+$	$(5/2)^+$	$(3/2)^+$
^{23}Na	11	12	14 fm^2	0.11	$(3/2)^+$	$(5/2)^+$	$(3/2)^+$
^{23}Mg	12	11			$(3/2)^+$	$(5/2)^+$	$(3/2)^+$

So far we have studied the motion of a single particle in a stationary deformed potential without regard to the motion of this well. The well is fixed in the nucleus. If the nucleus rotates, the potential rotates with it. In the previous section we have shown that the rotation of a deformed nucleus gives rise to a rotational band. Now the question arises: Is it correct to treat rotation and intrinsic motion separately, as was done in Eq. (18.21)? The separation is permissible if the motion of the particle in the deformed well is fast compared to the rotation of the well so that the particle traverses many orbits in one period of collective motion. In real nuclei, the condition is reasonably well satisfied because the rotational motion involves A nucleons and consequently is slower than the motion of the single valence nucleon. Nevertheless, for a realistic treatment, the effect of the rotational motion on the intrinsic level structure, given by the term H_p in Eq. (18.21), must be taken into account.[18,19]

After asserting that intrinsic and rotational motion are indeed independent to a good approximation, we can return to the interpretation of the spectra of deformed nuclei. Since the nucleus can rotate in any state of the deformed nucleus, each intrinsic level (Nilsson level) is the band head of a rotational band. In other words, a rotational band is built onto each intrinsic level. Figure 18.7 gives an example of three bands, built on three different Nilsson states.

18.5 Vibrational States in Spherical Nuclei

So far we have discussed two types of nuclear states, rotational and intrinsic. The occurrence of different types of excitations is not peculiar to nuclei; diatomic molecules were known long ago to display three different types of excitations,

[18]O. Nathan and S.G. Nilsson, in *Alpha-, Beta- and Gamma-Ray Spectroscopy*, Vol. 1 (K. Siegbahn, ed.), North-Holland, Amsterdam, 1965, p. 646.

[19]A.K. Kerman, *Kgl. Danske Videnskab. Selskab, Mat.-fys. Medd.* **30**, No. 15 (1956).

intrinsic (electronic), rotational, and vibrational.[20] In a first approximation, the wave function of a given state can be written as

$$|\text{total}\rangle = |\text{intrinsic}\rangle|\text{rotation}\rangle|\text{vibration}\rangle. \tag{18.34}$$

It turns out that nuclei are similar to molecules in that they, also, can have vibrational excitations.[6,21,22] In the present section, we shall describe some aspects of nuclear vibrations, restricting the treatment to spherical nuclei.

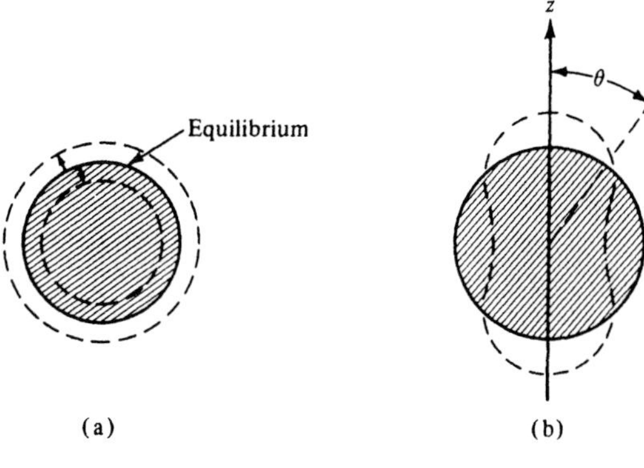

Figure 18.11: (a) Monopole vibration. (b) Quadrupole vibration, $l = 2, m = 0$.

The simplest vibration corresponds to a density fluctuation about an equilibrium value, as shown in Fig. 18.11(a). Since such a motion carries no angular momentum, it is called the *monopole* or *breathing* mode. Its isospin is $I = 0$. Although indications for this mode had occurred, definitive evidence did not become available until 1977.[23] The interest in this mode stems in part from its relationship to the incompressibility of nuclei, Eq. (16.21).

Another mode of motion, which can even occur for an incompressible system, corresponds to *shape oscillations*, without change of density. Such oscillations were first treated by Rayleigh[3], who observed: "The detached masses of liquid into which a jet is resolved do not at once assume and retain a spherical figure, but execute a series of vibrations, being alternately compressed and elongated in the direction of the axis of symmetry." The investigations of nuclear vibrations use much of the mathematical approach developed by Rayleigh, but, of course, the oscillations are quantized. Before describing shape oscillations, we shall briefly outline how permanent nuclear *deformations* are expressed mathematically. After Rayleigh, the surface of a figure of arbitrary shape can be expanded as

$$R = R_0 \left[1 + \sum_{l=0}^{\infty} \sum_{m=-l}^{l} \alpha_{lm} Y_l^m(\theta, \varphi) \right], \tag{18.35}$$

[20] G. Herzberg, *Molecular Spectra and Molecular Structure*, Van Nostrand Rinehold, New York, 1950; L. D. Landau and E. M. Lifshitz, *Quantum Mechanics*, transl. J.B. Sykes and J.S. Bell, 3d ed, Pergamon, Elmsford, N.Y., 1977, Chapters 11 and 13.

[21] N. Bohr and J. A. Wheeler, *Phys. Rev.* **56**, 426 (1939); D. L. Hill and J. A. Wheeler, *Phys. Rev.* **89**, 1102 (1953).

[22] A. Bohr, *Kgl. Danske Videnskab. Selskab, Mat.-fys. Medd.* **26**, No. 14 (1952).

[23] D. H. Youngblood et al., *Phys. Rev. Lett.* **39**, 1188 (1977).

where $Y_l^m(\theta,\varphi)$ are the spherical harmonics; θ and φ are polar angles with respect to an arbitrary axis and the α_{lm} are expansion coefficients. If the expansion coefficients are time-independent, Eq. (18.35) describes a permanent deformation of the nucleus. If α_{lm} is time dependent, then the term $l = 0$ describes the breathing mode. The term $l = 1$ corresponds to a displacement of the center-of-momentum and is not allowed, since no external force is acting on the system.[24] The term of interest here is $l = 2$, describing a *quadrupole deformation*. Since the salient features of nuclear collective vibrations appear in this mode, we restrict the following discussion to these terms. The nuclear radius then is written as

$$R(\theta,\varphi) = R_0\left[1 + \sum_{m=-2}^{2} \alpha_{2m} Y_2^m(\theta,\varphi)\right]. \tag{18.36}$$

The quadrupole deformation is determined by the five constants α_{2m}. For a mode with $\alpha_{2m} = 0$, for all $m \neq 0$, the radius is

$$R(\theta) = R_0\left[1 + \alpha_{20}\left(\frac{5}{16\pi}\right)^{1/2}(3\cos^2\theta - 1)\right]. \tag{18.37}$$

Such a deformation $(l = 2, m = 0)$ is shown in Fig. 18.11(b).

Equation (18.36) describes a quadrupole deformation if the coefficients α_{2m} are constants. Shape vibrations are expressed through the time dependence of the expansion coefficients. To write the relevant Hamiltonian, we note first that for small oscillations about an equilibrium position, the motion can be treated as harmonic. For such harmonic motion, we saw in Section 13.7 that the kinetic energy is given by $\frac{1}{2}mv^2 = \frac{1}{2}m\dot{x}^2$, the potential energy by $\frac{1}{2}m\omega^2 r^2$, and the Hamiltonian by $H = \frac{1}{2}m\dot{x}^2 + \frac{1}{2}m\omega^2 r^2$. In the present situation, the dynamical variable is the deviation of the radius vector from its equilibrium value. This deviation is given by α_{2m} so that the Hamiltonian for an oscillating liquid drop, for $l = 2$ and for small deformation, has the form[3] [21] [25]

$$H = \frac{1}{2}B\sum_m |\dot{\alpha}_{2m}|^2 + \frac{1}{2}C\sum_m |\alpha_{2m}|^2, \tag{18.38}$$

where B is the parameter corresponding to the mass and C is the potential energy parameter. H describes a five-dimensional harmonic oscillator because there are five independent variables α_{2m}. In analogy to Eq. (15.26), the energies of the quantized oscillator are given by

$$E_N = \left(N + \frac{5}{2}\right)\hbar\omega, \quad \hbar\omega = \left(\frac{C}{B}\right)^{1/2}. \tag{18.39}$$

[24] The dipole vibration of protons against neutrons is allowed, however, because it leaves the nuclear c.m. unaffected. The giant dipole resonance that occurs in nuclei at excitation energies between 10 and 20 MeV is explained as being due to such dipole vibrations, and it is particularly clearly observed in electromagnetic processes. See section 18.7.

[25] A detailed derivation of Eq. (18.38) is given by S. Wohlrab, in *Lehrbuch der Kernphysik*, Vol. II (G. Hertz, ed.), Verlag Werner, Dausien, 1961, p. 592.

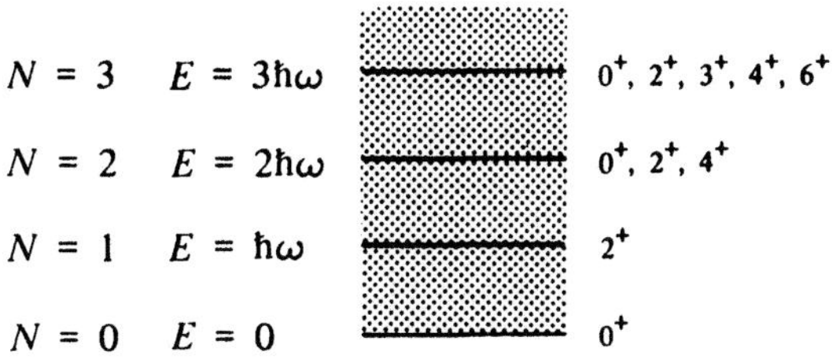

$N = 3 \quad E = 3\hbar\omega \quad 0^+, 2^+, 3^+, 4^+, 6^+$

$N = 2 \quad E = 2\hbar\omega \quad 0^+, 2^+, 4^+$

$N = 1 \quad E = \hbar\omega \quad 2^+$

$N = 0 \quad E = 0 \quad 0^+$

Figure 18.12: Vibrational states. The vibrational phonon carries an angular momentum 2 and positive parity. The states are characterized by the number, N, of phonons. The energy of the ground state has been set equal to zero.

The angular dependence of the shape oscillations is described by the spherical harmonics Y_2^m, and we know from Eq. (15.24) that these are eigenfunctions of the total angular momentum with quantum number $l = 2$. The vibration carries an angular momentum of 2 and positive parity.

Nuclear physicists have borrowed the expression *phonons* from their solid-state colleagues,[26] and the situation is described by saying that the phonon angular momentum is $2\hbar$, and that one phonon is present in the first excited state, two phonons in the second excited state, and so forth.

Since the ground states of even–even nuclei always have spin 0, the first excited vibrational states should have assignments 2^+. Two phonons have an energy $2\hbar\omega$ and they can couple to form states $0^+, 2^+$, and 4^+. The states with spin 1 and 3 are forbidden by the requirement that the wave function of two identical bosons must be symmetric under exchange. The expected spectrum is sketched in Fig. 18.12. Even–even nuclei near closed shells indeed show spectra with the characteristics predicted by the vibrational model.[27]

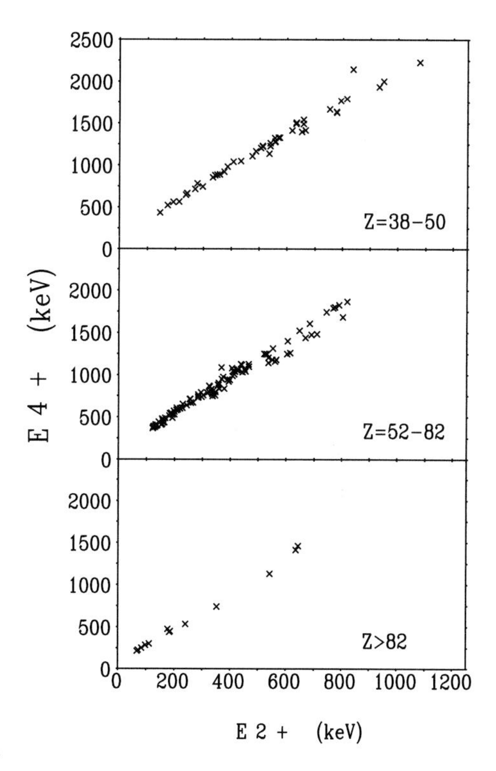

Figure 18.13: Plot of the excitation energy of the first 4^+ state versus the excitation energy of the first 2^+ state for a large range of nuclei [From ref. (28).]

[26] C. Kittel, *Introduction to Solid State Physics*, 6th ed., Wiley, New York, 1986, Chapters4 and 5; J.M. Ziman, *Electrons* and *Phonons*, Clarendon Press, Oxford University, 1960. J.A. Reisland, *The Physics of Phonons*, Wiley, New York, 1973.

If these states were describable as pure vibrations of a harmonic oscillator one would expect that the excitation energy of the second excited (4^+) state would be twice that of the first excited (2^+) state.

As shown in Fig. 18.13, measurements over a wide range of nuclei show a behavior that can be summarized by:

$$E_4 \approx 2E_2 + \text{constant} \tag{18.40}$$

where the constant varies for the different groups of nuclei shown in the figure.[(28)] The constant can be interpreted as an unharmonicity but this behavior is not fully understood.

In addition to the vibrational states described above, in which all nucleons tend to move together, there exist isospin 1 vibrations in which protons move against neutrons. An example is the so-called "scissors" mode in which the protons and neutrons have independent collective motions and move against each other in a scissors-like manner.[(29)]

18.6 The Interacting Boson Model

The interacting boson model (IBM) is an alternative to, and complementary description of, the collective model. Based on ideas of Iachello and Feshbach, it was first proposed in detail by Arima and Iachello in 1975.[(30)] Although the original model was based on symmetry considerations, the name derives from the fact that the model assumes pairs of like nucleons coupled to spins zero and two. Since for all but the lightest nuclei neutrons and protons are not in the same shell, the pairing between neutrons and protons tends to be much less important. We have already remarked on the short range residual interaction for like-nucleons in relative S-states. This residual force leads to pairing. The evidence for such pairing comes from the observation of an energy gap in the spectra of nuclei: The first intrinsic excitation of even–even heavy nuclei is at about 1 MeV (Fig. 17.3) whereas neighboring odd nuclei possess many levels below this energy. Even–even nuclei consequently exhibit an energy gap and this gap is taken as evidence for the pairing force:[(31)] Nucleons like to form pairs with angular momentum zero and the energy gap arises because to reach the first excited state requires a minimum energy corresponding to the break-up of such a pair. The pairing of nucleons bears a close resemblance

[27] G. Scharff-Goldhaber and J. Weneser, *Phys. Rev.* **98,** 212 (1955).
[28] R.F. Casten, N.V. Zamfir, D.S. Brenner, *Phys. Rev. Lett.* **71**, 227 (1993).
[29] E.B. Balbutsev, P. Schuck, nucl-th/0602031; D. Bohle et al., *Phys. Lett.* **137B,** 27 (1984); A. Faessler and R. Nojara, *Prog. Part. Nucl. Phys.* (A. Faessler, ed.) **19,** 167 (1987); I. Bauske et al., *Phys. Rev. Lett.* **71**, 975 (1993).
[30] A. Arima and F. Iachello, *Phys. Rev. Lett.* **35,** 1069 (1975); I. Talmi, *Comm. Nucl. Part. Phys.* **11,** 241 (1983); R. F. Casten, *Comm. Nucl. Part. Phys.* **12,** 119 (1984); A. E. L. Dieperink, *Comm. Nucl. Part. Phys.*, **14,** 25 (1985).
[31] A. Bohr, B. R. Mottelson, and D. Pines, *Phys. Rev.* **110,** 936 (1958).

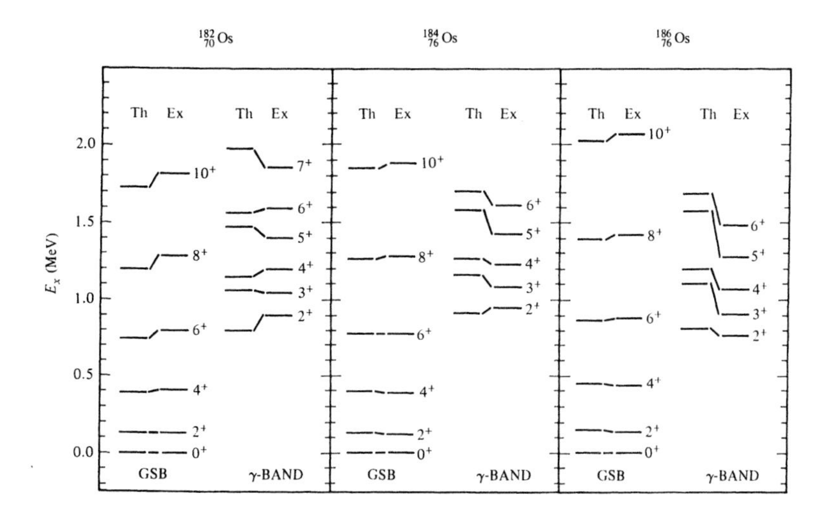

Figure 18.14: Energy spectra of even–even Os isotopes. The theoretical (th) rotational bands (GSB) and vibrational bands (γ-band) based on the ground state are compared to experiment (ex). [From W.-T. Chou, Wm. C. Harris, and O. Scholten, *Phys. Rev.* **C37,** 2834 (1988).]

to Cooper pairs[(32)] in superconductivity and it has been possible to use the tools and ideas developed to explain superconductivity[(33)] in nuclear physics. Arima and Iachello also take into account the somewhat weaker attraction for nucleons in a relative d-state. This inclusion can be related to the shell model and to the concept of seniority introduced by Racah.[(34)] In this scheme nucleons tend to pair to spin zero (seniority 0) and the next most likely pairing is to spin 2 (seniority 1).

In the IBM the paired particles in s- and d-states are treated as bosons and the bosonic degrees of freedom are able to describe well the spectra of even–even nuclei without invoking shape variables. The emphasis is on the dynamics of the bosons rather than on the shape variables of the collective model. Further, by incorporating s-bosons as well as d-bosons there are six degree of freedom to be compared to the five degrees of the collective model, represented by α_{2m} of Eq. (18.36). These features differentiate between the IBM and the collective models. By introducing also unpaired fermions, the model has been extended to odd–even nuclei. Thus, the IBM treats collective and pairing degrees of freedom on the same footing.

A state is described by fixed numbers of s-bosons (n_s) and d-bosons (n_d); the total number of bosons is $N = n_s + n_d$. In the more recent model, IBM2, neutron pairs and proton pairs are treated separately. For either neutrons or protons, the Hamiltonian thus consists of the kinetic energies of the bosons in s- and in d-states and the interactions between them. The connection to the collective model is obtained by considering the classical limit. A coherent state with n_s s- and

[32] L. N. Cooper, *Phys. Rev.* **104,** 1189 (1956).
[33] J. Bardeen, L. N. Cooper, and J. R. Schrieffer, *Phys. Rev.* **108,** 1175 (1957).
[34] G. Racah, *Phys. Rev.* **63,** 367 (1943); **76,** 1352 (1949).

n_d *d*-bosons can be shown to correspond to a collective state expressed in terms of the variables of the collective deformation. By minimizing the energy of the state with respect to these variables, one obtains the equilibrium deformation of a given nucleus and finds both spherical and deformed nuclei in the proper limits. In addition, the low-lying levels are found to correspond to those of the collective model. An example of some calculated low-lying rotational and vibrational spectra are compared to experiment in Fig. 18.14.

18.7 Highly Excited States; Giant Resonances

The last several decades have seen considerable growth in the study of highly excited states of nuclei, with particular attention focused on resonances and states of high angular momentum.[35] These states can be excited with reactions initiated by photons, electrons, pions, nucleons, and more massive projectiles.

To understand the nature of these states, residual forces between nucleons are considered. Although nucleons can be represented reasonably well as moving in an average (single particle) potential due to all other nucleons, there are important residual forces. We have already mentioned the short-range pairing force that is attractive and particularly strong for like nucleons in a relative *s*-state. The residual forces tend to parallel the free nucleon–nucleon force, but there are also residual effects due to long-range collective effects.

Resonances in the continuum can be studied with the help of high-resolution detectors. Breathing mode oscillations of angular momentum $L = 0$, dipole resonances ($L = 1$), quadrupole resonances ($L = 2$), octupole ($L = 3$), and higher L resonances, as well as resonances built on excited states have all been observed. Most of these resonances can occur with neutrons and protons oscillating together (isospin $I = 0$) or against each other ($I = 1$). The first resonance found, the electric dipole one, is an isovector mode with an energy of excitation given approximately by $E_1{}^* = 77\ A^{-1/3}$ MeV, the energy at which the strength of the 1^- excitation built on the ground state is concentrated. The resonance is observed in photoreactions such as (γ, n), or its inverse, neutron capture. It is called a "giant resonance," because the strength is many times that of a single particle excitation. The next resonance to be discovered was the giant quadrupole resonance of $I = 0$ with $E_2^* = 64\ A^{-1/3}$ MeV and a decay width that decreases with the mass number A.[36] It may appear odd that the giant quadrupole resonance lies at an excitation energy below that of the giant dipole, since the latter can be caused by moving a nucleon to the next higher unfilled shell, whereas the quadrupole requires the excitation through two major shells. Clearly, these are not simply single-nucleon excitations; strong residual forces and cooperative phenomena are involved. We can see the importance of residual forces and show the $A^{-1/3}$ dependence by treating

[35] G. F. Bertsch and R. A. Broglia, *Phys. Today* **39,** 44 (August 1986).
[36] M.B. Lewis and F.E. Bertrand, *Nucl. Phys.* **A196,** 337 (1972).

single-particle motion in a harmonic oscillator potential.

The degeneracy of a single particle level of energy $E = (N + \frac{3}{2})\hbar\omega$, is given by Eq. (15.30). For a nucleus with equal numbers of neutrons and protons, each of which can have spin up and spin down, the degeneracy is

$$\text{degeneracy} = 2(N+1)(N+2). \tag{18.41}$$

For a heavy nucleus with $A \gg 1$ or $N \gg 1$, and keeping only leading-order terms, we find

$$A \approx \sum_{N=0}^{N_{\max}} 2N^2 \approx \int_0^{N_m} 2N^2\,\mathrm{d}N = \tfrac{2}{3}\,N_m^3, \tag{18.42}$$

and energy levels are filled up to an energy E

$$\begin{aligned} E &= \hbar\omega \sum_{N=0}^{N_{\max}} (N + \tfrac{3}{2}) \times \text{degeneracy} \\ &\approx 2\hbar\omega \int_0^{N_m} \mathrm{d}N N^3 \\ &= \frac{1}{2}\,N_m^4 \hbar\omega = \tfrac{1}{2}\,(\tfrac{3}{2}A)^{4/3}\hbar\omega. \end{aligned} \tag{18.43}$$

The energy per nucleon in a harmonic oscillator can be written as $E/A = m\omega^2 R^2$. The total energy to leading order in A is therefore given by

$$E = Am\omega^2 R^2. \tag{18.44}$$

We identify R as the radius of a nucleus with uniform charge density, Eq. (6.30). By combining Eqs. (18.43) and (18.44), we obtain the A dependence of the harmonic oscillator level spacing as

$$\hbar\omega = \frac{5}{4}\left(\frac{3}{2}\right)^{1/3} \frac{\hbar^2}{mr_0^2} A^{-1/3} \approx 41A^{-1/3}\ \text{MeV}. \tag{18.45}$$

Thus, we reproduce the A-dependence found experimentally, but the predicted energy of the $L = 1$ resonance is almost a factor of two too low. This discrepancy shows the importance of residual interactions.

In addition to electric modes of excitation there are also magnetic ones. The "Gamow–Teller" resonance due to both spin and isospin oscillations, of $J^\pi = 1^+, I = 1$, and described by the operator $\vec{\sigma}\vec{\tau}$ has been observed cleanly in (p, n) reactions with protons of the order of 200 MeV[37] and in beta decays.[38]

[37] D.F. Barnum et al., *Phys. Rev. Lett.* **44,** 1751 (1980); C.D. Goodman, *Nucl. Phys.* **A374,** 241c, (1982), *Comm. Nucl. Part. Phys.* **10,** 117 (1981); G.F. Bertsch, *Comm. Nucl. Part. Phys.* **10,** 91 (1981), *Nucl. Phys.* **A354,** 157c (1981).

[38] Z.Q. Hu et al., *Phys. Rev. C* **62**, 064315 (2000).

Resonances can also be built on excited states, for instance single particle states.[39] Such resonances have been found in the deexcitation of states of high angular momentum formed in heavy ion reactions. Indeed, there has been considerable interest in the study of nuclei with very high spins, $J \gtrsim 30\hbar$.[40] Above angular momenta of this order of magnitude, pairing effects are destroyed by Coriolis forces and particles tend to align their angular momenta with collective axes of rotation.[40] Excited states of spins up to about $70\hbar$ have been observed by means of heavy ion fusion reactions, discussed in Chapter 16. If the spin becomes too high then the excited state is unstable against fission.[41] It appears that these high spin states are neither due to single particle nor collective motions, but rather a combination of these two, coupled together.

Of particular interest are the *yrast* levels.[42] An yrast level of a given nuclide, at a given angular momentum, is the level with least energy of that angular momentum.[43] The yrast line, connecting the yrast levels of a given nuclide, shows how the moment of inertia changes as the rotational angular velocity of nucleus varies.[44] For the highest angular momenta, the moment of inertia is close to that of a solid body.

The high spin states permit the study of nuclear matter when it is being subjected to enormous rotational forces. To understand some of the experimental results, we note that the angular velocity and the moment of inertia of an axially symmetric rotor with angular momentum $\hat{J} = \hbar[J(J+1)]^{1/2}$ are defined by[45]

$$\omega_{\text{rot}} = \frac{dE}{d\hat{J}} = \frac{dE}{\hbar d[J(J+1)]^{1/2}} \approx \frac{dE}{\hbar dJ}, \tag{18.46}$$

$$\mathcal{I} = \frac{\hat{J}}{\omega_{\text{rot}}} \approx \frac{\hbar J}{\omega_{\text{rot}}}. \tag{18.47}$$

These two definitions together give

$$\mathcal{I} \approx \hbar^2 J \frac{dJ}{dE}. \tag{18.48}$$

[39] K.A. Snover, *Comm. Nucl. Part. Phys.* **12,** 243 (1984); *Annu. Rev. Nucl. Part. Sci.* **36,** 545 (1986).

[40] F.S. Stephens, *Comm. Nucl. Part. Phys.* **6,** 173 (1976); R.M. Diamond and F.S. Stephens, *Annu. Rev. Nucl. Part. Sci.* **30,** 851 (1980); B.R. Mottelson and A. Bohr, *Nucl. Phys.* **A354,** 303c (1981).

[41] N. Bohr and F. Kalckar, *Kgl. Danske Videnskab Selskab, Mat-fys. Medd.* **14**, No. 10 (1937); S. Cohen, F. Plasil, and W.J. Swiatecki, *Ann. Phys. (New York)* **82**, 557 (1974).

[42] J. R. Grover, *Phys. Rev.* **157**, 832 (1967).

[43] The origin of the word "yrast" is given by Grover:[42]

The English language seems not to have a graceful superlative form for adjectives expressing rotation. Professor F. Ruplin (of the Germanic Languages Department of the State University of New York, Stony Brook) suggested the use of the Swedish adjective *yr* for designating these special levels. This word derives from the same Old Norse verb *hvirfla* (to whirl) as the English verb *whirl*, and forms the natural superlative, *yrast*. It can thus be understood to mean "whirlingest," although literally translated from Swedish it means "dizziest" or "most bewildered."

[44] A. Johnson, H. Ryde, and S. A. Hjorth, *Nucl. Phys.* **A179**, 753 (1972).

[45] Equations (18.47) and (18.48) are the rotational analogs of the relations $v = dE/dp$ and $m = p/v$.

The yrast line of a nucleus gives E as a function of J, as for instance shown for rotational families in Fig. 18.8. From such a plot, Eqs. (18.46) and (18.48) permit determination of $\omega_{\rm rot}$ and $\mathcal{I}$ for the yrast states. It has become customary to plot $2\mathcal{I}/\hbar^2$ against the square of the rotational energy, $(\hbar\omega_{\rm rot})^2$. The points on the plot are characterized by the values of the spin of the various yrast states. If nothing remarkable happens, then the plot will indicate a smooth increase of the rotational energy with J, and a smooth increase of the moment of inertia with rotational energy. Such a behavior is indeed observed for many nuclei.

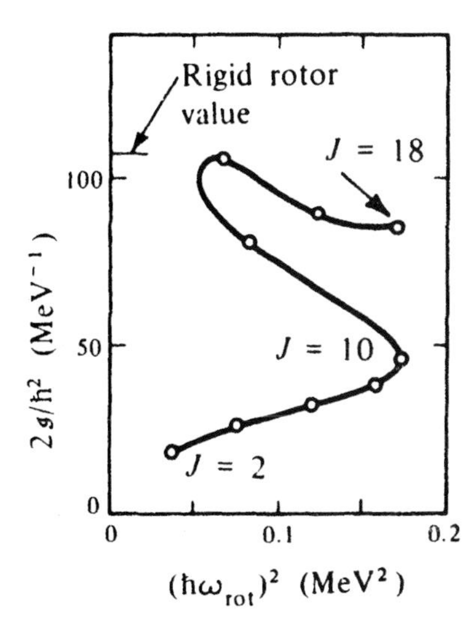

Figure 18.15: Plot of the nuclear moment of inertia as a function of the square of the angular frequency. The rigid rotor value was calculated for the nucleus in its ground state ($\omega = 0$). [After O. Taras et al., *Phys. Lett.* **41B**, 295 (1972).]

In some nuclides, however, a dramatic departure from a smooth picture has been discovered.(42) At some value of the spin J, the moment of inertia increases so rapidly that the rotational frequency actually decreases as higher spin states are reached. As an example, the yrast line for the even-spin states in ^{132}Ce are shown in Fig. 18.15. The yrast states up to $J = 18$ were found by using the reaction $^{16}\text{O} + {}^{120}\text{Sn} \rightarrow 4n + {}^{132}\text{Ce}$.(46) At $J = 10$, a backbending occurs and the rotational frequency at $J = 14$ is about the same as at $J = 2$! This backbending can be understood in the following manner. At low rotating frequencies the nucleus follows the yrast curve, as explained above. But as the rotational energy becomes higher the individual spins of nucleons tend to align themselves with the nuclear rotation. At some point pairs of nucleons with opposite spin are "broken" in favor of alignment with the nuclear rotation.(47)

18.8 Nuclear Models—Concluding Remarks

In the last three chapters we have discussed the simplest aspects of nuclear models. The spherical shell model is most successful near magic number nuclei, the collective model for nuclei far removed from shell closures. The transition from spherical to deformed nuclei can be understood in terms of the competition between the short-range pairing force and the longer range polarizing force. The latter is the force that

46 O. Taras et al., *Phys. Lett.* **41B**, 295 (1972).

47 The interesting interplay between the collective and single-particle contributions to the angular momentum has recently been studied by W.C. Ma et al., *Phys. Rev. C* **65**, 034312 (2002).

nucleons outside a closed shell exert on those inside the shell. For a single nucleon outside a closed shell the polarizing effect is too small to deform the core. When two nucleons are present outside the closed shell, two competing effects occur: The pairing force tends to keep the nucleus spherical, whereas the polarizing force tries to deform the nucleus. When only a few nucleons are present outside a closed shell, the pairing force wins out, but the polarizing effects become dominant as more and more nucleons are added. This feature is shown schematically in Fig. 18.16.

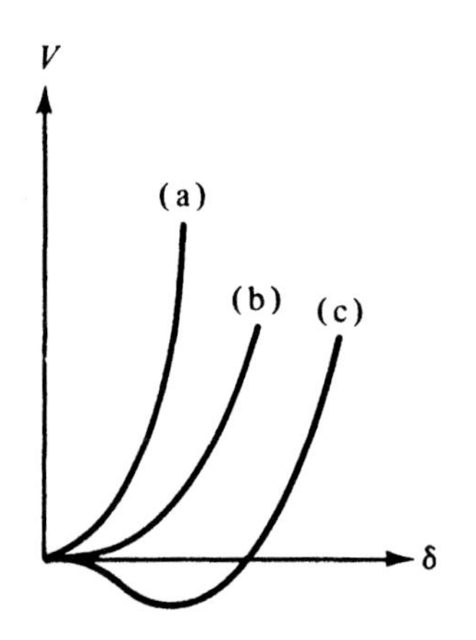

Figure 18.16: Potential energy surfaces as a function of deformation (see Section 18.1). The three curves are (a) for a closed shell nucleus, (b) for a nucleus near a closed shell and (c) for a nucleus far from a closed shell. A permanent deformation occurs in the last case.

The deformed shell or Nilsson model, Section 18.4, combines essential aspects of the two extremes. The Nilsson model shows that especially stable structures should occur for very anisotropic orbitals or large deformations. Nilsson energy levels, Fig. 18.17, show energy gaps when the ratio of major to minor axes are integers, such as 2:1, 3:1, and somewhat smaller gaps when the ratio is 3:2. These "superdeformed" shapes have been found at high angular momenta in heavy ion reactions through gamma-ray deexcitation studies in ^{152}Dy and ^{149}Gd; they correspond to yrast levels of 50–60$\hbar$.(48) Although the shell model and its extensions have given us considerable insights, what really is needed is a *microscopic* theory in which the features of the unified model are explained by the known properties of nuclear forces.

The nucleon density and the effective single particle potential that acts on a baryon can be investigated with *hypernuclei*.(49) In order to understand that we should first note that the time scale for nucleons to move accross the nucleus is $\sim 10^{-22}$ sec, which one gets by assuming nucleons are moving at $\sim 10\%$ of the speed of light and covering distances of $\sim 10^{-14}$ m. In hypernuclei nuclei, one or sometimes two neutrons are replaced by hyperons, mostly lambdas. The lambda is unaffected by the Pauli exclusion principle and can only decay weakly when bound in the nucleus so that it lives a long time ($\sim 10^{-10}$ sec) compared to the sampling time mentioned above. The potential that acts on the Λ^0 in a nucleus is related to that which acts

[48] R.V.F. Janssens and T.L. Khoo, *Annu. Rev. Nucl. Part. Sci.* **41**, 321 (1991); P.J. Nolan and P.J. Twin, *Annu. Rev. Nucl. Part. Sci.* **38**, 533 (1988).

[49] O. Hashimoto and H. Tamura, *Prog. Part. Nucl. Phys.* **57**, 564 (2006); A. Gal, *Adv. Nucl. Phys.*, (J.W. Negele and E. Vogt, eds.) **8**, 1 (1975); R.E. Chrien, *Annu. Rev. Nucl. Part. Sci.* **39**, 113 (1989).

[50] B. Holzenkamp, K. Holinde, and J. Speth, *Nucl. Phys.* **A500**, 485 (1989).

on a nucleon, but there are also differences. For instance, the spin–orbit potential is considerably weaker than that for nucleons. This difference can be understood easily on the basis of quark models,[(49)] but also with nucleons and mesons.[(50)] The study of hypernuclei has been pursued avidly. Excited as well as ground states have been observed. The hypernuclei are formed by (K^-, π^-) or (π^+, K^+) reactions. In the former case the reaction can proceed with small momentum transfer, so that the Λ^0 has a good chance of forming the ground state of a hypernucleus and remaining in the nucleus; the latter reactions tend to produce nuclei in states of higher angular momenta.

Other baryons that can be investigated in the nuclear medium are excited states of the nucleon, particularly the $\Delta(1232)$. Such nuclei are formed by scattering pions on a nuclear target at energies close to that of the $\Delta(1232)$. Investigations of Δ's in nuclei have been essential in understanding the scattering of pions from nuclei[(51)] and for probing the probability of finding Δ's in the wavefunction of the target nucleus.[(52)] The evidence points to a probability of at most a few percent.

A question that has received wide attention is that of the alteration of the properties of nucleons in nuclei. This topic was discussed in Section 6.10. Additional effects, particularly two-body correlations occur when two nucleons are close to each other in nuclei. Such short range correlations show up in the multiple scattering of high energy electrons and nucleons from nuclei, described in Section 6.11; they have also been studied in the double charge exchange reaction (π^+, π^-) to isobaric analogue states,[(53)] since two neutrons must be turned into protons in this process.

Finally, we have only talked about two-body forces between nucleons. However, meson-theoretical considerations suggest that three-body forces, even if weaker than two-body forces, should exist; such forces come into play only if *three* nucleons are close together. Evidence from ^{3}He and ^{3}H indicates that such forces do not play a large role in nuclei,[(54)] but more work is required to fully assess their importance.

[51]G.E. Brown, B.K. Jennings, and V.I. Rostokin, *Phys. Rep.* **50**, 227 (1979); A.W. Thomas and R.H. Landau, *Phys. Rep.* **58**, 122 (1980); E. Oset H. Toki, and W. Weise, *Phys. Rep.* **83**, 282 (1982); M. Rho, *Annu. Rev. Nucl. Part. Sci.* **37**, 531 (1984); W.R. Gibbs and B.F. Gibson, *Annu. Rev. Nucl. Part. Sci.* **37**, 411 (1987).

[52]H.J. Weber and H. Arenhövel, *Phys. Rep.* **36**, 277 (1978); B. ter Haar and R. Malfliet, *Phys. Rep.* **149**, 207 (1987); see also B. Julia-Diaz, T.-S.H. Lee, T. Sato, and L.C. Smith, nucl-th/0611033; O. Drechsel and L. Tiator, nucl-th0610112.

[53]W. R. Gibbs and B. F. Gibson, *Annu. Rev. Nucl. Part. Sci.* **37**, 411 (1987); E. Bleszynski, M. Bleszynski, and R. Glauber. *Phys. Rev. Lett.* **60**, 1483 (1988); O. Buss, L. Alvarez-Ruso, A.B. Larionov, and U. Mosel, *Phys. Rev. C* **74**, 044610 (2006).

[54]J.L. Friar, B.F. Gibson, and G.L. Payne, *Annu. Rev. Nucl. Part. Sci.* **34**, 403 (1984); *The Three-Body Force in the Three Nucleon System,* (B. L. Berman and B. F. Gibson, eds.) Springer Lecture Notes in Physics, 260, Springer, New York, 1986.

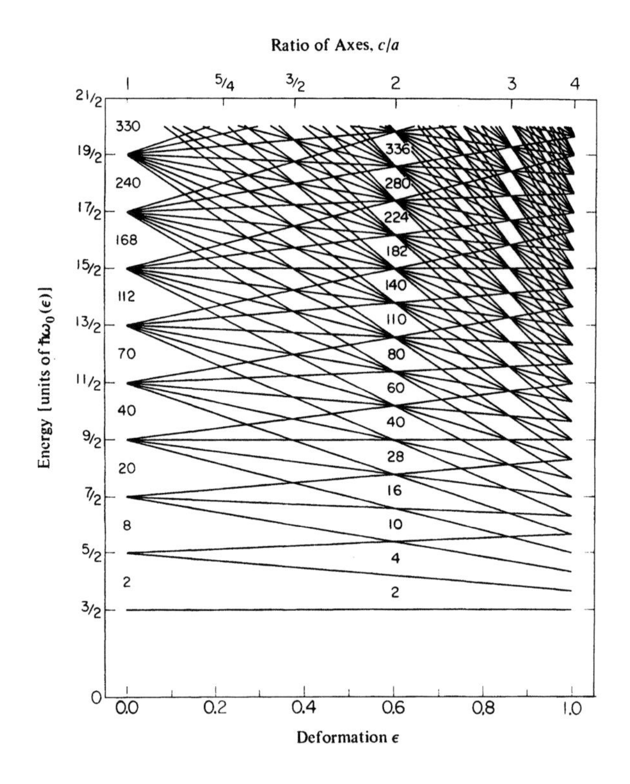

Figure 18.17: Energy levels for a harmonic oscillator potential with prolate spheroidal deformations. The particle numbers of closed shells are shown for a spherical potential and for an ellipsoid with a ratio of major to minor axes of 2. [Courtesy J. R. Nix and reproduced with permission from the *Annu. Rev. Nucl. Sci.* **22**, 65 (1972); © Annual Reviews, Inc.]

18.9 References

Classic treatments of most of the aspects described in this chapter are given in more detail in A. de Shalit and H. Feshbach, *Theoretical Nuclear Physics*, Vol. I: *Nuclear Structure*, Wiley, New York, 1974; J.D. Walecka, *Theoretical Nuclear and Subnuclear Physics*, World Sci. (2004); J. M. Arias, M. Lozano *An Advanced Course in Modern Nuclear Physics*, Springer (2001).

The authoritative work on the phenomenological description of the collective nuclear model is A. Bohr and B. R. Mottelson, *Nuclear Structure*, Vol. II, Benjamin, Reading, Mass. 1975.

Superdeformation and nuclear rotation at high spins are described in R.V.F. Janssens and T.L. Khoo, *Annu. Rev. Nuc. Part. Sci.* **41**, 321 (1991). The collective model, resonances and high spin states are reviewed in D. Ward and P. Fallon, *High Spin Properties of Atomic Nuclei* in *Adv, in Nucl. Phys.* **26**, 168 (2001); *Collective Phenomena in Atomic Nuclei*, (T. Engeland, J.E. Rekstad, and

J. S. Vaagen, eds.) World Sci., Teaneck, NJ, 1984; *Nuclear Structure 1985*, (R. Broglia, G. Hagemann, and B. Herskind, eds.) North-Holland, New York, 1985; *Nuclear Structure at High Spin, Excitation, and Momentum Transfer*, (H. Nann, ed.) AIP Proceedings **142**, Amer. Inst. Phys., New York, 1986.

Detailed comparisons between theoretical predictions and experimental data are given in Bohr and Mottelson, *Nuclear Structure*, Vol. II, and in B. R. Mottelson and S. G. Nilsson, *Kgl. Danske Videnskab. Sels-kab. Mat.-fys. Medd.* **1**, No. 8 (1959).

Giant resonances are discussed in G.F. Bertsch, *Phys. Today* **39**, 44 (August 1986); G.J. Warner in *Giant Multipole Resonances*, (F. E. Bertrand, ed.) Harwood Academic, New York, 1980. Magnetic resonances are reviewed in A. Arima et al., *Adv. Nucl. Phys.*, (J. W. Negele and E. Vogt, eds.) **18**, 1 (1987); E. Lipparini and S. Stringari, *Phys. Rep.* **175**, 104 (1989).

The IBM model is analyzed in numerous conference proceedings and review articles. They include A. Arima and F. Iachello, *Annu. Rev. Nucl. Part. Sci.* **31**, 75 (1981); F. Iachello and I. Talmi, *Rev. Mod. Phys.* **59**, 339 (1987); R. F. Casten and D. Warner, *Rev. Mod. Phys.* **60**, 389 (1988). There are also two books on the subject: F. Iachello and A. Arima, *The Interacting Boson Model*, Cambridge University Press, New York, 1987; D. Bonatsos, *Interacting Boson Models of Nuclear Structure*, Oxford University Press, New York, 1988.

The microscopic theory of nuclei is discussed in *Phys. Rep.* **6**, 214 (1973); S.O. Backman, G.E. Brown, and J.A. Niskanen, *Phys. Rep.* **124,** 1 (1985); *Microscopic Models* in *Nuclear Structure Physics*, (M. W. Guidry et al., eds.) World Scientific, Teaneck, NJ, 1989.

Pions in nuclei are discussed in detail in T. Ericson and W. Weise, *Pions and Nuclei*, Oxford University Press, New York, 1989.

A nice review of advances in models of nuclei is presented in *Hans Bethe and His Physics*, ed. G.E. Brown and C.-H. Lee, World Sci. (2006).

Problems

18.1. Find the expression for the energy of interaction between a system with quadrupole moment $\mathcal{Q}$ and an electric field $\boldsymbol{E}$ with field gradient $\boldsymbol{\nabla E}$.

18.2. The electric quadrupole moment of a nucleus can be determined by using atomic beams.

(a) Describe the principle underlying the method.

(b) Sketch the experimental apparatus.

(c) What are the main limitations and sources of error?

18.3. Repeat Problem 18.2 for the method using optical hyperfine structure.

18.4. Quadrupole moments can also be determined by using nuclear quadrupole resonance and the Mössbauer effect. Answer the questions posed in Problem 18.2 for these two methods.

18.5. Verify Eq. (18.2).

18.6. * The giant dipole resonance has a very different shape in spherical nuclei and in strongly deformed nuclei. Sketch typical resonances for the two cases. Explain the reason for the appearance of two peaks in deformed nuclei. How can the ground-state quadrupole moment be deduced from the positions of the two peaks? How are the relevant experiments performed? [F. W. K. Firk, *Annu. Rev. Nucl. Sci.* **20,** 39 (1970).]

18.7. How can the deformation of a nucleus be observed in an electron scattering experiment? [See, for instance, F. J. Uhrhane, J. S. McCarthy, and M. R. Yearian, *Phys. Rev. Lett.* **26,** 578 (1971).]

18.8. Prepare a $Z - N$ plot and indicate on this plot the regions where you expect spherical nuclei and where you expect large deformations. Plot the position of a few typical nuclides. [E. Marshalek, L. W. Person, and R. K. Sheline, *Rev. Mod. Phys.* **35,** 108 (1963).]

18.9. Verify Eq. (18.7).

18.10. Show that the expectation value of the quadrupole operator in states with spins 0 and $\frac{1}{2}$ vanishes.

18.11. Discuss the transition rates for electric quadrupole transitions in strongly deformed nuclei:

(a) Find a particular example and compare the observed half-life with the one predicted by a single-particle estimate.

(b) How can the observed discrepancy be explained?

18.12. Coulomb excitation. Discuss:

(a) The physical process of Coulomb excitation, and

(b) The experimental approach.

(c) What information can be extracted from Coulomb excitation?

(d) Sketch the information that supports the assumption of collective states in strongly deformed nuclei. [K. Alder and A. Winther, *Coulomb Excitation,* Academic Press, New York, 1966; K. Alder et al., *Rev. Mod. Phys.* **28,** 432 (1956).]

18.13. Verify the numbers in Table 18.1.

18.14. Compute the single-particle quadrupole moments for ^{7}Li, ^{25}Mg, and ^{167}Er. Compare with the observed values.

18.15. (a) Draw the energy levels of ^{166}Yb, ^{172}W, and ^{234}U. Compare the ratios $E_4/E_2, E_6/E_2$, and E_g/E_2 with the ones predicted on the basis of rotation of a spherical nucleus.

(b) Repeat part (a) for ^{106}Pd and ^{114}Cd. Compare with the predictions of the vibrational model.

18.16. Assume ^{170}Hf to be a rigid body (see Fig. 18.3). Calculate, very approximately, the centrifugal force in the state $J = 20$. What would happen to the nucleus if its mechanical properties were similar to those of steel? Support your conclusion with a crude calculation.

18.17. Verify the uncertainty relation equation (18.10).

18.18. Verify Eqs. (18.16) and (18.17).

18.19. Figure 18.5 shows the flow lines of particles for rigid and for irrotational motion in a rotating coordinate system. Draw the corresponding flow lines in a laboratory-fixed coordinate system.

18.20. Assume the moment of inertia, $\mathcal{I}$, in Eq. (18.14) to be a function of the energy E_J. Compute $\mathcal{I}(E_J)$ (in units of $\hbar^2$/MeV) for the rotational levels in ^{170}Hf, ^{184}Pt, and ^{238}U. Plot $\mathcal{I}(E_J)$ against E_J and show that a linear fit $\mathcal{I}_{\text{eff}} = c_1 + c_2 E_J$ reproduces the empirical data well.

18.21. Consider an even-even nucleus with equilibrium deformation δ_0 and spin $J = 0$ in its ground state. The energy in a state with spin J and deformation δ is the sum of a potential and a kinetic term,

$$E_J = a(\delta - \delta_0)^2 + \frac{\hbar^2}{2\mathcal{I}} J(J+1).$$

(a) Assume irrotational motion, $\mathcal{I} = b\delta^2$. Use the condition $(dE/d\delta) = 0$ to find the equation for the equilibrium deformation δ_{eq} in the state with spin J.

(b) Show for small deviations of the deformation from the ground-state deformation that the nucleus stretches and that the energies of the rotational states can be written as

$$E_J = AJ(J+1) + B[J(J+1)]^2.$$

(c) Use this form of E_J to fit the observed energy levels of ^{170}Hf by determining the constants A and B from the two lowest levels. Then check how well the computed energies agree with the observed ones up to $J = 20$.

18.22. Consider an axially symmetric deformed core plus one valence nucleon (Fig. 18.6). Why are J and K good quantum numbers, but not j?

18.23. Why are the states with odd J not excluded from the sequence (18.24)?

18.24. Discuss the rotational families of ^{249}Bk (Fig. 18.7):

(a) Check how well Eq. (18.23) fits the observed energy levels for each band.

(b) Show that K for each band can be found unambiguously from the lowest three levels of a band by using Eq. (18.25).

18.25. Compare the term H_c in Eq. (18.21) with the classical Coriolis force.

18.26. Use the slope of the trajectories in Fig. 18.8 and Eq. (18.23) for E_J to determine the moment of inertia as a function of J. Plot $\mathcal{I}$ against J for the three families. Is stretching apparent?

18.27. Find another example for rotational families and prepare a plot similar to Fig. 18.8.

18.28. Find the energy levels of the anharmonic oscillator, described by the potential

$$V = \frac{1}{2}\, m[\omega_\perp(x_1^2 + x_2^2) + \omega_3^2 x_3^2].$$

18.29. Describe the complete labeling of Nilsson levels.

18.30. Verify Eq. (18.30).

18.31. Justify that the rotational and the intrinsic motion in deformed nuclei can be separated by finding approximate values for the time of rotation and the time a single nucleon needs to traverse the nucleus.

18.32. * Discuss the level diagram of ^{165}Ho [M. E. Bunker and C. W. Reich, *Rev. Mod. Phys.* **43,** 348 (1971)]:

(a) Find the various band heads and their rotational spectra.

(b) Plot the bands in a Regge plot (Section 15.7).

(c) Use a Nilsson diagram to find the complete quantum number assignment for each band head.

18.33. Consider a completely asymmetric nucleus, with $\omega_1 > \omega_2 > \omega_3$. What is the spectrum of single particle levels in such a nucleus if $\omega_1/\omega_2/\omega_3 = \alpha/\beta/1$. (*Hint*: Use Cartesian coordinates.)

18.34. Compare molecular and nuclear spectra. Discuss the energies and energy ratios involved in the three types of excitations. Discuss the corresponding characteristic times. Sketch the essential aspects of the spectra.

18.35. Show that the term $l = 1$ in Eq. (18.35) corresponds to a translation of the nuclear c.m. Draw an example.

18.36. Find a relation between the coefficients α_{lm} and $\alpha^*_{l,-m}$ in Eq. (18.35) by using the reality of R and the properties of the Y_l^m.

18.37. Use Eq. (18.35) to draw a deformed nucleus described by $\alpha_{30} \neq 0$, all other $\alpha = 0$.

18.38. Verify the solution (18.39).

18.39. Show that for an incompressible irrotational nucleus the semiempirical mass formula gives for the coefficients B and C in Eq. (18.38)

$$B = \frac{3}{8\pi} AmR^2$$

$$C = \frac{1}{2\pi}\left(2a_s A^{2/3} - \frac{3}{5}\frac{Z^2e^2}{R}\right).$$

18.40. Show that vibrational motion implies the existence of excited vibrational states. (*Hint*: Consider the nuclear density and show that the density is always constant if only one state exists. Then consider a small admixture of an excited state.)

18.41. Discuss a plot of the energy ratio E_2/E_1 for even-even nuclei. Indicate where rotational and where vibrational spectra appear. Compare the corresponding excitation energies E_1.

18.42. Why can a state 3^+ turn up in the level $N = 3$, but not $N = 2$, in Fig. 18.12?

18.43. Consider nonazimuthally symmetric quadrupole deformations,

$$R = R_0\left(1 + \sum_m \alpha_{2m} Y_2^m\right)$$

$$\alpha_{20} = \beta\cos\gamma, \quad \alpha_{22} = \alpha_{2,-2} = \frac{1}{\sqrt{2}}\beta\sin\gamma.$$

(a) If $\gamma = 0$, what is $V(\beta)$ for a spherical harmonic oscillator?

(b) For a prolate nucleus and harmonic forces, what is $V(\beta)$?

(c) Consider harmonic γ vibrations for a prolate nucleus. What is the shape of the potential and the energy spectrum due to these vibrations?

18.44. What is the effect of an octupole term in Eq. (18.35) on

(a) The vibrational spectrum?

(b) Permanent deformations?

(c) The rotational spectrum?

18.45. Label the angular momenta and show the spacings of the first two excited states for nuclear octupole vibrations. (Take nuclear symmetries into account.)

18.46. The giant dipole resonances in even–even nuclei are $J^\pi = 1^-$ and have isospin $I = 1$. What is the reason for the absence of $1^-, I = 0$ modes?

18.47. What are the possible decay modes of a Λ^0 in a nucleus?

18.48. To check the accuracy of Eq. (18.43) carry out the exact sum for the energy of a nucleus with $Z = N = 64$, and compare the energy to that of Eq. (18.43).

18.49. Consider the effect of the exclusion principle in hypernuclei if the Λ^0 is considered to be a composite particle that dissolves into its constituent s, u, and d quarks, and nucleons into u and d quarks. Compare to the case when the Λ^0 and N retain their identities.

(a) What is the lowest baryon number and charge of the hypernucleus for which the exclusion principle could have an effect?

(b) How could you determine whether a Λ^0 in a low-lying energy level of a nucleus should be regarded as a composite object made up of quarks or a fundamental particle?

(c) Is experimental information available, and if so what does it indicate?

18.50. In the production of a hypernucleus with incident K^- and outgoing π^- what is the approximate incident momentum of the kaon required to produce a Λ^0 at rest, or to minimize the momentum transferred to the target nucleus? As a specific example, consider a ^{12}C target.

18.51. (a) In the case of a normal nucleus, ^{4}He is the one with a full S state. Is this also true if a quark basis is used, where the nucleons are made up of u and d quarks?

(b) If strange quarks are included, the color-singlet di-baryon with a full space–spin flavor S state is called H. What strong decays are allowed for the H if its mass is sufficiently large? What must be the upper limit of its mass (in MeV/c^2) if the H is to be stable for hadronic decays?

(c) If the mass is lower than the critical value of part (b), what would be the approximate expected lifetime of the H?

18.52. What reaction could be used to produce double hypernuclei, i.e., those with two lambdas replacing two neutrons?

Chapter 19

Nuclear and Particle Astrophysics

> The most incomprehensible thing about the world is that it is comprehensible.
>
> Albert Einstein

> The marriage between elementary particle physics and astrophysics is still fairly new. What will be born from this continued intimacy, while not foreseeable, is likely to be lively, entertaining, and perhaps even beautiful.
>
> M. A. Ruderman and W. A. Fowler[1]

For millenia, the stars, Sun, and Moon have fascinated humans; their properties have been subject to much speculation. Up to a short time ago, however, observation of the heavens was restricted to the very small optical window between about 400 and 800 nm, and mechanics was the branch of physics most intimately involved in astronomy. In the last century, the situation has changed dramatically and physics and astronomy have become much more closely intertwined. In this chapter, we shall sketch some of the areas in which subatomic physics and astrophysics are linked.

19.1 The Beginning of the Universe

> O God, I could be bounded in a nut shell and count
> myself a king of infinite space, were it not that I
> have bad dreams.
>
> Shakespeare, *Hamlet, Act II, Scene 2.*

In 1929 Hubble observed that well-known lines in the spectrum of gases (e.g. Hydrogen) from stars in galaxies were shifted towards the red. Hubble was able

[1]M. A. Ruderman and W. A. Fowler, "Elementary Particles," *Science, Technology and Society* (L. C. L. Yuan, ed.), Academic Press, New York, 1971, p. 72. Copyright © 1971 by Academic Press.

to show that, the more distant the galaxies were located, the larger the redshift. The shifts are due to the fact that galaxies are moving away from us and further galaxies are receding at higher speeds. This can be understood by considering a rubber sheet that is stretching at a constant rate. Any two points marked on the sheet move away from each other and the greater the distance between the points, the higher the speed. Like the sheet, the universe is expanding and any two points in the universe separated by a distance r are moving away from each other at a speed:

$$v = H_0 r \tag{19.1}$$

where $H_0 \sim 70$ km/s/Mparsec is called the Hubble constant (1 Mparsec $= 3.09 \times 10^{19}$ km.) Recent observations of distant supernovae[(2)] have yielded clear evidence that the constant is not really a constant, and that the universe's expansion is accelerating. We will consider this issue further below. Nevertheless, an expanding universe (as opposed to a stationary one) turns out to be a natural consequence of Einstein's equations of general relativity.

As the universe expands it necessarily cools down; consequently it is obvious that it must have been much hotter in its early stages. Gamow proposed[(3)] that the universe began as an extremely hot and highly compressed neutron ball, bathed in radiation. Its great internal energy meant that this primordial fireball expanded so rapidly that the proposed beginning is usually called the "big bang." The presently accepted "standard model" of the early universe differs from Gamow's picture, but it is still believed that the universe was hot and dense shortly after its beginning and has been expanding since. In the past several decades, the connection between subatomic physics and cosmology, the study of the evolution of the universe and its large scale structures, has become ever stronger. The resulting understanding is that the universe is $\sim$ 14 billion years old and has evolved through a set of well-defined periods and phases, separated by distinctive transitions. Table 19.1 lists some of the critical phase transitions. The universe may have begun as a singularity or as a vacuum fluctuation. It is difficult to consider times less than $\hbar c^{-2}\sqrt{G/\hbar c} \approx 10^{-43}$ sec, the so-called Planck time, because conventional concepts of space–time break down. In order to describe such times we would need a theory that combines gravity and quantum mechanics, which we presently don't have.

After the first phase transition, gravity became weaker and was no longer unified with the other forces. The universe is then believed to have entered the GUT (grand unified theory) era in which the electroweak and strong forces remained unified, as described in Chapter 14. At about 10^{-10} sec, when the temperature dropped to about 3×10^{15} K, a temperature that corresponds to about 100 GeV, the electromagnetic and weak forces separated into the two forces we have described in previous chapters. The weak force became short-ranged, whereas it had been of long

---

[2]S. Perlmutter et al., *Astrophys. J.* **517**, 565 (1999); A.G. Riess et al. *Astrophys. J.* **560**, 49 (2001) and references therein.

[3]G. Gamow, *Phys. Rev.* **70**, 572 (1946); *Rev. Modern Phys.* **21**, 367 (1949).

Table 19.1: SOME CRITICAL PHASES IN THE DEVELOPMENT OF THE UNIVERSE. The development is shown going backward in time. Numbers are approximate.

Age	Temperature/Energy (K)	(eV)	Transition	Era
1.4×10^{10} y	2.7	$\sim 10^{-4}$		Present, Stars
4×10^{5} y	3×10^{3}	$\sim 10^{-1}$	Plasma to atoms	Photon
3 min	10^{9}	$\sim 10^{5}$	Nucleosynthesis	Particle
10^{-6} sec	10^{12}	$\sim 10^{8}$	Quarks (hadronization)	Quark
10^{-10} sec	10^{15}	$\sim 10^{11}$	Weak and em forces unify	Electroweak
10^{-33} sec	10^{28}?	$\sim 10^{24}$	Inflation	Inflation
10^{-43} sec	10^{32}	$\sim 10^{28}$	All forces unify	SUSY, Planck
0			Vacuum to matter	

range, like the electromagnetic force, prior to this time, because the particles were all massless. The ratio of the number of baryons to photons acquired its final value of about 6×10^{-10}. At the next, QCD, phase transition, the quarks are combined into hadrons with their present masses. After this time the universe continued to be composed of elementary particles. The rate of a reaction is determined by the available energy or temperature. For instance, for energies larger than $2mc^2$ particle–antiparticle pair production can occur. When the production rate of a particle decreases and becomes negligible compared to the expansion rate, that particle is said to decouple or "freeze out." The thermal reaction rate depends on $e^{-E/kT}$, where E is the necessary energy, k the Boltzmann constant, and T the temperature. At temperatures of 3×10^{11} K, ($kT \sim 10^7$ eV), protons and neutrons remained in equilibrium because the reactions $\overline{\nu}p \leftrightarrow e^+n$ and $\nu n \leftrightarrow ep$ can proceed with ease in both directions. When the temperature fell to about 10^{10} K, ($kT \sim 10^6$ eV), at about 1 sec, the weak reactions were sufficiently weak that neutrinos decoupled and became free to roam the universe; so the reaction $\overline{\nu}p \leftrightarrow e^+n$ became negligible and the proton to neutron ratio increased to about 3:1 because then the $n - p$ mass difference was comparable to kT. As the temperature dropped further, e^+e^- creation became negligible, annihilation to photons continued to occur. The reaction $e^+e^- \to \nu\overline{\nu}$ was much slower; thus the photons gained energy and heated up compared to neutrinos. Consequently, 'relic' neutrinos are expected to have a smaller temperature ($\sim 1/3$) of that corresponding to 'relic' photons.

After about 3 minutes primordial nucleosynthesis began; it ceased at about 30 minutes, when the temperature had dropped to $\lesssim 4 \times 10^8$ K ($kT \sim 10^4$ eV). Not until about 4×10^5 y, when the temperature had dropped further to about 3000 K, ($kT \sim 1/4$ eV), did electrons and protons bind to form hydrogen atoms. At this time, the universe became "transparent" to light, and this is as far back as we can see by observing photons. The photons were consequently emitted with a blackbody radiation distribution with a characteristic temperature of ~ 3000 K. However, due to the expansion of the universe the radiation now has a characteristic temperature of ~ 2.7 K. Penzias and Wilson[4] first observed this radiation. Because of its characteristic temperature it is called the cosmic microwave background radiation (CMBR). In the early 1990's the COBE satellite mission[5] produced detailed measurements that showed excellent agreement with the expected blackbody spectrum and measured the temperature distribution over the whole sky. These measurements brought up the realization that it was possible to extract precise information on what the universe was like before galaxies existed. Recently the WMAP[6] mission has produced measurements with much improved precision and has measured the polarization of the CMBR as well. Now we know that the temperature of the CMBR is 2.725 ± 0.001 K and we infer that the age of the universe is $(1.37 \pm 0.02) \times 10^{10}$ y. The data however pose a problem. The temperature of the background radiation is the same in all directions to about 1 part in 10^5, once the motion of our galaxy is subtracted out.[7] The radiation observed from opposite directions was emitted when the source regions were separated by more than 90 times the distance that light could have traveled (horizon distance) since the beginning of the universe. The regions were, therefore, causally disconnected, and the isotropy is hard to understand. The observed isotropy is called the "horizon" problem. Another issue, called the "flatness" problem, is that of the various solutions of Einstein's equations, i.e., a convex, concave or flat universe, all evidence points to a flat one, i.e., Euclidean. The ratio, Ω, of the measured density of the universe to the critical density at which the geometry is precisely flat is known to be in the range $0.97 \lesssim \Omega \lesssim 1.12$.[6] Any deviation of Ω from unity tends to grow with time. The above limit on Ω implies that when the temperature T was of the order of 10^{28} K, Ω could not have differed from unity by more than 10^{-50}. How did this fine tuning come about?

A solution to these problems, dubbed *the inflationary scenario*, was proposed by Guth in 1980,[8], developed by Linde,[9] and Albrecht and Steinhardt,[10] and is continuing to be modified. The basic idea is that the potential energy of a scalar field (like the Higgs field described in Chapter 12) dominated the energy density of the early universe. This condition caused the expansion of the universe to accelerate, while the energy density remained approximately constant. The expansion was exponential, with the radius doubling in about 10^{-34} sec; the inflation continued until about 10^{-32} sec. Eventually, the scalar field "rolled" to the minimum of its

---

[4]A.A. Penzias and R.W. Wilson, *Astrophys. J.* **142**, 419 (1965); L. Page and D. Wilkinson, *Rev. Mod. Phys*, **71**, S173 (1999).

[5]John Mather and George Smoot were awarded the 2006 Nobel prize for this measurement.

[6]Wilkinson Microwave Anisotropy Probe mission; D.N. Spergel et al. astro-ph/0603449; see http://map.gsfc.nasa.gov/m_mm.html; M.D. Lemonick *Echo of the Big Bang*, Princeton University Press, Princeton (2005).

[7]A.S. Redhead et al., *Astrophys. J.* **346**, 566 (1989).

[8]A. Guth, *Phys. Rev.* **D23**, 347 (1981).

[9]A.D. Linde, *Phys. Lett.* **108B**, 389 (1982); A.D. Linde *Particle Physics and Inflationary Cosmology*, Harwood, Chur, Switzerland, 1990.

[10]A. Albrecht and P. Steinhardt, *Phys. Rev. Lett.* **48**, 1220 (1982); P.J. Steinhardt, *Comm. Nucl. Part. Phys.* **12**, 273 (1984).

potential and inflation terminated. By this time the radius had increased by a factor of $\sim 10^{50}$. Ω is driven to unity, since, like a balloon, the surface becomes flatter as it expands. Thus, the curvature became so small that it remains unimportant today. In this scenario, the universe begins from a much smaller ($\sim 10^{-50}$) region than without inflation, and the horizon and homogeneity problems are solved. The inflationary scenario does not tell us how the universe began, but it allows a wider variety of early conditions, including vacuum fluctuations and a beginning from "nothing."(11)

Fortunately, the uniform smoothness is not complete. Figure19.1 shows the distribution of temperatures across the sky measured by the WMAP collaboration. The temperature non-uniformities shown in Fig. 19.1, which are at the level of 1 part

Figure 19.1: Map of temperature differences across the whole sky. The darker regions are colder and the clearer regions are hotter. The white lines indicate the polarization direction. [From WMAP collaboration.(6)]

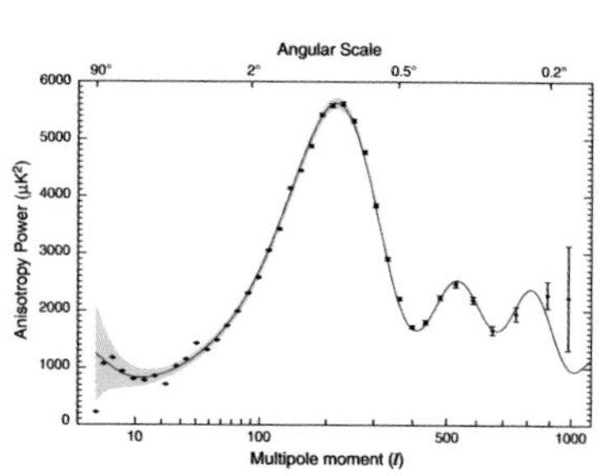

Figure 19.2: Power spectrum corresponding to the data shown in Fig. 19.1 (see Eq. 19.3.) The continuous line shows the best fit of a model that assumes Dark Matter to be cold (see text). [From WMAP collaboration.(6)]

in 10^5, are tell-tales of density non-uniformities, i.e. 'lumps' in the early universe matter distribution. In the inflation scenario these non-uniformities were quantum fluctuations that were amplified by the initial rapid expansion. These lumps are responsible for the subsequent gravitational aggregation that eventually produced galaxies and stars. Thus, the inflationary scenario explains several observations in a rather simple framework. The data of Fig. 19.1 is usually analyzed in terms of its spherical harmonics:

$$T(\boldsymbol{n}) = \sum_{l,m} a_{lm} Y_{lm}(\boldsymbol{n}), \tag{19.2}$$

where $\boldsymbol{n}$ is a unit vector indicating a direction in space. Fig. 19.2 shows the power spectrum:

$$C_l = \frac{1}{2l+1} \sum_m |a_{lm}|^2. \tag{19.3}$$

[11] A. H. Guth and P. J. Steinhardt, *Sci. Amer.* **250**, 116 (May 1984); S. Y. Pi, *Comm. Nucl. Part. Phys.* **14**, 273 (1984); K.A. Olive and D.N. Schramm, *Comm. Nucl. Part. Phys.* **15**, 69 (1985).

The peak in Fig. 19.2 indicates that on average the temperature fluctuations have a width of, and are separated by, ~ 1 degree in the sky.

Perhaps the most important present problem in cosmology which is closely connected to subatomic physics is that baryonic matter constitutes less than 4% of the mass required by the condition $\Omega = 1$; about 22% is *dark matter* and about 74% of it is *vacuum energy.*[12] The evidence for dark matter comes from the distribution of galaxies and clusters and their motions, from the study of stars, and from the expansion of the universe. The dark matter makes itself felt through its gravitational effects, but remains invisible to other probes. Weakly interacting massive particles (WIMPs) and axions (a particle that has been proposed to explain the smallness of CP-breaking in the strong interaction) are candidates to solve this problem.[13] Continuing efforts are being undertaken to search for this missing dark matter.[14] There are two important scenarios, named 'hot dark matter', where dark matter is assumed to have velocities close to the speed of light, and 'cold dark matter' that assumes very low velocities, similar to the baryonic matter. Fig. 19.2 shows that a model that assumes that dark matter is cold can fit the CMBR data very nicely, so this is presently a favored hypothesis (see problem 19.13.)

Vacuum energy became a possible explanation for the unanticipated accelerating expansion of the universe found by two separate investigative teams that were studying distant supernovae.[2] Supporting evidence comes from the studies of the CMBR and nucleosynthesis, which show that cold dark matter and baryons only account for $\sim 26\%$ of the mass required for a flat universe. The vacuum energy can be represented by a *cosmological constant*, introduced by Einstein in general relativity and dubbed by him as "my biggest mistake"![15] A deeper understanding of the birth of our universe and its transitional stages occurred with the development of grand unified theories or GUTs. These theories, which unify the electroweak and hadronic forces also predict baryonic decays, which together with CP or time reversal violation, are a possible scenario for understanding the particle over antiparticle excess and the ratio of baryons to photons (about 6×10^{-10}) in our universe. The conditions for obtaining an excess of baryons over antibaryons were stated succinctly by Sakharov.[16] They are: 1) CP nonconservation, 2) baryon nonconservation, and 3) nonequilibrium conditions. CP nonconservation permits a slight difference to develop between the number of baryons and antibaryons. As an example consider a particle X of mass $= 10^{14}$ GeV/c^2. If baryon and lepton numbers are not conserved exactly, X may decay to a quark and electron and $\overline{X}$ to a $\overline{q}$ and e^+. Above temperatures of 10^{14} GeV, decay and formation of X and $\overline{X}$ were in approximate

[12] W.L. Freedman and M.S. Turner, *Rev. Mod. Phys.* **75**, 1433(2003).

[13] K. van Bibber and L.J. Rosenberg, *Phys. Today* pg. 30, Aug. (2006).

[14] R.J. Gaitskell, *Annu. Rev. Nuc. Part. Sci.* **54**, 315 (2004); Sadoulet, *Rev. Mod. Phys.* **71**, S197 (1999)

[15] Einstein originally introduced the constant with the intention of precluding his equations from predicting an expanding universe, because at the time there was no evidence for the expansion.

[16] A.D. Sakharov, *Pis'ma Z. Eksp. Teor. Fiz.* **5**, 32 (1967); English Translation: *JETP Lett.* **5**, 24 (1967); L.B. Okun, Ya.B. Zeldovich, *Comments Nucl. Part. Phys.* **6**, 69 (1976).

equilibrium. As the temperature fell, only the decays could occur and an excess of quarks over antiquarks could develop if *CP* invariance does not hold and the partial decay rate of the X to a quark is slightly more rapid than that of the $\overline{X}$ to an antiquark. An excess of quarks over antiquarks of about 6×10^{-10} is sufficient to account for our present universe. The subsequent annihilation of quarks with antiquarks left the baryonic excess. This model is only one among several proposals to understand the baryon excess.

19.2 Primordial Nucleosynthesis

Primordial nucleosynthesis did not begin until the universe was more than several tenths sec old and had cooled, through expansion, to about 3×10^{10} K, ($kT \sim 10^6$ eV). Prior to this time, the temperature was so high that the light nuclei formed by nucleon and nuclear collisions broke up as soon as they were formed. At 3×10^{11} K, ($kT \sim 10^7$ eV), ^{4}He would remain bound, but the lighter nuclei would continue to break up, so that nucleosynthesis still could not begin. However, when about 1 min. later, owing to the expansion, the temperature had dropped somewhat below 10^{10} K, deuterons that formed in the capture reaction $np \to d\gamma$ remained stable. Further neutron and proton capture by deuterons led to ^{3}H and ^{3}He. The ^{3}H also beta decays to ^{3}He, which can capture a neutron to form ^{4}He, but this process is very slow compared to the formation of ^{4}He through direct neutron capture by ^{3}He or through the reaction $d\,^3\text{He} \to p\,^4\text{He}$. Capture of ^{3}H and ^{3}He by ^{4}He leads to small amounts of ^{7}Li and ^{7}Be. The latter beta decays to ^{7}Li, which is stable, although it can be destroyed by $p^7\text{Li} \to {}^4\text{He}^4\text{He}$. Other light nuclei may also be destroyed, e.g., $n\ {}^3\text{He} \to p\,^3\text{H}$. The amounts of ^{2}H, ^{3}H, ^{3}He, ^{4}He, and ^{7}Li produced primordially are, therefore, sensitive to the density of baryons, or the ratio of baryons to photons ($\sim 6 \times 10^{-10}$), as well as to the rate of expansion or cooling.

The competition between the rates of nuclear reactions and expansion determines the survival of a given nuclide. A plot of the abundances, by weight, of the light nuclei formed in the big bang to that of hydrogen is shown in Fig. 19.3.[(17)] The larger the density of baryons, or η in Fig. 19.3, the higher the rate of destruction of d, ^{3}H, and ^{3}He. Deuterium is the most sensitive *primordial baryometer* because, when incorporated in stars, the deuterium is quickly consumed by nuclear reactions.

The primordial production of heavier elements is stymied by the inability of neutron or proton capture on ^{4}He to lead to stable nuclei and by the slowness of other reactions. Neutron capture, for instance, leads to ^{5}He which is unstable and decays back to ^{4}He.

---

[17] PDG.

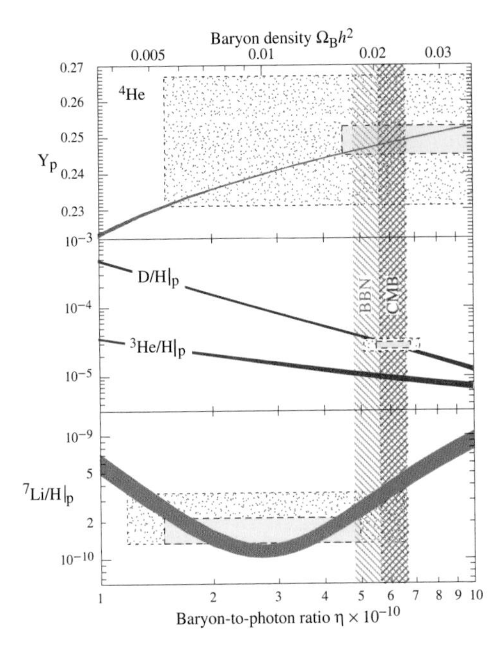

Figure 19.3: Abundances of ^{4}He, ^{2}H, ^{3}He, ^{7}Li as predicted by the standard model of big bang nucleosynthesis. The boxes correspond to the observed abundances (small boxes: $\pm 1\sigma$; large boxes: $\pm 2\sigma$.) The vertical line indicates the cosmic baryon density from CMBR. [From PDG.]

Further, the alpha-particle capture reaction,

$$^4\text{He}\ ^4\text{He} \longrightarrow\ ^8\text{Be}, \qquad (19.4)$$

leads to the highly unstable nuclide ^{8}Be which breaks up immediately into two alpha particles. When the temperature of the universe dropped to about 3×10^8 K, ($kT \sim 10^4$ eV), approximately half an hour after its birth, primordial nuclear synthesis ceased because the Coulomb barrier prevented further nuclear reactions. The abundances of the elements formed in the big bang were frozen, so that the presently observed abundances of the light elements d, ^{3}He, ^{4}He, and ^{7}Li still reflect this stage.

19.3 Stellar Energy and Nucleosynthesis

And God said, Let there be light; and there was light.

Genesis

The mechanism of energy production in the sun is understood and has been tested, and we shall discuss it as an example of stellar power sources. The construction of a terrestrial fusion reactor is difficult. The main difficulty is *containment*: A plasma with a temperature of about 10^8 K must be kept enclosed within a finite volume. Solid walls cannot withstand such a temperature and magnetic or laser confinement is used. The magnetic field volume must be relatively small (a few m^3), or power and construction costs become prohibitive. Instabilities plague all these confinement schemes, so the designer(s) of the Sun has chosen a simple but robust scheme: The "container" is huge, with a radius of about 7×10^8 m, with an outside temperature of about 6000 K, and with a central temperature of about 1.6×10^7 K, ($kT \sim 10^3$ eV). Fusion reactions then proceed at a much lower rate than that needed for terrestrial reactors. Nevertheless, total energy production is large because the volume is huge.

Before nuclear reactions were discovered, the energy production in the sun was unexplainable; no known source could provide sufficient energy, particularly because it was clear from geophysical studies that the sun must have had about the same temperature for at least 10^9 y. Among the first to recognize the nature of the energy-producing process was Eddington,[18] who showed that the fusion of four hydrogen atoms into one atom of He would release about 7 MeV/nucleon and thus provide millions of times more energy than a chemical reaction. However, one problem remained: Classically, fusion cannot occur because even at stellar temperatures protons do not have sufficient kinetic energy to overcome their mutual repulsion. Quantum mechanical tunneling, of course, permits reactions at much lower temperatures,[19] and specific reactions responsible for the stellar energy production were established.[20] The first sequence that was proposed is the carbon or *CNO cycle* in which a ^{12}C and $4p$ are transformed into an alpha particle and ^{12}C. The various steps in this cycle are:

$$\begin{aligned} {}^{12}\text{C}\,p &\longrightarrow {}^{13}\text{N}\,\gamma, \\ {}^{13}\text{N} &\longrightarrow {}^{13}\text{C}\,e^+\nu, \\ {}^{13}\text{C}\,p &\longrightarrow {}^{14}\text{N}\,\gamma, \\ {}^{14}\text{N}\,p &\longrightarrow {}^{15}\text{O}\,\gamma, \\ {}^{15}\text{O} &\longrightarrow {}^{15}\text{N}\,e^+\nu, \\ {}^{15}\text{N}\,p &\longrightarrow {}^{12}\text{C}\,{}^{4}\text{He}. \end{aligned} \tag{19.5}$$

In this sequence, ^{12}C acts as a catalyst; it undergoes changes but it is not used up, it appears again in the final state. Thus the overall reaction is

$$4p \longrightarrow {}^{4}\text{He} + 2e^+ + 2\nu + \text{photons}.$$

The total energy release in this reaction can easily be found with the known masses; it is

$$Q(4p \longrightarrow {}^{4}\text{He}) = 26.7\ \text{MeV}. \tag{19.6}$$

Of this energy, about 25 MeV heats the star, and the rest is carried off by the neutrinos.

The CNO cycle dominates in *hot* stars; in cooler stars, particularly in the sun, the *pp cycle* is much more important. The essential steps in the *pp* cycle are

$$\left.\begin{array}{r} pp \longrightarrow de^+\nu \\ \text{or} \\ ppe^- \longrightarrow d\nu \end{array}\right\} dp \longrightarrow {}^{3}\text{He}\,\gamma \tag{19.7}$$

[18] A. S. Eddington, *Brit. Assoc. Advan. Sci. Rep. Cardiff*, 1920. In this talk, Eddington also said: "If, indeed, the subatomic energy in the stars is being freely used to maintain their great furnaces, it seems to bring a little nearer to fulfillment our dream of controlling this latent power for the well-being of the human race—or for its suicide."

[19] R. Atkinson and F. Houtermans, *Z. Physik* **54**, 656 (1928).

[20] H. A. Bethe, *Phys. Rev.* **55**, 434 (1939); C. F. Weizsäcker, *Physik. Z.* **39**, 633 (1938); H. A. Bethe and C. L. Critchfield, *Phys. Rev.* **54**, 248 (1938).

$$\begin{aligned} &\text{and} \qquad {}^3\text{He}\,{}^3\text{He} \longrightarrow {}^4\text{He}\; 2p \\ &\text{or} \qquad {}^3\text{He}\,{}^4\text{He} \longrightarrow {}^7\text{Be}\,\gamma. \end{aligned} \tag{19.8}$$

In the first part of Eq. (19.8), the overall reaction $4p \rightarrow {}^4\text{He} + 2e^+ + 2\nu$ has already been achieved. In the second part, ^{7}Be has been formed, and it, in turn, leads to ^{4}He through two sequences:

$$\begin{aligned} &{}^7\text{Be}\, e^- \longrightarrow {}^7\text{Li}\, \nu; \quad {}^7\text{Li}\, p \longrightarrow 2{}^4\text{He} \\ \text{or} \quad &{}^7\text{Be}\, p \longrightarrow {}^8\text{B}\gamma; \quad {}^8\text{B} \longrightarrow {}^8\text{Be}^* e^+ \nu; \quad {}^8\text{Be}^* \longrightarrow 2{}^4\text{He}. \end{aligned} \tag{19.9}$$

The *pp* cycle has the same energy release, Eq. (19.6), as the CNO cycle. To compute the reaction rates, two very different input data are required. First, the temperature distribution in the interior of the sun must be known. The original work goes back to Eddington;[21], but over the years much improvement was achieved. At its center the temperature of the sun is about 16 million K ($kT \sim 10^3$ eV). Second, the cross sections for the reactions listed above must be known at temperatures of the order of 16 million K. This temperature corresponds to kinetic energies of only a few keV, and the relevant cross sections are extremely small. A glance at Eqs. (19.7)–(19.9) shows that two types of reactions are involved, hadronic and weak ones. All reactions where neutrinos are involved are weak. The mean life of the decay ${}^8\text{B} \rightarrow {}^8\text{Be}^* e^+ \nu$ has been measured. The two weak reactions in Eq. (19.7), however, are so slow that they cannot be measured in the laboratory; they must be computed using the weak Hamiltonian discussed in Chapter 11.[22] To find the cross sections for the hadronic reactions, values measured at higher energies are extrapolated down to a few keV.[23]

Both the stellar-structure and the nuclear-physics aspects of solar energy production thus appear to be understood. The confidence in our models of the sun was strengthened by the observation of neutrinos at the expected intensities.[24]

We described above how the lighter elements can be produced by primordial nucleosythesis but some of them may also be made in stars; for ^{4}He this production process leads to less than about 10% of the measured abundance. Deuterium cannot be made in any significant quantity in stars because it is converted to heavier nuclei at high densities. The lithium production in stars may be assisted by neutrino

[21] A. S. Eddington, *Internal Constitution of Stars*, Cambridge University Press, Cambridge, 1926.

[22] J. N. Bahcall, *Neutrino Astrophysics*, Cambridge University Press, New York, 1989.

[23] E.G. Adelberger et al., *Rev. Mod. Phys.* **70**, 1265 (1998).

[24] See e.g. http://nobelprize.org/physics/articles/bahcall/index.html.

interactions with ^{4}He; these reactions make ^{3}He, ^{3}H, protons and neutrons.[25] One of the successes of the standard model is its ability to predict the abundances of the lightest elements, even though they differ by nine orders of magnitude. On the other hand, the observed abundance of heavy elements cannot be explained through big-bang-synthesis.[26] Heavier elements were produced later, after stars had already been formed. Nucleosynthesis, the explanation of the abundance of nuclear species, thus becomes intimately involved with problems of stellar structure and evolution.

In a star, gravity pressure tends to decrease the star's volume, while the pressure of the hot gas inside tends to oppose this reduction. Pressure and temperature inside a star are immense. In the Sun, for instance, the pressure at the center is about 2×10^{10} bar (2×10^{15} Pa) and the temperature 16 M K. Under these circumstances, atoms will be almost completely ionized, resulting in a mixture of free electrons and bare nuclei. This mixture forms the "gas" mentioned above. The internal pressure is maintained by the nuclear reactions that provide the energy for the star's radiation. As long as these reactions proceed, gravitational and internal pressure balance and the star will be in equilibrium. What will happen, however, when the fuel is used up? Or to give one example, what will happen to our Sun when all hydrogen is used up and the *pp* cycle stops? At this point, the star will contract gravitationally and the central temperature and pressure increase. At some higher temperature new reactions occur, a new equilibrium will be reached, and new elements will be formed. There are thus alternate stages of nuclear burning and contraction. Burning may be quiescent as in the Sun or explosive as in supernovae,[27] and both are involved in the synthesis of heavier elements.

After the formation of ^{4}He, the next important step is the creation of ^{12}C. ^{8}Be, formed through the reaction Eq. (19.4), is unstable. Nevertheless, if the density of ^{4}He is very high, significant quantities of ^{8}Be are present in the equilibrium situation

$$^4\text{He}\,^4\text{He} \rightleftarrows {}^8\text{Be}^*.$$

Capture of an alpha particle can then occur,

$$^4\text{He}\,^8\text{Be}^* \longrightarrow {}^{12}\text{C}. \qquad (19.10)$$

This capture reaction is enhanced because the formation of ^{12}C proceeds mainly through a resonant capture to an excited state, ^{12}C*.[28]

The formation of ^{16}O occurrs mainly via *helium burning*,

$$^4\text{He}\,^{12}\text{C} \longrightarrow {}^{16}\text{O}\ \gamma. \qquad (19.11)$$

---

[25] S. E. Woosley et al., *Astrophys. J.* **356**, 272 (1990).

[26] C.E. Rolfs and W.S. Rodney, *Cauldrons in the Cosmos*, The University of Chicago Press, (1988).

[27] W.R. Hix et al., *Phys. Rev. Lett.* **91**, 201102 (2003) and references therein.

[28] It is interesting that this resonance was predicted by Hoyle as a possible solution to understand how ^{12}C could be formed in stars with the observed abundance. It was soon after found experimentally using the ^{14}N(d, α) reaction.

This sequence can be repeated up the ladder of elements. In addition (α, n) (α, p) and reactions with incoming neutrons and protons can form the elements that lie in between the alpha-like nuclides. When α-burning becomes insuficient the star compresses due to the gravitational pull and heats up until carbon burning occurs.

Such reactions,

$$
\begin{aligned}
{}^{12}\mathrm{C}\,{}^{12}\mathrm{C} &\longrightarrow {}^{20}\mathrm{Ne}\ \alpha \\
&\longrightarrow {}^{23}\mathrm{Na}\ p \qquad (19.12) \\
&\longrightarrow {}^{23}\mathrm{Mg}\ n
\end{aligned}
$$

require temperatures higher than about 10^9 K ($kT \sim 10^5$ eV). Such temperatures occur only in very heavy stars and carbon burning thus is believed to occur predominantly in massive, also sometimes called *exploding stars.* If it is assumed that the temperature in exploding stars is about 2×10^9 K, the abundance of elements produced appears to agree closely with observation, as is shown in Fig. 19.4.

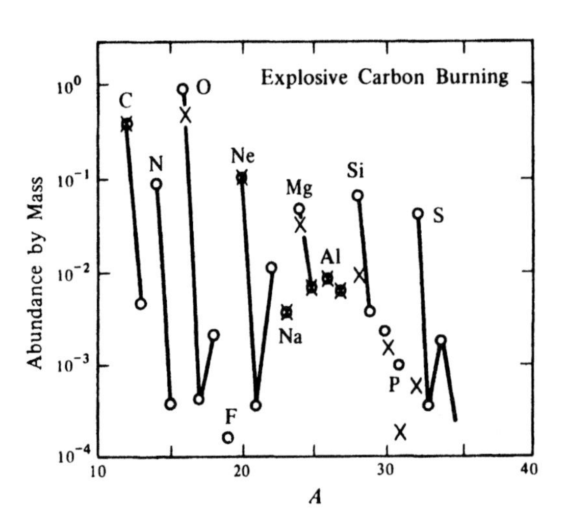

Figure 19.4: Products of carbon burning in an exploding star. Circles represent solar-system abundances, calculated abundances are shown as crosses. Solid lines connect all stable isotopes of a given element. The assumed peak temperature is 2×10^9K, the density 10^5 g/cm^3. [After W. D. Arnett and D. D. Clayton, *Nature* **227**, 780 (1970).]

The exact path that nucleosynthesis takes depends on the initial conditions and on whether there can be additional fresh hydrogen coming, e.g. from a companion young star, but in general the production proceeds toward the more stable nuclei, ending up in Fe.

As the formation of elements reaches iron, a new aspect appears. As Fig. 16.1 shows, the binding energy per nucleon reaches a maximum at the iron group. Beyond these elements, the binding energy per nucleon decreases. Hence the iron group cannot serve as fuel, and burning must cease once iron has been formed. This feature explains why elements centered around Fe are more abundant than others.

Most elements beyond the iron group are formed mainly through neutron capture reactions. There are two processes, a slow one, called s, and a rapid one, called r. The capture processes depend critically on the neutron flux. Figure 19.5 shows how the two processes may generate heavier nuclei with different relative abundances. With the beta decays of unstable nuclei, the proton number Z increases by one (as shown by the diagonal lines in Fig. 19.5.) In the s-process that takes place in stars like red giants, neutron captures are not very frequent because the fluxes are low. So once a capture occurs there is plenty of time for the beta decay to happen before

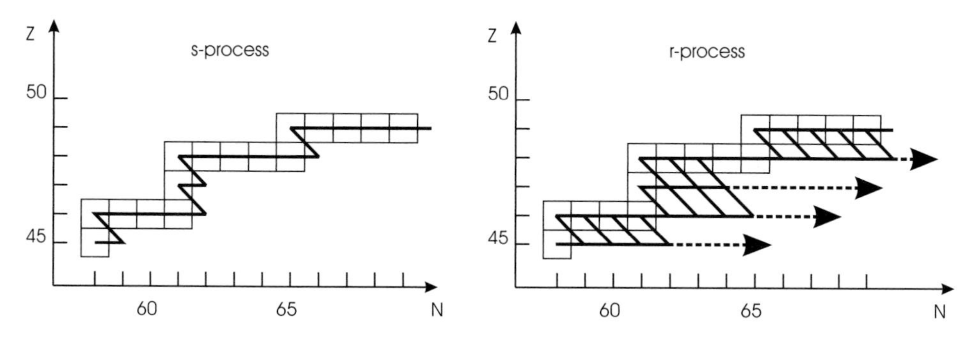

Figure 19.5: Example of isotope production as a result of the (left) s- and (right) r-process. In the parts of the s- and r-process shown, ^{103}Rh initally captures a neutron and all the other isotopes indicated with squares are produced. Although after beta-decay occurs the isotopes produced are the same, their relative abundance is very different.

another neutron can be captured. Thus, the s-process follows a path close to the stable nuclei and the path is determined by the relative beta-decay half lives. By contrast, if neutron fluxes are very high, as within a supernova,(29) the r-process can take place. Now neutron capture is much faster than beta decay (the roles of these two reactions are reversed from the s-process.) As before, the fast reaction (which is now neutron capture) determines the path of the nucleosynthesis. Nuclei in an intense neutron flux capture neutrons by the (n, γ) reaction, filling up the neutron shell until one approaches so close to the neutron drip line that the (γ, n) competes with (n, γ) reaction. This (n, γ) - (γ, n) reaction equilibrium determines the 'stable' nuclei in this explosion. These nuclei are very neutron rich and thus far from the valley of stability. As before, the slow reaction (which is now beta decay) determines the rate of the nucleosynthesis. When a beta decay occurs, a neutron changes into a proton, opening up a 'hole' in the neutron shell that can be quickly filled by neutron capture. In this way, through a succession of beta decays followed quickly by neutron capture, nuclei up to uranium can be synthesized. In fact, the r-process is the only way to produce elements like uranium or heavier ones. There is no s-process path through stable nuclei to uranium. Although the basic idea for the r-process is clear, the abundances that result of such a process are not possible to calculate presently. The main reason is that the masses of nuclei far from stability are not known and this has a strong impact on the capture cross sections. Fig. 19.6 shows a sketch of a nuclide chart showing that most of the nuclei that participate in the r-process have not yet been observed in the laboratory. In principle, the masses of these nuclei as well as the reactions involved can be studied with radioactive beams.(30,31)

We have described the neutron capture processes, but there are equivalent proton capture processes that take place in proton-rich environments. Explosive proton burning (the rp-process), for example, can take place when an old star (hot but out

[29] J.J. Cowan and F.-K. Thielemann, *Phys. Today* **57**, 47 (2004).

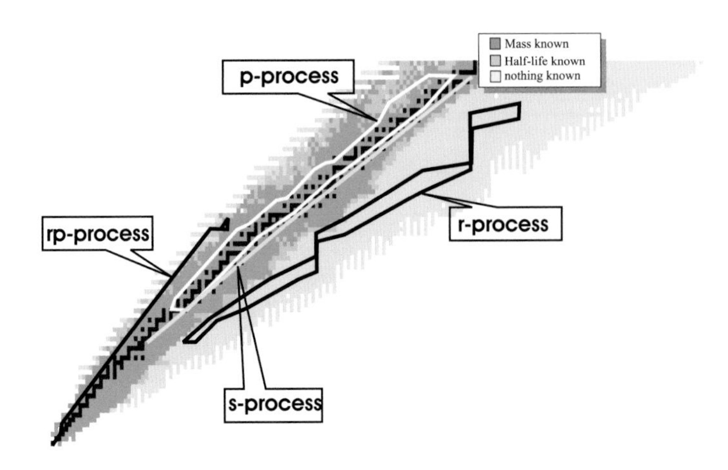

Figure 19.6: Sketch of nuclide chart showing several processes that contribute to nucleosynthesis. The nuclei in black are the stable ones. Of the predicted radioactive ones, about half remain unobserved. [Courtesy Hendrik Schatz.]

of hydrogen) can combine in a binary system with a younger star (full of hydrogen). Novae and x-ray bursts are probably generated in this way. We have only sketched the simplest ideas of nucleosynthesis. The correctness of these ideas can be examined only through detailed computations, involving nuclear physics and stellar evolution. Such investigations have arrived at encouraging results: Most of the salient abundance features can at least be qualitatively explained. However, much more needs to be understood and Fig. 19.6 gives an impression on how much more needs to be measured to get the nucleosynthesis on a secure footing.(30,31)

19.4 Stellar Collapse and Neutron Stars

In the previous section we have described the various burning processes in stars. These fusion reactions give rise to the elements; at the same time, they exhaust more and more of the nuclear fuel. What happens after the fuel is gone? According to present theory, the star can die one of four deaths; it can become a black hole, a white dwarf, a neutron star, or it can become completely disassembled. The ultimate fate is determined by the initial mass of the star. If this mass is less than about four solar masses, the star sheds mass until it becomes a white dwarf. If the initial mass is larger than about four solar masses, it may become a supernova which then results either in a neutron star, a black hole, or becomes completely

[30] D.F. Geesaman, C.K. Gelbke, R.V.F. Janssens, B.M. Sherrill, *Annu. Rev. Nuc. Part. Sci.* **56**, 53 (2006).
[31] M.S. Smith, K.E. Rehm, *Annu. Rev. Nuc. Part. Sci.* **51**, 91 (2001).

disassembled. Black holes contract forever and they approach, but never reach, a radius of roughly 3 km and a density exceeding 10^{16} g/cm^3. Neutron stars have a radius of about 10 km and a central density exceeding that of nuclear matter, about 10^{14} g/cm^3. We shall limit the discussion here to the formation(32) and properties(33) of neutron stars.

Neutron stars are believed to evolve from the gravitational collapse of stars more massive than about eight solar masses. Towards the end of their nuclear burning stages, stars have interior temperatures of about 8×10^9 K ($kT \sim 10^6$ eV) with central cores of about 1.5 solar masses composed primarily of iron. As discussed in Chapter 16, ^{56}Fe is the most stable nucleus at zero temperature and pressure. At the pressure, density, and temperature of the inner core, the atoms are fully ionized, and the freed electrons form a degenerate gas. The behavior of these electrons determines the further evolution of the star; the role of the electrons can be understood with the Fermi gas model treated in Section 16.2. We assume the electrons to form a gas of extremely relativistic free fermions, enclosed in a volume V. All available states up to the Fermi energy E_F are filled. This degenerate electron gas provides the pressure that balances the gravitational attraction. To compute the pressure, we first determine the total energy of n extremely relativistic electrons in a volume V, by following the steps from Eqs. (16.12) to (16.17), but using the extreme relativistic relation $E = pc$. The total energy of the electrons then becomes

$$E = \left(\frac{\pi^2}{4}\right)^{1/3} \hbar c \frac{n^{4/3}}{V^{1/3}}. \qquad (19.13)$$

The pressure due to this Fermi gas is

$$p = -\frac{\partial E}{\partial V} = \frac{1}{3}\left(\frac{\pi^2}{4}\right)^{1/3} \hbar c \left(\frac{n}{V}\right)^{4/3}, \qquad (19.14)$$

and this pressure balances the gravitational inward force, so that the core is in equilibrium.

The core loses electrons through capture by the iron, with the emission of neutrinos. When the mass of the core can no longer be supported by the electrons, the core collapses. The resulting gravitational energy is converted to heat and kinetic energy, the nuclei are stripped down to nucleons and the core density increases until it reaches densities above (of the order of twice) those of nuclear matter. At this point compression ceases because the nucleon gas provides the pressure required

[32] H. A. Bethe and G. Brown, *Sci. Amer.* **252**, 60 (May 1985); S. E. Woosley and T. A. Weaver, *Annu. Rev. Astronomy Astrophys.* **24**, 205 (1986).

[33] M. A. Ruderman, *Sci. Amer.* **224**, 24 (February 1971); G. Baym and C. Pethick, *Annu. Rev. Nucl. Sci.* **25**, 27 (1975); S. Tsuruta, *Comm. Astrophys.* **11**, 151 (1986); S. L. Shapiro and S. A. Teukolsky, *Black Holes, White Dwarfs, and Neutron Stars*, Ch. 9, John Wiley, New York, 1983; G. Baym, in *Encyclopedia of Physics*, 2nd. Edition, (R.G. Lerner and G. L. Trigg, eds.) VCH Publishers, Inc., New York, p. 809; D. Pines in *Proc. Landau Memorial Confer. on Frontiers of Physics*, (E. Gotsman and Y. Ne'eman, eds.) Pergamon, Elmsford, NY, 1989; C.J. Pethick and D.G. Ravenhall, *Annu. Rev. Nucl. Part. Sci.* **45**, 429 (1995).

to halt collapse. If, however, the star is too massive, $\gtrsim$ about 25 solar masses, the nucleon gas cannot supply enough pressure and gravitational collapse continues until a black hole forms. We do not consider this case. For smaller mass stars, when compression ceases, the core "bounces" back somewhat and outward pressure waves result that collect to form a shock wave. This shock wave will disrupt the surrounding star mantle and an explosion follows. A type II supernova is born. The enormous store of energy in the collapsed core, about 3×10^{53} ergs, is radiated in the form of neutrinos in approximately the next 10 seconds; a neutron star is left behind.

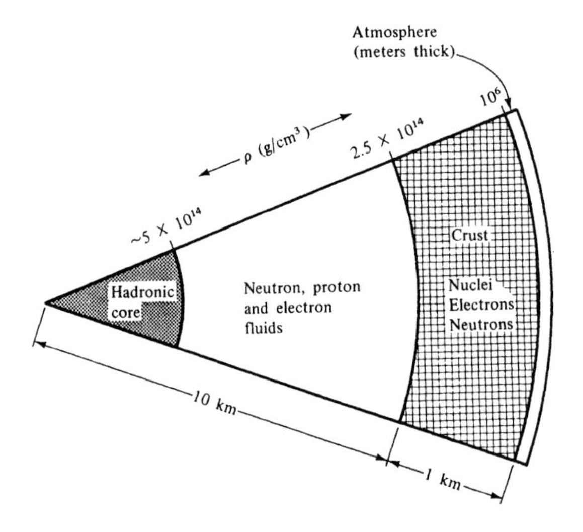

Figure 19.7: Cross section of a typical neutron star. The hadronic core could be quark matter or a pion condensate.

The neutrino emission is very efficient in cooling the remnant neutron star; initial rapid cooling occurs in a matter of seconds, and within a few days the internal temperature drops to about 10^{10} K, ($kT \sim 10^6$ eV), and keeps cooling. Neutrino emission continues for at least 10^3 years; photon emission only takes over as the dominant cooling mechanism when the remnant star reaches a temperature of about 10^8 K.(34)

The cross section of a typical neutron star is shown in Fig. 19.7. How did the star reach this terminal stage and why does it not collapse completely? The answer to these questions involves many fields, relativity, quantum theory, nuclear, particle, and solid state physics. Here we sketch some of the features of interest to subatomic physics.

Consider first *density and composition.* For a given neutron star mass, the radius and the density distribution can be computed. For a star with a radius of 10 km, the central density is of the order of 10^{14} to 10^{15} g/cm^3. The density thus increases from zero at the top of the atmosphere to a value larger than the density of nuclear matter at the center. From a knowledge of the density, the composition at a given depth can be inferred. The outermost layer is expected to be mainly ^{56}Fe, the end point of the thermonuclear burning process. Towards the interior the density increases and the Fermi energy becomes sufficiently high for electron capture processes to

[34] K. Nomoto and S. Tsuruta, *Astrophys. J. Lett.* **250**, L19 (1981); *Neutron Stars: Theory and Observation*, (J. Ventura and D. Pines, eds.), Kluwer, Dordrecht (1991); *The Lives of the Neutron Stars*, (M.A. Alpar et al., eds.) Kluwer, Dordrecht (1995); J.M. Lattimer and M. Prakash, *Science* **304**, 536 (2004).

occur, as in the formation process of the neutron star in the pre-supernova. At the increased pressure, more neutron rich nuclei are formed; the electron capture processes continue and at about 4×10^{11} g/cm^3, nuclei with 82 neutrons, such as ^{118}Kr, are most stable.(35,38) Ordinary krypton on Earth has $A = 84$. The most stable nuclides at high pressure thus are very neutron-rich. Under ordinary circumstances, such nuclides would decay by electron emission. However, at the pressure under discussion here, all available energy levels are already occupied by electrons and the Pauli principle prevents simple beta decay.

The last neutron in ^{118}Kr is barely bound. As the density increases beyond 4×10^{11} g/cm^3, the neutrons begin to leak out of the nuclei and form a degenerate liquid. As the pressure increases further, the nuclei in this *neutron drip regime* become more neutron rich and grow in size. At a density of about 2.5×10^{14} g/cm^3, they essentially touch, merge together, and form a continuous fluid of neutrons, protons, and electrons. Neutrons predominate and protons and electrons constitute only about 5% of the matter. Neutrons cannot decay to protons by simple beta decay because the decay electron would have an energy below the electron Fermi energy; the decay is thus forbidden by the Pauli principle.

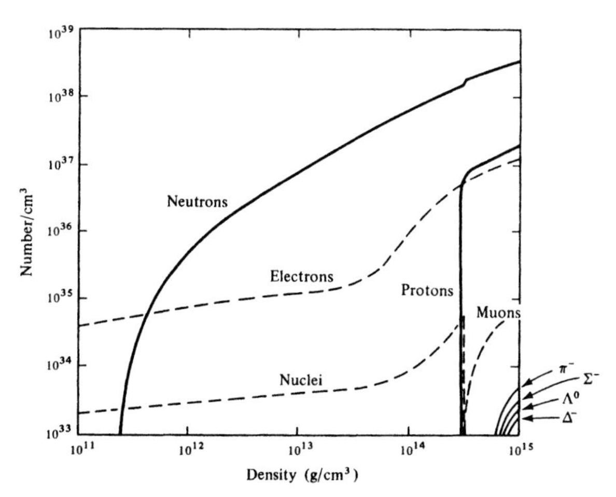

Figure 19.8: Composition of neutron-star matter as a function of the density. At higher densities, muons and strange particles appear. [Courtesy M. Ruderman.]

At still higher densities, it becomes energetically feasible to create more massive elementary particles through electron captures such as

$$e^{-}n \longrightarrow \nu\Sigma^{-};$$

these particles can again be stable because of the exclusion principle.(36) The number of constituents of matter as a function of density is shown in Fig. 19.8.(37)

We now turn again to the internal pressure in a neutron star. We have seen above that the degenerate electron gas provides pressure that prevents collapse at lower pressures. At higher pressure (or densities), complete collapse is prevented by a combination of two features, the repulsive core in the nucleon–nucleon force (Fig. 14.15), and the degeneracy energy of the neutrons. Fig. 19.8 indicates that neutrons pre-

[35] G. Baym, C. Pethick, and P. Sutherland, *Astrophys. J.* **170**, 299 (1971).

[36] V.R. Pandharipande, *Nucl. Phys.* **A178**, 123 (1971).

[37] For an updated version of the composition at densities larger than nuclear see T.Takatsuka and R. Tamagaki, *Prog. Theor. Phys.* **112**, 37 (2004).

dominate at high densities. They form a degenerate Fermi gas and the arguments leading to Eq. (19.14) can be repeated nonrelativistically. Again, as in Eq. (19.14), the degeneracy pressure increases with decreasing volume until it, together with the hard core repulsion, balances the gravitational attraction.

Neutron stars were predicted long ago,[38] but hope for their observation was small and they remained mythical objects for a long time. Their discovery was unexpected: In 1967, a strange new class of celestial objects was observed at the University of Cambridge.[39] The objects were point-like, definitely outside the solar system, and emitted periodic radio signals. They were nicknamed *pulsars*[40] and, despite the fact that the objects are not pulsating, but rotating, the name has been accepted. About 1500 pulsars are known;[41] each has its own characteristic signature. The pulsar periods range from a low of about 1 msec, and their periods lengthen in a very regular fashion; it is so regular, that some pulsars have been said to be the best clocks in the universe.

As suggested by Gold, a pulsar is a neutron star.[42] The pulsar period is associated with the rotational period of the neutron star: Because particles (and consequently x rays) are emitted preferentially along the magnetic axis the neutron star works as a lighthouse. The slow lengthening of their periods is caused by the loss of rotational energy. The rotational energy lost by the Crab pulsar, for instance, is of the same order as the total energy emitted by the nebula. The neutron star thus is the power source of the huge Crab nebula.

Pulsars have not only been observed as radio stars, but periodic light emission also has been seen. The periods, the slow-down rates, and the sudden changes in the period are being studied very carefully. Step by step, pulsars reveal properties of neutron stars. In an indirect way, astrophysicists and nuclear physicists have obtained access to a laboratory in which densities beyond 10^{15} g/cm^3 are available; the properties of nuclear matter can thus be studied in a beautiful combination of various disciplines.

19.5 Cosmic Rays

> The planetary system is a gigantic laboratory where nature has been performing an extensive high-energy physics experiment for billions of years.
>
> T. A. Kirsten and O. A. Schaeffer[43]

[38]W. Baade and F. Zwicky, *Proc. Nat. Acad. Sci. Amer.* **20**, 259 (1934); L.D. Landau, *Phys. Z. Sowiet* **1**, 285 (1932).

[39]A. Hewish, S. J. Bell, J. D. S. Pilkington, P. F. Scott, and R. A. Collins, *Nature* **217**, 709 (1968). A. Hewish, *Sci. Amer.* **219**, 25 (October 1968); J. P. Ostriker, *Sci. Amer.* **224**, 48 (January 1971).

[40]M. A. Ruderman and J. Shaham, *Commun. Astrophys.* **10**, 15 (1983); D. C. Backer, *Commun. Astrophys.* **10**, 23 (1983).

[41]R. Irion, *Science* **304**, 532 (2004); R.N. Manchester, *Science*, **304**, 542 (2004).

[42]T. Gold, *Nature* **218**, 731 (1968).

[43]T. A. Kirsten and O. A. Schaeffer, "Elementary Particles." *Science, Technology and Society* (L. C. L. Yuan, ed.), Academic Press, New York, 1971, p. 76. Copyright © 1971 by Academic Press.

We are constantly bombarded by energetic particles from outer space; about 1 charged particle/sec passes through every cm^2 of the Earth's surface. These "rays" were discovered by Victor Hess in 1912 by observing the ionization in an electrometer carried in a manned balloon; above 1000 m altitude, the intensity began to increase and it doubled by 4000 m.[44]

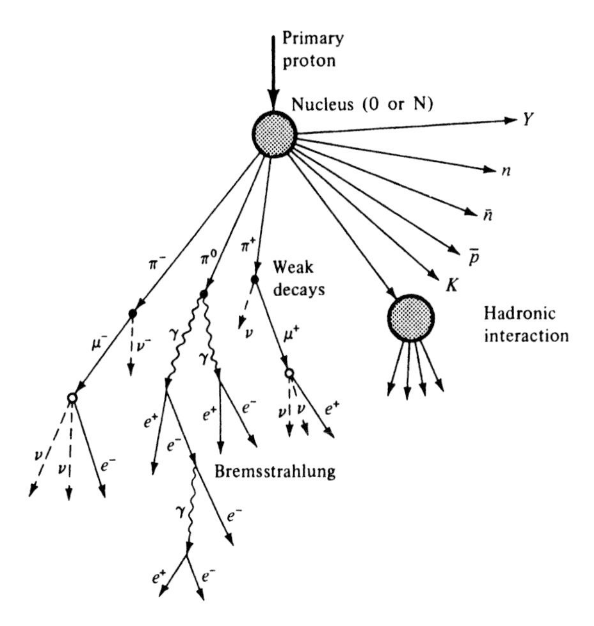

Figure 19.9: An incident high-energy proton strikes the top of the atmosphere and produces a cascade shower.

Since 1912, cosmic rays have been studied extensively; their composition, energy spectrum, spatial and temporal variation are being explored with ever-increasing sophistication, and many theories concerning their origin have been proposed. Cosmic rays are one of the main components of the galaxy. This assessment is based on the fact that the energy density of the cosmic rays in our galaxy, about 1 eV/cm^3, is of the same order of magnitude as the energy density of the magnetic field of the galaxy and of the thermal motion of the interstellar gas.

Cosmic rays have been observed and studied at various altitudes, in caverns deep underground, in mountaintop laboratories, with balloons at altitudes up to 40 km, with rockets, and with satellites. The radiation incident on the Earth's atmosphere consists of nuclei, electrons and positrons, photons and neutrinos. It is customary to call only the charged particles cosmic rays. X-ray, radio, and γ-ray astronomy have led to spectacular discoveries,[45] but we shall not treat these here. Consider first the fate of a cosmic ray proton of very high energy that strikes the top of the Earth's atmosphere. It interacts with an oxygen or nitrogen nucleus, and a cascade process is initiated. A simplified scheme is shown in Fig. 19.9. As discussed in Sections 14.7

[44] V. F. Hess, *Physik. Z.* **13,** 1084 (1912).

[45] E.M. Schlegel, *The Restless Universe, Understanding X-Ray Astronomy in the Age of Chandra and Newston,* Oxford Univ. Press, Oxford; 2002; *Exploring the Universe, A.Festschrift in Honor of R. Giacconi,* (H. Gursky, R. Ruffini and L. Stella,eds) World Sci., Singapore, 2000; B.F. Burke and F. Graham-Smith, *An Introduction to Radio Astronomy,* 2nd Edition, Cambridge Univ. Press, Cambridge, 1998.

and 6.11, the interaction will produce a large number of hadrons; pions predominate, but antinucleons, kaons, and hyperons also occur. These hadrons can again interact with oxygen or nitrogen nuclei; the unstable ones can also decay weakly. The decays result in electrons, muons, neutrinos, and photons (Chapter 11). The photons can produce pairs; the muons decay, but because of the time dilation (Eq. (1.9)), many penetrate into the Earth's solid mantle before doing so. Overall, a very-high energy proton can give rise to a large number of photons and leptons (Fig. 3.10); such a cosmic-ray shower can cover an area of many km^2 on the surface of the Earth.[(46)] In contrast, a photon produces a shower with very few muons. We shall not discuss the phenomena in the atmosphere further, but shall turn to the primary radiation.

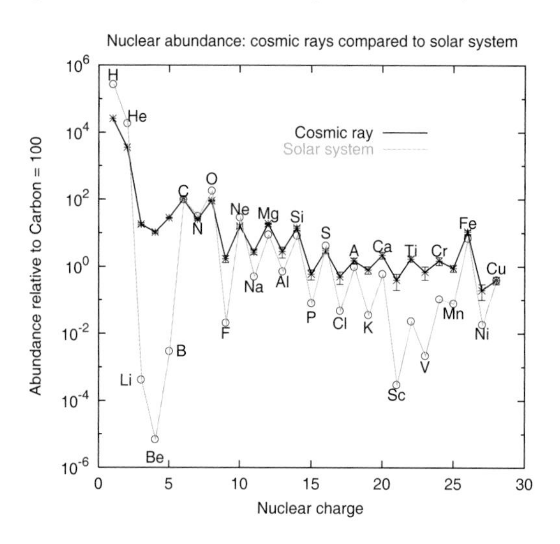

Figure 19.10: Composition of the nuclear component of the primary cosmic rays. Shown for comparison are the solar abundances. [From T.K. Gaisser and T. Stanev, *Nucl. Phys.* **A777**, 98 (2006).]

The *composition of the nuclear component* of the primary cosmic rays is shown in Fig. 19.10.[(47)] Also shown for comparison is the *universal* distribution of the elements observed in the solar atmosphere and in meteorites. A few remarkable facts emerge from a comparison of the cosmic-ray and the universal data: (1) The elements Li, Be, and B are about 10^5 times more abundant in cosmic rays than universally. (2) The ratio $^3He/^4He$ is about 300 times larger in cosmic rays. (3) Heavy nuclei are much more prevalent in cosmic rays.

The first two facts can be explained by assuming that the cosmic rays have traversed about several g/cm^2 of matter between their source and the top of the Earth's atmosphere. In such an amount of matter, nuclear reactions produce the observed distribution. Since the interstellar density is about 10^{-25} g/cm^3, the cosmic rays must have wandered around for 10^6–10^7 y. Two more facts have been established that may prove important for theories of the origin of cosmic rays: (4) So far, no antihadrons have been found in the primary cosmic rays.[(48)] (5) Electrons are about

[46] D.E. Nagle, T.K. Gaisser, and R.J. Protheroe, *Annu. Rev. Nuc. Part. Sci.* **38**, 609 (1988); M.V.S. Rao and B.V. Sreekantan, *Extensive Air Showers*, World Scientific, 1998.

[47] *Composition and Origin of Cosmic Rays*,(M. M. Shapiro, ed.) Reidel , Boston, 1982; J. A. Simpson, *Annu. Rev. Nuc. Part. Sci.* **33,** 323 (1983); N. Lund in *Cosmic Radiation in Contemporary Astrophysics*, (M. M. Shapiro, ed.) Reidel, Boston, 1986, p. 1.

[48] See e.g., M. Bongi et al, *IEEE Transc Nucl. Sci.* **51**, 854 (2004).

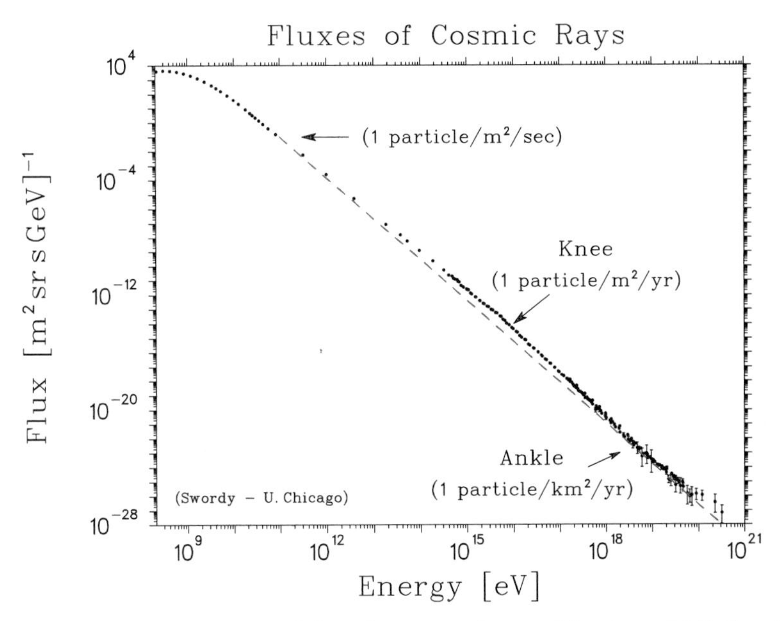

Figure 19.11: Energy spectrum of the primary cosmic rays. [Courtesy of S.P. Swordy.]

1% as abundant as nuclei in the same energy interval; positrons form about 10% of the electron component.

The energy spectrum, the number of primary particles as a function of energy, has been measured over an enormous range. For the nuclear component, it is shown in Fig. 19.11; the data extend over about 14 decades in energy and 32 decades in intensity. The highest observed energy is about 5×10^{20} eV or about 30 Joules.[(49)] Fig. 19.11 demonstrates that the cosmic-ray spectrum does not have a thermal shape; it is not exponential, but decays more slowly. A good fit to the data, except at the lowest energies, is[(50)]

$$I(E) \propto E^{-2.7}, \tag{19.15}$$

where $I(E)$ is the intensity of the nuclear component with energy E. At about 5×10^{15} eV there is a knee in the spectrum. The cause of the knee is thought to be due to either propagation effects or a new accelerating mechanism. Above this energy, the fit is about $I(E) \propto E^{-3}$ up to $\sim 10^{18}$ eV. Above this energy, the slope varies with energy. It is usually assumed that the cosmic ray particles above $\approx 10^{18}$ eV are extragalactic in origin because no galactic acceleration mechanism for these *ultra-high* energy cosmic rays has been found. There is an expected cut-off at about 6×10^{19} eV because of proton collisions with the cosmic microwave background (see Sec. 19.1) producing pions, so that the energy is degraded.[(51)] The electron

[49]see, e.g., M. Ahlers, A. Ringwald, and H. Tu *Astopart. Phys.* **24**, 438 (2006).
[50]J.W. Cronin, *Rev. Mod. Phys.* **71**, S165 (1999).
[51]K. Greisen, *Phys. Rev. Lett.* **16**, 748 (1966); G.T. Zatsepin and V.A. Kuszmin, *JETP Lett* **4**, 78 (1966).

spectrum is similar to Fig. 19.11 for energies above 1 GeV, but is somewhat steeper above 100 GeV because of the electromagnetic interaction during propagation. The electron spectrum thus provides a sensitive test of propagation models.[52]

Two more facts concerning the energy spectra are important for the discussion of the origin of cosmic rays. One is the isotropy of cosmic rays; the other is their constancy over a long period of time. Measurements in outer space indicate that the cosmic-ray flux is essentially isotropic for energies $\lesssim 10^{15}$ eV. The time dependence of the intensity over long periods has been studied by looking at the abundance of nuclides created in moon samples and meteorites. The cosmic ray intensity has been approximately constant over a period of about 10^9 y.

The experimental evidence discussed above implies that it is likely that there are several sources of cosmic rays.[53] For energies below $\approx 10^{15}$ eV the source should have the following properties: [50] The total produced energy must be of the order of 10^{49} ergs/y in our galaxy; the cosmic rays must be isotropic and constant during at least 10^9 y. The primary spectrum must include heavy elements up to about $Z = 100$ but less than about 1% antihadrons. For cosmic rays of *high energies* the sources are likely ultra-galactic.

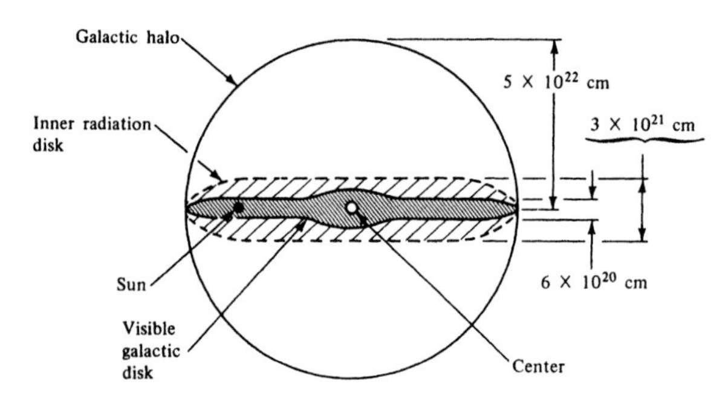

Figure 19.12: Sketch of cross section through our galaxy.

Where do cosmic rays originate? To sharpen this question, it helps to draw a cross section through our galaxy, as in Fig. 19.12. Cosmic rays can be produced in the inner radiation disk, in the galactic halo, or they can flow into the galaxy from the outside.[54] Most experts believe that the cosmic rays below 10^{18} eV originate in our galaxy.[54,55]

A favored hypothesis is that supernovae and neutron stars produce cosmic rays with the correct properties.[55] There is even some, not-universally accepted, evidence of the production of cosmic-ray nuclei in the expanding shock wave of a supernova explosion.[56] In our galaxy, a supernova appears about every 40 y and one supernova is believed to produce between 10^{51} and $10^{52.5}$ erg of energy. A recent bright extragalactic one observed in 1987, and called SN 1987a, has been studied

[52]D. Casadei and V. Bindi, *Astrophysical Journal* **612**, 262 (2004).
[53]See however A. Dar, A. De Rujula, hep-ph/0606199.
[54]T.K. Gaisser, T. Stanev, *Nucl. Phys.* **A777**, 98 (2006); V. Berezinsky, A. Gazizov, and S. Grigorieva *Phys. Rev. D* **74**, 043005 (2006).
[55]H. Bloemen in *Interstellar Processes*, (D.J. Hollenbach and H.A. Thronson, Jr, eds.) Reidel, Dordrecht, 1987, p. 143; P.K. MacKeown and T.C. Weeles, *Sci. Amer.* **252**, 60 (November 1985).
[56]R. Enomoto et al., *Nature*, **416**, 823 (2002); Y.M. Butt et al., *Nature* **418**, 499 (2002).

extensively.[57] The investigations of supernovae show that they can provide the energy required for cosmic rays, about 10^{49} ergs/y. However, supernova shock wave acceleration models have difficulty in accounting for particles above 10^{15} eV. Recent detection of cosmic rays from the binary system Cygnus X-3 and Hercules X-1 suggest that for energies beyond the knee most cosmic rays may come from pulsars or binary systems consisting of a neutron star and a giant companion.

It is possible that the sources emit cosmic rays with the energy spectrum Eq. (19.15). It is, however, also possible that nature uses the same technique as present day high-energy accelerators, acceleration in stages (Section 2.6). A mechanism for acceleration in interstellar space, collision of the particles with moving magnetic fields, has for instance been suggested by Fermi.[58] However, it is now more generally believed that the primary sources of cosmic rays are supernovae explosions and their remnants.[55,56]

For cosmic rays of highest energies no unique source has yet been identified. They could come from black holes or even from remnants of the early universe.[49]

19.6 Neutrino Astronomy and Cosmology

Classical astronomy is based on observations in the narrow band of visible light, from 400 to 800 nm. In the past few decades, this window has been enlarged enormously through radio and infrared astronomy on one side and through X-ray and gamma-ray astronomy on the other. The charged cosmic rays provide another extension. However, all these observations have one limitation in common: They cannot look at the inside of stars, because the radiations are absorbed in a relatively small amount of matter (Chapter 3). As a ballpark figure, it takes a photon $\sim 10^4 - 10^5$ years to come out from the center of the sun. Fortunately, there is one particle that escapes even from the inside of a very dense star, the *neutrino*; and neutrino astronomy,[22] even though extremely difficult, has become an irreplaceable tool in astrophysics. The properties that make the neutrino unique have already been treated in Sections 7.4 and 11.14:

1. The absorption of neutrinos and antineutrinos in matter is very small. For the absorption cross section, Eq. (11.78) gives

$$\sigma(\text{cm}^2) = 2.3 \times 10^{-44} \frac{p_e}{m_e c} \frac{E_e}{m_e c^2}, \tag{19.16}$$

   where p_e and E_e are momentum and energy of the final electron in the reaction $\nu N \to eN'$. With Eqs. (2.17) and (2.18) it is then found that the mean free path of a 1 MeV electron neutrino in water is about 10^{21} cm. It far exceeds the linear dimensions of stars, which range up to 10^{13} cm. (See also Fig. 1.1.)

[57] V. Trimble, *Rev. Mod. Phys.* **60**, 859 (1988); S. Woosley and T. Weaver, *Sci. Amer.* **261**, 32 (August 1989).

[58] E. Fermi, *Phys. Rev.* **75**, 12 (1949). (Reprinted in *Cosmic Rays, Selected Reprints*, American Institute of Physics, New York.)

2. Neutrinos and antineutrinos can be distinguished by their reactions with water. Although the neutrino luminosity at the Earth is dominated by the Sun, because of its closeness, it is believed that the primary galactic sources of neutrinos are supernovae and their remnants. In the cooling process of supernovae, neutrino-antineutrino pairs of all flavors are emitted through neutral current reactions such as $e^+e^- \rightarrow \nu\overline{\nu}$. In addition, electron neutrinos and antineutrinos are generated through charged current reactions in nuclei, e.g., $e^-p \rightarrow n\nu_e$. Electron antineutrinos, primarily, but also neutrinos were observed in connection with SN 1987a.

In addition to being a unique probe for getting information from the core of stars, neutrinos can be used to search for dark matter. In particular, if WIMPs existed, they would accumulate in the gravitational well of the Sun and at the center of our galaxy, where they would decay into neutrinos. These neutrinos could be observed on Earth and detectors like Super Kamiokande have put constraints[(59)] on these scenarios. Detectors such as AMANDA and ICECUBE could significantly improve the constraints in the future.

19.7 Leptogenesis as Basis for Baryon Excess

The baryon excess over antibaryons in the universe can also be explained via leptogenesis.[(60)]If neutrinos are Majorana particles, then their small mass could be explained by a see-saw mechanism. The latter postulates the existence of very heavy right-handed Majorana neutrinos, which could have masses as large as $M_{RH} \sim 10^{14} - 10^{16}$ GeV, of the order of the GUT (Grand Unified Theory) scale. The observable (left-handed) neutrinos end up having masses:

$$m_\nu \sim m_H^2/M_{RH} \tag{19.17}$$

where $m_H \sim 10^2$ GeV is the mass of the Higgs boson or of the electroweak scale. This yields $m_\nu \sim 10^{-2}$ eV so it is usually argued that this mechanism provides a natural explanation for the small masses of neutrinos.

The right-handed neutrinos decay into a positively charged Higgs and a negative lepton and also into a positively charged Higgs and a negative lepton. The rates need not be the same; thus there is CP violation as well as lepton non-conservation in the decay. If the temperature T is such that $kT \lesssim M_{RH}^< c^2$, where $M_{RH}^<$ is the lightest of the right-handed neutrinos, the decay would be out of equilibrium. Thus, the three Sakharov conditions are satisfied. There is a mechanism, called 'sphaleron' which preserves the difference between lepton and baryon number and is essentially an excursion over the temperature barrier.[(61)] This then leads to a

[59]S. Desai et al., *Phys. Rev. D* **70**, 083523 (2004).

[60]T. Yanagida, *Progr. Theor. Phys.* **64**, 1103 (1980); M. Fukugita and T. Yanagida, *Phys. Lett.* **B174**, 45 (1986); W. Buchmüller, R.D. Peccei, and T. Yanagida, *Annu. Rev. Nuc. Part. Sci.* **55**, 311 (2005); .

[61]V.A. Kuzmin, V.A. Rubakov and M.A. Shaposhnikov, *Phys. Lett.* **B155**, 36 (1985).

baryon excess and the numerical excess can explain the experimental value. Of course, many questions remain unanswered, but the mechanism solves two puzzles: the smallness of the neutrino masses and the baryon excess.

19.8 References

We have only scratched the surface of nuclear and particle astrophysics. For the reader who would like more information we suggest consulting the journals or books mentioned below. We have already referenced numerous reviews in the main body of the chapter. Some others are listed here.

Comments on Astrophysics, Annual Review of Nuclear and Particle Science, Annual Review of Astronomy and Astrophysics contain useful review articles.

There are a number of good texts on the present knowledge of astrophysics. A collection of articles that appeared in *Physics Today* are bound together in *Astrophysics Today*, (A. G. W. Cameron, ed.) Amer. Inst. Phys., New York, 1984. Another good book is M. Harwit, *Astrophysical Concepts*, 2nd. ed. Springer, New York, 1988; E. Choisson and S. McMillan, *Astronomy: A Beginner's Guide to the Universe*, Prentice-Hall, Upper Saddle River, NJ, 2001; S.N. Shore, *The Tapestry of Modern Astrophysics*, Wiley-Interscience, Hoboken, NJ, 2003; W. Hu and M. White, *The Cosmic Symphony, Sci. Amer.* **290**, 44, February 2004. Some more popular accounts include *Einstein's Universe, the Layperson's Guide*, Penguin Books, 2005; J. Silk, *On the Shores of the Unknown: A Short History of the Universe*, Cambridge University Press, (2005); B. Greene, *The Elegant Universe: Superstrings, Hidden Dimensions, and the Quest for the Ultimate Theory*, W.W. Norton, New York, 2003; B. Greene, *The Fabric of the Cosmos: Space, Time, and the Texture of Reality*, A. Knopf, New York, 2004; B.W. Carroll, A. Ostlie, *An Introduction to Modern Astrophysics*, Pearson, Addison-Wesley, San Francisco (2007).

Cosmic Rays *Genesis and Propagation of Cosmic Rays*, (M. M. Shapiro and J. P. Wefel, eds.) Reidel, Dordrecht, Holland (1988); P. Skokolsky, *Introduction to Ultrahigh Energy Cosmic Rays*, Addison-Wesley, Reading, Mass, 1989; M. W. Friedlander, *Cosmic Rays*, Harvard University Press, Cambridge, Mass, 1989; R. Schlickeiser, *Cosmic Ray Astrophysics*, Springer, New York, 2002; P.K.F. Grieder, *Cosmic Rays at Earth*, Elsevier, New York, 2001; T. Stanev, *High Energy Cosmic Rays*, Springer, New York, 2004; M.V.S. Rao and B.V. Skreekantan, *Extensive Air Showers*, World Scientific, Singapore, 1998; M.W. Friedlander, *A Thin Cosmic Rain*, Harvard University Press, Cambridge, MA, 2000; *Topics in Cosmic Ray Astrophysics*, (ed. M.A. DuVernois), Nova , Huntington, NY 2000.

X-Ray Astronomy *X-Ray Astronomy*, (R. Giacconi and G. Setti, eds.) Reidel, Boston, 1980; *X-Ray Astronomy with the Einstein Satellite*, (R. Giacconi, ed.) Reidel, Boston, 1981; C. L. Sarazin, *X-Ray Emission from Clusters of Galaxies*, Cambridge University Press, New York, 1988; *Exploring the Universe: A Festschrift for Riccardo Giacconi*, (M. Gursky R. Ruffini, and L. Stella, eds.) World Scientific, Singapore, 2000; E.M.Schliegel, *The Restless Universe: Understanding X-Ray Astronomy in the Age of Chandra and Newton*, Oxford University Press, New York, 2002; *Frontiers of X-Ray Astronomy*, (A.C. Fabian, K.A. Pounds, and R.D. Blandford, eds.), Cambridge University Press, 2004.

Radio Astronomy B.F. Burke and F. Graham-Smith, *An Introduction to Radio Astronomy*, Cambridge University Press, New York, 2002

Gamma-Ray Astronomy T. Weekes, *Very High Energy Gamma-Ray Astronomy*, Inst. Phys., Bristol and Philadelphia, 2003; F.A. Aharonian, *Very High Energy Gamma Radiation*, World Scientific, Singapore, 2004.

Neutrino Astronomy and Solar Neutrinos J.N. Bahcall, *Neutrino Astrophysics*, Cambridge University Press, New York, 1989; R. Davis, Jr., A.K. Mann, and L. Wolfenstein, *Annu. Rev. Nuc. Part. Sci.* **39**, 467 (1989); M. Stix, *The Sun: An Introduction*, 2nd Edition, Springer, New York, 2002; M.Fukugita and T. Yanagida, *Physics of Neutrinos and Applications to Astrophysics*, Springer, New York, 2003; R.N. Mohapatra and P.B. Pal, *Massive Neutrinos in Physics and Astrophysics*, World Scientific, River Edge, NJ, 2004; J.L. Tassoul and M. Tassoul, *A Concise History of Solar and Stellar Physics*, Princeton University Press, Princeton, NJ, 2004; A. Bhatnagar and W. Livingston, *Fundamentals of Solar Astronomy*, World Scientific, Hackensack, NJ, 2005.

Supernovae H.A. Bethe and G.E. Brown, *Sci. Amer.* **252**, 60 (May 1985) and *Nucl. Phys.* **429**, 527 (1984); S. A. Woosley and H.-T. Janka, *Nature Phys.* **1**, 147 (2005); J.A. Wheeler and R. F. Harkness, *Sci. Amer.* **257**, 50 (November 1987); R.W. Mayle, J. R. Wilson, and D. N. Schramm, *Astrophys. J.* **318**, 288 (1987); H.A. Bethe, *Annu. Rev. Nuc. Part. Sci.* **38**, 1 (1988); S. Woosley and T. Weaver, *Sci. Amer.* **261**, 32 (August 1989); W.D. Arnett et al., *Annu. Rev. Astronomy Astrophys.* **27**, 629 (1989). Some books are: *The Standard Model and Supernova 1987a*, (J. Tran Thanh Van, ed.) Edition Frontières, Gif-sur-Yvette, France, 1987; *Supernova Shells and Their Birth Events*, (W. Kundt, ed.) Springer Lecture Notes in Physics 316, Springer, New York, 1988; L.A. Marschall, *The Supernova Story*, Plenum, New York, 1988; *Supernova 1987a in the Large Magellanic Cloud.* (M. Kafatos and A.G. Michalitsiano, eds.) Cambridge University Press, New York, 1988; J.C. Wheeler, *Cosmic Catastrophes: Supernovae and Gamma-Ray Bursts and Adventures in Hyperspace*, Cambridge University Press, Cambridge, 2000; F.R. Stephenson and D.A. Green, *Historical Supernovae and their Remnants* Oxford University Press, New York, 2002; *Supernovae and Gamma-Ray Bursters*, (K.W. Weiler, ed.) Springer New York, 2003.

Neutron Stars G. Baym, nucl-th/0612021, Proceedings of Quark Confinement and the Hadron Spectrum VII, (2006); S.L. Shapiro and S.L. Teukolsky, *Black Holes, White Dwarfs, and Neutron Stars*, Wiley, New York, 1983; *The Origin and Evolution of Neutron Stars* (D.J. Helfand and J.H. Huang, eds) Reidel, Dordrecht, Holland, 1987; C.A. Pickover, *The Stars of Heaven*, Oxford, New York, 2001; A.M. Kaminker, *Physics of Neutron Stars*, Nova, New York, 1994; E. Choisson and S. McMillan, *Matter at High Density in Astrophysics: Compact Stars and the Equation of State*, (H. Riffert et al., eds.) Springer, New York, 1996; *Physics of Neutron Star Interiors*, (D. Blaschke, N.K. Glendenning, and A. Sedrakian, eds.), Springer, New York, 2001; R.P. Kishner, *The Extravagant Universe: Exploding Stars, Dark Energy, and the Accelerating Cosmos*, Princeton University Press, Princeton, 2002.

Nucleosynthesis F. Käppeler, F.-K. Thielemann, M. Wiescher, *Annu. Rev. Nuc. Part. Sci.* **48**, 175 (1998); C.E. Rolfs and W.S. Rodney, *Cauldrons in the Cosmos*, The University of Chicago Press, (1988); J. W. Truran, *Annu. Rev. Nuc. Part. Sci.* **34**, 53 (1984); J.H. Applegate, C.J. Hogan, and R.J. Scherrer, *Astrophys. J.* **329**, 572 (1988); D. Arnett, *Supernovae and Nucleosynthesis: An Investigation of the History of Matter, from the Big Bang to the Present*, Princeton University Press, Princeton, NJ, 1996; G. Wallerstein et al., *Rev. Mod. Phys.* **69**, 995 (1997); see also the list under "The Early Universe."

Dark Matter K. Freeman, G. McNamara, *In Search of Dark Matter*, Springer (2004); M.M. Waldrop, *Science* **234**, 152 (1987); V.R. Trimble, *Annu. Rev. Astronomy Astrophys.* **25**, 425 (1987); J.R. Primak, D. Seckel and B. Sadoulet, *Annu. Rev. Nuc. Part. Sci.* **38**, 751 (1988); *Dark Matter in the Universe*, (J. Bahcall, T. Piran, and S. Weinberg, eds) World Scientific., Teaneck, N.J., 1988; L.M. Krauss, *The Fifth Essence: The Search for Dark Matter*, Basic Books, New York, 1989; J.M. Overduin and P.S. Wesson, *Dark Sky, Dark Matter*, Bristol, Philadelphia, 2003.

Dark Energy N. Breton, J.L. Cervantes-Cota, M. Salgado (eds) *The Early Universe and Observational Cosmology*, Springer (2004); R.P. Kirshner *The Extravagant Universe : Exploding Stars, Dark Energy, and the Accelerating Cosmos*, Princeton University Press (2002); R.J.E. Peebles and B. Ratra, *Rev. Mod. Phys.* **75**, 559 (2003).

The Early Universe There are numerous books for the layman on this subject. Some good ones are S. Weinberg, *The First Three Minutes.* Basic Books, New York, 1977; H.R. Pagels, *Perfect Symmetry, The Search for the Beginning of Time*, Simon and Schuster, New York, 1985; L.M. Lederman and D.N. Schramm, *From Quarks to the Cosmos*, Scient. Amer. Lib., N. Y., 1989; J. Silk, *The Big Bang*, W.H. Freeman, New York, 2001; M. Mallory *Our Improbable Universe: A Physicist Considers How We Got Here*, Thunder's Mouth Press, New York, 2004; J. Silk, *On the Shores of* *the Unknown: A Short History of the Universe*, Cambridge, New Yrok, 2005. More serious reviews can be found in *Astronomy, Cosmology, and Fundamental Physics*, (M. Caffo et al., eds.) Kluiver, Dordrecht, 1989. A. D. Linde, *Particle Physics and Inflationary Cosmology*, Harwood Academic, New York, 1989; F. Lizhi, *Creation of the Universe*, World Scientific, Teaneck, NJ, 1989; S. G. Brush, *Rev. Mod. Phys.* **62**, 43 (1990); E.W. Kolb and M.S. Turner, *The Early Universe*, Addison-Wesley, Reading, MA, 1990; *The Scientific American Book of the Cosmos*, (D.H. Levy, ed.), St. Martin Press, New York, 2000; *The Early Universe and the Cosmic Microwave Background*, (N.G. Sànchez and Y.N. Prijskij, eds.), Dordrecht, Boston, 2003; G. Börner, *The Early Universe, Facts and Fiction*, Springer, New York, 2003; *The Early Universe and Observational Cosmology*, (N. Bretón , J.L. Cervantes-Cota, and M. Salgado, eds.) Springer, New York, 2004; *The Physics of the Early Universe*, (E. Papantonopoulos, ed.) Springer, New York, 2005; G. Veneziano, *The Myth of the Beginning of Time, Sci. Amer.* **290**, 54, (May 2004); G.D. Starkman and D.J. Schwarz, *Is the Universe Out of Tune?, Sci. Amer.* **293**, 48, (August 2005).

Problems

19.1. Show that the rate of a low energy reaction between two particles or nuclei of charges Z_1e and Z_2e depends exponentially on $Z_1Z_2e^2/\hbar v$, where v is the relative velocity of the two objects.

19.2. Discuss some of the difficulties of measuring cross sections relevant to nucleosynthesis and explain how some of them may be overcome.

19.3. Assume that the density distribution in the Sun (or a star) is given by $\rho = \rho_c[1-(r/R)^2]$, where ρ_c is the central density and R is the radius of the Sun.

(a) Evaluate the variation of the mass with radius by finding $dM(r)/dr$ and $M(r)$.

(b) Evaluate ρ_c in terms of the total mass M and radius R of the Sun.

19.4. List some reactions that can be used to test the solar energy cycle and explain the reason for your choices.

19.5. The total mass of a neutron star is limited by general relativity to be less than three solar masses or $\lesssim 6\times 10^{33}$ g. [M. Nauenberg and G. Chapline, *Astrophys. J.*, **179**, 277 (1973)]. Check whether the neutron star illustrated in Fig. 19.7 satisfies this criterion.

19.6. A star contains n_i particles/volume of type i which, at a temperature T, have an average velocity v_{ij} relative to particles of type j.

(a) What is the rate/volume for the reaction $i+j \rightarrow a+b$ at the temperature T? Express your answer in terms of the cross section σ_{ij} for the reaction, and assume that $i \neq j$.

(b) In a real star, the velocities follow a Maxwell–Boltzmann distribution. How is the answer to part (a) altered in this case?

19.7. How is the lifetime of the Z^0 affected by the number of neutrino families?

19.8. Use Hubble's law (Eq. 19.1) for the following:

(a) Determine the approximate age of the universe and compare to Table 19.1.

(b) In terms of the expansion of the universe, we can write Hubble's law as

$$\left.\frac{dR/dt}{R}\right|_{\text{now}} = H_0.$$

Under certain conditions, general relativity gives the relationship

$$8\pi G\rho R^2 = 3kc^2 + 3\left(\frac{dR}{dt}\right)^2,$$

where G is the gravitational constant, ρ is the mean density of the universe, and k is a constant. For a "flat" universe $k = 0$. Determine the critical density in terms of H_0 and G and evaluate it numerically.

19.9. What is the approximate minimum temperature necessary to allow pion production to occur?

19.10. What do you expect for the ratio of protons to neutrons at a temperature of 1.2×10^{11} K?

19.11. As an example of the effect of CP or time reversal (T) nonconservation on particle and antiparticle decays, show that the rates for

$$\Sigma^+ \longrightarrow p\pi^0 \quad \text{and} \quad \overline{\Sigma}^- \longrightarrow \overline{p}\pi^0$$

are not equal to each other unless CP or T holds.

19.12. (a) Show that the rotational speeds of stars in galaxies should be given by:

$$v(r) = \sqrt{\frac{GM(r)}{r}} \tag{19.18}$$

where $M(r)$ is the mass inside of a sphere of radius r.

(b) Speeds are observed to be roughly independent of r, and luminous matter (stars) seem to be near the center of the galaxy. What does this imply?

19.13. Explain qualitatively why the WMAP data is sensitive to whether one assumes that dark matter is cold versus hot. [Hint: consider the evolution of a spot with higher-than-average density and a universe made with cold dark matter (low velocity and only reacting to gravity), hot dark matter (e.g. neutrinos), gas (nuclei and electrons making a plasma), and photons.]

19.14. *Olber's paradox:* Consider a static infinite universe. Then consider a shell of radius r centered around the earth. While the observed brightness of the stars in this shell from earth would decrease like $1/r^2$, the number of stars would increase in the same proportion. Show that in a static infinite universe nights would be bright. Explain why this is not so.

Index